Lecture Notes
in Computational Science
and Engineering

2

Editors
M. Griebel, Bonn
D. E. Keyes, Norfolk
R. M. Nieminen, Espoo
D. Roose, Leuven
T. Schlick, New York

Springer
Berlin
Heidelberg
New York
Barcelona
Hong Kong
London
Milan
Paris
Singapore
Tokyo

Hans Petter Langtangen

Computational Partial Differential Equations

Numerical Methods and Diffpack Programming

 Springer

Author

Hans Petter Langtangen
Mechanics Division
Department of Mathematics
University of Oslo
Box 1053 Blindern
0136 Oslo, Norway
e-mail: hpl@math.uio.no

Cataloging-in-Publication Data applied for

Die Deutsche Bibliothek – CIP-Einheitsaufnahme

Langtangen, Hans Petter:
Computational partial differential equations: numerical methods and diffpack programming /
Hans Petter Langtangen. – Berlin; Heidelberg; New York; Barcelona; Hong Kong; London; Milan; Paris;
Singapore; Tokyo: Springer, 1999
(Lecture notes in computational science and engineering; 2)
ISBN 3-540-65274-4

Mathematics Subject Classification (1991):
primary: 65M60, 65M06, 68U20
secondary: 65F10, 65M12, 65N06, 65N30, 65Y99, 73C02, 73E60, 76D05, 76M10, 76M20

ISBN 3-540-65274-4 Springer-Verlag Berlin Heidelberg New York

Cover Design: Friedhelm Steinen-Broo, Estudio Calamar, Spain
Cover production: *design & production* GmbH, Heidelberg
Typeset by the author using a Springer TEX macro package
SPIN 10653075   46/3143 – 5 4 3 2 1 0 – Printed on acid-free paper

Preface

During the last decades there has been a tremendous advancement of computer hardware, numerical algorithms, and scientific software. Engineers and scientists are now equipped with tools that make it possible to explore real-world applications of high complexity by means of mathematical models and computer simulation. Experimentation based on numerical simulation has become fundamental in engineering and many of the traditional sciences. A common feature of mathematical models in physics, geology, astrophysics, mechanics, geophysics, as well as in most engineering disciplines, is the appearance of systems of partial differential equations (PDEs). This text aims at equipping the reader with tools and skills for formulating solution methods for PDEs and producing associated running code.

Successful problem solving by means of mathematical models in science and engineering often demands a synthesis of knowledge from several fields. Besides the physical application itself, one must master the tools of mathematical modeling, numerical methods, as well as software design and implementation. In addition, physical experiments or field measurements might play an important role in the derivation and the validation of models. This book is written in the spirit of computational sciences as inter-disciplinary activities. Although it would be attractive to integrate subjects like mathematics, physics, numerics, and software in book form, few readers would have the necessary broad background to approach such a text. We have therefore chosen to focus the present book on numerics and software, with some optional material on the physical background for models from fluid and solid mechanics.

The main goal of the text is to educate the reader in developing simulation programs for a range of different applications, using a common set of generic algorithms and software tools. This means that we mainly address readers who are or want to become professional *programmers* of numerical applications. As the resulting codes for solving PDEs tend to be very large and complicated, the implementational work is indeed non-trivial and time consuming. This fact calls for a careful choice of programming techniques.

During the 90s the software industry has experienced a change in programming technologies towards modern techniques such as object-oriented programming. In a number of contexts this has proved to increase the human efficiency of the software development and maintenance process considerably. The interest in these new techniques has grown significantly also in the numerical community. The software tools and programming style in this book reflect this modern trend. One of our main goals with the present text is in fact to explore the advantages of programming with objects in numerical contexts. The resulting programming is mainly on a high abstraction level, close to the numerical formulation of the PDE problem. We can do this because we build our PDE solvers on the Diffpack software.

Diffpack is a set of libraries containing building blocks in numerical methods for PDEs, for example, arrays, linear systems, linear and nonlinear solvers, grids, scalar and vector fields over grids, finite elements, and visualization support. Diffpack utilizes object-oriented programming techniques to a large extent and is coded in the C++ programming language. This means that we must write the PDE solvers in C++. If you do not already know C++, this text will motivate you to pick up the perhaps most popular programming language of the 90s. Most of the knowledge as a C++ programmer can be reused as a C, Java, or even Fortran 90/95 programmer as well, so we speak about a fortunate investment. You do not need to study a textbook on C++ before continuing with the present text, because experience shows that one can get started as a Diffpack programmer without any experience in C++ and learn the language gently as one proceeds with numerical algorithms, the software guide, and more advanced example codes. This book comes with a large number of complete Diffpack solvers for a range of PDE problems, and one can often adapt an existing solver to one's particular problem at hand.

In the past, Diffpack was distributed with a pure software guide as the only documentation. It soon became apparent that successful utilization of numerical software like Diffpack requires (i) a good understanding of the particular formulation of numerical methods that form the theoretical foundations of the package and (ii) a guide to the software tools, exemplified first in detail on simple problems and then extended, with small modifications, to more advanced engineering applications. The software guide can be significantly improved by providing precise references to the suitable *generic* description of various numerical algorithms. Although this generic view on methods is to a large extent available in the literature, the material is scattered around in textbooks, journal papers, and conference proceedings, mostly written for specialists. It was therefore advantageous to write down the most important numerical topics that must be mastered before one can develop PDE solvers in a programming environment like Diffpack, and tailor the exposition to a programmer.

Our decision to include brief material on the background for and derivation of a model helps the reader with physical knowledge and interest to more clearly see the link between our software-oriented numerical descriptions and comprehensive specialized text on the physics of a problem. Much of the literature on applications, and also on numerical analysis, works with formulations of equations that are not directly suited for numerical implementation. Our chapters on mechanical applications therefore emphasizes a combined physical, mathematical, and numerical framework that aids a flexible software implementation.

The present text has been used in a computationally-oriented PDE course at the University of Oslo. Experience from the course shows that whether the aim is to teach numerical methods or software issues, both subjects can benefit greatly from an integrated approach where theory, algorithms, programming, and experimentation are combined. The result is that people use less time to grasp the theory and much less time to produce running code than what we have experienced in the past.

Overview of the Text. Chapter 1 serves the purpose of getting the reader started with implementation of PDE simulators in Diffpack. The chapter also gives a brief introduction to mathematical modeling and numerical solution of PDEs. To keep the mathematics and numerics as simple as possible, we address the one-dimensional Poission equation and the one- and two-dimensional wave equation, discretized by finite differences. The material on finite difference algorithms is self-contained, but newcomers to the field of PDEs and finite differences may find it convenient to also consult a more comprehensive text like Strikwerda [111] or Tveito and Winther [120]. The next part of Chapter 1 motivates for the use of C++ and explains the basic concepts of object-oriented programming and numerical software engineering. The treatment of the programming techniques is quite brief and readers not fluent in C++ will need to access a textbook on C++ as they go through the more advanced parts of this book. Two recommended C++ books are Barton and Nackman[14] and Stroustrup [114]. Readers with minor experience with programming on beforehand will perhaps appreciate a more verbose C++ textbook, teaching both C++ and basic programming in general. Prata [91] is then a good choice.

An introduction to the finite element method is provided in Chapter 2. The exposition aims at giving the reader the proper view of the finite element method as a general computational algorithm for a wide range of problems. This view is important for the understanding and usage of the finite element toolbox in Diffpack. We start with the weighted residual method as a generic framework for solving stationary PDEs. Time-dependent equations are handled by first discretizing in time by finite differences and then applying the weighted residual method to the sequence of spatial problems. Various aspects of finite elements are introduced in detail for one-dimensional problems, but the formulation of the numerical tools is general so that the extension of the algorithms to multi-dimensional problems becomes trivial. We end Chapter 2 with a gentle treatment of the basic mathematics of variational formulations and present some of the main results regarding uniqueness, stability, best approximation principles, and error estimates. These topics build the basis for the final section on adaptive discretization of elliptic boundary-value problems and prepares the reader for accessing the large and important mathematically-oriented literature on the finite element method. Moreover, the abstract finite element framework forms the background for the treatment of the Conjugate Gradient-like methods in Appendix C.

Diffpack offers extensive support for solving PDEs by the finite element method, and Chapter 3 gives an introduction to Diffpack's finite element software tools. The description starts with a very simple program for the 2D

Poisson equation on the unit square. We then motivate strongly for making the solver more flexible. For instance, the solver should work in 1D, 2D, and 3D by parameterizing the number of space dimensions, arbitrary unstructured grids should be handled, input data should be given through a menu sytem, and the results should be visualized using various common tools, ranging from small public domain programs via Matlab to full visualization systems like IRIS Explorer. The increased flexibility enhances the user's productivity when performing extensive numerical experimentation and supports migration of the solver code towards much more complicated PDE problems. This is a philosphy that characterizes most programs associated with the book. Time-dependent PDE solvers are viewed as some minor extensions of stationary PDE solvers, and after having demonstrated how to solve a heat equation in 1D, 2D, and 3D, we end the chapter with a particularly efficient finite element solver for the standard multi-dimensional wave equation.

There is no separate chapter on tools for finite difference methods, but Chapter 1 and Appendices D.2 and D.7 provide fundamental examples on the implementation of explicit and implicit schemes for scalar PDEs, whereas Chapters 6.2 and 6.4 deal with systems of PDEs. These examples should be sufficient starting points for creating finite difference-based simulators with the aid of Diffpack.

Algorithms and software tools for nonlinear problems constitute the topic of Chapter 4. We describe the standard Successive Substitution method (Picard iteration) and the Newton-Raphson method for solving systems of nonlinear algebraic equations. We also apply these methods directly at the PDE level. Other topics covered are the group finite element method and continuation methods. Differences and similarities between the finite difference and finite element methods when applied to nonlinear problems are discussed. The exposition emphasizes that solution methods for nonlinear PDEs mainly involve solving a sequence of linear PDEs. As in the other parts of the book, the advocated view in the theory part is reflected in the usage of the software tools; the software example shows the few steps required to extend a simple heat equation solver from Chapter 3 to treat a nonlinear heat equation.

Simulation software for problems in fluid and solid mechanics are developed in Chapters 5–7 by combining and extending the numerical algorithms and software tools from Chapters 1–4.

Some elasticity and plasticity models are treated in Chapter 5. We first present the mathematical model, the finite element discretization, and the Diffpack implementation of a linear thermo-elasticity solver for 2D and 3D problems. Here we try to demonstrate the strong link between the vector PDE in elasticity and a scalar PDE like the Poisson equation, and how we can utilize this similarity to reuse software components from Chapter 3. Thereafter we discuss algorithms and implementations for an elasto-viscoplastic material model. In a certain sense, the plastic deformations can be computed as a sequence of elastic problems. This chapter also introduces a standard "engineering finite element notation" that simplifies the formulation of the elasto-viscoplastic solution algorithm and that prepares the reader for accessing the comprehensive engineering-oriented literature on the finite element method.

Chapter 6 is devoted to convection-diffusion problems, the shallow water equations, and the Navier-Stokes equations. The convection-diffusion PDE to be solved has principally the same complexity as the model PDEs in Chapters 3 and 4, but the implementation combines almost all Diffpack features introduced in previous chapters and thereby acts as a kind of summary and recommended composition of finite element software tools. The treatment of the shallow water equations is centered around finite difference methods on staggered grids and solution of systems of PDEs by operator splitting. Associated finite element techniques are also covered. A simple finite element solver for the incompressible Navier-Stokes equations, based on the penalty-function method, is thereafter described. This application is in some sense a nonlinear extension of the elasticity problem. Thus, there is a natural evolution of models and corresponding implementations from the Poisson or convection-diffusion equation via linear elasticity to a penalty-based Navier-Stokes solver. We also present a classical projection method for the incompressible Navier-Stokes equations along with its 3D finite difference implementation, including a multigrid solver for the pressure equation. The basic ideas of this fluid flow solver are extended to a finite element framework, with particular emphasis on efficient implementation, in the final section of Chapter 6.

Chapter 7 deals with coupled problems. A special fluid-structure interaction problem, so-called squeeze-film damping, is treated first. Thereafter we focus on non-Newtonian pipeflow coupled with heat transfer. In both applications we derive the mathematical model, present finite element-based numerical methods, and explain how one can implement a system of PDEs by assembling *independent solver classes for each scalar PDE* in the system. This is an important implementation technique that is supported by Diffpack and that significantly reduces the efforts required to develop and verify solvers for coupled systems of PDEs. The PDE systems in this chapter are highly nonlinear and extend the material from Chapter 4 with further examples on numerical methods and Diffpack programs for nonlinear PDEs.

The mathematical tools employed in this text do not require knowledge beyond multi-variable calculus and linear algebra. Throughout the book we make use of scaling of mathematical models, compact notations for PDEs and discrete equations, and tools for determining stability and accuracy of numerical approximations. These mathematical topics are covered in Appendix A. The treatment of accuracy and stability is centered around exact solutions of the difference equations and investigation of quantities like numerical dispersion relations. Discussion of dispersion relations helps the understanding of the nature of discretization errors and can be applied to both finite difference and finite element schemes in a simple and direct way.

Appendix B contains overview of the most important Diffpack functionality in tabular form, basic commands for operating Diffpack in the Unix and

Windows environments, and more detailed information about data storage and finite element programming. The final section in Appendix B demonstrates how to formulate algorithms and adjust the default "template" programs in Chapter 3 with the purpose of increasing the computational speed of Diffpack simulators.

Numerical solution of PDEs is very often centered around discretization techniques. That is, as soon as the inital PDE is reduced to a sequence of linear systems of algebraic equations, the problem is considered as "solved". This view is reflected in Diffpack simulators as well; calling up a linear system solver is trivial, and most of the programming work deals with input data, formulating the discrete equations, and setting the boundary conditions. Nevertheless, solution of linear systems can easily be the computational bottleneck and needs careful consideration. Diffpack offers access to several efficient iterative solution methods for linear systems, but proper use of these methods requires basic knowledge about the nature of the involved algorithms. This is the topic of Appendix C. We start with a brief introduction to classical iterative methods, like the Jacobi, Gauss-Seidel, SOR, and SSOR methods. Thereafter we outline the basic reasoning behind Conjugate Gradient-like methods, with strong emphasis on explaining the similarities between these methods for approximate solution of linear systems and the finite element method for approximate solution of PDEs. An overview of the most fundamental and generic preconditioning techniques is thereafter presented. The basic principles of multigrid and domain decomposition methods are also outlined.

Appendix D is devoted to software tools for preconditioned iterative solvers in Diffpack. These solvers can almost in a trivial way be combined with the simulators from the other parts of the book.

The text contains numerous *projects*, bringing together different aspects of computational science: physics/mechanics, mathematics, numerics, and software. Each project typically starts with defining the equations in a mathematical model. An optional step is to derive the model from basic continuum mechanics. The next step is to formulate a numerical method, and thereafter a Diffpack simulator is coded with the purpose of exploring the model. Throughout the text, particular emphasis is put on investigating the properties of models using experimental procedures. However, these experimental procedures must be assisted by a firm theoretical understanding of mathematical properties of similar, perhaps simplified, problems. Therefore most projects involve analysis of the mathematical and numerical model in simplified cases. The results from this analysis are also central for verifying the computer implementations. A fundamental property of all projects is that they can be carried out without working through the details of the physical derivation step.

What makes this book different from the many other texts on numerical solution of PDEs? First of all it is the interplay of models, generic algorithms, and software. The software part is built on modern concepts, such as object-oriented programming in C++, and the book is accompanied by the Diffpack tool and a large collection of ready-made simulators. The generic attitude to formulating algorithms and implementing them in reuseable software components explains why we can cover a broad application area, including heat transfer, fluid flow, and elasticity. If you do not find your application area in this book, it is likely that you will learn generic tools that can be applied to solve your problem at hand. Moreover, unified views on models, algorithms, and software tend – at least in this author's opinion – to increase the general understanding and intuition of mathematical modeling.

How to Use the Text. The text can be followed as is if the intention is to learn about numerical solution of PDEs and how to produce running software, with emphasis on modern implementational aspects. Readers having extensive previous experience with the theory and implementation of finite element methods, as well as C++ and object-oriented programming, might want to extract just the Diffpack specific parts of the book. A suitable start is Chapters 1.2.3, 1.3.3, 1.5.3, 1.5.5, 1.6.1, and 1.6.3, before studying the finite element software tools in 3 and 4.2. One can then continue with the application of interest. Some readers want to move to the application part as fast as possible, without first learning about numerous advanced Diffpack features. These readers can limit the study of Chapter 3 to sections 3.1, 3.2, and 3.9. Finally, readers who are mainly interested in numerical methods can use Chapters 1.2–1.3, 2, 4.1, and the theory parts of Chapters 5–7, combined with Appendix C for a course on numerical solution of PDEs with main emphasis on finite element methods. The software chapters can then act as optional add-on material for readers wanting to experiment with the methods.

Further Information. There is a web page associated with this book:

    http://www.diffpack.com/Book

In particular, a test version of the Diffpack software can be downloaded from this page. The page also encourages you to report errors and provide comments regarding the book and the software.

In parallel with writing the present text, the Diffpack software has undergone major restructuring and changes. Applications developed under Diffpack versions earlier than 3.0 will need to be updated. Fortunately, this update can be made (almost) automatic by the use of a certain script. The cited web page contains more information about the changes and the updating procedures.

Acknowledgements. Many people have contributed to the present book and the accompanying software. Appendix D is jointly written with Are Magnus Bruaset, who also developed the associated computer examples. All remarks regarding Diffpack under Windows NT/95, including Appendix B.2.2, are due to him. Chapter 2.10 is essentially a slight extension of a set of slides written by Aslak Tveito. Chapters 2.10.7 and 3.6, as well as the associated codes, were developed in collaboration with Klas Samuelsson. Harald Osnes revised the text and developed the associated solvers in Chapters 5.2 and 6.1, the program example in Chapter 6.2 was developed by Elizabeth Acklam, the software in Chapter 6.4 was made by Anders Jacobsen, and Otto Munthe developed the computer examples in Chapter 7.2. The illustration on the front cover originates from simulating the depolarization process of the electrical potential in the myocardium, using a simulator developed by Glenn Terje Lines and Vtk-based visualization tools made by Xing Cai.

The author is thankful for the many useful comments on the manuscript received from Elizabeth Acklam, Alfred Andersen, Are Magnus Bruaset, Erik Holm, Knut-Andreas Lie, Kent-Andre Mardal, Otto Munthe, Nigel Nunn, Atle Ommundsen, and Geir Pedersen.

During the nineties I have had the pleasure to experience a very creative and exciting collaboration with Are Magnus Bruaset and Aslak Tveito in the Diffpack project and related reseach and educational activites. The results from this work form the background and basis for the present book. All contributors to Diffpack are greatly acknowledged, and special thanks go to Xing Cai, Nigel Nunn, Klas Samuelsson, and Gerhard Zumbusch for their outstanding energy and creativity in the work with Diffpack. Nigel Nunn is particularly acknowledged for his extensive contributions to Diffpack version 3.0. Of importance to Diffpack is also the managing support from Morten Dæhlen and the financial support from the Research Council of Norway. Recently, the efforts and visions of Erlend Arge have given a new exciting dimension to the Diffpack development. Another significant collaborator to be mentioned is Geir Pedersen, who has shaped much of my view on numerics, modeling, and software development reflected in this book.

The encouragement and always prompt technical assistance from Martin Peters, Leonie Kunz, and Thanh-Ha LeThi at Springer-Verlag helped to bring the current book project to an end within reasonable time.

Finally, thanks to Liv, Mikkel, and Ingunn for all their support and for providing excellent working conditions during the writing of this book.

Oslo, December 1998   *Hans Petter Langtangen*

Table of Contents

List of Exercises

Chapter 1

Getting Started

This chapter introduces the reader to various basic aspects of numerical simulation, including derivation of partial differential equations, construction of finite difference schemes, implementation in Diffpack, and visualization of the simulation results. Particular attention is paid to fundamental concepts in Diffpack and C++ programming that will be needed for further reading of the book.

We start out with some trivial programs to make sure that C++ and Diffpack are correctly installed on your computer system. Thereafter we present the various aspects of the numerical simulation process illustrated for steady heat conduction, also with nonlinear effects, vibration of strings, and water surface waves. Although the finite element method will be a central numerical technique in most parts of the book, the examples in the present chapter apply finite differences on uniform grids to keep the numerics simple and enable focus on software issues. Topics like numerical stability and accuracy are illustrated through computational experiments, but readers wanting a theoretical treatment of these concepts can study Appendix A.4 in parallel with the present chapter.

The Diffpack-based simulation codes in Chapters 1.2 and 1.3 are made quite close to the ones you would create in plain Fortran 77 or C. However, throughout this text we advocate programming at a higher abstraction level than just manipulating plain arrays in the usual Fortran 77 and C style. Such high-level abstractions involve programming with *objects*. A brief introduction to programming with objects in the C++ language is provided by Chapter 1.5. Thereafter the examples from Chapters 1.2 and 1.3 are reprogrammed using high-level abstractions in Diffpack and a programming standard that we will make use of throughout the rest of the text.

All the source code files included or referred to in this book are located in subdirectories of $NOR/doc/Book, where $NOR is an environment variable that is available to any Diffpack user on Unix systems. *All code references are given relative to this path.* For example, when we state that the source code of an example is found in the directory src/fdm/Wave1D-func it actually means that the directory is

```
$NOR/doc/Book/src/fdm/Wave1D-func
```

Windows NT/95 Remark 1.1: Users on the Win32 platform will normally not make direct use of the NOR environment variable (if it is used, its value is obtained by the syntax %NOR%). Whenever $NOR occurs in the following discussions, simply replace it by the root directory for your Diffpack installation, e.g., C:\Program Files\WinDP\Src. Remember that Windows-specific utilities uses the backslash (\) rather than a slash (/) as delimiter in file paths.   ◇

Throughout this book, details dependending on your computer's operating system are primarily discussed in the context of Unix. However, the Diffpack software is also available for the Windows NT/95 platforms. To the extent that the discussion in the main text does not cover this platform, specific Win32 remarks are given. Further Win32-specific details can be found in Appendix B.2.2.

1.1 The First Diffpack Encounter

1.1.1 What is Diffpack?

Diffpack [34] is a sophisticated tool for developing numerical software, with main emphasis on numerical solution of partial differential equations. For a programmer, Diffpack acts as a numerical library consisting of C++ classes. We shall in this book try to demonstrate that programming with classes is easier and more flexible than shuffling data in and out of subroutines in the traditional Fortran or C way. The C++ classes can be combined into application codes to solve problems in diverse fields, including engineering, natural sciences, economics, and medicine. Since the application codes make extensive use of well-tested libraries and high-level abstractions, the time spent on writing and debugging code is significantly reduced compared with traditional software development in Fortran or C. This enables the computational scientist or engineer to concentrate more on modeling, algorithms, and numerical experimentation. Rapid prototyping, that is, spending only a couple of days on developing the first test version of a fairly general simulator for a system of partial differential equations, is one important application of Diffpack. Nevertheless, the same software development tools can be used to create specialized simulators for industrial use, where the development efforts span several man years.

Diffpack contains a large collection of useful abstractions, such as vectors, matrices, general multi-index arrays, strings, improved and simplified I/O, a menu system for getting data into programs, management of result files, couplings to visualization tools, representation of linear systems (particularly large sparse linear systems arising in finite element and finite difference methods), a large number of state-of-the-art iterative methods for sparse linear systems, solvers for nonlinear systems, finite difference and finite element grids, scalar and vector fields over grids, a collection of finite elements, various finite element algorithms and useful data structures, probability distributions, random number generators, solution methods for stochastic ordinary differential equations, support for random fields, adaptive grids, error estimation, multigrid methods, domain decomposition methods, generalized (mixed) finite element methods, parallel computing support, and numerous examples on simulators for problems like heat transfer, elasticity, and fluid flow[1]. Reference [23] gives an overview of Diffpack, but the package is being continuously further developed. Current projects at the time of this writing include a module for finite volume discretization, computational steering, and coupling of PDE solvers and optimization software.

Here is a list of some simulation codes that have been implemented using Diffpack.

- The Laplace, Poisson, Helmholtz, heat, and wave equations in general 1D/2D/3D geometries,
- structural analysis described by 2D/3D linear elasticity theory,
- compressible and incompressible 2D/3D Newtonian fluid flow,
- incompressible 2D non-Newtonian fluid flow between two plates, so-called Hele-Shaw flow for injection molding processes,
- Stefan problems in heat transfer,
- squeeze-film fluid-structure interaction,
- large plastic deformation in forming processes,
- optimal control and optimization problems in forming processes,
- electrical activity in the heart,
- deformation of tissue during surgery,
- shallow water waves and tsunami propagation, also with weakly dispersive and nonlinear effects,
- fully nonlinear 3D water waves,
- mild slope equations in the modeling of wave power plants,
- run-up of waves on beaches,
- solidification during aluminum casting,
- quasi 1D model for free surface flow of a non-Newtonian polymer,
- two- and three-phase flow in oil reservoirs,
- compositional flow in porous chemical reactors,
- poroelastic processes and earthquake analysis,
- heat and fluid flow in deformable rocks,
- continuous Markov processes, modeling e.g. random vibrations of simple structures,
- the 1D linear advection equation with random advection velocity,
- stochastic groundwater flow.

Recently, several of these applications have been adapted to parallel computers using an emerging method for reusing sequential Diffpack solvers in parallel computing environments [3,20,26].

Although usage of Diffpack requires programming in C++, only minor C++ knowledge is required to start using Diffpack, but experience with general programming is necessary.

---

[1] Not all of the functionality mentioned here is available in the public access version of Diffpack.

1.1.2   A Trivial C++ Program

Let us make a very simple C++ program that asks the user for a real number
r and then writes the value of $\sin r$ with some accompanying text. Make a
file `hello.cpp` with the following source code:

```
#include <iostream.h>    // make input/output functionality available
#include <math.h>        // make math functions available: e.g. sin(x)
int main ()              // function "main" is always the main program
{
  cout << "Hello world! Give a number: ";
  double r;  cin >> r;   // read number into double precision r
  double s = sin(r);     // declare s and initialize with sin(r)
  cout << "\nThe value of sin(" << r << ") is " << s << "\n";
  return 0;              // successful execution of the program
}
```

This is a pure C++ program – it does not make use of Diffpack. The C++
compilation procedure first calls the C preprocessor to process each line that
begins with a # character. For example, `#include <math.h>` means that the
preprocessor should copy the file `math.h`, found in some system directory,
e.g. `/usr/include` on Unix systems, into the program file `hello.cpp`. The
`#include` statements make various standard C++ functionality available to
the program: `math.h` gives access to mathematical functions, like the sine
function, and `iostream.h` enables input and output to the terminal screen.
Such files, ending with .h, are called *header files* in C and C++. It is not
necessary for the reader to locate and study the files `iostream.h` or `math.h`;
a C++ textbook [14,91,114] is the right source for learning about tools for
input/output in C++.

After the preprocessor has processed all lines starting with #, the compiler
starts compilation of the source code. This results in an object file `hello.o`.
Finally, the compiler calls a linker to link `hello.o` to the C++ libraries where
the code associated with `iostream.h` and `math.h` is located.

The variable `cout` represents standard output (the terminal screen) defined in `iostream.h`, whereas `cin` is standard input (the keyboard). The operators `<<` and `>>` direct texts and variables to `cout` or from `cin`. The symbol
`\n` signifies a newline character. A double precision variable in C and C++
has the name `double`, and C++ allows you to declare variables wherever you
want, provided that the variable is declared before it is used.

This `hello.cpp` program is compiled, linked, and run by the Unix commands

```
CC -c hello.cpp          # compile: make object file hello.o
CC -o app hello.o -lm    # link hello.o and the C/C++ math library
./app                    # run the program
```

Here, `CC` is the name of the C++ compiler. On some Unix systems, other
names are used for this compiler, e.g., `g++` or `xlC`.

We will refer to numerical computer programs as *applications*. This is
reflected in the name of the executable: `app`. The application is run by typing

app, or preferably[2] ./app. If the steps above work, we can proceed with your first Diffpack application.

Windows NT/95 Remark 1.2: Type the program into a file hello.cpp. Load one of the workspace definitions (files of type *.dsw) into Visual Studio and compile the program by use of the Build menu. Within the Visual Studio environment the resulting executable can be started from the same Build menu. Alternatively, the application can be run as any other Windows program simply by double-clicking the application icon. ◇

1.1.3 A Trivial Diffpack Program

We can write the previous hello.cpp program in the Diffpack environment, using some convenient Diffpack utilities that extend ordinary C++. This rewrite implies three modifications:

1. Diffpack applications must start with a call to the initDiffpack function,
2. cout and cin are replaced by the Diffpack equivalents s_o and s_i,
3. real variables in Diffpack are of type real[3],
4. Diffpack applications must be located in special directories.

However, before you can start programming with Diffpack, you need to customize your computer environment as explained in the paragraphs *Customizing Your Unix Environment* or *Customizing the Visual C++ Environment* in Appendix B.2.

To create a Diffpack application directory, type Mkdir hello, where hello is the name of the directory. The capital M in Mkdir is important; Mkdir is a Diffpack script (command) that runs the standard Unix command mkdir and sets up a correct makefile for compiling Diffpack applications (there is no need for the user to edit Makefile). Move to the subdirectory hello and prepare a file hello.cpp containing the following program:

```
#include <IsOs.h>   // Diffpack tools for input/output
#include <math.h>   // make math functions available: e.g. sin(x)
int main (int argc, const char* argv[])
{
  initDiffpack (argc, argv);   // should always be performed
  s_o << "Hello world! Give a number: ";
  real r;  s_i >> r;           // read real number into r
  real s = sin(r);
  s_o << "\nThe value of sin(" << r << ") is " << s << "\n";
  return 0;                    // successful execution of the program
}
```

---

[2] ./app means the app file in the current working directory (the dot . denotes the current directory in Unix). Depending on your $path variable, typing just app might start an app file in another directory.

[3] This allows transparent use of single or double precision.

The arguments `argc` and `argv` passed to `main` allow the user to give input data as a part of the execution command and will be used later. To compile and link this Diffpack application, type `Make` (capital `M` is again important). Instead of `Make`, you can type the name of GNU's make program on your computer.

Windows NT/95 Remark 1.3: Users on the Win32 platform will normally prefer to make new application directories by either copying the project folder for one of the supplied demo applications or by running the `AppWizard` in Visual C++. The steps for compilation was outlined in Remark 1.2 on page 5. Further information is available in Appendix B.2.2.   ◇

The `Make` command prints out the various steps of the compilation and linking phases. The result is an executable file with the name `app`. To run the program, just type `./app`. In the directory, you may notice that files like `SIMULATION.dp` and `SIMULATION.files` have been generated by `app`. You can ignore these files at the present stage.

For novice C++ programmers it can be confusing to detect whether a program statement refers to standard C++ or to Diffpack functionality. One solution to this problem is to learn the C++ language well; everything that is not recognized as pure C++ is then a part of Diffpack. Another solution is to prefix all Diffpack classes and functions with the namespace [114] identifier `dp::`. In the present book we have found the namespace prefix to fill up too much text space, and it is therefore dropped. We mention that all native C++ constructions can of course be used in Diffpack programs.

1.2   Steady 1D Heat Conduction

Our first simulation program deals with a simple differential equation, namely $-u''(x) = f(x)$ for $x \in (a, b)$. To ensure that the solution of this differential equation exists and is unique, we need to assign *boundary conditions* at $x = a$ and $x = b$. The conditions can be that either u or u' is known, but only one condition can be prescribed at each boundary point. That we actually need two boundary conditions becomes evident when solving the differential equation by integrating twice; for each integration we get an integration constant that must be determined by some additional conditions. The collection of a differential equation and boundary conditions make up a *boundary-value problem*. A one-dimensional (1D) boundary-value problem is often referred to as a *two-point boundary-value problem*.

We start by motivating our two-point boundary-value problem from an application in geology involving heat transfer in the continental crust. After having derived the differential equations and boundary conditions from physical principles, we outline the basic steps in a finite difference method for numerical solution of the problem. Thereafter, the numerical algorithm is implemented in a simulation program, using Diffpack tools.

1.2.1   The Physical and Mathematical Model

The differential equation $-u''(x) = f(x)$ arises in several physical contexts. Some examples are listed next.

- *Channel flow between two flat plates.* The function u is in this case the fluid velocity, $f(x) = \text{const}$ is the pressure gradient that "drives" the flow, and x is a coordinate normal to the planes $x = a, b$. Appropriate boundary conditions are that the fluid velocity coincides with the velocity of the plates, i.e.. $u(a) = u(b) = 0$ if the plates are at rest.
- *Deflection of a string under the load $f(x)$.* Here u is the displacement of a point on the string, and x is a coordinate running along the string. If the string is fixed at the ends, we have the boundary conditions $u(a) = u(b) = 0$.
- *Deflection of a beam under the load $f(x)$.* Now u is the bending moment (closely related to the stress) in a cross section of the beam, and x is a coordinate along the beam. For a simply supported beam we have the boundary conditions $u(a) = u(b) = 0$.
- *Heat conduction.* In this case u is the temperature, x is a coordinate along the direction of heat conduction, and $f(x)$ models heat generation, e.g., due to chemical reactions or radioactivity. The boundary conditions are of two types: The temperature u is known or u' (being proportional to the heat flux) is known.

We shall deal with the latter example in more detail to show how the differential equation arises from physical principles. Our application area is heat conduction in the continental crust as depicted in Figure 1.1 (from now on we choose $a = 0$). Heat is flowing through the crust with a velocity $q(x)$, referred to as the *heat flux*. The rocks that make up the mantle contain radioactive elements that release heat by an amount $s(x)$ per time and mass unit. We assume that the heat flow is steady in time. The first law of thermodynamics applied to this time-independent problem requires that the net flow of heat out of an arbitrary volume must be balanced by the total heat generation in that volume. This reflects that energy is conserved.

Considering the small box in Figure 1.1, with height h and widths w_y and w_z, the net outflow of heat is $(q(x + h/2) - q(x - h/2))w_y w_z$. The total heat generation is $\int_{\text{box}} s(x)dxdydz \approx s(x)hw_y w_z$. Putting these quantities equal to each other and dividing by the volume $hw_y w_z$ gives

$$\frac{q(x + \frac{h}{2}) - q(x - \frac{h}{2})}{h} = s(x). \tag{1.1}$$

Letting $h \to 0$, the left-hand side approaches $q'(x)$, which can easily be seen from a Taylor-series expansion of the two q terms around the point x,

$$\frac{q(x + \frac{h}{2}) - q(x - \frac{h}{2})}{h} = q'(x) + \frac{1}{24}q'''(x)h^2 + \cdots \tag{1.2}$$

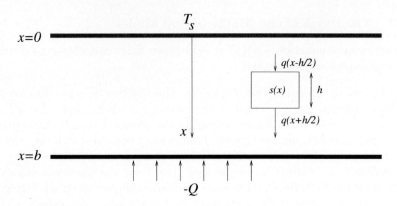

Fig. 1.1. Sketch of a simplified model for the heat conduction in the continental crust, with a surface temperature T_s and a heat flux $-Q$ from the mantle. The small box is used for deriving the governing differential equation.

In computer simulations, a finite h must be applied, and we then see that $q'(x)$ in a differential equation can be replaced by the finite difference on the left-hand side, resulting in an error whose leading term behaves as h^2 times the third derivative of u (which in practice means that halving h reduces the error in the approximation by a factor of four).

The boundary conditions are that q is known at the bottom: $q(b) = -Q$, while the temperature u equals T_s at the earth's surface: $u(0) = T_s$. The equation $q'(x) = s(x)$ is in this case incomplete for solving the problem; we need to relate q to u. This is normally done using *Fourier's law*, which states that heat flows from hot to cold regions: $q(x) = -\lambda u'(x)$. The quantity $\lambda > 0$ reflects the medium's ability to transport heat by conduction and varies with the rock type. Here we assume that the rock type, and thereby λ, vary with x only. Inserting $q = -\lambda u'$ in $q' = s$ gives the governing differential equation for u in this physical problem:

$$-\frac{d}{dx}\left(\lambda \frac{du}{dx}\right) = s(x), \quad u(0) = T_s, \quad -\lambda(b)u'(b) = -Q. \qquad (1.3)$$

Let us now for simplicity assume that λ is constant, although this is not a reasonable assumption since the continental crust is made up of many geological layers with different physical properties. Discretization methods that allow for a space-varying λ are dealt with in Chapter 1.2.6. Another assumption stems from the fact that heat generation due to the radioactive elements normally decreases exponentially with depth [119, Ch. 4] such that we can set $s(x) = R \exp(-x/L_R)$, where R is the generation at the earth's surface and L_R is the position where s is $1/e$ of its surface value.

The current model problem can easily be solved by straightforward integration, yielding a compact formula for the temperature u and its variation with x, λ, R, L_R, b, T_s, and Q. However, in most scientific and engineering problems the solution must be found using a computer. In the present case, we would end up investigating a function $u(x; \lambda, R, L_R, b, T_s, Q)$ through numerical simulations. Letting each of the six physical parameters λ, R, L_R, b, T_s, and Q vary among three values, results in a demand of $3^6 = 729$ computer experiments! *Scaling* constitutes a means for reducing the number of seemingly independent physical parameters in a problem. This consists in introducing new variables $\bar{x} = x/b$, $\bar{u} = \lambda(u - T_s)/(Qb)$, and $\bar{s}(\bar{x}) = s(b\bar{x})/R$, leading to

$$-\bar{u}''(\bar{x}) = \gamma e^{-\beta\bar{x}}, \quad \bar{u}(0) = 0, \ \bar{u}'(1) = 1,$$

where β and γ are dimensionless parameters: $\beta = b/L_R$ and $\gamma = bR/Q$. Observe that λ does not enter the scaled problem. At this stage it is common practice to drop the bars for notational convenience and just write

$$-u''(x) = \gamma e^{-\beta x}, \quad u(0) = 0, \ u'(1) = 1. \tag{1.4}$$

Appendix A.1 presents the details of scaling in general and this example in particular. To summarize, the model (1.3) with six physical parameters has been reduced to (1.4), containing only two dimensionless physical parameters.

The derivation of the governing differential equations was based on many simplifying assumptions. A more general model for heat conduction is listed in Project 2.6.1.

In the following we shall develop a Diffpack simulator for the slightly more general scaled boundary-value problem

$$-u''(x) = f(x), \quad 0 < x < 1, \tag{1.5}$$
$$u(0) = 0, \tag{1.6}$$
$$u'(1) = 1. \tag{1.7}$$

1.2.2 A Finite Difference Method

Application of the finite difference method to (1.5)–(1.7) consists of six basic steps:

1. The domain $(0, 1)$ is partitioned into $n-1$ cells $[x_i, x_{i+1}]$, $i = 1, \ldots, n-1$, with $x_1 = 0$ and $x_n = 1$. The points x_i will be referred to as *nodes* or *grid points*. The discrete representation of the domain is called a *mesh* or *grid*.
2. A numerical approximation u_i to the exact solution $u(x_i)$ is sought for $i = 1, \ldots, n$.

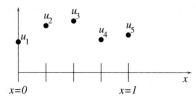

3. The derivatives in the differential equation are replaced by finite difference approximations. For example, we may set

$$u''(x_i) \approx \frac{u_{i+1} - 2u_i + u_{i-1}}{h^2},$$

where the cell length $h = x_{i+1} - x_i$ is assumed to be constant, $h = 1/(n-1)$. The error in this approximation to u'' is of order h^2.

4. The differential equation is required to be fulfilled at the nodes:

$$-u''(x_i) = f(x_i), \quad i = 1, \ldots, n.$$

5. The condition (1.6) is implemented by replacing the difference equation for $i = 1$ by $u_1 = 0$.

6. The condition (1.7) is implemented by using a centered finite difference approximation

$$\frac{u_{n+1} - u_{n-1}}{2h} = 1$$

in combination with the difference approximation to the differential equation for $i = n$ and eliminating the fictitious value u_{n+1}.

Carrying out steps 1-6 above with a constant cell length h results in the following algebraic equations:

$$u_1 = 0, \tag{1.8}$$
$$u_{i+1} - 2u_i + u_{i-1} = -h^2 f(x_i), \quad i = 2, \ldots, n-1, \tag{1.9}$$
$$2u_{n-1} - 2u_n = -2h - h^2 f(x_n). \tag{1.10}$$

Equations (1.8)–(1.10) constitute a coupled system of n linear algebraic equations for the n unknowns $u_1, u_2, \ldots, u_n$. One often refers to (1.8)–(1.10) as a *linear system*. We could of course eliminate (1.8) and obtain a linear system with $n - 1$ unknowns, but for compatibility with our treatment of finite element methods later, we keep u_1 as an unknown in the system.

Throughout this text, we shall refer to the underlying boundary-value problem, here (1.5)–(1.7), as the *continuous problem*. The discretized version, here (1.8)–(1.10), is then the corresponding *discrete problem*.

Example 1.1. The discrete equations can also be derived directly from the underlying physical principles. Considering the heat conduction problem from Chapter 1.2.1, energy conservation in a small box implies (1.1). This equation is to be combined with Fourier's law, $q = -\lambda u'$, which in discrete form reads

$$q(x + \frac{h}{2}) = -\lambda(x + \frac{h}{2})\frac{1}{h}\left(u((x + \frac{h}{2}) + \frac{h}{2}) - u((x + \frac{h}{2}) - \frac{h}{2})\right)$$
$$= -\lambda(x + \frac{h}{2})\frac{u(x+h) - u(x)}{h} = -\lambda_{i+\frac{1}{2}}\frac{u_{i+1} - u_i}{h}.$$

A similar formula can easily be derived for $q(x - h/2)$. Inserting these expressions in (1.1) yields

$$-\frac{1}{h}\left(\lambda_{i+\frac{1}{2}}\frac{u_{i+1} - u_i}{h} - \lambda_{i-\frac{1}{2}}\frac{u_i - u_{i-1}}{h}\right) = s_i\,. \tag{1.11}$$

When λ is constant, we recover (1.9). This example shows that a discrete model can be obtained directly from physics, without first taking the limit $h \to 0$ to obtain a differential equation and then discretizing this equation. The ideas here are fundamental to *finite volume methods*.   ◇

The discrete equations (1.8)–(1.10) can be written in matrix form $\boldsymbol{Au} = \boldsymbol{b}$, where $\boldsymbol{A}$ is an $n \times n$ matrix defined as

$$\boldsymbol{A} = \begin{pmatrix}
A_{1,1} & 0 & 0 & \cdots & \cdots & \cdots & \cdots & \cdots & 0 \\
A_{2,1} & A_{2,2} & A_{2,3} & \ddots & & & & & \vdots \\
0 & A_{3,2} & A_{3,3} & A_{3,4} & \ddots & & & & \vdots \\
\vdots & \ddots & & \ddots & \ddots & 0 & & & \vdots \\
\vdots & & \ddots & \ddots & \ddots & \ddots & \ddots & & \vdots \\
\vdots & & & 0 & A_{i,i-1} & A_{i,i} & A_{i,i+1} & \ddots & \vdots \\
\vdots & & & & \ddots & \ddots & \ddots & \ddots & 0 \\
\vdots & & & & & \ddots & \ddots & \ddots & A_{n-1,n} \\
0 & \cdots & \cdots & \cdots & \cdots & \cdots & 0 & A_{n,n-1} & A_{n,n}
\end{pmatrix}.$$

The entries $A_{i,j}$ are easily identified from the scheme (1.8)–(1.10):

$$A_{1,1} = 1, \tag{1.12}$$
$$A_{i,i-1} = 1, \quad i = 2,\dots,n-1, \tag{1.13}$$
$$A_{i,i} = -2, \quad i = 2,\dots,n, \tag{1.14}$$
$$A_{i,i+1} = 1, \quad i = 2,\dots,n-1, \tag{1.15}$$
$$A_{n,n-1} = 2\,. \tag{1.16}$$

The rest of the $A_{i,j}$ entries are zero. The unknown vector, i.e. the numerical solution, is $\boldsymbol{u} = (u_1, u_2, \dots, u_n)^T$. The right-hand side vector $\boldsymbol{b}$ is given by $\boldsymbol{b} = (b_1, b_2, \dots, b_n)^T$, where

$$b_1 = 0, \quad b_i = -h^2 f(x_i), \quad i = 2,\dots,n-1, \quad b_n = -2h - h^2 f(x_n)\,. \tag{1.17}$$

To solve a system $\boldsymbol{Au} = \boldsymbol{b}$, we can apply Gaussian elimination [32, Ch. 5.3.4]. The solution procedure consists first of an LU factorization of the coefficient matrix, that is, we compute factors $\boldsymbol{L}$ and $\boldsymbol{U}$ such that $\boldsymbol{A} = \boldsymbol{LU}$, where $\boldsymbol{L}$ is a lower triangular matrix and $\boldsymbol{U}$ is an upper triangular matrix.

The system $LUu = b$ can now be solved in two steps; first solve $Ly = b$ with respect to y (forward substitution) and then solve $Ux = y$ with respect to x (back substitution). The LU factorization and forward-back substitution is in general a slow method for solving linear systems arising from partial differential equations in more than one space dimension. More appropriate *iterative* solution methods are described in Appendix C. Nevertheless, in the present problem the coefficient matrix A is *tridiagonal*, and Gaussian elimination tailored to tridiagonal matrices constitutes an optimal solution method, see [40, Ch. 6.2.2] or [120, Ch. 2.2.3] for details regarding the algorithm. There is no particular need now for understanding the details of such algorithms, as we in the forthcoming programs will just call Diffpack library routines for carrying out the Gaussian elimination process.

1.2.3   Implementation in Diffpack

Constructing a Test Problem. Creating a simulator for (1.5)–(1.7) here means developing a computer program that generates and solves the discrete equations (1.8)–(1.10). The program will be implemented in Diffpack/C++, adopting a coding style close to plain Fortran 77 or C. Unfortunately, very few programmers are able to write an error-free program at first attempt. Syntax errors are found by the C++ compiler, but logical errors can only be removed by testing the code carefully. To this end, we need a suitable test problem. Let us choose $f(x) = -(\alpha+1)x^\alpha$, where $\alpha \in \mathbb{R}$ is a constant[4]. The analytical solution is found by integrating twice and applying the boundary conditions to determine the integration constants:

$$u(x) = \frac{1}{\alpha + 2}x^{\alpha+2}, \quad \alpha \neq -1, -2. \tag{1.18}$$

If we let $n = 2$, which means that we have only one cell, we can easily solve (1.8)–(1.10) by hand and find that

$$u_1 = 0, \quad u_2 = \frac{1-\alpha}{2},$$

with the error

$$u(1) - u_2 = \frac{1}{2}\alpha\frac{\alpha+1}{\alpha+2}.$$

We see that the solution becomes exact when $\alpha = 0$, i.e. for constant f, even when $h = 1$. It is easy to verify that the exact solution $u_i = \frac{1}{2}(x_i)^2$, $x_i = (i-1)h$, also fulfills the discrete equations (1.8)–(1.10) for $\alpha = 0$ and any $h > 0$. This result is very useful for verifying the simulation code.

---

[4] This f is not compatible with the requirements that $s' < 0$ in the heat conduction problem from Chapter 1.2.1, but the suggested f is convenient for debugging.

The Diffpack Code. We now present a simple Diffpack program for setting up the matrix system $Au = b$, according to the formulas in (1.12)–(1.17), solving the linear system, and writing the solution to file.

```
#include <Arrays_real.h> // for array functionality (and I/O)
#include <math.h>        // for the pow(.,.) function (x^alpha)

int main(int argc, const char* argv[])
{
  initDiffpack(argc, argv);
  s_o << "Give number of solution points: ";  // write to the screen
  int n;                     // declare an integer n (no of grid points)
  s_i >> n;                  // read n from s_i, i.e. the keyboard
  real h=1.0/(n-1);          // note: 1/(n-1) gives integer division (=0)
  Mat(real)      A(n,n);     // create an nxn matrix
  ArrayGen(real) b(n);       // create a vector of length n.
  ArrayGen(real) u(n);       // the grid point values
  s_o << "Give alpha: ";  real alpha;  s_i >> alpha;

  // --- Set up matrix A and vector b ---
  A.fill(0.0);               // set all entries in A equal to 0.0
  b.fill(0.0);               // set all entries in b equal to 0.0
  real x; int i;
  for (i = 1; i <= n; i++)                        // i++ means i=i+1
    {
      x = (i-1)*h;
      if (i == 1) {                               // does i equal 1?
        A(1,1) = 1;
        b(1) = 0;
      }
      else if (i > 1 && i < n) {                  // && means AND
        A(i,i-1) = 1;   A(i,i) = -2;   A(i,i+1) = 1;
        b(i) =    h*h*(alpha + 1)*pow(x,alpha);       // pow(a,b) is a^b
      }
      else if (i == n) {
        A(i,i-1) = 2;   A(i,i) = -2;
        b(i) = - 2*h + h*h*(alpha + 1)*pow(x,alpha);
      }
    }
  if (n <= 10) {
    A.print (s_o,"A matrix");        // print matrix to the screen
    b.print (s_o,"right-hand side"); // print vector to the screen
  }
  A.factLU();  A.forwBack(b,u);      // Gaussian elimination
  s_o << "\n\n x          numerical          error:\n";
  for (i = 1; i <= n; i++) {                      // \n is newline
    x = (i-1)*h;
    s_o << oform("%4.3f        %8.5f         %12.5e \n",
                 x,u(i),pow(x,alpha+2)/(alpha+2) - u(i));
  }
  // write results to the file "SIMULATION.res"
  Os file ("SIMULATION.res", NEWFILE);  // open file
  for (i = 1; i <= n; i++)
    file << (i-1)*h << "   " << u(i) << "\n";
  file->close();
}
```

If your Diffpack implementation applies true C++ templates, the notation Mat(real) and ArrayGen(real) should be replaced by the true template syntax Mat<real> and ArrayGen<real> in the source code. Throughout this text we shall stick to () rather than <> in parameterized types (templates).

The reader should first create a Diffpack directory and thereafter generate a file in this directory containing the statements above. The filename extension should be .cpp (which works fine on both Unix and Windows systems). The source code file can be obtained from the directory src/fdm/Heat1D-Mat, but we strongly encourage the reader to type the program by hand in an editor, because this will give valuable practical hands-on experience with the C++ syntax and probably some error messages from the compiler.

Compiling the Program. To compile this Diffpack application, one types Make as usual (see Chapter 1.1.3). However, on Unix systems we recommend compiling programs in the *emacs* editor, see page 541, as this brings you directly to the location in the source code where the compiler errors have occurred.

Windows NT/95 Remark 1.4: For instructions on how to compile a Diffpack program on Win32 platforms, see remark 1.2 on page 5 and Appendix B.2.2. Notice that by double-clicking the error messages displayed in the Build window the text editor in Visual Studio will load the file in question and put the cursor at the relevant line.   ◇

On Unix systems, Diffpack applications are compiled using makefiles [73, Ch. 7]. The name of the main makefile in an application directory is Makefile. This file should not be edited! Unless you are well experienced with advanced make utilities, you should not even look into this file. One can steer the compilation process by giving special options to Make. By default, Make turns on some internal time-consuming Diffpack safety checks to assist you in debugging the code. When the program is verified, you can turn off these safety checks and turn on the compiler's optimization features by giving the option MODE=opt to Make. By default, MODE equals nopt, which means no optimization. More information about useful Make commands are given in Chapter 3.5.4 and Appendix B.2.1.

Windows NT/95 Remark 1.5: Using Visual C++, each project can be compiled in at least two different modes, Release and Debug. These modes correspond to the opt and nopt modes on Unix. The resulting executables are placed in subdirectories named Release and Debug, respectively. To change between the different compilation modes, use the option Set Active Configuration on the Build menu.   ◇

Running the Program and Visualizing the Results. After having successfully compiled the program, there should be an executable file app that can be run. Try to give $n = 3$ and $\alpha = 0$ as input to validate that the numerical solution is exact even for such a coarse grid. Thereafter you can play with other values of n and α and observe the behavior of the numerical errors.

The `app` program generates a file `SIMULATION.res` containing two columns with the data pairs $(x, u(x))$. You can use almost any program for curve plotting to visualize the results. For example, here is a session with Gnuplot, which is a free program that works on Windows and Unix (type `gnuplot` to start the program):

```
gnuplot> plot 'SIMULATION.res' title 'u' with lines
gnuplot> exit
```

On Unix you can also use Plotmtv (`plotmtv SIMULATION.res`) or the Xmgr program (`xmgr SIMULATION.res`). Matlab is another alternative for visualization; just issue the command `curve` inside Matlab (this will run the Matlab script `curve.m` in the `Heat1D-Mat` directory). In later chapters (1.3.4, 1.5.5, 3.3, and 3.10.2) we describe several tools that make it easy to invoke various visualization programs with Diffpack-generated simulation data.

1.2.4 Dissection of the Program

Let us have a closer look at our program. The first line

```
#include <Arrays_real.h>
```

enables access to array classes in Diffpack[5]. This file includes several other files as well, such as `IsOs.h` for Diffpack's I/O functionality. Inclusion of `math.h` gives access to the `pow` function, which is needed in our code for evaluating expressions like x^α.

Flexible output formats are available through a Diffpack function `oform`, which combines numbers and text into a string. Numbers in the string are indicated by the special character `%` proceeded by formatting statements like those in C and Fortran. For example, `%4.3f` indicates a real number with 3 decimals written in a field of width 4. The variables to be written in the string are listed after the string. Obviously, the sequence of arguments is crucial. Other formatting commands include `%s` for a string (`char*`) and `%d` for an integer. Real numbers can also be written in "scientific notation", like 1.0453e-4, represented by the format `%12.4e`, which implies 4 decimals written in a field of width 12. Finally, `%g` denotes a real number written as compactly as possible. Actually, the `oform` function supports the same syntax as the standard `printf` function in C, so a textbook on the C programming language can give further information about the formatting strings in `oform`. Diffpack also has an `aform` function, which is identical to `oform`, except that it returns a `String` object rather than a `char*` pointer to a character array, making `aform` safer than `oform`. Class `String` is a Diffpack utility for representation and manipulation of strings in a program.

---

[5] A brief overview of some relevant array classes in Diffpack are given in Chapter 1.5.3.

At the end of the `main` function the reader should observe that there is no difference in syntax when writing to a file or to the screen, apart from the need to open and close files.

Exercise 1.1. Compilation in non-optimized mode turns on array index checking and thereby increases the CPU time. Choose $n = 500$ and compare the CPU time of the optimized and non-optimized versions of the program. (The CPU time used by a Diffpack simulator is written at the end of the file `SIMULATION.dp`.)   ◇

Exercise 1.2. Change the upper limit of a loop in the program to `i<=n+1`, compile the application in non-optimized mode, make a run, and observe the error message that is written when trying to access the invalid array entry `n+1`. What happens when you run the optimized version?   ◇

Exercise 1.3. Instead of writing the numerical error for every grid point, it is more convenient to report the error measure

$$e = \left(n^{-1} \sum_{i=1}^{n} (u(x_i) - u_i)^2 \right)^{\frac{1}{2}}, \tag{1.19}$$

where u_i and $u(x_i)$ denote the numerical and exact solution, respectively, at the point x_i. Modify the program accordingly.   ◇

1.2.5   Tridiagonal Matrices

Our program has one serious drawback. As seen from the discrete equations, the matrix A is tridiagonal, that is, each row has at most three nonzero entries. Matrices where most of the entries are zero are classified as *sparse* matrices. The efficiency of many numerical algorithms can be dramatically improved by utilizing the fact that the matrix is sparse. The current two-point boundary-value problem is a good example. Solving a system of n equations using Gaussian elimination (LU factorization) on a standard dense matrix, requires of the order n^3 arithmetic operations. However, if we know that the matrix is tridiagonal, the Gaussian elimination algorithm need only operate on the nonzero entries [40, Ch. 6.2.2] and the total work becomes of order n. In a problem with 1000 grid points this leads to an increase in the computational efficiency by a factor of order one million. Moreover, the memory requirement on the computer is reduced by the same factor.

The statement `Mat(real) A(n,n)` allocates memory for a full $n \times n$ matrix. To specify a tridiagonal matrix instead, simply change the declaration to `MatTri(real) A(n)`. The `MatTri(real)` matrix type is defined in `Arrays_real.h`. A `MatTri` object assumes indices of the form `A(i,k)` for `k=-1,0,1`, where `k=-1` corresponds to $A_{i,i-1}$, `k=0` refers to the main diagonal ($A_{i,i}$), and `k=1` corresponds to $A_{i,i+1}$. Besides tridiagonal matrices, Diffpack has many matrix formats that utilize the sparsity patterns usually encountered when solving partial differential equations numerically, see Appendix D.1.1.

Exercise 1.4. Replace the dense matrix `Mat` by a tridiagonal matrix `MatTri` in the test program. To this end, make a new Diffpack application directory, copy and edit the original test program[6], compile it in non-optimized mode and check that the numerical results remain unchanged. Determine the speed-up by using tridiagonal matrices instead of dense matrices when $n = 800$ and $n = 1600$. Use the optimized compilation mode for this test. (The CPU time of an execution appears in the file `SIMULATION.dp`.)   ◇

1.2.6   Variable Coefficients

A physically interesting extension of the model problem (1.20)–(1.22) is to allow for a variable coefficient $\lambda(x)$:

$$-\frac{d}{dx}\left(\lambda(x)\frac{du}{dx}\right) = f(x), \quad 0 < x < 1 \tag{1.20}$$

$$u(0) = 0, \tag{1.21}$$

$$u'(1) = 1. \tag{1.22}$$

Equation (1.20) naturally arose when modeling heat conduction in the continental crust in Chapter 1.2.1, because different geological layers normally have different heat conduction properties. A typical choice of $\lambda(x)$ is then a piecewise constant function, with possibly severe jumps at the interface between two geological layers. Also inside a layer λ may vary in space. Variable coefficients in differential equations appear in a variety of models for physical processes in *heterogeneous media*.

When discretizing $(\lambda u')'$ one must *avoid expanding the term* to $\lambda' u' + \lambda u''$. Instead, we discretize $(\lambda u')'$ in two steps. First, the "outer derivative" is discretized according to

$$\frac{d}{dx}\left(\lambda(x)\frac{du}{dx}\right)\bigg|_{x=x_i} \approx \frac{1}{h}\left(\lambda\frac{du}{dx}\bigg|_{x=x_{i+\frac{1}{2}}} - \lambda\frac{du}{dx}\bigg|_{x=x_{i-\frac{1}{2}}}\right).$$

Here, the points $x_{i\pm\frac{1}{2}}$ are $x = (i-1)h \pm h/2$, $i = 1,\ldots,n$, $h = 1/(n-1)$. The $\lambda u'$ term is discretized by a centered difference,

$$\lambda\frac{du}{dx}\bigg|_{x=x_{i+\frac{1}{2}}} \approx \lambda_{i+\frac{1}{2}}\frac{u_{i+1} - u_i}{h}.$$

Combining these basic ideas gives the discrete approximation to (1.20)–(1.22):

$$u_1 = 0, \tag{1.23}$$

$$\lambda_{i+\frac{1}{2}}(u_{i+1} - u_i) - \lambda_{i-\frac{1}{2}}(u_i - u_{i-1}) = -h^2 f(x_i), \tag{1.24}$$

$$i = 2,\ldots,n-1,$$

$$2\lambda_n(u_{n-1} - u_n) = -2h\lambda_{n+\frac{1}{2}} - h^2 f(x_n). \tag{1.25}$$

---

[6] You can find the answer to this exercise in `src/fdm/Heat1D-MatTri`.

In the latter equation, we have approximated $\lambda_{n-\frac{1}{2}} + \lambda_{n+\frac{1}{2}}$ by $2\lambda_n$. We still have the quantity $\lambda_{n+\frac{1}{2}}$, which needs a value of λ outside the domain[7].

Our particular discretization of $(\lambda u')'$, leading to (1.24), is strongly supported by Example 1.1 on page 10, where we actually derived (1.24) directly from physical principles.

A basic issue is the evaluation of $\lambda_{i+\frac{1}{2}}$. If $\lambda(x)$ is known as an explicit function, we can simply set $\lambda_{i+\frac{1}{2}} = \lambda(x_{i+\frac{1}{2}})$. However, in many cases λ is a discrete function or contains a discontinuity at $x_{i+\frac{1}{2}}$. It can then be necessary to express $\lambda_{i+\frac{1}{2}}$ in terms of λ_i and λ_{i+1}. The most obvious choice is the *arithmetic mean*

$$\lambda_{i+\frac{1}{2}} = \frac{1}{2}\left(\lambda_i + \lambda_{i+1}\right). \tag{1.26}$$

When λ exhibits severe jumps, the *harmonic mean* is often preferred [57, p. 227],

$$\frac{1}{\lambda_{i+\frac{1}{2}}} = \frac{1}{2}\left(\frac{1}{\lambda_i} + \frac{1}{\lambda_{i+1}}\right). \tag{1.27}$$

The *geometric mean* is also a possibility:

$$\lambda_{i+\frac{1}{2}} = \left(\lambda_i \lambda_{i+1}\right)^{1/2}. \tag{1.28}$$

The arithmetic, harmonic, and geometric means of more than two quantities follow from the obvious generalization of the preceding formulas.

Exercise 1.5. Extend the solver from Exercise 1.4 with variable coefficients, i.e., solve (1.20). The exact solution of (1.20) when $f = 0$ and $u(1) = 1$ is given by

$$u(x) = \left(\int_0^1 [\lambda(\tau)]^{-1} d\tau\right)^{-1} \int_0^x [\lambda(\tau)]^{-1} d\tau. \tag{1.29}$$

Test the program with a λ function for which you can evaluate the integrals analytically.   ◇

1.2.7   A Nonlinear Heat Conduction Problem

A Nonlinear Model. The mathematical model (1.20)–(1.22) was derived specifically in Chapter 1.2.1 for heat conduction in a geological medium. Of course, this model also applies to heat conduction in other media, for example, metals. Physical experiments show that the heat conduction properties of metals vary with the temperature. If the total temperature variation throughout the medium is significant, we may need to take the dependence of λ on u into account. The corresponding governing differential equation takes the form

$$-\frac{d}{dx}\left(\lambda(u)\frac{du}{dx}\right) = f(x), \quad 0 < x < 1. \tag{1.30}$$

---

[7] A specific example on evaluation of $\lambda_{n+\frac{1}{2}}$ appears later in Chapter 1.2.7.

Equation (1.30) contains products of the primary unknown u or its derivatives, which makes the equation *nonlinear* For example, if $\lambda(u) = 1 + u^2$ and we expand the factorized derivative, we end up with $u'' + u^2 u'' + 2uu'$ on the left-hand side. The terms $u^2 u''$ and $2uu'$ contain products of u or its derivatives and are hence nonlinear. We can of course apply the same discretization as in Chapter 1.2.6, but we end up with a nonlinear counterpart to (1.24); we cannot write the discrete equations on the form of a linear system, $\boldsymbol{Au} = \boldsymbol{b}$, because $\lambda_{i+\frac{1}{2}}$ and $\lambda_{i-\frac{1}{2}}$ in (1.24) now contain the unknown u itself. Instead, the discretization leads to a system of *nonlinear* algebraic equations. The next paragraph outlines a simple method for turning the nonlinear problem into a sequence of linear problems.

Numerical Algorithm. The most efficient numerical methods for dealing with nonlinear differential equations lead to somewhat technically complicated schemes, which are described in Chapter 4. However, we can devise a simple recipe for how to create a simulator that solves the nonlinear equation (1.30) by repeatedly solving linear problems of the same type as we covered in Chapter 1.2.6. The idea consists in starting with some guess of the solution, called u^0. Inserting u^0 in λ gives a differential equation $-(\lambda(u^0)u')' = f$, which is seen to be *linear* since $\lambda(u^0)$ is now a known function. This differential equation can be discretized by the techniques we used for (1.20) and leads to a linear system that we can solve with Gaussian elimination. Let us name the solution u^1. Of course, u^1 is not the correct solution of (1.30), because the coefficient $\lambda(u^0)$ was wrong, but hopefully u^1 is a better approximation than u^0 to the exact solution u. We can then repeat the process, that is, insert u^1 in λ and solve a linear differential equation for a new approximation u^2. The ideas can be summarized in the following iteration scheme:

$$-\frac{d}{dx}\left(\lambda(u^{k-1})\frac{du}{dx}^k\right) = f(x), \quad u^k(0) = 0, \quad \frac{du}{dx}^k(1) = 1, \tag{1.31}$$

for $k = 1, 2, 3, \ldots$ This simple iteration technique is called *Successive Substitutions* in Chapter 4.

The associated finite difference scheme is similar to (1.24), but $\lambda_{i+\frac{1}{2}}$ must now be expressed in terms of u_i and u_{i+1}, e.g.,

$$\lambda_{i+\frac{1}{2}} = \frac{1}{2}\left(\lambda(u_i) + \lambda(u_{i+1})\right).$$

This results in the following discrete version of (1.31):

$$\frac{1}{2}\left(\lambda(u_i^{k-1}) + \lambda(u_{i+1}^{k-1})\right)(u_{i+1}^k - u_i^k) - $$
$$\frac{1}{2}\left(\lambda(u_{i-1}^{k-1}) + \lambda(u_i^{k-1})\right)(u_i^k - u_{i-1}^k) = -h^2 f(x_i). \tag{1.32}$$

For the last point $i = n$, we get as in (1.25):

$$2\lambda(u_n^{k-1})(u_{n-1}^k - u_n^k) = -2h\lambda_{n+\frac{1}{2}} - h^2 f(x_n). \tag{1.33}$$

The quantity $\lambda_{n+\frac{1}{2}}$ can be approximated as usual by an average, $(\lambda(u_n^{k-1}) + \lambda(u_{n+1}^{k-1}))/2$, but this involves a quantity u_{n+1}^{k-1} outside the mesh. From the boundary condition at $x = 1$,

$$\frac{u_{n+1}^{k-1} - u_{n-1}^{k-1}}{2h} = 1, \quad k > 1,$$

we can approximate u_{n+1}^{k-1} by $u_{n-1}^{k-1} + 2h$, resulting in

$$\lambda_{n+\frac{1}{2}} = \frac{1}{2}\left(\lambda(u_n^{k-1}) + \lambda(u_{n-1}^{k-1} + 2h)\right).$$

To terminate the iteration, we need a stopping criterion, for example,

$$\sqrt{\sum_{j=1}^{n} |u_j^k - u_j^{k-1}|^2} \leq \epsilon,$$

where ϵ is a prescribed tolerance. Since the iteration procedure is not guaranteed to converge at all, we should also terminate the iteration after $k_{\max}$ iterations.

Implementation. The implementation can make use of the program from Chapter 1.2.3, or preferably the more efficient version from Chapter 1.2.5, with the following modifications:

1. We need to work with two arrays, `uk` and `ukm`, representing u_i^{k-1} and u_i^k, respectively, for $i = 1, \ldots, n$.
2. The array `ukm` must be initialized by the guess u^0.
3. Several additional variables are needed: an iteration counter, the maximum allowed number of iterations, help variables for storing evaluated values of λ, variables in the termination criterion etc.
4. The evaluation of λ is conveniently done in a separate function. For example, if $\lambda(u) = u^m$, where m is a user-given real number, $m \geq 0$, this function can take the form `real lambda(real u,real m) {return pow(u,m);}` in C++.
5. The formulas for the matrix entries and the right-hand side must be updated according to the present scheme.
6. The Gaussian elimination process is performed by calling `A.factLU()` and `A.forwBack(b,uk)`. However, the matrix object will not allow us to repeatedly perform the LU factorization unless we explicitly tell the matrix that the entries have been reinitialized and that the old factorization is overwritten. This is done by the call `A.resetFact()`, which must appear prior to `A.factLU()`.
7. The difference between the u_i^k and u_i^{k-1} values must be computed and reported.

8. We must set $u_i^{k-1} = u_i^k$, $i = 1, \ldots, n$, such that we are ready for a new iteration.

9. A while loop must be wrapped around the linear system generation and the actions 6, 7, and 8.

10. The error measure e in (1.19) can be reported, if an analytical solution to the problem is available.

Before starting the implementation, we should have a test example with exact solution. It appears that if $\lambda(u) = u^m$ and $f(x) = 0$, we can integrate (1.30) and apply $u(0) = 0$ and $u'(1) = 1$ to find that $u(x) = (m+1)x^{1/(m+1)}$.

The core part of the solver might look as follows.

```
int k = 0;                        // iteration counter
const int k_max = 200;            // max no of iterations
real lambda1, lambda2, lambda3;   // help variables
real udiff = INFINITY;            // udiff = ||uk - ukm||
const real epsilon = 0.0000001;   // tolerance in termination crit.

while (udiff > epsilon && k <= k_max)
  {
    k++;                          // increase iteration counter by 1
    A.fill(0.0); b.fill(0.0);     // initialize A and b

    for (i = 1; i <= n; i++)
      {
        if (i == 1) {
          A(1,0) = 1;
        }
        else if (i > 1 && i < n) {
          lambda1 = lambda(ukm(i-1), m);
          lambda2 = lambda(ukm(i), m);
          lambda3 = lambda(ukm(i+1), m);

          A(i,-1) =  0.5*(lambda1 + lambda2);
          A(i, 0) = -0.5*(lambda1 + 2*lambda2 + lambda3);
          A(i, 1) =  0.5*(lambda2 + lambda3);
        }
        else if (i == n) {
          A(i,-1) = 2*lambda(ukm(i), m);
          A(i, 0) = - A(i,-1);
          b(i) = -h*(lambda(ukm(i-1)+2*h,m)+lambda(ukm(i),m));
        }
      }
    A.resetFact();                     // ready for new factLU
    A.factLU();   A.forwBack(b,uk);    // Gaussian elimination

    // check termination criterion:
    udiff = 0;
    for (i = 1; i <= n; i++)
      udiff += sqr(uk(i) - ukm(i));
    udiff = sqrt(udiff);
    s_o << "iteration " << k << ":  udiff = " << udiff << "\n";

    ukm = uk;  // ready for next iteration
  }
```

The complete code is found in `src/fdm/Heat1Dn-MatTri`. Run this code with $m = 0$ and observe that the exact solution is obtained after the first iteration (`udiff` vanishes in the second iteration). The solution can be plotted as explained on page 14. Let us try $m = 5$ and $n = 10, 100, 1000$. The error is reduced when n is increased from 10 to 100, as expected, but with $n = 1000$ the program encounters fundamental problems with the solution of the linear systems. Lowering m to 4.2 results in a successful execution. With $m = 8$ and $n = 10$ the program terminates normally, but the nonlinear iteration process diverges, i.e., `udiff` becomes constant and the maximum number of iterations (200) is reached. Reducing m to 7 helps the nonlinear iteration process to converge, although at a very slow rate. From these examples we see that numerical solution of nonlinear differential equations can be a difficult task. We shall return to this example in Chapter 1.6.1 and perform more comprehensive experimentation and see how the numerical error depends on m, n, and alternative choices of the λ function. For that purpose, we need a more flexible program than what has been shown here.

Exercise 1.6. Assume that λ is a computationally expensive function to evaluate. Explain how one can reduce the number of λ evaluations in the code.
◇

Model Extensions. In the derivation of the heat conduction model in Chapters 1.2.1 we neglected two- and three-dimensional as well as time-dependent effects. Taking such effects into account results in a partial differential equation on the form

$$\beta \frac{\partial u}{\partial t} = \frac{\partial}{\partial x}\left(\lambda(u)\frac{\partial u}{\partial x}\right) + \frac{\partial}{\partial y}\left(\lambda(u)\frac{\partial u}{\partial y}\right) + \frac{\partial}{\partial z}\left(\lambda(u)\frac{\partial u}{\partial z}\right) + f(x, y, z). \quad (1.34)$$

This equation can be used to compute the temperature u in arbitrary three-dimensional solids. Later in the text we shall describe numerical methods and develop simulation software for such problems. Two-dimensional finite difference schemes will be explained already in Chapters 1.3.5 and 1.3.6, but we restrict the treatment of higher-dimensional finite difference methods to *explicit* difference schemes, where new values at a grid point are computed by an explicit analytical formula and not by solving a linear system like we encountered in the present 1D heat conduction model. Schemes which involve solution of coupled systems of algebraic equations are classified as implicit[8]. In 2D and 3D, implicit schemes lead to large algebraic systems of equations, and the solution of such systems frequently constitutes the computational bottleneck in simulation software. Fortunately, sophisticated methods tailored to the properties of linear systems arising from finite difference and

---

[8] This terminology is subject to debate; in Chapter 2.2.1 and (5.2) we apply explicit finite difference schemes in time, but the spatial discretization leads to systems of algebraic equations, i.e., the method is implicit according to our definition. However, many practitioners will consider such methods as explicit.

finite element methods can speed up the solution process, often in a dramatic way if we compare the efficiency with that of Gaussian elimination. The description of such specialized algorithms for linear systems is provided in Appendix C, whereas the implementational aspects of 2D and 3D implicit finite difference schemes are covered in Appendix D.

1.3 Simulation of Waves

The next introductory simulation example concerns vibrations of a string. This physical phenomenon can be modeled in terms of the partial differential equation (PDE)

$$\frac{\partial^2 u}{\partial t^2} = \gamma^2 \frac{\partial^2 u}{\partial x^2}, \quad x \in (a, b),$$

commonly known as the *wave equation*. Now the unknown function u depends on both x and t. The boundary conditions are similar to those in Chapter 1.2, i.e., u or $\partial u/\partial x$ must be prescribed at $x = a$ and $x = b$, with exactly one condition at each point. In addition, we need *initial conditions* for $u(x, 0)$ and $\partial u(x, 0)/\partial t$, $x \in [a, b]$. The collection of a partial differential equation with initial and boundary conditions is referred to as an *initial-boundary value problem*.

1.3.1 Modeling Vibrations of a String

While the first law of thermodynamics (energy balance) constituted the starting point for the heat conduction model in Chapter 1.2.1, the model for vibrating strings arises from Newton's second law. Consider a small part of the string as depicted in Figure 1.2. Newton's second law states that the total sum of forces on this string element balances the mass of the element times its acceleration. There is a tension force $\boldsymbol{T}(x, t)$ directed along the string (notice that $\boldsymbol{T}$ is a vector). Neglecting gravity and air resistance, which are usually much smaller forces than the string tension, the total sum of forces on the element becomes $\boldsymbol{T}(x + h/2) - \boldsymbol{T}(x - h/2)$. The mass per unit length is ϱ, and the length is Δs, leading to $\varrho \Delta s$ as the total mass of the string element. We assume that the element moves up and down in y direction only[9]. The position $\boldsymbol{r}$ of the element is given by $\boldsymbol{r} = x\mathbf{i} + u(x, t)\mathbf{j}$, where $\mathbf{i}$ and $\mathbf{j}$ are unit vectors along the x and y axis, respectively. The velocity $\boldsymbol{v}$ and the acceleration $\boldsymbol{a}$ of the element are then given by

$$\boldsymbol{v} = \frac{\partial \boldsymbol{r}}{\partial t} = \frac{\partial u}{\partial t}\mathbf{j}, \quad \boldsymbol{a} = \frac{\partial^2 \boldsymbol{r}}{\partial t^2} = \frac{\partial^2 u}{\partial t^2}\mathbf{j}.$$

---

[9] This is only an approximation, but the displacement in the x direction is very small, and the corresponding error will not dominate over the errors due to neglecting gravity or the approximations to come for small displacements in the y direction.

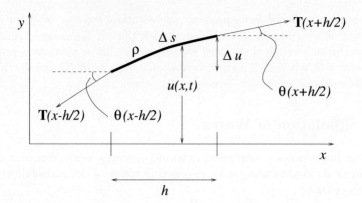

Fig. 1.2. Sketch of a part of a string, with length Δs (projected on to h and Δu in the x and y directions), tension $\boldsymbol{T}$, displacement $u(x,t)$, and density (mass per unit length) ϱ. The angle between the string and x axis is denoted by $\theta(x,t)$.

Newton's second law applied to the string element now takes the form

$$\boldsymbol{T}(x + \frac{h}{2}) - \boldsymbol{T}(x - \frac{h}{2}) = \varrho(x)\Delta s\frac{\partial^2}{\partial t^2}u(x,t)\mathbf{j} . \qquad (1.35)$$

This is a vector equation with two scalar component equations. The next step is to express $\boldsymbol{T}$ in terms of the unit vectors $\mathbf{i}$ and $\mathbf{j}$. From Figure 1.2 we see that

$$\boldsymbol{T}(x) = T(x)\cos\theta(x)\,\mathbf{i} + T(x)\sin\theta(x)\,\mathbf{j},$$

where T is the magnitude of $\boldsymbol{T}$. Each of the vector components in (1.35) must balance, leading to

$$T(x + \frac{h}{2})\cos\theta(x + \frac{h}{2}) - T(x - \frac{h}{2})\cos\theta(x - \frac{h}{2}) = 0, \qquad (1.36)$$

$$T(x + \frac{h}{2})\sin\theta(x + \frac{h}{2}) - T(x - \frac{h}{2})\sin\theta(x - \frac{h}{2}) = \varrho(x)\Delta s\frac{\partial^2 u}{\partial t^2} . \qquad (1.37)$$

Dividing (1.36) by h and taking the limit $h \to 0$, i.e.,

$$\lim_{h \to 0}\frac{1}{h}\left(T(x + \frac{h}{2})\cos\theta(x + \frac{h}{2}) - T(x - \frac{h}{2})\cos\theta(x - \frac{h}{2})\right) = 0,$$

leads to

$$\frac{\partial}{\partial x}\left(T\cos\theta\right) = 0 \qquad (1.38)$$

by a Taylor-series argument like we used in (1.2). The component (1.37), after division by h, has the limit equation

$$\frac{\partial}{\partial x}\left(T\sin\theta\right) = \varrho(x)\left(\lim_{h \to 0}\frac{\Delta s}{h}\right)\frac{\partial^2 u}{\partial t^2} . \qquad (1.39)$$

To estimate $\lim_{h \to 0} \Delta s/h$, we have from Figure 1.2 that

$$\Delta s^2 = h^2 + \Delta u^2 \quad \Rightarrow \quad \lim_{h \to 0} \frac{\Delta s}{h} = \sqrt{1 + \left(\frac{\partial u}{\partial x} \right)^2}. \tag{1.40}$$

Moreover,

$$\tan \theta = \frac{\partial u}{\partial x}, \quad \sin \theta = \frac{\tan \theta}{\sqrt{1 + \tan^2 \theta}} = \frac{\frac{\partial u}{\partial x}}{\sqrt{1 + \left(\frac{\partial u}{\partial x} \right)^2}}. \tag{1.41}$$

Inserting (1.40) and (1.41) in (1.39) leads to the following governing PDE for the string motion:

$$\varrho \left[1 + \left(\frac{\partial u}{\partial x} \right)^2 \right]^{\frac{1}{2}} \frac{\partial^2 u}{\partial t^2} = \frac{\partial}{\partial x} \left(T \left[1 + \left(\frac{\partial u}{\partial x} \right)^2 \right]^{-\frac{1}{2}} \frac{\partial u}{\partial x} \right). \tag{1.42}$$

This is a *nonlinear* PDE because the primary unknown u enters in products with itself or its derivatives[10].

Vibrations of a string are normally recognized as being small. Therefore we expect that $(\partial u / \partial x)^2$ is small compared with unity. This suggests that the bracket terms (square roots) can be approximated by 1. If we further assume that the tension T is constant[11] we obtain the *linear* wave equation for u:

$$\frac{\partial u^2}{\partial t^2} = c^2 \frac{\partial^2 u}{\partial x^2}, \tag{1.43}$$

where $c = \sqrt{T/\varrho}$, known as the phase velocity, reflects the speed of disturbances along the string. The boundary conditions follow from the fact that the string is fixed at the ends: $u(a,t) = u(b,t) = 0$. Initially, we may assume that the string is at rest, $\partial u / \partial t = 0$, with a prescribed shape $u(x,0) = I(x)$.

As in Chapter 1.2.1, it is advantageous to scale the model. The details of such a scaling are presented on page 495. The resulting scaled problem that we aim to solve numerically is listed next.

$$\frac{\partial^2 u}{\partial t^2} = \gamma^2 \frac{\partial^2 u}{\partial x^2}, \quad x \in (0,1), \quad t > 0, \tag{1.44}$$

$$u(x,0) = I(x), \quad x \in (0,1), \tag{1.45}$$

$$\frac{\partial}{\partial t} u(x,0) = 0, \quad x \in (0,1), \tag{1.46}$$

$$u(0,t) = 0, \quad t > 0, \tag{1.47}$$

$$u(1,t) = 0, \quad t > 0. \tag{1.48}$$

The γ parameter is dimensionless and equals unity, but we keep it in the PDE for labeling the spatial derivative term.

---

[10] This is perhaps more obvious if we expand the bracket terms in Taylor series.
[11] This is evident from (1.38), since $\cos \theta \approx 1$ for small displacements.

1.3.2 A Finite Difference Method

Finite difference discretization of the problem (1.44)–(1.48) starts with introducing a grid in space, $0 = x_1 < x_2 < \cdots < x_n = 1$, and in time, $0 = t_0 < t_1 < t_2 \cdots$. For simplicity, we assume constant grid spacings h and Δt such that $x_i = (i-1)h$ and $t_\ell = \ell \Delta t$. The governing equation (1.44) is to be satisfied at the discrete points (x_i, t_ℓ), $i = 1, \ldots, n$, $\ell = 0, 1, 2, \ldots$ At these points, we replace the derivatives by finite differences:

$$\frac{\partial^2}{\partial x^2} u(x_i, t_\ell) \approx \frac{u_{i-1}^\ell - 2u_i^\ell + u_{i+1}^\ell}{h^2}, \tag{1.49}$$

$$\frac{\partial^2}{\partial t^2} u(x_i, t_\ell) \approx \frac{u_i^{\ell-1} - 2u_i^\ell + u_i^{\ell+1}}{\Delta t^2}. \tag{1.50}$$

The errors in the approximations (1.49)–(1.50) are of order h^2 and Δt^2, respectively. The resulting finite difference equation can be solved with respect to $u_i^{\ell+1}$:

$$u_i^{\ell+1} = 2u_i^\ell - u_i^{\ell-1} + \gamma^2 \frac{\Delta t^2}{h^2} \left(u_{i-1}^\ell - 2u_i^\ell + u_{i+1}^\ell \right). \tag{1.51}$$

Assuming that all values at time levels $\ell - 1$ and ℓ are known, equation (1.51) yields an explicit updating formula for the new values $u_i^{\ell+1}$, $i = 2, \ldots, n-1$. There is no need to solve a coupled system of algebraic equations, like we had to in Chapter 1.2, and (1.51) is therefore referred to as an *explicit* finite difference scheme.

The boundary conditions (1.47)–(1.48) take the discrete form $u_1^\ell = u_n^\ell = 0$ and are used directly in (1.51) when $i = 2$ and $i = n - 1$. The initial condition (1.45) yields $u_i^0 = I(x_i)$, $i = 1, \ldots, n$. The other initial condition, given by equation (1.46), needs some more consideration. A finite difference approximation to $\partial u / \partial t$ at $t = 0$ gives

$$\frac{u_i^1 - u_i^{-1}}{2\Delta t} = 0 \quad \Rightarrow \quad u_i^{-1} = u_i^1. \tag{1.52}$$

Combining (1.52) with (1.51) for $\ell = 0$, results in a special formula for the first time step:

$$u_i^1 = u_i^0 + \gamma^2 \frac{\Delta t^2}{2h^2} \left(u_{i-1}^0 - 2u_i^0 + u_{i+1}^0 \right). \tag{1.53}$$

Apparently, we need two schemes in the program, one for the first time level and one for the other levels. However, we can in fact apply the general scheme (1.51) also for $\ell = 0$ to compute u_i^1, provided that u_i^{-1} has the value

$$u_i^{-1} = u_i^0 + \frac{1}{2}C^2(u_{i+1}^0 - 2u_i^0 + u_{i-1}^0). \tag{1.54}$$

We have here introduced the so-called Courant number: $C = \gamma \Delta t / h$. We shall use this latter approach in the simulation program. You can verify that inserting (1.54) in (1.51) yields (1.53).

After having derived the finite difference approximations to the PDE and the initial and boundary conditions, it is a good habit to summarize the complete scheme in algorithmic form:

Algorithm 1.1.

Explicit scheme for the 1D wave equation.

define u_i^+ , u_i and u_i^- to represent $u_i^{\ell+1}$, u_i^ℓ and $u_i^{\ell-1}$, respectively
SET THE INITIAL CONDITIONS:
$u_i = I(x_i)$,   for $i = 1, \ldots, n$
DEFINE THE VALUE OF THE ARTIFICIAL QUANTITY u_i^- :
$u_i^- = u_i + \frac{1}{2}C^2(u_{i+1} - 2u_i + u_{i-1})$ for $i = 2, \ldots, n - 1$
$t = 0$
while time $t \leq t_{\text{stop}}$
    $t \leftarrow t + \Delta t$
    UPDATE ALL INNER POINTS:
    $u_i^+ = 2u_i - u_i^- + C^2(u_{i+1} - 2u_i + u_{i-1})$ for $i = 2, \ldots, n - 1$
    INSERT BOUNDARY CONDITIONS:
    $u_1^+ = 0$, $u_n^+ = 0$
    INITIALIZE FOR NEXT STEP:
    $u_i^- = u_i$, $u_i = u_i^+$,   for $i = 1, \ldots, n$
    plot the solution $(u_i,\ i = 1, \ldots, n)$

1.3.3   Implementation

The finite difference scheme for the one-dimensional wave equation can easily be coded using basic array operations in almost any programming language. A particular advantage of using Diffpack for the present case is its handling of a large number of curve plots. If you want to make an animation of the string motion, you need perhaps several hundred plots. Each plot is typically a curve consisting of x and y points stored on files. In Diffpack there are tools for administering large amounts of curve plots and combining them into movies.

A simulation program for the vibrating string will now be presented in its complete form. The electronic version of the code can be found in `src/fdm/Wave1D-func/main.cpp`. As an extension of the programming style from Chapter 1.2, we now make use of *functions*. Functions in C++ (and C) that do not return any value are declared with `void` as "return" value in the function heading. Such functions hence correspond to subroutines in Fortran. Function arguments are listed with the type and the variable name, for example,

```
void setIC (real C, ArrayGen(real)& u0, ArrayGen(real)& um);
```

The argument `real C` leads to a local *copy* of the `C` value in the routine. Changing `C` inside `setIC` has no effect in the calling code[12]. The ampersand `&` means that the arrays `u0` and `um` are transferred to the `setIC` function by their addresses[13] only. This allows the function to alter the contents of `u0` and `um`, which is the purpose of the function. Moreover, transferring only the address is always important for efficiency if the vector is large, as we then avoid internal memory allocation and copying of the vector in the routine.

With this short introduction to functions in C++ and the previous program example for $-u'' = f$ fresh in mind, the source code below should be possible to understand. The fundamental quantities u^+, u, and u^- in the algorithm are represented by the arrays `up`, `u`, and `um` in the program.

```
#include <Arrays_real.h>
#include <CurvePlot.h>

// forward declarations:
// (we need to define function names and arguments before the
// functions can be called)
void timeLoop    (ArrayGen(real)& up, ArrayGen(real)& u,
                  ArrayGen(real)& um, real tstop, real C);
void setIC       (real C, ArrayGen(real)& u0, ArrayGen(real)& um);
void plotSolution (ArrayGen(real)& u, CurvePlotFile& plotfile,
                  real t, real C);

int main (int argc, const char* argv[])
{
  initDiffpack (argc, argv);
  s_o << "Give number of intervals in (0,1): ";
  int i; s_i >> i;  int n = i+1; // number of points;
  ArrayGen(real) up (n);  // u at time level l+1
  ArrayGen(real) u  (n);  // u at time level l
  ArrayGen(real) um (n);  // u at time level l-1
  s_o << "Give Courant number: ";  real C; s_i >> C;
  s_o << "Compute u(x,t) for t <= tstop, where tstop = ";
  real tstop; s_i >> tstop;

  timeLoop (up, u, um, tstop, C);   // finite difference scheme
  return 0;
}

void timeLoop (ArrayGen(real)& up, ArrayGen(real)& u,
               ArrayGen(real)& um, real tstop, real C)
{
  int  n = u.size();   // length of the vector u (no of grid points)
  real h = 1.0/(n-1);  // length of grid intervals
  real dt = C*h;       // time step, assumes unit wave velocity!!
  real t = 0;          // time
  CurvePlotFile plotfile(casename); // "databank" (file) with all plots

  setIC (C, u, um);                 // set initial conditions
```

---

[12] This is different from Fortran, where changes in any argument inside a subroutine are visible outside the routine.
[13] In C++ terminology, the term *references* is used rather than addresses.

```
    plotSolution (u, plotfile, t, C); // plot the initial displacement
    int  i;                          // loop counter over grid points
    int  step_no = 0;                // current step number

    while (t <= tstop)
      {
        t += dt;                     // increase time by the time step
        step_no++;                   // increase step number by 1

        // update inner points according to finite difference scheme:
        for (i = 2; i <= n-1; i++)
          up(i) = 2*u(i) - um(i) + sqr(C) * (u(i+1) - 2*u(i) + u(i-1));

        up(1) = 0;  up(n) = 0;   // update boundary points:
        um = u;  u = up;         // update data struct. for next step

        plotSolution (up, plotfile, t, C);
        if (step_no % 100 == 0) {  // write a message every 100th step:
          s_o << oform("time step %4d: u(x,t=%6.3f) is computed.\n",
                       step_no,t); // recall that \n is newline
          s_o.flush();             // flush forces immediate output
        }
      }
}

void setIC (real C, ArrayGen(real)& u0, ArrayGen(real)& um)
{
  int  n = u0.size();      // length of the vector u
  real x;                  // coordinate of a grid point
  real h = 1.0/(n-1);      // length of grid intervals
  real umax = 0.05;        // max string displacement
  int i;                   // loop counter over grid points
  for (i = 1; i <= n; i++) { // set the initial displacement u(x,0)
    x = (i-1)*h;
    if (x < 0.7)  u0(i) = (umax/0.7) * x;
    else          u0(i) = (umax/0.3) * (1 - x);
  }
  for (i = 2; i <= n-1; i++) // set the help variable um:
    um(i) = u0(i) + 0.5*sqr(C) * (u0(i+1) - 2*u0(i) + u0(i-1));
  um(1) = 0;  um(n) = 0;      // dummy values, not used in the scheme
}

void plotSolution (ArrayGen(real)& u, CurvePlotFile& plotfile,
                   real t, real C)
{
  int  n = u.size();        // the number of unknowns
  real h = 1.0/(n-1);       // length of grid intervals
  CurvePlot plot (plotfile);  // a single plot, tied to the "databank"
  plot.initPair ("displacement",        // plot title
                 oform("u(x,%.4f)",t),  // name of function
                 "x",                   // name of indep. var.
                 oform("C=%g, h=%g",C,h));  // comment
  for (int i = 1; i <= n; i++)          // add (x,y) data points
    plot.addPair (h*(i-1) /* x-value */, u(i) /* y value */);
  plot.finish();
}
```

Notice that the program defines the interfaces of the functions `timeLoop`, `setIC`, and `plotSolution` prior to the `main` function. These forward declarations could be avoided if we moved the `main` function to the end of the file. However, in this example we keep the Fortran-style approach of defining the main program at the beginning of the file.

Some C programmers will perhaps find the updating of the arrays for the next step a big strange; the statement `um=u` implies that we copy elements from u into um, while a more efficient way would be to switch the underlying array pointers. However, in all our introductory programs we shall pay more attention to safety than optimization as long as the constructions do not have severe performance penalties[14].

1.3.4 Visualizing the Results

Animation of the Wave Motion. Compile and run the program, using 20 intervals, unit Courant number, and `tstop` equal to 6. After the execution there should be a file `SIMULATION.map` in your directory. This file contains a "map" of all the 122 curves produced by the program. On Unix systems, you can now animate the computed string movement either using Gnuplot or Matlab. Animation in Gnuplot is most easily accomplished by the special Diffpack script `curveplotmovie`:

```
curveplotmovie gnuplot SIMULATION.map -0.1 0.1
```

The first argument is the map file, whereas the two next arguments are the minimum and maximum values of the y-values in the plot. You should see a nice movie on the screen. Alternatively, we may use Matlab for animation; simply replace `gnuplot` by `matlab`[15]. The command `curveplotmovie` is a simplified version of the much more powerful `curveplot` script that comes with Diffpack. With this latter script you can set various Gnuplot or Matlab options to customize the plot, select only a subset of all the curves to be included in the animation, and produce mpeg movies. A brief example appears later in this section, whereas Appendix B.4.1 documents the plotting features in more detail.

Setting the Casename. The stem of the files generated by the program (`SIMULATION` in the preceding example) can be changed by the user without out recompiling the program. If we want the name `C0.3`, e.g. indicating a simulation with $C = 0.3$, one types

```
./app --casename C0.3
```

---

[14] A profiling (see page 542) of the wave equation solver reveals that updating statements like `um=u` is not the most important place to start optimizing. Nevertheless, we address optimization and switching of array pointers on page 419.

[15] You need to issue an `exit` command to leave Matlab after the movie is shown.

The `--casename` option sets a casename for the simulation, and `SIMULATION` is the default casename. Anywhere in a Diffpack program you can access the casename of the current execution by the global variable `casename`. For example, in the demo program listed in the preceding section, we used `casename` to initialize the `plotfile` object, which then ensures that all the plot files have names containing `casename`.

Sometimes you will prefer to have all the casename-related files in a separate directory. The command-line option `--casedir mydir` to `app` creates a subdirectory `mydir`, which is used as current working directory during the simulation. Try, for example,

```
./app --casename CO.3 --casedir CO.3
```

To clean up all files generated by a computer experiment, you can use the Diffpack script `RmCase` with the casename as argument. Simply type `RmCase CO.3` in the present example. The `RmCase` script searches the current directory and all its subdirectories for files containing the stem `CO.3`, implying that files generated in the subdirectory `CO.3` will also be deleted.

Graphical Interface to Curve Plotting. There is a flexible Diffpack script, also with a graphical interface, that enables plotting of selected items from the collection of curves in the map file. First we describe the graphical interface called `curveplotgui`. Thereafter we explain how the plotting tool can be used without the graphical interface.

Let us generate a small number of curves, just for demonstration purposes. Run the program with casename `plotex`, Courant number 0.3, and `tstop` equal to 1.5. Then you will have a file `plotex.map`. Invoke the graphical curve plotting interface by typing `curveplotgui plotex.map`. The graphical menu allows you to select individual curves from the execution and plot them. Each curve is recognized by its three items: The plot title, the function or curve name, and the comment. You can now click on some individual curves, for example, those with function names $u(x,0.1650)$, $u(x,0.3600)$, and $u(x,0.5700)$. The chosen curves curves can be plotted using one of the listed plotting programs: Gnuplot, Xmgr, Matlab, and Plotmtv[16]. Figure 1.3 shows the result when Gnuplot is used. Clicking the Help button results in a description of the various features in `curveplotgui`. The fine details of a plot can be adjusted as outlined on page 560 under the option -c.

Windows NT/95 Remark 1.6: The Diffpack GUI on Win32 platforms supports functionality for selecting and plotting curves, much like `curveplotgui` tool.   ◇

Scripting Interfaces to Curve Plotting. The generation of the curve plot in Figure 1.3 can alternatively be made by the command

---

[16] Only the plotting program available on your computer system is listed.

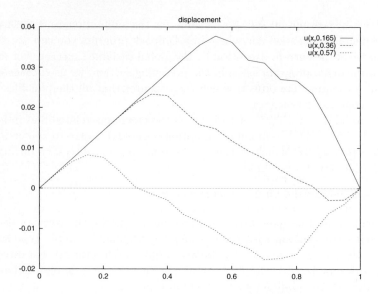

Fig. 1.3. Three snapshots of the waves on the string. The plot is made by Gnuplot.

```
curveplot gnuplot -f plotex.map -r '.' 'u\(x,0\.1650\)' '.'
    -r '.' 'u\(x,0\.3600\)' '.' -r '.' 'u\(x,0\.5700\)' '.'
    -ps myplot.ps
```

With identical syntax we can make plots with other plotting packages as well,
for example, Matlab, Plotmtv, or Xmgr; just replace gnuplot with matlab,
plotmtv, or xmgr. The -f option is used to give the name of mapfiles, and
the -r option is followed by three regular expressions [42,123] for the title, the function name, and the comment of the desired curves. Here, '.'
'u\(x,0\.3600\)' '.' means curves with any title ('.'), function name that
matches the string 'u(x,0.3600)', and any comment ('.'). It is necessary to pre-
cede the parentheses and the dots by backslashes, such as in u\(x,0\.3600\),
because (,), and . have special meanings in regular expressions.

To make an animation of the curves, we simply insert the -animate option
and specify that the regular expression is to match all titles, all functions
with name u, and all comments. In addition we need to fix the scale on the
y axis by giving the special Gnuplot command set yrange [-0.1:0.1]. This
command is transferred to Gnuplot through the -o option to curveplot. The
appropriate curveplot call then reads[17]

```
curveplot gnuplot -f plotex.map -r '.' 'u' '.' -animate
              -o 'set yrange [-0.1:0.1];'
```

---

[17] You can run the animation in slow motion by addition the command -fps 1
(giving one frame per second).

To make animations in Matlab instead, we just replace `gnuplot` by `matlab` (the specification of the axis is different in Matlab and Gnuplot so the `-o` must also be changed):

```
curveplot matlab  -f plotex.map -r '.' 'u' '.' -animate
          -o 'axis([0 1 -0.1 0.1]);' # [xmin xmax ymin ymax]
```

Sometimes it is convenient to store the animation in mpeg format. To this end, replace `-animate` by `-psanimate`. All the frames are then available in PostScript files `tmpdpc*.ps`, which the script converts to frames in an mpeg movie `movie.mpeg` by running a Diffpack script: `ps2mpeg tmpdpc*.ps`. Any mpeg player can be used to show the movie, try `mpeg_play movie.mpeg`. We refer to Appendix B.4 for additional information about curve plotting in Diffpack. For example, the movie can easily be equipped with your own drawings on a front page (see page 560). The moviemaking results in a lot of scratch files that should be removed using the Diffpack command

```
Clean .
```

See page 541 for more information about `Clean`, which is a useful script for cleaning directories in general.

Visualizing Numerical Errors. It is interesting to run the program again with another Courant number. It appears that $C = 1$ gives a numerical solution which is exact at all the grid points, regardless of the grid size! Trying $C > 1$ results in unstable solutions that have no physical relevance, see Figure 1.4 for an example. This illustrates that numerical solution methods may have a stability restriction on Δt. Here the stability criterion reads $C \leq 1$, implying that $\Delta t \leq h/\gamma$. If $C < 1$ the numerical solution will contain an error, visible as small-amplitude non-physical waves superimposed on the exact solution. You can easily see this effect by running the code with smaller Courant numbers. For example, give the same answers to the questions again, except 0.3 for the Courant number, plot the solution and compare the visible accuracy with the case $C = 1$. Appendix A.4 describes theoretical tools for investigating stability and accuracy of finite difference schemes. In particular, Example A.16 on page 516 applies these tools to our present simulation model.

Exercise 1.7. It can be of interest to make a plot that compares the numerical solution for $C = 0.25$ with the exact solution (the latter can be obtained by running the program with $C = 1$). Perform two simulations, one with $C = 1$ (casename `C1`) and one with $C = 0.25$ (casename `C0.25`). Of course, such comparisons only make sense if the curves $u(x, \cdot)$ for a particular time point are present in both the $C = 1$ and the $C = 0.25$ simulations. A plot of $u(x, t = 0.5)$ for the two choices of C (i.e. Δt) is enabled by the `curveplot` script with the file options `-f C1.map -f C0.25.map` and a regex option `-r '.' ',0\.5\)' '.'`. (Notice the backslash that precedes the dot; without the backslash the dot is interpreted as "any character", resulting in matches for, e.g., `0.025` and `0.075` as well.) ◇

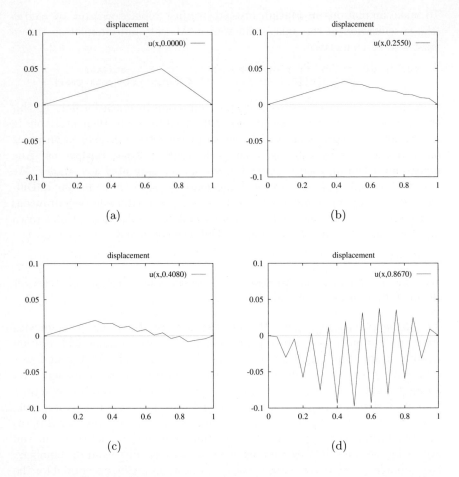

Fig. 1.4. Simulation of waves on a string with $C = 1.02$, i.e., slightly larger than the stability limit $C = 1$. (a) Initial condition; (b) small noise can be seen; (c) the noise has been amplified; (d) the noise has been further amplified and the solution is destroyed. (101 grid points were used in the simulation, and the particular plots shown here were generated by Gnuplot, using the special option -o 'set yrange [-0.1:0.1]; set data style lines;' to curveplot.)

As a prolongation of Exercise 1.7, we might want to show a movie of the $C = 0.25$ solution simultaneously with the exact solution. Noting that each frame in the C1 case corresponds to every four frame in the C0.25 case, we must first extract the lines in C0.25.map that match those of C1.map. To this end, we can use the Diffpack script ExtractLines (extracting every 4 line, starting with line 1):

```
ExtractLines C0.25.map 4 1 > C0.25x.map
```

Each frame in the resulting movie consists of two curves, where each curve must be specified with its own regex set:

```
curveplot gnuplot -f C1.map -r '.' 'u' 'C=1' -f C0.25x.map
-r '.' 'u' 'C=0\.25' -animate -o 'set yrange [-0.05:0.05];' -fps 1
```

A Tailored Scripting Interface to the Simulator. To increase the user's productivity in scientific-computing experiments, it is advantageous to automate the interaction with the computer. This can be done by creating a special-purpose *script* that in the present case runs the simulator based on some adjustable input data, displays an animation of the waves, and dumps a PostScript plot of the solution at a specified point in time. So in short, scripting saves you from boring and error-prone manual work with a computer code. Scripts can be written in a Unix shell (sh, bash, csh, ksh, or zsh) or in more advanced scripting languages such as Perl, Tcl, or Python. The latter three languages are rich and powerful, and the scripts run in Unix, Windows, and Macintosh environments. We shall base our example here on Perl [123].

First, we must decide upon the usage of the script. The simplest possibility is to use the command line (of the operating system) as interface to the script. Writing wave.pl 40 0.9 2 should then mean that the name of the script is wave.pl, the number grid cells equals 40, $C = 0.9$, and $t_{\text{stop}} = 2$. In Perl, the command-line arguments to the script are available in the built-in array ARGV, as the elements $ARGV[0], $ARGV[1] and so on. Variables in Perl are preceded by a $ sign, and you never need to specify the type of a variable. Strings are enclosed in double quotes, with the possibility of embedding variables directly in a string. For example, constructing a casename like C0.33, corresponding to a run with $C = 0.33$, is enabled by the string "C$C", where $C is a Perl variable for C. If we want to combine this into the filename C0.33.map, we need to write "C${C}.map", i.e., ${C} is a more accurate encapsulation of the variable than just $C. Embedding of variables in strings is heavily used in the forthcoming script.

Execution of operating system commands can be done via a call to the Perl function system. However, we recommend to restrict the use of system to running applications, like our simulator. Common operating system tasks, for example regarding manipulation of files and directories, are supported in Perl by special functions. For instance, deleting a file tmp.i is accomplished by the Perl call unlink("tmp.i"). This call will work on equally well on Unix as on Windows.

A particular advantage of scripts is that you do not need to compile them[18]. The reader is encouraged to study the following script and try to

---

[18] This is also the reason why scripts run slowly; they are hence not suitable for intensive numerical calculations.

grasp the very basics of the Perl language from this example. (Fortunately, most scripting languages have a syntax much inspired by C).

```perl
#!/usr/bin/perl
# script for automating the interaction with the simulator:
if ($#ARGV < 2) { # no of command line arguments = $#ARGV + 1
    # die writes the text and aborts execution
    die "Usage: wave.pl  ncells  Courant-no  tstop\n";
}
$ncells = $ARGV[0];   # 1st command-line arg is the no of intervals
$C      = $ARGV[1];   # 2nd command-line arg is the Courant no.
$tstop  = $ARGV[2];   # 3rd command-line arg is tstop
# write the parameters to a perl file HELPFILE with name $helpfile:
$helpfile = "tmp.i"; open (HELPFILE, ">$helpfile");
print HELPFILE " $ncells \n $C \n $tstop\n";
close (HELPFILE);
print "\n...running the simulator...\n";
system "./app --casename C$C < $helpfile";
unlink ("$helpfile");  # delete the tmp.i file
# make a curve plot at t=0.5 if possible:
$dt = $C * 1.0/$ncells;  # time step length
$stepno = int(0.5/$dt);
$curveplot_time = $dt * $stepno;  # possible t value
if ($curveplot_time <= $tstop) {
    $psfilename = "C${C}.t${curveplot_time}.ps";
    print "\n...curve plot at t=$curveplot_time in $psfilename\n";
    $regex = "-r '.' '${curveplot_time}' '.'";
    # alternative curvename regex: 'u\\(x,${curveplot_time}'
    # notice that a double backslash is needed (system eats one \ )
    system "curveplot gnuplot -f C${C}.map $regex -ps $psfilename";
}
print "\n...making the animation...\n";
system "curveplotmovie gnuplot C${C}.map -0.05 0.05";
```

Try the script with 30 0.99 4 as parameters and view the resulting file C0.99.t0.495.ps. Recall to clean up all files from the numerical experiment by the RmCase C0.99 command.

The wave.pl script can also quite easily be equipped with a graphical user interface. Check out the modified script wave-GUI.pl. To run it, just type wave-GUI.pl or perl wave-GUI.pl if your Perl installation has problems with including the Tk package. The usage of wave-GUI.pl is obvious and its comments should explain how a simple graphical user interface is realized by combining Perl and its bindings to the popular Tk widget set [109, Ch. 14].

We strongly encourage you to make scripting interfaces to a simulator when working with numerical experiments in the many projects and exercises throughout this book. Scripting interfaces constitute an indispensible tool for all kinds of research projects involving experimental scientific computing.

1.3.5 A 2D Wave Equation with Variable Wave Velocity

Two-dimensional waves in heterogeneous media can be modeled by the wave equation

$$\frac{\partial^2 u}{\partial t^2} = \nabla \cdot [\gamma^2 \nabla u]. \tag{1.55}$$

The wave velocity γ now varies with the properties of the medium and therefore becomes a function of the spatial variables: $\gamma = \gamma(\boldsymbol{x})$. For notational simplicity, we shall introduce $\lambda = \gamma^2$ in the following. A possible physical application of (1.55), which will be studied in Chapter 1.3.6, corresponds to large destructive ocean waves.

The variable-coefficient operator $\nabla \cdot \lambda(\boldsymbol{x})\nabla$, for some prescribed function $\lambda(\boldsymbol{x})$, appears frequently in the present text, simply because this is a common operator in a wide range of mathematical models. It is hence important to be familiar with the detailed meaning of this expression:

$$\nabla \cdot [\lambda(x,y)\nabla u] = \frac{\partial}{\partial x}\left(\lambda(x,y)\frac{\partial u}{\partial x}\right) + \frac{\partial}{\partial y}\left(\lambda(x,y)\frac{\partial u}{\partial y}\right). \tag{1.56}$$

In the case λ is constant, we achieve

$$\nabla \cdot [\lambda\nabla u] = \lambda\nabla \cdot [\nabla u] = \lambda\nabla^2 u = \lambda\left(\frac{\partial^2 u}{\partial x^2} + \frac{\partial^2 u}{\partial y^2}\right).$$

Let us show how straightforward it is to discretize (1.55) by combining elements from Chapters 1.2.6 and 1.3.2. To begin with, we assume that $u = 0$ on the boundary, u is prescribed as $I(x,y)$ at $t = 0$, with $\partial u/\partial t = 0$, and the domain is the rectangle $\Omega = (0, w_x) \times (0, w_y)$.

The 2D medium is given a uniform partition with n_x and n_y grid points in each of the two space directions x and y. The approximation to $u(x, y, t)$ at grid point (i, j) at time t_ℓ is denoted by $u_{i,j}^\ell$. We can set

$$x_i = (i - 1)\Delta x, \quad y_j = (j - 1)\Delta y, \quad t_\ell = \ell\Delta t,$$

where Δt is the constant time step and the constant space increments Δx and Δy are given by $\Delta x = w_x/(n_x - 1)$ and $\Delta y = w_y/(n_y - 1)$. The index i runs from 1 to n_x and j runs from 1 to n_y.

We see that (1.56) is essentially a sum of two "one-dimensional" terms, where each term can be discretized as in the left-hand side of (1.24). For example, the discrete version of the last term in (1.56) becomes

$$\frac{\partial}{\partial y}\left(\lambda\frac{\partial u}{\partial y}\right) \approx \frac{1}{\Delta y}\left(\lambda_{i,j+\frac{1}{2}}\left(\frac{u_{i,j+1}^\ell - u_{i,j}^\ell}{\Delta y}\right) - \lambda_{i,j-\frac{1}{2}}\left(\frac{u_{i,j}^\ell - u_{i,j-1}^\ell}{\Delta y}\right)\right).$$

For the time derivative in the wave equation we use the standard three-point approximation as explained in Chapter 1.3.2. Putting the elements together, we arrive at a finite difference method for the 2D wave equation with variable coefficients. Algorithm 1.2 lists the details and is a slight extension of Algorithm 1.1. The following abbreviation is used in the algorithm:

$$[\triangle u]_{i,j} \equiv \left(\frac{\Delta t}{\Delta x}\right)^2 (\lambda_{i+\frac{1}{2},j}(u_{i+1,j} - u_{i,j}) - \lambda_{i-\frac{1}{2},j}(u_{i,j} - u_{i-1,j})) +$$

$$\left(\frac{\Delta t}{\Delta y}\right)^2 (\lambda_{i,j+\frac{1}{2}}(u_{i,j+1} - u_{i,j}) - \lambda_{i,j-\frac{1}{2}}(u_{i,j} - u_{i,j-1})). \quad (1.57)$$

Algorithm 1.2.

Explicit scheme for the 2D wave equation with $u = 0$ on the boundary.

define $u^+_{i,j}$, $u_{i,j}$ and $u^-_{i,j}$ to represent $u^{\ell+1}_{i,j}$, $u^{\ell}_{i,j}$ and $u^{\ell-1}_{i,}$, resp.
define $[\triangle u]_{i,j}$ as in (1.57)
define $(i,j) \in \bar{\mathcal{I}}$ to be $i = 1, \ldots, n_x$, $j = 1, \ldots, n_y$
define $(i,j) \in \mathcal{I}$ to be $i = 2, \ldots, n_x - 1$, $j = 2, \ldots, n_y - 1$
set $u_{i,j} = 0$, $(i,j) \in \bar{\mathcal{I}}$
SET THE INITIAL CONDITIONS:
$u_{i,j} = I(x_i, y_j)$, $(i,j) \in \mathcal{I}$
DEFINE THE VALUE OF THE ARTIFICIAL QUANTITY $u^-_{i,j}$:
$u^-_{i,j} = u_{i,j} + \frac{1}{2}[\triangle u]_{i,j}$, $(i,j) \in \mathcal{I}$
$t = 0$
while time $t \leq t_{\text{stop}}$
    $t \leftarrow t + \Delta t$
    UPDATE ALL INNER POINTS:
    $u^+_{i,j} = 2u_{i,j} - u^-_{i,j} + [\triangle u]_{i,j}$, $(i,j) \in \mathcal{I}$
    INITIALIZE FOR NEXT STEP:
    $u^-_{i,j} = u_{i,j}$, $u_{i,j} = u^+_{i,j}$, $(i,j) \in \mathcal{I}$

Notice that we do not explicitly set $u_{i,j} = 0$ at the boundary. Instead, we set $u_{i,j} = 0$ initially and never touch the boundary values.

We remark that if λ is constant and $\Delta x = \Delta y = h$, the numerical scheme simplifies, and the Laplace term $\nabla \cdot [\lambda \nabla u] = \lambda \nabla^2 u$ takes the well-known discrete form

$$[\triangle u]_{i,j} = \lambda \left(\frac{\Delta t}{h}\right)^2 (-u_{i-1,j} - u_{i,j-1} - u_{i+1,j} - u_{i,j+1} + 4u_{i,j}) . \quad (1.58)$$

This formula for approximating $\lambda \nabla^2 u$ can be graphically exposed as in Figure 1.5. The circles denote the points in the grid that are used in the approximation, and the numbers reflect the weight of the point in the finite difference formula. One often refers to such a graphical representation as a *finite difference stencil* or a *computational molecule*. In the more general case when λ is not constant, the same stencil arises, except that the weights are different.

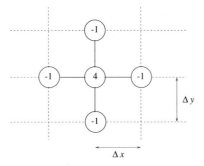

Fig. 1.5. Illustration of the finite difference stencil for approximating the Laplace operator $\nabla^2 u$ in a regular grid.

The implementation of the two-dimensional explicit finite difference scheme is easily accomplished by a slight extension of the one-dimensional wave equation solver. Now we need to declare 2D arrays, like

```
ArrayGen(real) u (nx,ny);
```

and use two nested loops, one for each index, for accessing all the spatial points at each time level. The entries in the ArrayGen object are stored columnwise (as in Fortran). Therefore, when running through the grid points in nested loops, the first index should have the fastest variation such that we run through the u(i,j) as they are stored in memory:

```
for (j = 1; j <= ny; j++)
  for (i = 1; i <= nx; i++)
    u(i,j) = ...
```

More details regarding implementation of the 2D wave equation are provided in the next section.

1.3.6   A Model for Water Waves

Some water wave phenomena are recognized as long waves in shallow water, meaning that the typical wave length is much larger than the depth. This feature simplifies models for water waves considerably and reduces the original 3D problem in the time-varying water volume to a 2D equation like (1.55)[19].

---

[19] The reduction to (1.55) also requires that nonlinear effects due to wave steepness or high amplitude/depth ratio can be neglected.

Shallow water models are used for simulating storm surges, tides, swells in coastal regions, and tsunamis[20].

When (1.55) is used to model water waves, the primary unknown $u(x, y, t)$ is the surface elevation, while the variable coefficient λ is related to the water depth: $\lambda = gH(x, y)$, where g is the acceleration of gravity and $H(x, y)$ is the still-water depth. A relevant boundary condition at the coastline is

$$\frac{\partial u}{\partial n} \equiv \nabla u \cdot \boldsymbol{n} = 0.$$

The notation $\frac{\partial u}{\partial n}$ for the derivative in the direction normal to the boundary is frequently used throughout this text. As initial condition we shall here take u as prescribed and $\partial u/\partial t = 0$. The assumptions behind the model (1.55) and more advanced wave models are discussed in Chapter 6.2.

The basic finite difference scheme for (1.55) was treated in Chapter 1.3.5. The extension here concerns the handling of the boundary condition $\partial u/\partial n = 0$, often referred to as a homogeneous Neumann condition. This boundary condition can be implemented in a way similar to what we did in Chapter 1.2.2, i.e., the boundary condition is discretized by a centered difference at the boundary. At the line $i = 1$ we then require

$$\frac{u_{2,j}^\ell - u_{0,j}^\ell}{\Delta x} = 0 \quad \Rightarrow \quad u_{0,j}^\ell = u_{2,j}^\ell.$$

Notice that this involves a fictitious value $u_{0,j}^\ell$ outside the grid. Using the discrete PDE at the same boundary point, with $u_{0,j}^\ell = u_{2,j}^\ell$ from the boundary condition, enables elimination of the fictitious value. The $[\triangle u]_{i,j}$ operator is then modified to

$$[\triangle u]_{1,j:i-1\rightarrow i+1} \equiv \left(\frac{\Delta t}{\Delta x}\right)^2 (\lambda_{1+\frac{1}{2},j}(u_{2,j} - u_{1,j}) - \lambda_{1-\frac{1}{2},j}(u_{1,j} - u_{2,j})) +$$

$$\left(\frac{\Delta t}{\Delta y}\right)^2 (\lambda_{1,j+\frac{1}{2}}(u_{1,j+1} - u_{1,j}) - \lambda_{1,j-\frac{1}{2}}(u_{1,j} - u_{1,j-1})), \; j \neq 1, n_y. \quad (1.59)$$

At the boundary $i = n_x$ we would then apply the modification $[\triangle u]_{n_x,j:i+1\rightarrow i-1}$. Similarly, for $j = 1$ and $j = n_y$ we replace the original $[\triangle u]_{i,j}$ operator by $[\triangle u]_{i,1:j-1\rightarrow j+1}$ and $[\triangle u]_{i,n_y:j+1\rightarrow j-1}$, respectively. The corner points of the grid require modification of both indices, for example, the $i = j = 1$ point leads to $[\triangle u]_{1,1:i-1\rightarrow i+1,j-1\rightarrow j+1}$.

Algorithm 1.3 precisely explains the updating of internal and boundary points in terms of a function WAVE(u^+, u, u^-, a, b, c). As a special case, the call WAVE$(u^+, u, u^-, 1, 1, 1)$ reproduces the original finite difference scheme with modifications due to homogeneous Neumann conditions $\partial u/\partial n = 0$.

---

[20] Tsunamis are destructive ocean waves generated by earthquakes, faulting, or slides.

Algorithm 1.3.

Basic finite difference updating formula for the 2D wave equation with $\frac{\partial u}{\partial n} = 0$ on the boundary.

define $[\triangle u]_{i,j}$ as in (1.57)
define $[\triangle u]_{1,j:i-1\to i+1}$, $[\triangle u]_{n_x,j:i+1\to i-1}$, $[\triangle u]_{i,1:j-1\to j+1}$,
   and $[\triangle u]_{i,n_y:j+1\to j-1}$ according to (1.59)
define $(i,j) \in \mathcal{I}$ to be $i = 2, \ldots, n_x - 1$, $j = 2, \ldots, n_y - 1$
define $u_{i,j}^+$, $u_{i,j}$ and $u_{i,j}^-$ to represent $u_{i,j}^{\ell+1}$, $u_{i,j}^{\ell}$ and $u_{i,j}^{\ell-1}$, resp.
function WAVE(u^+, u, u^-, a, b, c):

UPDATE ALL INNER POINTS:
$$u_{i,j}^+ = 2au_{i,j} - bu_{i,j}^- + c[\triangle u]_{i,j}, \quad (i,j) \in \mathcal{I}$$
UPDATE BOUNDARY POINTS:
$$i = 1; \; u_{i,j}^+ = 2au_{i,j} - bu_{i,j}^- + c[\triangle u]_{i,j:i-1\to i+1}, \; j = 2, \ldots, n_y - 1$$
$$i = n_x; \; u_{i,j}^+ = 2au_{i,j} - bu_{i,j}^- + c[\triangle u]_{i,j:i+1\to i-1}, \; j = 2, \ldots, n_y - 1$$
$$j = 1; \; u_{i,j}^+ = 2au_{i,j} - bu_{i,j}^- + c[\triangle u]_{i,j:j-1\to j+1}, \; i = 2, \ldots, n_x - 1$$
$$j = n_y; \; u_{i,j}^+ = 2au_{i,j} - bu_{i,j}^- + c[\triangle u]_{i,j:j-1\to j+1}, \; i = 2, \ldots, n_x - 1$$
UPDATE CORNER POINTS ON THE BOUNDARY:
$$i = 1, \, j = 1; \; u_{i,j}^+ = 2au_{i,j} - bu_{i,j}^- + c[\triangle u]_{i,j:i-1\to i+1,j-1\to j+1}$$
$$i = n_x, \, j = 1; \; u_{i,j}^+ = 2au_{i,j} - bu_{i,j}^- + c[\triangle u]_{i,j:i+1\to i-1,j-1\to j+1}$$
$$i = 1, \, j = n_y; \; u_{i,j}^+ = 2au_{i,j} - bu_{i,j}^- + c[\triangle u]_{i,j:i-1\to i+1,j+1\to j-1}$$
$$i = n_x, \, j = n_y; \; u_{i,j}^+ = 2au_{i,j} - bu_{i,j}^- + c[\triangle u]_{i,j:i+1\to i-1,j+1\to j-1}$$

In the implementation of Algorithm 1.3, two issues are important for computational efficiency: (i) the (i,j)-loop over the grid points is split into separate loops for the internal and boundary points, and (ii) the grid points should be visited in the sequence they are stored in memory. As an alternative to point (i), one could have one (i,j)-loop over all grid points and perform tests inside the loop whether a grid point is on the boundary or not. However, compilers have problems optimizing loops with if-tests, so such tests are better moved outside the loops. Regarding point (ii), it is important to know the underlying storage structure of the array entries.

Some readers experienced with finite difference programming would perhaps used an array for u that also includes the fictitious boundary points. Instead of using special finite difference stencils on the boundary, we then use the same stencil as in the interior. After the new values of u are computed, one needs to update the fictitious points, e.g., $u_{0,j} = u_{2,j}$ such that the discrete normal derivative vanishes. However, computing with fictitious points results in array structures that cannot directly be sent to visualization software, since they contain non-physical grid-point values, and this is the main reason for our choice of special stencils on the boundaries in Algorithm 1.3.

In the implementation of the WAVE function in Algorithm 1.3 it can be convenient to apply C macros for the operator $[\triangle u]_{i,j}$, for example,

```
#define LaplaceU(i,j, im1,ip1,jm1,jp1) \
  sqr(dt/dx)* \
  ( 0.5*(lambda(ip1,j )+lambda(i  ,j ))*(u(ip1,j )-u(i  ,j )) \
   -0.5*(lambda(i  ,j )+lambda(im1,j ))*(u(i  ,j )-u(im1,j )))\
 +sqr(dt/dy)* \
  ( 0.5*(lambda(i  ,jp1)+lambda(i  ,j ))*(u(i  ,jp1)-u(i  ,j )) \
   -0.5*(lambda(i  ,j )+lambda(i  ,jm1))*(u(i  ,j )-u(i  ,jm1)))
```

This statement defines a macro `LaplaceU` with six parameters. The C/C++
preprocessor will substitute each occurrence of expressions like

```
    LaplaceU(i,j, i-1,i+1,j-1,j+1)
```

by the defined formula involving `lambda` and `u`, where in this particular example `im1` is replaced by `i-1`, `ip1` by `i+1`, and so on. The `LaplaceU` macro saves
a lot of typing and makes the implementation of the scheme clearer:

```
// update inner points according to finite difference scheme:
for (j = 2; j <= ny-1; j++)
  for (i = 2; i <= nx-1; i++)
    up(i,j) = a*2*u(i,j) - b*um(i,j)
              + c*LaplaceU(i,j,i-1,i+1,j-1,j+1);

// update boundary points (modified finite difference schemes):
for (j = 2; j <= ny-1; j++) {
  // (run through j=const points in the sequence they are stored)
  i=1;
  up(i,j) = a*2*u(i,j)-b*um(i,j) + c*LaplaceU(i,j,i+1,i+1,j-1,j+1);
  i=nx;
  up(i,j) = a*2*u(i,j)-b*um(i,j) + c*LaplaceU(i,j,i-1,i-1,j-1,j+1);
}
j=1;
for (i = 2; i <= nx-1; i++)
  up(i,j) = a*2*u(i,j)-b*um(i,j) + c*LaplaceU(i,j,i-1,i+1,j+1,j+1);
j=ny;
for (i = 2; i <= nx-1; i++)
  up(i,j) = a*2*u(i,j)-b*um(i,j) + c*LaplaceU(i,j,i-1,i+1,j-1,j-1);
// corners:
i=1; j=1;
up(i,j) = a*2*u(i,j)-b*um(i,j) + c*LaplaceU(i,j,i+1,i+1,j+1,j+1);
i=nx; j=1;
up(i,j) = a*2*u(i,j)-b*um(i,j) + c*LaplaceU(i,j,i-1,i-1,j+1,j+1);
i=1; j=ny;
up(i,j) = a*2*u(i,j)-b*um(i,j) + c*LaplaceU(i,j,i+1,i+1,j-1,j-1);
i=nx; j=ny;
up(i,j) = a*2*u(i,j)-b*um(i,j) + c*LaplaceU(i,j,i-1,i-1,j-1,j-1);
```

We can now use the WAVE function to devise a compact description of
all the computational tasks for the discrete 2D wave equation with homogeneous Neumann conditions. The steps are listed in Algorithm 1.4, which has
been implemented in a simple demo program that can be found in the directory `src/fdm/Wave2D-func`. The implemented equations have been scaled with
the characteristic depth H_c as length scale and $H_c/\sqrt{gH_c}$ as time scale. We

remark that the explicit finite difference scheme is subject to stability restrictions in the same manner as the one-dimensional scheme. In two space dimensions the stability criterion becomes (see e.g. Example A.15 on page 515)

$$\Delta t \leq \left(\max_{(x,y)\in\Omega} \lambda(x,y) \right)^{-\frac{1}{2}} \left(\frac{1}{\Delta x^2} + \frac{1}{\Delta y^2} \right)^{-\frac{1}{2}}. \tag{1.60}$$

Algorithm 1.4.

Complete scheme for the 2D wave equation with $\frac{\partial u}{\partial n} = 0$ on the boundary.

define quantities in Algorithm 1.3
define $(i,j) \in \bar{\mathcal{I}}$ to be $i = 1, \ldots, n_x$, $j = 1, \ldots, n_y$
set $u_{i,j} = 0$, $(i,j) \in \bar{\mathcal{I}}$
SET THE INITIAL CONDITIONS:
$u_{i,j} = I(x_i, y_j)$, $(i,j) \in \mathcal{I}$
DEFINE THE VALUE OF THE ARTIFICIAL QUANTITY $u_{i,j}^-$:
WAVE($u^-, u, u^-, 0.5, 0, 1$)
$t = 0$
while time $t \leq t_{\text{stop}}$
    $t \leftarrow t + \Delta t$
    UPDATE ALL POINTS:
    WAVE($u^-, u, u^-, 1, 1, 1$)
    INITIALIZE FOR NEXT STEP:
    $u_{i,j}^- = u_{i,j}$, $u_{i,j} = u_{i,j}^+$, $(i,j) \in \mathcal{I}$

Let us apply the suggested algorithm to a physical problem involving ocean waves generated by an underwater earthquake. In this example the domain $\Omega = (s_x, s_x + w_x) \times (s_y, s_y + w_y)$ is a scaled segment of an ocean basin in which we want to study propagating wave patterns. The boundary condition $\partial u/\partial n = 0$ principally requires that the boundary $\partial\Omega$ of Ω really approximates coastlines. This is not the case in the present application; what we need is a type of boundary condition that transmits the waves through the boundaries without any reflection. Such conditions, often called radiation or open boundary conditions, constitute a difficult topic that we briefly address in Example A.21 on page 526. Generalization of the condition in that example to the present variable-coefficient 2D wave equation is not easy and clearly beyond the scope of this text. However, the nature of the wave equation allows us to use the currently non-physical boundary condition $\partial u/\partial n = 0$.

If the waves generated by the earthquake start from the interior of Ω, the impact of the boundary condition will not be visible before the waves hit the boundary. The condition $\partial u/\partial n = 0$ results in complete reflection of the waves, so when the reflected wave arrives at a point in the interior part of the domain, the value of u at this point is no longer physical. On the other hand, if we had applied a "wrong" boundary condition in the heat conduction problem in Chapter 1.2, the solution would be affected at all points in the domain. The different nature of the heat conduction PDE and the wave equation is fundamental for numerical solution techniques. Appendix A.5 deals with this topic, including the impact of changing boundary conditions in the wave equations. The results from Appendix A.5 should make the reader confident that the physically irrelevant $\partial u / \partial n = 0$ condition can be used for simulating earthquake-generated waves out in the open sea.

The impact of the earthquake on the sea water is typically modeled by an elevation of the bottom, but if the movement of the bottom is rapid compared with the time scale of wave propagation, which is usually the case, we may translate the bottom elevation into a corresponding initial surface elevation. Our initial conditions will hence be that $\partial u / \partial t = 0$ and $u(x, y, t) = I(x, y)$, ie., the surface is at rest with shape $I(x, y)$. A particular choice of I is the Gaussian bell function, centered at (x_u^c, y_u^c) with amplitude A_u and "standard deviation" σ_{ux} and σ_{uy} in the x and y directions:

$$I(x, y) = A_u \exp\left(-\frac{1}{2} \left(\frac{x - x_u^c}{\sigma_{ux}} \right)^2 - \frac{1}{2} \left(\frac{y - y_u^c}{\sigma_{uy}} \right)^2 \right). \qquad (1.61)$$

We shall here assume that the earthquake takes place in the vicinity of an underwater seamount, with the shape of a Gaussian bell function[21]:

$$H(x, y) = 1 - A_H \exp\left(-\frac{1}{2} \left(\frac{x - x_H^c}{\sigma_{Hx}} \right)^2 - \frac{1}{2} \left(\frac{y - y_H^c}{\sigma_{Hy}} \right)^2 \right). \qquad (1.62)$$

With the scaling used in the implemented equations, the scaled depth approaches unity far from the seamount. The characteristic wave length of the resulting waves should then be much larger than unity for the model to describe the wave propagation well.

For simulating and visualizing the wave motion we could extend the simple program in `src/fdm/Wave2D-func`. However, better flexibility both with respect to usage and modifications is obtained by using the high-level abstractions outlined in Chapter 1.6. It is not necessary to understand these programming techniques at the present stage when just playing around with the wave simulator.

In our first test, we try a circular bell shape for both H and the initial u, with center $y_H^c = y_u^c = x_H^c = x_u^c = 0$. The solution is then expected to be symmetric with respect to the lines $x = 0$ and $y = 0$. Therefore, we only need to simulate the wave motion for $x, y \geq 0$, thereby reducing the size of the domain and the computational work by a factor of four. The boundary condition at the symmetry lines is (also) $\partial u / \partial n = 0$. The recommended

---

[21] This simulation case is in fact inspired by the geometry of the Gorringe Bank southwest of Portugal. Severe ocean waves have been generated due to earthquakes in this region.

simulator is located in the directory src/fdm/Wave2D1. Here is an appropriate
execution command for that simulator:

```
./app --casename circles -nx 31 -ny 31 -sx 0 -wx 20 -sy 0 -wy 20
        -dt 0 -t 4 -A_H 0.7 -A_u 0.2 -xc_H 0 -yc_H 0 -xc_u 0 -yc_u 0
        -sigma_ux 2 -sigma_uy 2 -sigma_Hx 2 -sigma_Hy 2
```

The option -dt 0 causes the program to find the optimal value of Δt that
fulfills the stability criterion in each cell. The scripts u-*.pl take the output
from a run and visualizes the results. Try e.g. u-plotmtv.pl; some snapshots
of u and H are plotted[22] on the screen and an mpeg movie is made. After
the script has finished, use some mpeg player to visualize the animation of
the wave motion in movie.mpeg.

Let us choose $y_H^c = y_u^c$, such that $y = 0$ is still an expected symmetry
line, but now with the generated surface disturbance displaced slightly to the
right, e.g., $x_u^c = \sigma_{Hx}$. The modified command-line options become -nx 61
-sx -20 -wx 40 -xc_u 2. To make a visually attractive plot, one can simulate
the motion in the total physical domain and scale the initial amplitude A_u
such that the propagating waves exhibit significant visible displacement of
the surface[23], see Figure 1.6 for an example. Notice that the geometry in
Figure 1.6 is actually very thin; the z scale ranges from -1 to 0.2, whereas
the horizontal extent is $[-20, 20] \times [-20, 20]$. The waves are therefore long
compared with the depth, and our visualization of the geometry in a cube
can be misleading at first sight.

Exercise 1.8. Destructive surface waves can be generated from underwater
slides. Mathematically, we can model an underwater slide as a movement of
the bottom. Incorporation of a time-dependent (scaled) depth $H(x, y, t)$ in
the governing wave equation leads to an extra source term:

$$\frac{\partial^2 u}{\partial t^2} = \nabla \cdot [H(x, y, t)\nabla u] + \frac{\partial^2 H}{\partial t^2} . \tag{1.63}$$

In the finite difference scheme, we assume that H is available as an explicit
function and the second-order derivative can be approximated directly by

$$\left[\frac{\partial^2 H}{\partial t^2}\right]_{i,j}^{\ell} \approx \frac{1}{\Delta t^2} \left(H(x_i, y_j, t_{\ell+1}) - 2H(x_i, y_j, t_\ell) + H(x_i, y_j, t_{\ell-1})\right) .$$

Implement a simulation program for the equation (1.63) restricted to one
space dimension. Dump curve plots of $u(x, t)$ and $H(x, t)$ at each time level
such that we can make movies of the time-varying surface elevation *and* the
bottom (a suitable set of regular expressions for animations via curveplot is

---

[22] A brief guide to the plotting program Plotmtv, which is used by u-plotmtv.pl,
appears in Chapter 3.3.3.

[23] Observe that multiplying A_u by an arbitrary factor f just leads to multiplying
$u(x, y, t)$ by f as well.

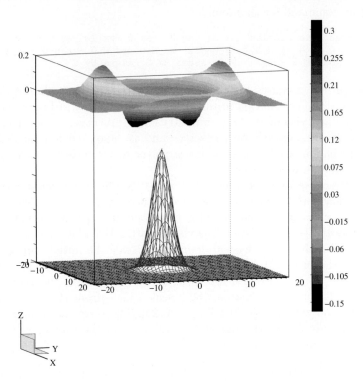

Fig. 1.6. Surface elevation (exaggerated scale) and bottom topography simulated by the `Wave2D1` solver on a 61 × 61 grid (plotted with Plotmtv).

-r '.' 'H.' '.' -r '. 'u' '.' provided the curve names of H and u contain H and u, respectively).

A specific function $H(x, y, t)$, modeling underwater slides in a fjord, may have the shape

$$H(x, t) = \Delta - \beta(x + \epsilon)(x + \epsilon - L)$$
$$- K \frac{1}{\sqrt{2\pi\gamma}} \exp\left(-\frac{1}{\gamma}\left[x - \frac{(L + \epsilon + 2)}{5} + ce^{\alpha t}\right]^2\right),$$

where Δ, ϵ, L, K, γ and α are constants that can be tuned to produce a particular slide. The function H is interpreted as the sum of a parabola and a bell-shaped curve, the latter moving with a velocity $\alpha c \exp(\alpha t)$. Choosing $\alpha < 0$ gives a retarding slide. A suitable choice of values are $\Delta = 0.2$, $\beta = 0.04$, $\epsilon = 0.5$, $L = 11$, $K = 0.7$, $\gamma = 0.7$, $c = 1.2$, and $\alpha = -0.3$.   ◇

1.4   Projects

1.4.1   A Uni-Directional Wave Equation

Mathematical Problem. In this project, we consider the uni-directional wave equation

$$\frac{\partial u}{\partial t} + \gamma \frac{\partial u}{\partial x} = 0, \tag{1.64}$$

where $u = u(x,t)$ and γ is a dimensionless number that equals unity (used for just labeling the $\partial u/\partial x$ term). The equation is to be solved for $x \in (0,1)$ and $t \in (0,T)$, with $u(0,t) = 1$ for $t \geq 0$ and $u(x,0) = 0$ for $x > 0$.

Physical Model. The mathematical problem may model one-dimensional advective transport of heat or of a contaminant. Choose an interpretation, present the full 3D model, demonstrate the simplifications, and introduce a suitable scaling.

Numerical Method. We solve (1.64) by an explicit upwind finite difference method:

$$u_i^{\ell+1} = (1 - C)u_i^\ell + Cu_{i-1}^\ell, \tag{1.65}$$

where $C = \gamma \Delta t/h$ is the Courant number. Explain how the formula (1.65) can be derived (hint: recollect terms and identify the difference approximations to the space and time derivatives). Formulate a numerical algorithm for the complete discrete problem, including initial and boundary values.

Analysis. Show that the analytical solution of (1.64) is in general $u(x,t) = f(x - \gamma t)$, where $f(x)$ is the initial condition. Specialize this solution to the current problem. Demonstrate that the analytical solution of (1.64) fulfills the discrete equations when $C = 1$, regardless of the mesh partition. As an optional analysis, derive a stability criterion for the numerical scheme and a measure of the error (Appendix A.4 gives suitable background material on stability and accuracy).

Implementation. The implementation should be as simple as possible; one can, for instance, simplify the program from Chapter 1.3. Demonstrate that the program reproduces the analytical solution of the continuous problem when $C = 1$. Provide plots of u when $C = 0.8, 1.0, 1.2$.

1.4.2   Centered Differences for a Boundary-Layer Problem

Mathematical Problem. In this project we shall work with the boundary-value problem

$$u'(x) = \epsilon u''(x), \ \ x \in (0,1), \ \ \ u(0) = 0, \ u(1) = 1, \tag{1.66}$$

where $\epsilon > 0$ is a prescribed constant. The physical relevance of this problem is somewhat limited, but one can think of (1.66) as the simplest possible model for an important phenomenon called boundary layers that appear in a wide range of fluid flow applications[24]. The boundary layer is a region of small extent, usually close to the boundary, where the solution changes rapidly. As will be evident from this project, simulation of boundary-layer phenomena may easily lead to *qualitatively* wrong results.

Numerical Method. A standard centered finite difference scheme for (1.66) takes the form

$$\frac{u_{i+1} - u_{i-1}}{2h} = \frac{\epsilon}{h^2}\left(u_{i-1} - 2u_i + u_{i+1}\right), \tag{1.67}$$

for $i = 2, \ldots, n-1$, $h = 1/(n-1)$, with $u_1 = 0$ and $u_n = 1$.

Implementation. Implement the scheme (1.67) in a Diffpack program based on the code from Chapter 1.2. The program should dump the u_i values for plotting and compute the error $e = \left(n^{-1}\sum_{k=1}^{\ell}(u(x_k) - u_k)^2\right)^{1/2}$, where u_k is the numerical solution at the grid point x_k, and $u(x_k)$ is the corresponding analytical solution of the PDE problem. Perform a partial verification of the program by letting $\epsilon \to \infty$ and observe that the limiting problem $u'' = 0$ is solved exactly (choose, e.g., $n = 4$).

Analysis. Find the analytical solution of the continuous problem and show that the solution $u(x)$ is monotone. Consider the limit $\epsilon \to 0$ in (1.67) when n is an even integer. Demonstrate that in this limit the following "saw-tooth" solution is possible:

$$u_1 = u_3 = u_5 = \cdots = u_{n-1} = 0, \quad u_2 = u_4 = \cdots = u_n = 1.$$

Find the analytical solution of the *discrete* equations (see Appendix A.4.4) and find a criterion, involving ϵ and h, such that the numerical solution is non-oscillatory, i.e., it exhibits the same qualitative features as the solution of the continuous problem. Find the truncation error of the scheme (1.67) (see Appendix A.4.9).

Computer Experiments. Demonstrate first what happens to the numerical solution, computed by the program, when $h = 2\epsilon - \delta, 2\epsilon, 2\epsilon + \delta$, where $\delta > 0$ is a small number. In addition, show a plot of the solution with $h \gg 2\epsilon$.

Set $\epsilon > h/2$ and make a table of e as a function of the grid spacing $h_k = 2^{-k}$, for $k = 4, 5, 6, 7$. Assuming in general a model $e(h) = Ch^r$ for the leading-order error term, we can estimate r from two values $e(h_k)$ and $e(h_{k+1})$ by

$$r = \frac{\ln(e(h_k)/e(h_{k+1}))}{\ln(h_k/h_{k+1})}.$$

---

[24] See [126, Ch. 3-6.1] for a fluid flow application where (1.66) occurs.

List r for $k = 4, 5, 6$ together with e in the table. How does the estimates compare with the expressions found for the truncation error? (Diffpack hackers might appreciate class `ErrorRate` for automated computation of r, see the man page for class `ErrorRate` or consult Chapter 3.5.8.)

1.4.3   Upwind Differences for a Boundary-Layer Problem

Mathematical Problem. This is a continuation of Project 1.4.2. We consider the same mathematical problem, but the finite difference discretization is different.

Numerical Method. The oscillating solutions that occurred in Project 1.4.2 when the cells were not sufficiently small, can be avoided by using so-called *upwind* differences [40, Ch. 9.1.2]. When $\epsilon > 0$, we use the approximation

$$u'(x_i) \approx \frac{u_i - u_{i-1}}{h}, \tag{1.68}$$

which has an error of order h. The unscaled differential equation corresponding to (1.66) reads $vu' = ku''$, and numerical instabilities occur when the term vu' dominates over ku''. Physically, ku'' models diffusion in positive and negative x direction, whereas vu' models transport of u in a flow with velocity v in positive x-direction (if $v > 0$). When forming the difference approximation to vu' at point x_i, one can apply the upstream or upwind value u_{i-1}, and avoid the downstream or downwind value u_{i+1}, to reflect transport of information from left to right also in the discrete version of vu'. Changing the sign of v (and thereby ϵ) means changing the direction of the flow and the upwind difference at x_i must then be based on $u_{i+1} - u_i$.

Implementation. Implement the upwind scheme in the same program as used for the centered scheme.

Analysis. Find the analytical solution of the *discrete* equations in this case and demonstrate that the numerical solution is monotone, like the solution of the continuous problem, regardless of the mesh size. Find also the truncation error of the upwind scheme.

Artificial Diffusion Interpretation. The upwind scheme may also arise from the following different reasoning: Add a diffusion term $\hat{\epsilon}u''$ to the original equation and discretize the modified equation by standard centered finite differences. Find the value of $\hat{\epsilon}$ that recovers the upwind scheme. In other words, this alternative view shows that adding artificial diffusion $\hat{\epsilon}u''$ can stabilize a scheme and produce the same effect as upwind differences.

Computer Experiments. Demonstrate that the upwind scheme, computed by the program, gives qualitatively correct solution profiles also in the case $h > 2\epsilon$. Produce a table of the error and convergence rate versus h, like in Project 1.4.2. Discuss the merits of the two schemes.

1.5   About Programming with Objects

The programming examples so far in the book reflect traditional scientific programming, where a problem is solved in terms of algorithms realized as subroutines. Data structures, usually in the form of arrays and scalar variables, are shuffled in and out of the subroutines. The use of C++ changes the syntax in comparison with Fortran 77 and C, but the programming concepts remain the same. The only slight advantage of using C++ over Fortran 77 is that the compiler checks the number and type of arguments when calling functions and that the validity of array indices can be controlled at run time. This removes many of the subtle errors in Fortran 77 programs. Since the way of thinking about programming is the same, experienced Fortran 77 programmers will quickly get used to the new C++ syntax after some hours with coding on their own.

Shuffling data in and out of subroutines is often referred to as procedural programming. This is an intuitive implementation technique that allows the programmer to get started and be productive at once. The problem is that PDE codes tend to be large, and procedural-oriented programming loses its human efficiency as code size increases. At some point, even small changes and extensions require substantial modification of existing code, with the danger of introducing errors in thoroughly debugged parts. Many experienced programmers will recognize this scenario, but take the unfortunate situation for granted. Redesign and reimplementation is a possible, but very expensive, recipe.

The basic problem is that procedural programming easily involves too many visible details. When the complexity of a program or problem grows, successful further management relies on introducing new high-level abstractions and hiding the details of more primitive parts. Researchers and practitioners in computer science have established software development techniques that help to increase the abstraction level. Such techniques have proven to be superior to procedural-oriented programming and result in significantly increased human efficiency associated with developing and maintaining computer codes. This may mean higher productivity for researchers who do programming of PDEs, with the opportunity to focus more on models and algorithms and less on debugging code.

Some of the keywords associated with successful modern software development techniques are object-oriented programming, object-based programming, user-defined (or abstract) data types, generic programming (templates), data encapsulation, and inheritance. We refer to Stroustrup [114, Ch. 2] for introductory examples on what these keywords mean. Unfortunately, such techniques have not been adopted in the scientific community until recently. One of the main reasons for the delay is obvious; the programming languages that supported the new techniques were too inefficient for number crunching until C++ became available.

Fortran 90 offers many of the features in C++, but lacks virtual functions (see Norton [83, Ch. 4] for a thorough discussion). Virtual functions are widely used in Diffpack for enhancing software flexibility. In fact, the use of the term object-oriented programming (OOP) normally implies utilization of virtual functions, and this is how we use the term in the present text. Several new languages with true support for OOP have become popular during the 90's. We can mention Java, Perl, Python, and Tcl[25]. These languages are in general too slow for number crunching, but they offer some very powerful features that can be useful when developing simulation and visualization software. For example, graphical user interfaces, combination of simulation and visualization codes, distribution of computational tasks over a network, text file processing, and operating system interactions are usually more conveniently handled by these languages than by C++. Java and Python have a syntax and class concept close to that of C++ and are hence quite easy to learn for a C++ programmer (or vice versa). The syntax and principles of Perl and Tcl require some time to get used to, but the languages are well worth a close study. Diffpack makes much use of Perl for non-numerical tasks.

Many students are now exposed to Java in their introductory programming course. Java coding implies object-oriented programming, because Java Java does not allow traditional procedural programming. A new generation of scientists and engineers will hence be trained in object-oriented concepts. This will definitely make an impact on scientific computing software and create demands for tools like Diffpack. Introductory Java programming is perhaps an optimal background for taking up C++ and Diffpack; the programming concepts are identical, the syntax is similar, and Diffpack supports many of the convenient features found in Java, e.g., garbage collection, platform-independent operating system interface, distibuted programming, and most important, large libraries of ready-made generic modules.

Software for PDEs, based on the OOP paradigm, has attracted significant attention in recent years, and several large-scale packages have emerged. Some of these are described in [10]. A collection of object-oriented numerics software in general can be found on the web page [84]. We remark that our discussion and emphasis in this text center around software tools *embedded in programming languages*, contrary to very user-friendly special-purpose interfaces or languages for solving PDEs, like Fastflo [52], FreeFEM [53,74], and the Matlab PDE toolbox [78]. Tools directly embedded in a standard computer language may take some time to master, but provide a very flexible and extensible problem solving environment.

The forthcoming sections outline the usage of C++ in numerical contexts and in Diffpack in particular. The material is meant as a gentle introduction to C++ programming and forms a basis for further reading of the book. However, we emphasize that the exposition is brief and by no means a substitute for a comprehensive textbook on C++, e.g. [14,91,114].

---

[25] Object orientation in Tcl is supported by a module called `[incr tcl]`.

1.5.1   Motivation for the Object Concept

Example: Computations with Dense Matrices. Suppose we want to compute
a matrix-matrix or matrix-vector product. In the mathematical language we
would express this as: Given $M \in \mathbb{R}^{p,q}$ and $B \in \mathbb{R}^{q,r}$, compute $C = MB$,
$C \in \mathbb{R}^{p,r}$. Similarly, the matrix-vector product is defined as: Given $x \in \mathbb{R}^q$
and $M \in \mathbb{R}^{p,q}$, compute $y = Mx$, $y \in \mathbb{R}^p$.

Let us express these computations in Fortran 77 (a C code will be quite
similar, at least conceptually). The relevant data structures are

```
integer p, q, r
double precision M(p,q), B(q,r), C(p,r)
double precision y(p), x(q)
```

Given these data items, we may simply call a function `prodm` for the matrix-
matrix product and another function `prodv` for the matrix-vector product:

```
call prodm (M, p, q, B, q, r, C)
call prodv (M, p, q, x, y)
```

This approach seems simple and straightforward, and Fortran 77 is often regarded as a convenient language for numerical computations involving arrays.
However, the Fortran 77 expressions for the products involve details about
the array sizes that are not explicitly needed in the mathematical formulations $C = MB$ and $y = Mx$. This observation is in contrast to the basic
strength of mathematics, namely the ability to define abstractions and hide
details. A more natural program syntax would be to declare the arrays and
then write the product computations using the arrays by themselves. We will
now indicate how we can achieve this goal in C++ through user-defined data
types. This means that we will create a new data type in C++, here called
`MatDense`, that is suitable for representation of the mathematical quantity
matrix.

User-Defined Data Types. Related to the example given above, what are the
software requirements for the representation of a dense matrix? Clearly, one
must be able to declare a `MatDense` of a particular size, reflecting the numbers
of rows and columns. Then one must be able to assign numbers to the matrix
entries. Furthermore, the matrix-matrix and matrix-vector products must be
implemented. Here is an example of the desired (and realizable) syntax:

```
// given integers p, q, j, k, r
MatDense M(p,q);        // declare a p times q matrix
M(j,k) = 3.54;          // assign a number to entry (j,k)

MatDense B(q,r), C(p,r);
Vector    x(q), y(p);   // vectors of length q and p
C=M*B;                  // matrix-matrix product
y=M*x;                  // matrix-vector product
```

Programmers of numerical methods will certainly agree that the example above demonstrates the desired syntax of a matrix type in an application code. We will now sketch how this syntax can be realized in terms of a new user-defined data type `MatDense`. This realization must be performed in a programming language. Using C++, user-defined data types like `MatDense` are implemented in terms of a *class*. Declaring a variable of class type `MatDense`, creates a matrix *object* that we can use for computations. Below we list a possible definition of the C++ class `MatDense`.

```
class MatDense
{
private:
  double** A;    // pointer to the matrix data
  int      m,n;  // A is an m times n matrix
public:
  //    --- mathematical interface ---
  MatDense (int p, int q);                    // create pxq matrix
  double& operator () (int i, int j);         // M(i,j)=4; s=M(k,l);
  void operator = (MatDense& B);              // M = B;
  void prod (MatDense& B, MatDense& result);  // M.prod(B,C); (C=M*B)
  void prod (Vector& x, Vector& result);      // M.prod(y,z); (z=M*y)
  MatDense  operator * (MatDense& B);         // C = M*B;
  Vector    operator * (Vector& y);           // z = M*y;
};
```

In this example, the type `Vector` refers to another class representing simple vectors in $\mathbb{R}^n$.

A class consists of *data* and *functions operating on the data*, which in either case are commonly referred to as *class members* in the C++ terminology. The data associated with a matrix typically consist of the entries and the size (the number of rows and columns) of the matrix. In the present case, we represent the matrix entries by a standard two-dimensional C/C++ array, i.e., as a `double` pointer to a block of memory. The number of rows and columns are naturally represented as integers using the built-in C/C++ type `int`. In most cases, the internal data representation is of no interest to a user of the matrix class. Hence, there are parts of a class that are *private*, meaning that these parts are invisible to the user. Other possibilities for representing the matrix entries could include the use of a linked list. To maintain full flexibility, it is therefore important that the user's application program is completely independent of the chosen storage scheme. We can obtain this by letting the user access the matrix through a set of functions specified by the programmer of the class. If the internal storage scheme is altered, only the contents of these functions are modified accordingly — without any changes to the argument sequences. That is, the syntax of any code that uses the `MatDense` class will be unaltered if the internal C/C++ array is replaced by a list structure. We shall later give examples on alternative internal storage schemes in a matrix class.

Member Functions and Efficiency. In the class `MatDense` we have introduced a subscripting operator for assigning and reading values of the matrix entries. This function is actually a redefinition of the parenthesis operator in C/C++,

```
double& operator () (int i, int j);
```

thus providing the common Fortran syntax `M(r,s)` when accessing the entry (r, s). The ampersand `&` denotes a *reference* to the matrix entry, which enables the calling code to alter the contents of the entry. One might argue that issuing a function call for each matrix look-up must be very inefficient. This is definitely true. To circumvent such problems, C++ allows functions to be *inlined.* That is, these functions are syntactically seen as ordinary functions, but the compiler will copy the body of inlined functions into the code rather than generating a function call. In this way, the subscripting operation is as efficient as a direct index look-up in a C array. The inline function can be equipped with a check on the array indices. Using the C/C++ preprocessor, we can automatically include the check in non-optimized code and completely remove it in optimized code. This use of inline functions enables the programmer to achieve the efficiency of pure C, while still having full control of the definition of the call syntax and the functionality of the index operator. We shall explain these concepts in detail on pages 61–62.

The matrix-matrix product is carried out by a member function called `prod`. The corresponding matrix-vector product function has also the name `prod`. This convenient naming scheme is available since C++ can distinguish the two `prod` functions due to different argument types. Fortran and C programmers are well aware of the problems of constructing names for a new function. In C++ one can usually limit the fantasy to the most obvious names, like `print`, `scan`, `prod`, `initialize` and so on, since the compiler will automatically use the class name and the argument types as a part of the name. Such *function overloading* reduces the number of function names employed by a package significantly, a feature that makes the software much easier to use.

The desired syntax for a matrix-matrix product, `C=M*B`, can be achieved by redefining the multiplication operator,

```
MatDense operator * (MatDense& B);
```

In this case, we get a matrix `B`, multiply it (from the right) with the matrix in the current `MatDense` object, and then return the answer as a new matrix. Similarly, one can redefine the addition operator to make it work for matrices,

```
MatDense operator + (MatDense& B);
```

Such constructs permit a compact and elegant syntax for compound expressions, e.g. `Q=M*B+P*R`, where `Q,M,B,P` and `R` are matrices. Unfortunately, in C++ such expressions can lead to loss of efficiency. This is what actually happens: `P` is multiplied by `R`, the result is stored in a temporary matrix

TMP1. Then M is multiplied by B and the result is stored in a temporary matrix TMP2. The TMP1 and TMP2 matrices are added and the result is stored in a temporary variable TMP3, which is finally assigned to Q. The temporary variables are created at run time, and the allocation of the associated data structures (plain C/C++ arrays in the implementation above) can be time consuming[26]. It should also be mentioned that the temporary objects allocated by the compiler can be a waste of storage space, since the compiler is unaware of whether algorithmic data items, say Q, can be used to hold the results of intermediate computations.

The prod function stores the result in a matrix C that the programmer supplies, while the operator* function allocates a matrix and returns it. Dynamic allocation is very convenient, but the Diffpack developers feel that full user control of matrix allocation is an important issue for efficient code. The compound expression Q=M*B+P*R is most effectively evaluated by a special function that implements the compound operator =*+*. This can be realized by creating a member function in class MatDense:

```
void add (MatDense& M, MatDense& B, MatDense& P, MatDense& R);
```

which computes the desired expression. Although one now sacrifices the attractive syntax Q=M*B+P*R, the add function has still lower complexity than the Fortran 77 counterpart. Inside the add function, as in all member functions, we have direct access to the underlying data structure and can utilize special low-level tricks to help compilers optimize the code.

Extension to Sparse Matrices. Let us consider numerical computations with *sparse* matrices (which is more relevant than dense matrices when solving partial differential equations). In the implementation it is then important to take advantage of the fact that only the nonzero entries of the sparse matrix need to be stored and used for computations. For example, the product Mx of a dense matrix $M \in \mathbb{R}^{n,n}$ and a vector $x \in \mathbb{R}^n$ requires in general n^2 multiplications and additions, while the work reduces to about $5n$ multiplications and additions when M stems from finite difference discretization of a 2D Laplace equation $-\nabla^2 u = f$; only about $5n$ entries in M are in this case different from zero. We shall show that the Fortran 77 application code for a sparse matrix-vector product grows in complexity, whereas the C++ version does not. From a mathematical point of view, the expression for the matrix-vector product is independent of whether the matrix is sparse or dense. Therefore, the interface of a matrix should be the same regardless of the storage format, but the internal data structure will be more complicated for a sparse matrix and functions like prod and add must be tailored to the internal storage structure for optimal efficiency.

---

[26] We remark that many of the user-friendly interactive matrix computation systems, like Matlab, S-Plus, Maple, and Mathematica, frequently have similar efficiency problems with compound expressions although it is seldom explicitly documented.

To exemplify, a well-known storage scheme referred to as Compressed Row Storage (see page 627 for an example) needs the following data structure,

```
class MatSparse
{
  private:
    double* A;      // long vector with the nonzero matrix entries
    int*    irow;   // indexing array
    int*    jcol;   // indexing array
    int     m, n;   // A is (logically) m times n
    int     nnz;    // number of nonzeroes
  public:
    // the same functions as in the example above
    // plus functionality for initializing the data structures
};
```

For a *user* of the class, the public functions that can be called are the same. Hence, an application code will not change when employing a sparse matrix instead of a dense matrix. As soon as we have declared a matrix, and given the information on whether it is dense or sparse, the storage details are inherent parts of the matrix representation and the programmer does not need to be concerned with this when using the matrix object.

Refining this approach, the dense and sparse matrix formats could be organized together with other matrix implementations in a class hierarchy. This means that we introduce a common *base class* Matrix that defines a generic interface to matrix operations. The base class does not contain any data; it just tells what you can do with a matrix. *Subclasses* of Matrix implement specific matrix formats, for example, class MatDense for dense matrices and MatSparse for sparse matrices. In addition, we would equip the matrix hierarchy with classes for tridiagonal matrices, banded matrices, diagonal matrices, and so on.

Among the operations on matrices that class Matrix defines, is the prod function for the matrix-vector product. Every subclass of Matrix provides its own implementation of the prod function. For example, class MatDense has a straightforward matrix-vector product loop, whereas class MatSparse offers a much more efficient prod function that utilizes the sparsity of the matrix. A key point is that prod is defined as a *virtual* function. This means that we can program with matrices in terms of a pointer or reference to Matrix, *without knowing whether the matrix is dense or sparse.* Say we have the reference Matrix& M and perform the matrix-vector product y=M*x by the call M.prod(x,y). Now, Matrix is no real matrix, and the reference M must actually refer to some concrete subclass object, e.g. of type MatSparse. C++ will then know at run time that M actually refers to a MatSparse object, and the M.prod(x,y) call is automatically invoked as a call to the tailored prod function in MatSparse. In other words, the matrix-vector product can always be written as M.prod(x,y) in C++, regardless of the type of matrix we work with. This way of "hiding" special matrix formats behind a common matrix interface (Matrix) simplifies the application code considerably and is a central issue in object-oriented programming.

At this point, it is appropriate to discuss how the extension from dense to sparse matrices affects the implementation in Fortran 77. We will then have to explicitly transfer the sparse matrix storage structure to all functions operating on the matrices. Both the name and the arguments of the matrix-vector product function reflect the matrix format. For example,

```
integer p, q, nnz
integer irow(p+1), jcol(nnz)
double precision M(nnz), x(q), y(p)
call prodvs (M, p, q, nnz, irow, jcol, x, y)
```

When writing numerical libraries, one often wants to provide generic solver routines that can work for any matrix format. This is much easier accomplished in C++ than in Fortran.

There is some overhead in a virtual function call, so when the virtual function has few arithmetic operations, the overhead can be noticeable. A striking example is the indexing function `operator()`. If this is implemented as a virtual function, indexing a matrix becomes very inefficient. Virtual functions must therefore be used with care. One could be skeptical in general to the overall efficiency of numerical codes based on C++ and object-oriented programming techniques, because there are numerous features in C++ that easily lead to slow code. The main rule proves to be that algorithms expected to carry out heavy computations should be implemented as member functions of the relevant classes using low-level C functionality. The sophisticated features of C++ are then used at higher abstraction levels for organizing calls to the low-level functions. Diffpack mainly works in this way. Experience has shown that such careful use of C++ makes the computational efficiency close to that of Fortran 77, see e.g. [9] for examples. There are also elements in C++ that allow construction of numerical codes with efficiency superior to that of Fortran 77 [121]. One can also combine Fortran and C++ by letting the C++ member functions call Fortran 77 for performing the CPU-intensive numerics.

1.5.2 Example: Implementation of a Vector Class in C++

The previous section motivated for the class concept and explained some of the basic structure and usage of C++ classes. Now we shall look at the inner details of a class. It would be natural to choose one of the matrix classes from the previous section, but it appears to be somewhat easier to explain the details of a simple class for vectors instead. Our prototype vector class is called MyVector. It turns out that the implementational details of class MyVector cover large parts of the C++ language. The goal for the reader should not be to *understand* C++ on the background of this example. The forthcoming material is instead meant to outline important features in C++, demonstrate how C++ objects behave, and indicate that C++ is a rich and complicated language that takes time to master. The inner details of class MyVector might be quite complicated for a novice C++ programmer, but the *usage* of the class in numerical computations is very simple, a fact that we have tried to demonstrate in Chapters 1.2 and 1.3. This observation should motivate utilization of software libraries like Diffpack; C++ programming related to application of ready-made classes can be straightforward, while implementation of the library classes can be quite tricky and requires extensive C++ knowledge and experience.

Specification of the Desired Functionality. Let us start with specifying the desired syntax and usage of the new type `MyVector`. Some useful features are listed next.

- Create vectors of arbitrary length n: `MyVector v(n);`
- Create a vector with zero length: `MyVector v;`
- Redimension a vector to length n: `v.redim(n);`
- Create a vector as a copy of another vector w: `MyVector v(w);`
- Set two vectors equal to each other: `w = v;`
- Take the inner product of two vectors: `double a = w.inner(v);`
  or alternatively `a = inner(w,v);`
- Write a vector to the screen: `v.print(cout);`
- Extract an entry: `double e = v(i);`
- Assign a number to an entry: `v(j) = e;`
- Extract the length of the vector: `int n = v.size();`

Each statement in the proposed syntax will correspond to calls to member functions in the `MyVector` class, thus giving us complete control of the behavior of the vector.

Definition of the Vector Class. A C++ class must first be defined, which means that we list the class name, the data members, and the member functions. The appropriate syntax is given next, with comments indicating how the public functions are used in application code.

```
class MyVector
{
private:
  double* A;                      // vector entries (C-array)
  int     length;
  void    allocate (int n);       // allocate memory, length=n
  void    deallocate();           // free memory
public:
  MyVector ();                    // MyVector v;
  MyVector (int n);               // MyVector v(n);
  MyVector (const MyVector& w);   // MyVector v(w);
  ~MyVector ();                   // clean up dynamic memory

  bool redim (int n);                      // v.redim(m);
  void operator= (const MyVector& w);      // v = w;
  double  operator() (int i) const;        // a = v(i);
  double& operator() (int i);              // v(i) = a;
```

```
  void print (ostream& o) const;          // v.print(cout);
  double inner (const MyVector& w) const; // a = v.inner(w);
  int size () const { return length; }    // n = v.size();
};
```

Dissection of the Vector Class. A standard one-dimensional C/C++ array `A` is used to store the vector entries. Or in other words, `A` points to a memory segment where all vector entries are stored consecutively. The entries are recovered as `A[0]`, `A[1]`, ..., `A[length-1]`. Notice that we also store the length of the vector as part of the vector object. Since the way we store the vector information technically in the computer is of no interest for application code programmers, the entries and the length are declared as private data members. Additional private information covers functions for allocating and deallocating dynamic memory, that is, the functionality that enables us to create vectors of a length specified at run time. The contents, or the *bodies*, of the `allocate` and `deallocate` functions show how we in C++ actually create and destroy dynamic memory segments:

```
void MyVector::allocate (int n)
{
  length = n;
  A = new double[n];  // create n doubles in memory
}
```

```
void MyVector::deallocate ()  { delete [] A; }
```

The full name of a function in a class is prefixed by the class name, as in `MyVector::allocate`. An important point at this stage is that member functions in a class can access and modify the private and public data members of the class as if they were global. In other words, `allocate` and `deallocate` have direct access to `A` and `length`. We should remark that functions operating directly on dynamic memory often lead to subtle errors in C and C++. As a Diffpack programmer, you will therefore never write code like the `allocate` and `deallocate` functions above. Instead, you will access ready-made Diffpack functions that perform dynamic memory operations in a safe (and thoroughly debugged) way[27].

   The functions in class `MyVector` that have the same name as the class itself, are called *constructors*. These functions define the actions to be taken when declaring an object of type `MyVector`. Writing just `MyVector v;` leads to a call to the `MyVector` function without arguments. In this case it is natural to perform no memory allocation and to initialize the pointer `A` to `NULL` (not pointing to anything) and set `length=0`. The body of this constructor can then take the following form:

---

[27] The Diffpack counterpart to `MyVector` is class `VecSimplest`. Looking at its `allocate` and `deallocate` functions, one will see that these are more complicated than the ones outlined here in the text.

```
MyVector::MyVector ()                    // MyVector v;
{ A = NULL; length = 0; }
```

The next constructor, `MyVector(int n)`, creates a vector of length n, with the declaration syntax `MyVector v(n)`;

```
MyVector::MyVector (int n)               // MyVector v(n);
{ allocate(n); }
```

Space for the vector entries is now allocated, but the entries are not initialized!

Another constructor, `MyVector(const MyVector& w)`, makes a copy of the vector w. Such copying of objects appear frequently in C++ code and the constructor is called a *copy constructor*. Inside this function we must distinguish between the supplied vector w and the vector in the class (i.e. v in the declaration `MyVector v(w)`). The latter is often referred to as the "this" object. A possible form of the copy constructor is

```
MyVector::MyVector (const MyVector& w)  // MyVector v(w);
{
  allocate (w.size());         // "this" object gets w's length
  *this = w;                   // call operator=
}
```

We first allocate memory for "this" vector, i.e. the A array in the class, based on the size of w. Then we want to set all entries in the A array equal to the entries in w. This is exactly what we perform in a statement like v = w, which inside the copy constructor takes the form *this = w. There is a hidden variable in every object, called this, which is a pointer to what we have called the "this" object. To extract the object itself from a pointer to an object, the pointer variable is prefixed by *, i.e., *this is the current ("this") object. Saying just A inside a class `MyVector` member function actually means this->A, or equivalently, (*this).A. Notice that in C++, the arrow -> is used to access a class member (data or function) from a *pointer* to an object, while the dot . is used to access the member from an *object*.

The behavior of the assignment operator in expressions like *this = w or v = w is defined by the `operator=` function in class `MyVector`:

```
void MyVector::operator= (const MyVector& w)     // v = w;
{
  redim (w.size());
  int i;
  for (i = 0; i < length; i++)  // C arrays start at 0: A[0],A[1],...
    A[i] = w.A[i];
}
```

The argument to this function is declared as const `MyVector&`. The const keyword means that w cannot be altered inside the function. If we try to change the length or contents of w, when w is declared as a const `MyVector` object, the compiler will issue an error message. The ampersand & signifies a *reference*. This means that we only transfer the address of w to the `operator=`

function. We could equally well have used a constant pointer, const MyVector*
w. Pointers and references both represent addresses of objects, but they lead
to slightly different syntax; with a pointer we would need to write w->A[i],
whereas a reference requires w.A[i].

The alternative to using a reference (or pointer) as argument to operator=
would be to write just MyVector w. In C++ this implies that we *take a copy*
of the supplied w and use this copy inside the operator= function. That is,
w inside the function and the supplied w in the calling statement (v=w) *are
two different objects*. This allows us to change w inside the function, but the
changes will not be visible outside the function since we work on a copy. The
copy is taken by calling the copy constructor (which again calls operator=
such that we end up with an infinite loop in the present case). If the vector is
long, taking a copy might imply significant (and unnecessary) computational
overhead. It is therefore much more efficient to just pass the address in form
of a reference. Objects with possibly large data structures should always be
passed by reference.

Inside a member function, we can always access the private data members
of another object of the same type, here exemplified by w.A in operator=.
The loop applies basic C indexing of arrays, and it is important to notice
that C arrays start at 0. In application codes working with an object of type
MyVector, we do not allow programmers to write v.A[i], because A is declared
as private. Instead they should index the vector via the operator() function,
i.e. as v(i). With this latter syntax we can also decide that MyVector objects
are indexed from 1, that is, v(1) is the first entry (technically equivalent to
v.A[0]). The definition of the operator() function can be like this:

```
// inline functions:
inline double  MyVector::operator() (int i) const    // a = v(i);
{ return A[i-1]; }

inline double& MyVector::operator() (int i)           // v(i) = a;
{ return A[i-1]; }
```

There are several details of this syntax that should be explained here. Calling
a function every time we want to access an entry in the vector implies a
significant overhead compared to the direct access w.A[i]. C++ therefore
offers functions to be *inlined*, which means that the compiler can copy the
body of the inline functions directly to the calling code. In other words, the
expression v(i) is converted by the compiler to w.A[i-1] and there is no
efficiency loss of indexing through the operator() function[28].

As we see, there are two versions of the operator() function. The first
one returns a double and has a const keyword after the list of arguments,

---

[28] This is not completely true; A[i-1] implies the overhead of a subtraction prior
to looking up the array entry. By a simple trick, namely setting A to A-1 in
allocate, A[i] will correspond directly to our definition of the entry v(i). This
trick is used in Diffpack arrays.

indicating that the function does not alter the data members of the class (otherwise the compiler issues an error message). We can only use such a function for *extracting* a certain entry in the vector, like in the statement `double a = v(j)`. However, we would also need the syntax `v(j)=a` for potentially *changing* a vector entry. The corresponding `operator()` function cannot be `const`. Moreover, it must return *access* to the relevant entry in the vector, that is, it must return the memory address of the entry, here represented as a `double` reference: `double&`. The contents of the two functions are the same, but the application areas differ.

One could ask the following question: Why not just provide *one* indexing function

```
double& operator() (int i) { return A[i-1]; }
```

This function can be used for setting `v(j)=a` and `a=v(j)`. However, if we want to set `a=v(j)` and `v` is a `const MyVector`, the compiler will not allow us to access `v(j)` because `operator()` is not a `const` function; only `const` member functions can be called for `const` objects. That is, to allow full use of `const`, both for documenting the code and increasing programming safety, we need two versions of `operator()`.

Checking the validity of array indices is central when debugging numerical codes. We can easily build an index check into the `operator()` function. It should be possible to turn off the check in optimized versions of the code and only use the check during program development. We therefore put the check inside preprocessor directives such that the preprocessor variable (or macro) `SAFETY_CHECKS` must be defined, usually as part of the compilation command, for the index check to be active:

```
inline double& MyVector::operator() (int i)
{
#ifdef SAFETY_CHECKS
  if (i < 1 || i > length)
    cout << "MyVector::operator(), illegal index, i=" << i;
#endif
  return A[i-1];
}
```

Diffpack makes heavy use of such tests, and `SAFETY_CHECKS` is automatically turned on when compiling Diffpack applications without optimization. The reader should realize that with the class concept in C++ one has full control of all aspects of the implementation and behavior of e.g. a vector class; the functionality is exactly how the programmer has defined it to be.

The `operator=` function could have been implemented using `operator()` indexing instead of the basic C/C++ indexing. The loop would then take form

```
for (i=1; i<=length; i++) (*this)(i) = w(i);
```

Observe here that (*this)(i) calls the non-const this->operator() function, while w(i) calls the const w.operator() function, because w is a const object according to the declaration const MyVector& w, and the "this" object is non-const when the function operator= is non-const. By inlining, the compiler should translate (*this)(i) = w(i) into A[i-1] = w.A[i-1].

A function for redimensioning the vector can be almost trivially constructed by some logic and calls to allocate and deallocate:

```
bool MyVector::redim (int n)
{
  if (length == n)
    return false;  // no need to allocate anything
  else {
    if (A != NULL) {
      // "this" object has already allocated memory
      deallocate();
    }
    allocate(n);
    return true;   // the length was changed
  }
}
```

On the background of the preceding text, the print and inner functions should be quite self-explanatory:

```
void MyVector::print (ostream& o) const           // v.print(cout);
{
  int i;
  for (i = 1; i <= length; i++)
    o << "(" << i << ")=" << (*this)(i) << '\n';
}
```

```
double MyVector::inner (const MyVector& w) const // a = v.inner(w);
{
  int i; double sum = 0;
  for (i = 0; i < length; i++)   sum += A[i]*w.A[i];
  // alternative: for (i=1; i<=length; i++)   sum += (*this)(i)*w(i);
  return sum;
}
```

Notice that both functions do not alter the "this" object, hence, they can be declared as const. One should always try to declare functions and function arguments as const if possible.

One might prefer the syntax a=inner(v,w) instead of a=v.inner(w). This is trivially accomplished by a global inline function:

```
inline double inner (const MyVector& v, const MyVector& w)
{ return v.inner(w); }
```

We saw that when an object of type MyVector is declared, one of the constructors are called. When the object is to be destroyed, C++ calls its *destructor*. The name of the destructor is the class name preceded by a tilde. The purpose of the function is to clean up dynamic memory:

```
MyVector::~MyVector () { deallocate(); }
```

Finally, we mention that the implementation of class MyVector does not make use of any Diffpack functionality; what we have seen is pure C++.

Splitting the Class Definition and Implementation. A simple demonstration program like the one in Chapter 1.2.3 fits easily into one single file. However, as a program grows in size, it should be partitioned into logical file components. Moreover, when programming in C++ it is usual to collect functions and data in classes. The program code of a class is usually broken up into a header file, with extension .h, and a source-code file, with extension .cpp. Normally, every class X is defined in a file called X.h, whereas all its member-function bodies are placed in a file X.cpp. All inline functions must be written out in the X.h file. The file X.cpp must include X.h in order to access the definition of class X.

The class definition in any header file X.h must be enclosed in a pair of #ifndef and #endif preprocessor directives to avoid multiple inclusions. As an example, consider MyVector.h:

```
#ifndef MyVector_IS_DEFINED
#define MyVector_IS_DEFINED

class MyVector
{
...
};
#endif
```

The first time the compiler sees this file, the preprocessor variable (or macro) MyVector_IS_DEFINED is not defined, making the first #ifndef test true. The variable is thereafter defined. If the compiler hits an inclusion of the header file MyVector.h again (in the same compilation), MyVector_IS_DEFINED is defined and the text between #ifndef and #endif is skipped. Sometimes one encounters compiler error messages complaining about multiple definitions of something. The source of such errors is usually that the header file is not surrounded by proper #ifndef and #endif directives.

Exercise 1.9. Type in the code of class MyVector in the files MyVector.h and MyVector.cpp. Recall that the bodies of inline functions must be placed in the header file MyVector.h. Make a main.cpp file with a simple demo program. ◇

Exercise 1.10. Extend class MyVector with a function for computing the Euclidean norm of the vector.   ◇

Exercise 1.11. Extend class MyVector with a scan(istream& i) function that reads a vector on the form n v_1 v_2 v_3 ... v_n, where n is the total number of entries and v_1, v_2, v_3 and so on are the entries in the vector. Reading an entry v(j) from an istream object i is accomplished by the statement i>>v(j).   ◇

Having seen some of the inner details of a vector class, we can turn to a simpler topic, namely the application of ready-made array classes in Diffpack.

1.5.3 Arrays in Diffpack

Arrays constitute a fundamental data structure in numerical applications. However, arrays are nothing else than sequences of memory locations used in a program to represent different mathematical quantities like vectors, matrices, sets, grids, and fields. In Chapter 1.5.2 we encapsulated a plain C/C++ array in a class for vectors, whereas we in Chapter 1.5.1 outlined the encapsulation of a C/C++ array in terms of a class for matrices. In Diffpack we emphasize programming with abstractions like vectors, matrices, grids, and fields as an alternative to just manipulating plain arrays, because such abstractions make the numerical code closer to the mathematical formulation of the problem and hide many disturbing book-keeping details. Especially in large numerical applications, a close relation between problem formulation and code is an important ingredient for increasing the human efficiency of software development and maintenance.

Basic Vector and Matrix Types. Diffpack programmers replace the typical Fortran 77 or C/C++ array of real-valued entries by the following objects:

- `Vec(real)`: vector v_i, $i = 1, \ldots, n$, of real numbers.
- `Mat(real)`: matrix $M_{i,j}$, $i = 1, \ldots, m$, $j = 1, \ldots, n$, of real numbers.
- `ArrayGen(real)`: a more general multi-index vector $u_{i_1 \cdots i_d}$ of real numbers, with general base indices $b_1, \ldots, b_d$, $i_j = b_j, \ldots, n_j$, $j = 1, \ldots, d$.
- `Ptv(real)`: spatial point x_i, $i = 1, \ldots, d$ (more efficient than `Vec` when $d \leq 3$).

These Diffpack types have been parameterized[29] such that `Vec(Complex)` is the corresponding vector with complex entries. However, `Vec(int)` does not make sense for reasons that we will explain later. In the following discussions we drop the parameterization and write only `Vec`, `Mat`, and so on. Notice that the base index of Diffpack arrays is fixed to unity in `Vec` and `Mat` and not zero, which is standard in C and C++. Unlike the C and C++ arrays, Diffpack arrays enable subscript checking. This is of course a very useful feature when debugging programs, and it can be turned off by simply compiling the application in optimized mode (`MODE=opt`).

---

[29] Those familiar with C++ and the use of templates may find the indicated syntax strange. For reasons of portability across Unix platforms, Diffpack does not yet use true C++ templates. Instead, a set of preprocessor macros is used to mimic the behavior of templates, see [63]. The source code can automatically be converted to a form using true templates, and this is done to generate the Windows version of Diffpack. If your Diffpack version applies true templates, simply look at, e.g., `Ptv(real)` as a notation for the source code type `Ptv<real>`.

From a mathematical viewpoint, the array classes Vec, Ptv, and ArrayGen are *vectors*, where the latter allows multiple indexing and an arbitrary base for each index. This increased generality has of course an efficiency penalty, and that is why Diffpack supports the simpler vector types Vec and Ptv in addition to ArrayGen. Choosing $d = 2$ in an ArrayGen vector gives a storage structure that seemingly coincides with Mat. However, Diffpack programmers should from the initial stage pay attention to the underlying mathematical quantity when they make a choice between ArrayGen with $d = 2$ and Mat. Consider, for example, a finite difference scheme that involves a discrete function $w_{i,j}$ over a two-dimensional uniform grid. In Fortran 77 or C, w would be represented by a two-dimensional array, but in Diffpack we have the choice of ArrayGen (with $d = 2$) or Mat. Since $w_{i,j}$ is actually a vector from a mathematical point of view, ArrayGen is the correct type. This becomes obvious if we consider the same mathematical problem in one or three space dimensions. We can then always use ArrayGen to store the grid point values and only change d according to the number of physical dimensions. Turning to another example, finite element methods usually work with discrete values in terms of a vector indexed from unity to the number of nodes, regardless of the number of space dimensions or the underlying grid. In this case, Vec is the most suitable representation of the vector of grid point values, since it is more efficient than ArrayGen.

Integer arrays in Diffpack have the names VecSimple(int), MatSimple(int), ArrayGenSimple(int), and Ptv(int). The word "Simple" is inserted because integer arrays cannot perform all the numerical computations that real and complex arrays can. The functionality of VecSimple is therefore a subset of the functionality of Vec, realized by inheritance in C++ [14, Ch. 9–11,13] and explained on page 69.

Syntax of Vectors and Matrices. In the following program we demonstrate the syntax of the Vec, Mat, ArrayGen, and Ptv types in Diffpack.

```
#include <ArrayGenSimple_int.h>   // some integer arrays in Diffpack
#include <Arrays_real.h>          // all real arrays in Diffpack
int main (int argc, const char* argv[])
{
  initDiffpack (argc, argv);
  int i,j,k,n,m,p; real r;
  n = m = 4;  p = 3;

  Vec(real) w(n);          // declare vector of length n
  w.redim (m);             // change length to m (no effect if n=m)
  i = w.size();            // read length of w vector
  w.fill (-3.14);          // set all entries of w equal to -3.14
  w = -3.14;               // alternative syntax to fill
  Vec(real) z;             // declare an empty vector (length 0)
  z.redim(w.size());
  z = w;                   // redim z and init z by w
  z(n-1) = w(1) - 4.3;     // this is how we index Vec objects
  z.print (s_o, "z");      // write z to standard output with a heading
```

```
z.printAscii(s_o,"z");// same output, for ascii format only
z.print (s_o);         // more compact output
Vec(real) q(z);        // make a vector identical to z
w.add(z,3.5,q);        // w = z + 3.5*q
w.print ("FILE=w.dat1", "w");        // print to file w.dat1
w.print (Os("w.dat2",NEWFILE),"w"); // alternative syntax
s_o << "\nRead a vector of length " << w.size() << ": ";
w.scan (s_i);          // read from cin (w must have correct length)
s_o->setRealFormat ("%6.2f");   // change output format
w.print (s_o, "w read from standard input");
w.scan ("FILE=w.dat1");          // read w from the file "w.dat1"
w.scan (Is("w.dat1",INFILE));    // read w from the file "w.dat1"
w.save("w.m","w");     // save w in Matlab format on file "w.m"
w.load("w.m","w");     // load w in Matlab format from file "w.m"
r = w.inner (z);       // compute r as the inner product of w and z
r = w.norm();          // compute r as the Eucledian norm of w
r = w.norm(Linf);      // L-infinity norm of w (=max(abs(w_i)))
r = w.norm(l2);        // discrete l2-norm (Eucledian norm)
r = w.norm(L2);        // continuous L2-norm (=l2-norm/sqrt(n))
r = w.norm(l1);        // discrete l1-norm
r = w.norm(L1);        // continuous L1-norm

Mat(real) B;           // declare an empty matrix (no data)
B.redim (n,m);         // redimension B to an n times m matrix
B = 1.0;               // set all entries to 1.0
B.fill (1.0);          // alternative syntax
B.prod (w, z);         // compute z=B*w
z.print (s_o, "z=B*w");
B.prod (w, z, TRANSPOSED); // compute z=(B-transposed) * x
z.print (s_o, "z=B^t*w");
Mat(real) C(n,n);
B.redim (n,n);  B = 0;  for (i = 1; i <= n; i++)  B(i,i) = i;
B.factLU ();           // overwrite B with its LU factorization
B.inverse (C);         // compute C as the inverse matrix of B
C.print (s_o, "C");    // print and scan work as for a Vec object
B.size (j,k);          // read dimensions j and k of B
i = B.rows();          // read the number of rows in B
j = B.columns();       // read the number of columns in B

ArrayGen(real) a;      // declare an empty vector (length=0)
a.redim (m);           // change length to m
a.redim (m,n,p);       // three-dimensional array
a.redim (m,n);         // a gets two indices: (1:m) and (1:n)
a.setBase(0,-1);       // a's indices become: (0:m-1) and (-1:n-2)
int i1, in, j1, jn;    // loop limits
a.getBase (i1, j1);    // get base indices
a.getMaxI (in, jn);    // get largest index in each array dimension
// typical loop:
for (i = i1; i <= in; i++)
  for (j = j1; j <= jn; j++)
    a(i,j) = i - 2*j;
// all Vec functionality is also available to ArrayGen, e.g.,
a.print(s_o,"a"); a=-3.15; r=a.norm();
// B.prod(...) can also take ArrayGen arguments instead of Vec

// integer array: VecSimple(int), ArrayGenSimple(int):
VecSimple(int) c;
```

```
    c.redim(11);   c.fill(2);   c.print(s_o,"v");
    // syntax is equal to Vec, but numerical computations
    // (inner, norm etc) are not allowed
    ArrayGenSimple(int) e(2,4); e.setBase(0,0);  // array e(0:1,0:3)
    e.fill(-1);   e(0,1) = 10;    e.print(s_o, "e");

    // effective class Ptv for an index or a point:
    Ptv(int) ind;            // empty object (length 0)
    ind.redim (3);           // ind is to be used as index in 3D
    ind(1)=2; ind(2)=m; ind(3)=n;
    a.redim (ind);           // redim to a 3-dim array of length 2*m*n
    ind(1)=1; ind(2)=1; ind(3)=2; // index
    a(ind) = 7.9;            // assignment
    ind.redim (4);           // makes it possible to work with 4-dim array
    // syntax is the same as that of Vec:
    ind.fill (3);   ind(2) = 5;
    // slightly different print functions:
    s_o << "\nind="; ind.print(s_o); s_o << " or "
        << ind.printAsIndex() << "\n";

    Ptv(real) x(3);          // 3D point
    Ptv(real) y(3);   x = 0.0;   y = 2.0;
    s_o << "||x-y||=" << x.distance(y) << "\n\n";
    return 0;
}
```

You can find the program above in src/start/arraysyntax/main.cpp. If you compile and run it (see Appendix B.2.1), you will observe that the header text in the matrix and vector output contains some special characters. The purpose of these is to indicate the array size and thereby help the scan functions reading the arrays back in memory again.

Sending Diffpack Arrays to C and Fortran Routines. Many programmers of numerical applications will need to access well-tested C or Fortran 77 software from their new C++ programs. The Diffpack vector types are easily sent to C or Fortran routines by calling the member function getPtr0, which returns a pointer to the underlying C array. This pointer points at the memory location containing the first entry in the array. Multi-dimensional ArrayGen-type objects employ the same storage structure as Fortran (i.e. the first index has the fastest variation). Sending, e.g., a two-dimensional ArrayGen object to a C function expecting a double** pointer causes problems, because the ArrayGen provides a single pointer only. Extra code must then be written for generating the double pointer. Another problem arises when Diffpack matrices (Mat) are to be communicated to Fortran, because the underlying C matrix (double pointer) employs the C convention for storing matrices: the first index has the slowest variation, i.e., the entries are stored row by row instead of column by column as in Fortran. Extra code for transposing arrays is needed in such cases. The matrix data to be sent to the Fortran function consists of a single pointer to the first entry, obtained by the member function getData in the matrix class.

The complete functionality of the Diffpack array classes is documented in the man pages (see Appendix B.3.1 for how to access Diffpack's man pages).

Organization of Vector and Matrix Types in Diffpack. We will now explain the class design of the array types in Diffpack. Preferably, the reader should be familiar with concepts such as inheritance and virtual functions. *The material on the design of array objects in Diffpack is not required for continuing with this text.* Hence, you can safely move on to Chapter 1.5.4, which is another optional section on object-orientation and C++.

The design of the array class hierarchies may seem complicated at first sight, but the reason for the design is flexibility, generality, and reliability. The vector hierarchy has a base class `VecSimplest(Type)` that is as simple as an array class can be; it is nothing but a plain C/C++ array of `Type` entries with a class interface. There are no requirements posed on the class type of the array entries (`Type`), besides that class `Type` must have a constructor without arguments. Class `VecSimplest` is similar to class `MyVector` in Chapter 1.5.2, except that the array entry is parameterized in terms of a class `Type`, and `operator=` and output functionality is removed to avoid the requirement that class `Type` has `operator=`, `operator<<`, and a copy constructor. Furthermore, the inner product function in `MyVector` has no meaning for a general entry type (class `Type` may fail to provide an `operator*` function).

A slightly more advanced class, `VecSimple`, is derived from `VecSimplest`, inheriting all the data and functions in `VecSimplest`. Class `VecSimple` can print and read vector entries and set two vectors equal to each other. If you use class `VecSimple` to create arrays, where each entry is an object of a class (`Type` in our parameterization), that class must implement a copy constructor, `operator=`, `operator<<`, and `operator>>`.

Class `VecSort` is a subclass of `VecSimple` and adds functionality for sorting vector entries. The sorting procedures require that the entry class (`Type`) has defined the operators <, >, <= and >=. If class `Type` also supports the operators +, -, *, and /, we can perform numerical computations with the entries. A class with many built-in numerical computations (like the norm, the inner product, etc.) is `Vec`. The split of the vector concept into the different layers `VecSimplest`, `VecSimple`, `VecSort`, and `Vec` is motivated by the diverse requirements of the functionality of class `Type`. We use inheritance for sharing code, such that procedures for indexing, allocation, and deallocation are only programmed and tested once.

The layered design of vector classes in Diffpack has the advantage that if you do not need sophisticated array functionality, you can use a very simple class, like `VecSimplest`. It guarantees optimal efficiency and has a code that is easy to understand. If the array entries are complicated objects, like full PDE simulators, you *must* use such a simple class, because it does not make sense to read, print, sort, or perform arithmetic operations on PDE simulators.

The `VecSimplest`, `VecSimple`, and `Vec` classes implement vectors with unit index base, contrary to the C/C++ convention where array indexing starts at zero. It is convenient to have an arbitrary base and to allow multiple indexing (one, two- and three-dimensional arrays). Such functionality could be incorporated into class `VecSimplest`. However, we think it is important that the simplest class is just a plain C/C++ array without additional data structures. The added functionality offering multiple indices and arbitrary base values is located in class `ArrayGenSimplest`. To increase the reliability, it can utilize the already developed vector data structure and indexing, enabled by deriving the class from `VecSimplest`. The idea that increased functionality should take place in subclasses is also realized for the `ArrayGen` classes: `ArrayGenSimple` can print its contents and set arrays equal to each other, while `ArrayGen` can perform all the numerics that `Vec` offers. This is simply achieved by deriving `ArrayGen` from `Vec`[30]. This design is easily accomplished by combining existing classes using multiple inheritance in C++. Notice that the numerical functionality of class `Vec(Type)` does not make sense when `Type` is `int`, which means that there are no `Vec(int)` or `ArrayGen(int)`.

A similar design is used for matrices as well. The most primitive dense matrix class is called `MatSimplest`, class `MatSimple` is an extension of functionality, while `Mat` is aimed at numerical computations. There are no `MatSort` or `ArrayGenSort` classes, but `ArrayGen` inherits sort functionality through its base class `Vec` (which has `VecSort` as base).

We have already pointed out that there is a clear difference between vectors and matrices from a mathematical point of view and that this difference is mirrored in all parts of the Diffpack software. Matrix classes for numerical computations can be accessed through a common interface specified by the base class `Matrix`. Various subclasses realize tridiagonal matrices (`MatTri`), banded matrices (`MatBand`), structured sparse matrices (`MatStructSparse`), arbitrary sparse matrices (`MatSparse`), diagonal matrices (`MatDiag`), etc. A similar generic interface for vectors is provided by class `Vector`. Its subclasses represent ordinary vectors with a single subscript (`Vec`), vectors that allow multiple indices (`ArrayGen`), and vectors with multiple indices and "ghost" (inactive) entries (`ArrayGenSel`). We refer to Appendix D.1 for more information about the design of vector, matrix, and linear system classes in Diffpack. Figure 1.7 shows the vector and matrix class hierarchies. A technical justification of the design is provided by the references [22,24].

We remark that there are two possible alternatives to array classes, namely sets and lists. These are well suited if the user does not know the length of the array in advance. After having used a set or a list to create a series of objects, it can be converted to an ordinary array (usually `VecSimple`). There are two set classes in Diffpack: `SetSimplest(Type)`, which can make sets of

---

[30] The observant reader will now claim that class `ArrayGen` gets one `VecSimplest` part from `Vec` and another `VecSimplest` part from `ArrayGenSimplest`. Thus a class `ArrayGen` object has two vectors! This could be true, but the effect is inconvenient and avoided by declaring `VecSimplest` to be a *virtual* base class when deriving `VecSimple` and `ArrayGenSimplest`. Consult a C++ textbook [14,114] for more information on virtual base classes.

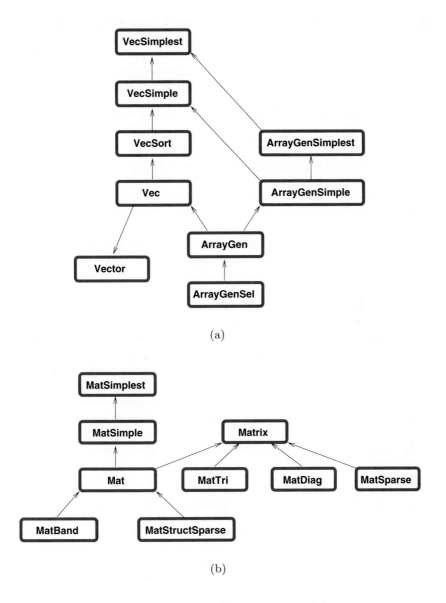

Fig. 1.7. Class hierarchy for (a) vectors and (b) matrices.

almost any type of objects, and SetOfNo(Type), which is aimed at set of numbers (Type must be int or real). For lists, there is a class List(Type,Item), where Type can be a user defined class, and Item reflects the type of each list item: an instance, a pointer, or a handle (handles are smart Diffpack pointers that can perform reference counting and some kind of garbage collection, see page 83). There is also a list of lists class: `Lists(Type,Item)`.

The most common use of the set classes is when a series of numbers or strings are to be read from input and the number of items is unknown. The `scan` function reads a set of numbers terminated by a semicolon, for example,

```
SetOfNo(int) nodes; String s="1 8 2 16 23;"; nodes.scan(s);
```

It is then easy to load the set into a vector:

```
VecSimple(int) v; nodes.convert2vector(v);
```

The `scan` function in `SetOfNo` objects can also read a string on the form `[0:10,2]`, which means all numbers from 0 to 10 in a step of 2, that is, $\{0, 2, 4, 6, 8, 10\}$.

The set and list classes in Diffpack are quite primitive and mostly suited for building up data structures of unknown sizes prior to converting them to Diffpack arrays. In the case you need to do advanced programming with sets or lists, we highly recommend to use the classes in the Standard Template Library (STL) [114] that ought to be a part of your C++ implementation.

Diffpack is centered around computation on arrays[31] rather than on sets or lists because of efficiency reasons. Algorithms for numerical solution of PDEs can usually be expressed as manipulation of (very) large arrays, and the natural "Fortran-like" loops over array indices are easily recognized by optimization modules in compilers. In other words, intensive numerical calculations should have a C++ implementation whose syntax triggers compilers to take advantage of the experience from four decades of optimizing Fortran codes. However, if we are not concerned about efficiency, implementations of many numerical methods, especially the finite element method, can be made very simple and elegant in terms of sets and lists.

1.5.4 Example: Design of an ODE Solver Environment

In this section we exemplify how advanced C++ features, such as inheritance and virtual functions, can be conveniently used in numerical contexts. Novice C++ programmers may find this section complicated; after all, we solve a simple mathematical problem using a fairly complicated C++ code. However, the basic principles demonstrated here extend to advanced and large numerical codes. The Diffpack libraries are founded on similar design principles, so to take real advantage of the Diffpack system, the contents of this section should be understood. You can probably jump to the next chapters and be successfully playing around with Diffpack without reading the forthcoming

---

[31] STL also contains vector classes and even an array type, `valarray`, aimed at numerical computations. The reason why Diffpack applies its own special array classes is mainly the need for high flexibility and efficiency at the same time. This has forced us into a collection of array classes with different properties.

sections, but as soon as you feel your understanding of C++ and Diffpack is too shallow to make satisfactory progress, it is a good point to study the current section and the rest of Chapter 1 in detail.

To focus on programming concepts, we shall work with numerical solution of ordinary differential equations (ODEs), because this keeps the mathematics simple. The ODE system is written as

$$\frac{dy_i}{dt} = f_i(y_1, \ldots, y_n, t), \quad y_i(0) = y_i^0, \quad i = 1, \ldots, n. \tag{1.69}$$

We assume that the reader is familiar with such problems and the common numerical methods for solving them, e.g., the forward Euler scheme and the fourth-order Runge-Kutta algorithm.

The most straightforward way to solve (1.69) is to hardcode the f_i functions and the numerical algorithm. Using the forward Euler or fourth-order Runge-Kutta method, this requires about an hour of work, including source code writing and testing. Why then spend time on advanced programming concepts? Although a simple algorithm is very easy to hardcode for a given problem, every new problem requires approximately the same lines of code. In such situations we should, as a fundamental principle, divide the code into general and problem-specific parts. The general parts can be collected in a numerical library and reused in other contexts, thus saving some coding and debugging when new applications are addressed. A quite practical reason for building libraries of ODE solvers is that many ODE systems require quite sophisticated solvers. Occasionally, the user must experiment with several solvers to determine an appropriate choice for the problem at hand. This means that our particular equation, i.e. the f_i functions, must be combined with several possibly complicated solution algorithms. Only the f_i functions should then be implemented by the user, and the numerics of ODE solvers should be offered by a general library. The current section outlines some basic ideas for designing and implementing such libraries.

General libraries for ODE solvers have been available in the scientific computing community for several decades. Fortran 77 libraries implement the algorithms in terms of subroutines. For example,

```
SUBROUTINE RK4(Y,T,F,WORK1,N,TSTEP,TOL1,TOL2,...)
```

is a typical interface. The Y variable is an array of the current y variable, T is the value of t, F is an external function defining f_i, WORK1 is a work array (recall that Fortran 77 cannot allocate work arrays at run time), TSTEP is the time step, TOL1, TOL2 and so on are various parameters for controlling the behavior of the algorithm. This is an obvious design and quite easy to use. However, we shall point out a couple of principal difficulties that we can handle more elegantly using object-oriented programming.

Let us be specific and treat a particular ODE:

$$\ddot{y} + c_1(\dot{y} + c_2\dot{y}|\dot{y}|) + c_3(y + c_4 y^3) = \sin \omega t. \tag{1.70}$$

To obtain a set of first order equations, which is assumed in both the mathematical and implementational framework, we rewrite the second-order equation as a system of two first-order equations, $\dot{y}_1 = y_2$, $\dot{y}_2 = \ddot{y}$, leading to

$$f_1 = y_2, \tag{1.71}$$
$$f_2 = -c_1(y_2 + c_2 y_2 |y_2|) - c_3(y_1 + c_4 y_1^3) + \sin \omega t. \tag{1.72}$$

It would have been convenient to let the user-defined function have y_i, c_i and ω as arguments. That is, the Fortran function F looks like this

```
SUBROUTINE F(YDOT,Y,T,C1,C2,C3,C4,OMEGA)
```

However, the call to F will then be problem dependent. Subroutines like RK4 force the F function to be the same in all problems:

```
SUBROUTINE F(YDOT,Y,T)
```

The parameters c_i and ω needed in the F function must be transferred by COMMON blocks, i.e., as global variables, frequently leading to side effects and bugs that are hard to find.

   In C++ we can improve the design by letting the user program with a generic solver of unknown type (i.e. there is no need to hardcode calls to a specific routine like RK4), and the particular type of solver can be chosen at run time. Class hierarchies and virtual functions are important tools in this respect. Furthermore, the code for the f_i functions should have a generic signature, such that it can be called from any solver, without requiring that the problem-dependent parameters $c_1, \ldots, c_4, \omega$ are global variables. This is enabled by letting the f_i functions be implemented in terms of a class instead of a function.

Basic Principles of the Software Design. A C++ tool for solving ODE systems may consist of two class hierarchies, one for the solution algorithms and one for the problem-dependent information (f_i). (We remark that the concepts outlined in this paragraph probably become clearer when we later discuss the corresponding source code in detail.) The solution algorithm hierarchy has a base class ODESolver that defines a virtual function advance, which advances the solution one time step. Specific algorithms are implemented in subclasses derived from class ODESolver. As examples, we will demonstrate the implementation of the advance function for the explicit Forward Euler integration method in a class ForwardEuler. The fourth-order Runge-Kutta scheme is realized as a class RungeKutta4. Although we only demonstrate the details of trivial solution methods, the reader should understand that it is in principle straightforward to equip our library with more sophisticated algorithms.

   Using an ODESolver pointer or reference you can call the virtual advance function and the program will at run time figure out *which* particular algorithm (e.g. Forward Euler or fourth-order Runge-Kutta) that is to be used.

This is an important feature made possible by object-oriented programming: In an application code we can work with an abstract solver ODESolver and ask it to solve our system without any knowledge of what type of algorithms that will be used or the implementational details of that algorithm. This concept is also very advantageous in more complicated settings, for instance, in numerical linear algebra and finite element methods.

The ODE solver needs access to the user's problem (f_i) through some general interface, such that the source code of the solution algorithm becomes independent of the particular problem being solved. To this end, we define a base class ODEProblem for the user's problem, with a virtual function equation that defines the f_i functions. Subclasses of ODESolver implement specific f_i functions. For example, we shall here make a subclass Oscillator that implements the virtual equation function corresponding to (1.70), or more precisely, (1.71)–(1.72). The ODEProblem class also contains a driver function, timeLoop, for calling the solver's advance function for $t = \Delta t, 2\Delta t, \ldots$ to some final time T. Since all problem classes involve common data, such as Δt, T, the name of the ODE solver etc., we can collect these data and functions reading the data from standard input in the base class ODEProblem. Every subclass automatically inherits access to the data and functions in the base class. In fact, the total functionality of the subclass consists of the features in the base class code plus the extra functionality declared in the subclass.

From the outline of the design it is clear that ODESolver must be able to access a general ODEProblem object and vica versa, that is, we need a two-way pointer between the base classes.

Implementation in C++. We are now in a position to specify how the various classes should look like in C++ code. The complete code is found in the directory src/start/ode. We have made some use of Diffpack, mostly the vector type Vec(real) and the general output or file class Os. A suitable definition of class ODESolver is listed next.

```
class ODESolver
{
protected:                // members only visible in subclasses
  ODEProblem* eqdef;      // definition of the ODE in user's class
public:                   // members visible also outside the class
  ODESolver (ODEProblem* eqdef_)
    { eqdef = eqdef_; }
  virtual ~ODESolver () {} // always needed, does nothing here...
  virtual void init();     // initialize solver data structures
  virtual void advance (Vec(real)& y, real& t, real& dt);
};
```

Notice that we have here used the keyword protected and not private. The former makes the members invisible for a user, but visible for the subclasses. The advance function has no meaning in class ODESolver, since this class does not represent any real ODE solver; it only defines a common interface to all ODE solvers.

The ODEProblem class can look like this:

```
class ODEProblem
{
protected:
  ODESolver* solver;   // some ODE solver
  Vec(real)  y, y0;    // solution (y) and initial condition (y0)
  real       t, dt, T; // time loop parameters
public:
  ODEProblem () {}
  virtual ~ODEProblem ();
  virtual void timeLoop ();
  virtual void equation (Vec(real)& f, const Vec(real)& y, real t);
  virtual int  size ();  // no of equations in the ODE system
  virtual void scan ();
  virtual void print (Os os);
};
```

The functions equation and size depend on the particular ODE problem being solved so the contents of these functions have meaning in a subclass and not here. The scan and print functions perform general read and write operations on the data in the class. These operations will be extended in subclasses, which will be evident later.

We can now demonstrate how the timeLoop function can be written in a general fashion:

```
void ODEProblem:: timeLoop ()
{
  Os outfile (aform("%s.y",casename.c_str()), NEWFILE);
  t = 0;   y = y0;
  outfile << t << " "; y.print(outfile); outfile << '\n';
  while (t <= T) {
    solver->advance (y, t, dt);  // update y, t, and dt
    outfile << t << " "; y.print(outfile); outfile << '\n';
  }
}
```

Let us now turn to class Oscillator, a subclass of ODEProblem that implements the specific system of ODEs (1.71)–(1.72):

```
class Oscillator : public ODEProblem
{
protected:
  real c1,c2,c3,c4,omega;  // problem dependent paramters
public:
  Oscillator () {}
  virtual void equation (Vec(real)& f, const Vec(real)& y, real t);
  virtual int  size () { return 2; }  // 2x2 system of ODEs
  virtual void scan ();
  virtual void print (Os os);
};
```

The first line tells that the class is derived from ODEProblem, which means that we inherit y, y0, t, dt, and T, as well as the various functions in ODEProblem.

The `timeLoop` function from `ODEProblem` is general and can be reused as it is in class `Oscillator`. Therefore, it does not explicitly appear in the class definition. The functions `equation` and `size` *must* be defined in `Oscillator`, whereas we need more functionality in `scan` and `print` than what `ODEProblem` provides so these must also be defined by class `Oscillator`. The contents of `scan` will be presented later. The most important function is probably `equation`, because it defines the problem to be solved:

```
void Oscillator::equation (Vec(real)& f, const Vec(real)& y, real t)
{
  f(1) = y(2);
  f(2) = -c1*(y(2)+c2*y(2)*abs(y(2))) - c3*(y(1)+c4*pow3(y(1)))
         + sin(omega*t);
}
```

Let us also show the code of a concrete ODE solver, e.g., the very simplest possible scheme, the forward Euler scheme

$$y_i^{\ell+1} = y_i^{\ell} + \Delta t f(y_1^{\ell}, \dots, y^{\ell}, n, t_m). \tag{1.73}$$

Here, $y_i^{\ell} = y_i(t_{\ell})$ and $t_{\ell} = \ell \Delta t$, where we for simplicity assume a constant step size Δt. The implementation takes place in a subclass of `ODESolver`:

```
class ForwardEuler : public ODESolver
{
  Vec(real) scratch1;     // needed in the algorithm
public:
  ForwardEuler (ODEProblem* eqdef_);
  virtual void init (); // for allocating scratch1
  virtual void advance (Vec(real)& y, real& t, real& dt);
};
```

The `advance` function[32] takes the following form:

```
void ForwardEuler:: advance (Vec(real)& y, real& t, real& dt)
{
  eqdef->equation (scratch1, y, t);  // evaluate scratch1 (as f)
  const int n = y.size();
  for (int i = 1; i <= n; i++)
    y(i) += dt * scratch1(i);
  t += dt;
}
```

A similar class, `RungeKutta4` for the standard fourth-order order Runge-Kutta method [92, Ch. 16.1], looks almost identically, except that it needs additional scratch vectors and has a more complicated `advance` function[33].

At this point, we should draw a diagram of the involved classes and their relations, see Figure 1.8. The figure caption refers to the useful "is a" and

---

[32] The observant reader will notice that `dt` is declared as a reference and can be altered by `advance`. This gives the possibility to incorporate adaptive time stepping in the solver.

[33] We refer to the `RungeKutta4.h` and `RungeKutta4.cpp` files for details.

"has a" relationships between classes. Since a subclass inherits everything in the base class, i.e., the base class is a subset of the subclass, we can say that, e.g., RungeKutta4 *is an* ODESolver. The converse is not true, because RungeKutta4 contains more than the ODESolver data and functions. It thus makes sense that a base class pointer ODESolver* can point to any subclass object, since any subclass is an ODESolver. If a class ODESolver has a pointer to another class ODEProblem, we say that ODESolver *has an* ODEProblem. The "is a" and "has a" relationships are used throughout the book in class diagrams.

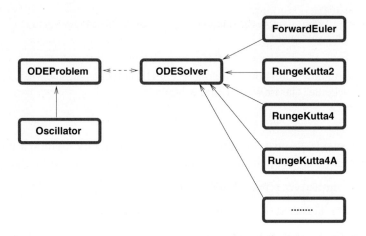

Fig. 1.8. A class hierarchy for ODE solvers and user problems. The solid line indicates class inheritance ("is a" relationship), while the dotted line indicates a pointer to the class being pointed at by the arrow ("has a" relationship).

We can now explain how the timeLoop function really works. At each time level, when we issue the call solver->advance, we actually call the advance function in some unknown *subclass* of ODESolver, where the desired solution scheme is implemented. The solver pointer points to this subclass, but inside the ODEProblem code we cannot see *which* subclass it points to; C++ keeps track of such information at run time. Writing general code and letting automatically generated code fill in problem-dependent details at run time is some of the "magic" that has made object-oriented programming so popular.

The advance function, e.g. ForwardEuler::advance, needs access to the f_i functions and obtains this access through an ODEProblem* pointer eqdef and the call eqdef->equation. The solver does not know which subclass of ODEProblem that eqdef actually points to, but again C++ handles this information when it is needed. In this way, the solver and the problem class are independent; all solvers can be combined with all problems and vica versa. Of course, in some problems certain solvers might be inappropriate for the problem at hand, so the user of such a flexible library must know the numerical limits of the software flexibility.

Having seen how the suggested class design solves the ODE problem, it remains to see how the `ODESolver*` and `ODEProblem*` pointers are initialized. At some place in the code we must say that the `solver` pointer should point to, e.g., a `ForwardEuler` solver, and at some place we must say that `eqdef` should point to an object of, e.g., class `Oscillator`. The Diffpack libraries collect related numerical methods in a class hierarchy, exactly as we have outlined for the ODE solvers here. To create a particular subclass object in such a hierarchy, we employ the following convention. Say the name of the base class is `X`. All the parameters that are needed for initializing objects in any subclass of `X` are collected in a *parameter object* of class `X_prm`. In addition, this class must contain the name of the subclass of `X` to be created. A function `X_prm::create` creates the actual subclass object. In the present example, we need the `ODEProblem*` pointer as input when initializing subclass objects of `ODESolver`. Class `ODESolver_prm` becomes quite simple:

```
class ODESolver_prm
{
public:
  String      method;    // name of subclass in ODESolver hierarchy
  ODEProblem* problem;   // pointer to user's problem class
  ODESolver*  create (); // create correct subclass of ODESolver
};
```

The `create` function must check the `method` string and allocate the proper subclass object in the `ODESolver` hierarchy:

```
ODESolver* ODESolver_prm:: create ()
{
  ODESolver* ptr = NULL;
  if (method == "ForwardEuler")
    ptr = new ForwardEuler (problem);
  else if (method == "RungeKutta4")
    ptr = new RungeKutta4 (problem);
  ...
  else
    errorFP("ODESolver_prm::create",
            "Method \"%s\" is not available",method.c_str());
  return ptr;
}
```

This is the only place in the code where we deal with specific subclass names of ODE solvers; in the rest of the code we only see ODE solvers as `ODESolver*` pointers. The specification of the particular problem subclass of `ODEProblem` to be used appears in the main function shown later.

In the `ODEProblem::scan` function we perform the necessary initialization of data structures (memory allocation), read input from the terminal screen, and allocate the proper solver object:

```
void ODEProblem:: scan ()
{
  const int n = size();  // call size in actual subclass
  y.redim(n);  y0.redim(n);
  s_o << "Give " << n << " initial conditions: ";
  y0.scan(s_i);
  s_o << "Give time step: ";    s_i >> dt;
  s_o << "Give final time T: "; s_i >> T;

  ODESolver_prm solver_prm;
  s_o << "Give name of ODE solver: ";
  s_i >> solver_prm.method;
  solver_prm.problem = this;
  solver = solver_prm.create();
  solver->init();
  // more reading in user's subclass
}
```

A particular problem class will usually need additional reading functionality, for example,

```
void Oscillator:: scan ()
{
  // first we need to do everything that ODEProblem::scan does:
  ODEProblem::scan();
  // additional reading here:
  s_o << "Give c1, c2, c3, c4, and omega: ";
  s_i >> c1 >> c2 >> c3 >> c4 >> omega;
  print(s_o);  // convenient check for the user
}
```

The `print` function can typically print the equation on the screen, together with other parameters of interest.

The library classes and our problem-dependent class `Oscillator` are now ready for application:

```
int main (int argc, const char* argv[])
{
  initDiffpack (argc, argv);
  Oscillator problem;
  problem.scan();       // read input data and initialize
  problem.timeLoop();   // solve problem
}
```

What Has Been Gained? The class hierarchies and all the virtual functions may be a disturbance to the basic numerics of the problem. However, our purpose here was to indicate how to implement a *flexible library*, and the basic ideas of the suggested code design carry over to numerous other numerical contexts. Also notice that although we can hardcode a solution algorithm for the problem (1.71)–(1.72) pretty quickly, it is even faster to just implement f_1 and f_2 and then call up a ready-made library for performing the numerics.

Diffpack is essentially a collection of libraries based on design principles similar to those used for the present ODE solver library. We emphasize that this ODE solver library is just a toy code and not a part of the Diffpack libraries.

Exercise 1.12. Modify class ODEProblem such that it stores all the t and $y_1(t)$ values in arrays of type Vec(real). Also improve the input in the scan functions by using the initFromCommandLineArg described on page 87 (these modifications prepare for the class in the next example).   ◇

Example 1.2. The modified ODEProblem class from Exercise 1.12 can be extended with functionality for interactive plotting in Matlab, such that we can observe the evolution of $y_1(t)$ *during the computations*. This is enabled by the MatlabEngine class in Diffpack [1]. Its usage is quite simple and briefly demonstrated on page 317, but now it should be sufficient to take a look at the source code of class ODEProblem in src/start/ode/matlab-movie. To compile the application, copy the *complete* ode directory to your local directory tree, go to the matlab-movie subdirectory and issue the Unix command Make MODE=opt CXXUF=-DMATLAB. This requires the availability of the Matlab library libmat.a on your computer system, see [1] for details. Run the code (./app) with the command-line options -c1 20 -c2 0.5 -c3 1 -c4 0.6 -omega 1.1. ◇

Exercise 1.13. Add a new solver to the ODESolver hierarchy, where you implement an *adaptive* Runge-Kutta method. See for example [92, Ch. 16.2] for a suitable adaptive algorithm. The whole adaptive step should appear within the advance function, that is, avance must administer a trial step, examination of the error, adjustment of Δt (dt) and so on. Additional data needed in an adaptive method (error tolerances, scaling of y_i etc.) can be stored in the solver class and set by command-line arguments (cf. Exercise 1.12).   ◇

Extensions. A real ODE solver library would contain sophisticated time stepping and error control. This will require many additional parameters. To accomplish such an extension, it would be wise to define a class ODESolver_prm that contains dt, tolerances, names of error control methods and so on. Class ODESolver should have an instance of this parameter class and it should be an argument to the constructor. Equipping class ODESolver_prm with a user-friendly interface (for example in terms of Diffpack's menu system), makes it easy for the user to set various algorithmic parameters at run time. Since the ODESolver_prm object is stored in the base class ODESolver, all the specific algorithms get access to all the common algorithmic parameters. Godess [86] is an example on a professional object-oriented library for solution ODEs, containing numerous advanced methods.

1.5.5 Abstractions for Grids and Fields

A nontrivial PDE simulator often counts more than 50,000 lines of code. In such software regimes a clear design closely related to the mathematical formulation is necessary to understand, extend, and maintain the source code. The key to obtaining an implementation that is close to the mathematical language, is to work with software abstractions, represented in terms of classes, that mirror the entities and concepts of mathematics.

In the mathematical formulation of a discrete method for PDEs we consider the unknown function u as a scalar *field*, defined over a *grid*. Many methods, like finite differences or finite elements, compute u only at the *grid points*, while the discrete mathematical field abstraction may imply some interpolation scheme for evaluating u at an arbitrary point in the domain. To reflect the mathematical quantities, we should develop our programs using grid and field objects. When programming simple 1D finite difference examples, it is not obvious that such abstractions represent an important improvement of the code. However, when working with more complicated problems, especially when using the finite element method, the field and grid abstractions really pay off in the sense that the user will soon realize that the code is more modular, reliable, and straightforward to extend.

As mentioned, the discrete counterpart of the primary unknown u of a PDE is viewed as a *field* in Diffpack. The finite difference field is a particular type of field which consists of a grid and an associated set of point values. The field class is called `FieldLattice`, and its grid is an object of the type `GridLattice`. The associated point values are represented by an object of the type `ArrayGenSel(real)`. We have already met class `ArrayGen` in Chapters 1.2 and 1.3. `ArrayGenSel` is a subclass (extension) of `ArrayGen` with useful additional functionality for ghost boundaries and inactive entries.

Simple Grids for Finite Difference Methods. The `GridLattice` class offers a lattice grid with uniform partition in d space dimensions. The constructor requires the number of space dimensions, while the grid must be initialized by the member functions. The easiest way to initialize the `GridLattice` object is to use its `scan` function, which builds the internal data structure on basis of an input string that compactly describes the domain's geometry and partitioning. The string

```
d=1 domain: [0,1], index [1:20]
```

initializes the 1D grid on the unit interval with 20 equally spaced grid points, using an index running from 1 to 20. To indicate the flexibility of class `GridLattice`, we outline the corresponding initialization string for a 3D grid[34]:

```
d=3 [0,1]x[-2,2]x[0,10] indices [1:20]x[-20:20]x[0:40]
```

Notice that the indices can start and stop at any integer value, but the shape of the domain must be a box[35]. Here is fragment of the `GridLattice` definition.

---

[34] Only the first = sign, the characters [ ] , : and the numbers are significant in the interpretation of the initialization string.

[35] When a `GridLattice` object is combined with a `FieldLattice` object, cells can be deactivated to approximate more complicated geometries.

```
class GridLattice
{
protected:
  // data that hold grid spacing, size of domain etc
public:
  GridLattice  (int nsd);   // give no of space dimensions
  int  getBase (int dir);   // base index for loops in dir direction
  int  getMaxI (int dir);   // max index for loops in dir direction
  real Delta   (int dir);   // grid spacing in dir direction (constant)
  void scan    (Is is);     // read init. string and initialize
  real getPt   (int dir, int i); // coord. of point i in dir dirction
};
```

We refer to the man page (see Appendix B.3.1) for the class for complete definition and explanation of all the member functions.

Fields over Lattice Grids. Typical scalar fields arising in finite difference methods can be viewed as a GridLattice object with an associated array containing the field values at the grid points. Such field abstractions are available as class FieldLattice in Diffpack. Two convenient functions are grid and values, which allow you to access the grid and to retrieve and store values associated with a given grid-point index. The Handle(X) construction used in the FieldLattice class can be read as X* and is explained below.

```
class FieldLattice
{
private:
  Handle(GridLattice)       grid; // pointer to a grid
  Handle(ArrayGenSel(real)) vec;  // pointer to vector of point values
public:
  FieldLattice (GridLattice& grid, const char* fieldname);
  GridLattice& grid ();              // return access to grid
  ArrayGenSel(real)& values (); // return access to point values
};

// given some 1D FieldLattice f
int i0 = f.grid().getBase(1);  // start index
int in = f.grid().getMaxI(1);  // stop index
for (int i = i0; i <= in; i++) // take the sine of the field:
  f.values()(i) = sin (f.values()(i));
```

The user should observe that we can easily create objects for higher-level abstractions by simply putting together more primitive objects. This principle demonstrates the strength of programming with objects – they let you build with walls and bricks rather than glue matches.

Handles (Smart Pointers). The constructor of FieldLattice requires a grid object. This is natural since the grid is a substantial part of the field abstraction. In our example problem (1.44)–(1.48), we need three fields on the same grid: up1, u, and um1. It would be a waste of memory to let all these fields have a local copy of the grid object. Instead we create *one* grid and let each of the fields *point to* the grid object. However, there is a serious problem with this approach. When `FieldLattice` depends on an external grid object, errors leading to unpredictable program behavior may occur if the grid object is deleted prematurely. Therefore, one should keep track of all the users of an object, through counting the number of pointers or references to the object. The object can only be deallocated when there are no other pointers or references to the object. This functionality is referred to as *reference counting*. In Diffpack, reference counting is offered by a type of smart pointer called *handle*.

A handle is basically a standard C/C++ pointer, but it can also keep track of other pointers to the object. Let `X` be the name of a class. The corresponding handle class then has the name `Handle(X)`. The interface to a handle is much like the interface to an ordinary pointer, for example, one can use the operator `->` in the same way. It is possible to declare an empty handle (implied by the constructor without arguments) that later can be initialized by calling the member function `rebind` as in the following examples.

```
Handle(X) x;            // create an empty handle (NULL pointer)

// later:
x.rebind (new X());     // x points to a new object
someFunc (x());         // send an x object (not the handle!)
someFunc (*x);          // alternative syntax
someFunc (x.getRef());  // yet another syntax for sending the object

// given Handle(X) y:
x.rebind (y());         // x points to y's object
x.rebind (*y);          // alternative syntax

*x = *y;                // copy y's object into x's object
x() = y();              // alternative syntax
x = y;                  // x points to y, same as x.rebind(y)
myFunc (x.getPtr());    // send X* pointer to myFunc(X*)
x.detach();             // x points to NULL and memory can be freed
```

It is a common error to write x=y when we actually want copy the contents of the object handled by x into the object handled by y, but x=y means copying of handles (pointers). The syntax x=y therefore leads to a warning message from Diffpack when running a program compiled without optimization. If one really means x=y, and not *x = *y, we recommend the programmer to write x.rebind(y). Memory is freed when a handle goes out of scope and there no other handles pointing to the object. In this way, handles offer a kind of garbage collection. With the use of x.detach(), which is identical to x.rebind(NULL), one can force the same actions as when the handle goes out of scope.

To access an object of class X by a handle, class X must be derived from class HandleId. The HandleId base class contains the necessary data structures for reference counting and for the correct dynamic allocation and deallocation of X objects administered by handles. If the application is compiled without optimization, class HandleId also contains various debug information.

Visualization of Fields. Visualization of field objects frequently requires that the field values and the grid are written to file in a format specific to the visualization program. A series of classes, having names starting with `SimRes2`, offer (static) functions for transforming Diffpack field objects to various file formats. As an example, we have the class `SimRes2mtv` for filtering Diffpack data to Plotmtv format. Just include `SimRes2mtv.h` and perform the call

```
SimRes2mtv::plotScalar (f, "myfield.res", ASCII, "%contstyle=3");
```

for dumping the field object `f` to the file `myfield.res` in ASCII Plotmtv format. The fourth argument specifies plotting program options that can be placed in the data file. Class `SimRes2vtk` has a similar function for dumping fields to file in Vtk format. Other supported formats include Matlab (`SimRes2matlab`), AVS (`SimRes2ucd`), IRIS Explorer (`SimRes2explorer`), and Gnuplot (`SimRes2gnuplot`). These classes work with both finite difference and finite element scalar and vector fields. There are numerous other useful functions in these classes for visualization of fields. We refer to the man pages for the classes and to Appendix B.4.2 for further information. Having the files with field data available in a particular format, Chapters 3.3 and 3.10.2 explain how to invoke the various visualization programs and produce plots for stationary and time-dependent problems.

1.6   Coding the PDE Simulator as a Class

In the next two sections we shall reimplement the user codes from Chapter 1.2 and 1.3 in a more advanced, flexible, and modular way. All routines that contribute to the solution of the discrete problem are collected in a class. Moreover, instead of working with plain array structures, we shall make use of high-level abstractions like *grids* and *fields*. The functionality of the new programs is exactly the same as in the original codes, but the program structure follows a Diffpack standard for PDE simulators. Consequently, the resulting software can easily be integrated with other parts of Diffpack. Of greater importance, however, is the fact that the more advanced implementation makes it easier to extend the code to problems in two and three dimensions or to other PDEs. The programming standard to be shown makes implementations of simple and complicated problems look quite similar. This also helps to ease the understanding of Diffpack simulators created by other programmers.

1.6.1   Steady 1D Heat Conduction Revisited

Let us reconsider the implementation of the numerical method (1.8)–(1.10) for the problem (1.5)–(1.7) that was covered in Chapter 1.2.3. We will develop a *class*, with name `Heat1D1`, that basically contains code pieces from the `main.cpp` file in `src/fdm/Heat1D-Mat`[36], but now split into several functions.

---

[36] We apply a tridiagonal matrix (`MatTri`) for efficiency, so class `Heat1D1` is actually a restructuring of the `main.cpp` file in `src/fdm/Heat1D-MatTri`.

The reader should be familiar with Chapter 1.5.5 as we program with the grid, field, and handle objects in the code that follows.

The Header File. Let us first take a look at the definition of class `Heat1D1` in the header file `Heat1D1.h`[37]:

```
#ifndef Heat1D1_h_IS_INCLUDED
#define Heat1D1_h_IS_INCLUDED
#include <Arrays_real.h>      // MatTri, ArrayGen and other arrays
#include <FieldLattice.h>     // includes GridLattice.h as well

class Heat1D1
{
protected:                    // data items visible in subclasses
  MatTri(real)        A;      // the coefficient matrix
  ArrayGen(real)      b;      // the right-hand side
  Handle(GridLattice) grid;   // 1D grid
  FieldLattice        u;      // the discrete solution
  real                alpha;  // parameter in the test problem
public:
  Heat1D1() {}
  ~Heat1D1() {}
  void scan ();               // read input, set size of A, b and u
  void solveProblem ();       // compute A, b; solve Au=b
  void resultReport ();       // write and plot results
};
#endif
```

Class `Heat1D1` is in fact a general outline of a Diffpack class for solving a differential equation. Data structures needed in the solution process are declared as *protected* variables, i.e., the data items are not accessible for users of the class, but visible in subclasses. The public interface should always consist of `scan`, `solveProblem`, and `resultReport`. We shall stick to these names since that will make it easy to couple the simulator to a menu system, parameter analysis, graphics, automatic report generation, and other powerful Diffpack features, as described in Chapter 3.

Elements of the Diffpack Programming Standard. Computer programs are much easier to understand and use if they are consistently written according to a programming standard. It is therefore appropriate to mention a few naming conventions from the Diffpack programming standard. Local variables in functions and classes have lower-case letters in their names, and words are separated by underscores, e.g., `local_counter`. Function names start with a lower-case letter, and words are separated by capitals, e.g., `someFunc`. Class (and enum) names start with a capital, and words are separated by capitals, e.g., `FieldLattice`. Macros (preprocessor variables) and enum values are written with upper-case letters, and words are separated by underscores, e.g., `SAFETY_CHECKS`. Additional information about Diffpack's programming standard is provided in the FAQ [63].

---

[37] The complete source code is found in `src/fdm/Heat1D1`.

Reading Data Into a Program. The example programs presented so far have read data from standard input (s_i). Much more user-friendly interfaces to simulation programs are available by use of the menu system as explained in Chapter 3.2.2. Nevertheless, here we cover a method that is simpler than using the menu system, but more flexible than asking questions in the terminal window. The method is particularly convenient if you build a tailored *graphical* user interface for the simulator.

Suppose we have two variables alpha and n that we want to read from the *command line*. That is, we want to execute the program like this

```
./app -a 0.1 -n 100
```

assigning 0.1 to alpha and 100 to n. This is achieved by the following statements:

```
initFromCommandLineArg ("-a", alpha, 0.0);
initFromCommandLineArg ("-n", n, 10);
```

The initFromCommandLineArg function can work with real, int, bool, and String variables. The first argument declares the option syntax on the command line (we could use, e.g., -alpha instead of -a), the second argument is the variable to be initialized, and the third argument reflects a default value that is used for the initialization if there are no tracks of the particular option on the command line[38].

The initFromCommandLineArg function has two optional arguments: a description string and a string specified valid answers. For example,

```
initFromCommandLineArg ("-a", alpha, 0.0, "f(x)=x^a", "R1[0:8]");
```

The first optional argument provides a description of what the -a option is used for. The second optional argument indicates that the answer consists of one real number (R1) in the interval $[0, 8]$. These optional parameters are useful for automatic generation of graphical user interfaces or LaTeX 2_ε manuals listing all the input data to the simulator. In case a string is initialized by the command-line option, the valid answer is one the form S (an arbitrary string) or S/choice1/choice2/choice3 (a string among the three listed choices). Integers are specified by the syntax I1[1:3] for an integer in the interval $[1, 3]$. If the interval of reals or integers is unknown, just R1 or I1 are suitable strings.

The Bodies of the Member Functions. Below we list the bodies of the member functions as they appear in the file Heat1D1.cpp.

---

[38] The command-line arguments to a C++ (or C) program are automatically transferred to the main function, and in Diffpack we process and store these arguments in the initialization function initDiffpack. Thereafter, the command-line arguments are available through the global variables (cl_argc and cl_argv) anywhere in a program.

```cpp
#include <Heat1D1.h>
#include <CurvePlot.h>

void Heat1D1:: scan ()
{
  int n;  // no of grid points in the domain (0,1)
  // read n from the command line, a la ./app -n 10
  initFromCommandLineArg ("-n", n, 5, "no of grid points");
  grid.rebind (new GridLattice(1));
  // GridLattice is initialized by a string
  // "d=1 domain=[0,1] index=[1:20]" :
  grid->scan (aform("d=1 domain=[0,1] index=[1:%d]",n));
  u.redim (*grid, "u");
  A.redim(n);             // set size (n rows) of tridiagonal matrix A
  b.redim(n);             // set size of vector b
  initFromCommandLineArg ("-a", alpha, 0.0, "a in rhs x^2");
}

void Heat1D1:: solveProblem ()
{
  // --- Set up matrix A and vector b ---
  A.fill(0.0);                   // set all entries in A equal to 0.0
  b.fill(0.0);                   // set all entries in b equal to 0.0
  const int  n = b.size();  // alternative: grid->getMaxI(1)
  const real h = grid->Delta(1);
  real x; int i;
  for (i = 1; i <= n; i++)
    {
      x = grid->getPt(1,i);
      if (i == 1) {
        A(1,0) = 1;
        b(1) = 0;
      }
      else if (i > 1 && i < n) {
        A(i,-1) = 1;   A(i,0) = -2;   A(i,1) = 1;
        b(i) = + h*h*(alpha + 1)*pow(x,alpha);
      }
      else if (i == n) {
        A(i,-1) = 2;   A(i,0) = -2;
        b(i) = - 2*h + h*h*(alpha + 1)*pow(x,alpha);
      }
    }
  A.factLU();  A.forwBack(b,u.values());   // Gaussian elimination
}

void Heat1D1:: resultReport ()
{
  // print numerical solution and compute the error:
  real x, uval; int i;
  s_o << "\n \n x              numerical            error:\n";
  const int n = grid->getMaxI(1);
  for (i = 1; i <= n; i++) {
    x = grid->getPt(1,i);
    uval = u.values()(i);
    s_o << oform("%4.3f          %8.5f          %12.5e \n",
                 x,uval,pow(x,alpha+2)/(alpha+2) - uval);
  }
```

```
// write results to the file "SIMULATION.res"
Os file ("SIMULATION.res", NEWFILE); // open file
for (i = 1; i <= n; i++)
  file << grid->getPt(1,i) << "   " << u.values()(i) << "\n";
file->close();
}
```

The main program looks like this:

```
#include <Heat1D1.h>

int main(int argc, const char* argv[])
{
  initDiffpack (argc, argv);
  Heat1D1 simulator;
  simulator.scan ();
  simulator.solveProblem ();
  simulator.resultReport ();
  return 0; // success
}
```

Some readers may find the step from the program in Chapter 1.2.3 to the class Heat1D1 listed above too large. An intermediate implementation, that introduces a class Heat1D0, but avoids working with fields and grids, is provided in the directory src/fdm/Heat1D0.

Exercise 1.14. Suppose we want to investigate the numerical error as a function of α and n in the Heat1D1 simulator. This can be accomplished by introducing an error field FieldLattice e in the class, where the nodal values e_i equal the difference between the numerical and exact u values. The L^2 norm $||e||_{L^2} = (\int_0^1 e^2 dx)^{\frac{1}{2}}$ of the error field e is then a suitable error measure, which can be approximately computed through the discrete L^2 norm (Euclidean norm) of the nodal values of e, enabled by the call e.values().norm(L2). Let us denote this error measure by ϵ. The main function can then be extended with a loop over n and α, where the α values are given as input and $n = 2^k \cdot 10$, $k = 1, 2, 3, 4, 5$. We can view the (n, α) pairs as points in a 2D lattice grid (GridLattice). Assigning a value of the error to each grid point then results in a 2D field (FieldLattice), which can be visualized to illustrate how the error depends on n and α. The specific error measure to be visualized equals ϵ/h^2 (ϵ/h^2 should be independent of h since we expect the error to behave as h^2). The visualization of this 2D error field can be performed as indicated on page 85. Carry out the necessary modifications of class Heat1D1 and run the experiment. You can optionally use the expression for the truncation error from Example A.17 on page 519 to explain the features in the plot.   ◇

Nonlinear 1D Heat Conduction Revisited. The solver from Chapter 1.2.7 can also benefit from a restructuring in terms of classes and application of grid and field abstractions. In parallel, it is desirable to enable visualization of the intermediate u^k fields during the nonlinear iterations and allow for different choices of $\lambda(u)$ functions. Visualization of u^k is easily accomplished by dumping the fields to a `CurvePlotFile` manager as we did in the vibrating string example in Chapter 1.3.

When it comes to different choices of the $\lambda(u)$ function, the most obvious implementation consists of a series of C++ functions taking u and perhaps other parameters as arguments. At run time we read the user's choice and apply an indicator in the solver to call the right λ function. For some of these functions we can also provide an exact solution, thus enabling calculation of the discretization error. This may give rise to a considerable collection of functions and, in general, many if-tests throughout the code. If an additional λ function is added, updates at scattered locations in the code are needed. Using classes, we can devise a more elegant implementation, which is explained next.

Suppose we create a base class which serves as interface to all information regarding a $\lambda(u)$ function, including evaluating λ, reading λ-specific parameters from the input, evaluating the corresponding analytical solution if it is available, and writing the formula for the λ function in various output from the solver:

```
class LambdaFunc : public HandleId
{
public:
  virtual real lambda (real u);
  virtual real exactSolution (real x);
  virtual void scan () {}       // read parameters in lambda func.
  virtual String formula ();   // LaTeX syntax for lambda func.
};
```

Specific λ functions are then realized as subclasses of `LambdaFunc`, for example,

```
class Lambda1 : public LambdaFunc
{
  real m;
public:
  Lambda1() {}
  virtual real lambda (real u)  { return pow(u,m); }
  virtual real exactSolution (real x) { return pow(x,1/(m+1)); }
  virtual void scan ()
    { initFromCommandLineArg("-m",m,0.0,"exponent m","R1[0:10]"); }
  virtual String formula () { return "u^m"; }
};
```

Notice that we include the parameter m as a member of the class. In this way, the `lambda` function always takes one argument (u) and all other data are local in the class and are initialized in the `scan` function. All calls to the λ function have the same argument list, here just the u variable. In fact, none of the λ function-specific parameters are visible in the simulator class.

In the solver class, here named `Heat1Dn1`, we include a handle to the base class `LambdaFunc`. The magic of object-oriented programming allows us to evaluate the λ function and the analytical solution everywhere in the program without knowing exactly which choice of λ that was given by the user at run time. A function represented in terms of a class is often referred to as a *functor* and is widely used in Diffpack.

The definition of class `Heat1Dn1` can look like this:

```
class Heat1Dn1
{
protected:                              // data items visible in subclasses
  MatTri(real)        A;               // the coefficient matrix
  ArrayGen(real)      b;               // the right-hand side
  Handle(GridLattice) grid;           // 1D grid
  FieldLattice        uk;             // the discrete solution u^k
  FieldLattice        ukm;            // the discrete solution u^{k-1}
  real                m;              // parameter in the test problem
  real                epsilon;        // tolerance in nonlinear iteration
  Handle(LambdaFunc)  lambda;        // specific lambda function
  CurvePlotFile       plotfile;      // for plotting results
public:
  Heat1Dn1() {}
 ~Heat1Dn1() {}
  void scan ();                       // read input, set size of A, b, and u
  void makeAndSolveLinearSystem();   // compute A, b; solve Au^k=b
  void solveProblem ();               // nonlinear iteration loop
  void resultReport ();               // write numerical error (if possible)
};
```

The `scan` function is similar to the `scan` function in the `Heat1D1` class. Therefore, we only show how to initialize the handle to the `LambdaFunc` hierarchy of objects, based on the user's input on the command line:

```
  String lambda_tp;
  initFromCommandLineArg
    ("-N", lambda_tp, "u^m", "lambda function", "S/u^m/(1+u)^m");
  // allow for numbers 1 and 2 as answers in addition to text:
  if (lambda_tp == "u^m" || lambda_tp == "1")
    lambda.rebind (new Lambda1());
  else if (lambda_tp == "(1+u)^m" || lambda_tp == "2")
    lambda.rebind (new Lambda2());
  else
    errorFP("Heat1Dn::scan","wrong -N %s option.",lambda_tp.c_str());
  lambda->scan();   // initialize function-specific parameters
```

In the rest of the solver code, we evaluate $\lambda(u)$ by a unified syntax. For example, $\lambda(u_{i-1}^k)$ is translated into

```
  lambda->lambda(ukm.values()(i-1));
```

while the exact solution can be evaluated by

```
  lambda->exactSolution(grid->getPt(1,i));
```

The exact solution is of course not available for all choices $\lambda(u)$. Subclasses of `LambdaFunc` can then just inherit the default `exactSolution` function in `LambdaFunc`, which returns a dummy argument. Look at `Heat1Dn1::resultReport` to see how we test on this dummy argument to determine if an exact solution is provided or not.

The lambda->formula() call is useful in output statements from the solver such that we can recognize which λ function that was used in the computations.

The splitting of the original code in src/fdm/Heat1Dn-MatTri into four different functions in a class should be a straightforward task after having seen the principle for the Heat1D1 solver. The complete source code of the nonlinear 1D heat equation solver is available in src/fdm/Heat1Dn1. Here we shall use this code to perform numerical experiments with the purpose of learning more about numerical solution of nonlinear differential equations.

1.6.2   Empirical Investigation of the Numerical Method

In this section we shall use computer simulations to investigate approximation and convergence properties of the numerical method for solving the nonlinear heat equation from Chapters 1.6.1 and 1.2.7:

$$\frac{d}{dx}\left(\lambda(u)\frac{du}{dx}\right) = 0, \quad 0 < x < 1,$$

now with boundary conditions $u(0) = 0$ and $u(1) = 1$.

To run the nonlinear heat equation solver Heat1Dn1 with, for example, $m = 4$ on a grid with 100 cells, you can type this command

```
./app --casename n100m4N1 -n 100 -m 4 -N 1 -e 1.0e-5
```

or you can use the graphical interface script gui.pl (type gui.pl or perl gui.pl). The -n option assigns the number of grid cells, -m is used for m, -N signifies the choice of lambda function, where 1 means $\lambda(u) = u^m$, and -e is used to set the tolerance in the termination criterion. The solver stores the u^k curves for $k = 1, 2, \ldots$ until convergence on files as we explained in Chapter 1.3.4. Animation of the iteration process is enabled by the Visualize button in the GUI, or with full control, through a command on the form

```
curveplot gnuplot -f n100m4N1.map -r '.' 'u' '.' -animate
                  -fps 1 -o 'set yrange [0:1.2];'
```

Just substitute -animate by -psanimate to make an mpeg movie[39]. If you prefer the animation facilities in Matlab instead, substitute gnuplot by matlab. Watching the movie of the iteration process, we observe that the first iterations exhibit a significant qualitative change in the solution from one iteration to the next, but the process converges. The error is about 0.008 in this particular run with $n = 100$ and $m = 4$. Repeating the experiment with $n = 200$

---

[39] Displaying the mpeg movie is most conveniently performed by manually marching through the frames.

reduces the error to about 0.005. This is less than what we would expect, because the difference approximations involved in the scheme are all of second order in h. The arithmetic averages used to compute $\lambda_{i+\frac{1}{2}}$ also have error terms proportional to h^2. We therefore expect the discretization error to behave as h^2, that is, doubling the number of grid points should reduce the error by a factor $1/4$. Let us look closer into this problem from an experimental point of view.

A common model for the discretization error is $e(h) = Ch^r$, where e is the error and C and r are constants that can be estimated either from theoretical considerations (see Appendix A.4 and in particular A.4.9) or from numerical experiments. The latter approach is simple and widely applicable and will be addressed in the following.

Having a sequence of grid spacings $h_1, h_2, h_3, \ldots$ and corresponding error measures $e_1, e_2, e_3, \ldots$, we have for two experiments s and $s+1$ that $e_s = Ch_s^r$ and $e_{s+1} = Ch_{s+1}^r$. Dividing these expressions eliminates C, and by solving with respect to r we get the estimate of the *convergence rate*

$$r = \frac{\ln(e(h_s)/e(h_{s+1}))}{\ln(h_s/h_{s+1}).} \tag{1.74}$$

This is a cheap alternative to the more general approach, which consists in estimating C and r from a least-squares fit to the complete collection of data points $(h_1, e_1), (h_2, e_2), \ldots$ However, if more than one discretization parameter appears in the problem, e.g., a spatial (h) and a temporal (Δt) parameter, a common model for the error reads $e(h, \Delta t) = C_1 h^r + C_2 \Delta t^s$, where C_1, C_2, r, and s must be determined by a nonlinear least-squares procedure.

When running the experiments, we must ensure that the ϵ tolerance in the termination criterion for the nonlinear iterations on page 20 is significantly less than the computed error e, otherwise we pollute the discretization error with errors from the nonlinear iteration method. A value of 10^{-8} might be suitable for our convergence studies. We remark that in practical engineering computations one should choose ϵ less, but not much less, than the typical level of discretization accuracy; a very low ϵ value will not contribute to more accurate results if the discretization error dominates anyway.

We can now run a series of grid sizes, e.g., $h_q = 0.2 \cdot 2^{-q}$ for $q = 1, 2, \ldots, 8$ and calculate r values from two successive experiments. Doing this by hand is a tedious job – it is typically work for the computer, conveniently done in a script that runs the program a specified number of times with appropriate input and calculates the corresponding r values. The writing of (Perl) scripts is introduced on page 35, but the script we need for the present purpose employs constructs that go beyond the introductory example on page 35. The name of our current script is `rate.pl` and you can run it like this: `rate.pl 0.1 7`, meaning that m is 0.1 and that 7 experiments are to be executed ($h_q = 0.2 \cdot 2^{-q}$ for $q = 1, \ldots, 7$).

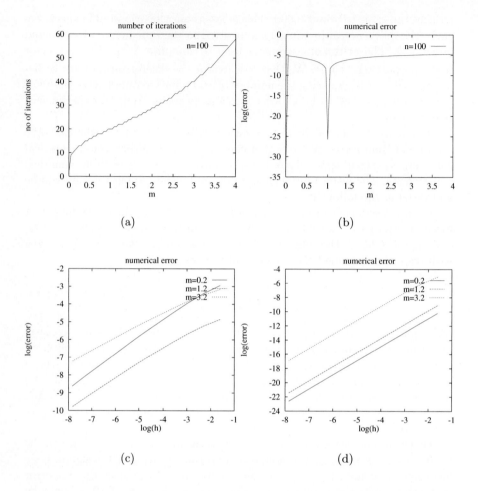

Fig. 1.9. The figures summarizes the numerical errors and the number of nonlinear iterations for a series of experiments (carried out by the `rate.pl` and `error.pl` scripts). (a) The number of iterations as function of m; (b) the numerical error as function of m; (c) numerical errors as functions of $\ln h$ for various m values; (d) as (c), but with $\lambda(u) = (1 + u)^m$. (100 grid points were used to produce (a) and (b), while $\lambda(u) = u^m$ was used in (a), (b), and (c).)

When $m \leq 1$ the rates r stabilizes quickly around unity, while the expected r value was 2. Increasing m leads to stabilization of r at lower values, e.g., $m = 3$ typically gives $e \sim h^{3/4}$. From a plot of the solution we recognize that the slope of u close to $x = 0$ is very steep; the value of $u'(0)$ is fact infinite according to the exact solution! We therefore need a small cell size h to capture the steep slope close to $x = 0$. Actually, the small cells are only needed in the vicinity of $x = 0$. The optimal grid should therefore employ a spatially varying cell size that *adapts* to the solution. Such adaptive grids are particularly attractive in the finite element method and the topic is covered in Chapters 2.10 and 3.6.

Let us try another λ function that avoids the infinite derivative of the solution at $x = 0$: $\lambda(u) = (1 + u)^m$. The analytical solution now becomes $u(x) = ((2^{m+1} - 1)x + 1)^{1/(m+1)} - 1$. This λ is implemented in terms of a class Lambda2 and referred to as the second choice of λ functions. The rate.pl script takes a third argument reflecting the type of λ function to be used (this argument is 1 by default). Try rate.pl 1.5 10 2 and observe that the rates quickly stabilize around the expected value 2.

With $m = 1.5$ the nonlinear iteration loop needs about 14 iterations to meet the termination criterion with $\epsilon = 10^{-9}$ (the ϵ value is automatically set by the rate.pl script). Increasing m to 4 increases the number of nonlinear iterations, now we typically need 25-30. It could be of interest to see how m affects the error and the number of iterations. The script error.pl runs a series of experiments with $m = 0, 0.05, 0.10, \ldots, 4$ and writes out the error and the number of iterations. Running error.pl 50 2 (50 grid points and $\lambda = (1 + u)^m$) reveals that the number of iterations is steadily increasing with m, while the error shows a peculiar behavior; it drops significantly in the vicinity of $m = 1$. Figure 1.9 summarizes a series of experiments with this nonlinear model.

The user is encouraged to play around with the rate.pl and error.pl scripts and investigate the impact of m and the choice of λ on the discretization accuracy and the number of nonlinear iterations.

1.6.3 Simulation of Waves Revisited

Algorithm 1.1 for the mathematical problem (1.44)–(1.48) can also be implemented using a simulator class, like Heat1D1, with grid and field abstractions. In addition, we need a library utility that holds the temporal grid.

Temporal Grid. The central parameters for time discretization are the time step size and the start and stop values of the time interval. These parameters are collected in a class TimePrm, which is useful in simulators for time-dependent problems. The initialization of a TimePrm object is performed by feeding a string, with the following syntax, to the TimePrm::scan function:

```
dt=0.05, t in [0,10]
```

In this example, the time interval is $[0, 10]$ and the constant time-step length is 0.05. We refer to the scan and timeLoop functions in the simulator below for examples on using the TimePrm class, as well as to the man page for that class.

A class version of the simulator for waves on a string appears in class Wave1D1. The corresponding source code files are collected in the directory src/fdm/Wave1D1. Here is the class definition, taken from Wave1D1.h:

```
class Wave1D1
{
  Handle(GridLattice) grid;       // lattice grid; here 1D grid
  FieldLattice        up;         // solution u at time level l+1
  FieldLattice        u;          // solution u at time level l
  FieldLattice        um;         // solution u at time level l-1
  TimePrm             tip;        // delta t, final t etc
  CurvePlotFile       plotfile;// for plotting results
  real                C;          // the Courant number

  void setIC ();              // set initial conditions
  void timeLoop ();           // perform time stepping
  void plotSolution ();       // make a curve plot of u
public:
  Wave1D1() {}               // no special construction is needed
  ~Wave1D1() {}              // no special destruction is needed
  void scan ();              // read input and initialize
  void solveProblem ();      // solve the problem
  void resultReport () {}    // not used here
};
```

The associated source code is found in a file Wave1D1.cpp.

```
#include <Wave1D1.h>
#include <SimRes2gnuplot.h>  // defines makeCurvePlot (used below)

void Wave1D1:: scan ()
{
  initFromCommandLineArg("-C",C,1.0,"Courant number","R1[0:1]");
  String g;
  initFromCommandLineArg("-g",g,"d=1 [0,1] [0:20]","grid","S");
  grid.rebind(new GridLattice(1)); grid->scan (g);
  real tstop;
  initFromCommandLineArg ("-t", tstop, 1.0, "tstop", "R1[0:10]");
  // construct the proper initialization string,
  // determine the time step from C, the grid size, and tstop:
  tip.scan (aform("dt=%g t in [0,%g]", C*(grid->Delta(1)), tstop));
  // (we assume unit wave velocity in the expression for the time step)

  up.redim(*grid,"up");  u.redim(*grid,"u");  um.redim(*grid,"um");
  plotfile.open (casename);
  // write input data for a check
  u.grid().print(s_o);  // could also just say grid->print(s_o)
  tip.print(s_o);
}

void Wave1D1:: setIC ()
{
  // set initial conditions on u and um
  const int  i0  = u.grid().getBase(1);   // start point index
  const int  n   = u.grid().getMaxI(1);   // end point index
  const real umax = 0.05;                 // max amplitude
  up.fill(0.0);                           // initialization of up
  u. fill(0.0);                           // initial displacement
  int i; real x;
  for (i = i0; i <= n; i++) {
    x = grid->getPt(1,i);  // get x coord of grid point no i
```

```
      if (x < 0.7)   u.values()(i) = (umax/0.7) * x;
      else           u.values()(i) = (umax/0.3) * (1 - x);
  }
  // initialization of um (the special formula)
  um.fill(0.0);
  for (i = i0+1; i <= n-1; i++) // set the help variable um:
    um.values()(i) = u.values()(i) + 0.5*sqr(C) *
              (u.values()(i+1) - 2*u.values()(i) + u.values()(i-1));
}

void Wave1D1:: timeLoop ()
{
  tip.initTimeLoop();
  setIC();

  const int i0 = u.grid().getBase(1);  // start index in x-dir.
  const int n  = u.grid().getMaxI(1);  // end   index in x-dir.
  // (could also used grid->getBase instead of u.grid().getBase)
  plotSolution ();  // plot initial condition
  int i;
  // useful abbreviations (also for efficiency):
  const ArrayGen(real)& U  = u. values(); // grab underlying array
  const ArrayGen(real)& Um = um.values();
        ArrayGen(real)& Up = up.values();

  while (!tip.finished()) {
    tip.increaseTime();

    for (i = i0+1; i <= n-1; i++)
      Up(i) = 2*U(i) - Um(i) + sqr(C) * (U(i+1) - 2*U(i) + U(i-1));

    Up(i0) = 0;  Up(n)  = 0; // insert boundary values
    um = u;      u = up;     // update data structures for next step
    plotSolution ();         // plot u, i.e. the new up
  }
}

void Wave1D1:: plotSolution ()
{
  // automatic dump of a curve plot of a 1D field:
  SimRes2gnuplot::makeCurvePlot          // (static) library routine
    (u,                                  // 1D field to be plotted
     plotfile,                           // curve plot manager
     "displacement",                     // plot title
     oform("u(x,%.4f)",tip.time()),      // name of function
     oform("C=%g, h=%g, dt=%g, t=%g",    // comment
         C, u.grid().Delta(1), tip.Delta(), tip.time()));
}

void Wave1D1:: solveProblem ()   { timeLoop(); }
```

Here is an example on execution the program:

```
    ./app -g 'd=1 [0,1] index: [0:100]' -t 10 -C 0.85
```

The quotes are important to ensure that the initialization command to the -g option is treated as a single compound string. The results of the simulation can be visualized as explained in Chapter 1.3.4.

Using Handles for the Field Objects. In general, we recommend to use handles instead of instances for potentially large data structures like fields, grids, and linear systems. Future integration of the simulator class with other simulator classes will then be safer and more flexible. Hereafter, the simulators in this text employ handles to a larger degree than what is strictly necessary for the problem at hand – the purpose is merely to stick to a programming standard. We here briefly outline the necessary modifications related to replacing a FieldLattice object by a Handle(FieldLattice). Initialization of a handle is accomplished by

```
u.rebind (new FieldLattice (*grid, "u"));
```

in for example the simulator's scan function. Because u is a handle, i.e. basically a pointer, one must access the member functions of FieldLattice with the arrow -> operator instead of the dot . operator. (This does not apply to u.rebind, because rebind is a member function of class Handle(FieldLattice), not of FieldLattice.) Another important difference is that when we wish to pass a field u as an argument to functions, we must write *u, u() or u.getRef() (all these expressions are equivalent). We must also be careful with the statement um1=u; this must read *um1=*u (or um1()=u()) and not um1=u when u and um1 are handles[40].

Exercise 1.15. Edit the class definition of Wave1D1 such that it uses the classes Handle(FieldLattice) and Handle(TimePrm) instead of just FieldLattice and TimePrm. Modify the member functions accordingly. (The answer can be found in the directory src/fdm/Wave1D2.)   ◇

Exercise 1.16. The waves on a piano string can be modeled by the wave equation (1.44) and no displacements at the ends ($u(0,t) = u(1,t) = 0$), but the initial condition should now model the impact of the hammer on the string. This can be accomplished by letting the string be at rest initially, $u(x,0) = 0$, with a prescribed velocity $\partial u/\partial t = v$, $v < 0$, at a small portion $[0.1, 0.2]$ of the string (the part of the string hit by the hammer). Modify Algorithm 1.1 such that it handles the new initial condition. Make a corresponding simulator and visualize waves on a piano string.   ◇

Exercise 1.17. From experience we know that the string motion will eventually die out, but this is not reflected in our governing PDE (1.44) because we neglected, e.g., air restistance in the derivation of the mathematical model. Including a damping force proportional to the velocity of the string, which is a reasonable model, results in the scaled PDE

$$\frac{\partial^2 u}{\partial t^2} + \beta \frac{\partial u}{\partial t} = \gamma^2 \frac{\partial^2 u}{\partial x^2},$$

---

[40] The um1=u assignment for handles means only that um1 points to the same data as u. In other words, *um1 and *u contain identical values and the original field values at the previous time level are lost (and deleted!) in the program.

where β is a dimensionless positive number and γ is a dimensionless constant (equal to unity) as before. The damping term can be approximated by a difference $(u_i^{\ell+1} - u_i^{\ell-1})/(2\Delta t)$. Modify Algorithm 1.1 to take damping into accound and develop a corresponding simulator. Make animations that demonstrate wave motion with damping.   ◇

Generation of Graphical User Interfaces. On page 35 we advocated the use of scripting interfaces to simulation and visualization and showed a simple example in Perl, the `wave.pl` script found in the directory `src/fdm/Wave1D-func`. This script gives you a quite flexible one-line command for running the simulator and producing a tailored visualization. Programmers in Unix environments find the command-line interface efficient and convenient, but a generation of computer users expect programs to have a graphical user interface (GUI). A Perl script like `wave.pl` can be equipped by calls to the Tk tool for generating windows, and an example is presented in `wave-GUI.pl`. Tk can also be used from Tcl and Python.

Since Diffpack simulators employing `initCommandLineArg` for fetching input data are quite standardized, a GUI for one application can with minor editing efforts be ported to another application. However, if such editing is simple, it can be automated. Assuming that you have run the application (successfully) once, the script `CloGUIO` automatically generates a GUI for your program[41], see Figure 1.10a for an example.

Go to the `Wave1D1` directory, run the program once (with e.g. the default values), and write `CloGUIO`. A graphical scripting interface `app.pl` is then automatically generated. Just type `app.pl` to see the interface. The grid, time interval, and Courant number from the last execution of the simulator appear as default values in the GUI. Clicking on Simulate runs the simulator with the one-line command as printed in the Messages window. The actions implied by clicking the Visualize button must be coded in the `app.pl` script explicitly. For example, the command

```
system "curveplot gnuplot -f $casename.map -r '.' '.' '.' -animate
        -o 'set yrange [-0.1:0.1];'"
```

can be placed in the `visualize` subroutine in `app.pl`. Clicking Visualize will then show a Gnuplot movie on the screen.

The `Wave1D1` is a toy program and not very exciting to play with. Closely related simulators, especially the `Wave1D3` solver in Example A.16 on page 516 and the `Wave1D4` and `Wave1D5` demos in Appendix A.5, are made for numerical experimentation, and graphical user interfaces for these simulators are useful, at least when demonstrating the programs. The GUI scripts associated with these simulators have all been generated automatically by `CloGUIO` and then

---

[41] The `initFromCommandLineArg` functions writes a special file with key information about each command-line option. The `CloGUIO` script uses this information to generate some appropriate sliders and buttons in a window.

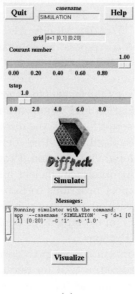

(a)

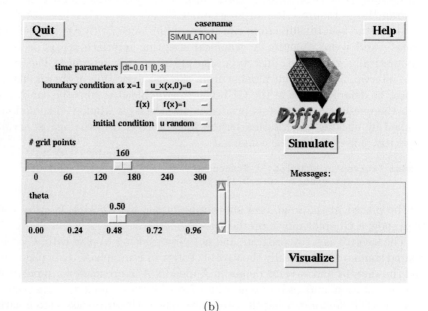

(b)

Fig. 1.10. Examples on automatically generated graphical user interfaces to command-line driven Diffpack simulators. In (a) we see the default GUI for the Wave1D1 solver as generated by the CloGUIO script, while (b) shows a slightly manually adjusted GUI for a heat equation solver in src/fdm/Parabolic1D1.

slightly adjusted (slider intervals or visualization commands might need some editing).

The 2D Wave Equation. It is natural to implement the algorithm for the 2D wave equation from Chapter 1.3.6 using fields and grid objects similar to what we did in the Wave1D1 simulator. To choose between different depth functions $H(x, y)$ (or $H(x, y, t)$ for modeling slide-generated waves) and initial elevations $I(x, y)$, we apply the same design as we explained for the $\lambda(u)$ coefficient in Chapter 1.6.1, i.e., we represent functions in terms of classes, termed functors. A suggested base class for such functors is

```
class WaveFunc : public HandleId
{
public:
  virtual real valuePt (real x, real y, real t = DUMMY);
  virtual void scan () {}     // read parameters in depth func.
  virtual String formula ();  // function label
};
```

In the wave simulator we include two Handle(WaveFunc) objects for evaluating the initial surface displacement and the depth, respectively. Evaluation of these functions at a point is performed by calling the virtual valuePt function, without knowing exactly which initial condition or depth function that the user has chosen on the input. Since there is some overhead associated with virtual function calls, we store the depth values at the grid points using a Handle(FieldLattice) object lambda and use lambda instead of the functor in the CPU-intensive statements of the numerical scheme. A typical code segment for the initialization of the λ values on the basis of a Handle(WaveFunc) object H might take this form:

```
int i, j; real x, y;
const int  nx = grid->getMaxI(1);
const int  ny = grid->getMaxI(2);
ArrayGen(real)& lambda_v = lambda->values();
for (j = 1; j <= ny; j++) {
  for (i = 1; i <= nx; i++) {
    x = grid->getPt(1,i);   y = grid->getPt(2,j);
    lambda_v(i,j) = 1.0 - H->valuePt(x,y);
  }
}
```

Considering the Gaussian bell functions (1.61) and (1.62), we can implement both types in the same subclass GaussianBell of WaveFunc if we parameterize the name of the function in the command-line arguments to avoid ambiguous commands. The idea and its associated string manipulations are documented in the source code in src/fdm/Wave2D1/Wave2D1.cpp.

The rest of the simulator code should be straightforward to understand provided that you have understood the underlying algorithm and its implementation in the more primitive version found in src/fdm/Wave2D-func, and that you are familiar with the present section and the Wave1D1 solver. The usage of the `Wave2D1` class was explained when performing the numerical experiments with 2D water waves in Chapter 1.3.6.

1.7 Projects

1.7.1 Transient Flow Between Moving Plates

Mathematical Problem. The model to be studied in the present project reads

$$\frac{\partial u}{\partial t} = \frac{\partial^2 u}{\partial x^2}, \quad x \in (0,1), \ t > 0, \tag{1.75}$$

$$u(0,t) = 0, \quad t > 0, \tag{1.76}$$

$$u(1,t) = 1, \quad t > 0, \tag{1.77}$$

$$u(x,0) = 0, \quad x \in [0,1]. \tag{1.78}$$

Physical Model. The initial-boundary value problem (1.75)–(1.78) models channel flow between two flat (infinite) plates $x = 0$ and $x = 1$, where the fluid is initially at rest and the plate $x = 1$ is given a sudden initial movement. The function $u(x,t)$ reflects the fluid velocity in direction parallel to the plates (normal to the x axis). For small t, only the part of the fluid close to the moving plate is set in significant motion, resulting in a thin boundary layer at $x = 1$. As time increases, the velocity approaches a linear variation with x (known as the stationary Couette flow profile). Equations (1.75)–(1.78) constitute a model for studying friction between moving surfaces separated by a thin fluid film.

Derive the model (1.75)–(1.78) from the incompressible Navier-Stokes equations and a suitable scaling. In addition, give the details of an alternative physical interpretation, where $u(x,t)$ is the temperature in a rod.

Numerical Method. We shall solve the problem (1.75)–(1.78) by three different finite difference methods. Let subscript i denote evaluation at a spatial grid point x_i, and let superscript ℓ denote evaluation at time level ℓ. The spatial and temporal grid increments are h and Δt, respectively.

- In the *explicit forward Euler scheme* the PDE (1.75) is approximated at the space-time index (i, ℓ) by a forward difference in time and a centered difference in space:

$$\left[\frac{\partial u}{\partial t}\right]_i^\ell \approx \frac{u_i^{\ell+1} - u_i^\ell}{\Delta t}, \quad \left[\frac{\partial^2 u}{\partial x^2}\right]_i^\ell \approx \frac{u_{i-1}^\ell - 2u_i^\ell + u_{i+1}^\ell}{h^2}.$$

These expressions lead to an explicit formula for updating u:

$$u_i^{\ell+1} = f(u_i^\ell, u_{i-1}^\ell, u_{i+1}^\ell).$$

– The *implicit backward Euler scheme* employs a backward difference in time, and the PDE is discretized at the space-time index $(i, \ell + 1)$.

$$\left[\frac{\partial u}{\partial t}\right]_i^{\ell+1} \approx \frac{u_i^{\ell+1} - u_i^{\ell}}{\Delta t}, \quad \left[\frac{\partial^2 u}{\partial x^2}\right]_i^{\ell+1} \approx \frac{u_{i-1}^{\ell+1} - 2u_i^{\ell+1} + u_{i+1}^{\ell+1}}{h^2}.$$

Combining these expressions leads to a tridiagonal matrix system for the new values $u_{i-1}^{\ell+1}$, $u_i^{\ell+1}$, and $u_{i+1}^{\ell+1}$. Because we have to solve a linear system to find the new values $u_i^{\ell+1}$, the scheme is classified as *implicit*.

– The *implicit Crank-Nicolson scheme* utilizes a centered difference in time, with a time average of the spatial derivative. The PDE is approximated at the space-time index $(i, \ell + 1/2)$.

$$\left[\frac{\partial u}{\partial t}\right]_i^{\ell+1/2} \approx \frac{u_i^{\ell+1} - u_i^{\ell}}{\Delta t},$$

$$\left[\frac{\partial^2 u}{\partial x^2}\right]_i^{\ell+1/2} \approx \frac{1}{2}\left(\frac{u_{i-1}^{\ell} - 2u_i^{\ell} + u_{i+1}^{\ell}}{h^2} + \frac{u_{i-1}^{\ell+1} - 2u_i^{\ell+1} + u_{i+1}^{\ell+1}}{h^2}\right).$$

These equations also lead to a tridiagonal matrix system for the new values $u_{i-1}^{\ell+1}$, $u_i^{\ell+1}$, and $u_{i+1}^{\ell+1}$. The scheme is therefore implicit.

Formulate complete computational algorithms (cf. Algorithm 1.1) for these three schemes.

Write the three schemes compactly using the notation in Appendix A.3. Outline how the schemes are generalized to a 3D version of the governing PDE (1.75), i.e., $\partial u/\partial t = \nabla^2 u$ on the unit cube.

Analysis. Find the truncation error of the three schemes and determine if the schemes are consistent. Investigate the stability properties. (Appendix A.4 covers truncation errors, consistency, and stability.) Find the analytical solution to the continuous problem (hint: introduce $v(x, t) = u(x, t) - x$ and expand $v(x, t)$ in a Fourier sine series $v = \sum_k a_k(t) \sin \pi k x$; details are found in [126, Ch. 3-5.2]).

Implementation. Implement the three schemes for (1.75)–(1.78) in the same computer program. A mixture of the basic programs from Chapters 1.2 and 1.3, or their more advanced counterparts in Chapter 1.6, might be a suitable starting point. Compute and visualize the error at three time points, $t_I < t_{II} < t_{III}$, where $u(\cdot, t_I)$ exhibits a thin boundary layer close to $x = 1$, $u(\cdot, t_{II})$ is smoother, but still significantly curved, whereas $u(\cdot, t_{III})$ is almost linear, i.e., close to the stationary state.

Computer Experiments. For each of the cases $h = 1/10$ and $h = 1/100$, run the three schemes with Δt corresponding to the stability limit dictated by the explicit scheme. Compare these solutions with the analytical solution of the continuous problem at $t = t_I, t_{II}, t_{III}$ with respect to the error measure e explained in Project 1.4.2. Which of these schemes will you classify as "best"?

Extension: Higher-Order Forward and Backward Schemes. The forward and backward Euler schemes for equation (1.75) typically lead to discretization errors of order Δt. To obtain an error of order Δt^2, but still enable an explicit time marching, we can apply a three-point, one-sided difference approximation:

$$\left[\frac{\partial u}{\partial t}\right]_i^\ell \approx au_i^{\ell+2} + bu_i^{\ell+1} + cu_i^\ell, \tag{1.79}$$

where a, b, and c are constants to be determined. Make Taylor-series expansions of $u_i^{\ell+2}$, $u_i^{\ell+1}$, and u_i^ℓ around time level ℓ and insert these in (1.79). Find equations for a, b, and c such that $au_i^{\ell+2} + bu_i^{\ell+1} + cu_i^\ell$ approximates a first derivative. The leading term in the error should be of order Δt^2. Find a corresponding difference formula for a backward scheme, that is, a three-level, one-sided difference approximation to $\partial u/\partial t$ at time $\ell + 2$, based on the values $u_i^{\ell+2}$, $u_i^{\ell+1}$, and u_i^ℓ. Formulate three-level forward and backward schemes for the problem (1.75)–(1.78). (At the first time level one can use the Crank-Nicolson scheme.)

Find the truncation error of the forward and backward three-level schemes. Deduce the stability properties of the schemes.

Implement the three-level time schemes in the program and use numerical experiments to determine the feasibility of the potentially improved schemes.

1.7.2   Transient Channel Flow

Mathematical Problem. This project applies the numerical methods from Project 1.7.1 to a slightly different problem:

$$\frac{\partial u}{\partial t} = \frac{\partial^2 u}{\partial x^2} + \beta(t), \quad x \in (0,1),\ t > 0, \tag{1.80}$$

$$u(0,t) = 0, \quad t > 0, \tag{1.81}$$

$$u(1,t) = 0, \quad t > 0, \tag{1.82}$$

$$u(x,0) = 0, \quad x \in [0,1]. \tag{1.83}$$

Physical Model. The physical problem modeled by (1.80)–(1.83) concerns flow in a straight channel with fixed walls, such that $u(0,t) = u(1,t) = 0$, but with a time-dependent pressure gradient, giving rise to a time-varying source term $\beta(t)$ in the governing PDE. While the moving wall caused the fluid flow in Project 1.7.1, the pressure gradient is now the driving force.

Derive the model (1.80)–(1.83) from the incompressible Navier-Stokes equations and a suitable scaling.

Numerical Method. The numerical schemes from Project 1.7.1 can be straightforwardly applied to (1.80)–(1.83). In the forward Euler method, the PDE is approximated at time level ℓ, and the source term β is then to be evaluated as $\beta(\ell\Delta t)$, here noted by β^ℓ as usual. In the backward Euler scheme, we employ the value $\beta^{\ell+1}$, whereas in the Crank-Nicolson scheme β is to be evaluated at level $\ell + 1/2$, that is, as $\beta((\ell + \frac{1}{2})\Delta t)$ or as an average of β^ℓ and $\beta^{\ell+1}$.

Implementation. Implement the three schemes in a program (see the guidelines in Project 1.7.1). It is of interest to make animations of u and β simultaneously, so one needs to dump curve plots of u and β as functions of x for each time step and also find a suitable scaling of β in the plots.

To verify the implementation, one can construct a simple solution $u(x, t)$ of a slightly different problem. For example, one can seek u on the form $f(t) + \text{const} \cdot (1 - x^2)$, calculate $f(t)$ in terms of $\beta(t)$, and adjust the boundary and initial conditions accordingly. Choosing a constant β should lead to a solution which is exactly reproduced by the schemes.

Computer Experiments. Produce a set of movies of u and β corresponding to $\beta(t) = C_1 \sin^{2n} \omega t$ and $\beta(t) = C_2 \sin \omega t$, where you vary the parameters $n \in \mathbb{N}^+$ and $C_1, C_2, \omega \in \mathbb{R}$. Run the simulations until the velocity appears to be periodic in time.

Remark. The model (1.80)–(1.83) can easily be extended to flow in a straight pipe with circular cross section by introducing a radial coordinate r instead of x. This implies replacing $\frac{\partial^2 u}{\partial x^2}$ in (1.80) by $\frac{1}{r} \frac{d}{dr} \left(r \frac{\partial u}{\partial r} \right)$ and removing the condition (1.81). The corresponding finite difference equations can utilize the approximations from Chapter 1.2.6, but the point $r = 0$ gives rise to difficulties. An appropriate set of difference equations is in fact most readily obtained by applying a finite element method to the problem in radial coordinates, see Project 2.6.2.

1.7.3   Coupled Heat and Fluid Flow

Mathematical Problem. This project is a continuation of Project 1.7.2, but we now consider a nonlinear version of (1.80) coupled with a heat equation.

$$\frac{\partial u}{\partial t} = \alpha \frac{\partial}{\partial x} \left(m(T) \left| \frac{\partial u}{\partial x} \right|^{n-1} \frac{\partial u}{\partial x} \right) + \beta(t), \quad x \in (0, 1), \ t > 0, \ (1.84)$$

$$\frac{\partial T}{\partial t} = \gamma \frac{\partial^2 T}{\partial x^2} + m(T) \left| \frac{\partial u}{\partial x} \right|^{n+1}, \quad x \in (0, 1), \ t > 0, \quad (1.85)$$

$$u(0, t) = 0, \quad t > 0, \quad (1.86)$$

$$u(1, t) = 0, \quad t > 0, \quad (1.87)$$

$$u(x, 0) = 0, \quad x \in [0, 1], \quad (1.88)$$

$$T(0, t) = 0, \quad t > 0, \quad (1.89)$$

$$T(1, t) = 0, \quad t > 0, \quad (1.90)$$

$$T(x, 0) = 0, \quad x \in [0, 1]. \quad (1.91)$$

This is a *coupled system of nonlinear PDEs*. The parameters $\alpha, \gamma, n \in \mathbb{R}^d$ are dimensionless numbers, while the function $m(T)$ is in general nonlinear and can be taken as $\exp(-\tau T)$, where τ is another dimensionless constant.

Physical Model. The apparent viscosity of highly viscous fluids, like metals under extrusion or liquid plastics, often varies with the velocity gradient and the temperature. Furthermore, the internal friction caused by viscosity may generate significant heat, which in turn influences the viscosity. Chapter 7.2.1 derives a set of PDEs and corresponding boundary conditions for coupled heat and fluid flow in straight pipes with arbitrary cross section. Simplifying the equations in Chapter 7.2.1 to channel flow, and introducing a suitable scaling, results in the system (1.84)–(1.91). The viscosity is reflected by the function $m(T)|\partial u/\partial x|^{n-1}$. The last term in (1.85) models the heat generation by internal friction in the fluid. We notice that the coefficients in the PDE depend on both u and T, making the two PDEs fully coupled.

Numerical Method. Construction of stable and robust methods for numerical solution of the time-dependent nonlinear system of PDEs (1.84)–(1.85) is a challenging task. Chapters 4.1 and 7.2 present quite comprehensive solution methods for such problems, but in this project we shall formulate a special time discretization that leaves us with two standard linear problems of the same type as in Project 1.7.2 at each time level. We discretize both (1.84) and (1.85) by a backward Euler scheme. However, in the evaluation of the nonlinear coefficients we approximate u and T by values at the previous time level:

$$\frac{u^\ell - u^{\ell-1}}{\Delta t} = \alpha \frac{\partial}{\partial x}\left(m(T^{\ell-1}) \left|\frac{\partial u}{\partial x}^{\ell-1}\right|^{n-1} \frac{\partial u^\ell}{\partial x}\right) + \beta(t^\ell), \qquad (1.92)$$

$$\frac{T^\ell - T^{\ell-1}}{\Delta t} = \gamma \frac{\partial^2 T^\ell}{\partial x^2} + m(T^{\ell-1})\left|\frac{\partial u^\ell}{\partial x}\right|^{n+1}, \qquad (1.93)$$

for $x \in (0,1)$ and $t > 0$. Notice now that (1.92) is a linear equation in u^ℓ and that (1.93) is linear in T^ℓ if we have already solved (1.92) for u^ℓ. In other words, the particular time discretization technique turns the original coupled nonlinear system of PDEs into two linear PDEs that can be solved in sequence at each time level. The penalty for this great simplification is less stability and robustness of the simulations in comparison with the more sophisticated techniques in Chapters 4.1 and 7.2.

Implementation. Extend the program from Project 1.7.2 to allow for two tridiagonal linear systems at each time level. To partially verify the implementation, set $m(T) = 1$ and $n = 1$, let β be constant, and calculate the analytical solution of the PDEs as $t \to \infty$ (i.e. $\partial u/\partial t, \partial T/\partial t \to 0$), and check that the program reproduces this solution. Another stationary solution can be obtained for $m(T) = 1$ and arbitrary n.

Computer Experiments. Repeat the experiments from Project 1.7.2, but make animations that show the time evolution of u, β, and T.

1.7.4   Difference Schemes for Transport Equations

Mathematical Problem. This project is an extension of Project 1.4.1 and concerns numerical experimentation with different finite difference schemes for the one-dimensional transport equation (1.64) and its nonlinear extension

$$\frac{\partial u}{\partial t} + \frac{\partial}{\partial x} f(u) = 0, \tag{1.94}$$

with $u(x, 0)$ prescribed.

Physical Model. Equations on the form (1.94) are often referred to as *hyperbolic conservation laws* and appear frequently in gas dynamics and multiphase porous media flow. We shall consider three specific choices of $f(u)$:

1. The linear form $f(u) = \gamma u$ reproduces Equation (1.64).
2. The choice $f(u) = u^2/2$ leads to Burgers' equation, which is a prototype equation mirroring the nonlinear acceleration terms in fluid flow models.
3. The Buckley-Leverett equation, modeling 1D flow of oil and water in a porous medium [6], corresponds to taking

$$f(u) = \frac{u^2}{u^2 + \mu(1 - u)^2},$$

where μ is a constant reflecting the ratio of the viscosity of oil and water. The unknown $u(x, t)$ is the saturation of water ($u \in [0, 1]$).

Numerical Methods. There exists a large number of difference schemes for hyperbolic equations like (1.94). Let u_j^ℓ be the numerical approximation to the analytical solution u at grid point $x_j = (j - 1)h$ and time point $\ell \Delta t$. The following list of schemes is taken from LeVeque [67], and the reader is referred to that text for a comprehensive treatment of finite difference methods for hyperbolic conservation laws.

Backward Euler scheme:

$$u_j^{\ell+1} = u_j^\ell - \frac{\Delta t}{2h} \left(f(u_{j+1}^{\ell+1}) - f(u_{j-1}^{\ell+1}) \right) \tag{1.95}$$

Explicit upwind scheme:

$$u_j^{\ell+1} = u_j^\ell - \frac{\Delta t}{h} \left(f(u_j^\ell) - f(u_{j-1}^\ell) \right), \quad f'(u) \geq 0 \tag{1.96}$$

Lax-Friedrichs scheme:

$$u_j^{\ell+1} = \frac{1}{2} \left(u_{j-1}^\ell + u_{j+1}^\ell \right) - \frac{\Delta t}{2h} \left(f(u_{j+1}^\ell) - f(u_{j-1}^\ell) \right) \tag{1.97}$$

Leap-Frog scheme:

$$u_j^{\ell+1} = u_j^{\ell-1} - \frac{2\Delta t}{2h} \left(f(u_{j+1}^\ell) - f(u_{j-1}^\ell) \right) \qquad (1.98)$$

Richtmyer's two-step Lax-Wendroff method:

$$u_{j+\frac{1}{2}}^{\ell+\frac{1}{2}} = \frac{1}{2} \left(u_j^\ell + u_{j+1}^\ell \right) - \frac{\Delta t}{2h} \left(f(u_{j+1}^\ell) - f(u_j^\ell) \right), \qquad (1.99)$$

$$u_j^{\ell+1} = u_j^\ell - \frac{\Delta t}{h} \left(f(u_{j+\frac{1}{2}}^{\ell+\frac{1}{2}}) - f(u_{j-\frac{1}{2}}^{\ell+\frac{1}{2}}) \right) \qquad (1.100)$$

MacCormack's method:

$$u_j^* = u_j^\ell - \frac{\Delta t}{h} \left(f(u_{j+1}^\ell) - f(u_j^\ell) \right), \qquad (1.101)$$

$$u_j^{\ell+1} = \frac{1}{2} \left(u_j^\ell + u_j^* \right) - \frac{\Delta t}{2h} \left(f(u_j^*) - f(u_{j-1}^*) \right), \qquad (1.102)$$

All methods, except the backward Euler scheme, are explicit. The stability criterion for the explicit finite difference methods is that the Courant number, given by $C = \max |f'(u(x,t))|\Delta t/h$, must fulfill $C \leq 1$ (referred to as the CFL condition).

Implementation. Implement the listed schemes and flux functions in the same program such that the user can combine any of the schemes with any of the flux functions at run time. As initial data we set $u = 0$ for $x > 0$ and $u = 1$ for $x = 0$. The grid covers $x \in [0,1]$. (To start the Leap-Frog scheme, one can use one of the other schemes.) The numerical solution must be plotted at each time level. In the linear case $f(u) = \gamma u$, one should also produce plots of the simple exact solution. Use the linear case to partially verify the program. To facilitate the implementation, one can, e.g., simplify the program from Chapter 1.6.3.

Make a script (see page 35) that (i) reads the name of the scheme, the type of flux function, the number of grid points, and value of the Courant number, (ii) runs the simulator, (iii) animates the evolution of the numerical solution, and (iv) produces a PostScript plot of the numerical solution when the front is located in the middle of the domain. In the case $f(u) = \gamma u$, the animation and the plot should include the exact solution.

Analysis. We now consider the linear case $f(u) = \gamma u$. Try to write as many of the schemes as possible in the finite difference operator notation from Appendix A.3. Calculate the truncation error of each scheme. Also analyze a selected set of schemes by finding and discussing numerical dispersion relations as explained in Appendices A.4.5–A.4.8 and try to use this analysis to explain some of the visual numerical effects that can be observed when playing with the computer code.

Chapter 2

Introduction to Finite Element Discretization

The finite element method is a flexible numerical approach for solving partial differential equations. One of the most attractive features of the method is the straightforward handling of geometrically complicated domains. It is also easy to construct higher-order approximations. The present chapter gives an introduction to the basic ideas of finite elements and associated computational algorithms. No previous knowledge of the method is assumed.

First, we present the key ideas of a discretization framework called the weighted residual method, where the finite element method arises as a special case. Particular emphasis is put on the reasoning behind the derivation of discrete equations and especially the handling of various boundary conditions. Our formulation of the discretization procedure attempts to give the reader the proper background for understanding how to operate the finite element toolbox in Diffpack.

The finite element tools in Diffpack allow the user to concentrate on specifying the weighted residual statement (also referred to as the discrete weak formulation) and the essential boundary conditions. Element-by-element assembly, numerical integration over elements, etc. are automated procedures. In the present chapter we will, however, explain all details of the finite element algorithms in 1D examples and show how the algorithms are coded at a fairly low level using only straightforward array manipulations. Thereby, the reader should gain a thorough understanding of how the methods work and hopefully realize how these algorithms can, at least in principle, easily be extended to treat complicated multi-dimensional PDE problems. Advanced generalized versions of the algorithms are available in Diffpack, and we focus on their usage in later chapters.

After the algorithmic aspects of the finite element method are introduced, we turn to variational forms and a more precise mathematical formulation of continuous and discrete PDE problems. This framework allows derivation of generic properties of the finite element method, such as existence and uniqueness of the solution, stability estimates, best-approximation properties, error estimates, and adaptive discretizations.

There are numerous textbooks on finite elements, emphasizing different aspects of the method. Some texts are written in an engineering style with special focus on structural analysis, where the method can be derived directly from physical considerations. Other texts are written in an abstract mathematical framework and emphasize the method as an optimal approach for solving certain classes of PDEs. The treatment of the finite element method in this book is mainly intuitive and informal with weight on generic algorithmic building blocks that apply to a wide range of PDEs. The emphasis on detailed hand calculations of 1D problems is not only motivated on pedagogical grounds – of even more importance is the need for hand calculations of element matrices and vectors when debugging finite element codes.

As we will demonstrate in later chapters, the combination of the generic discretization principles formulated in the present chapter and the Diffpack software provides a flexible workplace for experimenting with finite element methods in quite complicated scientific and engineering applications.

2.1 Weighted Residual Methods

2.1.1 Basic Principles

While the main idea of the finite difference method is to replace derivatives in a partial differential equation by difference approximations, the main idea of the finite element and related methods is to seek an approximation

$$\hat{u} = \sum_{j=1}^{M} u_j N_j(\boldsymbol{x})$$

to the unknown function $u(\boldsymbol{x})$. The sum in $\hat{u}$ involves prescribed functions $N_j(\boldsymbol{x})$ and unknown coefficients u_j. The functions N_j are often referred to as *basis functions* or *trial functions*. In the finite element community, the word *shape functions* is frequently used. Throughout this book we will use the term basis functions.

The ultimate aim is to construct a method for computing u_j such that the error $u - \hat{u}$ is minimized. For some special problems it is possible to minimize a problem-dependent norm of the error, $||u - \hat{u}||$, without knowing the exact solution u (see Theorems 2.13 and 2.14 in Chapter 2.10), but in general we must rely on seemingly less attractive strategies. Although the true error $u-\hat{u}$ is unknown, the error in the PDE, arising from inserting $\hat{u}$ instead of u, is easy to measure and work with. Let $\mathcal{L}(u(\boldsymbol{x})) = 0$, $\boldsymbol{x} \in \Omega$, denote the PDE, where $\mathcal{L}$ is some differential operator. If we insert the approximation $\hat{u}$ in the PDE, we generally have that $\mathcal{L}(\hat{u}) \neq 0$. The error in the equation, $R = \mathcal{L}(\hat{u})$, is termed the *residual*. What we hope, is that a small residual implies a small error $u - \hat{u}$. For a typical stationary PDE, like the Poisson equation, we shall actually in Chapter 2.10.7 derive a bound on the error in terms of the residual. Similar results have been established for several prototype PDEs.

Let us now formulate some procedures for determining u_j. The M equations we need to determine the M parameters $u_1, \ldots, u_M$ can be obtained by forcing the residual

$$R(u_1, \ldots, u_M; \boldsymbol{x}) = \mathcal{L}(\hat{u})$$

to be small in different senses. Notice that R varies in space so we need to minimize some averages of R.

The Least-Squares Method. In this method we minimize the average square of the residual $\int_\Omega R^2 d\Omega$ with respect to $u_1, \ldots, u_M$. This results in M algebraic equations,

$$\frac{\partial}{\partial u_i} \int_\Omega R^2 d\Omega = \int_\Omega 2R \frac{\partial R}{\partial u_i} d\Omega = 0, \quad i = 1, \ldots, M. \tag{2.1}$$

The Weighted Residual Method. The idea in this approach is to find $u_1, \ldots, u_M$ such that the weighted mean of R over Ω vanishes for M linearly independent weighting functions W_i:

$$\int_\Omega RW_i d\Omega = 0, \quad i = 1, \ldots, M. \tag{2.2}$$

The least-squares method is hence a weighted residual method with weighting functions $W_i = 2\partial R/\partial u_i$. Choosing various weighting functions gives rise to different methods that will be outlined in the following. The W_i functions are also often referred to as *test* functions.

Remark. In the weighted residual formulation, the PDE is to be fulfilled in an average sense: $\int_\Omega \mathcal{L}(\hat{u}) W_i d\Omega = 0$. This is often called a *discrete weak formulation, weighted residual formulation,* or *discrete variational formulation.* One can also derive a corresponding weak formulation of the *continuous* problem, as will be explained in Chapter 2.10.1. The associated solution u is then a *weak solution* of the problem. When the PDE $\mathcal{L}(u) = 0$ is taken to hold in a pointwise sense (i.e. at each point $\boldsymbol{x} \in \Omega$), we speak about the *classical solution* u to the problem. We refer to [98, Ch. 8.2] for precise definitions of classical, weak, and strong solutions to PDEs. In this text we will sometimes use the term *weak formulation* (or just *weak form*) when we actually mean the *discrete weak formulation.* However, the precise meaning should be evident from the context.

The Collocation Method. Let $W_i = \delta(\boldsymbol{x} - \boldsymbol{x}^{[i]})$, where $\delta(\boldsymbol{x} - \boldsymbol{x}^{[i]})$ is the Dirac delta function that vanishes when $\boldsymbol{x} \neq \boldsymbol{x}^{[i]}$ and has the property

$$\int_\Omega f(\boldsymbol{x}) \delta(\boldsymbol{x} - \boldsymbol{x}^{[i]}) d\Omega = f(\boldsymbol{x}^{[i]})$$

for an arbitrary function f. Application of these weighting functions results in discrete equations on the form

$$R(u_1, \ldots, u_M; \boldsymbol{x}^{[i]}) = 0 \quad \text{or} \quad \mathcal{L}(\hat{u}(\boldsymbol{x}^{[i]})) = 0, \quad i = 1, \ldots, M. \tag{2.3}$$

That is, the partial differential equation is required to be satisfied at M *collocation points* $\boldsymbol{x}^{[i]}$. Equivalently, the residual is forced to vanish at M distinct points. Observe that the collocation method is closely related to the finite difference method. In fact, we can view the finite difference method as a collocation method, where the derivatives at each collocation point are replaced by finite difference approximations.

The Subdomain Collocation Method. By dividing Ω into M subdomains Ω_i, such that $\Omega = \cup_{i=1}^{M}\Omega_i$, one can introduce a weighting function defined by $W_i(\boldsymbol{x}) = 1$ if $\boldsymbol{x} \in \Omega_i$ and $W_i = 0$ otherwise. The discrete equations corresponding to this subdomain collocation method read

$$\int_{\Omega_i} \mathcal{L}(\hat{u})d\Omega = 0, \quad i = 1, \ldots, M. \tag{2.4}$$

2.1.2   Example: A 1D Poisson Equation

Let us apply the methods from the previous section to the specific problem $\mathcal{L}(u) = u''(x) + f(x)$, $\Omega = (0, 1)$, and $u(0) = u(1) = 0$. For the discretization we set

$$u(x) \approx \hat{u}(x) = \sum_{j=1}^{M} u_j N_j(x)$$

into the differential equation and get

$$R = f(x) + \sum_{j=1}^{M} u_j N_j''(x).$$

Applying the least-squares method or the method of weighted residuals, yields a linear system for $u_1, \ldots, u_M$. Nothing in these methods so far deals with the boundary conditions. It is therefore necessary to assure that $\hat{u}$ fulfills the prescribed boundary values, e.g., by letting $N_j(0) = N_j(1) = 0$ for $j = 1, \ldots, M$. A least-squares approach results in (2.1). We then need the derivative

$$\frac{\partial R}{\partial u_i} = \sum_{j=1}^{M} \frac{\partial}{\partial u_i} u_j N_j''(x) = N_i''(x).$$

The least-squares equations become

$$\int_0^1 \left(f(x) + \sum_{j=1}^{M} u_j N_j''(x) \right) N_i''(x)dx = 0, \quad i = 1, \ldots, M.$$

This is actually a linear system for $u_1, \ldots, u_M$, a fact that is easier to realize if we move the sum outside the integral sign and place the fN_i'' term on the right-hand side:

$$-\sum_{j=1}^{M}\left(\int_0^1 N_i''(x)N_j''(x)dx\right)u_j = \int_0^1 f(x)N_i''(x)dx, \quad i=1,\ldots,M. \quad (2.5)$$

In matrix form this linear system reads $\boldsymbol{Au} = \boldsymbol{b}$, with the matrix $\boldsymbol{A}$ having entries $A_{i,j} = -\int_0^1 N_i''N_j''dx$ and the vector $\boldsymbol{b}$ having entries $b_i = \int_0^1 fN_i''dx$.

The method of weighted residuals leads to a similar linear system:

$$-\sum_{j=1}^{M}\left(\int_0^1 W_i(x)N_j''(x)dx\right)u_j = \int_0^1 f(x)W_i(x)dx, \quad i=1,\ldots,M. \quad (2.6)$$

The most common choice of W_i in the method of weighted residuals is to let $W_i = N_i$. This choice is referred to as *Galerkin's method*. The weighted residual method with $W_i \neq N_i$ is also called a Petrov-Galerkin formulation.

The collocation method gives

$$-\sum_{j=1}^{M} N_j''(x^{[i]})u_j = f(x^{[i]}), \quad i=1,\ldots,M. \quad (2.7)$$

Equally spaced collocation points is a simple choice:

$$x^{[i]} = (i-1)/(M-1), \quad i=1,\ldots,M.$$

The subdomain collocation method on equal-sized subdomains results in

$$-\sum_{j=1}^{M}\int_{(i-1)/M}^{i/M} N_j''(x)dx\, u_j = \int_{(i-1)/M}^{i/M} f(x)dx, \quad i=1,\ldots,M. \quad (2.8)$$

Let us investigate some choices of the $N_i(x)$ functions. Two possible families of functions fulfilling the requirement $N_i(0) = N_i(1) = 0$ are

$$N_i(x) = \sin i\pi x, \quad i=1,\ldots,M, \quad (2.9)$$
$$N_i(x) = x^i(1-x), \quad i=1,\ldots,M. \quad (2.10)$$

All the mentioned discretization approaches give rise to a linear system

$$\sum_{j=1}^{M} A_{i,j}u_j = b_i, \quad i=1,\ldots,M.$$

With Galerkin's method and the choice $N_i = \sin i\pi x$, $A_{i,j}$ becomes *diagonal* because

$$\int_0^1 \sin i\pi x \sin j\pi x\, dx = \begin{cases} \frac{1}{2} & i=j \\ 0, & i \neq j \end{cases}. \quad (2.11)$$

The diagonal coefficient matrix in the linear system enables a close-form solution for u_i:

$$u_i = \frac{2}{\pi^2 i^2} \int_0^1 f(x) \sin i\pi x \, dx. \tag{2.12}$$

The reader should show that the same result arises from the least-squares method as well.

Ill-Conditioning. Let us choose $f(x) = -1$, which leads to $u(x) = x(x - 1)/2$. We see that with $N_i = x^i(1-x)$, $\hat{u} = \sum_j u_j N_j$ is now capable of reproducing the exact solution. Is the method intelligent enough to always compute $u_1 = -1/2$ and $u_i = 0$ for $i > 1$ in this case? The answer is yes; if the expansion $\sum_j u_j N_j$ contains the analytical solution, Galerkin's method will automatically extract that solution. One can exemplify this principle by computing u_i for, e.g. $M = 8$, using symbolic manipulation software like Maple. However, if we implement the expressions in an ordinary computer program with fixed-length floating point arithmetic, we can only expect $u_i \approx 0$ for $i > 1$. Computations with finite arithmetic can be simulated in Maple, and with 6 significant digits we get the following solutions:

M	$(u_1, \ldots, u_M)$
2	(-.50000, .00001)
4	(-.50001, .00039, -.00079, .00048)
6	(-.50072, .01296, -.07323, .17561, -.18727 , .073122)
8	(-.49977, -.00510, .01485, .11669, -.73884, 1.56298, -1.4438, .49399)

These results demonstrate a fundamental shortcoming of the method: It does not converge – the error increases when we add more terms to the expansion $\sum_j u_j N_j$. The reason is that the coefficient matrix in finite arithmetic becomes ill-conditioned as M grows. The solution procedure then becomes sensitive to round-off errors. The ill-conditioning stems from almost linear dependence among the $N_i = x^i(1-x)$ functions when i is large, see Figure 2.1. To cure the problem, one should choose basis functions that are as orthogonal as possible. The basis functions $N_i = \sin i\pi x$ are exactly orthogonal and give a stable approximate solution for any odd M:

$$\hat{u} = \sum_{j=1}^{(M+1)/2} \frac{4}{(2j-1)^3 \pi^3} \sin\left[(2j-1)\pi x\right] .$$

We need to let $M \to \infty$ to reproduce the analytical solution. For practical purposes only a few terms are needed to obtain an approximation whose error is within machine precision.

Fourier Series. Let us briefly demonstrate that the least-squares method and Galerkin's method can be used to derive the well-known Fourier (sine) series.

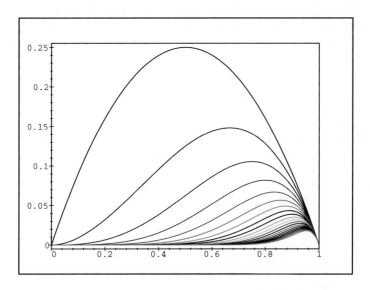

Fig. 2.1. Plot of the functions $x^i(1-x)$ for $i = 1, 2, \ldots, 20$. The height of the graph is reduced with increasing i. For large i, we see that the graphs are very close to each other, demonstrating almost linear dependence of the functions.

We seek an approximate solution of the equation

$$u(x) = f(x), \quad x \in [0, 1],$$

where $u(x)$ is on the form $\hat{u}(x) = \sum_j u_j N_j(x)$. The method of weighted residuals with $N_i = W_i = \sin i\pi x$ gives

$$\sum_{j=1}^{M} \left(\int_0^1 \sin i\pi x \sin j\pi x \, dx \right) u_j = \int_0^1 f(x) \sin i\pi x \, dx, \quad i = 1, \ldots, M. \quad (2.13)$$

Due to the orthogonality of the sine functions, we can explicitly calculate that

$$u_i = 2 \int_0^1 f(x) \sin i\pi x dx, \quad i = 1, \ldots, M, \quad (2.14)$$

which is recognized as the coefficients in a truncated Fourier sine series with M terms. Using the least-squares method, we see that $\partial R / \partial u_j = N_j$, such that this method becomes identical to Galerkin's method in the present problem. When $M \to \infty$, the expression for $\hat{u}$ recovers the standard Fourier sine series of $f(x)$ on $[0, 1]$.

Exercise 2.1. Choose $N_1 = 1$, $N_{2i} = \sin i\pi x$, and $N_{2i+1} = \cos i\pi x$, $i = 1, \ldots$ Show that both the least-squares method and Galerkin's method applied to the equation $u(x) = f(x)$ reproduce the standard sine-cosine Fourier series of $f(x)$ on $[0, 1]$. $\diamond$

2.1.3   Treatment of Boundary Conditions

When u vanishes on the boundary $\partial\Omega$ of Ω, we can require $N_i = 0$ on $\partial\Omega$ such that the approximate solution $\hat{u}$ fulfills the boundary conditions. In the case where $u = \tilde{\psi}(\boldsymbol{x}) \neq 0$ on $\partial\Omega$, one can introduce a function ψ that equals $\tilde{\psi}$ at $\partial\Omega$ and use the expansion

$$\hat{u}(\boldsymbol{x}) = \psi(\boldsymbol{x}) + \sum_{j=1}^{M} u_j N_j(\boldsymbol{x}) \,. \tag{2.15}$$

With this approach we still demand $N_i = 0$ on $\partial\Omega$. Note that ψ in (2.15) is not determined uniquely; it can be replaced by another function that equals the correct boundary value $\tilde{\psi}$ on $\partial\Omega$.

As an example, consider a boundary-value problem on $(0, 1)$ where the unknown function $u(x)$ has the boundary conditions $u(0) = U_L$ and $u(1) = U_R$. The ψ function must then fulfill these boundary conditions, but it can be arbitrary in the interior of $(0, 1)$. Some possible choices are, e.g.,

$$\psi(x) = x^i U_R + (1 - x)^j U_L \,, \quad i, j > 0 \,.$$

Boundary conditions involving derivatives can be conveniently treated by integration by parts. The principle is most easily explained through an example. Let us consider the boundary-value problem

$$-u''(x) = f(x), \ 0 < x < 1, \quad u(0) = 1, \ u'(1) = \beta \,.$$

We introduce an expansion $u \approx \hat{u} = 1 + \sum_{j=1}^{M} u_j N_j(x)$, with $N_j(0) = 0$, that fulfills the condition $\hat{u}(0) = 1$. Using weighting functions $W_i = N_i$ in the weighted residual method, leads to

$$-\sum_{j=1}^{M} \left(\int_0^1 N_i N_j'' dx \right) u_j = \int_0^1 f N_i dx, \quad i = 1, \ldots, M \,.$$

Integrating by parts on the left-hand side results in

$$\sum_{j=1}^{M} \left(\int_0^1 N_i' N_j' dx \right) u_j - N_i(1)\hat{u}'(1) + N_i(0)\hat{u}'(0) = \int_0^1 f N_i dx, \quad i = 1, \ldots, M \,.$$

Recall now that $N_i(0) = 0$. Moreover, it is natural to require that $\hat{u}'(1) = \beta$ such that we end up with the linear system

$$\sum_{j=1}^{M} \left(\int_0^1 N_i' N_j' dx \right) u_j = \int_0^1 f N_i dx + N_i(1)\beta, \quad i = 1, \dots, M . \qquad (2.16)$$

The integration by parts introduced a term in the weighted integral formulation that could be conveniently used for inserting the boundary condition $u'(1) = \beta$. We also recognize that the coefficient matrix has become symmetric. The most important achievement is, however, that derivatives of only first order appear in the formulation. This allows simple piecewise polynomials as basis functions, which is a fundamental choice of N_i in the finite element method.

The ideas above carry over to multi-dimensional problems. Consider for example

$$-\nabla \cdot [k(\boldsymbol{x})\nabla u(\boldsymbol{x})] = f(\boldsymbol{x}), \quad \boldsymbol{x} \in \Omega, \qquad (2.17)$$

$$-k(\boldsymbol{x})\frac{\partial u}{\partial n} = g(\boldsymbol{x}), \quad \boldsymbol{x} \in \partial\Omega_N, \qquad (2.18)$$

$$u(\boldsymbol{x}) = \psi(\boldsymbol{x}), \quad \boldsymbol{x} \in \partial\Omega_E . \qquad (2.19)$$

Here, k, f, and g are prescribed functions and u is unknown. The boundary $\partial\Omega$ of the domain Ω is partitioned into two non-overlapping parts, $\partial\Omega_N$, where a condition on the normal derivative applies, and $\partial\Omega_E$, where a condition on the function itself applies. We can write $\partial\Omega = \partial\Omega_N \cup \partial\Omega_E$, $\partial\Omega_N \cap \partial\Omega_E = \emptyset$. This boundary-value problem appears in a wide range of physical applications, including fluid flow, heat conduction, electromagnetism, and elasticity. A porous media flow interpretation is briefly explained on page 170.

To solve the boundary-value problem (2.17)–(2.19), we might use the expansion $u \approx \hat{u} = \psi + \sum_j N_j u_j$, the weighted integral method, and integration by parts. Green's lemma is an important tool to accomplish the latter step:

$$-\int_\Omega \nabla \cdot [k\nabla u] W_i d\Omega = \int_\Omega k\nabla u \cdot \nabla W_i d\Omega - \int_{\partial\Omega} W_i k \frac{\partial u}{\partial n} d\Gamma . \qquad (2.20)$$

The N_i functions must vanish on $\partial\Omega_E$ to ensure that $\hat{u} = \psi$ is fulfilled here. A weighted residual approach with $W_i = N_i$ (Galerkin's method) then leads to the linear system

$$\sum_{j=1}^{M} \left(\int_\Omega k(\boldsymbol{x}) \nabla N_i \cdot \nabla N_j \, d\Omega \right) u_j = \int_\Omega f(\boldsymbol{x}) N_i d\Omega - \int_{\partial\Omega_N} g(\boldsymbol{x}) N_i d\Gamma \qquad (2.21)$$

for $i = 1, \ldots, M$. Some readers will claim that this simple incorporation of the flux condition $-k\partial u/\partial n = g$ arose with a great amount of luck. It turns out, however, that integration by parts usually results in boundary terms that correspond to common physical flux-type conditions.

To devise a principally more general method for incorporating boundary conditions, we can proceed as follows. Given a partial differential equation on the form $\mathcal{L}(u) = 0$ in Ω and a boundary condition $\mathcal{B}(u) = 0$ on $\partial\Omega_N$, where $\mathcal{B}$ is some possible differential operator on the part $\partial\Omega_N$ of the boundary, we insert the approximate solution and demand

$$\int_\Omega \mathcal{L}(\hat{u})W_i d\Omega + \int_{\partial\Omega_N} \mathcal{B}(\hat{u})\overline{W}_i d\Gamma = 0.$$

Here we have introduced two sets of weighting functions, W_i and $\overline{W}_i$, $i = 1, \ldots, M$. As we shall see, W_i and $\overline{W}_i$ are closely related. A requirement is that $W_i = 0$ on the part $\partial\Omega_E = \partial\Omega \backslash \partial\Omega_N$ of the boundary where u is prescribed. The expansion of u reads $u \approx \hat{u} = \psi + \sum_{j=1}^M u_j N_j$, with $N_i = 0$ on $\partial\Omega_E$.

Let us apply this general set-up to the problem (2.17)–(2.19). We demand the total weighted residual to vanish for M linearly independent weighting functions W_i and $\overline{W}_i$

$$-\int_\Omega \left(\nabla \cdot [k\nabla\hat{u}] + f\right) W_i d\Omega - \int_{\Omega_N} \left(k\frac{\partial\hat{u}}{\partial n} + g\right)\overline{W}_i d\Gamma = 0, \quad i = 1, \ldots, M.$$

(2.22)

Application of Green's lemma (2.20) gives

$$\int_\Omega k\nabla\hat{u} \cdot \nabla W_i \, d\Omega - \int_{\partial\Omega} k\frac{\partial\hat{u}}{\partial n}W_i d\Gamma - \int_\Omega fW_i d\Omega - \int_{\Omega_N} \left(k\frac{\partial\hat{u}}{\partial n} + g\right)\overline{W}_i d\Gamma = 0.$$

(2.23)

Let us now choose $W_i = N_i$ and $\overline{W}_i = -N_i$. The boundary terms involving $k\partial\hat{u}/\partial n$ then cancel, and because $N_i = 0$ on $\partial\Omega_E$, we are left with equation (2.21). The present approach is advantageous when more complicated derivative boundary conditions appear in the model.

The reader should notice the following striking fact: When $g = 0$ in (2.18) there is no sign of the derivative boundary condition in the final weak formulation (2.21)!

In general, we have seen that the weighted residual method with integration by parts admits a natural mechanism for incorporating derivative boundary conditions involving $-k\partial u/\partial n$. This type of boundary condition is therefore called a *natural boundary condition* for the operator $\nabla \cdot [k\nabla u]$. We have implemented the other condition $u = \psi$ by demanding the approximate solution $\hat{u}$ to fulfill $\hat{u} = \psi$ at all points on the boundary. This condition is therefore termed an *essential boundary condition*. More generally, if we have a differential operator of order $2m$, the boundary conditions that involve derivatives of order less than m are essential boundary conditions, whereas those of order greater than or equal to m constitute natural boundary conditions [94, p. 217]. The essential conditions must be enforced on the approximate solution $\hat{u}$, and the natural conditions are typically incorporated in boundary terms of the weighted residual statement.

Exercise 2.2. The boundary-value problem (2.17)–(2.19) arises in numerous physical contexts. Go through a series of applications, mention the interpretation of u, f, and k, and point out relevant types of boundary conditions. $\diamond$

Example 2.1. We shall now try to derive an approximate solution of the problem (2.17)–(2.19) when $\Omega = (0,1) \times (0,1)$, $k(\boldsymbol{x}) = 1$, $f = 2$, $u = 0$ at $\partial\Omega_E$, and $\partial\Omega_N = \emptyset$. A product $\sin i\pi x \sin j\pi y$ could be used as basis functions,

$$\hat{u} = \sum_{i=1}^{n_x} \sum_{j=1}^{n_y} u_{i,j} N_{i,j}, \quad N_{i,j} = \sin i\pi x \sin j\pi y,$$

where (i,j) is now a double index. As weighting functions we can choose $N_{k,\ell}$. For implementation in a computer program it may, however, be necessary to transform the double indices to single indices prior to calling linear algebra software to solve the algebraic equation system. Our general set-up of the methods in this text also requires single-indexed basis functions and unknown parameters. To switch to a single index, we can define

$$N_{(j-1)n_x+i}(x,y) = \sin i\pi x \sin j\pi y, \quad i = 1, \ldots, n_x, \; j = 1, \ldots, n_y, \; M = n_x n_y.$$

Nevertheless, for analytical calculations it is more convenient to work with double indices. Galerkin's method can then be formulated as

$$\sum_{k=1}^{n_x} \sum_{j=1}^{n_y} \int_0^1 \int_0^1 \nabla N_{k,\ell} \cdot \nabla N_{i,j} \, dx dy = 2 \int_0^1 \int_0^1 N_{k,\ell} \, dx dy,$$

for $k, \ell = 1, \ldots, n$. This is a linear system that can also be written on the form

$$\sum_i \sum_j A_{i,j,k,\ell} u_{i,j} = b_{k,\ell}.$$

Inserting the form of $N_{i,j}$ and using the fact that (2.11) also holds for cosine functions, gives

$$A_{i,j,k,\ell} = \int_0^1 \int_0^1 (ik\pi^2 \cos i\pi x \sin j\pi y \cos k\pi x \sin \ell\pi y +$$
$$j\ell\pi^2 \sin i\pi x \cos j\pi y \sin k\pi x \cos \ell\pi y) dx dy$$

$$= ik\pi^2 \int_0^1 \cos i\pi x \cos k\pi x \, dx \int_0^1 \sin j\pi y \sin \ell\pi y \, dy +$$

$$j\ell\pi^2 \int_0^1 \sin i\pi x \sin k\pi x \, dx \int_0^1 \cos j\pi y \cos \ell\pi y \, dy$$

$$= ik\pi^2 \frac{1}{2}\delta_{ik}\frac{1}{2}\delta_{j\ell} + j\ell\pi^2 \frac{1}{2}\delta_{ik}\frac{1}{2}\delta_{j\ell}$$

$$= \pi^2(ik + j\ell)\frac{1}{4}\delta_{ik}\delta_{j\ell}.$$

The notation δ_{ik} means the Kronecker delta, which equals unity when $i = k$ and zero otherwise. This results shows that $A_{i,j,k,\ell}$ vanishes unless $i = k$ and $\ell = j$, that is, the coefficient matrix in the linear system is *diagonal*:

$$\sum_i \sum_j A_{i,j,k,\ell} u_{i,j} = A_{k,\ell,k,\ell} u_{k,\ell} = b_{k,\ell},$$

with $A_{k,\ell,k\ell} = \pi^2(k^2 + \ell^2)/4$. The right-hand side becomes

$$b_{k,\ell} = \int_0^1 \int_0^1 2 \sin k\pi x \sin \ell\pi y \, dxdy = \frac{8}{k\ell\pi^2} \begin{cases} 1, \ k \text{ and } \ell \text{ odd} \\ 0, \text{ otherwise} \end{cases}$$

Switching from k and ℓ to i and j, the final results can then be written as

$$\hat{u}(x,y) = \sum_{i=1}^{(n_x+1)/2} \sum_{j=1}^{(n_y+1)/2} \alpha_{ij} \sin\left[(2i-1)\pi x\right] \sin\left[(2j-1)\pi y\right],$$

where

$$\alpha_{ij} = 32\left[((2i-1)^2 + (2j-1)^2)(2i-1)(2j-1)\pi^4\right]^{-1}.$$

$\diamond$

2.2   Time Dependent Problems

The weighted residual method and its variants (collocation, subdomain collocation, least squares) are usually thought of as procedures for discretization in space. Of course, these methods can be used in time as well, but it is more common to solve time dependent partial differential equations by finite difference approximation of time derivative terms, combined with some weighted residual method in space.

2.2.1 A Wave Equation

As an example, consider the following initial-boundary value problem involving the wave equation:

$$\frac{\partial^2 u}{\partial t^2} = \nabla \cdot [c^2 \nabla u], \quad \boldsymbol{x} \in \Omega, \ t > 0, \tag{2.24}$$

$$u(\boldsymbol{x}, 0) = f(\boldsymbol{x}), \quad \boldsymbol{x} \in \Omega, \tag{2.25}$$

$$\frac{\partial}{\partial t} u(\boldsymbol{x}, 0) = 0, \quad \boldsymbol{x} \in \Omega, \tag{2.26}$$

$$\frac{\partial u}{\partial n} = 0, \quad \boldsymbol{x} \in \partial\Omega, \ t > 0. \tag{2.27}$$

This model can describe, for instance, long ocean waves as explained in Chapter 1.3.6.

We introduce a grid in time, with points $t_\ell = \ell \Delta t$, $\ell = 0, 1, 2, \ldots$. A time-discrete function $u^\ell(\boldsymbol{x}) = u(\boldsymbol{x}, t_\ell)$, or more compactly written as u^ℓ, can then be defined. We approximate the time derivative by a centered (second-order accurate) finite difference formula:

$$\frac{\partial^2}{\partial t^2} u(\boldsymbol{x}, t_\ell) \approx \frac{u^{\ell-1} - 2u^\ell + u^{\ell+1}}{\Delta t^2}. \tag{2.28}$$

This results in a problem for $u^\ell(\boldsymbol{x})$ that is discrete in time, but continuous in space:

$$u^{\ell+1} = 2u^\ell - u^{\ell-1} + \Delta t^2 \nabla \cdot [c^2 \nabla u^\ell]. \tag{2.29}$$

The initial condition $\partial u/\partial t = 0$ can be approximated by a centered (second-order accurate) finite difference to yield $u^1(\boldsymbol{x}) = u^{-1}(\boldsymbol{x})$. The fictitious quantity u^{-1} is eliminated by using (2.29) for $\ell = 0$. This yields a modification of (2.29) for $\ell = 0$, that is, a special formula for u^1:

$$u^1 = u^0 + \frac{1}{2} \Delta t^2 \nabla \cdot [c^2 \nabla u^0]. \tag{2.30}$$

At time levels $\ell \geq 2$ one can of course apply (2.29) directly. We can alternatively adopt the strategy from Chapter 1.3.2, where we apply (2.29) as it stands for $\ell \geq 0$, but with a special form of the artificial u^{-1}, cf. (1.54),

$$u^{-1} = u^0 + \frac{1}{2} \Delta t^2 \nabla \cdot [c^2 \nabla u^0].$$

This form is convenient from an implementational point view, as we can use the same updating formula (2.29) for all time steps.

Exercise 2.3. Derive (2.30) by the following reasoning. First, express u^1 as a three-term Taylor series around $t = 0$. Then insert the initial $\partial u/\partial t$ value and replace the second-order time derivative of u by $c^2 \nabla^2 u$ (from the PDE).
◇

The time discretization yields the following sequence of purely spatial problems:

$$u^0 = f(\boldsymbol{x}), \quad \boldsymbol{x} \in \Omega, \tag{2.31}$$

$$u^{-1} = u^0 + \frac{1}{2}\Delta t^2 \nabla \cdot [c^2 \nabla u^0], \quad \boldsymbol{x} \in \Omega, \tag{2.32}$$

$$u^{\ell+1} = 2u^\ell - u^{\ell-1} + \Delta t^2 \nabla \cdot [c^2 \nabla u^\ell], \quad \boldsymbol{x} \in \Omega, \ \ell = 0, 1, 2, \ldots, \tag{2.33}$$

$$\frac{\partial u^\ell}{\partial n} = 0, \quad \boldsymbol{x} \in \partial\Omega, \ \ell = 1, 2, 3, \ldots \tag{2.34}$$

A suitable expansion for $u^\ell(\boldsymbol{x})$ can look like

$$u^\ell(\boldsymbol{x}) \approx \hat{u}^\ell = \sum_{j=1}^M u_j^\ell N_j(\boldsymbol{x}), \quad \ell = -1, 0, 1, 2, \ldots, \tag{2.35}$$

where u_j^ℓ are constants to be determined by the method. Applying Galerkin's method ($W_i = N_i$) to (2.31)–(2.33), and integrating second-order derivatives by parts, one obtains a discrete problem in both space and time:

$$\sum_{j=1}^M M_{i,j} u_j^0 = \int_\Omega f(\boldsymbol{x}) N_i d\Omega,$$

$$\sum_{j=1}^M M_{i,j} u_j^{-1} = \int_\Omega \left[\hat{u}^0 N_i - \frac{1}{2}(c\Delta t)^2 \nabla N_i \cdot \nabla \hat{u}^0 \right] d\Omega + \frac{1}{2}\Delta t^2 \int_{\partial\Omega} c^2 \frac{\partial u^0}{\partial n} N_i d\Gamma,$$

$$\sum_{j=1}^M M_{i,j} u_j^{\ell+1} = \int_\Omega \left[\left(2\hat{u}^\ell - \hat{u}^{\ell-1}\right) N_i - (c\Delta t)^2 \nabla N_i \cdot \nabla \hat{u}^\ell \right] d\Omega.$$

The matrix

$$M_{i,j} = \int_\Omega N_i N_j d\Omega \tag{2.36}$$

is often called the *mass matrix* (the symbols M and $M_{i,j}$ in this chapter should not be confused!). Observe that in this method, the weighted residual approach has been used for approximating the spatial parts of the initial conditions (2.31)–(2.32) as well as the time-discrete equation (2.33).

The reader should notice that although we use the same time discretization as in a standard explicit finite difference method for the wave equation (cf. Chapters 1.3.2 and 1.3.6), Galerkin's method leads to coupled systems of algebraic equations. In this sense, the time discretization is implicit[1]. At each time level we must solve a linear system with the mass matrix as coefficient matrix. This is a serious disadvantage in 2D and 3D problems, because the computational labor increases significantly in comparison with an explicit

---

[1] See the footnote on page 22 for a remark on the terminology.

scheme. This could only be acceptable if there were a corresponding increase in accuracy, which is not the case. However, in Chapters 2.4 and 2.7.3 we show how finite element methods for the wave equation can be made as fast as the standard explicit finite difference schemes.

Exercise 2.4. Restrict the initial-boundary value problem (2.24)–(2.27) to one space dimension. Choose $N_i = \cos i\pi x$ and deduce an *explicit* updating formula for $u_i^{\ell+1}$. This choice of N_i yields a *spectral method* for the wave equation. Demonstrate how the formula for $u_j^{\ell+1}$ is simplified when the wave velocity c is constant.   ◇

2.2.2   A Heat Equation

The previous example demonstrated discretization of a PDE with a time derivative of second order. Now we consider the heat (or diffusion) equation, which has a first-order time derivative:

$$\frac{\partial u}{\partial t} = \nabla \cdot (\lambda \nabla u), \quad \boldsymbol{x} \in \Omega, \ t > 0, \tag{2.37}$$

$$u(\boldsymbol{x}, 0) = f(\boldsymbol{x}), \quad \boldsymbol{x} \in \Omega, \tag{2.38}$$

$$-\lambda \frac{\partial u}{\partial n} = g(\boldsymbol{x}, t), \quad \boldsymbol{x} \in \partial\Omega, \ t > 0. \tag{2.39}$$

A widely used finite difference scheme for first-order equations is the so-called θ-rule. It approximates an equation

$$\frac{\partial u}{\partial t} = G$$

by

$$\frac{u^\ell - u^{\ell-1}}{\Delta t} = \theta G^\ell + (1 - \theta)G^{\ell-1}, \quad 0 \le \theta \le 1. \tag{2.40}$$

For $\theta = 0$ we get the *forward Euler scheme*, $\theta = 1$ gives the *backward Euler scheme*, and $\theta = 1/2$ corresponds to the *Crank-Nicolson scheme*, also referred to as the *mid-point rule*. The forward Euler scheme may be subject to severe stability restrictions, cf. (A.35), but it can lead to explicit equations and thereby avoid solution of large coupled systems of equations. The backward Euler and Crank-Nicolson schemes are unconditionally stable. For $\theta \ne 1/2$ the error in the time approximation is of order Δt, while the choice $\theta = 1/2$ gives one order higher accuracy.

Applying the θ-rule to the heat equation results in the following sequence of spatial problems:

$$u^0 = f(\boldsymbol{x}), \quad \boldsymbol{x} \in \Omega, \tag{2.41}$$

$$\frac{u^\ell - u^{\ell-1}}{\Delta t} = \theta \nabla \cdot (\lambda \nabla u^\ell) + (1 - \theta)\nabla \cdot (\lambda \nabla u^{\ell-1}), \quad \boldsymbol{x} \in \Omega, \tag{2.42}$$

$$-\lambda \frac{\partial u^\ell}{\partial n} = g(\boldsymbol{x}, t_\ell), \quad \boldsymbol{x} \in \partial\Omega, \tag{2.43}$$

for $\ell = 1, 2, \ldots$ Discretizing these equations by the method of weighted residuals, with $W_i = N_i$, gives

$$\sum_{j=1}^{M} M_{i,j} u_j^0 = \int_{\Omega} f(\boldsymbol{x}) N_i d\Omega, \tag{2.44}$$

$$\sum_{j=1}^{M} \left(M_{i,j} + K_{i,j} \right) u_j^\ell = \int_{\Omega} \hat{u}^{\ell-1} N_i d\Omega - (1-\theta)\Delta t \int_{\Omega} \lambda \nabla N_i \cdot \nabla \hat{u}^{\ell-1} d\Omega -$$

$$\theta \Delta t \int_{\partial\Omega} N_i g(\boldsymbol{x}, t_\ell) d\Gamma - (1-\theta)\Delta t \int_{\partial\Omega} N_i g(\boldsymbol{x}, t_{\ell-1}) d\Gamma, \tag{2.45}$$

for $\ell \geq 1$. We have, as in the preceding section, $\hat{u}^\ell = \sum_{j=1}^{M} u_j^\ell N_j(\boldsymbol{x})$. The matrix $M_{i,j}$ is the mass matrix (2.36), whereas

$$K_{i,j} = \theta \Delta t \int_{\Omega} \lambda \nabla N_i \cdot \nabla N_j \, d\Omega \,. \tag{2.46}$$

Equation (2.45) is valid for $\ell \geq 1$, but for $\ell = 1$ the last boundary integrals must involve $\partial f / \partial n$ or $\partial u^0 / \partial n$ rather than $g(\boldsymbol{x}, 0)$.

Exercise 2.5. Formulate a discrete version of the PDE

$$\frac{\partial u}{\partial t} = \nabla^2 u + f(\boldsymbol{x}, t) u,$$

using the θ-rule in time and a Galerkin method in space. $\diamond$

Exercise 2.6. Consider the heat equation problem (2.37)–(2.39). Instead of first discretizing in time by a finite difference method, we first apply the weighted residual method in space, with u approximated by

$$\hat{u}(\boldsymbol{x}, t) = \sum_{j=1}^{M} u_j(t) N_j(\boldsymbol{x}) \,.$$

Show that this approach yields a system of first-order *ordinary differential equations* (ODEs):

$$\sum_{j=1}^{M} M_{i,j} u_j(0) = b_i, \tag{2.47}$$

$$\sum_{j=1}^{M} M_{i,j} \dot{u}_j(t) + \sum_{j=1}^{M} K_{i,j} u_j(t) = c_i(t), \tag{2.48}$$

for suitable matrices $M_{i,j}$ and $K_{i,j}$ and vectors b_i and c_i. Some readers might prefer to see (2.47)–(2.48) in vector notation:

$$\boldsymbol{M} \boldsymbol{u}(0) = \boldsymbol{b}, \quad \boldsymbol{M} \dot{\boldsymbol{u}}(t) + \boldsymbol{K} \boldsymbol{u}(t) = \boldsymbol{c}(t) \,. \tag{2.49}$$

Apply the θ-rule to this system and compare the resulting equations with (2.44) and (2.45). Suggest some other schemes for (2.49) by consulting literature on numerical solution of ODEs, e.g. [92, Ch. 16.6-16.7] (we remark that one usually needs to apply methods for *stiff* ODEs).   ◇

Remark. Transforming the PDE, by finite element discretization in space, to a system of ODEs as we demonstrate in Exercise 2.6, is a strategy advocated by many leading experts[2]. Obviously, this strategy gives immediate access to the many sophisticated time-discretization methods developed in the ODE world. Gresho and Sani [44] explains how to take advantage of such methods when solving PDEs. Fortunately, the difference between starting with discretization in time or space soon becomes transparent after some experience with finite elements and transient PDEs.

2.3   Finite Elements in One Space Dimension

When choosing specific functions N_i in a weighted residual method, we need to have $N_i = 0$ at the part of the boundary where essential conditions are prescribed. This can be a difficult task if the boundary does not have a very simple shape. Furthermore, exact or approximate orthogonality of the functions N_i is important for numerical stability when solving the linear system arising from the discretization method. Both these facts complicate the application of the weighted residual method in problems with nontrivial geometry. Finite elements constitute a means for constructing N_i functions that are "nearly orthogonal" and that are very flexible with respect to handling essential boundary conditions, regardless of the geometric shape of the domain.

2.3.1   Piecewise Polynomials

The finite element choice of N_i consists of three fundamental ideas:

1. divide the domain into non-overlapping *elements*,
2. let N_i be a simple polynomial over each element,
3. construct the global N_i as a piecewise polynomial that vanishes over most of the elements, except for a local patch of elements.

For example, in 1D we can sketch a possible choice of N_i as in Figure 2.2, where N_i is linear over each element, but $N_i \neq 0$ only over a patch of two elements. It is not required in general that N_i is a *continuous* piecewise polynomial, but many physical applications involve second-order differential operators, and we then need the integral of the square of the first derivative of

---

[2] See e.g. [44, p. 232] for a discussion.

N_i in the weighted residual statement. To ensure a finite integral, it is natural to let N_i be continuous[3].

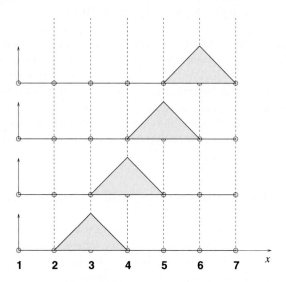

Fig. 2.2. Sketch of piecewise linear N_i functions in 1D for a few elements. The numbers along the x axis refer to the nodes in the grid. The graphs thus illustrate the basis functions $N_3(x)$, $N_4(x)$, $N_5(x)$, and $N_6(x)$.

Let us write down a more formal specification of the N_i functions in Figure 2.2. Assume that $\Omega = [0, 1]$, and divide this domain into m non-overlapping *elements* $\Omega_1, \ldots, \Omega_m$. Throughout Ω there is a set of points $x^{[i]}$, called *nodes*, $i = 1, \ldots, n$. One possible choice of the nodes is to place them at the boundary of each element, that is, $\Omega_e = [x^{[e]}, x^{[e+1]}]$, $x^{[e]} < x^{[e+1]}$ (with $n = m + 1$ and $i = 1, \ldots, m$). This is what we have in Figure 2.2.

One of the strengths of the finite element method is the flexibility in the choice of elements. In regions where the solution is rapidly varying, one can have small elements. The smoother parts of the solution can have an associated grid with larger elements and perhaps high polynomial degree of N_i. This flexibility is particularly important in 2D and 3D.

We want N_i to have two properties:

1. N_i is a polynomial over each element, uniquely determined by its values at the nodes in the element.
2. $N_i(x^{[j]}) = \delta_{ij}$.

---

[3] Using the weighted residual method locally on each element and considering the jumps of N_i between the elements allow discontinuous N_i functions also for PDEs of second order.

Here, δ_{ij} is the Kronecker delta, which equals unity when $i = j$ and vanishes otherwise. Property 2 has the nice implication that

$$\hat{u}(x^{[i]}) = \sum_j u_j N_j(x^{[i]}) = \sum_j u_j \delta_{ij} = u_i,$$

or in other words, u_i is the value of $\hat{u}$ at node no. i. This simple interpretation of the coefficients u_i is convenient in practical computations. The discrete unknowns in the finite element method are then values of the unknown function at grid points, like in the finite difference method. When each basis function is piecewise linear, the $\hat{u}(x)$ function will also be piecewise linear, as we illustrate here:

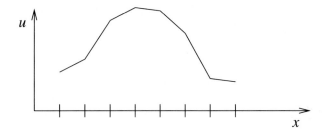

Let us look at the construction of $N_4(x)$ in Figure 2.2. We must have $N_4 = 1$ at node number 4 ($x^{[4]}$), and at all other nodes $N_4 = 0$. In addition, $N_4(x)$ must be a polynomial over each element, uniquely determined by its nodal values. Since there are two nodes per element, N_4 restricted to an element can fulfill two conditions and can therefore at most be a linear function. The only piecewise linear function that equals unity at $x = x^{[4]}$ and zero at all other nodes, is the $N_4(x)$ function depicted in Figure 2.2.

It is natural to construct higher-order polynomials for N_i. For example, let each element have three nodes, two on the boundaries and one at the mid point:

$$\Omega_e = [x^{[2(e-1)+1]}, x^{[2(e-1)+3]}], \quad x^{[2(e-1)+2]} = \frac{1}{2}\left(x^{[2(e-1)+1]} + x^{[2(e-1)+3]}\right),$$

for $e = 1, \ldots, m$. Moreover, the number of nodes is now $n = 2m + 1$. We must of course demand $x^{[i]} < x^{[i+1]}$, $i = 1, \ldots, n-1$, for the definition of Ω_e to make sense. With three nodes per element, $N_i(x)$ can be uniquely determined as a quadratic polynomial. We encourage the reader to make a sketch of a domain consisting of three quadratic elements. Number the elements and the nodes from left to right and draw the piecewise quadratic functions $N_3(x)$ and $N_4(x)$. When i corresponds to an internal node, $N_i(x) \neq 0$ only over the element that contains this node. From the sketch the reader should realize that there are basically two sets of N_i functions in this case: those that correspond to internal nodes and those that correspond to the boundary nodes (interfaces) of the elements.

One can easily extend the ideas above and construct piecewise cubic polynomials, with four nodes per element. Alternatively, one can have only two nodes (at the boundary of each element), and let N_i be a cubic polynomial that is uniquely determined by values of N_i and dN_i/dx at the two nodes. This construction ensures continuity of the derivative of N_i and is advantageous when solving fourth-order problems (e.g. bending of elastic beams).

One often refers to piecewise linear basis functions as *linear elements*. Similarly, piecewise quadratic basis functions are frequently denoted as *quadratic elements*.

2.3.2 Handling of Essential Boundary Conditions

Let us apply the piecewise linear basis functions to a simple 1D boundary-value problem

$$-u''(x) = f(x), \ 0 < x < 1, \quad u(0) = u_L, \ u(1) = u_R. \qquad (2.50)$$

We divide the domain $\Omega = [0, 1]$ into m elements of equal length h, $\Omega_e = [(e-1)h, eh]$, with $h = 1/m$, and $e = 1, \dots, m$. The nodes are located at $x^{[i]} = (i-1)h$, $i = 1, \dots, n = m + 1$. See Figure 2.2 for a sketch of the situation.

Each N_i function is piecewise linear, and $N_i(x^{[j]}) = \delta_{ij}$. The latter property makes it easy to fulfill essential boundary conditions; we can simply choose

$$\psi(x) = u_L N_1(x) + u_R N_n(x)$$

and use the expansion

$$u(x) \approx \hat{u}(x) = \psi(x) + \sum_{j=2}^{n-1} u_j N_j(x).$$

The reader should check in detail that $\hat{u}$ equals u_L and u_R at $x = 0$ and $x = 1$, respectively. A weighted residual formulation can be obtained by inserting the expression for $\hat{u}$ in the differential equation, multiplying by $W_i = N_i$, $i = 2, \dots, n-1$, and integrating over $[0, 1]$. Note that we need only $n - 2$ weighting functions because there are only $n - 2$ unknown parameters $(u_2, \dots, u_{n-1})$ to solve for. The second-order derivative is integrated by parts, resulting in the discrete formulation

$$\sum_{j=2}^{n-1} \left(\int_0^1 N_i' N_j' dx \right) u_j = \int_0^1 f N_i dx, \quad i = 2, \dots, n-1.$$

When programming this, it will be convenient to renumber the unknowns from 1 to $n - 2$. In other words, we introduce a numbering of the unknowns in the linear system that is different from the numbering of the nodes. However, this renumbering is actually not necessary: By introducing essential boundary conditions directly in the linear system instead of in the expansion, we can work with $u_1, \ldots, u_n$ as unknowns through the whole computational procedure. We skip the function ψ in the expansion and work directly with

$$u \approx \hat{u} = \sum_{j=1}^{n} u_j N_j$$

for all types of essential conditions. The linear system now has the complete set of parameters $(u_1, \ldots, u_n)$ as unknowns. To enforce an essential boundary condition at a node j, we simply replace equation no. j by the boundary condition equation $u_j = \psi(x^{[j]})$, i.e., in our case $u_1 = u_L$ or $u_n = u_R$. This is a general procedure that applies to all type of finite element functions as long as $N_i(x^{[j]}) = \delta_{ij}$ holds, such that u_j has the interpretation of being the value of $\hat{u}$ at node j. In particular, the approach can be used in 2D and 3D.

2.3.3   Direct Computation of the Linear System

Let us calculate explicit expressions for the n discrete equations arising from a Galerkin finite element method applied to (2.50). With piecewise linear N_i functions, we can deduce from the simple sketch in Figure 2.3 that the matrix entry $A_{i,j} = \int_0^1 N_i' N_j' dx$ is different from zero *if and only if i and j are nodes belonging to the same element*. In other words, $A_{i,j} \neq 0$ only

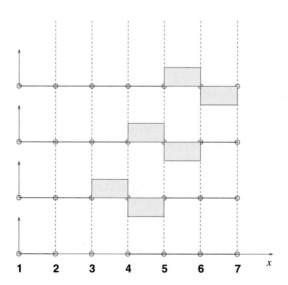

Fig. 2.3. Sketch of piecewise constant $N_i'(x)$ functions i 1D for a few elements. See Figure 2.2 on page 126 for more information.

for $j = i - 1, i, i + 1$. Assuming constant element size h (this is done only for simplicity – it is straightforward to handle varying element size as will be shown later), it is easy to realize that $N_i' = \pm 1/h$. By inspection of Figure 2.3 we can calculate

$$A_{i,i-1} = \int_0^1 N_{i-1}' N_i' dx = -\frac{1}{h}$$

$$A_{i,i} = \int_0^1 N_i' N_i' dx = \frac{2}{h}$$

$$A_{i,i+1} = \int_0^1 N_i' N_{i+1}' dx = -\frac{1}{h}$$

for $i = 2, \ldots, n-1$. The end-point contributions are $A_{1,1} = 1/h$, $A_{1,2} = -1/h$, $A_{n,n-1} = -1/h$, and $A_{n,n} = 1/h$.

The expression for the right-hand side vector of the linear system reads

$$b_i = \int_0^1 f(x) N_i(x) dx \,. \tag{2.51}$$

In order to evaluate this expression, an explicit formula for $N_i(x)$ is needed:

$$N_i(x) = \begin{cases} 0, & x \le x^{[i-1]} \\ h^{-1}(x - x^{[i-1]}), & x^{[i-1]} \le x \le x^{[i]} \\ -h^{-1}(x - x^{[i+1]}), & x^{[i]} \le x \le x^{[i+1]} \\ 0, & x \ge x^{[i+1]} \end{cases} \tag{2.52}$$

for $1 < i < n$. The formulas for $N_1(x)$ and $N_n(x)$ are a slight modification of (2.52). The integral (2.51) is now split into an integral over $[x^{[i-1]}, x^{[i]}]$ and another integral over $[x^{[i]}, x^{[i+1]}]$. Only in certain cases these integrals can be evaluated analytically. A general approach is therefore to integrate (2.51) numerically.

A convenient numerical integration scheme is the trapezoidal rule, with the nodal points as integration points:

$$\int_0^1 f(x) N_i dx \approx \frac{h}{2} f(x_1) N_i(x_1) + \sum_{j=2}^{n-1} h f(x_j) N_i(x_j) + \frac{h}{2} f(x_n) N_i(x_n) \,.$$

Note that for a fixed i, only one integration point, namely $x^{[i]}$, will contribute to the integral, because $N_i(x^{[j]}) = \delta_{ij}$. This leads to a right-hand side $b_i = f(x^{[i]})h$, $i = 2, \ldots, n-1$, like in the finite difference method. At the end points we get $b_1 = f(0)h/2$ and $b_n = f(1)h/2$. Other numerical integration schemes with sampling points also inside the elements will lead to a b_i value that is a weighted mean of f in the neighborhood of $x = x^{[i]}$.

We enforce the boundary conditions by replacing the equations for $i = 1$ and $i = n$ by the conditions themselves: $u_1 = u_L$ and $u_n = u_R$. The complete linear system can be written on the form

$$u_1 = u_L, \tag{2.53}$$

$$-\frac{1}{h}u_{i-1} + \frac{2}{h}u_i - \frac{1}{h}u_{i+1} = f(x^{[i]})h, \quad i = 2, \ldots, n-1, \tag{2.54}$$

$$u_n = u_R, \tag{2.55}$$

with $x^{[i]} = (i-1)h$, $i = 2, \ldots, n-1$. These equations are identical to those arising from the finite difference method when a centered (second-order accurate) difference is used to approximate $u''(x)$. The finite element method gives discrete equations that in general differ from those of the finite difference method. This is also the case in the present example if analytical integration or more accurate integration rules are used for the term on the right-hand side.

2.3.4   Element-By-Element Formulation

The calculation of the linear system above was straightforward, but the technique is not feasible for multi-dimensional problems with geometrically complicated domains. Therefore we will from now on focus on a more general computational algorithm that can easily be applied to all kinds of finite element calculations. The understanding of this general algorithm is fundamental for the understanding the problem specification required by generic software tools like Diffpack.

Background. The general algorithm for constructing the linear system arising in the finite element method, works in an element-by-element fashion. The fundamental idea is to write

$$A_{i,j} = \int_0^1 N_i' N_j' dx = \sum_{e=1}^m A_{i,j}^{(e)} \quad \text{where} \quad A_{i,j}^{(e)} = \int_{\Omega_e} N_i' N_j' dx, \tag{2.56}$$

and

$$b_i = \int_0^1 f N_i dx = \sum_{e=1}^m b_i^{(e)} \quad \text{where} \quad b_i^{(e)} = \int_{\Omega_e} f N_i dx. \tag{2.57}$$

Now, $A_{i,j}^{(e)}$ is different from zero if and only if i and j are nodes in element no. e, because in the case i (or j) corresponds to a node that does not belong to the present element, N_i' (or N_j') will be zero. Similarly, $b_i^{(e)}$ is different from zero if and only if i is a node in element no. e. It is therefore necessary to consider only the local i and j numbers in the element when computing $A_{i,j}^{(e)}$ and $b_i^{(e)}$. For example, i and j are the two node numbers in element no. e in the case of linear elements. This motivates the concept of a local node numbering in each element. At present, we restrict the presentation to linear elements for simplicity. If $\Omega_e = [x^{[e]}, x^{[e+1]}]$, we refer to $x^{[e]}$ as local node no. 1 and $x^{[e+1]}$ as local node no. 2. There is clearly a mapping $i = q(e, r)$ that gives the global node number i corresponding to local node number r in element no. e. For the present linear elements we have $q(e, r) = e - 1 + r$, $r = 1, 2$. In multi-dimensional problems with nontrivial domain geometry, the q function has no simple analytical expression and is known only in form of a table.

Local Coordinates. We collect the nonzero contributions to $A_{i,j}^{(e)}$ in a 2×2 *element matrix* $\tilde{A}_{r,s}^{(e)}$, where $r, s = 1, 2$ are local node numbers. An *element vector* $\tilde{b}_r^{(e)}$, $r = 1, 2$, is also introduced. The element matrix and vector involve integrals over element no. e. It will be convenient, at least in nontrivial multi-dimensional problems, to map the physical element onto a reference element of fixed size. In 1D we map $\Omega_e = [x^{[e]}, x^{[e+1]}]$ onto $[-1, 1]$. The new coordinate $\xi \in [-1, 1]$ is related to x through $x = x^{(e)}(\xi)$, where

$$x^{(e)}(\xi) = \frac{1}{2}\left(x^{[e]} + x^{[e+1]}\right) + \xi\frac{1}{2}\left(x^{[e+1]} - x^{[e]}\right). \tag{2.58}$$

We now need expressions for the basis functions in local coordinates in order to compute the integrals. Let us denote the basis functions in local coordinates as $\tilde{N}_r(\xi)$. It is easy to construct these functions, because we know that $\tilde{N}_1(-1) = \tilde{N}_2(1) = 1$, $\tilde{N}_1(1) = \tilde{N}_2(-1) = 0$, and that the functions must be linear. The unique choice is then

$$\tilde{N}_1(\xi) = \frac{1}{2}(1 - \xi), \quad \tilde{N}_2(\xi) = \frac{1}{2}(1 + \xi). \tag{2.59}$$

We have in general $\tilde{N}_r(\xi) = N_i(x^{(e)}(\xi))$, $i = q(e, r)$, when $x \in \Omega_e$.

The expansion of $\hat{u}$ over the reference element is now $\hat{u} = \sum_{s=1}^{2} \tilde{N}_s(\xi)\tilde{u}_s$, $\xi \in [-1, 1]$, provided $x^{(e)}(\xi) \in \Omega_e$. The parameters $\tilde{u}_1$ and $\tilde{u}_2$ are the values of $\hat{u}$ at local nodes 1 and 2 in the element; in general we have $\tilde{u}_r = u_{q(e,r)}$.

Notice that the expressions for $\tilde{N}_r(\xi)$ can be used for *all* the element matrices and vectors, whereas the formulas for the basis functions in physical coordinates depend on the element's shape and size. Basis functions for multi-dimensional finite elements are frequently defined in local coordinates only. The general mapping between local and global (or physical) coordinates is taken as

$$x^{(e)}(\xi) = \sum_{r=1}^{n_e} \tilde{N}_r(\xi)x^{[q(e,r)]}, \tag{2.60}$$

where n_e is the number of nodes in the element ($n_e = 2$ for linear elements). This formula coincides with the relation (2.58) for the special case of linear elements. When the $\tilde{N}_r$ functions are used for both interpolating the unknown function and for mapping the reference element to global coordinates, the element is referred to as an *isoparametric element*. The associated mapping (2.60) is called the *isoparametric mapping*.

Transforming Derivatives and Integrals. Changing coordinates from x to ξ affects both the integral and the derivatives. The Jacobian matrix $J = dx/d\xi$ of the mapping $x = x^{(e)}(\xi)$ is fundamental when transforming the integral and the derivatives in the integrand. The derivative transforms according to

$$\frac{dN_i}{dx} = \frac{d\tilde{N}_r}{d\xi}\frac{d\xi}{dx} = J^{-1}\frac{d\tilde{N}_r}{d\xi}, \quad i = q(e, r).$$

The integral then becomes

$$\int\limits_{x^{[e]}}^{x^{[e+1]}} N_i'(x)N_j'(x)dx = \int\limits_{-1}^{1} J^{-1}\frac{d\tilde{N}_r(\xi)}{d\xi}J^{-1}\frac{d\tilde{N}_s(\xi)}{d\xi}\det J d\xi,$$

with $i = q(e, r)$ and $j = q(e, s)$. From (2.60) we have $J = (x^{[e+1]} - x^{[e]})/2$, or $J = h/2$ if the the element size is constant. Although J is a scalar in the integral above, the formula carries over to 2D and 3D when $x^{(e)}(\boldsymbol{\xi})$ is a vector mapping. Then J is a $d \times d$ matrix and $J_{i,j} = \partial x_j/\partial \xi_i$ in d space dimensions. The gradient operator is transformed according to the formula $\nabla N_i = J^{-1} \cdot \tilde{\nabla} Nr$, where $\nabla_\xi = (\partial/\partial\xi_1, \dots, \partial/\partial\xi_d)^T$ (and $i = q(e, r)$ as usual).

Here is an example of transforming the integral of $\nabla N_i \cdot \nabla N_j$ over a 2D element Ω_e to the corresponding reference element $\tilde{\Omega}$, which is taken to be the square $[-1, 1] \times [-1, 1]$:

$$\int\limits_{\Omega_e} \nabla N_i \cdot \nabla N_j dx_1 dx_2 = \int\limits_{-1}^{1}\int\limits_{-1}^{1} J^{-1}\nabla_\xi \tilde{N}_r \cdot J^{-1}\nabla_\xi \tilde{N}_s \det J \, d\xi_1 d\xi_2.$$

Notice that $J^{-1}\nabla_\xi \tilde{N}_j$ is a matrix-vector product, resulting in a vector, such that we end up with the inner product of two vectors and hence a scalar integrand.

The formulation of finite element problems for implementation in Diffpack makes use of the weak formulation restricted to the reference element $\tilde{\Omega}$, but we omit the explicit transformation of derivatives and the explicit shape of the reference element. This means that we write

$$\int\limits_{\Omega_e} \nabla N_i \cdot \nabla N_j d\Omega = \int\limits_{\tilde{\Omega}} \nabla \tilde{N}_r \cdot \nabla \tilde{N}_s \det J \, d\xi_1 \cdots d\xi_d.$$

We also often use i and j as indices on the right-hand side, $i, j = 1, \dots, n_e$ (this prevents us of course from using i and j in the left-hand side expression).

Diffpack provides tools for evaluating $\tilde{N}_r$, $\nabla\tilde{N}_r$, and $\det J$ at a point in $\tilde{\Omega}$. A Laplace term in the PDE is then reflected in the program code through the integrand

$$\nabla \tilde{N}_r \cdot \nabla \tilde{N}_s \det J \tag{2.61}$$

times a weight in the numerical integration rule used for evaluating integrals.

The Element Matrix and Vector in the Model Problem. We can now write the explicit expressions for the element matrix and vector in our model problem (2.50):

$$\tilde{A}_{r,s}^{(e)} = \int_{-1}^{1} \frac{2}{h} N_r'(\xi) \frac{2}{h} N_s'(\xi) \frac{h}{2} d\xi, \quad r, s = 1, 2, \tag{2.62}$$

$$\tilde{b}_r^{(e)} = \int_{-1}^{1} f(x^{(e)}(\xi)) \tilde{N}_r(\xi) \frac{h}{2} d\xi. \tag{2.63}$$

Inserting the formulas for basis functions in local coordinates and carrying out the integration for each pair (r, s), $r = 1, 2$, results in an element matrix

$$\left\{ \tilde{A}_{r,s}^{(e)} \right\} = \frac{1}{h} \begin{pmatrix} 1 & -1 \\ -1 & 1 \end{pmatrix}. \tag{2.64}$$

The element vector calculated by the numerical integration rule

$$\int_{-1}^{1} g(\xi) d\xi \approx g(-1) + g(1)$$

reads

$$\left\{ \tilde{b}_r^{(e)} \right\} = \frac{h}{2} \begin{pmatrix} f(x^{(e)}(-1)) \\ f(x^{(e)}(1)) \end{pmatrix}. \tag{2.65}$$

Incorporating Essential Boundary Conditions. Recall from Chapters 2.3.2 and 2.3.3 that the essential boundary conditions correspond to known values of some of the nodal values $(u_1, \ldots, u_n)$. Instead of modifying the global linear system, we can insert the conditions at the element level. In element no. 1, local node no. 1 is subjected to an essential boundary condition. Hence, we try to modify the element matrix and vector such that the first equation in the local 2×2 linear system reflects the boundary condition $\tilde{u}_1 = u_L$. The modification of (2.64) and (2.65) can be carried out by replacing the first row in the element matrix by zeroes, except on the main diagonal, where we insert the value 1. The first entry in the element vector is overwritten with the boundary value u_L. It follows that the modified element linear system becomes

$$\begin{pmatrix} 1 & 0 \\ -\frac{1}{h} & \frac{1}{h} \end{pmatrix} \begin{pmatrix} \tilde{u}_1 \\ \tilde{u}_2 \end{pmatrix} = \begin{pmatrix} u_L \\ \frac{h}{2} f(x^{(1)}(1)) \end{pmatrix}. \tag{2.66}$$

The original element matrix and the corresponding global matrix were symmetric. This is an important property to preserve when it comes to storage requirements and efficient algorithms for solving linear systems. Unfortunately, our modifications for incorporating the essential boundary conditions destroyed the symmetry of the element matrix, with the consequence that also the global matrix will become nonsymmetric. Nevertheless, it is possible to incorporate the condition $\tilde{u}_1 = u_L$ and still preserve symmetry. This requires the subtraction of column 1 in the element matrix, times the boundary value u_L, from the right-hand side. The general algorithm, where an arbitrary nodal value $\tilde{u}_k$ is prescribed as u_L, reads:

$$\tilde{b}_r^{(e)} \leftarrow \tilde{b}_r^{(e)} - u_L \tilde{A}_{r,k}^{(e)}, \quad r = 1, \ldots, n_e, \tag{2.67}$$

$$\tilde{A}_{r,k}^{(e)} \leftarrow 0, \quad r = 1, \ldots, n_e, \tag{2.68}$$

$$\tilde{A}_{k,r}^{(e)} \leftarrow 0, \quad r = 1, \ldots, n_e, \tag{2.69}$$

$$\tilde{A}_{k,k}^{(e)} \leftarrow 1, \tag{2.70}$$

$$\tilde{b}_k^{(e)} \leftarrow u_L. \tag{2.71}$$

Using this general algorithm, the element equations in our example become

$$\begin{pmatrix} 1 & 0 \\ 0 & \frac{1}{h} \end{pmatrix} \begin{pmatrix} \tilde{u}_1 \\ \tilde{u}_2 \end{pmatrix} = \begin{pmatrix} u_L \\ \frac{h}{2} f(x^{(1)}(1)) + \frac{1}{h} u_L \end{pmatrix}. \tag{2.72}$$

The corresponding procedure can be used for the element matrix and vector for element no. m. In that case, $\tilde{u}_2 = u_R$ and we need to modify row 2 and column 2 in the element matrix. The result is

$$\begin{pmatrix} \frac{1}{h} & 0 \\ 0 & 1 \end{pmatrix} \begin{pmatrix} \tilde{u}_1 \\ \tilde{u}_2 \end{pmatrix} = \begin{pmatrix} \frac{h}{2} f(x^{(m)}(-1)) + \frac{1}{h} u_R \\ u_R \end{pmatrix}. \tag{2.73}$$

Numerical Integration. Although many integrals can be computed by analytical means in 1D and over geometrically simple 2D and 3D domains, integrals in finite element computations must normally be computed numerically. This is conveniently performed in local coordinates. The general numerical integration rule in 1D has the form

$$\int_{-1}^{1} g(\xi) d\xi \approx \sum_{k=1}^{n_I} g(\xi_k) w_k,$$

where ξ_k is a numerical integration point (sampling point) and w_k is a weight. The accuracy of the rule is determined by the number of integration points n_I and the location ξ_k of the points. Most of the rules used in finite element computations are constructed to integrate a polynomial of a given order p exactly. Some widely used rules are listed in table 2.1. See references [57], [115], or [131] for more complete information on integration rules.

Finite element procedures on uniform grids with element length h result in an error $||u - \hat{u}||$ that usually goes to zero like h^r, where r depends on the norm $|| \cdot ||$ and the type of finite elements being used. Numerical integration introduces additional errors that might destroy the convergence rate. However, it can be shown that the value of r is retained by using Gauss-Legendre

Table 2.1. Numerical integration (quadrature) rules for integrals on $[-1, 1]$ that integrate a polynomial of degree p exactly. The Gauss-Lobatto rules with 2 and 3 points coincide with the well known trapezoidal and Simpson's rules, respectively.

name	n_I	p	weights	points
Gauss-Legendre	1	1	(1)	(0)
Gauss-Legendre	2	3	$(1, 1)$	$(-1/\sqrt{3}, 1/\sqrt{3})$
Gauss-Legendre	3	5	$(5/9, 8/9, 5/9)$	$(-\sqrt{3/5}, 0, \sqrt{3/5})$
Gauss-Lobatto	2	2	$(1, 1)$	$(-1, 1)$
Gauss-Lobatto	3	3	$(1/3, 4/3, 1/3)$	$(-1, 0, 1)$

rules with q points if the finite element basis functions are polynomials of degree $q - 1$. In 1D we should therefore use the two-point rule for linear elements and the three-point rule for quadratic elements.

Assembly of Element Matrices and Vectors. We recall the splitting of the global coefficient matrix in a sum of element matrices as indicated in (2.56). How do we actually add $\tilde{A}^{(e)}$ to $A_{i,j}$? The entry (r, s) in the element matrix corresponds to the coupling of local nodes r and s in element no. e. Hence, this entry is a contribution to the coupling of the global nodes i and j, where $i = q(e, r)$ and $j = q(e, s)$. We can then formulate the algorithm for updating $A_{i,j}$ with the entries in $\tilde{A}_{r,s}^{(e)}$:

$$A_{q(e,r),q(e,s)} \leftarrow A_{q(e,r),q(e,s)} + \tilde{A}_{r,s}^{(e)}. \tag{2.74}$$

The right-hand side is updated in a similar way:

$$b_{q(e,r)} \leftarrow b_{q(e,r)} + \tilde{b}_r^{(e)}. \tag{2.75}$$

As usual, r and s runs from 1 to the number of nodes in the element (n_e). The process of adding element contributions to the global linear system is called *assembly of element matrices and vectors*.

Figure 2.4 shows how the 2×2 element matrices in our 1D model problem are added to the global matrix. Carrying out this process, we get the following

element matrices   **global matrix**

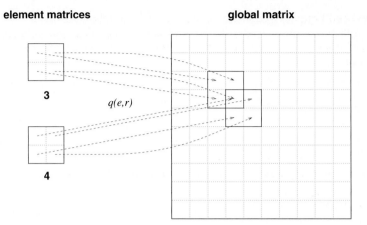

Fig. 2.4. Illustration of matrix assembly: element matrices, corresponding to element no. 3 and 4 in a 1D problem with linear elements, are added to the global system matrix.

global system, provided h is constant.

$$
\begin{pmatrix}
1 & 0 & 0 & \cdots & & & & \cdots & 0 \\
0 & \frac{2}{h} & -\frac{1}{h} & \ddots & & & \ddots & \ddots & \ddots \\
0 & -\frac{1}{h} & \frac{2}{h} & -\frac{1}{h} & \ddots & & & \ddots & \ddots \\
\vdots & \ddots & & \ddots & \ddots & \ddots & & & \ddots \\
\vdots & & \ddots & \ddots & \ddots & \ddots & & & \vdots \\
\vdots & & & 0 & -\frac{1}{h} & \frac{2}{h} & -\frac{1}{h} & 0 & \vdots \\
\vdots & & \ddots & & & \ddots & \ddots & \ddots & \vdots \\
\vdots & \ddots & \ddots & \ddots & & \ddots & -\frac{1}{h} & \frac{2}{h} & 0 \\
0 & \cdots & & \cdots & & \cdots & 0 & 0 & 1
\end{pmatrix}
\begin{pmatrix}
u_1 \\ u_2 \\ \vdots \\ \vdots \\ \vdots \\ \vdots \\ \vdots \\ u_{n-1} \\ u_n
\end{pmatrix}
=
\begin{pmatrix}
b_1 \\ b_2 \\ \vdots \\ \vdots \\ \vdots \\ \vdots \\ \vdots \\ b_{n-1} \\ b_n
\end{pmatrix},
$$

where $b_1 = u_L$, $b_2 = hf(x^{[2]}) + u_L/h$, $b_i = hf(x^{[i]})$, $i = 3, \ldots, n-2$, $b_{n-1} = hf(x^{[n-1]}) + u_R/h$, and $b_n = u_R$. Notice that in the assembly process we use the relation $f(x^{(e)}(-1)) = f(x^{(e-1)}(1)) = f(x^{[e]})$.

Non-Constant Element Size. Elements of unequal lengths are trivially incorporated in the elementwise formulation presented above. The important observation to make is that the element matrix and vector involve only the length of the current element. All we have to do is to replace h by $h_e = x^{[e+1]} - x^{[e]}$ in the expressions for the element matrix and vector. In case of nonsymmetric modifications of the element matrices and vectors due to essential boundary conditions, one obtains the following linear system corresponding to $-u'' = f$, $u(0) = u_L$, and $u(1) = u_R$, on a non-uniform grid:

$$u_1 = u_L, \tag{2.76}$$

$$-\frac{1}{h_{i-1}} u_{i-1} + \left(\frac{1}{h_{i-1}} + \frac{1}{h_i} \right) u_i - \frac{1}{h_i} u_{i+1} = \frac{1}{2} (h_{i-1} + h_i) f(x^{[i]}), \tag{2.77}$$

$$i = 2, \ldots, n-1,$$

$$u_n = u_R. \tag{2.78}$$

2.3.5 Extending the Concepts to Quadratic Elements

The finite element examples so far have been restricted to linear elements in one space dimension. However, we have tried to parameterize the number of nodes and integration points in an element etc. such that it shall be easy to adapt the procedures to more complicated situations.

Consider quadratic elements as an example. We then have three nodes per element. In the reference element, node 1, 2, and 3 are given as $\xi = -1$, $\xi = 0$, and $\xi = 1$, respectively. The local basis functions are quadratic polynomials fulfilling the condition $\tilde{N}_r = \delta_{rs}$ at local node no. s.

$$\tilde{N}_1(\xi) = \frac{1}{2} \xi(\xi - 1), \tag{2.79}$$

$$\tilde{N}_2(\xi) = (1 + \xi)(1 - \xi), \tag{2.80}$$

$$\tilde{N}_3(\xi) = \frac{1}{2} \xi(1 + \xi). \tag{2.81}$$

The reader should sketch these three functions in $[-1, 1]$. The mapping from x to ξ is in general given by (2.60), but if local node no. 2 is located at the mid point of the element, (2.60) reduces to the linear form (2.58). With the formulas for $\tilde{N}_r$ at hand, we can run through the algorithm from Chapter 2.3.6 and solve 1D differential equations using quadratic elements.

Example 2.2. Applying the Galerkin method to the equation $-u'' = 1$, $x \in (0, 1)$, and dividing $(0, 1)$ into quadratic elements, leads to the following element matrix and vector, using exact integration,

$$\frac{1}{3h_e} \begin{pmatrix} 7 & -8 & 1 \\ -8 & 16 & -8 \\ 1 & -8 & 7 \end{pmatrix}, \quad \frac{h_e}{6} \begin{pmatrix} 1 \\ 4 \\ 1 \end{pmatrix}.$$

As usual, h_e is the physical length of the element. See for instance [37, Ch. 8.3] or [16, Ch. 2.9] for more details regarding hand calculations with quadratic basis functions.   ◇

Exercise 2.7. Consider the problem (1.5)–(1.7) with $f(x) = -(\alpha + 1)x^\alpha$. Introduce a uniform finite element mesh consisting of m linear elements and

$n = m + 1$ nodes. Use the Galerkin finite element method with exact evaluation of all integrals. Compute the global linear system directly by the method in Chapter 2.3.3. Compute also element matrices and vectors and assemble the contributions. Compare the discrete finite element equations with those obtained by the finite difference method. Formulate the discrete finite element equations also for a nonuniform mesh. What is the truncation error of the method on a uniform grid? (Appendix A.4.9 gives an introduction to the truncation error concept.)   ⋄

Exercise 2.8. This is a continuation of Exercise 2.7. Now we shall make use of a uniform finite element mesh consisting of m *quadratic* elements and $n = 2m+1$ nodes. Use the Galerkin finite element method with (i) exact evaluation of all integrals and (ii) three-point Gauss-Lobatto (i.e. Simpson's) rule for approximating integrals over one element. Compute the element matrices and vectors. The results from Example 2.2 can be helpful. Also notice that the mapping (2.60), which is needed for the integral of $\int_{-1}^{1} f(x(\xi))\tilde{N}_i d\xi$, reduces to the linear form (2.58). This also implies that $J = h/2$. Assemble the contributions from each element, and write each equation in the linear system. Equation no. i can be of two types, depending on whether node no. i is located in the middle of an element or at the end points. Explain how u_i can be eliminated from the linear system when i is a node in the middle of the element. Alternatively, the mid-point values can be eliminated at the element level, yielding a 2×2 element matrix. Set up the resulting set of difference equations involving only the same nodes as in linear elements, that is, the end points of the elements. Suggest how the resulting equations can be implemented in the Diffpack class `Heat1D1` from Chapter 1.6.1.   ⋄

Exercise 2.9. This exercise concerns the two-point boundary-value problem

$$-(\lambda(x)u'(x))' = f(x), \quad x \in (0, 1), \quad u(0) = u_L, \ u'(1) = \gamma.$$

Calculate the element matrices and vectors, as well as the global linear system, corresponding to a uniform grid with linear elements. All integrals should be evaluated by the trapezoidal rule. Then extend the calculations to quadratic elements, with a three-point Gauss-Lobatto (Simpson's) rule for integration. Compare the representation of λ in the finite element equations with the representation of λ in the finite difference method (see Chapter 1.2.6).   ⋄

2.3.6   Summary of the Elementwise Algorithm

Algorithm 2.1 summarizes the element-by-element construction of the global linear system, where all evaluations of the basis functions are performed in local coordinates and all integrals are computed by numerical integration. The formulation of this finite element algorithm is general enough to make it relevant for a wide range of boundary-value problems (although formulas associated with our model problem (2.50) are used in the algorithm as examples on expressions when sampling the integrands in the weighted residual formulation).

Algorithm 2.1.

Finite element assembly algorithm.

INITIALIZE GLOBAL LINEAR SYSTEM:
set $A_{i,j} = 0$ for $i, j = 1, \ldots, n$
set $b_i = 0$ for $i = 1, \ldots, n$
LOOP OVER ALL ELEMENTS:
for $e = 1, \ldots, m$
    set $\tilde{A}_{r,s}^{(e)} = 0$, $r, s = 1, \ldots, n_e$
    set $\tilde{b}_r^{(e)} = 0$, $r = 1, \ldots, n_e$
    LOOP OVER NUMERICAL INTEGRATION POINTS:
    for $k = 1, \ldots, n_I$
        evaluate $\tilde{N}_r$, $d\tilde{N}_r/d\xi$, $d\tilde{N}_r/dx$, and J at $\xi = \xi_k$
        ADD CONTRIBUTION TO ELEMENT MATRIX AND VECTOR FROM
        THE CURRENT INTEGRATION POINT:
        for $r = 1, \ldots, n_e$
            for $s = 1, \ldots, n_e$
$$\tilde{A}_{r,s}^{(e)} \leftarrow \tilde{A}_{r,s}^{(e)} + \frac{d\tilde{N}_r}{dx} \frac{\tilde{N}_s}{dx} \det J \, w_k$$
$$\tilde{b}_r^{(e)} \leftarrow \tilde{b}_r^{(e)} + f(x^{(e)}(\xi_k)) N_r \det J \, w_k$$
    INCORPORATE ESSENTIAL BOUNDARY CONDITIONS:
    for $r = 1, \ldots, n_e$
        if node r has an essential boundary condition then
            modify $\tilde{A}_{r,s}^{(e)}$ and $\tilde{b}_r^{(e)}$ due to this condition
    ASSEMBLE ELEMENT MATRIX AND VECTOR:
    for $r = 1, \ldots, n_e$
        for $s = 1, \ldots, n_e$
$$A_{q(e,r),q(e,s)} \leftarrow A_{q(e,r),q(e,s)} + \tilde{A}_{r,s}^{(e)}$$
$$b_{q(e,r)} \leftarrow b_{q(e,r)} + \tilde{b}_r^{(e)}$$

The nomenclature for Algorithm 2.1 reads as follows: n is the number of nodes, m is the number of elements, n_e is the number of nodes in an element, n_I is the number of numerical integration points, ξ_k is the numerical integration points, w_k is the numerical integration weights, $q(e, r)$ is the global node number corresponding to local node r in element e, $A_{i,j}$ is the global coefficient matrix, $\tilde{A}_{r,s}^{(e)}$ is the element matrix, b_i is the global right-hand side in the linear system, and $\tilde{b}_r^{(e)}$ is the element vector.

2.4   Example: A 1D Wave Equation

In this section we shall investigate the application and performance of finite element methods in a problem involving the wave equation:

$$\frac{\partial^2 u}{\partial t^2} = c^2 \frac{\partial^2 u}{\partial x^2}, \quad x \in (0,1), \quad t > 0, \tag{2.82}$$

$$u(x,0) = f(x), \quad x \in [0,1], \tag{2.83}$$

$$\frac{\partial}{\partial t} u(x,0) = 0, \quad x \in [0,1], \tag{2.84}$$

$$\frac{\partial}{\partial x} u(0,t) = 0, \quad t > 0, \tag{2.85}$$

$$\frac{\partial}{\partial x} u(1,t) = 0, \quad t > 0. \tag{2.86}$$

2.4.1   The Finite Element Equations

Discretizing the PDE (2.82) in time by second-order finite differences yields a sequence of spatial problems. Each spatial problem is then discretized by the finite element method, using the approximation $\hat{u}^\ell = \sum_{j=1} u_j^\ell N_j(x)$ to the unknown $u(x,t)$ at time level ℓ. See Chapter 2.2 for a derivation of the finite element formulation. The discrete equations can be written on the form

$$\sum_{j=1}^{n} M_{i,j} u_j^0 = \int_0^1 f(x) N_i dx,$$

$$\sum_{j=1}^{n} M_{i,j} u_j^{-1} = \int_0^1 \left[\hat{u}^0 N_i - \frac{1}{2}(c\Delta t)^2 \frac{dN_i}{dx} \frac{d\hat{u}^0}{dx} \right] dx +$$

$$\frac{1}{2}(c\Delta t)^2 \left(\frac{d\hat{u}^0}{dx} N_i \Big|_{x=1} - \frac{d\hat{u}^0}{dx} N_i \Big|_{x=0} \right),$$

$$\sum_{j=1}^{n} M_{i,j} u_j^{\ell+1} = \int_0^1 \left[\left(2\hat{u}^\ell - \hat{u}^{\ell-1} \right) N_i - (c\Delta t)^2 \frac{dN_i}{dx} \frac{d\hat{u}^\ell}{dx} \right] dx, \quad \ell \geq 0,$$

where $M_{i,j} = \int_0^1 N_i N_j dx$ is the mass matrix. In the following, we omit the boundary terms by demanding $f' = 0$, and thereby approximately $d\hat{u}^0/dx = 0$, at $x = 0,1$, just for making the expressions more compact. Notice that with $f'(0) \neq 0$ or $f'(1) \neq 0$ we also get an extra boundary term (from the integration by parts) in the equation for $u_j^{\ell+1}$ when $\ell = 0$.

Sometimes it is convenient to express the right-hand side of the linear system as a matrix-vector product. This is particularly the case for analytical work and formulation of fast computer implementations in many problems. To this end, we simply insert $\hat{u}^\ell = \sum_{j=1}^{n} u_j^\ell N_j$ on the right-hand side, move the sum outside the integrals and write the terms as matrix-vector products.

The result becomes

$$\sum_{j=1}^{n} M_{i,j} u_j^0 = \int_0^1 f(x) N_i dx, \tag{2.87}$$

$$\sum_{j=1}^{n} M_{i,j} u_j^{-1} = \sum_{j=1}^{n} M_{i,j} u_j^0 - \frac{1}{2}(c\Delta t)^2 \sum_{j=1}^{n} K_{i,j} u_j^0, \tag{2.88}$$

$$\sum_{j=1}^{n} M_{i,j} u_j^{\ell+1} = 2 \sum_{j=1}^{n} M_{i,j} u_j^\ell - \sum_{j=1}^{n} M_{i,j} u_j^{\ell-1} - (c\Delta t)^2 \sum_{j=1}^{n} K_{i,j} u_j^\ell, \tag{2.89}$$

with $K_{i,j} = \int_0^1 N_i'(x) N_j'(x) dx$.

2.4.2   Interpretation of the Discrete Equations

We shall interpret the various terms in the difference equations and thereby see how the finite element formulation above relates to certain finite difference schemes. For this purpose, it will be convenient to assemble the $M_{i,j}$ and $K_{i,j}$ matrices separately, since they correspond to different terms in the original PDE. Concentrating on the general scheme (2.89) for $\ell \geq 0$, we can easily calculate the element mass and stiffness matrices:

$$\left\{ \tilde{M}_{r,s}^{(e)} \right\} = \frac{h}{6} \begin{pmatrix} 2 & 1 \\ 1 & 2 \end{pmatrix}, \quad \left\{ \tilde{K}_{r,s}^{(e)} \right\} = \frac{1}{h} \begin{pmatrix} 1 & -1 \\ -1 & 1 \end{pmatrix}. \tag{2.90}$$

Assembling a typical row in the global mass matrix results in the only nonzero entries $M_{i-1,i} = M_{i+1,i} = h/6$ and $M_{i,i} = 4h/6$. We can then evaluate, for a fixed $1 < i < n$:

$$\sum_j M_{i,j} u_j^{\ell+1} = \frac{h}{6} \left(u_{i-1}^{\ell+1} + 4u_i^{\ell+1} + u_{i+1}^{\ell+1} \right). \tag{2.91}$$

This is the contribution from the mass matrix terms to the difference equations corresponding to (2.89). The finite difference method leads to a single term $u_i^{\ell+1}$ only, which would be $hu_i^{\ell+1}$ in the finite element context since we integrate and thereby get the factor h. We can rewrite the right-hand side of (2.91) as

$$h \left(u_i^{\ell+1} + \frac{1}{6} \left(u_{i-1}^{\ell+1} - 2u_i^{\ell+1} + u_{i+1}^{\ell+1} \right) \right).$$

Using the difference operator notation from Appendix A.3, this can be written as

$$h[u + \frac{h^2}{6} \delta_x \delta_x u]_i^{\ell+1}.$$

Expressing discrete finite element equations in terms of finite difference operators has several important aspects: (i) the notation becomes more compact,

(ii) the interpretation of finite element equations as finite difference approximations to derivatives becomes evident, and (iii) the tools for analyzing stability and accuracy of finite difference schemes can immediately be applied to finite element schemes.

The contributions from $\sum_j K_{i,j} u_j^\ell$ yield a standard centered finite difference formula: $-h[\delta_x \delta_x u]_i^\ell$. The scheme according to (2.89) can then be written, after division by h,

$$[\delta_t \delta_t (u + \frac{h^2}{6} \delta_x \delta_x u) = c^2 \delta_x \delta_x u]_i^\ell . \tag{2.92}$$

Compared with the standard finite difference scheme for the same wave equation, see (1.51) on page 26, we observe that the finite element method introduces an extra term $[\frac{1}{6} h^2 \delta_t \delta_t \delta_x \delta_x u]_i^\ell$, arising from the time derivative $\partial^2 u / \partial t^2$ in the PDE. The extra term looks like a kind of diffusion, but the effect of the new term is to alter the velocity of a wave component, not its amplitude. This is hence a *dispersion* term in the discrete equations. On the other hand, in PDEs with first-order time derivatives the extra term takes the form $\frac{1}{6} h^2 [\delta_x \delta_x u]_i^\ell$ (see Exercise 2.10) and can in such cases represent a negative diffusion, which can have a destabilizing effect on the numerical scheme.

To explicitly demonstrate that our finite element scheme requires solution of coupled equations, we may write out the $\delta_t \delta_t$ operator, yielding

$$[u + \frac{h^2}{6} \delta_x \delta_x u]_i^{\ell+1} = 2[u + \frac{h^2}{6} \delta_x \delta_x u]_i^\ell - [u + \frac{h^2}{6} \delta_x \delta_x u]_i^{\ell-1} + (c\Delta t)^2 [\delta_x \delta_x u]_i^\ell . \tag{2.93}$$

It is now evident that we have to solve a tridiagonal matrix system with coefficient matrix corresponding to the difference operator $(1 + \frac{1}{6} h^2 \delta_x \delta_x)$, that is, the coefficient matrix is the mass matrix.

Lumped (Diagonal) Mass Matrix. To obtain an explicit scheme, it is common to *lump* the mass matrix such that it becomes diagonal. The original mass matrix is referred to as the *consistent mass matrix*, whereas the diagonal version is commonly named the *lumped mass matrix*. The simplest lumping method is to approximate $M_{i,j}$ by

$$M_{i,j}^{(L)} = \begin{cases} \sum_k M_{i,k}, & i = j, \\ 0, & i \neq j \end{cases} \tag{2.94}$$

This is often referred to as the *row-sum technique*. Another lumping strategy is to use nodal-point integration. In that case, the basis functions are always evaluated at a node, resulting in the value 0 or 1. At an arbitrary node k we then have $N_i(x^{[k]}) N_j(x^{[k]}) = \delta_{ik} \delta_{jk}$, that is, we get nonzero contributions only when $i = j$, which means that the matrix becomes diagonal. Both the row-sum technique and nodal-point integration give the same results for the lumped version of the mass matrix when using linear elements, which is $\{M_{i,j}^{(L)}\} = (h/2) \mathrm{diag}(1,1)$. The reader should verify that the finite element scheme now coincides with the standard finite difference scheme (1.51). This means that mass lumping removes the extra term $\frac{1}{6}h^2[\delta_t\delta_t\delta_x\delta_x u]_i^\ell$ in (2.92).

We remark that the row-sum technique and nodal-point integration do not work for all element types. For example, a 6-node quadratic triangular element then leads to vanishing diagonal terms M_{ii} for nodes i corresponding to the corner nodes of the elements. The right-hand side also vanishes in this case. More sophisticated techniques (e.g. the moving least squares method briefly touched on page 251 are then required.

2.4.3   Accuracy and Stability

Since we have transformed the finite element equation to a set of difference equations, we can use the tools for analyzing finite difference schemes also in a finite element context. For example, the accuracy can be measured by calculating the truncation error. Alternatively, we can use the concept of numerical dispersion relations to obtain an exact analytical solution of the difference equations, and using this solution, accuracy and stability can be investigated. These techniques are introduced in detail in Appendix A.4. The reader should be familiar with this material before proceeding with the application of the techniques to the present wave equation problem.

With the aid of Table A.3 we can compute the contribution to the truncation error from the various terms in (2.92). The $[\delta_t\delta_t u]_i^\ell$ term gives a contribution

$$\left(\frac{\partial^2 u}{\partial t^2}\right)_i^\ell + \frac{\Delta t^2}{12}\left(\frac{\partial^4 u}{\partial t^4}\right)_i^\ell + \mathcal{O}(\Delta t^4). \tag{2.95}$$

The $[\delta_x\delta_x u]_i^\ell$ term gives a similar contribution to the truncation error:

$$\left(\frac{\partial^2 u}{\partial x^2}\right)_i^\ell + \frac{h^2}{12}\left(\frac{\partial^4 u}{\partial x^4}\right)_i^\ell + \mathcal{O}(h^4). \tag{2.96}$$

The additional term $\frac{h^2}{6}[\delta_t\delta_t\delta_x\delta_x u]_i^\ell$ generated by the consistent mass matrix results in

$$\frac{h^2}{6}\left(\frac{\partial^2}{\partial x^2}(\delta_t\delta_t u)\right)_i^\ell + \mathcal{O}(h^4)$$

$$= \frac{h^2}{6}\left(\frac{\partial^4 u}{\partial x^2\partial t^2}\right)_i^\ell + \mathcal{O}(h^2\Delta t^2) + \mathcal{O}(h^4) + \mathcal{O}(\Delta t^4).$$

Summarizing the various contributions, the truncation error becomes

$$\tau = \frac{\Delta t^2}{12}\left(\frac{\partial^4 u}{\partial t^4}\right)_i^\ell - c^2\frac{h^2}{12}\left(\frac{\partial^4 u}{\partial x^4}\right)_i^\ell + \frac{h^2}{6}\left(\frac{\partial^4 u}{\partial x^2\partial t^2}\right)_i^\ell + \mathcal{O}(h^2\Delta t^2, \Delta t^4, h^4).$$

Let us then turn the attention to numerical dispersion relations as a measure of the accuracy and stability. We know from Appendix A.4.3 that the analytical dispersion relation reads $\omega = \pm ck$. The corresponding numerical dispersion relation is found by inserting $u_j^\ell = A \exp\left(i(kjh - \tilde\omega\ell\Delta t)\right)$ into the scheme (2.92) and using Table A.1, yielding

$$-\frac{4}{\Delta t^2}\sin^2\frac{\tilde\omega\Delta t}{2} + \frac{h^2}{6}\frac{4}{\Delta t^2}\sin^2\frac{\tilde\omega\Delta t}{2}\frac{4}{h^2}\sin^2\frac{kh}{2} = -c^2\frac{4}{h^2}\sin^2\frac{kh}{2},$$

which can be written

$$\sin^2\frac{\tilde\omega\Delta t}{2} = \frac{c^2\Delta t^2}{h^2}\left(1 - \frac{2}{3}\sin^2\frac{kh}{2}\right)^{-1}\sin^2\frac{kh}{2}.$$

Following the reasoning in Appendix A.4.7, a complex $\tilde\omega$ must be avoided since this will always give rise to instability. This means that the amplitude of the $\sin^2 kh/2$ term must be equal to or less than unity:

$$\frac{c^2\Delta t^2}{h^2}\left(1 - \frac{2}{3}\sin^2\frac{kh}{2}\right)^{-1} \le 1.$$

The worst case arises when the sine function equals unity, leading to $C \le 1/\sqrt{3}$ as the stability criterion. This criterion is more severe than the one $(C \le 1)$ we have met in the finite difference scheme, which is equivalent to the finite element scheme with lumped mass matrix. To summarize, *lumping the mass matrix improves stability and makes the scheme explicit.*

Remark. When solving other types of PDEs, in particular the convection-diffusion equation, see e.g. (6.1), lumping the mass matrix can reduce the numerical accuracy significantly. Gresho and Sani [44, Ch. 2] present a comprehensive treatment of this subject.

To calculate the error in the numerical dispersion relation associated with our discretization of the wave equation, we can solve for $\tilde\omega$ and make a Taylor-series expansion in terms of h and Δt:

$$E_\omega(k, h, \Delta t) \equiv \omega(k) - \tilde\omega(k; h, \Delta t) = \frac{1}{24}ck^3\left(h^2 + c^2\Delta t^2\right) + \mathcal{O}(h^2\Delta t^2, h^4, \Delta t^4).$$

If we compare this result with the similar result on page 516, we see that there is no cancellation of error terms now for the maximum C $(1/\sqrt{3})$ value. The relative error in phase velocity is shown in Figure 2.5. Comparison with Figure A.2 on page 517 reveals that lumping the mass matrix in fact improves the accuracy when solving the wave equation by linear finite elements. For a uni-directional wave equation $\partial u/\partial t + c\partial u/\partial x = 0$, lumping the mass matrix *reduces* the accuracy significantly, see [44, Ch. 2] for a thorough study.

Exercise 2.10. Apply the steps of the calculations and analysis in Chapter 2.4 to a one-dimensional heat equation.   ◇

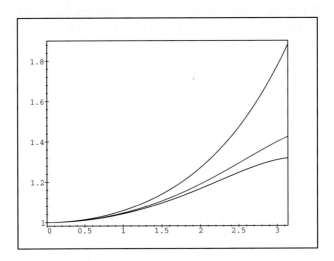

Fig. 2.5. Error in numerical phase velocity, normalized by the exact phase velocity, as a function of $p = kh$. The wave 1D equation is discretized by linear finite elements, keeping the mass matrix consistent. The different curves correspond to different Courant numbers: the stability limit $C = 1/\sqrt{3}$ (top curve), $C = 0.3$, and $C = 0.1$ (bottom curve). The corresponding results when using a lumped mass matrix are shown in Figure A.2.

Exercise 2.11. Lumping the mass matrix might seem like a trick with little physical or mathematical justification. Nevertheless, by approximating a continuous medium by a set of discrete particles, we can justify the lumped version of the mass matrix. For example, waves on a string lead to the unscaled PDE

$$\varrho(x)\frac{\partial^2 u}{\partial t^2} = T\frac{\partial^2 u}{\partial x^2},$$

where $u(x, t)$ is the string displacement, $\varrho(x)$ is the mass per unit length of the string, i.e., a measure of the density, and T is the tension of the string. The equation is derived in Chapter 1.3.1. The term $\varrho(x)\partial^2 u/\partial t^2$ reflects mass (per unit length) times the acceleration, and usually ϱ is constant. In the Galerkin finite element formulation this term gives rise to an integral on the form $\int \varrho N_i N_j dx$. Instead of having the mass of the string uniformly distributed (constant ϱ), we can think of lumping the mass at the nodes. That is, the string consists of particles with mass m_i, located at each node i and connected with massless springs. Such a model is explained in detail by Spiegel [107, Problems 8.29-32, 8.79-80]. In a continuum setting we can then model $\varrho(x)$

in terms of Dirac delta functions $\delta(x)$:

$$\varrho(x) = \sum_{j=1}^{n} m_j \delta(x - x^{[j]}).$$

An important property of the delta function is that $\int_{-\epsilon}^{\epsilon} \delta(x)dx = 1$ for any $\epsilon > 0$. Moreover, $\delta(x) = 0$ for all $x \neq 0$ (the value of $\delta(0)$ is hence infinite). We see that $\int \varrho(x)dx = \sum_j m_j$ which is the total mass of the spring. Assume now that the density is constant, equal to ϱ_0, and that we distribute the mass of each element equally between its two nodes. Then $m_i = \varrho_0 h$ for $1 < i < n$, while at the end points we have $m_1 = m_n = \varrho_0 h/2$. Use the lumped model representation $\varrho(x)$ in the mass matrix term $\int \varrho(x)N_i N_j dx$ and demonstrate that exact integration then results in a diagonal mass matrix. Also compute the corresponding mass matrix term $\int \varrho_0 N_i N_j dx$, using standard lumping in terms of the row-sum technique or nodal-point numerical integration. Show that the different approaches yield the same diagonal mass matrix.

The mass matrix often arise from terms in the PDE corresponding to the acceleration term in Newton's second law, and the reasoning in this exercise can then be applied to justify the lumping technique from a physical point of view.   ⋄

Exercise 2.12. We consider the mathematical problem (2.82)–(2.86) solved by quadratic elements in space and a higher-order difference scheme in time. For the spatial discretization we apply a Galerkin finite element method with quadratic elements. Establish the 3×3 element matrix corresponding to the terms $\partial^2 u/\partial t^2$ and $\partial^2 u/\partial x^2$. Show that the internal node of an element is only coupled to the other nodes in that element. This means that the internal node can be eliminated from the element equations. The associated procedure is called *static condensation* and is frequently used in many finite element contexts. The first step in the static condensation method is to write the element matrix system in block form,

$$\begin{pmatrix} A_{E,E} & A_{E,I} \\ A_{I,E} & A_{I,I} \end{pmatrix} \begin{pmatrix} u_E \\ u_I \end{pmatrix} = \begin{pmatrix} b_E \\ b_I \end{pmatrix}. \tag{2.97}$$

The subscript E refers to *external nodes*, i.e., the nodes that are coupled to other elements, whereas the subscript I identifies contributions from internal nodes. Eliminating u_I (which will not interact with contributions from other elements), yields

$$u_I = A_{I,I}^{-1}(b_I - A_{I,E}u_E).$$

Inserting this expression in the system (2.97) yields a modified equation system at the element level. Go through this procedure for the quadratic 1D element and set up the resulting 2×2 system. Assemble the element matrices and vectors and identify a difference approximation to the second order spatial derivative.

For the discretization in time we shall use two methods: (i) standard three-point finite difference approximation and (ii) the standard three-point approximation combined with correction terms (FD-corr). In the latter case we add a new term R to the partial differential equation. By performing a standard truncation error analysis, we can choose R such that terms in the truncation error expression cancel. This will result in a higher-order scheme in time. Carry out the procedure by considering the time-discrete equation

$$\frac{1}{\Delta t^2} \left(u^{\ell+1} - 2u^{\ell} + u^{\ell-1} \right) = c^2 \frac{\partial^2 u^{\ell}}{\partial x^2} + R.$$

Finally, write up the complete discrete equations. How can the accuracy of this method be established?   ◇

2.5   Naive Implementation

In this section, we present a simple implementation of the finite element algorithm from Chapter 2.3.6. The program is written in C++ using Diffpack arrays, but it can be straightforwardly translated into any other computer language[4]. The nested loops of the finite element algorithm can of course be coded directly in a `main` program, but to increase the modularity of the code, we try to break up the algorithm into some basic pieces, where each piece is coded as a function. Other element types, integration rules, or equations can then easily be incorporated by switching functions. The mathematical symbols used in the algorithm on page 140 have been replaced by somewhat more self-explanatory C++ variables. Many of the function and variable names in the example program are also found in the general finite element toolbox in Diffpack.

Here is a table of the basic mathematical symbols and the corresponding C++ names used in the code.

n	`nno`	number of nodes
m	`nel`	number of elements
n_e	`nne`	number of nodes in an element
n_I	`no_itg_pts`	number of numerical integration points
ξ_k	`num_itg_points`	numerical integration points
w_k	`num_itg_weights`	numerical integration weights
$q(e,r)$	`nodel(e,r)`	global node number of local node r in elm. e
$A_{i,j}$	`A`	global coefficient matrix
$\tilde{A}_{r,s}^{(e)}$	`elm_matrix(r,s)`	element matrix
b_i	`b`	global right-hand side in the linear system
$\tilde{b}_r^{(e)}$	`elm_vector(r)`	element vector
x_i	`coor(i)`	nodal coordinates

---

[4] To this end, class variables become global variables and class functions become ordinary global functions.

For a specific implementation example, we choose linear elements, $f(x) = \beta = $ constant, $u_L = 0$, and $u_R = 1$ in the 1D problem (2.50) on page 128. The most important functions in our program are listed next.

scan: read n and β, allocate vectors, matrices etc., call initGrid.
initGrid: compute x_i and $q(e, r)$, i.e., the finite element grid.
makeSystem: calculate the linear system.
solve: solve the linear system by Gaussian elimination.
calcElmMatVec: compute element matrix and vector for an element.
integrands: evaluate the integrands of the weighted residual statement (this
    is the part of the program that really depends on the PDE being solved!).
N: evaluate the basisfunctions in local coordinates.
dN: evaluate the derivatives of the basisfunctions in local coordinates.

The program is, as usual in Diffpack, realized as a class. The source code can be found in src/fem/MyFirstFEM. Here is the definition of the solver class.

```
class MyFirstFEM
{
  Vec(real) u;              // solution to be found
  Vec(real) b;              // right-hand side of linear system
  Mat(real) A;              // global coefficient matrix

  int       nel;            // number of elements
  int       nne;            // number of nodes in an element
  int       nno;            // number of nodes
  int       no_itg_pts;     // number of integration points
  real      beta;           // equation parameter

  Vec(real) coor;           // nodal coordinates
  MatSimple(int) nodel;     // element connectivity array
  Vec(real) num_itg_weights; // weights in numerical integration rule
  Vec(real) num_itg_points;  // points  in numerical integration rule

  // compute value of basis func. no. i at local point xi:
  real N  (int i, real xi);
  // compute derivative of basis func. no. i at local point xi:
  real dN (int i, real xi);

  void initGrid ();      // set up coor and nodel in 1D
  void makeSystem ();    // the assembly process
  void solve ();         // solve linear system by Gaussian elim.
  // compute the elemental matrix and vector:
  void calcElmMatVec (Mat(real)& m, Vec(real)& v, int e);
  // evaluate the integrand in the weighted residual statement:
  void integrands (Mat(real)& m, Vec(real)& v, int e, int p);

public:
  void scan ();          // read no of elements, initialize
  void solveProblem ();  // main administering function of the class
  void resultReport ();  // compare with analytical solution
};
```

The implementation of the member functions of this class is located in a file MyFirstFEM.cpp:

```
#include <MyFirstFEM.h>

void MyFirstFEM:: scan()
{
  initFromCommandLineArg("-nel",  nel,  10);
  initFromCommandLineArg("-beta", beta, 0.1);
  nne = 2;  nno = nel+1;
  A.redim (nno,nno);  b.redim (nno);  u.redim (nno);
  no_itg_pts = 2;
  num_itg_weights.redim (no_itg_pts);
  num_itg_points. redim (no_itg_pts);
  if (no_itg_pts == 2) {   // two-point Gauss rule:
    num_itg_points(1) = -1/sqrt(3);  num_itg_weights(1) = 1;
    num_itg_points(2) =  1/sqrt(3);  num_itg_weights(2) = 1;
  }
  initGrid();
}

void MyFirstFEM:: initGrid ()  // compute coor and nodel
{
  coor.redim (nno);
  real h = 1.0/nel;
  coor(1) = 0;
  for (int j = 2; j <= nno; j++)
    coor(j) = coor(j-1) + h;    // const partition, 2 nodes per element

  nodel.redim (nel, nne);
  for (int e = 1; e <= nel; e++)
    for (int i = 1; i <= nne; i++)
      nodel(e, i) = e*(nne-1)+i-1;
}

void MyFirstFEM:: solveProblem ()  { makeSystem();  solve(); }

void MyFirstFEM:: makeSystem ()
{
  A.fill(0.0);  b.fill(0.0);
  Mat(real) elm_matrix (nne,nne);   Vec(real) elm_vector (nne);
  for (int e = 1; e <= nel; e++) {  // element-by-element loop
    elm_matrix.fill(0.0);  elm_vector.fill(0.0);
    calcElmMatVec(elm_matrix, elm_vector, e);

    if (e==1)    // enforce the boundary condition u(0)=0
      { elm_matrix(1,1)=1; elm_matrix(1,2)=0; elm_vector(1)=0; }
    if (e==nel) // enforce the boundary condition u(1)=1
      { elm_matrix(2,2)=1; elm_matrix(2,1)=0; elm_vector(2)=1; }

    // assemble local contributions into A and b
    for (int r=1; r <= nne; r++) {
      for (int s=1; s <= nne; s++)
        A(nodel(e,r),nodel(e,s)) += elm_matrix(r,s);
      b(nodel(e,r)) += elm_vector(r);
    }
  }
}

void MyFirstFEM:: solve ()
```

```
{
  LinEqSystemStd eq_system(A, u, b);    // make linear system Au=b
  GaussElim gauss;                      // Gaussian elim. solver
  gauss.solve (eq_system);
  // could also used A.factLU() and A.forwBack(b,u)
}

void MyFirstFEM:: calcElmMatVec
  (Mat(real)& elm_matrix, Vec(real)& elm_vector, int e)
{
  // numerical integration over the element:
  for (int p = 1; p <= no_itg_pts; p++)
    integrands (elm_matrix, elm_vector, e, p);
}

void MyFirstFEM:: integrands
  (Mat(real)& elm_matrix, Vec(real)& elm_vector, int e, int p)
{
  real h = coor(e+1)-coor(e);                // length of this element
  real detJxW = h/2 * num_itg_weights(p); // Jacobian * weight
  int r,s;
  for (r = 1; r <= nne; r++) {
    for (s = 1; s <= nne; s++)
      elm_matrix(r,s) += dN(s, num_itg_points(p))*2/h
                      * dN(r, num_itg_points(p))*2/h * detJxW;
    elm_vector(r) += beta * N(r, num_itg_points(p)) * detJxW;
  }
}

real MyFirstFEM:: N (int i, real xi)
{
  if (i==1)   return 0.5*(1-xi);
  else        return 0.5*(1+xi);
}

real MyFirstFEM:: dN (int i, real /*xi*/)
{
  if (i==1)   return -0.5;
  else        return +0.5;
}

void MyFirstFEM:: resultReport ()
{
  Vec(real) u_exact (nno);
  int i;
  for(i = 1; i <= nno; i++)
    u_exact(i) = -0.5*beta*coor(i)*coor(i) + (0.5*beta + 1)*coor(i);
  s_o << "\n \n x              numerical        exact:           "
      << "difference:\n";   // \n is newline
  for (i = 1; i <= nno; i++) {
    s_o << oform("%4.3f         %8.6f          %8.6f          %8.6f \n",
              coor(i), u(i), u_exact(i), u_exact(i)-u(i) );
  }
}
```

The reader should realize that the coefficient matrix is represented by a
Mat(real) object, that is, a dense matrix. We know that with 1D elements,

the global coefficient matrix is tridiagonal, and Diffpack's tridiagonal matrix object, MatTri(real) should be used (see Chapter 1.2.5).

Extension of the code to quadratic elements in 1D will involve a pentadiagonal matrix. The relevant Diffpack matrix is then a banded matrix, represented by class MatBand(real). Of course, a special case of MatBand(real) is a tridiagonal matrix. In Chapter 3 we present finite element programs where the particular matrix format and solution procedure for the linear system can be flexibly chosen at run time. Explicit appearance of the matrix format (Mat vs. MatTri or MatBand) in the code is then avoided.

The main program associated with class MyFirstFEM looks like this:

```
#include <MyFirstFEM.h>
int main (int argc, const char* argv[])
{
  initDiffpack (argc, argv);
  MyFirstFEM problem;
  problem.scan(); problem.solveProblem(); problem.resultReport();
  return 0;   // success
}
```

It is easy to verify that a simulation with three elements and $\beta = 0.1$, i.e., ./app -nel 3 -beta 0.1, results in numerical values that coincide with the exact solution, as expected (see page 12).

Exercise 2.13. Modify class MyFirstFEM such that it employs a tridiagonal global coefficient matrix. Then implement the problem from Project 1.4.2:

$$u'(x) = \epsilon u''(x), \ x \in (0,1), \ \epsilon > 0, \quad u(0) = 0, \ u(1) = 1, \tag{2.98}$$

using linear elements on a *nonuniform* grid. Assume that one first generates a uniform grid and then moves a node x_i to a new position x_i^* according to a function $\mu(x)$: $x_i^* = \mu(x_i)$. From the results of Project 1.4.2 we know that the gradients are large in the vicinity of $x = 1$. Suggest a function $\mu(x)$ that concentrates the nodes in this critical region and test if a nonuniform mesh can cure the oscillations that appear when (2.98) is solved on a uniform grid with $h/\epsilon > 2$. See also Chapter 2.9 for various other approaches that stabilize the finite element solution of (2.98) when $h/\epsilon > 2$.   ◇

Exercise 2.14. Go through the source code of class MyFirstFEM and figure out the necessary extensions that are required for solving the PDE $-u'' + u = 1$, $u(0) = u(1) = 0$, using quadratic elements.   ◇

While solving Exercise 2.14, the reader will notice that extending the code in class MyFirstFEM to handle quadratic elements require modifications that are unfortunately scattered all over the code. When we think of programs for multi-dimensional time-dependent and nonlinear PDEs combined with flexible choice of finite elements, grids, numerical integration rules etc., scattered editing of a code for adapting it to a different problem or method is neither efficient nor reliable. The main purpose of program systems like Diffpack is to provide a programming environment for nontrivial applications where only the strongly problem-dependent parts of the problem at hand are visible in the application code. Ingredients that are the same from problem to problem are programmed and tested once and thereafter made available in generic libraries.

Weakly Imposed Essential Conditions. We shall now investigate an alternative method of imposing boundary conditions. Let the model problem be

$$-u''(x) = 0, \quad u(0) = 0, \ u(1) = 1.$$

The finite element approximation $\hat{u}$ to u reads as usual $\hat{u} = \sum_{j=1}^{n} u_j N_j(x)$. We introduce a weighted residual formulation where we add a boundary term consisting of a (large) parameter λ times the difference between u and the essential boundary conditions, in general integrated along the boundary:

$$-\int_0^1 \hat{u}''(x)W_i dx + \lambda\left((\hat{u} - 1)W_i|_{x=1} + (\hat{u} - 0)W_i|_{x=0}\right) = 0,$$

for $i = 1, \ldots, n$. This approach is often referred to as incorporation of boundary conditions by *penalization*. Using $W_i = N_i$ and assuming that $N_i(x^{[j]}) = \delta_{ij}$, we get

$$\sum_{j=1}^{n} \left(\int_0^1 N_i' N_j' dx\right) u_j + \lambda \delta_{in} u_n + \lambda \delta_{i1} u_1 = \lambda \delta_{in}, \quad i = 1, \ldots, n.$$

Straightforward computations result in the following system:

$$(\frac{1}{h} + \lambda)u_1 - \frac{1}{h}u_2 = 0, \tag{2.99}$$

$$\frac{1}{h}(-u_{i-1} + 2u_i - u_{i-1}) = 0, \ i = 2, \ldots, n-1, \tag{2.100}$$

$$-\frac{1}{h}u_{n-1} + (\frac{1}{h} + \lambda)u_n = \lambda. \tag{2.101}$$

Let us try to find an analytical solution of these discrete equations. Inserting $u_i = Q^i$ in (2.100) gives a double root $Q = 1$. A candidate for solution is then $u_i = A + iB$. The constants A and B can be determined from (2.99) and (2.101). This gives

$$u_i = \frac{1}{\lambda + 2} + \frac{\lambda}{\lambda + 2}(i - 1)h.$$

Since the exact solution of the continuous problem is $u(x_i) = (i - 1)h$, we realize that the numerical approximation becomes correct in the limit $\lambda \to \infty$ for any h. From the discrete equations we also see that $\lambda \to \infty$ recovers the equations corresponding to our general method for incorporating essential boundary conditions.

Exercise 2.15. Formulate the penalization approach for a 3D Poisson equation problem with $u = g$ on the boundary. Apply a Gauss-Lobatto (nodal-point) integration rule for the new boundary term. Explain that the method can be implemented by first computing the element matrix and vector without paying attention to boundary conditions and then adding λ on the main diagonal and the right-hand side for each node that is located at the boundary.   ⋄

The advantage of the penalization method is that the handling of essential boundary conditions becomes simpler from an implementational point of view. Constraints on the solution, such as periodic boundary conditions $u(0) - u(1) = 0$, are also easy to implement. The disadvantage is that the conditions are only fulfilled approximately and that a large value of λ leads to ill-conditioning of the coefficient matrix. Nevertheless, the penalization approach is widely used in finite element software (but not in Diffpack), see for example [106, Ch. 3.8], [5, Ch. 20.5], and [30, Ch. 9].

Instead of letting $\lambda \to \infty$, one can enforce the boundary condition as an extra constraint $\int_{\partial\Omega}(u - g)d\Gamma = 0$ and view λ as a Lagrange multiplier [93, p. 227].

2.6   Projects

2.6.1   Heat Conduction with Cooling Law

Mathematical Problem. We consider a two-point boundary-value problem on $(0, 1)$ with a special boundary condition known as a cooling law or Robin condition at $x = 1$:

$$(ku')' = 0, \ x \in (0,1), \quad u(0) = 1, \ -k(1)u'(1) = \beta(u(1) - U_S). \quad (2.102)$$

The quantity $k > 0$ is a given function of x, whereas β and U_S are prescribed constants.

Physical Model. The problem (2.102) can be derived from a more general model for steady heat conduction,

$$\varrho C_p \frac{\partial u}{\partial t} = \nabla \cdot (k \nabla u) + f(\boldsymbol{x}, t), \quad \boldsymbol{x} \in \Omega \subset \mathbb{R}^d, \ t > 0, \quad (2.103)$$

$$u(\boldsymbol{x}, 0) = g_I(\boldsymbol{x}), \quad \boldsymbol{x} \in \Omega, \quad (2.104)$$

$$u = g_D(\boldsymbol{x}, t), \quad \boldsymbol{x} \in \partial\Omega_D, \quad (2.105)$$

$$-k \frac{\partial u}{\partial n} = g_N(\boldsymbol{x}, t), \quad \boldsymbol{x} \in \partial\Omega_N, \quad (2.106)$$

$$-k \frac{\partial u}{\partial n} = g_T(u - U_S), \quad \boldsymbol{x} \in \partial\Omega_R. \quad (2.107)$$

Here, ϱ is the density, C_p is the heat capacity, u is the temperature, k is the heat conduction coefficient, f denotes external heat sources, (2.104) specifies the initial condition, (2.105) models parts of the boundary where the temperature is controlled, (2.106) models boundaries with a known heat flux, and (2.107) is the so-called *Newton's cooling law*, modeling heat transfer from the medium (Ω) to its surroundings. In (2.107), g_T is a heat transfer coefficient and U_S is the temperature in the surroundings. At each point on the boundary, only one of the conditions (2.105)–(2.107) applies.

We shall consider stationary heat conduction, without heat sources, in a cylindrical rod, $0 \leq x \leq L$, $y^2 + z^2 \leq a^2$. At $x = 0$, u is fixed at u_L. At the outer boundary $y^2 + z^2 = a^2$ the rod is insulated, i.e., the heat flux is zero, and at $x = L$ the cooling law (2.107) applies. If u_L, g_T, and U_S do not vary with y or z, the initial and boundary conditions suggest that u is independent of y and z, thus reducing the original 3D problem to a 1D model. Set up the resulting stationary 1D model, perform a suitable scaling (use u_L as temperature scale), and derive (2.102).

Numerical Method. Start with formulating a Galerkin procedure, with $u \approx \hat{u}(x) = \psi + \sum_{j=1}^{M} u_j N_j(x)$, for (2.102). The function ψ is used to incorporate the essential boundary condition. Moreover, $N_j(0) = 0$.

Analysis. Explain how the cooling law $-ku'(1) = \beta(u(1) - U_S)$ contributes to the coefficient matrix and the right-hand side in the linear system for $u_1, \ldots, u_M$. Restrict the choice of N_i to typical finite element basis functions and discuss how this simplifies the boundary terms. Set up the resulting discrete equations in the case k is constant and we use linear elements on a uniform mesh. Then explain how to deal with the boundary conditions using Algorithm 2.1. Set up the corresponding discrete equations. How are these equations changed if we apply quadratic elements (see Example 2.2 on page 138)? Finally, suggest a finite difference method for the problem and compare the discrete equations of the finite element and difference methods.

Derive a closed-form expression for the analytical solution of the problem. (Hint: Integrate (2.102) directly to $u = 1 + C \int_0^x [k(\tau)]^{-1} d\tau$ and determine the integration constant C from the condition at $x = 1$.) Choose a specific k function and specialize the analytical solution in this case.

Implementation. Modify class `MyFirstFEM` from Chapter 2.5, or a more advanced solver from Chapter 3, to handle the present problem using linear elements. Partially verify the implementation by comparing intermediate numerical results against the hand-calculated numerical expressions for the element matrices and vectors in the case k is constant.

2.6.2   Retardation of a Well-Bore

Mathematical Problem. The current project involves a simplified version of the incompressible Navier-Stokes equations. The simplification results in the following set of equations:

$$\frac{\partial u}{\partial t} = \frac{\partial}{\partial r}\left(\frac{\partial u}{\partial r} - \frac{u}{r}\right) + \frac{2}{r}\left(\frac{\partial u}{\partial r} - \frac{u}{r}\right), \quad r \in (\alpha, 1),\ t > 0, \quad (2.108)$$

$$\frac{\partial u}{\partial t} = \beta\left(\frac{\partial u}{\partial r} - \frac{u}{r}\right), \quad r = \alpha,\ t > 0, \quad (2.109)$$

$$u(1, t) = 0, \quad t > 0, \quad (2.110)$$

$$u(r, 0) = \alpha \frac{r^{-1} - r}{1 - \alpha^2}, \quad r \in [\alpha, 1], \quad (2.111)$$

$$\frac{\partial p}{\partial r} = \frac{u^2}{r}, \quad r \in (\alpha, 1),\ t > 0. \quad (2.112)$$

Equations (2.108)–(2.111) constitute an initial-boundary value problem for $u(r, t)$, while the function $p(r, t)$ can be found from (2.112) when $u(r, t)$ is known. The parameter β is a dimensionless constant. Notice that a time derivative enters the boundary condition at $r = \alpha$, cf. (2.109).

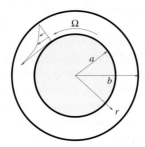

Fig. 2.6. Flow of a viscous fluid between one rotating and one fixed cylinder.

Physical Model. We consider flow between two concentric cylinder (often referred to as circular Couette flow), see Figure 2.6. The inner cylinder $0 \leq r \leq a$ models a well-bore rotating with constant angular velocity Ω, while the outer cylinder $r \geq b$ models the fixed wall of the well. At $t = 0$, the external forces causing the steady rotation of the bore are turned off, with the result that the friction in the fluid will eventually bring the inner cylinder to rest. The mathematical model (2.108)–(2.111) can be used to simulate the retardation process and determine the time it takes to stop the cylinder. The function $u(r, t)$ is the fluid velocity in angular direction and $p(r, t)$ is the fluid pressure.

The fluid motion is assumed to be governed by the incompressible Navier-Stokes equations, while the relation between the torque on the inner cylinder and the time rate of change of its angular momentum determines the governing equation for the well-bore. After a scaling, (2.108) and (2.112) correspond to the simplified Navier-Stokes equation and (2.109) stems from the rigid-body motion of the well-bore. Steady state rotation of the inner cylinder results in a fluid velocity according to (2.112), in scaled form, which acts as initial condition for the retardation problem. Derive (2.108)–(2.112) in detail and perform a suitable scaling[5].

Numerical Method. The problem (2.108)–(2.111) is to be solved by a standard Galerkin finite element method in space and a θ-rule in time. When formulating the Galerkin equations in radial coordinates, one should recall that the one-dimensional problem is actually a simplification of PDEs for the corresponding 3D axisymmetric problem. If we integrate over the 3D axisymmetric domain, with unit length in the direction z of the cylinder axis, but utilize the property that $u \approx \hat{u} = \sum_j u_j N_j(r)$ only varies with r, we typically get

$$
\int_a^b \int_0^{2\pi} \int_0^1 \mathcal{L}(\hat{u}) N_i(r) dz d\theta dr = 2\pi \int_a^b \mathcal{L}(\hat{u}) N_i(r) r dr \ .
$$

Here (r, θ, z) are the cylindrical coordinates, and $\mathcal{L}$ is a differential operator.

Set up the global system of discrete equations to be solved at each time level. Allow for arbitrary grid spacing in r direction and arbitrary time step Δt. The element choice can be restricted to the linear type.

Implementation. After having found the difference equations arising from the finite element method, one can create a finite difference-like solver[6] based on these discrete equations.

Computer Experiments. Let T be dimensionless retardation time of the well-bore, defined by $u(\alpha, T) \leq 10^{-5}$. Set up a set of experiments and determine T as a function of β.

2.7 Higher-Dimensional Finite Elements

Figure 2.7 demonstrates how finite elements typically look like in two space dimensions. The elements are of triangular or rectangular shape, with possibly curved sides. This gives great flexibility in discretizing domains of complicated shape. An important requirement of the subdivision is that a node on the boundary between two elements must be a node in both elements. For example, the mesh of rectangular elements as shown in Figure 2.8 is not a proper finite element mesh. However, by dividing the left-most element into two new elements, a legal mesh is obtained.

---

[5] The characteristic length is chosen as b, the characteristic velocity is $a\Omega$, the characteristic pressure is $\varrho a^2 \Omega^2$, where ϱ is the density of the fluid, and the characteristic time equals $\varrho b^2 / \mu$, μ being the viscosity of the fluid.

[6] The program in `src/fdm/Parabolic1D` is an appropriate starting point.

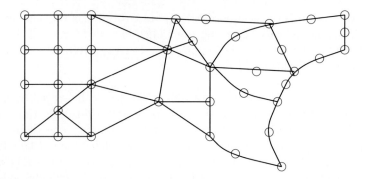

Fig. 2.7. Example on combining various 2D finite elements for discretizing a non-trivial geometry. The circles mark the nodes.

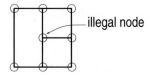

Fig. 2.8. Improper construction of a finite element mesh.

Our formulation of the finite element method in the previous sections makes it easy to define two- and three-dimensional elements and use these in Algorithm 2.1 from page 140. The elements are conveniently defined in a local coordinate system and then mapped by the isoparametric transformation (2.60) to the global physical coordinate system. The local coordinates are denoted by $(\xi_1, \ldots, \xi_d)$ or $\boldsymbol{\xi}$, while the corresponding global coordinates read $(x_1, \ldots, x_d)$ or $\boldsymbol{x}$.

2.7.1   The Bilinear Element and Generalizations

Basic Constructions. An obvious generalization of the one-dimensional linear element is the two-dimensional rectangle $[-1, 1] \times [-1, 1]$ in local coordinates, with nodes at the corners, see Figure 2.9. Since there are four nodes, and therefore four constraints of the type $\tilde{N}_r(\boldsymbol{\xi}^{[s]}) = \delta_{rs}$, we need four parameters in the local basis functions. This implies that the two-dimensional polynomial over a rectangle must be bilinear,

$$\tilde{N}_r(\xi_1, \xi_2) = a_r + b_r \xi_1 + c_r \xi_2 + d_r \xi_1 \xi_2, \quad r = 1, 2, 3, 4.$$

The term $\xi_1 \xi_2$ gives rise to the name *bilinear*.

The reader can now set up a 4×4 linear system for a_r, b_r, c_r, and d_r, for a given r, based on the conditions that $\tilde{N}_r = \delta_{rs}$ at node s, $s = 1, 2, 3, 4$. A worked example involving this process is given in Chapter 2.7.2.

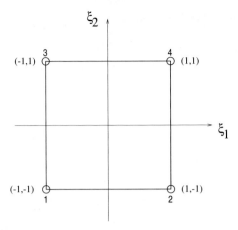

Fig. 2.9. Local numbering of nodes in the 2D bilinear element.

Applying the standard isoparametric transformation (2.60) to the bilinear element results in an element in global coordinates with four straight sides as depicted in Figure 2.10.

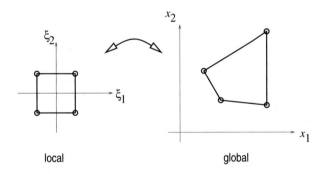

local   global

Fig. 2.10. Sketch of a two-dimensional rectangular element with 4 nodes and bilinear functions in its reference domain. The isoparametric mapping (2.60) results in an element, in global coordinates, that has 4 straight sides.

Tensor-Product Generalization of 1D Elements. When constructing higher order elements, the linear systems for determining the coefficients in expressions for $\tilde{N}_r$ become tedious to solve analytically. It is therefore advantageous to use simpler procedures for determining $\tilde{N}_r$. One such procedure is based on so-called *tensor-product* generalization of 1D elements. We can write the expressions for the bilinear basis functions $\tilde{N}_r(\xi_1, \xi_2)$ in terms of the *one-*

dimensional linear $\tilde{N}_r(\xi)$ functions:

$$\tilde{N}_1(\xi_1, \xi_2) = \tilde{N}_1(\xi_1)\tilde{N}_1(\xi_2), \tag{2.113}$$

$$\tilde{N}_2(\xi_1, \xi_2) = \tilde{N}_2(\xi_1)\tilde{N}_1(\xi_2), \tag{2.114}$$

$$\tilde{N}_3(\xi_1, \xi_2) = \tilde{N}_1(\xi_1)\tilde{N}_2(\xi_2), \tag{2.115}$$

$$\tilde{N}_4(\xi_1, \xi_2) = \tilde{N}_2(\xi_1)\tilde{N}_2(\xi_2). \tag{2.116}$$

The point now is to observe that the functions $\tilde{N}_1, \ldots, \tilde{N}_4$ can be viewed as tensor products of the linear 1D basis functions. The tensor (or dyadic) product ab of two vectors $a = (a_1, a_2)^T$ and $b = (b_1, b_2)^T$ is

$$\begin{pmatrix} a_1 \\ a_2 \end{pmatrix} \begin{pmatrix} b_1 & b_2 \end{pmatrix} = \begin{pmatrix} a_1 b_1 & a_1 b_2 \\ a_2 b_1 & a_2 b_2 \end{pmatrix}.$$

If we use indicial notation, two vectors a_i and b_i simply form the tensor product $a_i b_j$. The construction can be generalized to n vectors $a_i^{(1)}, a_i^{(2)}, \ldots, a_i^{(n)}$ whose tensor product becomes $a_{i_1}^{(1)} a_{i_2}^{(2)} \cdots a_{i_n}^{(n)}$, where $(i_1, i_2, \ldots, i_n)$ is an n-tuple index.

Forming the vector $v(\xi) = (\tilde{N}_1(\xi), \tilde{N}_2(\xi))^T$, the bilinear functions appear as the tensor product $v(\xi_1) v(\xi_2)^T$. Alternatively, we may use indicial notation and write the basis functions as $\tilde{N}_p(\xi_1)\tilde{N}_q(\xi_2)$, $p, q = 1, 2$. The single-index numbering of the 2D function $\tilde{N}_r(\xi_1, \xi_2)$ is then "columnwise" in the tensor product $v(\xi_1)v(\xi_2)^T$ or $\tilde{N}_p(\xi_1)\tilde{N}_q(\xi_2)$, where columnwise numbering corresponds to forming the vector $(A_{1,1}, A_{1,2}, A_{2,1}, A_{2,2})$ from a 2×2 matrix $A_{i,j}$, or equivalently, the vector index r equals $(p-1)2 + q$.

The advantage of the tensor-product formalism is that it becomes very easy to construct higher order elements and elements in higher space dimensions. Basis functions in three-dimensional elements $[-1, 1] \times [-1, 1] \times [-1, 1]$ can be generated from tensor products $\tilde{N}_i(\xi_1)\tilde{N}_j(\xi_2)\tilde{N}_k(\xi_3)$ of one-dimensional basis functions $N_i(\xi)$. Using linear one-dimensional basis functions results in the 3D *trilinear* element (each polynomial $\tilde{N}_r(\xi_1, \xi_2, \xi_3)$ is trilinear). Moreover, tensor products of 1D quadratic basis functions immediately yield biquadratic (2D) and triquadratic (3D) elements. The tensor-product construction can easily be applied for defining elements for PDEs in an arbitrary number of space dimensions. Such higher-dimensional PDEs arise, for example, in stochastic models from economics and engineering.

Numerical integration over the bilinear rectangle element is accomplished by a tensor product of the one-dimensional rules. For example, the relevant Gauss-Legendre rule with two points $(-1/\sqrt{3}, 1/\sqrt{3})$ can be used in a tensor product representing the four integration points in the rectangle. The corresponding vector, using a columnwise ordering, reads

$$\left((-1/\sqrt{3}, -1/\sqrt{3}), (1/\sqrt{3}, -1/\sqrt{3}), (-1/\sqrt{3}, 1/\sqrt{3}), (1/\sqrt{3}, 1/\sqrt{3}) \right).$$

A similar procedure is used for the weights. Numerical integration rules for other elements that arise from tensor-product generalization of 1D elements, is similarly constructed from tensor products of 1D rules. Notice that multiple indices arising from tensor-product constructions must be converted to single indices, such that the single-index based Algorithm 2.1 remains applicable. It is then just a matter of adjusting n_e, n_I, the basis functions, and the numerical integration points and weights when applying the algorithm to problems involving higher-dimensional elements.

Applying the general isoparametric mapping (2.60) and transforming the bilinear element to global coordinates, results in an element shape as depicted in Figure 2.10. The sides of the element are still straight. The angles must be less than π for the mapping (2.60) to be well defined, and angles close to π may lead to inaccurate numerical results.

2.7.2   The Linear Triangle

Local Coordinates. Linear polynomials in two variables have three coefficients. This means that a finite element with linear basis functions in 2D must have three nodes and hence be a triangle. Local coordinates for triangles are usually not varying in the interval $[-1, 1]$, but in $[0, 1]$. One can define local node no. 1 to be the point $(1, 0)$, local node no. 2 is then $(0, 1)$, whereas local node no. 3 is $(0, 0)$, see Figure 2.11.

Let us demonstrate the procedure for computing, e.g., $\tilde{N}_2(\xi_1, \xi_2)$ in detail. We know from the general principles that (i) $\tilde{N}_2(\xi_1, \xi_2)$ must be a polynomial, and that (ii) $\tilde{N}_2 = \delta_{ij}$ at local node no. j. Since condition (ii) gives three constraints (from three local nodes), $\tilde{N}_2$ can only have three free parameters, that is, $\tilde{N}_2$ must be a linear function:

$$\tilde{N}_2(\xi_1, \xi_2) = \alpha_2 + \beta_2 \xi_1 + \gamma_2 \xi_2.$$

The three constraints read

$$\tilde{N}_2(1, 0) = \alpha_2 + \beta_2 = 0,$$
$$\tilde{N}_2(0, 1) = \alpha_2 + \gamma_2 = 1,$$
$$\tilde{N}_2(0, 0) = \alpha_2 = 0.$$

This is a linear system in the three unknowns α_2, β_2, and γ_2. The solution is trivial to find: $\alpha_2 = \beta_2 = 0$, $\gamma_2 = 1$. Hence, $\tilde{N}_2(\xi_1, \xi_2) = \xi_2$. Following this procedure for $\tilde{N}_1$ and $\tilde{N}_3$ as well, one finds the local trial functions

$$\tilde{N}_1(\xi_1, \xi_2) = \xi_1, \qquad (2.117)$$
$$\tilde{N}_2(\xi_1, \xi_2) = \xi_2, \qquad (2.118)$$
$$\tilde{N}_3(\xi_1, \xi_2) = 1 - \xi_1 - \xi_2. \qquad (2.119)$$

The reference element can be mapped by (2.60) onto a general triangle with straight sides as depicted in Figure 2.11. A typical basis function in global coordinates is shown in Figure 2.12.

Notice that tensor-product generalizations of expressions for 1D elements do not make sense for a triangle, since its geometry is not a tensor product of an interval. Basis functions and numerical integration rules must therefore be especially constructed for triangles. We do not present appropriate integration rules here, but refer to more comprehensive textbooks on the finite element method, for instance [130,131]. The important information is that such rules exist and can be formulated in our general form as n_I 2D points with n_I corresponding weights.

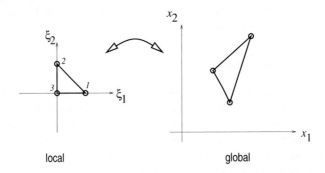

Fig. 2.11. Sketch of a two-dimensional triangular element with 3 nodes and linear functions in its reference domain. The isoparametric mapping (2.60) results in an element, in global coordinates, that has 3 straight sides.

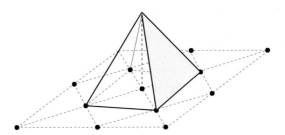

Fig. 2.12. Sketch of a typical basis function over a patch of linear triangular elements.

Global Coordinates. The advantage of working in local coordinates is hardly present when using linear finite elements in 1D. The same is true for multidimensional linear basis functions. The main reason for this is that there are simple formulas, expressed directly in global coordinates, for most of the common discrete terms that arise in PDE applications. These formulas will be given below.

Let the three nodes, with local numbers i, j, and k, of a triangle have global coordinates $(x_1^{[i]}, x_2^{[i]})$, $(x_1^{[j]}, x_2^{[j]})$, and $(x_1^{[k]}, x_2^{[k]})$. The linear basis functions are then

$$N_i(x_1, x_2) = \frac{1}{2\Delta}\left(\alpha_i + \beta_i x_1 + \gamma_i x_2\right), \tag{2.120}$$

$$\alpha_i = x_1^{[j]} x_2^{[k]} - x_1^{[k]} x_2^{[j]}, \tag{2.121}$$

$$\beta_i = x_2^{[j]} - x_2^{[k]}, \tag{2.122}$$

$$\gamma_i = x_1^{[k]} - x_1^{[j]}, \tag{2.123}$$

$$2\Delta = \det \begin{pmatrix} 1 & x_1^{[i]} & x_2^{[i]} \\ 1 & x_1^{[j]} & x_2^{[j]} \\ 1 & x_1^{[k]} & x_2^{[k]} \end{pmatrix}, \quad \Delta = \text{area of the element}. \tag{2.124}$$

A very useful result exists for the integral of products of basis functions over a linear triangular finite element Ω_e:

$$\int_{\Omega_e} N_i^p N_j^q N_k^r dx_1 dx_2 = \frac{p!q!r!}{(p+q+r+2)!} 2\Delta, \quad p, q, r \in \mathbb{N}. \tag{2.125}$$

2.7.3   Example: A 2D Wave Equation

We shall now demonstrate how the formulas (2.120)–(2.125) can be used for hand-calculation of element matrices and vectors for a two-dimensional problem. As example, we consider the 2D wave equation

$$\frac{\partial^2 u}{\partial t^2} = c^2 \nabla^2 u$$

with appropriate initial and boundary conditions. Chapter 2.2.1 deals with the finite element formulation of this problem. At the element level, we have the following central equation:

$$\sum_{s=1}^{3} \tilde{M}_{r,s}^{(e)} \tilde{u}_s^{\ell+1} = 2\sum_{s=1}^{3} \tilde{M}_{r,s}^{(e)} \tilde{u}_s^{\ell} - \sum_{s=1}^{3} \tilde{M}_{r,s}^{(e)} \tilde{u}_s^{\ell-1}$$

$$-c^2 \Delta t^2 \sum_{s=1}^{3} \int_{\Omega_e} \nabla N_r \cdot \nabla N_s dx_1 dx_2 \, \tilde{u}_s^{\ell}, \quad r = 1, 2, 3,$$

where

$$\tilde{M}_{r,s}^{(e)} = \int_{\Omega_e} N_r N_s dx_1 dx_2$$

is an element mass matrix that can be lumped by the row-sum technique or nodal-point integration (see Chapter 2.4). All the integrals above are straightforwardly calculated analytically using the preceding formulas. First, we consider the terms arising from the time derivative,

$$\int\limits_{\Omega_e} N_r N_s dx_1 dx_2 = \left\{ \begin{array}{l} \Delta/12, \ r \neq s \\ \Delta/6, \ r = s \end{array} \right.$$

The element matrix associated with the ∇^2 operator becomes

$$\int\limits_{\Omega_e} \nabla N_r \cdot \nabla N_s dx_1 dx_2 = \int\limits_{\Omega_e} \left(\frac{\partial N_r}{\partial x_1} \frac{\partial N_s}{\partial x_1} + \frac{\partial N_r}{\partial x_2} \frac{\partial N_s}{\partial x_2} \right) dx_1 dx_2$$

$$= (\beta_r \beta_s + \gamma_r \gamma_s) \frac{1}{4\Delta^2} \int\limits_{\Omega_e} dx_1 dx_2$$

$$= \frac{1}{4\Delta} (\beta_r \beta_s + \gamma_r \gamma_s) \ .$$

Given the coordinates of the nodal-points, it is now easy to evaluate the element matrix and vector. On a regular grid, it is sufficient to study the patch of triangles illustrated in Figure 2.13. Element numbers in that figure

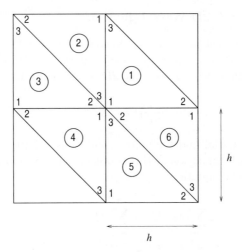

Fig. 2.13. Patch of six triangles contributing to the discrete equation that corresponds to the common node in the middle.

are surrounded by circles, and local nodal numbers appear in the corners of each element. There are basically two types of elements, exemplified by element no. 1 and 2. With our special local numbering in the elements, we find that the element matrices are the same for both types of elements.

$$\{\beta_r \beta_s\} = h^2 \begin{pmatrix} 1 & -1 & 0 \\ -1 & 1 & 0 \\ 0 & 0 & 0 \end{pmatrix}, \quad \{\gamma_i \gamma_j\} = h^2 \begin{pmatrix} 1 & 0 & -1 \\ 0 & 0 & 0 \\ -1 & 0 & 1 \end{pmatrix}.$$

Noting that $\Delta = h^2/2$, we get the following element matrices associated with the ∇^2 operator and the $\tilde{M}_{r,s}^{(e)}$ term, respectively:

$$\frac{1}{2} \begin{pmatrix} 2 & -1 & -1 \\ -1 & 1 & 0 \\ -1 & 0 & 1 \end{pmatrix}, \quad \frac{h^2}{24} \begin{pmatrix} 2 & 1 & 1 \\ 1 & 2 & 1 \\ 1 & 1 & 2 \end{pmatrix}.$$

Let us assemble the equations for the mid node of the patch. To easily interpret the resulting discrete equations as a finite difference scheme, we assign the index pair (i, j) to the mid-mode. Local node no. 1 in element no. 3 then has index pair $(i-1, j)$ and so on. All elements that contain the node (i, j) will contribute to equation no. (i, j), i.e., we must assemble the matrices from elements 1–6. Let $(r, s; e)$ denote entry (r, s) in the element matrix from element no. e. Equation (i, j) can be written as

$$c_{i,j-1} u_{i,j-1} + c_{i+1,j-1} u_{i+1,j-1} + c_{i-1,j} u_{i-1,j} + c_{i,j} u_{i,j} +$$
$$c_{i+1,j} u_{i+1,j} + c_{i-1,j+1} u_{i-1,j+1} + c_{i,j+1} u_{i,j+1} = 0,$$

where the coefficients equal

$$c_{i,j-1} = (1, 3; 4) + (3, 1; 5)$$
$$c_{i+1,j-1} = (3, 2; 5) + (2, 3; 6)$$
$$c_{i-1,j} = (1, 2; 4) + (2, 1; 3)$$
$$c_{i,j} = (1, 1; 1) + (3, 3; 2) + (2, 2; 3) + (1, 1; 4) + (3, 3; 5) + (2, 2; 6)$$
$$c_{i+1,j} = (1, 2; 1) + (2, 1; 6)$$
$$c_{i-1,j+1} = (2, 3; 3) + (3, 2; 2)$$
$$c_{i,j+1} = (3, 1; 2) + (1, 3; 1)$$

For the ∇^2 operator term we find the following contribution to equation no. (i, j):

$$-u_{i,j-1} - u_{i-1,j} - u_{i+1,j} - u_{i,j+1} + 4u_{i,j}.$$

We observe that this is the standard 5-point finite difference stencil (modulo a factor h^{-2}) for the Laplace operator. For the mass matrix term we get this contribution to equation no. (i, j):

$$\frac{h^2}{12} (u_{i,j-1} + u_{i+1,j-1} + u_{i-1,j} + u_{i+1,j} + u_{i-1,j+1} + u_{i,j+1}) + \frac{h^2}{2} u_{i,j}.$$

By lumping the mass matrix, the element mass matrix becomes $\frac{1}{6} h^2 \text{diag}(1, 1, 1)$, which assembles to one term $h^2 u_{i,j}$. The reader is strongly encouraged to work through the details of this section as this will improve the understanding of the finite element method.

2.7.4   Other Two-Dimensional Element Types

The examples so far should illustrate the fairly straightforward definition of basis functions in local coordinates and the flexibility of the element shape in global coordinates that results from the isoparametric mapping (2.60). Higher-order elements are easily defined in local coordinates and enable elements with curved sides in global coordinates. For example, a 9-node reference element has 9 parameters that can be used to define a biquadratic polynomial in ξ_1 and ξ_2. The basis functions can easily be constructed from a tensor product of the 1D quadratic functions. Figure 2.14 depicts this 2D element. Each straight side in the reference element is mapped by a biquadratic function through (2.60) and will result in a parabola in the physical domain (global coordinates). This gives a high degree of flexibility in fitting the elements to curved boundaries. Moreover, the higher polynomial degree usually gives higher accuracy for a fixed number of unknowns, compared with bilinear or linear elements.

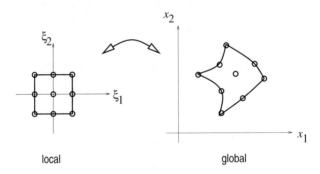

local   global

Fig. 2.14. Sketch of a two-dimensional rectangular element with 9 nodes and biquadratic functions in its reference domain. The isoparametric mapping (2.60) results in an element, in global coordinates, that has 4 curved sides. Each side has the shape of a parabola.

Exercise 2.16. Construct the expressions for the basis functions in the 9-node biquadratic element: $\tilde{N}_1(\xi_1, \xi_2), \ldots, \tilde{N}_9(\xi_1, \xi_2)$. Hint: Use the tensor products of quadratic 1D basis functions. Derive also an appropriate numerical integration rule.   ◇

The element in Figure 2.14 has 9 nodes, but the internal node is not used for geometric flexibility and can be removed. The resulting 8-node element has basis functions without the $\xi_1^2 \xi_2^2$ term. The sides in the physical domain are still of parabolic shape. Figure 2.15 depicts this element, which is popular in structural analysis.

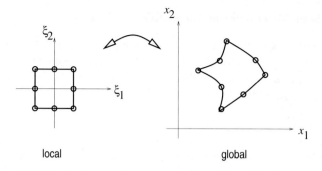

Fig. 2.15. Sketch of a two-dimensional rectangular element with 8 nodes and quadratic functions in its reference domain. The isoparametric mapping (2.60) results in an element, in global coordinates, that has 4 curved sides. Each side has the shape of a parabola.

We can also define quadratic elements of triangular shape. For example, extending the linear triangular element with nodes on each side results in 6 parameters (nodes) which can be used to fit quadratic basis functions. Figure 2.16 shows this element. Each straight side of the reference element can be mapped to a parabola in the physical domain. Together with the triangular shape, this gives a very high degree of flexibility for partitioning a geometrically complicated domain into triangles with curved sides. See [131, Ch. 7.8] for useful formulas regarding triangular elements with quadratic basis functions.

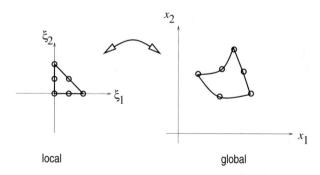

Fig. 2.16. Sketch of a two-dimensional triangular element with 6 nodes and quadratic functions in its reference domain. The isoparametric mapping (2.60) results in an element, in global coordinates, that has 3 curved sides. Each side has the shape of a parabola.

2.7.5   Three-Dimensional Elements

It should be obvious that the ideas presented so far can be carried over to three dimensions. The trilinear element, with eight nodes at the corners of a cube $[-1, 1]^3$, was briefly mentioned in Chapter 2.7.1. The isoparametric mapping (2.60) will then map the cube to a deformed cube with plane sides in the physical domain.

Triquadratic basis functions are enabled by having a cube $[-1, 1]^3$ with 27 $(3 \times 3 \times 3)$ nodes. The basis functions can be constructed from the three-factor tensor product of 1D quadratic functions, similarly to the trilinear functions. Appropriate integration rules are also constructed from three-factor tensor products of 1D rules. The isoparametric mapping now consists of second-order polynomials so the shape of the sides in the physical domain is described by quadratic functions. Other element shapes in 3D, like tetrahedra or prisms, are also straightforwardly constructed.

The flexibility of tetrahedral elements are often required when gridding complex 3D geometries. Linear basis functions over a tetrahedron frequently make it possible to compute element matrices and vectors analytically. One can hence avoid numerical integration and thereby improve the efficiency of simulation codes (see Appendix B.6.3). We therefore list closed-form expressions for the linear basis functions over a tetrahedron, together with a useful integration formula.

Let the local nodes have the numbers i, j, m, and p. The ordering of these nodes most follow a right-hand rule in the sense that the first three nodes are numbered in an anti-clockwise manner when viewed from the last one. The local nodes are chosen as $(1, 0, 0)$, $(0, 1, 0)$, $(0, 0, 1)$, and $(0, 0, 0)$, with the corresponding numbering 1-4. The expressions for $\tilde{N}_i$ in local coordinates are then trivial:

$$\tilde{N}_1(\xi_1, \xi_2, \xi_3) = \xi_1, \tag{2.126}$$

$$\tilde{N}_2(\xi_1, \xi_2, \xi_3) = \xi_2, \tag{2.127}$$

$$\tilde{N}_3(\xi_1, \xi_2, \xi_3) = \xi_3, \tag{2.128}$$

$$\tilde{N}_4(\xi_1, \xi_2, \xi_3) = 1 - \xi_1 - \xi_2 - \xi_3. \tag{2.129}$$

As for linear triangles, the formulas for hand calculation of element matrices and vectors are given in global coordinates. Let the global coordinates of node i be given as $(x_1^{[i]}, x_2^{[i]}, x_3^{[i]})$. The basis functions in global coordinates become

$$N_i(x_1, x_2, x_3) = \frac{1}{6V} \left(\alpha_i + \beta_i x_1 + \gamma_i x_2 + \delta_i x_3 \right), \tag{2.130}$$

where the coefficients α_i, β_i, γ_i, δ_i, and the element volume V are given as

$$\alpha_i = \det \begin{pmatrix} x_1^{[j]} & x_2^{[j]} & x_3^{[j]} \\ x_1^{[m]} & x_2^{[m]} & x_3^{[m]} \\ x_1^{[p]} & x_2^{[p]} & x_3^{[p]} \end{pmatrix}, \quad \beta_i = \det \begin{pmatrix} 1 & x_2^{[j]} & x_3^{[j]} \\ 1 & x_2^{[m]} & x_3^{[m]} \\ 1 & x_2^{[p]} & x_3^{[p]} \end{pmatrix},$$

$$\gamma_i = \det \begin{pmatrix} x_1^{[j]} & 1 & x_3^{[j]} \\ x_1^{[m]} & 1 & x_3^{[m]} \\ x_1^{[p]} & 1 & x_3^{[p]} \end{pmatrix}, \quad \delta_i = \det \begin{pmatrix} x_1^{[j]} & x_2^{[j]} & 1 \\ x_1^{[m]} & x_2^{[m]} & 1 \\ x_1^{[p]} & x_2^{[p]} & 1 \end{pmatrix},$$

$$6V = \det \begin{pmatrix} 1 & x_1^{[i]} & x_2^{[i]} & x_3^{[i]} \\ 1 & x_1^{[j]} & x_2^{[j]} & x_3^{[j]} \\ 1 & x_1^{[m]} & x_2^{[m]} & x_3^{[m]} \\ 1 & x_1^{[p]} & x_2^{[p]} & x_3^{[p]} \end{pmatrix}.$$

Integrals are easily computed by the formula

$$\int_{\Omega_e} N_i^q N_j^r N_m^s N_p^t dx_1 dx_2 dx_3 = \frac{q!r!s!t!}{(q+r+s+t+3)!} 6V, \qquad (2.131)$$

with $q, r, s, t \in \mathbb{N}$. See [131, Ch. 7.12] for appropriate formulas regarding quadratic tetrahedral elements.

Remark: Simplified Notation for the Rest of the Book. So far in this introduction to the finite element method, it has been important to distinguish between local and global quantities, and we used a tilde to mark the element-level quantities. For the rest of the book we will, however, mostly drop the tilde and use N_i, $A_{i,j}^{(e)}$, $b_i^{(e)}$ etc. as symbols also at the element level. It will be apparent from the context whether a quantity refers to the local or global level.

2.8   Calculation of derivatives

Having computed a finite element field $\hat{u} = \sum_{j=1}^{n} u_j N_j(\boldsymbol{x})$, it is trivial to calculate the derivatives,

$$\nabla \hat{u} = \sum_{j=1}^{n} u_j \nabla N_j(\boldsymbol{x}).$$

Such derivatives are ingredients in formulas for derived quantities, like flux, stress, velocity etc., which can often be of more physical importance than the primary unknown u. However, there is a fundamental problem with calculating derivatives of finite element fields: Typical finite element functions N_j have discontinuous derivatives, which implies that the values of the derivatives at the nodes or other points at the element boundaries are not uniquely defined. To obtain continuous derivatives, some smoothing procedure can be applied.

2.8.1   Global Least-Squares Smoothing

Our interest now concerns the derivative q of a finite element field,

$$q = \sum_{j=1}^{n} u_j \frac{\partial N_j}{\partial x_1}.$$

The aim of the smoothing procedure to be described here is to approximate q by a continuous finite element field $\hat{q}$,

$$\hat{q} = \sum_{j=1}^{n} q_j N_j(\boldsymbol{x}).$$

To this end, we can solve the equation $\hat{q} = q$ approximately, using a least-squares method or a Galerkin approach. Both procedures lead to the linear system

$$\sum_{j=1}^{n} M_{i,j} q_j = b_i, \quad i = 1, \ldots, n, \tag{2.132}$$

with

$$M_{i,j} = \int_{\Omega} N_i N_j d\Omega, \quad b_i = \int_{\Omega} N_i q \, d\Omega. \tag{2.133}$$

As usual, it is customary to lump the mass matrix $M_{i,j}$ to increase the efficiency of the solution process.

It turns out that the derivatives are most accurate at certain points inside an element. These points correspond to the integration points of a Gauss-Legendre rule of one order lower than what is the required rule for integrating the mass matrix (and preserving the convergence rate expected from the order of the finite element polynomials). This is commonly referred to as a *reduced Gauss-Legendre rule*. The associated integration points are referred to as *reduced integration points*. For multilinear basis functions (linear in 1D, bilinear in 2D, and trilinear in 3D), a Gauss-Legendre rule with 2^d points is the standard rule. The reduced rule has hence $(2 - 1)^d = 1$ point, which means that the derivatives of multilinear finite element fields have optimal accuracy at the centroid of each element. The reduced rule should be applied for the computation of b_i such that we sample q at the point(s) with optimal accuracy.

2.8.2 Flux Computations in Heterogeneous Media

In many situations we are not primarily interested in the solution of a PDE, but the derivative of the solution multiplied by a variable coefficient. We shall here use the term *flux* for such quantities. Consider, for example, the multi-dimensional PDE

$$-\nabla \cdot [\lambda(\boldsymbol{x})\nabla u(\boldsymbol{x})] = 0, \quad \boldsymbol{x} \in \Omega \subset \mathbb{R}^d, \tag{2.134}$$

with appropriate boundary conditions. The associated flux is here the vector $-\lambda\nabla u$. In solid mechanics (Chapter 5) the stress represents the counterpart to the flux $-\lambda\nabla u$.

It can be motivating to have in mind a specific physical application where the flux is of primary interest. One such application is porous media flow, where (2.134) is the governing PDE and the flux $\boldsymbol{v} = -\lambda\nabla u$ is the velocity of the fluid.

Porous Media Flow. A porous medium, like sand, soil, or rock, contains a geometrically complicated network of pores where a fluid can flow under the action of pressure and gravity forces. The mathematical model for the flow of a single fluid in a porous medium consists of basically two equations, one reflecting mass conservation and one reflecting Newton's second law. Let $\boldsymbol{v}(\boldsymbol{x}, t)$ be the velocity of the fluid averaged over a large number of pores in the medium. If the fluid is considered as incompressible, which is often relevant for flow of water, oil, and air in the small pores, the mass conservation equation reads $\nabla \cdot \boldsymbol{v} = 0$. Newton's second law takes the form $\nabla p + (\mu/K)\boldsymbol{v} - \varrho\mathbf{g} = 0$, where p is the fluid pressure, μ is the viscosity coefficient, K (called the permeability) is a quantity that reflects the medium's ability to transport the fluid through the porous network, ϱ is the fluid density, and $\mathbf{g}$ is the acceleration of gravity. This version of Newton's second law is usually called *Darcy's law* and states that the pressure forces ∇p balance viscous forces in the pores ($\mu\boldsymbol{v}/K$) and gravity ($\varrho\mathbf{g}$). We can eliminate $\boldsymbol{v}$ by solving Darcy's law with respect to $\boldsymbol{v}$ and inserting $\boldsymbol{v}$ in the mass conservation equation $\nabla \cdot \boldsymbol{v} = 0$, yielding the PDE

$$\nabla \cdot \left[\frac{K}{\mu} \left(\nabla p + \varrho\mathbf{g}\right) \right] = 0 \,.$$

With constant $\varrho\mathbf{g}$ we get (2.134), where λ equals K/μ and u is the pressure. For simplicity, we drop the gravity term $\varrho\mathbf{g}$ in the forthcoming discussion. The associated flux $\boldsymbol{v} = -\lambda\nabla u$ is then the fluid velocity.

Porous media are often of geological nature, and the spatial variations of the physical properties can therefore be significant. This is particularly true for the permeability K and thereby λ (the fluid property μ can be regarded as constant). Geological media are also often layered, and the physical parameters normally change rapidly between two layers. This leads to discontinuous coefficients, and especially the permeability can exhibit jumps of several orders of magnitude.

As a remark, we mention that the d-dimensional version of the heat conduction problem from Chapter 1.2.1 also takes the form (2.134), now with λ as the heat conduction coefficient, u as the temperature, and $-\lambda\nabla u$ as the heat flow (heat flux). The heat conduction properties will then change in a discontinuous way between the geological layers.

One can show that $-\lambda\nabla u \cdot \boldsymbol{n}$ is continuous across any surface S with outward unit normal $\boldsymbol{n}$. That means that if λ has large jumps, ∇u becomes discontinuous. In the discrete case, $-\lambda\nabla\hat{u} \cdot \boldsymbol{n}$ is continuous in a weak sense. We refer to [44, Ch. 2.2.1] for details on these issues. The following exercise demonstrates that the finite element method is capable, at least in a simple 1D example, of computing an exact flux even when λ is discontinuous, provided that the jump in λ appears on the element boundaries.

Exercise 2.17. Consider the equation $(\lambda u')' = 0$ in $\Omega = (0, 1)$ with $u(0) = 1$ and $u(1) = 0$. The function $\lambda(x)$ equals λ_1 for $x \in \Omega_1 = (0, 0.5)$ and λ_2

for $x \in \Omega_2 = [0, 5, 1)$. Solve the problem first analytically by integrating the PDE independently in Ω_1 and Ω_2 and using the boundary conditions and the requirement of continuous u and $-\lambda u'$ at $x = 0.5$. Then let each of Ω_1 and Ω_2 consist of a single linear element. Calculate the two element matrices, assemble the system for the unknown value u_2 at $x = 0.5$, solve for u_2, and show that u_2 coincides with the solution of the continuous problem. Also show that the flux from the finite element computations coincides with the exact flux.   ◇

In the case λ varies continuously, ∇u is also continuous, and if Γ then intersects an element the overall accuracy of the method will be preserved. However, when λ is discontinuous, there is no way we can force a discontinuity of ∇u in the interior of an element. The possible inaccuracy of letting the jumps in λ appear in the interior of an element can be explored by the experimental procedure to be presented in the following, involving a 1D Poisson equation. Exercise 3.12 on page 322 experiments with a 2D time-dependent heat equation and the impact of a discontinuous coefficient.

Experimental Analysis of a 1D Case. To gain insight into various aspects of flux computations in a finite element setting, we can study a simple 1D problem,

$$-(k(x)u')' = f_0, \quad x \in (0, 1), \quad u(0) = 1, \ u(1) = 0, \tag{2.135}$$

where f_0 is a constant. Our aim is to investigate the effect of jumps in k. To this end, we choose

$$k(x) = \begin{cases} 1, & x < \gamma \\ k_0, & x \geq \gamma \end{cases} \tag{2.136}$$

for some $\gamma \in (0, 1)$. The corresponding solution $u(x)$ can be found by direct integration of (2.135) and determining the integration constants from the boundary conditions, leading to

$$u(x; x < \gamma) = 1 - \frac{f_0}{2}x^2 - \frac{x}{\gamma + \frac{1-\gamma}{k_0}}\left[1 - \frac{f_0}{2}\left(\gamma^2 + \frac{1-\gamma^2}{k_0}\right)\right] \tag{2.137}$$

$$u(x; x \geq \gamma) = 1 - \frac{f_0}{2}\left(\gamma^2 + \frac{x^2 - \gamma^2}{k_0}\right)$$

$$-\frac{\gamma + \frac{x-\gamma}{k_0}}{\gamma + \frac{1-\gamma}{k_0}}\left[1 - \frac{f_0}{2}\left(\gamma^2 + \frac{1-\gamma^2}{k_0}\right)\right]. \tag{2.138}$$

The associated flux is

$$q = -ku' = f_0x + \frac{1}{\gamma + \frac{1-\gamma}{k_0}}\left[1 - \frac{f_0}{2}\left(\gamma^2 + \frac{1-\gamma^2}{k_0}\right)\right]. \tag{2.139}$$

For the discretization, we divide $(0, 1)$ into $2m$ linear elements and let $\gamma = \frac{1}{2} + \epsilon/h$, with $0 \leq \epsilon < 1$ and h as the constant element length. The jump in

k is then located in element no. $m + 1$, either at its left boundary ($\epsilon = 0$) or in the interior ($0 < \epsilon < 1$). If the jump appears at the element boundary (i.e. $\epsilon = 0$), one can quite easily prove that linear elements are able to predict the jump correctly, but the details are left for the reader as an extension of Exercise 2.17.

To investigate what happens when $f_0 \neq 0$ or $\epsilon > 0$, it might be instructive to create a simulator for the current problem. Such a simulator is located in the directory src/fem/Poisson2. Let us look at the quality of the solution with 10 elements and $\gamma = 0.50, 0.51$. The script gui-sec2.8.2.pl provides a simple graphical user interface to the current test problem, where the user can adjust γ, k_0, f_0, the element type, and the grid. Alternatively, the test case can be run directly from the command line:

```
./app --iscl --class Poi2flux --gamma '{ 0.5 & 0.51 }' --k0 10
      --gridfile 'P=PreproBox | d=1 [0,1] | d=1 e=ElmB2n1D [10] [1]'
      --f0 0 --casename flux1
netscape flux1-report.html
```

The precise meaning of the command-line options is explained in Chapter 3, but for now it suffices to know that --gamma, --k0, and --f0 assign values to γ, f_0, and k_0, whereas linear elements are specified by ElmB2n1D (quadratic elements would be ElmB3n2D), and the number of divisions between nodes[7] is 10 ([10]). As you can see, the simulator makes a web page with plots of u, the error in u, as well as various types of numerical computations of the flux $-ku'$. An example on plotting the latter quantity is given in Figure 2.17.

In the case $\gamma = 0.5$, the error in u and $-ku'$ is zero (at least within machine precision), whereas for $\gamma = 0.51$, the error is piecewise linear in u. You can check how the error is reduced by decreasing h. Simply replace 10 (in [10]) by

```
{ 10 & 20 & 40 & 80 & 160 }
```

and set --gamma 0.51 (the same settings apply in the graphical user interface as well). This input implies that we investigate the case $\gamma = 0.51$ for $h = 1/10, 1/20, 1/40, 1/80, 1/160$. Run the application and reload the flux1-report.html into the web browser. You will see that the error is always piecewise linear with three regions, where the mid region consists of element no. $m + 1$. However, it appears that refining the mesh does not lead to a monotone reduction in the error. We can investigate this phenomenon further by choosing four quadratic elements (8 divisions between nodes) and three γ values: $\gamma = 0.5, 0.51, 0.55$. The execution command becomes

```
./app --iscl --class Poi2flux --gamma '{ 0.5 & 0.51 & 0.55 }'
      --k0 10 --f0 0 --gridfile 'P=PreproBox | d=1 [0,1] |
      d=1 e=ElmB3n1D [8] [1]' --casename flux3
```

---

[7] Switching to quadratic elements, ElmB3n2D, but keeping the number of divisions between nodes constant at 10, then gives 5 elements. In this way the number of nodes is kept constant when the element type changes, which is convenient in convergence studies.

If you prefer the graphical user interface, choose a 3-node element and use the syntax for the three γ values as in the above command when filling out the gamma entry. Again we obtain exact results when the jump occurs at $x = 0.5$, i.e., at the boundary between two elements. Figure 2.17 dis-

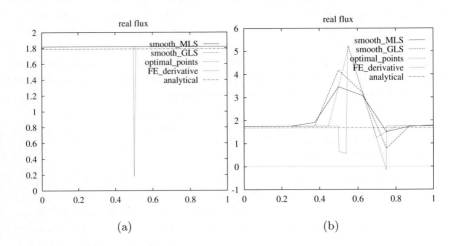

Fig. 2.17. The flux $-ku'$ computed by various numerical methods (see the text for explanations), using four quadratic elements, $k_0 = 10$, and $f_0 = 0$. (a) $\gamma = 0.51$; (b) $\gamma = 0.55$.

plays a plot of the flux $-ku'$ computed by various methods for the cases $\gamma = 0.51$ and $\gamma = 0.55$. The "smooth_GLS" curve applies the global least-squares smoothing technique, based on sampling $-ku'$ at the reduced Gauss points. The curve marked with "smooth_MLS" corresponds to a more sophisticated, but also more computationally expensive, smoothing technique, called moving least-squares smoothing (see page 251). The curve labeled with "optimal_points" consists of a straight line drawn between the optimal sampling (reduced Gauss) points. "FE_derivative" means that we compute $-ku' = -k \sum_j u_j N_j'(x)$ at a dense collection of discrete points inside each element.

We realize from Figure 2.17 that for $\gamma = 0.51$ the smoothed curves deviate slightly from the analytical flux value throughout the domain, whereas pointwise evaluation of $-ku' = -k \sum_j u_j N_j'(x)$ lead to a large error in the small region $[0.50, 0.51]$. However, when the point of discontinuity in k is moved more to the interior of the element ($\gamma = 0.55$), the errors in all methods increase significantly. The global flux is fortunately not much disturbed by the discontinuity; the dominant errors are localized to the element where the jump occurs and its neighbors. When performing convergence studies, the relative location of the jump in k inside the element (i.e. the ϵ value) will change, so reducing the element size may lead to an increase in the error. However, the main trend over many successively refined grids is that the error decreases. Uniform refinement is actually a waste of computational resources as the nature of this problem calls for refining only the element containing the jump in k. This leads to *adaptive* finite element methods, which is treated in Chapters 2.10 and 3.6, with Exercise 3.5 being particularly relevant for the present case study.

The lessons learned from these experiments are that severe inaccuracy may occur if variable coefficients exhibit jumps in the interior of elements, but the inaccuracy is a local phenomenon. The reader is encouraged to play around with the solver and the HTML report to learn more about finite element computations with discontinuous coefficients.

Remark. Our preceding exposition of finite element flux computations focuses at $-\lambda\nabla\hat{u}$ as a *field* throughout the domain Ω. In many applications one is mainly interested in the normal component $-\lambda\nabla\hat{u}\cdot\boldsymbol{n}$ *at the boundary* or at an internal surface. The simple flux computation method that we have outlined here can then be significantly improved by defining a separate finite element problem for the flux at the boundary or internal surface. We refer to [44, Ch. 4] for details on this numerical approach as well as for computations of derivatives of finite element fields in general.

2.9 Convection-Diffusion Equations

The finite element methods from the previous sections are readily applied to the convection-diffusion equation $\boldsymbol{v}\cdot\nabla u = k\nabla^2 u$, where $\boldsymbol{v}$ is a vector. In two space dimensions, the explicit form of this equation reads

$$v_x\frac{\partial u}{\partial x} + v_y\frac{\partial u}{\partial y} = k\left(\frac{\partial^2 u}{\partial x^2} + \frac{\partial^2 u}{\partial y^2}\right),$$

if $\boldsymbol{v} = (v_x, v_y)^T$. When solving such equations by the Galerkin method, the solutions may be polluted by nonphysical oscillations. This calls for a modification of the Galerkin approach. The present section briefly explains some of the problems with Galerkin methods for convection-diffusion problems and then outlines some strategies for obtaining more qualitatively correct solutions. The purpose is, as usual in this book, to introduce the reader to the fundamental ideas such that further information from the specialized literature is easier to process. Fortunately, the book by Morton [81] gives an excellent survey of finite element, difference, and volume methods for convection-diffusion problems. The particular subject of advanced finite element methods for convection-diffusion problems is presented, at an introductory level, by Eriksson et al. [36]. The recent text by Gresho and Sani [44] contains the most comprehensive treatment of finite element discretization of convection-diffusion PDEs at the time of this writing and constitutes an important source for gaining a thorough understanding of the topic.

We consider the following initial-boundary value problem:

$$\beta\frac{\partial u}{\partial t} + \boldsymbol{v}\cdot\nabla u = k\nabla^2 u + f, \quad \boldsymbol{x}\in\Omega,\ t>0, \tag{2.140}$$

$$u(\boldsymbol{x},0) = I(\boldsymbol{x}), \quad \boldsymbol{x}\in\Omega, \tag{2.141}$$

$$u = g(\boldsymbol{x}), \quad \boldsymbol{x}\in\partial\Omega_E, \tag{2.142}$$

$$-k\frac{\partial u}{\partial n} = 0, \quad \boldsymbol{x}\in\partial\Omega_N. \tag{2.143}$$

Here, $u(\boldsymbol{x},t)$ is the primary unknown, β is a parameter that is either zero or unity, with the purpose of turning the time derivative on or off in the problem, $\boldsymbol{v}$ is a prescribed velocity field, k is a constant diffusion coefficient, $f(\boldsymbol{x},t)$ is a prescribed function, and I and g are known functions. The complete boundary $\partial\Omega = \partial\Omega_E\cup\partial\Omega_N$, where $\partial\Omega_E$ covers at least the inflow boundary where $\boldsymbol{v}\cdot\boldsymbol{n}<0$, $\boldsymbol{n}$ being the outward unit normal vector to the boundary. The boundary $\partial\Omega_N$ is often an outflow boundary $(\boldsymbol{v}\cdot\boldsymbol{n}>0)$ where we can assume constant behavior outside the computational domain (i.e. a homogeneous Neumann condition applies).

The dimensionless form of the convection-diffusion equation is derived in Appendix A.1, and the important physical parameter of the problem is the Peclet number $\mathrm{Pe} = UL/k$, where U is a characteristic size of $\boldsymbol{v}$, and L is a characteristic length in Ω. In Appendix A.1 we show that the Peclet number measures the relative importance of the convective and the diffusive terms. Looking at convection and diffusion in an element, it is more appropriate to choose the element size h^* as the characteristic length. The relative importance of convection and diffusion in an element is hence described by the *local mesh Peclet number* $\mathrm{Pe}_\Delta = Uh^*/k$. The quality of the results produced by a Galerkin method depends strongly on the size of Pe_Δ.

2.9.1   A One-Dimensional Model Problem

Let us restrict the attention to a stationary 1D problem on $\Omega = [0, L]$ with constant $\boldsymbol{v} = (U, 0, 0)^T$ and Dirichlet conditions $u(0) = 0$ and $u(L) = \alpha$. The equation to be solved is then, in dimensionless variables,

$$u'(x) = \epsilon u''(x), \quad u(0) = 0,\ u(1) = 0, \tag{2.144}$$

where $\epsilon^{-1} = \mathrm{Pe} = UL/k$. The dimensionless primary unknown u equals the original u function divided by its typical scale, α, but we have as usual dropped special labeling of dimensionless quantities.

Finite difference methods for the problem (2.144) represent the subject of Projects 1.4.2 and 1.4.3. We strongly recommend the reader to go through these projects before proceeding. The main results from Projects 1.4.2 and 1.4.3 can be summarized as follows:

1. Centered (second-order accurate) finite differences for (2.144), written in the notation of Appendix A.3,

$$[\delta_{2x} u = \epsilon \delta_x \delta_x u]_i, \qquad (2.145)$$

leads to nonphysical oscillations in the solution if $h/\epsilon > 2$. The quantity $\mathrm{Pe}_\Delta = h/\epsilon$ is the local mesh Peclet number[8].

2. An upwind one-sided finite difference for the convection term stabilizes the method,

$$[\delta_x^- u = \epsilon \delta_x \delta_x u]_i. \qquad (2.146)$$

Qualitatively correct solutions are now achieved for all values of the local mesh Peclet number. Formally, the upwind scheme has a truncation error of first order in h, whereas the centered scheme has second-order accuracy.

3. The upwind scheme (2.146) can alternatively be obtained by adding artificial diffusion of size $h/2$ in dimensionless variables ($U h^*/2$ in the unscaled problem) and discretizing the equations by centered differences:

$$[\delta_{2x} u = (\epsilon + \frac{h}{2}) \delta_x \delta_x u]_i.$$

Discretizing (2.144) by a Galerkin finite element method with linear basis functions on a uniform grid, results in the centered difference scheme (2.145). The finite element method therefore faces exactly the same problems as the centered finite difference approach. Unfortunately, upwind differences are not trivially constructed via finite elements, but the artificial diffusion technique can easily be applied in a finite element context.

A popular way of constructing upwind differences is to perturb the weighting functions for the convective term. Taking

$$W_i(x) = N_i(x) + \tau N_i'(x)$$

leads to the following modified representation of the convection term in the weighted residual statement:

$$\int_0^1 u' W_i dx = \int_0^1 u' N_i dx + \int_0^1 \tau N_i' u' dx.$$

The last term, arising from the perturbation of the weighting function, can be interpreted as a Galerkin formulation of an extra diffusion term $\tau u''$ in the governing PDE. We are then left with a Galerkin formulation of a differential equation with artificial diffusion. Knowing that Galerkin's method and linear elements lead to centered difference approximations of the derivatives, the resulting equations become identical to those arising from an upwind finite difference scheme or centered finite differences for a differential equation with

---

[8] With m cells, we have $h^* = L/m$, $h = 1/m$, and therefore $\mathrm{Pe}_\Delta = h\epsilon^{-1} = hULk^{-1} = m^{-1}LUk^{-1} = h^*Uk^{-1}$.

artificial diffusion. To get the equivalence with the upwind scheme, we must take $\tau = h/2$.

We shall now explain that our 1D model problem can in fact be solved exactly by a modified finite difference or finite element method. The computer experiments from Projects 1.4.2 and 1.4.3 reveal that the upwind scheme has too much diffusion for intermediate mesh Peclet numbers, whereas the centered scheme has too little diffusion in the same parameter region. An obvious improvement is to introduced a difference approximation to u' that is a weighted sum of an upwind and a centered difference:

$$[u']_i \approx [\theta \delta_x^- u + (1 - \theta)\delta_{2x} u]_i, \quad 0 \le \theta \le 1.$$

Letting $\theta = \theta(\mathrm{Pe}_\Delta)$, with

$$\lim_{\mathrm{Pe}_\Delta \to 0} \theta = 0, \quad \lim_{\mathrm{Pe}_\Delta \to \infty} \theta = 1,$$

should give the desired qualitative behavior of this new difference approximation.

We can find an analytical solution to the modified difference scheme

$$[\theta \delta_x^- u + (1 - \theta)\delta_{2x} u = \epsilon \delta_x \delta_x u]_i, \tag{2.147}$$

using the methods of Appendix A.4.4. Inserting $u_i = \gamma^i$ gives a quadratic equation for the constant γ. One root is equal to unity, while the other can be forced to equal the analytical behavior: $u \sim \exp(x/\epsilon) = (\exp(\mathrm{Pe}_\Delta))^i$, i.e., $\gamma = \exp(\mathrm{Pe}_\Delta)$. Using the full expression for the second root leads to

$$\theta(\mathrm{Pe}_\Delta) = \coth \frac{\mathrm{Pe}_\Delta}{2} - \frac{2}{\mathrm{Pe}_\Delta}. \tag{2.148}$$

Reordering (2.147), we can rewrite the difference scheme as

$$[\delta_{2x} u = \epsilon \delta_x \delta_x u + \tau_o \delta_x \delta_x u]_i, \quad \tau_o = \frac{h}{2}\theta(\mathrm{Pe}_\Delta). \tag{2.149}$$

This is again a centered difference approximation to a modified PDE, where the modification consists in adding an artificial diffusion term $\tau_o u''$. With the optimal value τ_o of the artificial diffusion coefficient, the solution of the discrete equations becomes exact at the grid points in a uniform mesh.

It should be obvious how to construct a finite element method that also gives the exact solution at the nodes; τ_o is simply used either in the expression for the perturbed weighting function or in an artificial diffusion term. Notice that the formulas above are not restricted to the dimensionless form of our model problem. If we work with $Uu' = ku''$, we simply use (2.149) with $\mathrm{Pe}_\Delta = Uh^*/k$.

One can also use other types of perturbed weighting functions. In fact, there are numerous numerical approaches to our model problem that recover the exact solution at the grid points. See Morton [81] for further material about this topic.

2.9.2   Multi-Dimensional Equations

So far we have analyzed and cured the finite element method by (i) recognizing that the Galerkin approach and linear elements are equivalent to centered finite difference approximations to derivatives in 1D problems and by (ii) directly using results from the improvements of the centered finite difference method. When moving to multi-dimensional problems in $\mathbb{R}^d$, transferring finite difference technology to the finite element world is less attractive, because multi-dimensional upwind finite difference schemes are known to contain too much diffusion. Artificial diffusion is in fact only needed in the direction of $\boldsymbol{v}$, i.e. in the *streamline direction*. Straightforward upwind differences lead to significant crosswind diffusion (normal to the streamline), and this is undesired from a qualitative point of view.

Diffusion in a particular direction is enabled by working with a tensor diffusion coefficient k_{ij}. Choosing $k_{ij} = \tau v_i v_j / ||\boldsymbol{v}||^2$ gives a diffusive flux

$$q_i = \sum_{j=1}^{d} k_{ij} \frac{\partial u}{\partial x_j} = \alpha v_i, \quad \alpha = \frac{\tau}{||\boldsymbol{v}||^2} \sum_{j=1}^{d} v_j \frac{\partial u}{\partial x_j} = \tau \frac{\boldsymbol{v} \cdot \nabla u}{||\boldsymbol{v}||^2}.$$

Or in other words, q_i is proportional to v_i, which means that all diffusion is directed along the streamlines as required. Notice that in the multi-dimensional expressions we work with quantities with dimension, i.e., our model equation is $\boldsymbol{v} \cdot \nabla u = k \nabla^2 u$.

Writing $\tau^* = \tau / ||\boldsymbol{v}||^2$, we can now add the streamline-diffusion term

$$\sum_{i=1}^{d} \sum_{j=1}^{d} \frac{\partial}{\partial x_i} \left(\tau^* v_i v_j \frac{\partial u}{\partial x_j} \right) \tag{2.150}$$

to the multi-dimensional convection-diffusion equation and apply a standard Galerkin finite element method. Alternatively, we can perturb the weighting function for the convective term,

$$W_i = N_i + \tau^* \sum_{j=1}^{d} v_j \frac{\partial N_i}{\partial x_j} = N_i + \tau^* \boldsymbol{v} \cdot \nabla N_i.$$

The product of the convective term and the perturbation above yields a term that coincides with what we get from a standard Galerkin technique applied to the term (2.150). The particular perturbation of the weighting functions is therefore equivalent to adding a streamline-diffusion term to the original PDE.

Another justification for the weighting function perturbation $\tau \boldsymbol{v} \cdot \nabla N_i$ follows by looking at a least-squares formulation of the reduced equation $\boldsymbol{v} \cdot \nabla u = 0$:

$$\int_{\Omega} \boldsymbol{v} \cdot \nabla u \; \boldsymbol{v} \cdot \nabla N_i \, d\Omega = 0,$$

which is a weighted residual formulation of $\boldsymbol{v} \cdot \nabla u = 0$ with weighting function $\boldsymbol{v} \cdot \nabla N_i$.

Various choices of τ^* appear in the literature. In the next formulas, h^* represents a characteristic element length. A simple choice [81, Ch. 5.5] is $\tau = \max(\|\boldsymbol{v}\| h^* - k, 0)$. An obvious extrapolation of the optimal 1D results gives $\tau^* = \tau_o$ with $\mathrm{Pe}_\Delta = \|\boldsymbol{v}\| h^*/k$ [54]. Claes Johnson and co-workers have developed streamline-diffusion methods with parameters calculated from precise error analysis [36, Ch. 18]. Application of their methods to the current problem involves perturbed weighting functions $\tau^* \boldsymbol{v} \cdot \nabla N_i$ *and* an artificial diffusion term $\hat{\epsilon} \nabla^2 u$.

So far we have applied the perturbed weighting functions to the convective term only. A strong property of the weighted residual method is that the analytical solution of the problem is also always a solution of the weighted residual statement, simply because $R = 0$ (exact solution of the PDE) fulfills $\int_\Omega R W_i d\Omega = 0$. If we apply different weighting functions to different terms in the PDE, we can no longer factor out a common weighting function. In other words, $R = 0$ is no longer a solution of the weighted residual statement. This motivates for applying the same weighting function $W_i \neq N_i$ to all the terms in the PDE. The literature on numerical methods for convection-dominated transport refers to this approach as a consistent *Petrov-Galerkin* formulation. In our model problem this leads to

$$\sum_{j=1}^{n} \int_\Omega \left(W_i \boldsymbol{v} \cdot \nabla N_j + k \nabla N_i \cdot \nabla N_j + \tau^* \boldsymbol{v} \cdot \nabla N_i k \nabla^2 N_j \right) d\Omega \, u_j = \cdots,$$

where the dots on the right-hand side indicate possible surface integrals arising from integration by parts of the term $N_i k \nabla^2 u$. The term $\tau^* \boldsymbol{v} \cdot \nabla N_i k \nabla^2 N_j$ poses some problems since integration by parts cannot remove the second-order derivatives. The term vanishes in the interior of linear elements, and it has been common in the literature to neglect it also for multi-linear elements. In other words, the consistent Petrov-Galerkin formulation leads to the same results as we obtained by applying the perturbed weighting function to the convection term only, provided we use linear or multi-linear elements.

Figure 3.19a on page 303 shows the stabilizing effect of a Petrov-Galerkin formulation in a challenging 2D convection-dominated transport problem.

2.9.3   Time-Dependent Problems

Petrov-Galerkin Methods. Most of the preceding methods can be directly applied to the time-dependent version of the model problem (2.140)–(2.143) (see e.g. Project 1.4.1 for an example of upwind differencing in a pure convection problem). The common methodology nowadays is to use consistent Petrov-Galerkin formulations, where the perturbed weighting functions also affects the time-derivative term. The optimal value of τ^* might, however, be different in time-dependent problems. One early suggestion was $\tau^* = \Delta t/2$, which is justified in the next paragraph.

The Lax-Wendroff and Taylor-Galerkin Schemes. The Lax-Wendroff finite difference method has long been popular for solving convection-dominated problems with smooth solutions. The basic idea of the method is simple. Having a solution u^ℓ at time level ℓ (continuous or discrete), find a new solution at the next time level from a forward Taylor-series expansion to second order:

$$u^{\ell+1} = u^\ell + \Delta t \left[\frac{\partial u}{\partial t}\right]^\ell + \frac{1}{2}\Delta t^2 \left[\frac{\partial^2 u}{\partial t^2}\right]^\ell.$$

Then we use the PDE to replace the time derivatives by spatial derivatives and discretize the spatial derivatives by centered (second-order accurate) finite difference approximations. As an example, consider the PDE

$$\frac{\partial u}{\partial t} + U\frac{\partial u}{\partial x} = 0. \tag{2.151}$$

It follows that

$$\frac{\partial}{\partial t} = -U\frac{\partial}{\partial x}.$$

Using this in the Taylor-series expansion results in

$$u^{\ell+1} = u^\ell - U\Delta t \left[\frac{\partial u}{\partial x}\right]^\ell + \frac{1}{2}U^2\Delta t^2 \left[\frac{\partial^2 u}{\partial x^2}\right]^\ell, \tag{2.152}$$

which can be interpreted as

$$\left[\delta_t^+ u = -U\frac{\partial u}{\partial x} + \frac{1}{2}U^2\Delta t\frac{\partial^2 u}{\partial x^2}\right]^\ell. \tag{2.153}$$

We immediately observe that (2.153) is a forward temporal scheme for the original equation with an additional artificial diffusion term $\frac{1}{2}U^2\Delta t\partial^2 u/\partial x^2$.

The original Lax-Wendroff method approximates the spatial derivatives in (2.153) by centered differences, resulting in the scheme

$$[\delta_t^+ u + U\delta_{2x}u = \frac{1}{2}U^2\Delta t\delta_x\delta_x u]_i^\ell. \tag{2.154}$$

One can equally well apply a Galerkin finite element method to (2.152). Galerkin's method with linear elements results in centered difference approximations to the spatial derivatives and recovers (2.154). This finite element approach is usually referred to as the Taylor-Galerkin method in the literature.

In multi-dimensional problems, e.g.,

$$\frac{\partial u}{\partial t} + \boldsymbol{v}\cdot\nabla u = 0,$$

we have

$$\frac{\partial}{\partial t} = -\boldsymbol{v}\cdot\nabla \frac{\partial^2}{\partial t^2} = \boldsymbol{v}\cdot\nabla(\boldsymbol{v}\cdot\nabla).$$

For a divergence-free velocity field, $\nabla \cdot \boldsymbol{v} = 0$, which appears if the convection-diffusion process takes place in incompressible fluid flow, we can rewrite the second-order derivative in time such that it takes the form of anisotropic diffusion:

$$\frac{\partial^2}{\partial t^2} = \nabla \cdot \boldsymbol{v}\boldsymbol{v}\nabla = \sum_{r=1}^{d}\sum_{s=1}^{d} \frac{\partial}{\partial x_r}\left(v_r v_s \frac{\partial}{\partial x_s}\right).$$

Inserting this in a three-term temporal Taylor-series expansion gives

$$[\delta_t^+ u + \boldsymbol{v} \cdot \nabla u = \frac{1}{2}\Delta t \nabla \cdot (\boldsymbol{v}\boldsymbol{v}\nabla u)]^\ell.$$

The artificial anisotropic diffusion term has the same form as the streamline-diffusion term introduced in the stationary case, but now $\tau^* = \Delta t/2$. This suggests the usage of $\tau^* = \Delta t/2$ in Petrov-Galerkin formulations for transient problems.

Other Methods for Unsteady Convection-Dominated Problems. During the last two decades, numerous successful methods have been developed for accurate solution of convection-dominated transport, see Morton [81, Ch. 7]. Advanced techniques, combining space-time finite elements and streamline diffusion are well explained in [36, Ch. 19]. See also [40, Ch. 9-10] and [39] for overview of many successful methods.

The brief presentation of various strategies for handling convection terms in the finite element method indicates that streamline diffusion is a fundamental concept. Some popular standard choices of streamline diffusion-based methods are supported in Diffpack, see Chapter 3.8.

Exercise 2.18. Consider the model problem (2.151). A popular finite difference upwind scheme for this equation reads

$$[\delta_t^+ u + U\delta_x^- u = 0]_i^\ell.$$

The stability condition is given as $\Delta t \leq h/U$, and the optimal choice is to use the largest possible Δt value (this recovers the exact solution at the nodes as was proved in Project 1.4.1). An interesting question is how we can construct finite element methods that are mathematically equivalent to the upwind scheme above. Follow three alternative strategies: (i) add a suitable artificial diffusion term in the equation and apply Galerkin's method, (ii) devise an inconsistent Petrov-Galerkin formulation (perturbed weighting function on the convective term only), and (iii) use a Taylor-Galerkin approach. Emphasize the value of "free parameters" like τ in each case. Use linear elements.   ◇

2.10   Analysis of the Finite Element Method

The numerical properties of finite element methods can be established by applying techniques from finite difference analysis to the difference equations arising from a particular finite element discretization. Such an approach was demonstrated in Chapter 2.4.3, and the methodology is generally applicable, although it can become quite tedious to derive difference schemes for 2D and 3D problems, especially if quadratic elements are involved.

Despite the obvious idea of reusing finite difference analysis in a finite element context, the literature on finite element analysis employs almost exclusively a mathematical framework based on functional analysis for investigating the properties of the methods. The results from these theories are powerful; one can prove existence and uniqueness of the solutions u and u_h to the continuous and discrete[9] problem, one can derive bounds on u and u_h in terms of coefficients in the PDE, and one can derive general bounds on the numerical error $u - u_h$ in different norms. Tools from this analysis are fundamental for constructing *adaptive* finite element discretizations, where the numerical error can be controlled and the element size can be distributed in an optimal way throughout the mesh. The theoretical results from this type of finite element analysis are applicable to a wide range of PDEs discretized by various types of elements in any number of space dimensions.

The mathematical analysis literature on the finite element method is comprehensive, and finite element practitioners will occasionally need to exploit parts of this literature. Unfortunately, the literature frequently employs advanced mathematical tools and is mainly written by and for mathematicians. There is hence a need for a gentle introduction to the subject that can be understood on basis of straight calculus and linear algebra. The present exposition guides the reader through the basic concepts, some fundamental results, and some tools used to prove the results. With this knowledge, it should be easier for researchers in computational sciences to proceed with the many excellent books on mathematical analysis of the finite element method, for example, Brenner and Scott [17], Ciarlet [28], Glowinski [43], Johnson [59], Quarteroni and Valli [93], or Reddy [94]. These books contain a more comprehensive and precise treatment of the various mathematical topics than what we aim at in the following.

---

[9] We have previously used $\hat{u}$ for the discrete finite element solution. The literature on mathematical analysis of the finite element method normally applies the symbol u_h for the discrete solution, and this notation will be adopted in the present section.

2.10.1   Weak Formulations

Galerkin's Method Revisited. We recall from Chapter 2.1 that Galerkin's method for a partial differential equation $\mathcal{L}(u) = 0$ in a domain Ω reads

$$\int_{\Omega} \mathcal{L}(u_h) N_i d\Omega = 0, \quad i = 1, \ldots, M \tag{2.155}$$

where

$$u_h = \psi + \sum_{j=1}^{M} u_j N_j,$$

and N_i are linearly independent functions that vanish on the part of the boundary $\partial\Omega$ where essential boundary conditions are prescribed. The N_i functions then span a vector space V_h with basis

$$\mathcal{B} = \{N_1, \ldots, N_M\}.$$

The subscript h in V_h indicates that the vector space has finite dimension ($M = \dim V_h$), that is, we are dealing with a discrete formulation.

Let us define the inner product

$$(u, v) = \int_{\Omega} uv \, d\Omega. \tag{2.156}$$

The weighted residual statement (2.155) can now be viewed as an inner product of the residual and the weighting function, i.e., Galerkin's method can be expressed as

find $u_h - \psi \in V_h$ such that $(\mathcal{L}(u_h), v) = 0 \quad \forall v \in V_h$.

In the rest of this chapter we will for simplicity drop the ψ function and assume homogeneous essential boundary conditions. See [93, p. 166] for how nonhomogeneous Dirichlet conditions can be incorporated in the results we derive in the following.

Normally, one performs an integration by parts if $\mathcal{L}$ is a second- or higher-order differential operator. Let

$$a(u_h, v) = L(v)$$

denote the equation that arises from an integration by parts of $(\mathcal{L}(u_h), v) = 0$. For example, if $\mathcal{L}(u_h) = u_h''(x) + f(x)$ on $(0, 1)$, $a(u_h, v) = \int_0^1 u_h' v' dx$, and $L(v) = \int_0^1 f v dx$. The Galerkin method can now be formulated as follows:

find $u_h \in V_h$ such that $a(u_h, v) = L(v) \quad \forall v \in V_h$.

Intuitively, we expect that $u_h \to u$ as $M \to \infty$ and $V_h \to V$. The limiting space V, containing the solution u of the continuous problem, has of course infinite dimension. We therefore anticipate that the corresponding continuous problem can be expressed as

$$\text{find } u \in V \text{ such that } a(u, v) = L(v) \quad \forall v \in V. \tag{2.157}$$

Instead of having a partial differential equation that is fulfilled at every point in the domain Ω, we have an integral statement that is supposed to be fulfilled for an infinite number of *test functions* v. The integral statement is a kind of average and is hence weaker than the pointwise requirement of fulfilling a partial differential equation, motivating the term *weak formulation* for the statement (2.157). Other frequently used terms are *variational formulation* or *variational problem*. Sometimes we will use the term *weak form* or *variational form* for the equation $a(u, v) = L(v)$.

Example 2.3. The weak form $a(u, v) = L(v)$ of the continuous problem is normally obtained by multiplying the PDE by a test function $v \in V$, integrating over the domain Ω, and performing integration by parts (of second-derivative terms). As an example, consider

$$-\nabla \cdot [\lambda(\boldsymbol{x})\nabla u] = f(\boldsymbol{x}), \quad \boldsymbol{x} \in \Omega, \tag{2.158}$$

$$u = 0, \quad \boldsymbol{x} \in \partial\Omega_E, \tag{2.159}$$

$$-\lambda\frac{\partial u}{\partial n} = \beta(u - U_s), \quad \boldsymbol{x} \in \partial\Omega_N. \tag{2.160}$$

The expressions for $a(u, v)$ and $L(v)$ then becomes

$$a(u, v) = \int_\Omega \lambda\nabla u \cdot \nabla v \, d\Omega + \int_{\partial\Omega_N} \beta uvd\Gamma, \tag{2.161}$$

$$L(v) = \int_\Omega fvd\Omega + \int_{\partial\Omega_N} \beta U_svd\Gamma. \tag{2.162}$$

The functions in the space V must vanish on $\partial\Omega_E$.   ◇

The integration by parts leads to a statement $a(u, v) = L(v)$ where only the first-order derivatives of u and v enters. The partial differential equation, on the other hand, requires second-order derivatives of u to exist, but no requirements on the derivatives of v. The space V is therefore "larger" than the space containing the analytical solution of the partial differential equation, because it has less restrictions on the regularity of its member functions.

Weak Formulation with Temporal Derivatives. In time-dependent problems we can establish a weak formulation in space. For example, consider the PDE

$$\frac{\partial^2 u}{\partial t^2} = \nabla \cdot [\lambda\nabla u], \quad \boldsymbol{x} \in \Omega,$$

with u known on the complete boundary $\partial\Omega$. An appropriate weak form then involves the sum of two bilinear forms,

$$(\frac{\partial^2 u}{\partial t^2}, v) + a(u, v),$$

with $(\cdot, \cdot)$ defined in (2.156) and $a(u, v) = \int_\Omega \lambda \nabla u \cdot \nabla v \, d\Omega$.

Discretization. Given a continuous weak formulation,

$$\text{find } u \in V \text{ such that } a(u, v) = L(v) \quad \forall v \in V,$$

we can derive a corresponding discrete form by introducing a finite dimensional subspace $V_h \subset V$ and state

$$\text{find } u_h \in V_h \text{ such that } a(u_h, v) = L(v) \quad \forall v \in V_h.$$

Some Remarks. The procedure of first deriving a continuous weak formulation and then discretizing the problem by restricting the weak formulation to a finite dimensional function space is similar to the method of weighted residuals, but the actions are performed in a different order. The weighted residual method is more intuitive and therefore easier to extend to new situations. That is why the finite element formulations in the application parts of this book are based on the weighted residual method. Discretizing a continuous weak formulation gives the same system of algebraic equations as one obtains from the weighted residual method.

2.10.2   Variational Problems

By requiring some properties of $a(u, v)$ and $L(v)$, we can derive useful general results regarding the continuous and the discrete problem. Let us first list some definitions.

- A *linear form* $L(v)$ on a linear space V is a mapping $L : V \to \mathbb{R}$ such that
$$L(\alpha v + \beta w) = \alpha L(v) + \beta L(w).$$

- A *bilinear form* $a(u, v)$ on a linear space V is a mapping $a : V \times V \to \mathbb{R}$ such that $a(u, v)$ is linear in both arguments:
$$a(\alpha u + \beta w, v) = \alpha a(u, v) + \beta a(w, v), \quad a(u, \alpha v + \beta w) = \alpha a(u, v) + \beta a(u, w).$$

- The bilinear form $a(u, v)$ is *symmetric* if $a(u, v) = a(v, u)$.
- A (real) *inner product*, denoted by $(\cdot, \cdot)$, is a symmetric bilinear form on a linear space V that satisfies
$$(v, v) \geq 0 \; \forall v \in V \text{ and } (v, v) = 0 \Leftrightarrow v = 0.$$

– A linear space V together with an inner product defined on it is called an *inner-product space*. This space has an associated norm $||v||_V = \sqrt{(v, v)}$.

In the presentation and derivation of the mathematical results, the following properties of $a(u, v)$ and $L(v)$ are fundamental:

1. $L(v)$ is a linear form on V.
2. $a(u, v)$ is a bilinear form on $V \times V$.
3. $L(v)$ is bounded (or continuous) if there exists a positive constant c_0 such that
$$|L(v)| \leq c_0 ||v||_V \quad \forall v \in V.$$
4. $a(u, v)$ is bounded (or continuous) if there exists a positive constant c_1 such that
$$|a(u, v)| \leq c_1 ||u||_V ||v||_V \quad \forall u, v \in V.$$
5. $a(u, v)$ is V-elliptic (or coercive) on V if there exist a positive constant c_2 such that
$$a(v, v) \geq c_2 ||v||_V^2 \quad \forall v \in V.$$
6. $a(u, v)$ is symmetric: $a(u, v) = a(v, u)$.

These six items will later be referred to as properties 1-6.

Remark. Property 3 is classified as a continuity requirement of $L(v)$, and this may seem a bit strange. Note, however, that since L is linear we have

$$|L(v) - L(w)| = |L(v - w)| \leq c_0 ||v - w||_V.$$

Hence, if $w \to v$ it follows that that $L(v) \to L(w)$, i.e., L is continuous. A similar argument can be used to justify the formulation of property 4.

Some Hilbert Spaces with Associated Inner Products and Norms. Hilbert spaces play a central role in weak formulations of boundary-value problems. We refer to Brenner and Scott [17] for information about relevant Hilbert spaces for finite element problems. Here, we merely take the intuitive approach and say that Hilbert spaces are function spaces where certain integrals of the derivatives of the functions exist. For example, when solving an equation containing the ∇^2 operator, we need to ensure that $\int \nabla v \cdot \nabla v \, d\Omega < \infty$. This places a restriction on the type of functions v that V can contain.

Let Ω be a bounded and smooth domain in $\mathbb{R}^d$, and define $L^2(\Omega)$ to be the set of square-integrable functions on Ω, that is,

$$L^2(\Omega) = \{v \mid \int_\Omega v^2 d\Omega < \infty\}. \tag{2.163}$$

The space $V = L^2(\Omega)$ is equipped with the inner product

$$(u, v)_{L^2(\Omega)} = \int_\Omega uv \, d\Omega \tag{2.164}$$

and the associated norm

$$||v||_{L^2(\Omega)} = (v,v)_{L^2(\Omega)}^{\frac{1}{2}}.$$

A common Hilbert space in the analysis of second-order PDEs is $V = H^1(\Omega)$, which is the subset of functions in $L^2(\Omega)$ whose first derivatives are square integrable:

$$H^1(\Omega) = \left\{ v \in L^2(\Omega) \mid \frac{\partial v}{\partial x_i} \in L^2(\Omega) \text{ for } i = 1,\ldots,d \right\}. \tag{2.165}$$

The inner product is given by

$$(u,v)_{H^1(\Omega)} = \int_\Omega (uv + \nabla u \cdot \nabla v)\, d\Omega$$

$$= (u,v)_{L^2(\Omega)} + \sum_{r=1}^d (\frac{\partial u}{\partial x_r}, \frac{\partial v}{\partial x_r})_{L^2(\Omega)}. \tag{2.166}$$

The norm is, as usual, defined in terms of the inner product,

$$||v||_{H^1(\Omega)} = (v,v)_{H^1(\Omega)}^{\frac{1}{2}}.$$

Sometimes we work with boundary-value problems where the unknown $u = 0$ on a part $\partial\Omega_E$ of the boundary. We must then restrict all functions in V to vanish on $\partial\Omega_E$. The appropriate subspace of $H^1(\Omega)$ is then

$$H_0^1(\Omega) = \left\{ v \in H^1(\Omega) \mid v = 0 \text{ on } \partial\Omega_E \right\}. \tag{2.167}$$

The norm and inner product are inherited from $H^1(\Omega)$. Occasionally we also need a Hilbert space with square-integrable second-order derivatives,

$$H^2(\Omega) = \left\{ v \mid v \in L^2(\Omega),\ \frac{\partial v}{\partial x_i} \in L^2(\Omega),\ \frac{\partial^2 v}{\partial x_i \partial x_j} \in L^2(\Omega) \right\}, \tag{2.168}$$

for $i,j = 1,\ldots,n$. The associated inner product reads

$$(u,v)_{H^2(\Omega)} = (u,v)_{H^1(\Omega)} + \sum_{r=1}^d \sum_{s=1}^d (\frac{\partial^2 u}{\partial x_r \partial x_s}, \frac{\partial^2 v}{\partial x_r \partial x_s})_{L^2(\Omega)}$$

and the norm is

$$||v||_{H^2(\Omega)} = (v,v)_{H^2(\Omega)}^{\frac{1}{2}}.$$

We shall in the following also make use of so-called *semi-norms*

$$|v|_{H^1(\Omega)} = \left(\int_\Omega \nabla v \cdot \nabla v\, d\Omega \right)^{\frac{1}{2}} = ||\nabla v||_{L^2(\Omega)},$$

$$|v|_{H^2(\Omega)} = \left(\int_\Omega \sum_{r=1}^d \sum_{s=1}^d \left(\frac{\partial^2 v}{\partial x_r \partial x_s} \right)^2 d\Omega \right)^{\frac{1}{2}}.$$

The semi-norm $|\cdot|_{H^k(\Omega)}$ measures the L^2 norm of the partial derivatives of order k. Since we can have $|v|_{H^k(\Omega)} = 0$ even if $v \neq 0$, e.g. when $v \equiv 1$, $|\cdot|_{H^k(\Omega)}$ is not a proper norm.

2.10.3   Results for Continuous Problems

In the following, we list some basic theorems that are valid for a large class of stationary PDEs[10]. The theorems assume in general the existence of a Hilbert space V, a bilinear form $a(v, w)$ with $v, w \in V$, a linear functional $L(v)$ with $v \in V$, and that the bilinear form and the linear functional fulfill properties 1-6. Sometimes not all six properties are required for a theorem to hold. Property 6, the symmetry of a, can in particular be relaxed.

Theorem 2.4. Existence and uniqueness of the continuous problem (the Lax-Milgram Theorem). *There exists a unique $u \in V$ such that $a(u, v) = L(v)$ $\forall v \in V$.*

Proof. The existence part of this theorem follows from Riesz representation theorem, see [17, p. 60]. We shall here only prove uniqueness. Assume that we have two solutions $u_1, u_2 \in V$, that is,

$$a(u_1, v) = L(v) \quad \forall v \in V,$$
$$a(u_2, v) = L(v) \quad \forall v \in V.$$

Subtracting these equalities yields

$$a(u_1 - u_2, v) = 0 \quad \forall v \in V.$$

Choosing $v = u_1 - u_2 \in V$, it follows from property 5 (the V-ellipticity of a) that

$$0 = a(u_1 - u_2, u_1 - u_2) \geq c_2 ||u_1 - u_2||_V \geq 0,$$

which implies $u_1 = u_2$. □

Theorem 2.5. Stability of the continuous problem. *The solution $u \in V$, fulfilling $a(u, v) = L(v) \ \forall v \in V$, obeys the stability estimate*

$$||u||_V \leq \frac{c_0}{c_2}. \tag{2.169}$$

Proof. Since $u \in V$ and $a(u, v) = L(v) \ \forall v \in V$, we can choose $v = u$ and use properties 3 and 5 to get

$$c_2 ||u||_V^2 \leq a(u, u) = L(u) \leq c_0 ||u||_V,$$

and then divide by $c_2 ||u||_V$. □

Exercise 2.20 on page 194 demonstrates how (2.169) can be used to predict the stability of u due to perturbations of the input data to the PDE problem.

We remark that Theorems 2.4 and 2.5 do not require symmetry of $a(u, v)$ (property 6), but this is central for the next result.

---

[10] This class of PDEs is often referred to as *elliptic* PDEs, here recognized by fulfilling the requirement of V-ellipticity. See also Appendix A.5 for information on the nature of elliptic PDEs.

Theorem 2.6. Equivalent minimization problem (continuous case). *The weak formulation: find $u \in V$ such that $a(u, v) = L(v) \ \forall v \in V$, is equivalent with the minimization problem: Find $u \in V$ such that*

$$J(u) \leq J(v) \quad \forall v \in V, \quad J(v) = \frac{1}{2}a(v, v) - L(v).$$

That is, u minimizes the functional $J(v)$.

Proof. We first prove that if $u \in V$ satisfies $a(u, v) = L(v) \ \forall v \in V$, then $J(u) \leq J(v) \ \forall v \in V$. Define $w = v - u \in V$. Using the definition of J, we get

$$\begin{aligned}
J(v) &= J(u + w) \\
&= \frac{1}{2}a(u + w, u + w) - L(u + w) \\
&= \frac{1}{2}a(u, u) - L(u) + \frac{1}{2}a(w, w) + \underbrace{a(u, w) - L(w)}_{=0} \\
&= J(u) + \frac{1}{2}a(w, w) \geq J(u).
\end{aligned}$$

That is, $J(u) \leq J(v) \ \forall v \in V$. The next task is to prove that the minimization problem implies $a(u, v) = L(v) \ \forall v \in V$. Suppose $u \in V$ satisfies

$$J(u) \leq J(v) \quad \forall v \in V,$$

and define the function $g(\epsilon) = J(u + \epsilon w)$ for an arbitrary $w \in V$. Since J attains its minimum at u, we have $g'(0) = 0$. Using the definition of g,

$$g(\epsilon) = \frac{1}{2}a(u, u) + \epsilon a(u, w) + \frac{1}{2}\epsilon^2 a(w, w) - L(u) - \epsilon L(w)$$

and

$$g'(\epsilon) = a(u, w) + \epsilon a(w, w) - L(w).$$

The requirement $g'(0) = 0$ then gives $a(u, w) = L(w)$. Since w was arbitrarily chosen, we have that $a(u, w) = L(w) \ \forall w \in V$. $\qquad\square$

In some literature, the minimization problem is referred to as a *variational problem*. We will, however, use this term exclusively for the problem $a(u, v) = L(v) \ \forall v \in V$. The term *variational principle* is also used for the minimization problem and should cause no confusion. Formulation of a mathematical model as a minimization problem is often closely related to physical principles, where the motion of a continuum is such that the potential or kinetic energy, or a combination of them, is minimized.

Example 2.7. Irrotational fluid flow is recognized by $\nabla \times \boldsymbol{v} = 0$, where $\boldsymbol{v}$ is the velocity field in the fluid. This is a reasonable assumption in many flow cases where viscous effects can be neglected. The property $\nabla \times \boldsymbol{v} = 0$ implies in general that $\boldsymbol{v} = \nabla\phi$, where ϕ is a *velocity potential*. Mass conservation of an incompressible fluid leads to the requirement that $\nabla \cdot \boldsymbol{v} = 0$. Inserting $\boldsymbol{v} = \nabla\phi$ in the latter equation yields the governing equation $\nabla^2\phi = 0$ for ϕ in a fluid domain Ω. An appropriate boundary condition is $\boldsymbol{v} \cdot \boldsymbol{n} = 0$, which means no flow through boundaries ($\boldsymbol{n}$ is a normal vector to the boundary). This implies $\partial\phi/\partial n = 0$ at $\partial\Omega$. (The solution of the boundary-value problem is not unique so we need an additional condition, for example, $\int_\Omega \phi\, d\Omega = 0$.)

We can easily establish that $a(u,v) = \int_\Omega \nabla u \cdot \nabla v\, d\Omega$ and $L(v) = 0$, with the associated Hilbert space $H^1(\Omega)$. Provided that properties 1-6 are fulfilled (and this can be shown to be the case [43, p. 335]), solving the PDE for ϕ is equivalent to minimizing $J(v) = \frac{1}{2}a(v,v) = \int_\Omega |\nabla v|^2 d\Omega$. The physical interpretation of ∇v is the velocity $\boldsymbol{v}$. Forming the expression for the kinetic energy of a flow, $\int_\Omega \frac{1}{2}\varrho v^2 d\Omega = \text{const} \int_\Omega |\nabla\phi|^2 d\Omega$, where ϱ is the constant density of the fluid, we have justified the following physical principle: *The motion of an irrotational fluid in a domain Ω is such that the kinetic energy $\int_\Omega \frac{1}{2}\varrho v^2 d\Omega$ is minimized.* $\diamond$

Exercise 2.19. Find expressions for $a(u,v)$, $L(v)$, and $J(v)$ in the case where $-(ku')' = f$ on $(0,1)$ with $u(0) = 0$ and $u'(1) = 1$. (Just assume that properties 1-6 are fulfilled.) $\diamond$

Theorem 2.8. The energy norm. *The bilinear form $a(u,v)$ defines the energy norm*

$$||v||_a = a(v,v)^{\frac{1}{2}} \quad v \in V. \tag{2.170}$$

Moreover, $a(u,v)$ is an inner product. The energy norm is equivalent to the V norm,

$$\sqrt{c_2}||v||_V \le ||v||_a \le \sqrt{c_1}||v||_V \quad \forall v \in V. \tag{2.171}$$

Proof. Provided properties 2, 4, 5, and 6 are fulfilled, $a(u,v)$ fulfills the requirement of being an inner product with $a(v,v)^{\frac{1}{2}}$ as the associated norm. Properties 4 and 5 give

$$c_2||v||_V^2 \le |a(v,v)| \le c_1||v||_V^2,$$

which leads to (2.171). $\square$

Some Boundary-Value Problems that Fit into the Framework. The purpose now is to verify that properties 1-6 are fulfilled in some examples involving common boundary-value problems. To this end, we need two inequalities, the *Cauchy-Schwartz inequality*,

$$|(v,w)_{L^2(\Omega)}| \le ||v||_{L^2(\Omega)}||w||_{L^2(\Omega)}, \quad v,w \in L^2(\Omega), \tag{2.172}$$

$$|(v,w)_{H^1(\Omega)}| \le ||v||_{H^1(\Omega)}||w||_{H^1(\Omega)}, \quad v,w \in H^1(\Omega), \tag{2.173}$$

and *Poincaré's inequality*,

$$||v||_{L^2(\Omega)}^2 \le C_\Omega ||\nabla v||_{L^2(\Omega)}^2, \quad 0 < C_\Omega < \infty \text{ and } v \in H_0^1(\Omega). \tag{2.174}$$

Poincaré's inequality also holds for $v \in H^1(\Omega)$ provided $\int_\Omega v\, d\Omega = 0$.

Example 2.9. We consider

$$-u''(x) + u(x) = f(x), \quad x \in \Omega = (0,1), \tag{2.175}$$
$$u'(0) = u'(1) = 0. \tag{2.176}$$

The appropriate function space is

$$V = H^1(\Omega) = \left\{ v \mid \int_0^1 [(v(x))^2 + (v'(x))^2] \, dx < \infty \right\}.$$

Unless otherwise stated, *we shall assume here and in the rest of this chapter that* $f \in L^2(\Omega)$. Multiplying the differential equation by a test function $v \in H^1(\Omega)$ and integrating over the domain, using integration by parts, lead to

$$a(u,v) = \int_0^1 [u'(x)v'(x) + u(x)v(x)] \, dx,$$

$$L(v) = \int_0^1 f(x)v(x)dx.$$

The next step is to check the validity of properties 1-6.

1. $L(v)$ is a linear form because

$$L(\alpha v + \beta w) = \int_0^1 f(\alpha v + \beta w)dx \tag{2.177}$$

$$= \alpha \int_0^1 fvdx + \beta \int_0^1 fwdx = \alpha L(v) + \beta L(w). \tag{2.178}$$

2. Using the same technique as we applied for showing property 1, it follows that $a(u,v)$ is a bilinear form because of the linearity of the integral operator.
3. To show that $L(v)$ is bounded, we use the Cauchy-Schwartz inequality and the fact that $||v||_{L^2(\Omega)} \leq ||v||_{H^1(\Omega)}$, cf. (2.166),

$$|L(v)| = |\int_0^1 fvdx|$$

$$= |(f,v)_{L^2(\Omega)}| \leq ||f||_{L^2(\Omega)} ||v||_{L^2(\Omega)} \leq ||f||_{L^2(\Omega)} ||v||_{H^1(\Omega)}.$$

Hence, we can choose $c_0 = ||f||_{L^2(\Omega)}$. We now see why it is natural to demand $f \in L^2(\Omega)$.

4. The bound on $a(u,v)$ follows from the observation that $a(u,v)$ equals the inner product on $H^1(\Omega)$, and application of Cauchy-Schwartz' inequality:

$$a(v,w) = (v,w)_{H^1(\Omega)} \leq ||v||_{H^1(\Omega)} ||w||_{H^1(\Omega)},$$

which means that $c_1 = 1$.

5. The V-ellipticity is trivial to show, since $a(v, w) = (v, w)_{H^1(\Omega)}$ implies $a(v, v) = ||v||_{H^1(\Omega)}$ and therefore $c_2 = 1$.
6. Finally, $a(v, w)$ is symmetric since $vw = wv$ and $v'w' = w'v'$ in the integral expression for $a(v, w)$.

Having proved that properties 1-6 are fulfilled, we know that there exists a unique solution $u \in H^1(\Omega)$ of the variational problem $a(u, v) = L(v) \; \forall v \in H^1(\Omega)$. Moreover, u satisfies the bound $||u||_{H^1(\Omega)} \leq ||f||_{L^2(\Omega)}$.   ◇

Example 2.10. This example concerns $-u''(x) = f(x)$ in $\Omega = (0, 1)$, with $u(0) = u(1) = 0$. The relevant space is now

$$V = H_0^1(\Omega) = \left\{ v \in H^1(\Omega) \,|\, v(0) = v(1) = 0 \right\} .$$

We multiply the PDE by $v \in H_0^1(\Omega)$, integrate over $(0, 1)$, and apply integration by parts, leading to

$$a(u, v) = \int_0^1 u'(x)v'(x)dx, \qquad L(v) = \int_0^1 f(x)v(x)dx .$$

To check properties 1-3, we proceed as in Example 2.9. Property 4 follows from the Cauchy-Schwartz inequality (this time in $L^2(\Omega)$) and the fact that $||v'||_{L^2(\Omega)} \leq ||v||_{H^1(\Omega)}$:

$$|a(v, w)| = |(v', w')_{L^2(\Omega)}| \leq ||v'||_{L^2(\Omega)}||w'||_{L^2(\Omega)} \leq ||v||_{H^1(\Omega)}||w||_{H^1(\Omega)} .$$

Hence, $c_1 = 1$. Property 5, the V-ellipticity, follows from Poincaré's inequality,

$$||v||_{H^1(\Omega)}^2 = \int_0^1 (v^2 + (v')^2)dx \leq (C_\Omega + 1) \int_0^1 (v')^2 dx = (C_\Omega + 1)a(v, v) .$$

This means that $c_2 = 1/(C_\Omega + 1)$. The final property regarding symmetry is obviously fulfilled. Therefore, there exists a unique solution u of this variational problem, satisfying the bound $||u||_{H^1(\Omega)} \leq (1 + C_\Omega)||f||_{L^2(\Omega)}$.   ◇

Example 2.11. Consider the multi-dimensional boundary-value problem

$$-\nabla^2 u + u = f, \quad \boldsymbol{x} \in \Omega \subset \mathbb{R}^d, \tag{2.179}$$

$$\frac{\partial u}{\partial n} = 0, \quad \boldsymbol{x} \in \partial\Omega . \tag{2.180}$$

The appropriate space is $V = H^1(\Omega)$ as defined in (2.165). Multiplying the PDE by a test function $v \in H^1(\Omega)$ and integrating over Ω, with the aid of integration by parts, yield

$$a(u, v) = \int_\Omega (\nabla u \cdot \nabla v + uv)\, d\Omega, \qquad L(v) = \int_\Omega fv\, d\Omega .$$

Using the ideas of Example 2.9, it should be trivial to show that $L(v)$ is linear, $a(u,v)$ is bilinear and symmetric, and $L(v)$ is bounded with $c_0 = 1$. In the present example we observe that $a(v,w) = (v,w)_{H^1(\Omega)}$. Property 4, the continuity of the bilinear form, then follows from Cauchy-Schwartz' inequality,

$$|a(v,w)| = |(v,w)_{H^1(\Omega)}| \leq ||v||_{H^1(\Omega)}||w||_{H^1(\Omega)},$$

which means that $c_1 = 1$. The V-ellipticity (property 5) is trivially fulfilled with $c_2 = 1$ since $a(v,w) = (u,v)_{H^1(\Omega)}$. We have hence shown the fulfillment of properties 1-6, and there exists a unique solution u to this variational problem, with the stability property $||u||_{H^1(\Omega)} \leq ||f||_{L^2(\Omega)}$.   ◇

Example 2.12. Our final example is the multi-dimensional analog to Example 2.10, namely $-\nabla^2 u = f$ in $\Omega \subset \mathbb{R}^d$, with $u = 0$ on the boundary $\partial\Omega$. The appropriate Hilbert space is

$$V = H_0^1(\Omega) = \left\{ v \in H^1(\Omega) \,|\, v = 0 \text{ on } \partial\Omega \right\}.$$

Recall that the inner product and the norm are the same as for $H^1(\Omega)$. The bilinear and linear forms now become

$$a(u,v) = \int_\Omega \nabla u \cdot \nabla v \, d\Omega, \qquad L(v) = \int_\Omega f v \, d\Omega.$$

The boundary term vanishes since $v \in H_0^1(\Omega)$. Properties 1-3 and 6 are trivially shown as in the preceding examples. Continuity of $a(u,v)$ follows the proof of property 4 in Example 2.10:

$$|a(v,w)| = \left| \int_\Omega \nabla v \cdot \nabla w \, d\Omega \right| \leq \left| \int_\Omega (\nabla v \cdot \nabla w + vw) \, d\Omega \right|$$
$$= |(v,w)_{H^1(\Omega)}| \leq ||v||_{H^1(\Omega)}||w||_{H^1(\Omega)},$$

that is, $c_1 = 1$. Property 5 (the V-ellipticity) is shown by using Poincaré's inequality in the same way as we did in Example 2.10. Again, properties 1-6 are fulfilled, and we can apply the general results regarding the continuous problem as well as the forthcoming results for the corresponding discrete problem.   ◇

Exercise 2.20. Suppose the f function in Example 2.12 on page 194 is given a perturbation ϵ (e.g. caused by round-off or measurement errors). Use the general stability property of Theorem 2.5 and the specific expressions for c_0 and c_2 in the present problem to establish that the corresponding perturbation in u, $||\Delta u||_{H^1(\Omega)}$, is bounded by $||\epsilon||_{L^2(\Omega)}$.   ◇

Exercise 2.21. Consider the problem in Example 2.12. Since properties 1-6 are fulfilled, we know from Theorem 2.8 that $a(v,v)^{1/2}$ is a norm. However, choosing e.g. $v = 1$ results in $a(v,v) = 0$ for $v \neq 0$. How can then $a(v,v)^{1/2}$ be a *norm*?   ◇

Exercise 2.22. Extend Example 2.12 to the PDE $-\nabla \cdot (\lambda \nabla u) + \alpha u = f$, where α and λ obey $0 \leq \alpha \leq A$ and $\lambda_{\min} \leq \lambda \leq \lambda_{\max}$.   ◇

2.10.4   Results for Discrete Problems

The discrete version of the continuous problem is obtained by introducing a finite dimensional subspace $V_h \subset V$ to obtain the weak formulation in V_h instead of V:

$$\text{find } u_h \in V_h \text{ such that } a(u_h, v) = L(v) \quad \forall v \in V_h. \tag{2.181}$$

For example, a basis for V_h can exist of all piecewise linear functions associated with a finite element mesh. The formulation (2.181) is a Galerkin method for the underlying boundary-value problem.

The next two approximation results are fundamental for deriving error estimates for various choices of approximation spaces V_h (i.e. finite elements).

Theorem 2.13. Best approximation property in the energy norm. *Let $u \in V$ be the solution of $a(u, v) = L(v) \ \forall v \in V$, and let $u_h \in V_h$ be the solution of $a(u_h, v) = L(v) \ \forall v \in V_h$. Then u_h is the best approximation to $u \in V$ among all $v \in V_h$, measured in the energy norm:*

$$||u - u_h||_a \leq ||u - v||_a \quad \forall v \in V_h. \tag{2.182}$$

Proof. The statement $a(u, v) = L(v) \ \forall v \in V$ is also valid when $v \in V_h \subset V$:

$$a(u, v) = L(v) \quad \forall v \in V_h.$$

Subtracting this equation and the discrete equation that determines u_h: $a(u_h, v) = L(v) \ \forall v \in V_h$, gives $a(u - u_h, v) = L(v) - L(v) = 0$, which means that the error $e = u - u_h$ is orthogonal to the space V_h with respect to the inner product $a(u, v)$. That is,

$$a(e, v) = 0 \quad \forall v \in V_h. \tag{2.183}$$

We now want to show that this orthogonality implies the best approximation property. Pick an arbitrary $v \in V_h$ and define $w = u_h - v \in V_h$.

$$\begin{aligned}
a(u - v, u - v) &= a(u - u_h + u_h - v, u - u_h + u_h - v)\\
&= a(u - u_h + w, u - u_h + w)\\
&= a(e, e) + 2\underbrace{a(e, w)}_{=0} + a(w, w) \geq a(e, e).
\end{aligned}$$

Hence,

$$a(e, e) \leq a(u - v, u - v) \quad \forall v \in V_h,$$

which is another way of writing (2.182).   □

Roughly speaking, Galerkin's method finds the "best" solution u_h among all finite element functions $v = \sum_j v_j N_j$ associated with a particular grid. This is indeed a remarkable property.

Theorem 2.14. Best approximation property in the V norm. *Let $u \in V$ be the solution of $a(u,v) = L(v) \ \forall v \in V$, and let $u_h \in V_h$ be the solution of $a(u_h, v) = L(v) \ \forall v \in V_h$. If $a(u,v)$ is symmetric, then*

$$||u - u_h||_V \leq \left(\frac{c_1}{c_2}\right)^{\frac{1}{2}} ||u - v||_V \quad \forall v \in V_h . \tag{2.184}$$

In the case $a(u,v)$ is nonsymmetric the constant in (2.184) is replaced by c_1/c_2.

Proof. The result (2.184) follows from the best approximation property of u_h in the energy norm, as given by equation (2.182), and the equivalence of the a norm and the V norm, equation (2.171). For any $v \in V_h$ we have

$$||u - u_h||_V \leq \frac{1}{\sqrt{c_2}}||u - u_h||_a \leq \frac{1}{\sqrt{c_2}}||u - v||_a \leq \left(\frac{c_1}{c_2}\right)^{\frac{1}{2}} ||u - v||_V .$$

If $a(u,v)$ is nonsymmetric, we cannot use (2.171) and (2.182). Nevertheless, the following argument does not make use of any symmetry property of $a(u,v)$. Pick a $w \in V_h$ and define $v = u_h - w \in V_h$. We then have

$$c_2||v||_V^2 \leq a(e,e) = a(e,e) + a(e,w) = a(e, e+w)$$
$$= a(e, u - (u_h - w)) = a(e, u - v) \leq c_1||e||_V||u - v||_V .$$

Notice that we can add $a(e,w)$ $(=0)$ for any $w \in V_h$ since (2.183) does not require symmetry of $a(u,v)$. Dividing by $||e||_V$ we have

$$||u - u_h|| \leq \frac{c_1}{c_2}||u - v||_V \quad \forall v \in V_h .$$

$\square$

Nonsymmetric Bilinear Forms. More general boundary-value problems than what we have considered in Examples 2.9–2.12 can also be shown to fit into the framework, but the proofs are occasionally more technical. The quite general PDE

$$\boldsymbol{v} \cdot \nabla u = \lambda \nabla^2 u + \alpha u + f$$

with Dirichlet and Neumann boundary conditions can be shown to fulfill properties 1-5, see [43, App. I] or [93, Ch. 6]. The term $\boldsymbol{v} \cdot \nabla u$ makes $a(u,v)$ nonsymmetric, which means that error estimates based on the results in Theorem 2.14 typically contain the constant c_1/c_2. One can show [93, p. 169] that $c_2 = \lambda$ and

$$c_1 = \lambda + \sqrt{C_\Omega}||\boldsymbol{v}||_{L^\infty(\Omega)} + C_\Omega||\alpha||_{L^\infty(\Omega)},$$

where $||v||_{L^\infty(\Omega)} = \sup\{|v(\boldsymbol{x})| \mid \boldsymbol{x} \in \Omega\}$. This means that if the size of $\boldsymbol{v}$ is much larger than λ, the constant c_1/c_2 also becomes large, resulting in a poor bound on $||e||_V$. This effect was demonstrated in simple 1D examples in Chapter 2.9. Figure 3.19a on page 303 shows a 2D example where the finite element approximations by the Galerkin method are very inaccurate in a test case with $||\boldsymbol{v}||_{L^\infty(\Omega)}/\lambda = 10^3$.

The Galerkin Equations. Let the finite dimensional space V_h be spanned by linearly independent functions $N_1, \ldots, N_n$:

$$V_h = \text{span}\{N_1, \ldots, N_n\}, \quad \dim V_h = n\,.$$

The N_i functions are then basis functions for the space V_h, and any element $v \in V_h$ can be expressed as a linear combination of the basis functions: $v = \sum_{j=1}^{n} u_j N_j$. The Galerkin method (2.181) is equivalent to

$$a(u_h, N_i) = L(N_i), \quad i = 1, \ldots, n\,.$$

Inserting $u_h = \sum_j u_j N_j$ gives the linear system

$$\sum_{j=1} A_{i,j} u_j = b_i, \quad A_{i,j} = a(N_j, N_i), \ b_i = L(N_i), \quad (2.185)$$

for $i = 1, \ldots, n$. Very efficient solution methods exist for linear systems where the coefficient matrix is *symmetric* and *positive definite*. Let us show that $A_{i,j}$ have these two important properties. Symmetry is obvious if property 6 is fulfilled. To show that $A_{i,j}$ is positive definite, we first recall that an $n \times n$ matrix $\boldsymbol{A}$ is positive definite if $\boldsymbol{v}^T \boldsymbol{A} \boldsymbol{v} > 0$ for all nonzero $\boldsymbol{v} \in \mathbb{R}^n$. Write $\boldsymbol{v} = (v_1, \ldots, v_n)^T$ and $v = \sum_{j=1}^{n} v_j N_j$. Then,

$$\boldsymbol{v}^T \boldsymbol{A} \boldsymbol{v} = \sum_{i=1}^{n} \sum_{j=1}^{n} v_i A_{i,j} v_j = \sum_{i=1}^{n} \sum_{j=1}^{n} v_i a(N_i, N_j) v_j,$$

$$= a\left(\sum_{i=1}^{n} v_i N_i, \sum_{j=1}^{n} v_j N_j\right) = a(v, v) \geq c_2 \|v\|_V > 0,$$

for any nonzero $\boldsymbol{v}$. In other words, the positive definiteness of the coefficient matrix stems from the V-ellipticity (property 4).

We can now easily establish a discrete counterpart to the Lax-Milgram Theorem (Theorem 2.4).

Theorem 2.15. Existence and uniqueness of the discrete problem. *There exists a unique solution vector* $(u_1, \ldots, u_n)^T$ *of the linear system (2.185).*

Proof. A symmetric and positive definite matrix is nonsingular. Hence, the solution of the corresponding linear system exists and is unique. □

We also have a discrete counterpart to the stability estimate in Theorem 2.5.

Theorem 2.16. Stability of the discrete problem. *The solution* $u_h = \sum_j u_j N_j$ *of the discrete problem fulfills the stability estimate* $\|u_h\|_V \leq c_0/c_2$.

Proof. Since $V_h \subset V$ the result follows immediately from Theorem 2.5. Alternatively, we can repeat that proof by using properties 3 and 5, but now with $v = u_h \in V_h$. □

Theorem 2.17. Equivalent minimization problem (discrete case). *The linear system (2.185) is equivalent to the minimization problem: Find $\boldsymbol{u} = (u_1, \ldots, u_n)^T$ such that*

$$J_h(\boldsymbol{u}) \leq J_h(\boldsymbol{v}) \quad \forall \boldsymbol{v} \in \mathbb{R}^d,$$

where J_h is defined as

$$J_h(\boldsymbol{v}) = \frac{1}{2} \sum_{r=1}^{n} \sum_{s=1^n} v_r A_{r,s} v_s - \sum_{r=1}^{n} b_r v_r, \ \ \boldsymbol{v} = (v_1, \ldots, v_n)^T,$$

$$= \frac{1}{2} \boldsymbol{v}^T \boldsymbol{A} \boldsymbol{v} - \boldsymbol{b}^T \boldsymbol{v}.$$

That is, $\boldsymbol{u}$ minimizes the quadratic form $J_h(\boldsymbol{v})$.

Proof. We first prove that if $\boldsymbol{u} \in \mathbb{R}^d$ satisfies $\boldsymbol{A}\boldsymbol{u} = \boldsymbol{b}$, then $J_h(\boldsymbol{u}) \leq J(\boldsymbol{v})$ for an arbitrary $\boldsymbol{v} \in \mathbb{R}^d$. Define $\boldsymbol{w} = \boldsymbol{v} - \boldsymbol{u}$. Using the definition of J_h, we get

$$J_h(\boldsymbol{v}) = J(\boldsymbol{u} + \boldsymbol{w})$$

$$= \frac{1}{2}(\boldsymbol{u} + \boldsymbol{w})^T \boldsymbol{A}(\boldsymbol{u} + \boldsymbol{w}) - \boldsymbol{b}^T(\boldsymbol{u} + \boldsymbol{w})$$

$$= \frac{1}{2}\boldsymbol{u}^T \boldsymbol{A} \boldsymbol{u} - \boldsymbol{b}^T \boldsymbol{u} + \frac{1}{2}\boldsymbol{w}^T \boldsymbol{A} \boldsymbol{w} + \frac{1}{2}\boldsymbol{u}^T \boldsymbol{A} \boldsymbol{w} + \frac{1}{2}\boldsymbol{w}^T \boldsymbol{A} \boldsymbol{u} - \boldsymbol{b}^T \boldsymbol{w}$$

$$= J_h(\boldsymbol{u}) + \frac{1}{2}\boldsymbol{w}^T \boldsymbol{A} \boldsymbol{w} + \boldsymbol{w}^T \underbrace{(\boldsymbol{A}\boldsymbol{u} - \boldsymbol{b})}_{=0}$$

$$= J_h(\boldsymbol{u}) + \frac{1}{2}\boldsymbol{w}^T \boldsymbol{A} \boldsymbol{w} \geq J(\boldsymbol{u}).$$

Here we used that $\boldsymbol{u}^T \boldsymbol{A} \boldsymbol{w} = \boldsymbol{w}^T \boldsymbol{A} \boldsymbol{u}$, which is true when $\boldsymbol{A}$ is symmetric and positive definite. Next, we prove that the minimization problem implies $\boldsymbol{A}\boldsymbol{u} = \boldsymbol{b}$. We shall make use of the Kronecker delta δ_{rs} and the fact that $\sum_r A_{r,s} \delta_{ri} = A_{i,s}$ (see Appendix A.2). Since $J_h(\boldsymbol{v})$ is a function of n variables, the minimum of J_h can be calculated by requiring $\partial J_h / \partial v_i$ to vanish.

$$\frac{\partial J_h}{\partial v_i} = \frac{\partial}{\partial v_i} \left(\frac{1}{2} \sum_{r=1}^{n} \sum_{s=1}^{n} v_r A_{r,s} v_s - \sum_{r=1}^{n} b_r v_r \right)$$

$$= \frac{1}{2} \sum_{r=1}^{n} \sum_{s=1}^{n} \frac{\partial v_r}{\partial v_i} A_{r,s} v_s + \frac{1}{2} \sum_{r=1}^{n} \sum_{s=1}^{n} v_r A_{r,s} \frac{\partial v_s}{\partial v_i} - \sum_{r=1}^{n} b_r \frac{\partial v_r}{\partial v_i}$$

$$= \frac{1}{2} \sum_{r} \delta_{ri} \sum_{s} A_{r,s} v_s + \frac{1}{2} \sum_{r} v_r \sum_{s} A_{r,s} \delta_{si} - \sum_{r} b_r \delta_{ri}$$

$$= \frac{1}{2} \sum_{s} A_{i,s} v_s + \frac{1}{2} \sum_{r} \underbrace{A_{r,i}}_{=A_{i,r}} v_r - b_i$$

$$= \sum_{j=1}^{n} A_{i,j} v_j - b_i = 0.$$

□

The reader should notice the need for symmetry of $\boldsymbol{A}$ in the proof above. The symmetry of $\boldsymbol{A}$ is directly related to the symmetry of the underlying bilinear form $a(u, v)$ and thereby the properties of the differential operators in the PDE.

Exercise 2.23. Discretize the functional $J(v) = \frac{1}{2}a(v, v) - L(v)$, using $v = \sum_j v_j N_j \in V_h$, and show that this leads to a quadratic form $J_h(\boldsymbol{v}) = \frac{1}{2}\boldsymbol{v}^T \boldsymbol{A}\boldsymbol{v} - \boldsymbol{b}^T \boldsymbol{v}$, $\boldsymbol{v} = (v_1, \ldots, v_n)^T$, $A_{i,j} = a(N_i, N_j)$, $b_i = L(N_i)$. Minimization of the quadratic form implies $\boldsymbol{A}\boldsymbol{u} = \boldsymbol{b}$. In other words, discretization of the functional $J(v)$ also leads to the discrete equations (2.185). This is called *Rayleigh–Ritz'* method. It is frequently used in problems where there exists a variational principle, such as minimization of some type of energy. One can then discretize the variational principle directly instead of first deriving the corresponding PDE and then formulate a discrete variational form.   ◇

Interpolation. The standard *interpolant* $I_h v$ of a function $v \in V$ is a finite element function on the mesh that equals $\boldsymbol{v}$ at the nodal points. Let $\boldsymbol{x}^{[i]}$ be the coordinates of node no. i in a finite element mesh. Then $I_h v(\boldsymbol{x}^{[i]}) = v(\boldsymbol{x}^{[i]})$ for $i = 1, \ldots, n$. The finite element representation of $I_h v$ is hence $I_h v = \sum_{j=1}^n N_j v(\boldsymbol{x}^{[j]})$.

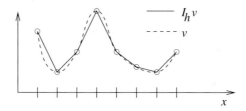

By means of I_h we can take "any" function $v \in V$ and transform it to a function $I_h v \in V_h$. We shall use this property to estimate the right-hand side of the error estimate (2.184).

Some basic results regarding the accuracy of interpolation are given next. These results are fundamental, together with the best approximation properties, for deriving error estimates in Chapters 2.10.5 and 2.10.7.

Suppose $v \in H^s(\Omega)$. Then there exists a constant C, independent of h and v, such that

$$||v - I_h v||_{L^2(\Omega)} \le Ch^s |v|_{H^s(\Omega)}, \quad (2.186)$$

$$|\nabla v - \nabla(I_h v)|_{H^1(\Omega)} = ||\nabla v - \nabla(I_h v)||_{L^2(\Omega)} \le Ch^{s-1} |v|_{H^s(\Omega)}. \quad (2.187)$$

In our context, s is an integer that equals 1 or 2. We refer to, e.g. Ciarlet [29] for derivation of (2.186)–(2.187) and many more general results regarding finite element interpolation. In finite element grids where the element size

h varies significantly, it is convenient to have h inside the norm [36]. For example, the following estimate is needed in Chapter 2.10.7:

$$||h^{-1}(v - I_h v)||_{L^2(\Omega)} \leq C||\nabla v||_{L^2(\Omega)} = C|v|_{H^1(\Omega)}. \qquad (2.188)$$

The constants C in (2.186)–(2.188) depend on properties of the mesh, as explained next.

We restrict the attention to finite element grids consisting of linear triangles. For each triangle K, we let h_K be the diameter of K, here defined as the longest side of K. Moreover, we define ϱ_K as the diameter of the circle inscribed in K. The largest element size h is taken as the maximum of h_K over all triangles. The ratio ϱ_K/h_K is a measure of the smallest angle in the triangle. Because thin triangles are unfavorable for the accuracy of the interpolation operator, we shall demand that ϱ_K/h_K is bounded from below by a constant β, reflecting the smallest angle among all elements in the mesh. The concept of β can be easily generalized to tetrahedral elements. The important information now is that the constants C in (2.186)–(2.188) depend on β; reducing β increases C and thereby the interpolation error [59]. This will have direct influence on the accuracy of the solution, as the forthcoming Theorems 2.18 and 2.19 predict.

2.10.5   A Priori Error Estimates

We can combine the general approximation result (2.184) with the interpolation results (2.186)–(2.187) to derive estimates for the discretization error in the finite element method. For simplicity the attention is limited to the Poisson problem from Example 2.12, discretized by linear triangular or tetrahedral elements.

Theorem 2.18. Error estimates for the derivatives. *Suppose $u \in H^2(\Omega)$ is the solution to $-\nabla^2 u = f$ in $\Omega \in \mathbb{R}^d$, with $u = 0$ on $\partial\Omega$, and that this problem is discretized by the Galerkin method, utilizing linear elements. Let u_h be the solution of the discrete problem. Then there is a finite constant C, independent of the element size h, such that*

$$|u - u_h|_{H^1(\Omega)} = ||\nabla u - \nabla u_h|| \leq Ch|u|_{H^2(\Omega)} \leq Ch||u||_{H^2(\Omega)} \quad (2.189)$$
$$||u - u_h||_{H^1(\Omega)} \leq Ch|u|_{H^2(\Omega)}. \qquad (2.190)$$

Proof. To prove (2.189), we start with the best approximation property (2.182):

$$||u - u_h||_a \leq ||u - v||_a \quad \forall v \in V_h.$$

Since

$$||v||_a = a(v, v)^{\frac{1}{2}} = \left(\int_\Omega \nabla v \cdot \nabla v \, d\Omega \right)^{\frac{1}{2}} = ||\nabla v||_{L^2(\Omega)},$$

we have

$$||\nabla u - \nabla u_h||_{L^2(\Omega)} \leq ||\nabla u - \nabla v||_{L^2(\Omega)} \quad \forall v \in V_h \,.$$

We can now choose $v = I_h u \in V_h$, since we have a bound on $||\nabla u - \nabla(I_h u)||_{L^2(\Omega)}$ from (2.187) in terms of the element size h:

$$||\nabla u - \nabla u_h||_{L^2(\Omega)} \leq ||\nabla u - \nabla(I_h u)||_{L^2(\Omega)} \leq Ch||u||_{H^2(\Omega)} \,.$$

The estimate (2.190) is shown similarly, but we start with the best approximation property (2.184). Choosing $v = I_h u \in V_h$ and using (2.186),

$$||u - u_h||_{H^1(\Omega)} \leq \left(\frac{c_1}{c_2}\right)^{\frac{1}{2}} ||u - I_h u||_{H^1(\Omega)}$$

$$= \left(\frac{c_1}{c_2}\right)^{\frac{1}{2}} \left(||u - I_h u||_{L^2(\Omega)} + ||\nabla u - \nabla(I_h u)||_{L^2(\Omega)}\right)$$

$$\leq Ch|u|_{H^2(\Omega)},$$

where C absorbs $\sqrt{c_1/c_2}$ and the constants from each of the estimates (2.186) and (2.187). $\qquad\square$

The estimate (2.189) can be generalized to finite element basis functions of order s:

$$||\nabla u - \nabla u_h|| \leq Ch^s||u||_{H^{s+1}(\Omega)} \,.$$

A standard finite difference scheme for Poisson's equation with homogeneous Dirichlet conditions gives a truncation error of second-order in the grid increments. From Chapter 2.7.3 we know that linear triangles reproduce the standard 5-point finite difference representation of $\nabla^2 u$ in 2D. Therefore, we would expect the error to be of order h^2, because the truncation error is $\mathcal{O}(h^2)$, but the error estimate (2.190) is only of first order in h. The explanation is that the H^1 norm in (2.190) also contains terms with the *derivatives* of $u - u_h$. According to (2.189), these derivatives have only first-order accuracy. A more optimal estimate would involve only the L^2 norm of $u - u_h$. Such an estimate can indeed be derived, but the proof is more technical than what we have seen so far, and a few additional requirements must be fulfilled.

Theorem 2.19. Error estimate in $L^2(\Omega)$. *Suppose we solve the same problem as in Theorem 2.18. Provided that Ω is a convex bounded domain, with a polygonal boundary, and $f \in L^2(\Omega)$, there exists a constant C, independent of the element size h, such that*

$$||u - u_h||_{L^2(\Omega)} \leq Ch^2|u|_{H^2(\Omega)} \,. \tag{2.191}$$

Proof. We start by considering the problem $-\nabla^2 w = e$ in Ω, with $w = 0$ on $\partial\Omega$. The function e is the error $u - u_h$. For this problem we have $a(w, v) = \int_\Omega \nabla w \cdot \nabla v \, d\Omega$ and $L(v) = \int_\Omega ev \, d\Omega$. Hence, by choosing $v = e$ we get

$$a(w, e) = L(e) = ||e||_{L^2(\Omega)}^2 \,.$$

We know from (2.183) that e is orthogonal to V_h,

$$a(e, v) = a(u - u_h, v) = 0 \quad \forall v \in V_h,$$

which gives

$$||e||^2_{L^2(\Omega)} = a(e, w) = a(e, w) - a(e, I_h w) = a(e, w - I_h w),$$

since $I_h w \in V_h$ implies $a(e, I_h w) = 0$. Cauchy-Schwartz' inequality can be used to bound

$$a(e, w - I_h w) = \int_\Omega \nabla e \cdot (\nabla w - \nabla (I_h w)) d\Omega \leq ||\nabla e||_{L^2(\Omega)} ||\nabla w - \nabla (I_h w)||_{L^2(\Omega)} .$$

The second factor can be related to the interpolation error through (2.187),

$$||\nabla w - \nabla (I_h w)||_{L^2(\Omega)} \leq Ch|w|_{H^2(\Omega)},$$

which means that

$$||e||^2_{L^2(\Omega)} \leq ||\nabla e||_{L^2(\Omega)} Ch|w|_{H^2(\Omega)} .$$

A bound for the $|w|_{H^2(\Omega)}$ norm in terms of $||e||_{L^2(\Omega)}$ can be obtained from a regularity result for the Poisson equation $-\nabla^2 u = f$ in Ω with $u = 0$ on $\partial \Omega$. The result tells that

$$||u||_{H^2(\Omega)} \leq \hat{C}||f||_{L^2(\Omega)}$$

for some finite constant $\hat{C}$ if Ω is a convex polygonal domain and $f \in L^2(\Omega)$. With this result we immediately have that

$$|w|_{H^2(\Omega)} \leq ||w||_{H^2(\Omega)} \leq \hat{C}||e||_{L^2(\Omega)} .$$

Summarizing so far,

$$||e||^2_{L^2(\Omega)} \leq ||\nabla e||_{L^2(\Omega)} C\hat{C}h||e||_{L^2(\Omega)},$$

or by dividing by $||e||_{L^2(\Omega)}$ and writing $\nabla e = \nabla u - \nabla u_h$,

$$||e||_{L^2(\Omega)} \leq \tilde{C}h||\nabla u - \nabla u_h||_{L^2(\Omega)} .$$

Using (2.187), we can bound the right-hand side, which gives the desired result:

$$||e||_{L^2(\Omega)} \leq Ch^2|u|_{H^2(\Omega)},$$

Now, C is a new constant that has absorbed various other constants, including the interpolation constants that depend on the smallest angle in the mesh.

$\square$

Remark. Since we already know that the finite difference method and the finite element method with linear elements are equivalent for the Poisson equation on a uniform mesh (cf. Chapter 2.7.3), we can use (2.191) to estimate the *error*, and not only the truncation error, of the corresponding finite difference scheme.

2.10.6   Numerical Experiments

The Model Problem. It is interesting to see if the general theorems above are in accordance with computational results in specific model problems. Of particular interest is the convergence rate of the error as predicted by Theorems 2.18 and 2.19.

The following d-dimensional model problem is considered:

$$-\nabla^2 u + \beta u = f, \quad \boldsymbol{x} \in \Omega = (0,1)^d, \tag{2.192}$$

$$u = 0, \quad x_1 = 0, \ x_1 = 1, \tag{2.193}$$

$$\frac{\partial u}{\partial n} = 0, \quad x_k = 0, \ x_k = 1, \ k = 2, \ldots, d. \tag{2.194}$$

With

$$f(x_1, \ldots, x_d) = (\pi^2 d + \beta) \sin \pi x_1 \prod_{j=2}^{d} \cos \pi x_j,$$

the exact solution becomes

$$u(x_1, \ldots, x_d) = \sin \pi x_1 \prod_{j=2}^{d} \cos \pi x_j.$$

In this particular test example we shall examine the following norms of the error $e = u - u_h$:

$$\|e\|_{L^1(\Omega)} = \int_{\Omega} |e| d\Omega,$$

$$\|e\|_{L^2(\Omega)} = \left(\int_{\Omega} e^2 d\Omega \right)^{\frac{1}{2}},$$

$$\|e\|_{L^\infty(\Omega)} = \sup \{|e(\boldsymbol{x})| \mid \boldsymbol{x} \in \Omega\},$$

$$\|e\|_{H^1(\Omega)} = \left(\int_{\Omega} \left(e^2 + \nabla e \cdot \nabla e \right) d\Omega \right)^{\frac{1}{2}}$$

$$\|e\|_a = \left(\int_{\Omega} \left(\beta e^2 + \nabla e \cdot \nabla e \right) d\Omega \right)^{\frac{1}{2}}.$$

We refer to these five norms as the L^1, L^2, L^∞, H^1, and energy norm, respectively. When $\beta \geq 0$, one can show that properties 1-6 are fulfilled and that the theorems in the previous sections hold. Our perhaps most interesting result, the estimate of the L^2 error in Theorem 2.19, was restricted to linear elements and the Poisson equation with Dirichlet boundary conditions. We therefore want to check if the same conclusions apply to our model problem, and what type of results we achieve for *quadrilateral* elements of first and *second* order. In addition, we want to see the effect of negative β values.

The Simulator. We have developed a Diffpack simulator for the current problem. The simulator class is called `Poi2estimates` and is a part of the `Poisson2` solver, which is thoroughly explained in Chapter 3.5. The reader does not need to look up the reference to various software tools used in this simulator; the purpose now is just to apply a ready-made program. Diffpack offers some convenient tools for estimating errors and corresponding convergence rates, thus making it very simple to develop a flexible solver for problems like (2.192)–(2.194) and compute quantities that are central in the mathematical theory of finite elements.

Using Diffpack's multiple loop functionality (explained in Chapter 3.4.2), we can easily set up[11] a simulation where we run through a sequence of grids, with element size $h = 0.25 \cdot 2^{-k}$, for $k = 0, 1, 2, 3, 4$, using 4-node bilinear and 9-node biquadratic elements, and varying β among -10, 0, and 1000. Convergence rates are estimated from two successive experiments as explained in Chapter 3.5.8 or in Project 1.4.2.

We should remark that when $\beta = -10$, the coefficient matrix in the linear system is not positive definite, thus requiring a solver that handles such type of matrices. The `Symmlq` solver in Diffpack, combined with SSOR preconditioning, worked satisfactorily for this and even less favorable negative values of β. The Conjugate Gradient method with MILU preconditioning is a good choice when $\beta \geq 0$. A critical point when using iterative solvers in numerical experiments for estimating convergence rates, is the choice of termination criterion for the linear solver; we must ensure that the errors due to approximate solution of the linear systems are negligible compared with discretization errors. For the experiments here we stopped the iterations when the initial residual was reduced by a factor of 10^{-10}. Solvers and termination criteria for linear systems are explained in Appendix C.

Computational Results. Tables 2.2 and 2.3 present convergence rates for different β values and choice of elements[12]. The convergence rates for the bilinear element are in accordance with the theory of linear elements, that is, the L^2 norm of the error has second-order convergence, whereas the H^1 norm of the error, containing derivatives of u, has first-order convergence. Table 2.2 also shows that the L^1 norm and the maximum error (L^∞) give second-order convergence, and that the energy norm rates are close to those of the H^1 norm.

---

[11] The appropriate input file for this experiment is found in the directory `src/fem/Poisson2/Verify/femtheory.i`.

[12] We remark that the tables were automatically generated by the simulation code.

Table 2.3 shows that biquadratic elements give convergence rates that are one order higher than in the bilinear case. The reader is encouraged to learn about the simulator in Chapter 3 and set up test cases involving triangular elements of first and second order. One can also verify that the estimated convergence rates are valid in any number of space dimensions.

The overall conclusion is that the theory in this chapter is in accordance with at least one real-world example, and as far as discretization errors are concerned, it seems that the convergence rates hold also in problems not covered by our exposition of the theory, e.g. the case $\beta = -10$ (cf. Exercise 2.22).

Table 2.2. Convergence rates of errors in the model problem (2.192)–(2.194). Assuming that the error e behaves like $e = Ch^r$, where h is the distance between two nodes in x-direction, we estimate r from two successive experiments, see (1.74) on page 93. The column heading L^1 indicates the rate r associated with the L^1 norm of the error, with similar interpretation of the other headings (a denotes the energy norm). The simulations are based on uniform grids with 4-node bilinear elements.

β	h	L^1	L^2	L^∞	a	H^1
-10	1.2500e-01	1.95	1.92	1.86	0.96	1.02
-10	6.2500e-02	1.99	1.98	1.96	0.99	1.01
-10	3.1250e-02	2.00	1.99	1.99	1.00	1.00
-10	1.5625e-02	2.00	2.00	2.00	1.00	1.00
0	1.2500e-01	2.04	1.99	1.88	0.99	0.99
0	6.2500e-02	2.01	2.00	1.97	1.00	1.00
0	3.1250e-02	2.00	2.00	1.99	1.00	1.00
0	1.5625e-02	2.00	2.00	2.00	1.00	1.00
1000	1.2500e-01	3.04	3.02	2.90	1.10	1.02
1000	6.2500e-02	2.96	2.94	2.83	1.01	1.00
1000	3.1250e-02	2.83	2.79	2.65	1.00	1.00
1000	1.5625e-02	2.55	2.49	2.41	1.00	1.00

2.10.7   Adaptive Finite Element Methods

Computing the solution to a desired level of accuracy and at the same time minimizing the computational resources, is the ultimate goal of numerical simulation. Adaptive finite element methods constitute a means for reaching this goal. To measure the accuracy, we need estimates of the error in the numerical solution. For this purpose, *a priori* error estimates of the type we have met in Theorems 2.18 and 2.19 can be applied. The error bounds involve, unfortunately, the *unknown* solution u of the continuous problem. Alternative *a posteriori* error estimates can be derived, where the error bounds involve the computed solution u_h.

Table 2.3. See the caption of Table 2.2 for details. The difference here is that the simulations make use of 9-node biquadratic elements.

β	h	L^1	L^2	L^∞	a	H^1
-10	1.2500e-01	3.19	3.04	3.05	1.96	1.99
-10	6.2500e-02	3.06	3.02	3.06	1.99	2.00
-10	3.1250e-02	3.02	3.00	3.06	2.00	2.00
-10	1.5625e-02	3.01	3.00	3.04	2.00	2.00
0	1.2500e-01	2.99	2.88	2.65	1.98	1.98
0	6.2500e-02	3.00	2.97	3.02	2.00	2.00
0	3.1250e-02	3.00	2.99	3.03	2.00	2.00
0	1.5625e-02	3.00	3.00	3.02	2.00	2.00
1000	1.2500e-01	2.67	2.55	2.42	2.32	2.05
1000	6.2500e-02	2.91	2.88	2.94	2.29	2.02
1000	3.1250e-02	2.98	2.98	2.97	2.12	2.00
1000	1.5625e-02	3.00	3.00	3.00	2.03	2.00

Basic Ideas of Adaptive Algorithms. Suppose we can compute the exact error e_K in element K. If the overall target error for the computation is ϵ, we want that $\sum_K e_K \leq \epsilon$. Having a finite element mesh, we would like to construct a new mesh, with the minimal number of nodes, such that $\sum_K e_K$ meets the target error. This is actually a constrained nonlinear optimization problem. To facilitate its solution, we can introduce an iteration over successively finer grids. First, we need to *estimate* or bound the true error. The error bound in such estimates normally have the form $\mathcal{E} = \sqrt{\sum_K \mathcal{E}_K^2}$, where $\mathcal{E}_K^2$ is the contribution from element K. The estimates are only valid as the sum, but it is common to use the local components $\mathcal{E}_K^2$ of the sum as indicators for the refinement of individual elements. That means that we locally employ estimates of the form $e_K^2 \leq \mathcal{E}_K^2$. Of course, $\mathcal{E}_K$ should be a computationally attractive formula.

We let $\mathcal{T}^j$ be the mesh in iteration j, that is, after j refinements of the initial mesh. Furthermore, let $m(\mathcal{T}^j)$ be the number of elements in $\mathcal{T}^j$. We can then apply the values of $\mathcal{E}_K$ in the current mesh $\mathcal{T}^j$ to select the elements to be refined. Applying a mesh refinement algorithm yields a new mesh $\mathcal{T}^{j+1}$. The selection of elements to be refined is often based on the principle that the error should be uniformly distributed throughout the mesh. Thus, we aim at having $\mathcal{E}_K^2 \leq \epsilon^2/m_{\text{opt}}$, where m_{opt} is the number of elements in the final (optimal) mesh. A natural consequence is that an element K is marked for refinement if $\mathcal{E}_K^2 > \epsilon^2/m(\mathcal{T}^j)$. The iteration is stopped when the total estimated error $\mathcal{E}$ is less than the target error ϵ, or when the number of elements in the current mesh, $m(\mathcal{T}^j)$, exceeds a prescribed maximum value m_{max}. For practical computations we often choose $\epsilon = \eta \|u_h\|_a$, where η is a given tolerance for the *relative* error in the global energy norm, if the energy norm is used for the estimate; otherwise we use alternative quantities to make ϵ dimensionless. Another popular refinement strategy is to refine a given percentage of the elements with the highest $\mathcal{E}_K$ values. Algorithm 2.2 lists the basic steps in adaptive finite element computations.

Algorithm 2.2.

Adaptive finite element computation.

choose initial mesh $\mathcal{T}^0$
for $j = 0, 1, 2, \ldots$ until $\mathcal{E} \leq \eta \|u_h\|_a$ or $m(\mathcal{T}^j) \leq m_{\max}$
    compute u_h using current mesh $\mathcal{T}^j$
    compute estimator $\mathcal{E}_K$ in each element
    if the total error $\mathcal{E} = \sqrt{\sum_K \mathcal{E}_K^2} > \eta \|u_h\|_a$ then
        refine the elements K for which $\mathcal{E}_K > \eta \|u_h\|_a / \sqrt{m(\mathcal{T}^j)}$

Sophisticated local mesh refinement algorithms are needed to carry out the subdivision of elements into new elements and construct a new finite element grid. The plain application of such algorithms can be quite user-friendly as we demonstrate in Chapter 3.6.

The algorithm above involves *refinement* only. If $\mathcal{E}_K$ is less than the target value in the element, one could instead *coarsen* the element, that is, merge it with some of its neighbors. Especially in time-dependent problems this can be an economical strategy.

A Priori Error Estimates and Adaptive Mesh Refinement. An error estimate like (2.189),

$$|u - u_h|_{H^1(\Omega)} = \|\nabla u - \nabla u_h\| \leq Ch|u|_{H^2(\Omega)}$$

might act as a starting point for defining $\mathcal{E}$ as $Ch|u|_{H^2(\Omega)}$. The estimated error in element K could then be taken as $\mathcal{E}_K = Ch|u|_{H^2(K)}$. Notice that this formula involves the unknown *global* solution u. To get an approximation of this quantity, we can attempt to numerically integrate the second-order derivatives of u_h over the element. This requires numerical approximation of the second-order derivatives based on ∇u_h in a patch of neighboring elements. However, it would be better to work with an estimate $\mathcal{E}_K$ that involves the approximation u_h and only its derivatives in element K. Certain a posteriori error estimates meet these demands.

An A Posteriori Error Estimate Based on the Residual. We consider the Dirichlet problem

$$-\nabla^2 u = f \quad \text{in } \Omega, \text{ with } u = 0 \text{ on } \partial\Omega. \tag{2.195}$$

From Example 2.12 we know that $a(u, v) = \int_\Omega \nabla u \cdot \nabla v \, d\Omega$, $L(v) = \int_\Omega fv d\Omega$, the relevant space V is $H_0^1(\Omega)$, and properties 1-6 are fulfilled. Moreover, we have a priori error estimates from Theorems 2.18 and 2.19.

Theorem 2.20. A posteriori error estimate in energy norm. *The numerical error $e = u - u_h$ in the finite element solution of the problem (2.195) fulfills*

$$||e||_a \leq \mathcal{E},$$

where the error estimator $\mathcal{E}$ is defined by

$$\mathcal{E}^2 = \sum_K \left(\alpha \int_K h^2 R(u_h)^2 d\Omega + \beta \int_{\partial K} h_S [\boldsymbol{n} \cdot \nabla u_h]^2 d\Gamma \right).$$

Here, K is an element, $R(u_h) = |f + \nabla^2 u_h|$ is the residual in an element, the notation $[g]$ denotes the jump of a quantity g over a side, h is the size of the element, h_S is the size of a side, $\boldsymbol{n}$ is the outward unit normal of a side of the element, and α and β are interpolation constants that depend on the element type. The integral over ∂K is omitted on boundaries with Dirichlet values (u prescribed and $e = 0$).

Proof. We start with

$$||e||_a^2 = a(e, e) = a(u, e) - a(u_h, e) = L(e) - a(u_h, e).$$

We can add $a(u_h, v) - L(v)$ ($= 0$) for any $v \in V_h$ and obtain

$$||e||_a^2 = L(e - v) - a(u_h, e - v).$$

The purpose now is to integrate the bilinear term by parts:

$$a(u_h, e - v) = \int_\Omega \nabla u_h \cdot \nabla (e - v) d\Omega$$
$$= -\sum_K \int_K (e - v) \nabla^2 u_h \, d\Omega + \sum_K \int_{\partial K} (e - v) \frac{\partial u_h}{\partial n} d\Gamma,$$

where K denotes an element. The normal derivative of the solution, $\partial u / \partial n$, is continuous over the element boundaries, but $\partial u_h / \partial n$ is in general *not* continuous across elements; with linear triangles we know that the derivatives of u_h are constant in each element and thus discontinuous over the element boundaries. For each common boundary S between two elements we can introduce the *jump* in $\partial u_h / \partial n$ by the notation $[\boldsymbol{n} \cdot \nabla u_h]$. Assuming that $e - v$ is continuous across S, we can write

$$\sum_K \int_{\partial K} (e - v) \frac{\partial u_h}{\partial n} d\Gamma = \sum_S \int_S [\boldsymbol{n} \cdot \nabla u_h](e - v) d\Gamma.$$

In the computation of these quantities it is convenient to have the S-integral along each side of an element. This can be accomplished by distributing the jump $[\boldsymbol{n} \cdot \nabla u_h]$ equally between the two side integrals in the two elements that share side S:

$$\sum_S \int_S [\boldsymbol{n} \cdot \nabla u_h](e - v) d\Gamma = \sum_K \int_{\partial K} \frac{1}{2} [\boldsymbol{n} \cdot \nabla u_h](e - v) d\Gamma.$$

We can then summarize,

$$||e||_a^2 = \sum_K \int_K (f + \nabla^2 u_h)(e - v)d\Omega + \sum_K \int_{\partial K} \frac{1}{2}[\boldsymbol{n} \cdot \nabla u_h](e - v)d\Gamma.$$

The next step is to estimate the size of the integrals. The standard procedure is to choose $v = I_h e$, because $e - v = e - I_h e$ can then be estimated using interpolation results. Notice that $I_h e$ is continuous across element boundaries, so that $e - v = e - I_h e$ is also continuous, which was a requirement when forming the jump expression above. Let

$$I_K = |\sum_K \int_K (f + \nabla^2 u_h)(e - I_h e)d\Omega|.$$

A slight rewrite gives

$$I_K = |\sum_K \int_K \left(h(f + \nabla^2 u_h)\right)\left(h^{-1}(e - I_h e)\right)d\Omega|.$$

Introducing the residual $R(u_h) = f + \nabla^2 u_h$ on a triangle K, applying the Cauchy-Schwarz inequality, then using the estimate (2.188), we can bound I_K:

$$I_K \leq \left(\sum_K ||hR(u_h)||_{L^2(K)}^2\right)^{\frac{1}{2}} ||h^{-1}(e - I_h e)||_{L^2(\Omega)}$$

$$\leq C \left(\sum_K ||hR(u_h)||_{L^2(K)}^2\right)^{\frac{1}{2}} ||\nabla e||_{L^2(\Omega)}.$$

The side integrals of the jumps can be estimated in a similar fashion [35,36], but the steps are more technical (involving trace inequalities). The final result takes the form

$$||e||_a^2 \leq \sum_K \left(\alpha \int_K h^2 R(u_h)^2 d\Omega + \beta \int_{\partial K} h_S[\boldsymbol{n} \cdot \nabla u_h]^2 d\Gamma\right).$$

□

Theorem 2.20 is valid in any number of space dimensions and higher-order elements. A possible choice of h is $\Delta^{1/d}$, where $\Delta = \int_K d\Omega$ and d is the number of space dimensions. Examples on calibrated values for α and β are $\alpha = 0.1$ and $\beta = 0.15$ (in the case of linear elements).

Considering a PDE $-\nabla \cdot [\lambda \nabla u] = f$ with $u = 0$ on the boundary and with a variable coefficient $\lambda(\boldsymbol{x})$, possibly with large discontinuities (located at the element boundaries), the error estimator $\mathcal{E}$ can be generalized to

$$\mathcal{E}^2 = \sum_K \left(\alpha \int_K \lambda^{-1}h^2 R(u_h)^2 d\Omega + \beta \int_{\partial K} h_S(\lambda^- + \lambda^+)^{-1}[\lambda \boldsymbol{n} \cdot \nabla u_h]^2 d\Gamma\right).$$

$$(2.196)$$

Now, $R(u_h) = |f + \nabla \cdot (\lambda \nabla u_h)|$. Furthermore, λ^+ and λ^- are the limiting values of λ at the common side in two neighboring elements.

The ZZ Estimator. The ZZ (Zienkiewicz-Zhu) estimator [131,132] has become a popular error indicator for lower-order elements. Suppose the model equation is $-\nabla \cdot (\lambda \nabla u) = f$. The energy norm is then

$$||v||_a^2 = a(v,v) = \int_\Omega \lambda \nabla v \cdot \nabla v \, d\Omega = ||\boldsymbol{p}||_{L^2(\Omega),\lambda^{-1}}^2, \quad \boldsymbol{p} = -\lambda \nabla v.$$

Here,

$$||\boldsymbol{p}||_{L^2(\Omega),k}^2 \equiv \int_\Omega k\boldsymbol{p} \cdot \boldsymbol{p} \, d\Omega$$

is a weighted L^2-norm of the vector $\boldsymbol{p}$. The original idea was to postulate that

$$||e||_a^2 \le C \sum_K \int_K ||\boldsymbol{q}^* - \boldsymbol{q}_h||_{L^2(K),\lambda^{-1}}^2, \qquad (2.197)$$

where C is a global constant, $\boldsymbol{q}_h = -\lambda \nabla u_h$ is the computed flux in an element, and $\boldsymbol{q}^*$ is a more accurate flux. The corresponding error indicator becomes $\mathcal{E}_K^2 = ||\boldsymbol{q}^* - \boldsymbol{q}_h||_{L^2(K),\lambda^{-1}}^2$. The concept of a flux is not limited to the $\nabla \cdot \lambda \nabla$ operator. For example, in elasticity $\boldsymbol{q}$ is typically the stress ($\boldsymbol{\sigma}$ in the notation of Chapter 5.1.3).

A very simple choice of $\boldsymbol{q}^*$ is the smoothed version of $\boldsymbol{q}_h$, using the Galerkin, least-squares, or L^2 projection methods in Chapter 2.8.1. A better estimator arises when $\boldsymbol{q}^*$ is based on the so-called super convergent patch recovery procedure [12,133]. This method consists basically of fitting a local polynomial approximation of $\boldsymbol{q}^*$ to a set of discrete $\boldsymbol{q}_h$ values in a patch of elements. The ZZ estimator is related to certain residual estimators, see the references in [12], but its attractive feature is the potential applicability to widely different PDE problems. Unfortunately, the quality of the ZZ estimator for measuring the true error can vary greatly, see [12] and [113].

Remarks. For wide application of adaptive finite elements based on a posteriori error estimation one will need estimates in several norms, because different applications demand monitoring the error in different norms. The residual-based error estimators above can be generalized to other norms, for example, the L^∞ norm [35]. However, with the present lack of precise a posteriori error estimates in most challenging physical applications, one is often left with more intuitive approaches to adaptive computations, like refining the grid where the gradients are large or where one knows from physical insight that the mesh should be dense (e.g. close to a singularity or a moving front). In the case $\mathcal{E}_K$ measures the error, which means that we have estimated numerical values for the interpolation constants in the expressions for $\mathcal{E}_K$, the adaptive procedure gives the possibility for *error control*, that is, the solution u_h is computed with a measureable accuracy $\mathcal{E}$. When we lack precise estimates of the interpolation constants or when we lack the precise form of the error estimator itself in a given problem, we can still use the structure of an existing error estimator as a candidate formula for $\mathcal{E}_K$. This time $\mathcal{E}_K$ does not reflect the true level of accuracy in the element, but we may use the relative sizes of the $\mathcal{E}_K$ values for marking the elements to be refined. Quite simple, but physically reasonable principles, can be used to define working $\mathcal{E}_K$ formulas. For example, $\mathcal{E}_K$ can be the local mesh Peclet or Reynolds number. The rationale for this is to obtain a mesh where the local mesh Peclet or Reynolds number is uniformly distributed, cf. Chapter 2.9. Other alternatives consist in letting $\mathcal{E}_K$ be the inverse distance to a singularity or to a steep front in u_h.

There are basically four categories of refinement methods:

1. h-refinement, consisting of subdividing elements,
2. p-refinement, where one locally increases the polynomial order p of the basis functions N_j in $u_h = \sum_j u_j N_j$,
3. r-refinement, where the nodal positions are moved,
4. hp-refinement, which is a combination of h- and p-refinement.

Our present exposition has been restricted to h-refinement.

Some recommended literature for further introduction to a posteriori error estimators and adaptive finite element algorithms is Eriksson et al. [35,36] and Verfürth [122]. Babuska et al. [12] and Strouboulis and Haque [112,113] provide an overview and evaluation of a wide range of error estimators for the Poisson equation (and to a less extent also the equations of linear elasticity), with comprehensive references to the specialized literature. These references focus mainly on h-refinement methods. Szabo and Babuska [115] treat p-refinement in detail, whereas an overview of theoretical and implementational issues of hp-refinement methods is provided in papers from Oden's group [33,85].

Chapter 3

Programming of Finite Element Solvers

The present chapter explains the usage of the finite element toolbox in Diffpack. The model problems are kept quite simple in order to concentrate on programming details. In Chapters 3.1–1.7 we deal with the Poisson equation. A trivial extension to the convection-diffusion equation is exemplified in Chapter 3.8. Chapters 3.9 and 3.10 demonstrate that only minor extensions of the Poisson equation solvers are needed to handle time-dependent problems, in this case the heat equation. Finally, Chapter 3.11 brings in some new tools for particularly efficient solution of PDEs with time-independent coefficients. The required background for working with this chapter is knowledge of the finite element method (at least the material corresponding to Chapters 2.1–2.7) and basic concepts in C++ and Diffpack (at least Chapters 1.1–1.3, 1.5.1–1.5.3, and 1.5.5; Chapter 1.6 is also useful).

Only linear problems are treated in the present chapter. Finite element solvers for nonlinear PDEs constitute the topic of Chapter 4.2, where we reuse the tools from linear problems and basically only add some functionality for administering an outer iteration loop. The application of these tools to problems in fluid and solid mechanics is the subject of Chapters 5–7.

The present chapter is comprehensive, and if the aim is to quickly get an overview of how Diffpack can handle applications involving nonlinear time-dependent problems, it is suggested to read Chapters 3.1, 3.2, 3.5.10, and 3.9, before moving on to Chapter 4.2. However, the human efficiency of scientific computing projects that last for months (or years) can be significantly improved by adopting the many tools and techniques that fill up the present chapter. We recommend to view the introductory solvers `Poisson0` and `Poisson1` in Chapters 3.1 and 3.2 as pedagogical steps towards a finite element program, the `Poisson2` solver in Chapter 3.5, that has sufficient flexibility for use in research. A similar refinement of the solvers appear in the time-dependent examples as well.

3.1 A Simple Program for the Poisson Equation

We shall first deal with a scalar stationary PDE:

$$-\nabla \cdot [k(\boldsymbol{x})\nabla u(\boldsymbol{x})] = f(\boldsymbol{x}), \quad \boldsymbol{x} \in \Omega \subset \mathbb{R}^d, \tag{3.1}$$

$$u(\boldsymbol{x}) = g(\boldsymbol{x}), \quad \boldsymbol{x} \in \partial\Omega_E, \tag{3.2}$$

where $f(\boldsymbol{x})$, $g(\mathbf{x})$, and $k(\boldsymbol{x})$ are given functions, $u(\boldsymbol{x})$ is the primary unknown, and $\partial\Omega_E$ is the complete boundary[1] of the domain Ω. Our aim is to create a short program, utilizing Diffpack's finite element toolbox, to solve this problem numerically.

3.1.1   Discretization

We assume that the reader is capable of formulating a finite element method for the boundary-value problem (3.1)–(3.2). Therefore, we here simply state the element matrix and vector contributions in local coordinates, denoted by $\tilde{A}_{i,j}^{(e)}$ and $\tilde{b}_i^{(e)}$, respectively:

$$\tilde{A}_{i,j}^{(e)} = \int_{\tilde{\Omega}} k\nabla N_i \cdot \nabla N_j \det J \, d\xi_1 \cdots d\xi_d, \tag{3.3}$$

$$\tilde{b}_i^{(e)} = \int_{\tilde{\Omega}} f N_i \det J \, d\xi_1 \cdots d\xi_d. \tag{3.4}$$

Here, $\tilde{\Omega}$ is a reference element in local coordinates $\boldsymbol{\xi} = (\xi_1, \ldots \xi_d)$, N_i denotes a finite element basis function, and J is the Jacobian matrix of the mapping between the coordinates in the reference and physical domains. The integrals are usually computed by numerical integration, resulting in expressions on the form

$$\tilde{A}_{i,j}^{(e)} \approx \sum_{k=1}^{n_I} I_{i,j}(\boldsymbol{\xi}_k) w_k, \tag{3.5}$$

$$\tilde{b}_i^{(e)} \approx \sum_{k=1}^{n_I} K_i(\boldsymbol{\xi}_k) w_k. \tag{3.6}$$

The symbol $I_{i,j}(\boldsymbol{\xi}_k)$ represents the integrand in (3.3), evaluated at the numerical integration point $\boldsymbol{\xi}_k$, with a corresponding interpretation of $K_i(\boldsymbol{\xi}_k)$. Moreover, w_k is the weight at the kth integration point. When solving a problem by means of Diffpack, it is only necessary to implement the expressions $I_{i,j}(\boldsymbol{\xi}_k) w_k$ and $K_i(\boldsymbol{\xi}_k) w_k$:

$$I_{i,j}(\boldsymbol{\xi}_k) w_k = [k\nabla N_i \cdot \nabla N_j \det J]_{\boldsymbol{\xi}=\boldsymbol{\xi}_k} w_k, \tag{3.7}$$

$$K_i(\boldsymbol{\xi}_k) w_k = [f N_i \det J]_{\boldsymbol{\xi}=\boldsymbol{\xi}_k} w_k. \tag{3.8}$$

The notation $[\,]_{\boldsymbol{\xi}=\boldsymbol{\xi}_k}$ means that the expression inside the brackets is to be evaluated at integration point no. k in the reference element.

The implementation of (3.7) and (3.8) is performed in a routine

---

[1] As in Chapter 2, we use the subscript E to indicate essential boundary conditions.

```
void integrands (ElmMatVec& elmat, const FiniteElement& fe);
```

The object `fe` contains N_i, ∇N_i, $\det J \cdot w_k$ and other useful quantities, such as the global coordinates $\boldsymbol{x}$ of the current integration point. The user must provide functions for $f(\boldsymbol{x})$ and $k(\boldsymbol{x})$. The purpose of `integrands` is then to add the contributions $I_{i,j} w_k$ and $K_i w_k$ to the element matrix, represented by the `Mat(real)` structure `elmat.A`, and the element vector, represented by the `Vec(real)` structure `elmat.b`. In case of a 2D problem, $\boldsymbol{x} = (x, y)$ and

$$\nabla N_i \cdot \nabla N_j = \frac{\partial N_i}{\partial x} \frac{\partial N_j}{\partial x} + \frac{\partial N_i}{\partial y} \frac{\partial N_j}{\partial y},$$

the "heart" of a Diffpack program for solving (3.1)–(3.2) is typically coded like this in the `integrands` routine:

```
elmat.A(i,j) += k(x,y)*(fe.dN(i,1)*fe.dN(j,1)
                      + fe.dN(i,2)*fe.dN(j,2))*fe.detJxW();
elmat.b(i)   += f(x,y)*fe.N(i)*fe.detJxW();
```

This outline of `integrands` demonstrates the close connection between the Diffpack C++ code and the numerical expressions.

3.1.2 Basic Parts of a Simulator Class

In order to solve the boundary-value problem numerically by the finite element method, the user must supply a problem-dependent class containing information about the PDE, the boundary conditions, the discretization, data structures for grids and fields etc. This class is often referred to as the *solver class* or the *simulator class*. We have previously introduced this concept in Chapter 1.6, but then in the context of finite difference methods.

The simulator class must be a subclass of `FEM` when we use the finite element method for the spatial discretization. The predefined library class `FEM` contains data and default versions of algorithms that are frequently used in finite element programs. At the moment it is enough to know that the finite element "engine" is administered by class `FEM` and that some of the functions in class `FEM` need to have information about the boundary-value problem to be solved. This information is provided by functions in the simulator class.

Let the name of the simulator class be `Poisson0`. The class should provide the following main functions.

- A routine `scan` which reads information about the finite element partition of the computational domain. For simplicity, we restrict the shape of the domain to be the unit square, $\Omega = (0, 1) \times (0, 1)$. A preprocessor (grid generator) is used to generate the finite element mesh. In the `scan` function we also dynamically allocate some large objects (fields and linear systems/solvers) whose sizes depend on the grid.
- A routine `fillEssBC` which sets the essential boundary condition $u = 0$ on $\partial \Omega_E$.

– A routine `integrands` which samples the integrands in the element matrix and vector at a numerical integration point. In other words, the $I_{i,j}w_k$ and $K_i w_k$ expressions (3.7)–(3.8) are evaluated as briefly described above. The assembly process and the numerical integration are administered automatically by a function `makeSystem` in class `FEM`.

– A main routine `solveProblem` which calls the routines `scan`, `fillEssBC`, `makeSystem` (in `FEM`) and `LinEqAdmFE::solve` (a function for solving linear systems).

– A routine `resultReport` for automated reporting of selected results, e.g., the differences between the numerical and the analytical solution in a test case.

The data members in class `Poisson0` will typically be

– a finite element grid object of type `GridFE`,
– a finite element field object of type `FieldFE` for representing the scalar field u over the grid,
– an object of type `LinEqAdmFE` containing the linear system in the current problem and various solvers for linear systems,
– a degree of freedom handler object of type `DegFreeFE` that transforms the field values of u into a vector of unknowns in the linear system[2].

The definition of class `Poisson0` is placed in a header file `Poisson0.h`, which is listed below[3].

```
// Simple FEM solver for the 2D Poisson equation
#ifndef Poisson0_h_IS_INCLUDED
#define Poisson0_h_IS_INCLUDED

#include <FEM.h>            // FEM algorithms, class FieldFE, GridFE
#include <DegFreeFE.h>      // degree of freedom book-keeping
#include <LinEqAdmFE.h>     // linear systems: storage and solution

class Poisson0 : public FEM
{
protected:
  // general data:
  Handle(GridFE)       grid;    // pointer to a finite element grid
  Handle(DegFreeFE)    dof;     // trivial book-keeping for a scalar PDE
  Handle(FieldFE)      u;       // finite element field, primary unknown
  Vec(real)            linsol;  // solution of the linear system
  Handle(LinEqAdmFE)   lineq;   // linear system: storage and solution

  void fillEssBC ();            // set boundary conditions u=g
  virtual void integrands      // evaluate weak form in the FEM-equations
    (ElmMatVec& elmat, const FiniteElement& fe);
public:
```

---

[2] This transformation is simply the identity mapping in the present example. However, when solving systems of PDEs the transformation might be complicated.

[3] The `Handle(X)` construction can be read as a pointer declaration `X*` and is explained right after the class definition.

```
  Poisson0 ();
 ~Poisson0 () {}

  void scan ();            // read and initialize data
  void solveProblem ();    // main driver routine
  void resultReport ();    // write comparison with analytical sol.

  real f(real x, real y);  // source term in the PDE
  real k(real x, real y);  // coefficient in the PDE
  real g(real x, real y);  // essential boundary conditions
};
#endif
```

Let us for convenience explain most of the details of this header file, despite the fact that many of the topics are dealt with in Chapter 1. The `#include` lines have the effect of making the classes `FEM` (and implicitly `FieldFE` and `GridFE`), `DegFreeFE`, and `LinEqAdmFE` available.

In Diffpack we try to avoid primitive C/C++ pointers. Instead we use a Diffpack tool called *handle*. Actually, a handle is a normal C++ pointer with a few additional "intelligent" features. The main reason for using handles is that they make the management of dynamic memory simple and reliable. Handles were introduced in Chapter 1.6, but we briefly repeat some of the most important syntax here for immediate reference. Statements of the type

```
  Handle(X) x;
```

declare a handle x to an object of class X. This handle is initially empty, which actually means that it is a null pointer. To let a handle point to an object of class X, we use the `rebind` function:

```
  x.rebind (new X(...));
```

where the dots indicate possible arguments to the X constructor. When using handles, the programmer can create objects where desired and forget about deallocation: *The handle will automatically delete the objects when they are no longer in use.* Having a `Handle(X) x`, `*x` or `x()` denotes the X object itself (one can also write `x.getRef()` as an alternative). The `*x` (or `x()`) construction must be used in calls to functions requiring an `X&` argument.

We now turn to the contents of the functions in class `Poisson0`. The functions are written for a 2D problem, but the Diffpack tools for finite element programming make it easy to develop code that can run unaltered in 1D, 2D, and 3D (and even in higher dimensions). The statements that must be altered to obtain *one* code that runs both in 1D, 2D, and 3D, are marked with the comment `2D specific`. We will modify these statements in a more advanced version of class `Poisson0` in order to obtain "dimension-independent" code.

The algorithms of the various functions are presented first. Thereafter we show the corresponding Diffpack code.

The `scan` function:

1. Read information about the number of nodes in the x- and y directions.
2. Read information about the element type.
3. Construct input strings to the box preprocessor, specifying the domain, the partition, and the element type.
4. Call the box preprocessor to generate the grid.
5. Set `u` to point to a new finite element field over the grid.
6. Set `dof` to point to a new `DegFreeFE` object.
7. Set `lineq` to point to a new `LinEqAdmFE` object.
8. Redimension the unknown vector in the linear system and attach it to `lineq`.

The code might look like this:

```
void Poisson0:: scan ()
{
  // extract input from the command line:
  int nx, ny;  // number of nodes in x- and y-direction
  initFromCommandLineArg ("-nx", nx, 6);  // read nx, default: nx=6
  initFromCommandLineArg ("-ny", ny, 6);
  String elm_tp;
  initFromCommandLineArg ("-elm", elm_tp, "ElmB4n2D");

  // the box preprocessor requires input on the form (example):
  // geometry:   d=2 [0,1]x[0,1]
  // partition:  d=2 elm= ElmB4n2D  div=[4,4] grading=[1,1]
  String geometry  = "d=2 [0,1]x[0,1]";          // 2D specific
  String partition = aform("d=2 elm=%s div:[%d,%d] grading:[1,1]",
                           elm_tp.c_str(),nx-1,ny-1); // 2D specific

  grid.rebind (new GridFE()); // make an empty grid
  PreproBox p;                      // preprocessor for box-shaped domains
  p.geometryBox() .scan (geometry);   // initialize the geometry
  p.partitionBox().scan (partition);  // initialize the partition
  p.generateMesh (*grid);           // run the preprocessor

  u.rebind (new FieldFE (*grid,"u"));     // allocate, set name="u"
  dof.rebind (new DegFreeFE (*grid, 1)); // 1 unknown per node
  lineq.rebind (new LinEqAdmFE());       // Ax=b system and solvers
  linsol.redim (grid->getNoNodes());     // redimension linsol
  lineq->attach (linsol);                // use linsol as x in Ax=b
  // banded Gaussian elimination is the default solver in lineq
}
```

A significant portion of this function concerns string manipulation. The statements should be self-explanatory and demonstrate how one can flexibly construct strings in Diffpack. Functions like `aform` and `initFromCommandLineArg` are covered in Chapters 1.2 and 1.6. More information on grid generation and variations of the input to the box preprocessor is given in the report [64], but the syntax used in the geometry and partition strings in `scan` is explained in detail on page 228.

The `fillEssBC` function:
1. Initialize assignment of essential boundary conditions.
2. for $i = 1$ to the number of nodes in the grid:
   3. if this node is on the boundary
      4. set the essential boundary condition.

```
void Poisson0:: fillEssBC ()
{
  dof->initEssBC ();                      // init for assignment below
  const int nno = grid->getNoNodes();     // no of nodes
  Ptv(real) x;                            // a nodal point
  for (int i = 1; i <= nno; i++) {
    // is node i subjected to any boundary indicator?
    if (grid->boNode (i)) {
      x = grid->getCoor (i);              // extract coor. of node i
      dof->fillEssBC (i, g(x(1),x(2)));   // u=g at boundary nodes
    }
  }
  dof->printEssBC (s_o, 2);               // debug output
}
```

The `Ptv(real)` class is in principle a vector like `ArrayGen` and `Vec` (or rather `VecSimple`), but the `Ptv` type is optimized for vectors of length 1, 2, and 3, representing spatial points. The syntax of basic operations coincides with that of other vector classes in Diffpack. See Chapter 1.5.3 for more information about Diffpack arrays.

The implementation of other types of boundary conditions, e.g., different Dirichlet conditions or prescribed flux $-k\partial u/\partial n$, is treated in Chapters 3.5.1 and 3.5.2.

The `integrands` function:
1. Evaluate the Jacobian determinant times the integration weight.
2. Find the global coordinates of the current integration point.
3. Evaluate the f and k functions at this global point.
4. for $i = 1$ to the number of basis functions (element nodes)
   5. for $j = 1$ to the number of basis functions
      6. Add the appropriate value to the element matrix.
   7. Add the appropriate value to the element vector.

```
void Poisson0::integrands (ElmMatVec& elmat,const FiniteElement& fe)
{
  // find the global coord. xy of the current integration point:
  Ptv(real) xy = fe.getGlobalEvalPt();
  const real x = xy(1);   const real y = xy(2);   // 2D specific
  const real f_value = f(x,y);                    // 2D specific
  const real k_value = k(x,y);                    // 2D specific
  int i,j;
  const int nbf = fe.getNoBasisFunc(); // = no of nodes in element
  const real detJxW = fe.detJxW();     // Jacobian * intgr. weight
```

```
for (i = 1; i <= nbf; i++) {
  for (j = 1; j <= nbf; j++)
    elmat.A(i,j) += k_value*(fe.dN(i,1)*fe.dN(j,1) // 2D specific
                            + fe.dN(i,2)*fe.dN(j,2))*detJxW;
  elmat.b(i) += fe.N(i)*f_value*detJxW;
  }
}
```

We use declarations like `const int nbf` to indicate that that `nbf` is a constant, i.e., it gets its value in the declaration statement and will not be changed later. Looking at the argument list to `integrands`, we see that `fe` is `const` and therefore an input parameter, whereas `elmat` is an output parameter since it is not `const` and can hence be changed by the routine. Comprehensive use of `const` is an important aspect of documenting C++ codes. See Chapter 1.5.2 for a quick introduction to the `const` concept.

Looking at the `integrands` function, we see that it is not coded in an optimal way. Symmetry of the element matrix should be exploited, and we should move invariant arithmetic expressions outside the loops. Nevertheless, since we want a close relationship between the `integrands` expressions and the numerical formulas, and easy extension to PDEs leading to nonsymmetric matrices, we shall present compact and quite general `integrands` functions throughout this book. The reader should know that after a finite element solver has been thoroughly verified, the programmer can use the techniques from Appendix B.6 to improve the computational performance.

The `solveProblem` function:
1. Call `fillEssBC` to assign boundary conditions.
2. Call `makeSystem` (inherited from `FEM`) to generate the linear system.
3. Initialize the values of `linsol`.
4. Call `lineq->solve` to solve the linear system of equations.
5. Load the solution (`linsol`) into the field `u`.

Point 3 deserves perhaps a comment. Normally, iterative solution methods are used for the linear systems (see Appendices C and D). The current contents of the solution vector we attached to `lineq` (i.e. `linsol`) constitute the initial guess for the iterative solver. Therefore, it is crucial to initialize `linsol` prior to calling `lineq->solve`. Here we shall just set all the values of `linsol` to zero. In the present case, where we use a direct solver, banded Gaussian elimination, and there is no use for the start vector, performing the initialization is a good habit and makes the program ready for efficient iterative solution methods.

```
void Poisson0:: solveProblem () // main routine of class Poisson0
{
  fillEssBC ();                    // set essential boundary conditions
  makeSystem (*dof, *lineq);       // assembly algorithm from class FEM
  linsol.fill (0.0);               // init start vector (good habit)
  lineq->solve();                  // solve linear system
  dof->vec2field (linsol, *u);     // load solution (linsol) into u
}
```

Since we view u as the basic quantity we want to compute, we should load the solution of the linear system into u. In the present simple problem, the `linsol` vector and the nodal-values vector are identical and we could have written[4] `u->values()=linsol`. However, in more complicated problems, the loading of the linear system solution into a field representation may involve nontrivial bookkeeping and data shuffling. It is the responsibility of the `DegFreeFE` object `dof` to control the mapping between the field representation of the unknown functions and the vector of degrees of freedom in the linear system. The call to `dof->vec2field` will always work and is introduced here as yet another good habit.

The `resultReport` function is so simple that no further details are necessary at this stage.

Remark. In general, it is convenient to store the nodal values and the linear system unknowns in separate vectors like we have outlined (`u->values()` and `linsol`). However, for the present stationary scalar PDE, we could save the memory associated with `linsol` by letting the linear system operate directory on the nodal-values vector. That is, we could remove the `linsol` vector and associated statemements, attach `u->values()` to the `lineq` object, and initialize `u->values()` prior to calling the linear solver. It is a good exercise for novice C++ and Diffpack programmers to perform this modification to the `Poisson0` code and check that the numerical results remain unaltered.

The reader who is interested in what is going on inside Diffpack's finite element engine, and especially in the `makeSystem` function, can consult Appendix B.5.1.

As a specific example for testing the program, we choose

$$f(x, y) = -2x(x - 1) - 2y(y - 1), \quad g(x, y) = 0, \quad k(x, y) = 1.$$

It is then easy to verify that $u(x, y) = x(x - 1)y(y - 1)$ is the analytical solution of (3.1)–(3.2). The f, g, and k functions take the simple forms

```
real Poisson0:: f (real x, real y)           // 2D specific
{ return -2*x*(x-1)-2*y*(y-1); }
```

---

[4] The basic principles, content, syntax of class `FieldFE` is similar to those of class `FieldLattice` (see Chapter 1.5.5). For example, `FieldFE` objects have a function `values` for returning access to `Vec(real)` object containing the nodal point values of the field.

```
real Poisson0:: g (real /*x*/, real /*y*/)   // 2D specific
{ return 0; }

real Poisson0:: k (real /*x*/, real /*y*/)   // 2D specific
{ return 1; }
```

The arguments in the g and k function might look strange. Since neither x nor y is used in g and k, compilers will flag these arguments as "not used" and issue a warning if we do not omit the argument names or enclose them in C-style comments /* */. Notice that there are two types of comments in C++: // for comment on the rest of the current line and /* */ for arbitrary comments. We recommend applying // because this allows later use of /* */ to leave out large code sections even if they contain // comments.

Some readers may find it more appropriate to implement f, g, and k in (3.1) as global functions instead of member functions in Poisson0. Nevertheless, we claim that it is advantageous that problem-dependent explicit functions are made member functions of the problem class. The main reason is that these functions may need access to problem-dependent parameters that are data items in the class. With a separate global function, these parameters must be transferred as additional arguments or as global variables[5]. In the present example, the f, g, and k functions are so simple that we do not need the flexibility of having them as class members; we do it just to conform to a common Diffpack programming standard.

Finally, we list the small main program for the present model problem:

```
#include <Poisson0.h>
int main (int argc, const char* argv[])
{
  initDiffpack (argc, argv);
  Poisson0 simulator;          // make a simulator object
  simulator.scan ();           // read and initialize data
  simulator.solveProblem ();   // main routine: compute the solution
  simulator.resultReport ();   // compare with exact solution
  return 0;
}
```

The complete code for the Poisson0 solver is found in src/fem/Poisson0. Copy that directory to your local directory tree and compile the application (see Appendix B.2 for details). Run the program without giving any command-line arguments. This results in the default mesh with 5×5 bilinear elements and 6×6 nodes. The output of the program should conform to the file Verify/test1.r.

We do not recommend using the present program to solve similar problems, because there are too many restrictions in the code, which will soon require significant additional programming when attacking more complicated stationary PDEs. The program given in the next section is a much better starting point.

---

[5] One can alternatively use the functor concept for f, g, and k, as explained in Chapter 1.6.1.

3.2   Increasing the Flexibility

One of the strengths of Diffpack is that a few minor changes of the simple program like the one in the previous section, can increase the flexibility of the program dramatically. The gain is a simulator that makes you more productive when exploring mathematical models through numerical experiments. Another advantage of increasing the flexibility is that the code becomes easier to reuse in related problems. Our idea of this tutorial on finite element software tools is to introduce more general classes and concepts than what is strictly needed to solve the particular PDE at hand. In this way we develop a working habit of generalizing the problem formulations and its solution methods such that the implementation techniques as well as the code itself can be reused in a wide range of problems. You will experience that the stationary template solver `Poisson1` presented in this section, and in particular the extended solver `Poisson2` to be presented in Chapter 3.5, can be applied with very small adjustments in many problems involving scalar stationary PDEs. Although you might see alternative ways of programming a particular feature, we encourage you to stick to our suggested source code constructions, because these are carefully chosen to enable easy transition from simple Poisson equation solvers to simulators for complicated mathematical models.

In the `Poisson1` solver we have focused at the following extensions of class `Poisson0`:

– making the simulator valid for 1D, 2D, and 3D problems,
– using a menu system for convenient assignment of input parameters,
– storing the solution and other fields for later visualization,
– enabling access to some simple grid generation tools,
– allowing automatic execution of several combinations of input parameters, e.g., for studying parameter sensitivity or convergence,
– computing the error field and various associated norms.

The source code for the `Poisson1` solver is found in `src/fem/Poisson1`. You should have the source code at hand before continuing with the forthcoming material.

3.2.1   A Generalized Model Problem

We extend our test problem from the last section to d space dimensions, where d is a parameter that can be chosen at run time. The program can be debugged for $d = 1$, and in principle a 2D, 3D, and even higher-dimensional solver is thereafter available.

Let us choose $f(\boldsymbol{x})$, $g(\boldsymbol{x})$, and $k(\boldsymbol{x})$ in the boundary-value problem (3.1)–(3.2) such that

$$u(x_1, \ldots, x_d) = \sum_{j=1}^{d} A_j \sin(j\pi x_j).$$

Then we can take $g(\boldsymbol{x})$ as $u(\boldsymbol{x})$, set $k = 1$, and let the right-hand side be given by $f(\boldsymbol{x}) = \sum_{j=1}^{d} j^2 \pi^2 A_j \sin(j\pi x_j)$. The parameters $A_1, \ldots, A_d$ are arbitrary and can be specified by the user of the program.

By using `Ptv(real)` and `Ptv(int)` vectors for spatial points and indices, respectively, it is easy to write code that can work in any number of space dimensions. Typically, a point $\boldsymbol{x} = (x_1, \ldots, x_d)$ or a set of parameters like $A_1, \ldots, A_d$ are represented by `Ptv(real)` objects. Recall that the statements originally restricted to the two-dimensional problem of Chapter 3.1 were marked with special comments in the source code of class `Poisson0`. Generalizing these statements is a matter of introducing `Ptv` objects.

3.2.2   Using the Menu System

In Diffpack we usually apply a menu system for feeding data into the program. This saves considerable coding effort and offers enhanced flexibility.

Running an Application with the Menu System. Suppose you are in the `Poisson1` directory and have the executable `app` available. Writing just `./app` on the command line leads to a prompt, and the program waits for input data to the menu system. Commands to the menu system may be conveniently placed in a file. The file `test1.i` in the `Verify` subdirectory is an example on a very simple set of menu commands[6]:

```
set gridfile = P=PreproBox| d=2 [0,1]x[0,1]| d=2 e=ElmB4n2D [5,5] [1,1]
set A parameters = 2 1
ok
```

To run the the `Poisson1` solver with `test1.i` as input, simply write

```
./app --casename g5x5 < Verify/test1.i
```

The results that are printed on the screen can be compared with reference results in `Verify/test1.r`. In the present example, the casename `g5x5` reflects a simulation with a 5×5 grid (cf. the `gridfile` item in `test1.i`). All files generated by the Diffpack libraries during the execution will have the stem `g5x5` in their names such that they are easy to locate, move, or delete.

The forthcoming paragraphs explain how to program with the menu system and access various types of user interfaces.

Overview of Programming with the Menu System. We need to perform the following modifications of class `Poisson0` in order to utilize the Diffpack menu system:

– The various menu items must be defined in a function
```
void Poisson1::define (MenuSystem& menu, int level = MAIN);
```

---

[6] The syntax of the answer to the `gridfile` item is explained in Chapter 3.2.3.

– The menu answers must be read from the menu system and loaded into the local data structures in the simulator using the function
```
void Poisson1::scan ()
```
– The function
```
void Poisson1::adm (MenuSystem& menu)
```
administers the calls to `define` and `scan` in addition to prompting the user for input data.

– The `main` program must initialize the global menu object right after having called `initDiffpack`:
```
global_menu.init ("Flexible Poisson simulator","Poisson1");
```
– The execution of the solver can in `main` be performed by
```
global_menu.multipleLoop (simulator);
```
where `simulator` is a `Poisson1` object. The `multipleLoop` function in the menu system administers the call to `adm` for setting up the menu and initializing the simulator, the call to `solveProblem` for computing a numerical solution, and the call to `resultReport` for reporting key results from the simulation.

Details of Menu System Programming. Here is an example of typical statements in the `define` and `scan` functions.

```
void Poisson1:: define (MenuSystem& menu, int level)
{
  menu.addItem (level,            // menu level: level+1 is a submenu
                "gridfile",        // menu command/name
                "file or preprocessor command",  // help/description
                "mygrid.grid"); // default answer

  menu.addItem (level,            // menu level: level+1 is a submenu
                "A parameters", // menu command/name
                "A_1 A_2 ... in f expression", // help/description
                "3 2 1");        // default answer
  // submenus:
  LinEqAdmFE::defineStatic (menu, level+1);   // linear solver
  FEM::defineStatic (menu, level+1);          // num. integr. rule etc
  SaveSimRes::defineStatic (menu, level+1);   // storage of fields
}
```

A fundamental feature of programming with the menu system can here be seen; we add menu items related to problem-dependent data in detail, whereas menus regarding modules in the Diffpack libraries are automatically generated by the `defineStatic` (or `define`) function of library classes. By specifying the argument `level+1`, we force the menus for linear solver storage and solution, numerical integration rules, and storage of fields on file to appear as submenus.

The reading of menu items is performed in the `scan` function.

```
void Poisson1:: scan ()
{
  MenuSystem& menu = SimCase::getMenuSystem();
  String gridfile = menu.get ("gridfile");
  ...
  A.scan (menu.get ("A parameters"));    // A is a Ptv(real)
  lineq->scan (menu);                    // LinEqAdmFE reads the menu
}
```

The first statement provides access to the menu system data base[7]. The menu.get call returns a String object. Sometimes this is sufficient; one either expects a string like gridfile or one can pass the menu answer string to an object's scan function[8]. In other occasions one needs to read an int or a real from the menu. This is exemplified as follows:

```
int  n = menu.get ("number of nodes").getInt();
real r = menu.get ("diffusion parameter").getReal();
bool b = menu.get ("plot solution").getBool();
```

There are only two menu commands in the test1.i file. The menu items that are requested in scan functions, but not specified in the input file, will get their default values as answers. For example, all parameters related to linear system storage and solution remain at their default values when using test1.i as input.

User Interfaces to the Menu System. The menu system can be operated in several ways. You can put command statements in a file, you can use command-line options, or you can use a graphical user interface. A particular menu item is recognized by its command name, so when you are using files, you will typically write instructions like

```
set A parameters = 1.0 2.3  ! A_1 and A_2 in a 2D problem
```

When you use the instruction set A parameters =, the rest of the line is treated as the answer. The *first* = sign always divides the line into the command name part (before) and the answer part (after). All text after the exclamation mark ! is treated as comments.

The graphical menu interface offers in the present case a label A parameters and a text field where you can fill in the answer. Alternatively, you can assign input data through command-line arguments; writing --A_parameters '1.0 2.3' on the command line initializes A to $(1.0, 2.3)$. Sufficient information on how to interact with the menu system through the graphical interface or command-line options appears in Chapter 3.4.1.

---

[7] We do not need to write the prefix SimCase:: – this is done to explicitly state where the function is defined, since class SimCase is a base class of FEM and therefore does not explicitly appear in the class definition of finite element solvers like Poisson1.

[8] In the A.scan call, the String object returned from menu.get is automatically converted to an Is object that can be used for reading in Ptv(real)::scan(Is), see Appendix B.3.3.

Static Define Functions. A part of the Diffpack programming standard is that each class defines its own menu items. In other words, most library classes that require input data, have `define` and `scan` functions that work with a `MenuSystem` object. Sometimes the programmer wants to define the menu items of an object that is not yet created. This is the case in class `Poisson1`; when `define` is executed, the `lineq` handle is empty and there is no `LinEqAdmFE` object. The `defineStatic` function is always a *static* version [14, Ch. 6.9] of `define` that can be called without having created an object of that class. The `define` function in class `Poisson1` demonstrates a call statement to the `defineStatic` function in class `LinEqAdmFE`. In C++, one can call a static function `f()` in a class `X` by saying `X::f()`. This means that you do not need an instance `x` of `X` to perform the call (i.e. `x.f()`). The `FEM::defineStatic` call could equally well be `FEM::define`, since `FEM` is a base class of `Poisson1` and the object is fully created when issuing the call[9]. Most library classes have a static function `defineStatic` in addition to `define`.

Unfortunately, there is not yet any support for automatic detection of incompatible parameter settings. The menu system is therefore aimed at users who have a certain knowledge of Diffpack, the mathematical and numerical methods that are applied, and the particular application program. However, the flexibility of the menu system allows you to build your own user interface on top of Diffpack, e.g., with the aid of scripting languages like Tcl/Tk, Perl, or Python. An example on an interface is provided in Chapter 3.10.5. With such interfaces tailored to a particular application, you can build in intelligent choices of default values and check for compatibility of the input parameters. You can also integrate grid generation and visualization tools and create a special-purpose problem solving environment.

With the `test1.i` input file listed previously, we rely on default values for most of the menu items. Some basic questions immediately arise:

1. How can we provide our own set of default values?
2. How can we obtain documentation of the actual default values employed for all the menu answers that we do not explicitly set in the input file? For example, what type of matrix storage and linear system solver is actually implied by the default values? (You should not be surprised if Gaussian elimination on banded matrices is the answer!)
3. How can one construct a more comprehensive set of menu commands than what is exemplified in the `test1.i` file above?
4. How can we operate the graphical interface to the menu system, where we more easily can see the available menu commands?

All these questions are covered in Chapter 3.4.1, but to get started with the menu system, the self-explanatory examples provided in the present section are sufficient.

---

[9] Notice that writing only `define` implies a call to `Poisson1::define` (i.e. the `define` function of the current object). The prefix `FEM::` is therefore required to access a function with the same name in the base class.

3.2.3   Creating the Grid Object

The `gridfile` item on the menu enables reading a ready-made grid on file[10] or running a preprocessor for generating a new grid. Having extracted the menu answer and stored it in the string `gridfile`, the grid object is initialized by calling `readOrMakeGrid`:

```
readOrMakeGrid (*grid, gridfile);
```

If `gridfile` contains a string on the form

```
P=pre | geometry-string | partition-string
```

the preprocessor `pre`, usually a class in the Diffpack libraries, will be called to generate the mesh, based on information about the geometry of the domain and the partition. Strings not following the syntax above are assumed to contain the name of a grid file, which is then read by the `readOrMakeGrid` function (using `GridFE::scan`).

The Box Preprocessor. The simplest preprocessor is class `PreproBox`, which generates box-shaped grids (interval in 1D, rectangle in 2D, and box in 3D). For example,

```
P=PreproBox | d=2 [0,1]x[0,1] | d=2 e=ElmB4n2D div=[6,8] g=[1,1]
```

generates a 2D grid (`d=2`) over the unit square $[0,1] \times [0,1]$, with bilinear elements (`ElmB4n2D`), and 6×8 *intervals between nodes* in the x and y direction. The last command, `g=[1,1]`, indicates a grading or stretch factor. A unit value leads to a uniform mesh. The "variable names" `d`, `e`, `div`, and `g` given in the string above can be arbitrary; they are only meant to enhance the readability.

The grading is mathematically performed as follows. First, the box pre-processors generates a uniform mesh on the hypercube

$$[\mu_1^-, \mu_1^+] \times \cdots \times [\mu_x^-, \mu_d^+].$$

Then each node with coordinates $(z_1, \ldots, z_d)$ in this mesh is mapped onto a new point $(x_1, \ldots, x_d)$ in the same domain according to a formula

$$x_j = G(z_j; \alpha_j, \mu_j^-, \mu_j^+), \quad j = 1, \ldots, d.$$

The parameters $\alpha_1, \ldots, \alpha_d$ are the *grading parameters* in the grid specification string. For example, if $\alpha_1 = -0.8$ and $\alpha_2 = 1.6$ in a 2D problem, we could write `g=[-0.8,1.6]` in the grid specification. The value $\alpha_i = 1$ corresponds to the identity mapping. When $\alpha_i > 0$ we use the mapping function

$$G(z; \alpha, a, c) = \frac{1}{2} \text{sign}\{2z - (c+a)\}(c-a) \left| \frac{2z - (a+c)}{c-a} \right|^{1/\alpha} + \frac{1}{2}(c+a). \tag{3.9}$$

---

[10] The format must correspond to that produced by the `GridFE::print` function. There are filters from other grid formats to Diffpack's format.

This results in a grading towards the mid point ($0 < \alpha < 1$) or the end points ($\alpha > 1$) of the interval $[a, c]$. Grading towards only one side of an interval is offered by setting $\alpha_i < 0$, which implies the mapping function

$$G(z; \alpha, a, c) = \left(\frac{z - a}{c - a}\right)^{1/|\alpha|} (c - a) + a. \tag{3.10}$$

Here is a guide to the choice of α_1 for grading the nodal coordinates when $x_1 \in [a, c]$:

1. $\alpha_1 > 1$ gives symmetric refinements towards the ends of the interval, $x_1 = a$ and $x_1 = c$,
2. $0 < \alpha_1 < 1$ gives symmetric refinements towards the center of the interval, $x_1 = (a + c)/2$,
3. $-1 < \alpha_1 < 0$ gives refinements towards the left boundary, $x_1 = a$,
4. $\alpha_1 < -1$ gives refinements towards the right boundary, $x_1 = c$.

Changing the Element Type. Up to now we have used bilinear basis functions defined on quadrilateral elements. This is just one of many possible choices of elements and basis functions in Diffpack. The names of the elements reflect the element type in terms of the geometrical shape, the number of nodes, and the number of space dimensions: `ElmB2n1D` (linear 2-node 1D element), `ElmB3n1D` (quadratic 3-node 1D element), `ElmT3n2D` (linear 3-node 2D triangle), `ElmT6n2D` (quadratic 6-node 2D triangle), `ElmB8n2D` (quadratic 8-node 2D brick element), `ElmB9n2D` (biquadratic 9-node 2D brick element), `ElmB8n3D` (trilinear 8-node brick element), `ElmB20n3D` (quadratic 20-node 3D brick element), `ElmB27n3D` (triquadratic 27-node brick element), `ElmBT6n3D` (3D 6-node prism element), plus more specialized elements. The letter `B` indicates box-shaped elements, whereas `T` denotes triangles or tetrahedra.

To change the element type, simply edit the partition part of the `gridfile` answer on the input menu. Recall that the partition information is given in terms of *intervals between nodes* in each space direction. Therefore, the string `[6,6]` means 7×7 nodes regardless of whether the element type is `ElmB4n2D` (36 elements), `ElmB9n2D` (9 elements), `ElmT3n2D` (72 elements), or `ElmT6n2D` (18 elements). In this way it is trivial to keep the partition constant and experiment with different types of elements. Notice that for quadratic elements, the number of intervals between nodes must be even since there are always two intervals between nodes along each side of the element.

Playing Around with the Box Preprocessor. A Diffpack preprocessor can either be run as a result of the `gridfile` argument to the `readOrMakeGrid` function or we can run a stand-alone Diffpack program called `makegrid`. The initialization string to the `readOrMakeGrid` function and the `makegrid` program are quite similar. The particular example on usage of the box preprocessor above has the following syntax if we use `makegrid` instead of `readOrMakeGrid`.

```
makegrid --iscl -m PreproBox -g 'd=2 [0,1]x[0,1]'
         -p 'd=2 e=ElmB4n2D div=[6,8] g=[1,1]' --casename t1
```

The grid is now stored in the file `t1.grid` in the `GridFE` file format. We can plot the grid using the Plotmtv program and the Diffpack script `plotmtvgrid`:

```
plotmtvgrid t1
```

The effect of various choices of the grading parameters is illustrated by running the Perl script `boxplay.pl` in `src/bin`; it takes α_1 and α_2 as arguments, runs `makegrid`, and plots the resulting grid.

Generation of grids over more complicated geometries is addressed in Chapter 3.5.3.

3.3   Some Visualization Tools

The `Poisson1` simulator has functionality for storing computed fields on file, using the `SaveSimRes` class. The storage format of the fields is specific to Diffpack and referred to as the *simres* format. Some information about this format is given in Appendix B.4.2. To invoke a plotting program, we need to export the simres field format to a file format required by the plotting program. This is accomplished by using a *filter*. Filters exist for many popular plotting systems, e.g., Vtk, Matlab, AVS, IRIS Explorer, Plotmtv, and Gnuplot. Below we explain some visualization processes for stationary scalar and vector fields. Time-dependent fields are covered in Chapter 3.9.

3.3.1   Storing Fields for Later Visualization

To implement the `SaveSimRes` tool in a simulator class, one typically includes a data member `database` of type `Handle(SaveSimRes)` in the class. This data member is allocated and initialized in the `scan` function. To dump a field (represented by a handle u) to the database, one simply writes

```
database->dump(*u);
```

The `database` member can be used for storing complete fields of several types, including `FieldFE` and `FieldLattice` objects, fields along lines through the domain, and time series of fields at some selected spatial points.

3.3.2   Filtering Simres Data

Invoking Filters from the Command Line. Let us assume that you have run a simulation with casename `mycase`, using the `Poisson1` code. The simulation has resulted in the computation of three fields, u, `error`, and `flux`, which are stored in a simres database with name `mycase`. The field u is the solution of the PDE, `error` is the numerical error in u, and `flux` is the vector field $-k\nabla u$.

To export the u field to a specific visualization tool `xxx`, one must first run a filter

```
simres2xxx -f mycase -s -n u -a
```

to produce a file `mycase.u.xxx`. This file can be directly loaded into the visualization tool. The options of the `simres2xxx` filter have the following meaning: `-f mycase` specifies the name of the simres dataset, which is usually the casename of the run of current interest, `-s` denotes filtering of a *scalar* field, `-n u` informs the filter to search for a field with name `u`, and `-a` indicates ASCII format of the resulting `mycase.u.xxx` file (most plotting programs can also handle a binary file format, which is then produced by the `-b` option to `simres2xxx`). There is also a `-t` option for giving a set of time points at which the field `u` is to be extracted (see page 313), but in stationary problems we simply omit this option. Running `simres2xxx` without any options gives a description of the usage, including examples.

Vector fields are filtered in the same way as scalar fields, the only difference being a flag `-v` instead of `-s` to the `simres2xxx` filter:

```
simres2xxx -f mycase -v -n flux -a
```

After a simulation is run, you can get a list of all the fields that are stored on file in the simres format by looking at the file `.mycase.simres`. One will usually consult this file, either directly or through or a nice browser, for an overview of the database prior to running specific filters and invoking visualization systems. Each scalar field, including components of vector fields, in a simres database is associated with a number. This number can be used directly as (scalar) field specification in the `simres2xxx` command. If it is of interest to plot the second component of a vector field and the field number of this component is 3, one can specify the field identification by the `-r` option:

```
simres2xxx -f mycase -r 3 -s -a
```

In fact, the argument to `-r` can be a general Diffpack set (`SetOfNo::scan` syntax, see page 72): `-r '3 5 8 9;'` results in four data files, one for each of the fields with numbers 3, 5, 8, and 9.

Windows NT/95 Remark 3.1: When running a Diffpack application which utilizes the graphical user interface on Win32 platforms, there is no need to run the `simres2xxx` filters as stand-alone applications. This GUI incorporates a browser for simulation results capable of immediate rendering of the data (actually using the same code as `simres2vtk`). From the visualization options available in this GUI the user can output the computed results in different formats, such as the Vtk or VRML formats as well as the UCD format supported in AVS and IRIS Explorer. See Appendix B.2.2 for further details. The material in Chapters 3.3.3–3.3.7 is therefore of less interest to users on Win32 platforms.   ◇

A Graphical User Interface. The simres filters are often embedded in scripts to ease coupled simulation and visualization when solving a specific problem, and in such cases the command-line driven interfaces shown previously are a requirement. However, for novice users it might be more comfortable to invoke a graphical user interface. Typing `simresgui mycase` starts a simple GUI with a list of the available fields for visualization in the `mycase` dataset and a multiple-choice button for the visualization program that is requested. One can simply click on the fields to be visualized and then click Export the chosen fields to run the filter corresponding to the chosen visualization program. The name of the resulting files to be loaded into the visualization tool is written in the message window.

Test Problem. The forthcoming sections illustrate various visualization programs and refer to a simulation case `g10x10` with the `Poisson1` simulator for exemplifying the techniques. In the `Poisson1` directory, the test data are generated like this:

```
./app --casename g10x10 < Verify/testplot2D.i
```

You are then ready to explore different visualization tools.

3.3.3   Visualizing Diffpack Data in Plotmtv

The Plotmtv visualization program is free and easy to use, but not as sophisticated as Vtk, IRIS Explorer, or AVS. For 2D finite element fields, however, Plotmtv can do a pretty good job.

Windows NT/95 Remark 3.2:   The Plotmtv program is based on X-Windows and is thus available only on Unix platforms. However, the Win32 version of Diffpack is capable of generating Plotmtv data files for later transfer to a Unix machine.   ◇

Typing

```
simres2mtv -f g10x10 -s -n u -a
```

gives the output

```
scalar "u" (stationary) in dataset "g10x10" found -> g10x10.u.mtv
```

which means that the search for a scalar (`-s`) field with name u (`-n u`) in the dataset with casename `g10x10` (`-f g10x10`) was successful. The field is available in ASCII format (`-a`) on the file `g10x10.u.mtv`. You can now invoke Plotmtv with this file,

```
plotmtv g10x10.u.mtv
```

to obtain a visualization of $u(x, y)$ as in Figure 3.1a. Click on the 2D Plot field in the plot window to get a 3D view of the surface $z = u(x, y)$, see Figure 3.1b. With Left, Right, Up, and Down you can view the 3D surface from various positions. To store the plot in PostScript format on a file, click on Print to File. By default the name of the PostScript file is `dataplot.ps`. If you want a PostScript plot with color and without the date, you can invoke Plotmtv with some additional command-line options:

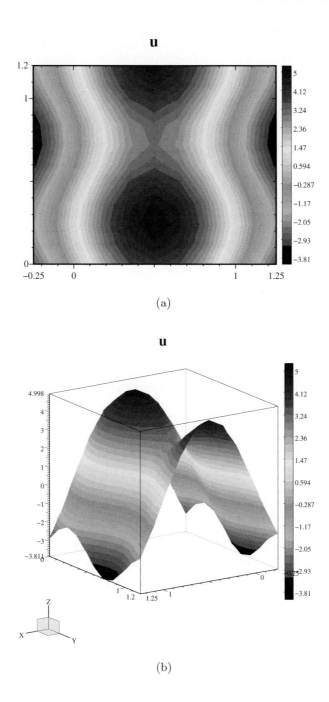

Fig. 3.1. Plot of $u(x, y)$ (computed by the `Poisson1` solver with the `testplot2D.i` input file) using Plotmtv. (a) 2D filled contours; (b) 3D elevated filled contours.

```
plotmtv -nodate -color g10x10.u.mtv
```

The Diffpack script `plotmtvps` runs Plotmtv in the background and produces a PostScript plot without any user interaction. Try

```
plotmtvps g10x10.u.ps -3d g10x10.u.mtv
```

The `-3d` Plotmtv option has the same effect as toggling the 2D Plot button.

Vector fields are filtered and plotted in the same way as scalar fields. The `Poisson1` simulator computes $-k\nabla u$ and stores the vector field in the simres database under the name `flux` (the name is evident from the `.g10x10.simres` file or the source code). Write

```
simres2mtv -f g10x10 -v -n flux -a
```

to filter $-k\nabla u$ to Plotmtv format. Invoke Plotmtv with the `g10x10.flux.mtv` file to see an arrow plot of the vector field.

All the plotting commands above and the next ones to be presented, are available in a demo script `src/fem/Poisson1/plot.pl`.

Setting Specific Plotting Options. When using Plotmtv, specifications regarding the plot must normally be placed in the datafile. The reader is encouraged to invoke an editor with the file `g10x10.u.mtv`. At the top of the file you will see several lines starting with %, indicating that these lines contain Plotmtv commands. Everything after # is considered as a comment. The `simres2mtv` filter writes several such comments to indicate alternative useful Plotmtv commands. For example, changing the value of `contstyle` from 2 to 1 results in a contour-line plot. The parameter `nsteps` controls the number of contour lines (or the number of colors when `contstyle=2`). Figure 3.2 shows the resulting plot with 15 contour lines. A wireframe plot is enabled by `contstyle=3`. From the comments in the datafile it should be clear how to change the title, comment, and labels in the plot. For further information about possible commands in the datafile, we refer to the Plotmtv manual.

Command-Line Customization of the Plot. Setting Plotmtv commands by editing the datafile is often inconvenient. The commands can alternatively be inserted by an extra option (`-o`) to the `simres2mtv` filter. For example,

```
simres2mtv -f g10x10 -s -n error
           -o '%contstyle=1 nsteps=20 subtitle="10x10 mesh"'
```

yields a plot with 20 contour lines of the error field. When you perform a large number of computer experiments, it is important to write scripts that automate the process of running the simulator and producing plots. Customized `simres2mtv` commands, using the `-o` option, are then very useful[11].

---

[11] Diffpack also offers tools for automatic generation of reports, containing plots, from a set of computer experiments, see Chapter 3.5.5.

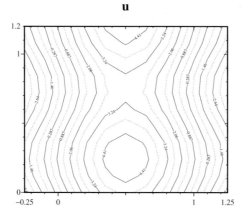

Fig. 3.2. Plot of $u(x, y)$ (computed by the `Poisson1` solver with the `testplot2D.i` input file) using Plotmtv with `contstyle=1` and `nsteps=15`.

Adjusting the Length of Vector Arrows. When plotting vector fields with Plotmtv one often wants to adjust the size of the arrows. This is accomplished by the `vscale` option in the Plotmtv datafile. The default vector scale factor, as computed by Plotmtv, is displayed in the output when Plotmtv is started. With the `g10x10` data for `flux`, the factor is approximately 0.0025. Larger arrows are obtained by increasing this value, for example by the command

```
% vscale = 0.008
```

in the `g10x10.flux.mtv` datafile or by the -o option to the `simres2mtv` filter:

```
simres2mtv -f g10x10 -v -n flux -a -o '%vscale=0.008'
```

The reader is encouraged to try this modification and see the effect on the plot.

Field Plot with Grid Overlay. Occasionally you want to plot the grid as well as the field. This is easily accomplished by adding the option -G 0 to the `simres2mtv` program[12]. For example,

```
simres2mtv -f g10x10 -s -n error -a -G 0
plotmtv g10x10.error.mtv &
plotmtv -plotall g10x10.error.mtv &
```

The first Plotmtv command results in two plots, one for the field and one for the grid. With the -plotall option the field and the grid are visualized in the same plot. The option -G starts the `drawgrid` program inside `simres2mtv`. The

---

[12] -G 0 is a special `simres2mtv` option; other visualization systems may have built-in support for plotting the grid.

value 0 indicates plotting of the grid (values greater than zero imply plotting boundary indicators). The string proceeding -G is actually any valid option to drawgrid. Just start drawgrid without arguments to get a brief description of what that program can do. Plotting the boundaries, but not the grid, is enabled by -G 99. Try this in combination with vector field plot of $-k\nabla u$, and zoom out to really see the boundary. The correct statements can be found at the end of the plot.pl script.

Plotting of 1D and 3D Fields. The Plotmtv program has limited capabilities of visualizing 3D scalar and vector fields. The only visualization technique available for 3D scalar fields is colored contours at the visible sides of the domain, provided the domain is a 3D box. Three-dimensional vector fields can be plotted for an arbitrary domain in terms of 3D arrows at the nodal points. However, such arrow plots are only useful if the number of arrows is small. Run the Poisson1 simulator with the input file Verify/testplot3D.i to produce some 3D scalar and vector field data. Suppose the casename of this run is g3D. The following session exemplifies the 3D plotting capabilities in Plotmtv:

```
simres2mtv -f g3D -s -n u -a
plotmtv -3d g3D.u.mtv
simres2mtv -f g3D -v -n flux  # can add -o '%vscale=0.002'
plotmtv -3d g3D.flux.mtv
```

During execution of the simulator, the SaveSimRes class issues a warning; the amount of field data is quite large so one should switch to binary storage format in the simres database. This is enabled by the menu commands

```
sub SaveSimRes
set field storage format = BINARY ! store fields in binary format
set grid storage format  = BINARY ! store grid in binary format
ok
```

One-dimensional fields can also be handled by Plotmtv and the simres2mtv filter. Run the simulator with the file Verify/testplot1D.i as input and g1D as casename. Then execute

```
simres2mtv -f g1D -s -n u -a
plotmtv g1D.u.mtv
```

Visualizing Diffpack Data in Gnuplot. Although Gnuplot is primarily aimed at plotting curves, $y = f(x)$, it has some capabilities of visualizing two-dimensional scalar fields defined on uniform grids. You can try the following interactive plotting session:

```
unix> simres2gnuplot -f g10x10 -s -n u -a
unix> gnuplot
gnuplot> set hidden3d; set parametric
gnuplot> splot "g10x10.u.gnu" with lines
gnuplot> exit
```

Gnuplot supports surface and primitive contour-line plots, with possible output in PostScript format. The plotting session can be run in the background by putting the commands in a file. We refer to the Gnuplot manual for documentation of the available commands.

3.3.4   Visualizing Diffpack Data in Matlab

To use Matlab for visualizing the results from a Diffpack application, we must first filter the simres data to Matlab format by executing the `simres2matlab` filter. For example,

```
simres2matlab -f g10x10 -s -n u -a
```

This results in a file `g10x10.u.mlb` with the grid and field data, plus a file `dpd.m` containing a Matlab script for plotting the Diffpack data in `g10x10.u.mlb`. Start Matlab and type `dpgui` to load Diffpack's graphical user interface for Matlab. Then run the script `dpd.m`:

```
>> dpgui
>> dpd
```

As the Diffpack data have been loaded into Matlab variables, you have now all Matlab commands *and* a GUI available at your disposal for further development of the plot.

Take a look at the `dpd.m` file and see the names of the Matlab variables holding the grid and field data. You can adjust axis, the title of the plot, the colormap, the viewing angle etc. either through the GUI, by direct Matlab commands in the terminal window, or by editing the script `dpd.m`.

Some Examples on Using the GUI. For the novice Matlab user, the GUI gives immediate access to a range of visualization features. Let us demonstrate some of the options. Changing the title is enabled by double clicking on the title string. You can now edit the title string. To insert a text in the plot, choose the Text/New item on the pulldown menu, click at the desired text location (`text` is now visible), double click on `text`, and edit the text string. The Axis menu enables the user to change between 2D and 3D plot (2D, 3D), insert an axis grid in the plot (Grid), change scales (Auto/Freeze, X Opts, Y Opts, Z Opts), define aspect ratio of the axis (Aspect), zoom (Zoom), and – perhaps most important – change the viewing angle (Viewer). Try the latter option and adjust the sliders to see the effect on the plot. The color map can be changed from the Figure/ColorMap menu. Finally, we can print the plot to file in PostScript format. Choose Save As on the File menu and then, for example, EPS Color. In a window you can fill in the desired filename. You can also generate the PostScript file from the Matlab prompt:

```
>> print -depsc 'myplot.eps';
```

Changing the Plot Type. In the present example, the underlying grid is a lattice and the default Matlab plot type is `mesh`. The field values are stored in the Matlab variable `v1`. We can easily change the plot type, try for instance

```
>> surf(v1);
>> contour(v1);
```

The selection of plot types is more limited when the finite element grid is not a lattice, but Matlab is capable of visualizing fields over any 2D geometry. We refer to the Matlab documentation for further information about the graphical capabilities. Writing just `simres2matlab` lists some extra options to the filter.

Diffpack can open a run-time communication with Matlab that can be used for computation or visualization in Matlab. See page 317 for information about this feature.

3.3.5   Visualizing Diffpack Data in Vtk

The Vtk (Visualization Toolkit) system is a comprehensive C++ library for sophisticated visualization of scientific data. The software is available in the public domain. The theory of visualization techniques in Vtk, along with introductory instructions on how to utilize the software, is available in a book [102].

Since Vtk is basically a library, one needs to write some sort of user interface to the library in order to visualize a specific set of data. For this purpose we have developed a general Tcl/Tk-based interface that allows you to explore many of the methods in Vtk for visualizing scalar and vector fields. Simply type `vtkviz` to start the interface[13].

We exemplify the features of `vtkviz` using the `g10x10` data. As usual, the simres data must first be filtered to an appropriate Vtk-readable format:

```
simres2vtk -f g10x10 -s -n u -a
```

This results in a file `g10x10.u.vtk`. Start `vtkviz` and choose Input Vtk Data from the File menu. Specify the file `g10x10.u.vtk` either by typing the name or clicking on the list of files. Type 'r' (reset) to scale the plot properly. Another plot, with colored grid lines, is enabled by typing 'w' on the keyboard. Type 's' to get filled colors back again. Elevated surfaces can be obtained by tilting the plot with the left mouse button and setting the Z scale on the front panel to a suitable value, say 0.1 for the present data.

To save the plot to file, choose Save PostScript from the File menu and fill in the filename `g10x10.u.vtk.ps`. Move the file dialog window such that it

---

[13] You need to have Vtk properly installed on your system. Moreover, the Diffpack libraries must be compiled with the particular make option `VTK_GRAPHICS=on`. The Tcl/Tk script launched by `vtkviz` is found in the file `$NOR/dp/etc/graphics/vtk/Examiner.tcl`.

does not hide parts of the plot. Press Save and use a PostScript previewer, e.g. ghostview, to examine the file g10x10.u.vtk.ps. The default background color is black, but a white background is more suitable for figures that are to be included in documents. Choose Background Color from the Options menu and adjust the red, green, and blue colors to their maximum value of 255 (implying a white background). Press Apply, save the plot in PostScript format again, and check the result.

You can play around with the View options to see various plotting capabilities. First, try Grid Outline On, which adds the edges of a grid box. With the right mouse button you can zoom in and out, whereas the button in the middle displaces the plot. Contour lines are enabled by the Contour Lines/Isosurfaces option. A second window pops up where you can fill in the number of contours and the maximum and minimum values. Setting the Z scale to zero and clicking on View from above xy-plane results in a traditional contour plot.

We can also plot the vector field $-k\nabla u$ in Vtk. Start with filtering the data,

```
simres2vtk -f g10x10 -n flux -v -a
```

to get the file g10x10.flux.vtk and then load this file into the graphical Vtk interface. The length of the arrows are adjusted on the front panel, whereas the view options work as in the scalar case. Choosing contour lines results in a plot of the magnitude of the vectors.

Vtk is well suited for visualization of three-dimensional data. Some sample data can be generated by running the Poisson1 simulator with the input file Verify/testplot3D.i. Filter the simres data for u and flux to Vtk format, load the resulting files into the vtkviz interface, and try some of the viewing options.

There is a help menu in the graphical interface to Vtk, and we refer to the items on this menu for additional information on the options that are currently available.

The strong side of Vtk is that it is a programmable system. Together with our Tcl/Tk script you can quite easily build a visualization environment to meet your own specific needs. You can also call Vtk directly from C++ in the simulator and thereby create an intimate coupling of simulation and visualization.

Windows NT/95 Remark 3.3: As pointed out in Remark 3.1 on page 231, Vtk is used for the visualization part of Diffpack's graphical user interface on Win32 platforms. For instructions on how to operate this visualization resource, see Appendix B.2.2. ⋄

3.3.6   Visualizing Diffpack Data in IRIS Explorer

The visualization system IRIS Explorer is normally operated through a user-friendly graphical interface, where the flow of data from files through visualization algorithms to render windows is programmed in terms of visual symbols, making up a *map*. To help novice IRIS Explorer users, we have made six maps for visualizing Diffpack fields. As usual, the simres files must be filtered to the appropriate format:

```
simres2explorer -f g10x10 -s -n u -a
```

This command generates a file g11x11.u.exp. Start IRIS Explorer by typing explorer and load the map view.2Dlat.map. This and other maps are found in the directory $NOR/dp/etc/graphics/explorer. The map automatically opens up a dialog box and makes a list of all the *.exp files in the current directory. Double-click on g11x11.u.exp and the render window is filled with a colored contour surface plot. The graphical interface allows you to change between discrete points, wire frames, or solid surfaces as visualization techniques. To get an elevated surface, simply adjust the scale field on the interface. The help button in the upper right corner of the main window lists information about the various parts of the user interface.

The view.2Dlat.map map is tailored to 2D fields over uniform grids consisting of bilinear or linear elements, i.e., lattice grids. Fields over 3D lattice grids can be visualized using the map view.3Dlat.map. Unstructured grids demand the maps view.2Ducd.map and view.3Ducd.map (ucd stands for unstructured cell data). Moreover, visualization of vector fields is offered by the maps view.veclat.map for 2D/3D vector fields over lattice grids and view.vecucd.map for 2D/3D vector fields over arbitrary unstructured grids. It makes sense to distinguish between lattice grids and unstructured grids, because the former class of grids allows a wider collection of visualization techniques.

Experienced IRIS Explorer users should note that the collected user interface in our maps can be ungrouped back into individual IRIS Explorer modules, so that user customization, such as reconnecting modules or including new modules, can easily be accomplished. It is also easy to visualize Diffpack fields simultaneously with other forms of data. For instance, when a user is interested in visualizing a Diffpack 2D FieldLattice together with a geometry object stored in an Inventor data file, the user should first use the map view.2Dlat.map to visualize the FieldLattice data. Then a "Read-Geom" module that reads the Inventor data file can be included in the map. Finally, connecting the output from "ReadGeom" to the "Input" port of the collected user interface "View2Dlat" will produce the desired visualization. This examples shows that the supplied maps may serve as a starting point for creating an IRIX Explorer environment according to your own needs.

3.3.7 Plotting Fields Along Lines Through the Domain

Although you can certainly hardcode some simple plotting statements inside class Poisson1 tailored to your favorite visualization tool, we strongly recommend using the SaveSimRes class and the dump function. There are numerous useful options automatically offered by class SaveSimRes: plot of fields along lines through the domain, time series plot at specified spatial points, and binary storage of data. And most important, when the simulation is finished, you are free to choose among many visualization programs.

Let us demonstrate how you can create a curve plot of a field along a line. Looking at the `Verify/testplot2D.i` input file, you can see the following lines:

```
sub SaveSimRes
set line1: start  = (-0.25,0) ! plot u along a line from start to stop
set line1: stop   = (1.25,1.2)
ok
```

These commands define a plot of u along a straight line from the start to the stop point as assigned on the menu. In the simulator, the plotting of u along lines is activated by the call `database->lineCurves(*u)`. At present, you can define a maximum of three such lines through the grid.

The curve plots of fields along lines are added to a curve plot database and can be visualized using Diffpack tools like `curveplotgui` or `curveplot`. We refer to Chapter 1.3.3 for a short introduction and to Appendix B.4.1 for more information about curve plot databases and associated tools. Here we just give some alternative Unix commands for visualizing the curve plot of u:

```
curveplot plotmtv -f .g10x10.curve.map -r '.' 'u' 'line1'
curveplot xmgr    -f .g10x10.curve.map -r '.' 'u' 'line1'
curveplot gnuplot -f .g10x10.curve.map -r '.' 'u' 'line1'
curveplot gnuplot -f .g10x10.curve.map -r '.' 'u' 'line1' -ps u1.ps
```

The x axis in the plot always reflects one of the x, y, or z coordinates of the points along the line. The reader is encouraged to define a `line2` in the input file and visualize u along this line as well.

3.4 Some Useful Diffpack Features

3.4.1 The Menu System

A Graphical User Interface. Instead of feeding the program from an input file, we can use the Tcl/Tk-based graphical user interface to Diffpack's menu system. To see this facility in action, the reader may try to run the program with the following command-line options[14]

```
./app --GUI --Default Verify/test1.i
```

---

[14] To activate the GUI, it may be necessary to compile the application using `Make MODE=opt GUIMENU=tcl`. The `GUIMENU` option might be turned on automatically for all applications in the file `$NOR/bt/src/MakeFlags`. If the application does not link successfully, the Diffpack library `bt2` must be compiled with the make option `GUIMENU=tcl`.

where the option `--Default` (or just `-D`) is used to set default values of selected menu items based on standard menu commands in a file. In this case we use the file `Verify/test1.i` for providing default answers. You can of course omit the `--Default` option; in that case the graphical menu will contain the default values as they are hard-coded in the `Poisson1::define` function and in the `define` functions of the classes in the Diffpack libraries. The `--GUI` option tells Diffpack to create a graphical window with the menu, see Figure 3.3 for an example[15]. Just point in the relevant window and write or select the menu answer. When all menu items are set as desired, click on Run to continue with the numerical simulation.

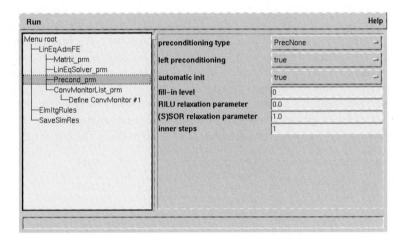

Fig. 3.3. Example on the Tcl/Tk graphical interface to the menu system. Shown here is a submenu for preconditioning techniques (see Appendix C.3).

Windows NT/95 Remark 3.4: When using Diffpack on a Win32 platform, the menu system is embedded in a flexible graphical user interface that can easily be put on top of any Diffpack simulator. This GUI also provides a browser for computational results and integrated visualization, see Appendix B.2.2 for further details.   ◇

Menu Answers on the Command Line. There is also a command-line mode of the menu system, indicated by the option `--iscl` (input source: command line). For example,

```
./app --iscl --A_parameters '1.2 -4.6' --Default Verify/test1.i
```

---

[15] The graphical interface to the menu system is programmed in C++ bindings to the Tcl, Tk, and Tix tools, a fact that makes it easy to customize the interface (e.g. include buttons for visualization).

In this case, all menu items are given default values according to the file test1.i, and we override the $A_1, \ldots, A_d$ parameters by an explicit command-line option. Notice that the command-line option has a prefix -- and a name equal to the menu command, but with blanks replaced by underscores. Also note that the answer 1.0 2.3 must be enclosed in quotes; the menu system assumes that the answer is a single string.

We also mention that there are overloaded versions of MenuSystem::addItem where you are allowed to specify your own name for the command-line options as well as giving a description of valid answers (see page 282). You can also couple a menu item to callback functionality (see the man page for class MenuSystem).

Example on More Comprehensive Input Files. Increasing the number of nodes beyond a few thousand results in such large linear systems that the default solver, which is Gaussian elimination on banded matrices, becomes too slow. More efficient *iterative* solvers should then be invoked. Useful theoretical background for iterative solvers is provided in Appendix C, while Appendices D.3–D.6 give information on the Diffpack tools for iterative solvers that the user can play with through the menu system in a finite element simulator. The input file Verify/testsolvers.i exemplifies the necessary commands for specifying a useful all-round solver for the present model problem (having symmetric positive definite coefficient matrix): the Conjugate Gradient method with relaxed incomplete LU factorization as preconditioner and a sparse matrix storage scheme.

```
set gridfile =P=PreproBox|d=2 [0,1]x[0,1]|d=2 e=ElmB4n2D [20,20] [1,1]
set A parameters = 2 1
sub LinEqAdmFE                        ! submenu for linear systems/solvers
 sub Matrix_prm                       ! submenu for matrix storage formats
  set matrix type = MatSparse         ! matrix storage format = sparse matrix
  !set matrix type = MatBand          ! banded matrix
  ok                                  ! quit the submenu
 sub LinEqSolver_prm                  ! submenu for linear solvers
  set basic method = ConjGrad         ! Conjugate Gradient iterative method
  !set basic method = GaussElim       ! Gaussian elimination
  ok                                  ! quit the LinEqSolver_prm  menu
 sub Precond_prm                          ! invoke preconditioner submenu
  set preconditioning type = PrecRILU  ! choose RILU preconditioning
  set RILU relaxation parameter = 0.0  ! 0 gives ILU, 1 gives MILU
  ok                                      ! quit Precond_prm  submenu
 ok                                   ! quit LinEqAdmFE submenu
ok                                    ! quit main menu and continue execution
```

From the comments one can see how to edit the file in order to explicitly specify Gaussian elimination on a banded matrix as equation solver.

Exercise 3.1. Set up a series of numerical experiments for the Poisson1 solver where you investigate the relative efficiency of banded Gaussian elimination versus the preconditioned Conjugate Gradient method as a function of the number of nodes. Find the break-even point for each grid. Remember to compile the application in optimized mode. To measure the CPU time, you can either (i) apply the command-line option `--verbose 1`, (ii) use an operating system functionality (like the `time` command on Unix systems), (iii) examine the end of the `casename.dp` file, or (iv) employ Diffpack's `CPUclock` class inside the code (cf. the man page for `CPUclock`). The latter approach, or the option `--verbose 1`, allows you to monitor the CPU time consumption in the `lineq->solve()` operation only. The results from this exercise should motivate you for studying Appendix C!   ◇

Automatic Generation of an Input File. The `InputMenuFile` script offers automatic generation of an input file to the menu system. The application must be run at least once before you can try

```
InputMenuFile > all.i
```

The answers in `all.i` reflect the answers in the most recent run of the application. All menu items that are available in the application appear in this file, but you can usually rely on the automatically generated values for all items that you do not understand. Explanations of the different commands can be obtained by generating the input data manual as explained next.

Automatically Generated Manual for Input Data. When you use the Diffpack menu system it is easy to generate a nicely formatted report describing all menu items or a table containing the input data used for the most recent run of the program. After a successful execution of the program, write

```
InputManualInLaTeX
```

and you will get a message indicating that two new files have been produced: `manual.tex` and `table.tex`. Run LATEX 2_ε on the file `manual.tex` and look at the result. The other file, `table.tex`, contains a table of the input data. You can remove the comment sign from the first five lines and the last line in `table.tex`, or you can run `InputManualInLaTeX -et` to produce a `table.tex` file ready for compilation by LATEX 2_ε.

3.4.2   Multiple Loops

The menu system enables the user to assign multiple values to the menu items and run the program repeatedly for all combinations of these answers. The feature is most easily explained through an example:

```
set gridfile = P=PreproBox| d=1 [0,1]| d=1 elm={ ElmB2n1D & ElmB3n1D }
              div={ [4] & [8] & [16] & [32] } grading=[1]
set A parameters = { 2 1 & 10 1 }
```

The strings separated by & inside brackets {} are interpreted as multiple values. In this example, we give two element types, four divisions, and two sets of (A_1, A_2) parameters, which results in $2 \times 4 \times 2 = 16$ combinations of the input data. If we use `MenuSystem::multipleLoop` to manage the execution of the solver, this function will automatically set up a loop over the 16 combinations and call `scan`, `solveProblem`, and `resultReport` inside the loop. Each time the `scan` function is called, a new combination of the menu answers are made available in the menu system.

We will refer to the automatically computed loop over multiple sets of input data as a *multiple loop*. Examples on input files that result in multiple loops can be found in `Verify/testdDelms.i`, where d can be 1, 2, or 3.

Inside the multiple loop, the casename will be automatically modified to `casename_mX`, where X is the value of the iteration counter in the loop. Suppose the original casename[16] in the current example is `gb`. After the simulation is finished, you will see several files, including

```
gb.dp gb.files gb_m01.ml gb_m02.ml ... gb_m16.ml
```

The `gb.files` file contains a list over all files generated by the Diffpack libraries in this simulation. The `gb_m02.ml` file contains the particular values of the varied menu answers in run no. 2 in the multiple loop. Looking at the files starting with a dot in the name, we see, e.g.,

```
.gb_m01.field .gb_m02.field ... .gb_m16.field
```

containing the field values of the fields that were dumped in the various runs. To plot the results from run no. 9 one can execute a filter `simres2xxx` with the option `-f gb_m09` for specifying the name of the simres dataset.

The execution time might be substantial when running multiple loops, so it is important to compile the program in optimization mode and to carefully count the number of multiple answers.

In Chapter 3.5.5 we equip the simulator with a nice report for presenting the results. Such automatic generation of result reports in combination with multiple loops are very useful tools in experimental scientific computing.

3.4.3 Computing Numerical Errors

Numerical studies often involve estimation of discretization errors and how they, hopefully, decrease with decreasing element size. To simplify such studies, a special class `ErrorNorms` has been made. The `ErrorNorms` class contains some static functions for computing the L^1, L^2, and L^∞ norms of the error $e = u - \hat{u}$, that is, the difference between the analytical (u) and the numerical ($\hat{u}$) solution over a grid. See page 203 for precise definitions of these norms. The appropriate call for calculating the L^1, L^2, and L^∞ norms of the error is

---

[16] The original casename is always available in the global string `casename_orig`, while the modified casename is stored in the global string `casename`.

```
ErrorNorms:: Lnorm (uanal, *u, DUMMY, L1_err, L2_err, Linf_err,
                    GAUSS_POINTS);
```

The variable `uanal` represents a *functor* implementing the analytical solution (see Chapter 3.4.4), and `L1_err`, `L2_err`, `Linf_err` are `real` variables containing estimated norms at the return of the `Lnorm` function. The `Lnorm` function is static, which means that we do not need an object to call the member function, but you must prefix the call by `ErrorNorms::`. The argument `DUMMY` is just a dummy constant for the point of time when the `uanal` functor is to be evaluated. Since we have a stationary problem, this argument is of no interest[17].

The computation of the error norms is performed by numerical integration over each element. The final argument to `Lnorm`, `GAUSS_POINTS`, reflects that Gauss-Legendre quadrature is to be used in the integration. The argument `GAUSS_POINTS` can be replaced by `NODAL_POINTS` which means that the nodes are used as integration points. The error norm `Linf_error` will then coincide with the L^∞- or max-norm of the nodal values vector. Occasionally one experiences super-convergence of the solution at the nodal points, and error norms based solely on nodal points can then be somewhat misleading.

Besides the norms of the error, it can be of interest to examine the error field itself. This is easily accomplished by declaring a `Handle(FieldFE) error` data member in the header file and create the field like we create u. When the solution is computed and available in u, we can compute the error field by the call

```
ErrorNorms::errorField (uanal, *u, DUMMY, *error);
```

Again, the `DUMMY` constant refers to the point of time used to evaluate the analytical solution `uanal`. We can store the error field for later visualization by simply calling `database->dump(*error)`.

Remark. Class `ErrorNorms` can also compare a field to a reference field, made in another simulation, for instance, with a much finer mesh. There are also other norms available, such as the H^1 and (user-defined) energy norms that appear frequently in theoretical finite element error estimates (see Chapter 2.10.6 for an example on various error-norm computations).

3.4.4 Functors

The `ErrorNorms::Lnorm` routine, which is supposed to work with a function reflecting the analytical solutions of a problem, must be implemented with a variable for this function. This requires that all analytical solution functions have the same signature, i.e., the same arguments and return values. Class `ErrorNorms` assumes that the analytical solution function takes a spatial

---

[17] See page 308 for how to deal with this argument in time-dependent problems.

point (Ptv(real)) and a time parameter (real) as arguments and returns a
real variable. In C and C++ we can easily introduce a function pointer [14,
Ch. 17.1] to represent such a function in the argument list of other functions:

```
typedef real (*analsol)(const Ptv(real)& x, real time);
```

Now, analsol is a name for all such functions, and one can, inside a function
that takes an analsol argument, say analsol(x,t) when it is desired to evaluate the analytical solution. If we try to implement a function in accordance
with analsol for our present analytical solution, we run into a problem. The
analytical solution needs the $A_1, \ldots, A_d$ parameters as well as the space-time
point. This is of course a common situation in C and Fortran as well, and the
standard solution is to introduce global variables that represent $A_1, \ldots, A_d$.
However, this is not good programming practice. In C++ we can devise an
elegant solution to the problem.

Instead of using a pure function, we create a class that contains user-
specific data and a virtual function with a fixed signature. This implementation of a function is called a *functor*[18] and we touched this idea in Chapter 1.6.1. The function pointer is then replaced by a base class pointer to
a hierarchy of such functors. The ErrorNorms functions can then take the
base class pointer as argument and thereafter access the analytical solution
through the base class pointer and an associated virtual function.

Functors representing functions of space and time are derived from class
FieldFunc in Diffpack. In the present example, the data members of the functor are either copies of $A_1, \ldots, A_d$ from the simulator or a pointer to the
simulator class, such that we can access $A_1, \ldots, A_d$. A pointer to the simulator class is safest and most flexible; with only one pointer we can access
all problem dependent parameters. The virtual function in class FieldFunc
that evaluates the function of space and time is called valuePt. Inside class
ErrorNorms one typically says f.valuePt(x), where f is a FieldFunc reference
and x a spatial point. Here is an example on how our analytical solution
function can be implemented as a functor:

```
class Ell1AnalSol : public FieldFunc
{
  Poisson1* data;        // access to simulator
public:
  Ell1AnalSol (Poisson1* data_) { data = data_; }
  virtual real valuePt (const Ptv(real)& x, real /*t*/ = DUMMY);
};

real Ell1AnalSol:: valuePt (const Ptv(real)& x, real /*t*/)
{
```

---

[18] The Standard Template Library (STL) in C++, as well as the book by Barton
and Nackman, use the term *function objects* instead of functors and let the corresponding function in the function object have the name operator(), see [82,
p. 78] and [14, Ch. 17.3]. However, Diffpack functors are used beyond the scope
of function objects and therefore follow other naming conventions.

```
const int nsd = x.size();
real sum = 0;
for (int j = 1; j <= nsd; j++)
    sum += data->A(j)*sin(j*M_PI*x(j)); // M_PI=3.14...see math.h
return sum;
}
```

Observe that the code is short and that much more complicated analytical expressions, involving perhaps dozens of physical parameters from the simulator class, can be trivially implemented in the same framework. Hopefully, we have illustrated that every time you need a function as argument to a function, you should use a functor.

Alternatively, we can often make a quicker implementation of the analytical solution functor by letting the `valuePt` function be a member function of class `Poisson1`. This means that `Poisson1` must also be derived from `FieldFunc`. Then there is no need for communication between the functor and the simulator, because the functor *is* the simulator. This approach will be demonstrated in the `Poisson2` solver (see class `Poisson2anal` and its subclasses). In general, a separate functor class gives the clearest code – letting the simulator itself be the functor has more the flavor of a quick hack.

3.4.5   Computing Derivatives of Finite Element Fields

The derivatives of a finite element field $\sum_{j=1}^{n} u_j N_j(\boldsymbol{x})$ are directly calculated as

$$\frac{\partial u}{\partial x_k} = \sum_{j=1}^{n} u_j \frac{\partial N_j}{\partial x_k}, \quad k = 1, \ldots, d.$$

The corresponding software tools in Diffpack work in terms of a `FieldFE` object, containing the u_j values, the grid, and the basis function information, and a `FiniteElement` object, containing the values of N_j and its derivatives at a point inside an element. The field value is computed by `u->valueFEM(fe)`, if `u` is a `FieldFE` handle and `fe` is a `FiniteElement` object. The `fe` object must be initialized, that is, evaluated at a point in an element. This is automatically done by many of the library routines, which means that in functions like `integrands` and `derivedQuantitiesAtItgPt` (described later) the `fe` object is ready to use. Manual initialization of a `FiniteElement` object at a point is explained in the man page for that class.

The values of the derivatives of a field handle `Handle(FieldFE)` `u` are computed by `u->derivativeFEM(gr,fe)`, where `gr` is of type `Ptv(real)`, representing the gradient of u as a vector in $\mathbb{R}^d$. The `valueFEM` and `derivativeFEM` functions are frequently used in `integrands` functions, which is exemplified in Chapters 3.9 and 4.2.

If you do not know the element number and the local element coordinates of the point where the derivatives are to be evaluated, you can simply call `u->derivativePt(x)`, which returns a `Ptv(real)` containing the gradient

at the arbitrary spatial point x. This function is dramatically slower than
FieldFE::derivativeFEM.

An inconvenient fact is that the derivatives of standard finite elements
are continuous inside an element, but discontinuous on the boundary between the elements. Hence, the derivatives become multiple valued at a node
that is shared among several elements. This calls for some smoothing procedure and a continuous finite element field representation of the derivatives,
especially if the derivatives are to be visualized. Smoothing of derivatives
of finite element fields is briefly covered in Chapter 2.8. A popular smoothing method for a possibly discontinuous field q is based on a least-squares
or Galerkin formulation, or equivalently an L^2 projection, of the equation
$w = q$, where w is a smooth finite element approximation to q. This leads to
a linear system with the mass matrix $\int_\Omega N_i N_j d\Omega$ as coefficient matrix and a
right-hand side

$$\int_\Omega q N_i d\Omega . \tag{3.11}$$

The mass matrix is usually lumped to simplify and speed up the computations. Diffpack tools for smoothing derivatives are described next.

Flux Computation. The simplest way of computing a smooth `flux` field $-k\nabla u$
is to call FEM::makeFlux like this:

```
FEM::makeFlux (*flux, *u);
```

where the scalar field u is a Handle(FieldFE) object and the vector field flux
is a Handle(FieldsFE) object. The makeFlux function needs information about
the coefficient k in the PDE and assumes that the solver, being a subclass of
FEM, implements a virtual function

```
real k (const FiniteElement& fe, real time = DUMMY);
```

representing the k function. Notice that the first argument to k is not a
spatial point, which could be expected, but a FiniteElement object. This
gives greater flexibility and efficiency in more complicated problem settings.
The time parameter is of course only relevant in time-dependent problems
and is set to a dummy value if it is redundant.

Sometimes the coefficient k in the PDE is represented as a finite element
field kf. One can then skip the implementation of the k function and instead
call an overloaded makeFlux function that makes use of the kf field directly:

```
FEM::makeFlux (*flux, *u, kf);
```

This call is valid for other field representations of k as well, e.g., kf can be
a constant field or a functor. In Chapter 3.11.4 we provide more information
about the general field concept and related tools in Diffpack.

The flux field $-k\nabla u$ is computed by the following algorithm. First, the
derivatives are computed at the optimal sampling points in each element.

Recall that the optimal sampling points for $\partial N_j/\partial x_k$ coincide with the *reduced Gauss-Legendre integration points*. For example, when working with bilinear elements we normally employ a 2×2 Gauss-Legendre rule, and the reduced rule has one integration point (the centroid). The flux computation proceeds with multiplying the gradient by k at each sampling point, which yields a set of discrete values of $-k\nabla u$. We then compute the integral in (3.11) using a reduced Gauss-Legendre rule that samples discrete values of $q = -k\partial u/\partial x_r$, $r = 1, \ldots, d$. Multiplying the right-hand side vector (3.11) by the inverse lumped mass matrix yields a smooth representation of $-k\partial u/\partial x_r$. The d components of the flux are represented by a handle `flux` to a `FieldsFE` object. Class `FieldsFE` is just an array of handles to `FieldFE` objects so that vector fields in Diffpack can reuse all functionality for scalar fields. We might evaluate, for instance, the second component of the `flux` vector at node 1 using the syntax `flux()(2).values()(1)`. The syntax `flux()` is equivalent to `*flux`, see page 84, and we could have written `(*flux)(2)` instead of `flux()(2)`. Evaluation at a non-nodal point is enabled by `flux()(2).valueFEM` or `flux()(2).valuePt`.

At this point we make a digression and comment on the C++ technicalities of the statement `flux()(2).values()(1)`. This is actually a quite complicated compound C++ statement: `flux()` calls `Handle(FieldsFE)::operator()` and returns a `FieldsFE&` reference, which is used to call the subscripting operator `FieldsFE::operator(int)`, returning access to a `FieldFE` object (second component in the vector field), which is used to call `FieldFE::values()`, returning access to a `Vec(real)` object containing the nodal values, which is finally indexed by the `Vec(real)::operator(int)` function (or more precisely, this is an inherited function from the base class `VecSimplest(real)`). Such compound statements demonstrate the flexibility and compactness of the C++ language and the Diffpack library design. Unfortunately, few compilers are able to optimize such compound statements. If you have computationally intensive loops containing this type of statements, it is wise to extract a reference to the `Vec(real)` object before the loop starts and then index this vector directly. In fact, optimal efficiency could be guaranteed by extracting the underlying C array in `Vec(real): real* v = flux()(2).values().getPtr1()`. (A loop over `v` would now be as efficient as pure low-level C code can be.)

Smoothing of a General Field. Class `FEM` contains several ready-made functions, having the name `smoothField`, for smoothing a possibly discontinuous scalar field. The field can be any subclass of `Field` (see Chapter 3.11.4) or a functor representing the integrand qN_i in (3.11) (see Appendix B.5.2). Some versions of the `smoothFields` function can smooth possibly discontinuous vector fields of type `FieldsPiWisConst` (fields that are constant over each element) or `FieldsFEatItgPt` (fields consisting of discrete values at integration points in each element).

Suppose you want to smooth some expression containing derivatives (flux, stress, strain etc.) of finite element fields. We first compute the expression at discrete points, e.g., the reduced Gauss-Legendre points, and save the values in a `FieldsFEatItgPt` object. Thereafter we can smooth the computed values by calling a suitable function in class `FEM`. Here is an example:

```
// Aim: compute Handle(FieldsFE) smooth_vec
FieldsFEatItgPt vec;
vec.derivedQuantitiesAtItgPt (*this, *grid, nfields, GAUSS_POINTS, -1);
FEM::smoothFields (*smooth_vec, vec);
```

The `nfields` parameter reflects the number of components in the vector field `vec`. The final argument `-1` indicates *reduced* integration, using a Gauss-Legendre rule (indicated by `GAUSS_POINTS`). Putting the last two arguments equal to `NODAL_POINTS` and `0` results in evaluating the vector field at the nodes in each element. The `vec.derivedQuantitiesAtItgPt` function assumes the existence of a subclass of `FEM` with a virtual function `derivedQuantitiesAtItgPt` for evaluating the vector field at an integration point in an element. Such a function must therefore be implemented in the solver class[19]. We refer to the `Elasticity1` solver in Chapter 5.1 for an example on the implementation of the `derivedQuantitiesAtItgPt` function.

Moving Least-Squares Smoothing. When using the `makeFlux` function or the `FieldsFEatItgPt` class, one can choose between the simple global Galerkin (or least-squares or L^2 projection) smoothing method and a method known as *moving least-squares smoothing* [116]. The latter approach consists in fitting a linear or quadratic polynomial to the discrete values in a patch of elements and then interpolating the polynomial for finding nodal point values of a finite element field. We use a variant of the moving least-squares method where we fit a polynomial to the *derivatives*. The derivatives at the reduced integration points are computed from a standard finite element formula. Class `MovingLS` supports the moving least-squares method in Diffpack.

There is an enum type in class `FEM` for distinguishing between different smoothing methods. The global Galerkin-based smoothing method has the name `FEM::GLOBAL_LS`, whereas the moving least-squares method bears the name `FEM::MOVING_LS`. The enum values enter the final argument to `makeFlux` (which is `FEM::GLOBAL_LS` by default). Hence, to apply moving least-square smoothing of the flux, one can just call

```
makeFlux (*flux, *u, FEM::MOVING_LS);
```

The `Poi2flux` class, which is a part of the `Poisson2` solver, contains a test problem where both the global least-squares and the moving least-squares smoothing methods are compared (see Chapter 2.8.2).

---

[19] `derivedQuantitiesAtItgPt` is a virtual function specified by class `FEM`, similar to `integrands` and `k`.

3.4.6   Specializing Code in Subclass Solvers

Suppose you want to solve the Poisson equation, but with slightly different boundary values and right-hand side. As an example, consider the classical *torsion* problem [56]: $\nabla^2 u = -2$ in a 2D domain Ω, with $u = 0$ on the boundary. The most straightforward way to develop a solver for this problem is to take a copy of the class `Poisson1` files and edit the `f` and `g` functions. The disadvantage with such an approach is that you get two almost identical versions of the source code. In case you later improve the efficiency of the original `Poisson1` class, the optimizations will not available to the torsion problem solver. Fortunately, using inheritance and virtual functions in C++, the torsion problem solver can be implemented in a subclass `Poisson1T` of `Poisson1` such that the `Poisson1` files are reused without any physical editing, but the `Poisson1T` subclass must provide new definitions of the functions `f` and `g`. In other words, the subclass reimplements only the parts of the base class that cannot be completely reused.

Creating a Subclass Solver. The `Poisson1T` torsion solver can use everything in class `Poisson1`, except the `f` and `g` functions and the error computation in `Poisson1::resultReport`[20]. A fundamental quantity in the torsion problem is $\|\nabla u\|$, which reflects the size of the stress vector (u is in fact only a "stress potential" and not of physical significance). We therefore need to compute $\|\nabla u\|$, a quantity not provided by class `Poisson1`. The `solveProblem` would normally be a natural place to perform the $\|\nabla u\|$ computation.

Letting `Poisson1T` be a subclass of `Poisson1` means that it inherits all functionality in class `Poisson1`. The special features of `Poisson1T` is that it needs to reimplement are the `f`, `g`, `resultReport`, and `solveProblem` functions. The torsion problem demands new expressions in `f` and `g`, `resultReport` can be empty, and `solveProblem` should carry out the steps in `Poisson1::solveProblem` and in addition compute the magnitude of the `flux` field ∇u. Because the reimplemented functions are *virtual*, the `fillEssBC` and `integrands` functions in `Poisson1` will *automatically* call the version of `f` and `g` in class `Poisson1T`. This is in fact the central idea of object-oriented programming. In a similar manner, `MenuSystem::multipleLoop` will also automatically call the modified `solveProblem` and `resultReport` functions. The magic behind this behavior of C++ is explained in more detail on page 278.

Class `Poisson1T` can now be declared as follows[21]:

---

[20] Except for very simple geometries, the analytical solution of the torsion problem is not available.

[21] Some C++ standards warn against defining virtual functions as inline, that is, including the function body in the header file, because C++ does not know at compile time which virtual function that will be called and inlining becomes impossible. Nevertheless, including the function body in the header file saves some typing and contributes to quick documentation of the class.

```
class Poisson1T : public Poisson1
{
public:
  virtual void resultReport () {}
  virtual void solveProblem ();
  virtual real f(const Ptv(real)& /*x*/) { return -2; }
  virtual real g(const Ptv(real)& /*x*/) { return  0; }
};
```

The only function not provided in the header file is `solveProblem`, which can be implemented like this:

```
void Poisson1T:: solveProblem ()
{
  // we need everything that Poisson1::solveProblem does:
  Poisson1::solveProblem ();

  Handle(FieldFE) stress = new FieldFE (*grid, "stress");
  flux->magnitude (*stress);
  database->dump (*stress);    // enable plotting of stress
}
```

We also need a new main function, similar to that used for the `Poisson1` simulator, but with the name `Poisson1` substituted by `Poisson1T`. As we see, the code in class `Poisson1T` is very short.

Applications with Source Code in Two Directories. The `Poisson1T` simulator needs to have its own `main` function and therefore needs to be located in its own directory, but we also need the `Poisson1` files when compiling `Poisson1T`. There are two ways to solve this problem *without copying the* `Poisson1` *files*: (i) we can make links to the `Poisson1` files in the `Poisson1T` application directory, or (ii) we can instruct the make program to also compile files in the `Poisson1` directory when compiling the `Poisson1T` solver. The latter technique is the easiest and safest. By default, the `Make` command copies and links all C++ files that it can find in the current directory. If we run the Diffpack script `AddMakeSrc` with `dir` as argument, `Make` will also compile all the C++ files in the `dir` directory, except for the file `main.cpp`, and link the corresponding object files to the current application. In the present case we have installed the `Poisson1T` in the subdirectory `torsion` of the `Poisson1` application directory and issued the command `AddMakeSrc ..` to extend the set of files that `Make` pays attention to (recall that `..` denotes the parent directory, here `Poisson1`).

For the reader with knowledge of makefiles, we remark that the `AddMakeSrc` command adds some instructions to the makefile. It is a basic principle in Diffpack that one should never edit the `Makefile` file, but perform customization of makefiles in the `.cmake2` and `.cmake1` files that are found in every Diffpack application directory. `AddMakeSrc` therefore appends make commands at the end of `.cmake2`.

Windows NT/95 Remark 3.5: When using Visual C++, the way to handle multiple source code directories is to interactively add the files to the project definition.   ◇

The source code of the torsion simulator is found in the subdirectory `torsion` of `src/fem/Poisson1`. Run the application and plot the solution, for example, by the following set of commands:

```
./app --iscl --gridfile 'P=PreproBox | d=2 [-0.25,1.25]x[0,1.2] |
                         d=2 e=ElmB4n2D [10,10] [1,1]'
simres2mtv -f SIMULATION -s -n u -a
simres2mtv -f SIMULATION -s -n stress -a
plotmtv SIMULATION.u.mtv SIMULATION.stress.mtv
```

There are contents of class `Poisson1`, e.g. the `A` parameters and their associated menu item, that are not relevant for the torsion problem. Having one class aimed at a general Poisson problem, one subclass for the original test problem, and another subclass for the torsion problem, is a better design of the code. That is, we put problem-dependent data and functions like `A`, `f`, `g`, and `k` in the subclasses. Such a design is realized in the `Poisson2` class and explained in detail in Chapter 3.5.6.

3.5   Introducing More Flexibility

The purpose of this section is to add more flexibility to our Poisson equation solver and thereby demonstrate the usage of many general and useful Diffpack tools. As a slight redesign of class `Poisson1` is advantageous, we replace class `Poisson1` by a new class `Poisson2` with several specialized subclasses. The extensions in class `Poisson2` cover

- inclusion of a boundary term (line/surface integral) in the weighted residual statement for incorporating Neumann and Robin conditions,
- a report module for automatic report generation in ASCII, LATEX 2_ε, and HTML format,
- a simulator hierarchy where closely related problems can share as much common code as possible, and where subclasses can specialize more general base class solvers,
- permanent debug output that can be turned on and off at compile time,
- empirical estimation of convergence rates.

We shall also discuss general Diffpack tools for setting boundary conditions and generating finite element grids. As mentioned previously in this chapter, class `Poisson2` is the recommended Diffpack template program for scalar stationary problems. Users who want even more flexibility can consult the `CdBase` hierarchy of convection-diffusion solvers in Chapter 6.1.

At this stage, some readers experienced with the details of finite element programming in general might appreciate having a brief overview of the more primitive objects that are used by the high-level FEM, FieldFE, and LinEqAdmFE classes. This is provided in Appendix B.5.3.

The target boundary-value problem in Poisson2 is quite general:

$$-\nabla \cdot [k(\boldsymbol{x})\nabla u(\boldsymbol{x})] = f(\boldsymbol{x}), \quad \boldsymbol{x} \in \Omega \subset \mathbb{R}^d, \tag{3.12}$$

$$u = D_1, \quad \boldsymbol{x} \in \partial\Omega_{E_1}, \tag{3.13}$$

$$u = D_2, \quad \boldsymbol{x} \in \partial\Omega_{E_2}, \tag{3.14}$$

$$u(\boldsymbol{x}) = g(\boldsymbol{x}), \quad \boldsymbol{x} \in \partial\Omega_{E_3}, \tag{3.15}$$

$$-k\frac{\partial u}{\partial n} = 0, \quad \boldsymbol{x} \in \partial\Omega_N, \tag{3.16}$$

$$-k\frac{\partial u}{\partial n} = \alpha u - U_0, \quad \boldsymbol{x} \in \partial\Omega_R. \tag{3.17}$$

Here, D_1, D_2, α, and u_0 are constants, while f, k, and g are prescribed functions. Moreover,

$$\partial\Omega = \partial\Omega_{E_1} \cup \partial\Omega_{E_2} \cup \partial\Omega_{E_3} \cup \partial\Omega_N \cup \partial\Omega_R,$$

with $\partial\Omega_{E_1}, \partial\Omega_{E_2}, \partial\Omega_{E_3}, \partial\Omega_N, \partial\Omega_R$ being non-overlapping, but some of these boundary parts can be empty.

We recommend to have a source code listing of the files in the directory src/fem/Poisson2 available while reading the forthcoming sections.

3.5.1 Setting Boundary Condition Information in the Grid

The Poisson0 and Poisson1 solvers employed the same boundary condition over the whole boundary. Normally, different conditions are applied at different parts of the boundary. The present section explains Diffpack tools for flexible assignment of various boundary conditions.

Boundary conditions in a finite element simulator are of two types, either natural or essential conditions. The former type of conditions involve integrals along (parts of) the boundary, whereas essential conditions involve manipulation of linear systems and specification of nodal values. Implementation of the boundary conditions requires knowledge of (i) the sides of an element that are subjected to a certain type of natural conditions, and (ii) the nodes in the mesh that are subjected to a particular type of essential condition.

The Concept of Boundary Indicators. Boundary information is in Diffpack represented by a concept called *boundary indicators*. A set of q binary-valued boundary indicators is introduced, their values being *on* or *off*. The q boundary indicators are defined at all the nodes in the mesh, with values that can vary among the nodes. For example, at one node, boundary indicators 1, 2, and q can be *on*, the others being *off*, whereas at another node, only indicators 1 and $q - 1$ can be *on*. If boundary indicator i is *on* at a node, we say that the node is marked with indicator i.

Usually, a boundary indicator is associated with a particular boundary condition. As an example, indicator 1 can be used to mark nodes subjected to a homogeneous Dirichlet condition $u = 0$. To find the nodes at which we should implement $u = 0$, we simply run through all nodes and check if a node is marked with boundary indicator 1. If so, special actions for assigning an essential condition must be taken. Any `GridFE` object has a function `bool boNode(n,i)` to check if node `n` is marked with boundary indicator `i`.

The boundary indicators make it very simple to implement essential boundary conditions. When it comes to natural conditions, we need to check if a side in an element has a boundary indicator. We define a side in an element as marked with a certain boundary indicator if all nodes on that sides are marked with the particular indicator. The function `bool boSide(s,i)` in class `FiniteElement` is used to check if side `s` in an element is marked with boundary indicator `i`.

The boundary indicators are given logical names. These names have no direct use in the libraries, but in some circumstances they might help to make the output from a program more readable.

Default Boundary Indicators. The box preprocessor (`PreproBox`) produces grids that have the shape of a box in $\mathbb{R}^d$. As a default set of boundary indicators for the box domain, it is natural to mark each of the $2d$ sides of the box with an indicator. The side-numbering convention is that side $i = 1, \ldots, d$ has its normal vector directed along the x_i axis, whereas the side $i = d + 1, \ldots, 2d$ is recognized by having its normal pointing in the negative direction of the x_{i-d} axis. For example, in a 2D rectangle-shaped domain $\Omega = (a_1, a_2) \times (b_1, b_2)$, side 1 corresponds to $x = a_2$ and is marked with boundary indicator 1. Side 2 is given by $y = b_2$ and marked with indicator 2. Indicator 3 marks side 3, $x = a_1$, while the final indicator 4 marks side 4, $y = b_1$[22]. The mesh generation tools introduced in Chapter 3.5.3 produce other default settings of boundary indicators, simply because the domain shapes may differ from a box.

The default boundary indicators might be appropriate even if they do not correspond directly to the set of boundary conditions. For example, in the `Poisson0` and `Poisson1` solvers we had the condition $u = 0$ on the whole boundary. A single boundary indicator marking all nodes on the boundary would be a natural choice in this case, but the 2D grid was generated by the box preprocessor, with default boundary indicators for each of the four sides. Nevertheless, checking if a node is subject to $u = 0$, that is, if a node is on the boundary, is straightforwardly accomplished by checking if *any*

---

[22] The same side numbering convention also applies to the sides in box-shaped finite elements, see the FAQ [63].

boundary indicator is turned *on* at a node. This is exactly what the function `GridFE::boNode(i)` does, whereas `GridFE::boNode(i,j)` checks if node i is subjected to the specific indicator j.

Redefining Boundary Indicators. Because the default boundary indicators are seldom appropriate, Diffpack supports a mapping from one set of indicators to another. Suppose we have a 2D grid over the unit square, with four default boundary indicators, marking each of the sides. On $x = 0, 1$, we have $u = 0$, while $\partial u/\partial n = 0$ on the two other sides. We can then either use one indicator in total, marking the essential condition $u = 0$ only, or we can use two indicators, one for each condition, although the natural Neumann condition does not require any actions in the code. It is recommended to have all points on the boundary marked with at least one boundary indicator, because these indicators are often used for more than just assigning boundary conditions; they are also convenient for visualizing the domain. We therefore introduce two indicators in this problem. The indicators can be assigned logical names, which are here chosen as `u=0` (indicator 1) and `Neumann0` (indicator 2). We assume that the grid is generated by the `PreproBox` tool, leading to four default indicators. The new indicator 1 is made up of the old indicators 1 (side 1, $x = 1$) and 3 (side 3, $x = 0$), whereas the new indicator 2 is the union of the old indicators 2 ($y = 1$) and 4 ($y = 0$). The mapping onto our new set of indicators is accomplished by the statement

```
grid->redefineBoInds ("n=2 names= u=0 Neumann0 1=(1 3) 2=(2 4)");
```

Two new indicators are here introduced (`n=2`), their names being `u=0` and `Neumann0`. The new indicator 1 consists of the old indicators 1 and 3, and the new indicator 2 is made up of the old indicators 2 and 4.

The boundary indicator mapping string is conveniently given on the menu and thereafter fed into the `redefineBoInds` function. The `Poisson2` class offers this functionality.

We also mention that the `makegrid` utility met on page 229 enables redefinition of boundary indicators, as well as extending boundary indicators and adding material definitions, according to the forthcoming descriptions, on a menu. Generation and manipulation of grids are in this case performed with `makegrid` outside the simulator, which then only needs to load the grid file.

Extending Boundary Indicators. Sometimes the mapping of default boundary indicators onto a new set does not give enough flexibility in defining suitable indicators in a problem. The `GridFE::addBoIndNodes` functions makes it possible to mark new nodes with one of the existing boundary indicators. Suppose we solve an equation on the unit square and want to mark the boundary segments $y = 0$, for $x > 1/2$, and $x = 1$ with indicator 1 and the rest of the boundary with indicator 2. If the grid is generated by a box preprocessor, we first map the four default indicators onto two new indicators, with indicator

1 preserved and indicator 2 equal to the old indicators 2, 3, and 4. Thereafter we extend indicator 1 to also cover nodes on $y = 0$ for $x > 1/2$. These
statements perform the tasks:

```
grid->redefineBoInds ("n=2 names= bo1 bo2 1=(1) 2=(2 3 4)");
grid->addBoIndNodes  ("n=1 b1=[0.5,1]x[0,0]");
```

The last statement has the following interpretation. We want to extend *one*
boundary indicator (n=1) with new nodes, and that is boundary indicator
1 (the number following b in b1, here 1, specifies the boundary indicator in
question). All nodes inside the prescribed domain $[0.5, 1] \times [0, 0]$, which is an
interval in this case, will be marked with indicator 1. The observant reader
will now claim that the nodes on $y = 0$, $x > 1/2$, have *two* indicators, 1 and
2, turned on. This calls for very careful coding of the fillEssBC function since
the DegFreeFE object can only hold a single essential boundary condition at
each node. A safe strategy is therefore to remove indicator 2 from $y = 0$,
$x > 1/2$, enabled by calling the addBoIndNodes function again, but this time
with a third argument that equals false or OFF for removing nodes or true/ON
for adding nodes (default):

```
grid->addBoIndNodes ("n=1 b2=[0.5,1]x[0,0]", OFF);
```

Again, it can be convenient to read the string argument to addBoIndNodes
from the menu, like we do in the Poisson2 class[23].

When addBoIndNodes determines which nodes that lie inside the given
domain, it uses a tolerance. The numerical value of the tolerance is 10^{-6} by
default, but it can be changed to, e.g., 10^{-4} by the option

```
--tolerance 1.0e-4
```

on the command line when executing the program. This means that when we
specify a domain with extent $[0, 0]$ in the x_2 direction, the code will actually
work with an interval $[-10^{-4}, 10^{-4}]$. The tolerance is used for many other
purposes as well, e.g., in comparison operators, see page 549.

Sometimes one wants to mark nodes with a new indicator, but restrict the
attention to nodes that are already at the boundary (i.e. marked). This is enabled by putting an optional capital B in front of the hypercube specifications
in the string argument to addBoIndNodes:

```
grid->addBoIndNodes("n=2 b1=B[0.5,1]x[0,0] b3=B[0.1,0.2]x[0.5,1]");
```

The B option is useful when marking new nodes on curved boundaries, see
Chapter 5.1.5 for an example.

---

[23] We use two menu items, one for adding indicators and one for removing indicators.

Boundary Indicator Convention in a Simulator. The boundary-value problem (3.12)–(3.17) operates with five parts of the boundary. It is therefore natural to introduce a convention for the numbering of boundary indicators in the `Poisson2` solver. We have that boundary indicator 1 represents $\partial\Omega_{E_1}$, indicator 2 represents $\partial\Omega_{E_2}$, indicator 3 represents $\partial\Omega_{E_3}$, indicator 4 represents $\partial\Omega_N$, whereas indicator 5 represents $\partial\Omega_R$. The following code, taken from `Poisson2::fillEssBC`, demonstrates how the boundary indicator numbering convention directly affects the values of u at the boundary:

```
dof->initEssBC ();                    // init for assignment below
const int nno = grid->getNoNodes() ;  // no of nodes
Ptv(real) x;                          // a nodal point
for (int i = 1; i <= nno; i++) {
  // is node i subjected to any Dirichlet value boundary indicator?
  if (grid->boNode (i, 1))
    dof->fillEssBC (i, dirichlet_val1);
  if (grid->boNode (i, 2))
    dof->fillEssBC (i, dirichlet_val2);
  if (grid->boNode (i, 3)) {
    x = grid->getCoor (i);            // extract coord. of node i
    dof->fillEssBC (i, g(x));
  }
}
```

In a particular problem we then need to map the default indicators, set by the preprocessor, to the appropriate set of five indicators used by the `Poisson2` functions. If we fail to map the default boundary indicators, we might actually end up using other boundary conditions than we intend to in a simulation. However, when generating the grid by more flexible preprocessors than what we normally apply in this text, the correct boundary indicators are set during the grid generation [64].

Other Applications of Indicators. Boundary indicators are in Diffpack implemented using class `Indicators`. This class is quite general and allows the programmer to work with an arbitrary number of indicators. Class `GridFE` has an instance of `Indicators` for marking elements, besides the `Indicators` object used for the boundary indicators. This enables easy specification of overlapping subdomains of the grid as required in certain domain decomposition methods (the elements marked with a particular indicator make up one subdomain).

Class `GridFE` allows the user to have several `Indicators` objects for various sets of boundary indicators. To use a particular set, one just attaches the relevant `Indicators` object to the grid object. See the man page for class `GridFE` and Chapter 7.2 (in particular page 484) for more information about this feature.

The Material Concept. Many applications involve a PDE defined over a domain containing different materials with different physical properties. To handle such cases, we must be able to partition the domain into a non-overlapping set of "materials". Some grid generation tools [64] allow flexible definition of materials, but one can also define box-shaped material subdomains, even on a one-material grid, through the function `GridFE::addMaterial`. Here is an example:

```
grid->addMaterial ("mat=2 [0,0.2]x[0.5,1]");
```

All elements whose centroids are inside the subdomain $[0, 0.2] \times [0.5, 1]$ will be marked with material number 2 (`mat=2`). Full flexibility in defining materials is offered by the `setMaterialType` function in class `GridFE`, which takes the element number and its corresponding material number as paramters. By default, all elements have material number 1. Checking the material number of an element is done by the function

```
int GridFE::getMaterialType (int element) const;
```

It is now easy to extend our Poisson equation solvers `Poisson1` or `Poisson2` to cover a two-material domain with two corresponding values of $k(\boldsymbol{x})$. Simply define a second material and implement a k function on the form

```
real Poisson1::k (const FiniteElement& fe, real t)
{
  if (grid->getMaterialType (fe.getElmNo()) == 1)
    return k_val1;
  else
    return k_val2;
}
```

We remark that a function `k(Ptv(real)& x)` would be inconvenient now, since this would require locating the element containing the point x, which is a time-consuming process. In general we find it more flexible to let functions like k and f take a `FiniteElement` argument instead of a spatial point.

A plot of the partitioning of the domain into material regions is enabled by calling `database->dumpMaterials(*grid)` and then filtering the simres field `materials` to an appropriate plotting program format.

3.5.2 Line and Surface Integrals

Consider a Poisson equation $-\nabla \cdot [k\nabla u] = f$ with a Neumann boundary condition,

$$-k\frac{\partial u}{\partial n} = p,$$

or a Robin condition like in (3.17),

$$-k\frac{\partial u}{\partial n} = \alpha u - U_0,$$

where k, p, U_0, and α are prescribed quantities.

In a typical Galerkin finite element formulation of this problem, the Neumann or Robin condition is incorporated through a boundary integral term:

$$\int_{\partial\Omega_N} N_i\, p\, ds \quad \text{or} \quad \int_{\partial\Omega_R} N_i \left(\alpha \sum_j N_j u_j - U_0 \right) ds,$$

over the part $\partial\Omega_N$ or $\partial\Omega_R$ of the boundary where the condition applies. This boundary term is a surface integral in 3D, a line integral in 2D, or just two point evaluations in 1D. We notice that the Neumann condition is a special case of the Robin condition with $\alpha = 0$. It is therefore sufficient to show how one implements the Robin condition. Since the unknown parameters u_j enter the Robin condition, we get contributions to both the element matrix $(p\alpha N_i N_j)$ and to the right-hand side $(N_i p U_0)$. The Neumann condition gives no contribution to the element matrix. From an implementational point of view, integration over sides is very similar to integration over the element's interior. This means that we implement the boundary terms in an `integrands`-like function in the simulator. The details of such an implementation are explained next.

In the previous program examples we have silently applied many functions provided by class `FEM`. The virtual `calcElmMatVec` function for calculating the element matrix and vector is an example on a function whose default implementation in class `FEM` often suffices[24]

```
void FEM::calcElmMatVec
   (int elm_no, ElmMatVec& elmat, FiniteElement& fe)
{
   fe.refill (elm_no, this);       // init for this element
   fe.setLocalEvalPts (itg_rules); // init num.itg. points
   numItgOverElm (elmat, fe);      // call integrands for each pt
}
```

However, to incorporate the Robin condition, we must extend `calcElmMatVec` with an integration loop over the relevant sides. We do this by redefining `calcElmMatVec` in our solver class:

```
void Poisson2::calcElmMatVec
   (int elm_no, ElmMatVec& elmat, FiniteElement& fe)
{
   FEM::calcElmMatVec (elm_no, elmat, fe);   // volume integral

   int nsides = fe.getNoSides();
   for (int s = 1; s <= nsides; s++)   {
      if (fe.boSide (s, 5))   // Robin condition (ind. 5) on side s?
         numItgOverSide (s, 5, elmat, fe);     // surface integral
   }
}
```

---

[24] The `itg_rules` variable in `calcElmMatVec` is of type `ElmItgRules` and is a data member of class `FEM`. The `itg_rules` object contains numerical integration rules for finite elements in the reference coordinate system.

According to the numbering convention of boundary indicators adopted in class `Poisson2`, indicator 5 marks the boundary where the Robin condition applies. The function `numItgOverSide` is similar to `numItgOverElm` and is inherited from class `FEM`. The nice thing about C++ and object-oriented programming is that if you find these inherited functions inappropriate, you are free to rewrite them in your problem class. Hence, you can inherit general algorithms and use them for quickly establishing a prototype solver. Later, you can in a step-wise manner rewrite these general algorithms and adapt them to the special features of your problem.

The `numItgOverSide` function runs through a loop over the integration points on the current side. For each point the function calls `integrands4side`, which is similar to `integrands`, but used for computing surface or line integrals. The implementation also works for the boundary terms in 1D problems. Here is a suitable implementation of `integrands4side` (the parameters α and U_0 are named `robin_u` and `robin_U0` in the `Poisson2` code):

```
void Poisson2:: integrands4side
  (int /*side*/, int boind, ElmMatVec& elmat, const FiniteElement& fe)
{
  const int nbf = fe.getNoBasisFunc();
  real detSideJxW = fe.detSideJxW();
  int i,j;

  if (boind == 5) {
    for (i = 1; i <= nbf; i++) {
      elmat.b(i) += fe.N(i)*robin_U0*detSideJxW;
      for (j = 1; j <= nbf; j++)
        elmat.A(i,j) += robin_u*fe.N(i)*fe.N(j)*detSideJxW;
    }
  }
}
```

If we have several different integrals over this element side, e.g. arising from different values of U_0 on different sides, one can insert `if-else` clauses in `integrands4side` that test for the side number, the boundary indicator number, or other parameters managed by the simulator class.

To summarize, surface integral terms are straightforwardly implemented by the following two steps.

1. Rewrite the `calcElmMatVec` function in the problem class by first performing the standard integration over the interior of the element and then check each of the sides whether it involves an integral over that side.

2. Implement the integrands of the side integrals in the function `integrands4side` in the problem class.

A subclass `Poi2Robin` of `Poisson2` implements a special test case for verifying the implementation of the Robin condition (see also page 276).

3.5.3   Simple Mesh Generation Tools

In Chapter 3.2.3 we explained the box preprocessor for generating grids over box-shaped domains in an arbitrary number of space dimensions. Diffpack also has somewhat more advanced mesh generation tools, or offers interfaces to external grid generation packages, as described in the report [64]. Since one often encounters domains with fairly simple geometry (box, triangle, disk) in academic and educational work, Diffpack has an easy-to-use interface to grid generation for some common domain shapes. This interface is itself a preprocessor in Diffpack, called `PreproStdGeom` (preprocessor for standard geometries). We shall explain the usage of this facility next. The reader is encouraged to try the examples on the `Poisson1` or `Poi2sinesum` simulators[25].

Box-Shaped Domains in 1D, 2D, and 3D. To generate a grid over a box-shaped domain in 1D, 2D, 3D, or higher dimensions, we send a string with the following syntax as argument to the `readOrMakeGrid` function:

```
P=PreproStdGeom | BOX geometry | partition
```

The `geometry` and `partition` strings are the same as for the `PreproBox` pre-processor. For example,

```
P=PreproStdGeom | BOX d=2 [0,2]x[1,3]|d=2 e=ElmB4n2D [10,28] [1,1.8]
```

generates a 2D grid over $[0,2] \times [1,3]$, using bilinear elements (`ElmB4n2D`), with 10 elements in the x_1 direction and 28 elements in the x_2 direction. The gradings are set to 1 for the x_1 direction (uniform partition) and 1.8 for the x_2 direction (stretching towards $x_2 = 1, 3$). A uniform 3D grid on the unit box is generated by the following command:

```
P=PreproStdGeom | BOX d=3 [0,1]x[0,1]x[0,1] |
                  d=3 e=ElmB27n3D [4,4,4] [1,1,1]
```

Here, we have $2 \times 2 \times 2$ elements since the division 4 in each space direction reflects the *number of intervals between nodes*. (Recall that with second-order elements, like `ElmB27n3D`, there are three nodes, i.e. two intervals between nodes, per element in each space direction.)

Rectangle with Triangulation. For some academic test purposes it can be convenient to apply a triangulation algorithm to generate the mesh over a rectangle. The command

```
P=PreproStdGeom | BOXT d=2 [0,1]x[-1,-0.7] | e=ElmT3n2D nel=200
```

---

[25] The geometries `BOXT`, `TRIANGLE`, or `DISK` require the Geompack software [58] or the Triangle software [104] to be installed properly with Diffpack.

generates an unstructured triangular mesh on $[0,1] \times [-1, -0.7]$. The man page for class `PreproStdGeom` gives information on the default preprocessor that the current Diffpack version applies for generating the mesh. The requested preprocessor type can be enclosed in parenthesis after the keyword `BOXT`, e.g., `BOXT(Geompack)` for using the Geompack software by Joe [58] or `BOXT(Triangle)` for specifying the Triangle package by Shewchuk [104]. Figure 3.4 depicts a possible mesh. The `nel` variable specifies the desired number of elements in the triangulation. The names `d`, `e`, and `nel` are (as usual) not significant; only the = signs proceeding the names are really used when interpreting the string. The boundary indicators associated with the `BOXT` domain shape are the same as those produced by default by the `PreproBox` preprocessor.

The other keyword indicators for geometry, like `TRIANGLE`, can also be equipped with a specification of the preprocessor to be used. However, some geometries are linked to special preprocessors (for example, we try to use `PreproBox` wherever possible).

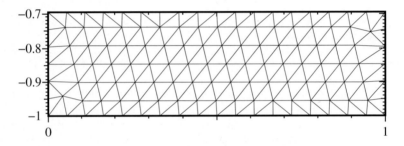

Fig. 3.4. Grid generated by the `PreproStdGeom` preprocessor, using the instruction `BOXT d=2 [0,1]x[-1,-0.7] | e=ElmT3n2D nel=200`.

Triangular domain. The following command generates a domain with the shape of a triangle:

```
P=PreproStdGeom | TRIANGLE (0,0) (3,0) (0,1) | e=ElmT3n2D nel=50
```

The vertices of the triangle become $(0,0)$, $(3,0)$, and $(0,1)$, the element type is restricted to `ElmT3n2D`, and the meaning of the `nel` parameter is the same as for the `BOXT` domain shape. Three boundary indicators are set by default, one for each side. Indicator no. 1 marks side 1, which is defined as the side between vertex 1 and 2 (here: $(0,0)$ and $(1,0)$). Indicator no. 2 and 3 mark side 2 (vertex 2 and 3) and 3 (vertex 3 and 1). The particular command given here results in the mesh in Figure 3.5.

Disk. A grid over a circular disk with radius 2.5 can be generated by

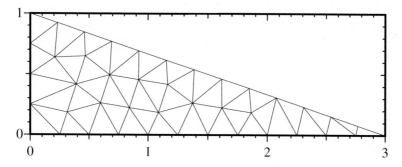

Fig. 3.5. Grid generated by the `PreproStdGeom` preprocessor, using the instruction `TRIANGLE (0,0) (3,0) (0,1) | e=ElmT3n2D nel=50`.

```
P=PreproStdGeom | DISK r=0.8 degrees=360 |
                e=ElmT3n2D nel=1000 resol=100
```

Here, `nel` is as usual the expected number of elements, `degrees` is explained below, and `resol` is the number of vertices used for specifying the outer boundary polygon of the disk. Only triangular elements of `ElmT3n2D` type are allowed. Increasing the value of `resol` reduces the error in approximating a circle with a polygon, but more elements are then required in the interior to avoid undesired element shapes. A warning is issued if the mismatch between `nel` and `resol` is large. The suggested number of elements can roughly be computed by

$$2\pi r^2 d \frac{D}{360} \left(\frac{2\pi r D}{360(n-1)} \right)^{-2},$$

where D is the `degrees` parameter, n is the `resol` parameter, and r is the radius of the disk. An example is presented in Figure 3.6a.

Mesh generation in a fraction of a disk is enabled by the `degrees` parameter. Setting e.g. `degrees=90` gives a quarter of a disk. Note that `degrees` is given in degrees (0 to 360). When `degrees` is less than 360, there are two boundary indicators. The first one marks the outer boundary, while the second one marks the cut. In case of a complete disk (`degrees` equals 360), there is of course only one boundary indicator by default.

Disk with Hole. A finite element mesh in a disk of radius `b`, with an inner hole of radius `a`, can be generated by the command

```
P=PreproStdGeom | DISK_WITH_HOLE a=0.5 b=0.8 degrees=120 |
                 d=2 e=ElmB9n2D [10,20] [1,1]
```

The opening of the disk is here 120 degrees, and the mesh consists of 5×10 second-order elements with $(10 + 1) \times (20 + 1) = 231$ nodes. The syntax of the partition part is identical to that of the box preprocessor. The boundary

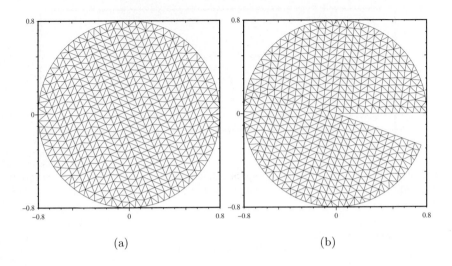

(a)   (b)

Fig. 3.6. Grid generated by the `PreproStdGeom` preprocessor, using (a) the instruction `DISK r=0.8 degrees=360 | e=ElmT3n2D nel=1000 resol=100` and (b) the same instruction, but with `degrees=340`.

indicators are also the same, one for each side. Figure 3.7 depicts an example of the grid.

In polar coordinates (r, θ) the hollow disk domain is $(a, b) \times (0, \theta_0)$, where θ_0 is the `degrees` parameter. Boundary indicator 1 marks the side $r = b$, indicator 2 marks $\theta = \theta_0$, indicator 3 marks $r = a$, and indicator 4 marks $\theta = 0$. The mesh is easily generated from a rectangular mesh on the unit square, mapped by the transformation

$$x = \varrho \cos \eta \theta_0, \quad y = \varrho \sin \eta \theta_0, \quad \varrho = \xi(b - a) + a,$$

where $(\xi, \eta) \in [0, 1] \times [0, 1]$. We remark that only a fraction of a disk can be generated, i.e., `degrees=360` is not legal (two of the boundaries in the original square would then coincide).

Box with Box-Shaped Hole. The command

```
P=PreproStdGeom | BOX_WITH_BOX_HOLE
   d=2 [0,1]x[0,0.4]-[0,0.4]x[0,0.2] | d=2 e=ElmB4n2D [20,20] [1,1]
```

specifies a rectangle $(0, 1) \times (0, 0.4)$, where a rectangular region $(0, 0.4) \times (0, 0.2)$ is removed. The syntax of the partition string is the same as for the box preprocessor. The boundary indicators for the original domain $(0, 1) \times (0, 0.4)$ are as for the box preprocessor, and the new boundary due to removal of a rectangle is marked by an additional indicator, which is then indicator no. 5 in the present 2D example. An example is presented in Figure 3.8. The syntax for three-dimensional problems should be evident.

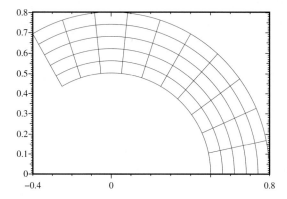

Fig. 3.7. Grid generated by the `PreproStdGeom` preprocessor, using the instruction `DISK_WITH_HOLE a=0.5 b=0.8 degrees=120 | d=2 e=ElmB9n2D [10,20] [1,1]`.

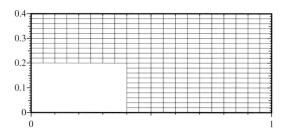

Fig. 3.8. Grid generated by the `PreproStdGeom` preprocessor, using the instruction `BOX_WITH_BOX_HOLE d=2 [0,1]x[0,0.4] - [0,0.4]x[0,0.2] | d=2 e=ElmB4n2D [20,20] [1,1]`.

Box with Elliptic Hole. The command

```
P=PreproStdGeom | BOX_WITH_ELLIPTIC_HOLE(box)
       a=2 b=1 c=3 d=3.5 deg=90 | d=2 e=ElmB4n2D [14,20] [-0.7,1]
```

specifies a quarter of a rectangle $(-a - c, a + c) \times (-b - d, b + d)$, with an elliptic hole $(x/a)^2 + (y/b)^2 = 1$, see Figure 3.9a. The syntax of the partition string is the same as for the box preprocessor. A `deg` parameter less than 90 (the value is arbitrary) specifies that $1/8$ of the total enclosing rectangle is to be meshed, `deg=90` specifies $1/4$ of the total rectangle, while `deg=180` means half of the total rectangle with a semi-ellipse as hole. The numbering of the boundary indicators depend on the `deg` parameter, see Figure 3.9b.

Sphere with Hole. The command

```
P=PreproStdGeom | SPHERE_WITH_HOLE a=2.5 b=6 theta=45 phi=45 |
                    d=3 e=ElmB8n3D [8,6,6] [1,1,1]
```

is similar to the command for creating a mesh in a hollow disk, but generates a grid in a part of a hollow sphere (a complete sphere or semi-sphere with

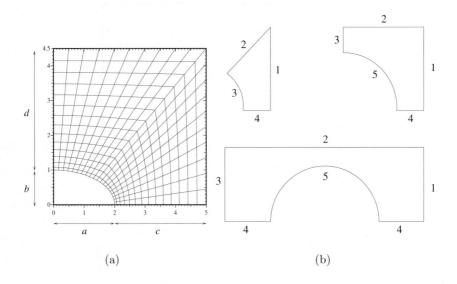

Fig. 3.9. (a) Grid generated by the `PreproStdGeom` preprocessor, using the instruction   `BOX_WITH_ELLIPTIC_HOLE a=2 b=1 c=3 d=3.5 deg=90 | d=2 e=ElmB4n2D [14,20] [-0.7,1]`; (b) numbering of boundary indicators for `deg<90` (top left), `deg=90` (top right), and `deg=180` (bottom).

a hole is not possible). The inner and outer radius of the hollow sphere are a and b, while `phi` and `theta` denote the opening of the meshed part of the sphere, expressed in the angular spherical coordinates (ϕ, θ), where ϕ is the angle between the radius vector and the z axis, and θ is the angle between the x axis and the projection of the radius vector onto the xy plane. The mesh is realized by transforming a grid over the unit cube onto the hollow sphere, as we described in detail for the `DISK_WITH_HOLE` feature. The boundary indicators are as follows: 1 marks the outer boundary $r = b$, 2 marks $\theta = \theta_0$, 3 marks $\phi = 0$, 4 marks the inner boundary $r = a$, 5 marks $\theta = 0$, and 6 marks $\phi = \phi_0$.

On page 455 we exemplify a geometry `BOX_WITH_BELL`. Class `PreproStdGeom` is occasionally equipped with new geometries – check the man page for the last updates.

More Flexible Tools for Grid Generation. Nontrivial geometries require the use of more advanced grid generation software, e.g., the super element preprocessor in Diffpack [64] or packages like Geompack [58], Triangle [104], or GRUMMP [46]. All of these can be accessed from Diffpack, either directly in a simulator or through the `makegrid` utility. Suppose you have used the Triangle preprocessor to generate a finite element grid stored in the files `mygrid.node` and `mygrid.ele`. The filter `triangle2dp mygrid` can then transform this grid to

Diffpack's grid format, resulting in the file `mygrid.grid`, which can be loaded into, e.g., a `Poisson1` or `Poisson2` simulator. Alternatively, the `makegrid` command

```
makegrid --iscl -m PreproTriangle -g 'file=tg' -p 'options=-pIq'
         --casename mygrid --plot ON
```

feeds Triangle with the options `-pqI` and a description of the boundary polygon in the file `tg.poly` (written according to Triangle's file format). This results is a Diffpack grid file `mygrid.grid` and a series of Plotmtv plots (due to the `--plot` option) of the mesh and the boundary indicators. Similar functionality exists for the GRUMMP and Geompack preprocessors. We refer to the report [64] for further information.

If your favorite grid generation software is not directly supported by Diffpack it is normally a few hours job to write a filter that transforms a grid file format to Diffpack's `GridFE` format.

3.5.4 Debugging

The first versions of a Diffpack simulator will most likely halt or abort due to error messages from Diffpack or from the operating system. Correcting such errors is usually quite easy, at least if you use the right tools and know the C++ language well. Computational errors are more difficult to correct and require a thorough understanding of the mathematics and numerics of the problem. Below we list some debugging hints that have proven to be valuable when developing Diffpack applications. Additional information is provided in the FAQ [63].

Array Indices Out of Bounds and Null Pointers. Before you even think about debugging, make sure that you run the program in non-optimized mode. The non-optimized mode forces checks on index bounds in arrays and on null pointers in `Handle` objects – the sources of the most common errors among programmers.

Handling Diffpack Error Messages. When a Diffpack error message is issued it can be difficult to locate where the error actually occurred in the program. Examining the sequence of function calls leading to the erroneous state is often helpful for finding the origin of the error. To this end, C++ exceptions must be caught inside the libraries, which is ensured by the `--exceptions` 0 command-line option to your program (this is the default behavior). If you then also supply the option `--verbose 1` when rerunning the program, the execution will be terminated, and the program will dump a core file. With the aid of a debugger this core file can recover the stack of function calls, supplied with corresponding line numbers in source code files (if your program was compiled in non-optimized mode). The output from the verbose error message briefly describes how to invoke a debugger and issuing a suitable command for displaying the call stack.

If the program terminates with a *segmentation fault* or a *bus error*, a core file is usually dumped. The error is then easily located by invoking a debugger and examining the call stack in the same way as we described for the error messages.

Debug Output. It is often convenient to insert output statements in the program for debugging, but include these in preprocessor directives such that the statements can be activated or deactivated at compile time. Keeping the output statements in the code allows for easy reactivation in future debugging sessions. Here are four examples:

```
#ifdef DP_DEBUG
  s_o << "Inside loop, i=" << i << " r=" << r << "\n";
#endif
// equivalent statement:
  DBP0(oform("Inside loop, i=%d, r=%g",i,r));

#if DP_DEBUG >= 2
  s_o << "Inside loop, i=" << i << " r=" << r << "\n";
#endif
// equivalent statement:
  DBP2(oform("Inside loop, i=%d, r=%g",i,r));
```

In the first example, DP_DEBUG must be *defined*, but its value is not used. The second example contains code that is activated only if DP_DEBUG has an integer value larger than or equal to 2. The DBP (DeBug Print) macros are shorthand for the same output: DBP is always active, DBP0 requires DP_DEBUG to be defined, while DBPX is active of DP_DEBUG equals the integer X. An advantage of the DBP macros is that they write the filename and the line number in addition to the string given as argument to the macros. The compiler option -DDP_DEBUG defines the preprocessor variable DP_DEBUG on Unix systems, whereas -DDP_DEBUG=2 assigns the value 2 (see page 541). Examples on practical use of DP_DEBUG are given in, e.g., the functions calcElmMatVec and solveProblem in class Poisson2.

Manual Program Trace. The Diffpack macro DPTRACE(msg) is convenient for tracing the program flow. The macro is typically inserted in the beginning of central functions:

```
void MySimulator::fillEssBC ()
{
  DPTRACE("MySimulator::fillEssBC");
  ...
}
```

When the program is compiled with the DP_DEBUG preprocessor variable defined, DPTRACE will print a message, containing the string argument to the macro, at the beginning *and at the end* of that function. If DP_DEBUG is not defined, DPTRACE results in empty code so there is no overhead in having this built-in trace of function calls.

Exercise 3.2. Compile the files in the Poisson2 application such that all the debug output and the trace information is activated (DP_DEBUG=2). Run a test case using the Verify/test1.i file as input to see what the debug information looks like.   ◇

Invoking Debuggers. As an alternative to inserting output statements for debugging in the code, you can invoke a debugger. This allows you to stop at specified locations in the program and examine variables. The ability of debuggers to examine C++ objects that are built of layers of other objects vary greatly, and only the most sophisticated debuggers eliminate the need for inserting your own output statements in the code. Here we shall briefly mention the usage of the widely distributed GNU debugger, called gdb [73, Ch. 6], on Unix systems. To apply such a debugger, you must compile the program in non-optimized mode; the compiler will then also generate the necessary debugging information. Thereafter you can invoke the debugger by the command gdb app. The file .gdb.demo in the Poisson2 directory contains a self-explanatory demonstration of a debug session with gdb. Readers who are working in Unix environments are strongly encouraged to go through this session and become familiar with the basic usage of gdb.

Windows NT/95 Remark 3.6: Using Visual C++, fatal error situations are best handled by using the debugger integrated in Visual Studio. This is a very powerful tool that gives complete source code access and the possibility of setting breakpoints and examining data interactively.   ◇

Problems with the Menu System. Errors in operating the menu system can often be found by inserting a help command before each ok command. This will print the status of the current submenu, including the most recent answers. In this way you can control that the right menu commands are issued. Sometimes it might be necessary to give menu commands manually in an interactive program session (cut and paste from the input file) to see exactly how and where errors occur.

Using the Same Code for 1D, 2D, and 3D Problems. Parameterizing the number of space dimensions, like we demonstrate in the Poisson1 and Poisson2 classes, helps the debugging of 2D and 3D problems immensely; experience shows that careful testing of the 1D case removes most of the errors in the code.

Evaluating Intermediate Results. Verification of intermediate computational results requires a thorough understanding of the mathematical problem and the numerical method. Try to find a simplified test case, work through the details of the numerical algorithm by hand, and compare hand-calculations with intermediate output from the program. Class `Poisson2` has debug output statements for the essential boundary conditions, the element matrices and vectors, the global linear system after assembly, and the solution. Our emphasis on calculating the discrete (finite difference-like) equations in Chapter 2 is mainly motivated by this debugging method. Using the methods of Appendix A.4, one can also often find an exact analytical solution of the discrete equations. In this way one can construct test cases with, e.g., two elements and hand-calculated results for the element vectors and matrices as well as the global linear system and its exact solution.

3.5.5   Automatic Report Generation

Exploration of mathematical models through computer experiments requires careful conduction of the experiments. We have already mentioned Diffpack's support for running all combinations of multiple-valued menu answers, the so-called multiple loop functionality, in Chapter 3.4.2. Such extensive experiments produce lots of results, and it is a nontrivial problem to extract the most relevant results for scientific or engineering interpretation. When exploring a mathematical model using a Diffpack simulator, you can increase your own efficiency significantly by having the simulator generate nicely formatted reports containing important input and output data together with relevant visualization. Class `MultipleReporter` enables Diffpack programs to write reports using statements similar to those of $\text{\LaTeX}\,2_\varepsilon$. The final report comes in three formats, ASCII, $\text{\LaTeX}\,2_\varepsilon$, and HTML.

The basic idea of report writing in Diffpack is that each class is responsible for writing its part of a report, which means that many library classes have functionality for writing information to a `MultipleReporter` object. The simulator holds the `MultipleReporter` object and calls up various classes to fill the report.

The size of reports containing numerous runs usually becomes inconveniently large. The `MultipleReporter` class can therefore also produce a *summary report* that enables a quick overview of the various runs in a multiple loop. For each run, just a few selected key parameters are visible. If you want more information about a particular run, you can look up the relevant part of the full report. The HTML version of the summary report offers of course a link to the detailed information, such that it is easy to jump back and forth between the summary report and the full report.

Having a simulator like class `Poisson2` (or `Poisson1`), it is trivial to generate code for automatic report generation. Just execute a Diffpack script: `MkReport -s Poisson2`. The arguments tell that class `Poisson2` is a simulator for a *stationary* (`-s`) problem. Many newcomers to Diffpack and C++ find it convenient to collect the additional code for automatic report generation in a subclass of the simulator (that is not necessary; it is just a way of separating the original simulator from the new report generation functions). The `MkReport` scripts generates appropriate code and files for a subclass `ReportPoisson2`. Substituting `Poisson2` by `ReportPoisson2` in the `main` file should make the simulator with report functionality ready for execution.

In problems where variable names deviate from that in class `Poisson2` you must be prepared to edit the report class. Fortunately, editing almost working code is much easier than writing the statements from scratch. The `MkReport` script assumes the existence of a `Handle(FieldFE)` u for the primary unknown, a `Handle(GridFE)` grid for the grid, and a `Handle(LinEqAdmFE)` lineq for the linear system tools. The application developer is encouraged to add new report writing statements by looking at the existing code and consulting the man page for class `MultipleReporter`. Class `SimReport` contains useful functions for including graphics.

The `ReportPoisson2` class contains more functions than what is strictly necessary for our present `Poisson2` solver. The splitting of operations into several functions provides the necessary flexibility when combining different simulator classes for solving systems of PDEs. Each PDE solver can then easily generate its part of the report for the whole system.

The reader is encouraged to study the `ReportPoisson2` class. The man page for the base class `FEM` and its base class `SimCase`, together with the man page for class `SimReport`, should provide the necessary information for customizing the report generation code to suit your own needs.

The following command runs a possible test case[26].

```
./app --class Poi2flux --casename arg1 < Verify/test4b.i
```

When the execution is finished, the program should have generated the files `arg1-summary.*`, containing summary reports, and `arg1-report-*`, containing full reports. The HTML format is the favorite of most users for browsing the results. Enter Netscape or any other browser with the report filename as "address". On the Unix command line you can typically write

```
netscape arg1-summary.html &
```

Clicking on "run-2" leads you to the part of the full report where the information about the second run in the multiple loop is provided.

Windows NT/95 Remark 3.7:   On Win32 platforms you can browse the HTML reports by any available web browser, However, you can also start such browsing from within the graphical user interface. This feature only works with Internet Explorer version 3.01 or newer. If you prefer the Netscape browser, you will have to load the HTML report manually outside the Diffpack application.   ◇

The report writing functions are called by `MenuSystem::multipleLoop`. Before the loop starts, `openReport` is called. By default `openReport` is an empty

---

[26] `Poi2flux` is a subclass, implementing a special test problem, in the `Poisson2` solver, see page 275 and Chapter 2.8.2.

function, but if it is defined in the simulator, it will typically open the report files and call `writeHeadings` for defining the row and column headings of the tables in the summary report. Inside the multiple loop, the functions `scan`, `solveProblem`, and `resultReport` are called in that order. After exit of the multiple loop, the `closeReport` function is called. Hence, if you want to postprocess data from several runs, you should do this in `closeReport`. The `resultReport` function contains calls to `writeResults` for writing summary report items and `writeExtendedResults` for writing the full report in the current run. Many Diffpack classes have their own `writeResults` and `writeExtendedResults` functions which can be called from the simulator's version of these functions. This is exemplified in the source code for class `ReportPoisson2`.

Sometimes you want to run a quick test problem and not wait for the time-consuming generation of plots in the report. The option `--nographics` to `app` turns off the plotting, but makes the report. The graphics can be generated later as we shall explain. The option `--noreport` is even more efficient; it avoids all calls to report generating functions and is very useful during debugging for speeding up the executions.

If you encounter problems with the creation of graphics during run time, this is usually due to missing programs. Error messages on the form "last Unix command .... failed" are then common. If you encounter such problems, rerun the case using the command-line options `--nounix --nographics`. This will put all Unix and plotting commands in scripts that can be executed after the simulation is finished. Of course, there will be no pictures in the reports before these scripts are run successfully. The try

```
SIMULATION.unix; SIMULATION.makegraphics
```

if `SIMULATION` is the casename of the run. During the execution of these scripts one can easier detect and explain error messages from the operating system.

3.5.6 Specializing Code in Subclass Solvers

Motivation. A flexible Poisson equation solver, like the one in class `Poisson1` or in `Poisson2`, can be reused in many problems. The most straightforward way to adapt the solver to a new problem consists in copying and editing the source code. We warned against this approach in Chapter 3.4.6 because it leads to multiple versions of the program; improvements of the original Poisson solver are not automatically available to all the derived versions of the code.

Suppose you incorporate efficient multigrid methods for solving linear systems in class `Poisson2`. If customization of the Poisson solver in various problems are done along the lines of Chapter 3.4.6, i.e., customization takes place in small subclasses, any improvements of the original solver are immediately available to the subclass solvers after a recompilation. Ideally, all common parts of Poisson equation simulators should be collected in one class.

The details that differ from problem to problem, like f, g, k, and the analytical solution (if it is available), should be implemented separately from the general solver. Such a software design is typical for object-oriented programming. We will now extend the ideas from Chapter 3.4.6 and create subclass solvers, derived from class Poisson2, which deal with specialized problems.

Basic Principles of the Design. Looking at class Poisson1, we see that it contains information about the details of a special test problem, namely the A parameters and the analytical solution. In class Poisson2, we have moved such problem-specific data and functionality out of the class, such that the class is better suited for all types of Poisson equation problems. The particular test problem in class Poisson1, as well as other test cases with analytical solutions, can be implemented in the Poisson2 framework by the following procedure. We derive a subclass Poisson2anal from Poisson2, or rather from ReportPoisson2, to inherit the report generation tools as well. The purpose of class Poisson2anal is to adapt the general Poisson2 solver to Poisson equation problems where the analytical solution is known. In such problems we want to compute the error field, various norms of the error, and estimate the convergence rate from a series of experiments. Moreover, the results about the errors are to be included in the automatically generated reports. We therefore need some extra functionality in class Poisson2anal, e.g., an ErrorRate object as well as calls to Lnorm and errorField in the ErrorNorms class, as explained in Chapters 3.4.3 and 3.5.8. This functionality is common to all problems where the analytical solution is available. Class Poisson2anal can therefore act as base class for subclass solvers that implement specific choices of the f, g, k, and valuePt (the analytical solution) functions. A sketch of the Poisson2 class hierarchy is displayed in Figure 3.10.

Using class derivation, we can specialize the original Poisson2 solver without touching the source code of the Poisson2 files. Moreover, we only program the *differences* from the base class in a subclass, i.e., we add data members and redefine some of the virtual functions. This leads to very short code in the subclass solvers. Moreover, any bug corrections or computational improvements of class Poisson2 are automatically available to these subclass solvers.

A List of the Subclass Solvers. Several special cases of the boundary-value problem (3.12)–(3.17) are implemented as subclasses of Poisson2anal:

- Class Poi2sinesum implements the test case from Chapter 3.2.1. Its subclass Poi2randgrid offers random displaced nodes for investigating the impact of distorted elements on the accuracy of the solution.
- The Poi2disk solver deals with the Laplace equation in a hollow disk (2D) or sphere (3D), with $u = 0$ on the inner boundary and $u = 1$ on the outer boundary.
- Class Poi2flux implements $-\nabla \cdot [k\nabla u] = f_0$ on a grid $[0,1] \times [a,b] \times [c,d]$, where f_0 is constant, a, b, c, and d are arbitrary, $k = 1$ for $x_1 \in [0, \gamma]$ and

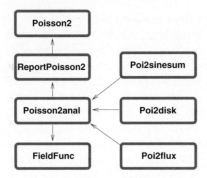

Fig. 3.10. The main parts of the `Poisson2` hierarchy. Arrows denote inheritance. Class `Poisson2` is the general solver for the problem (3.12)–(3.17), `ReportPoisson2` extends the solver with automatic report generation, the `Poisson2anal` class is a base for specialized solvers where the analytical solution is known and the numerical errors can be computed. Examples on some specialized solvers are the classes `Poi2sinesum`, `Poi2disk`, and `Poi2flux`.

$k = k_0$ for $x_1 \in (\gamma, 1]$, $u = 1$ at $x_1 = 0$, $u = 0$ at $x_1 = 1$, and $\partial u / \partial n = 0$ at all other boundaries. The solution is then only varying with x_1. The purpose of this test case is to investigate the effect of jumps in k on u and the flux $-k\nabla u$. Numerical experiments using this solver appear in Chapter 2.8.2.

– Class `Poi2Robin` solves a two-point boundary-value problem $u'' = 0$ on $(0, 1)$ with $u(0) = 0$ and $-u'(1) = 2.5u(1) - 4$, but the actual computational domain is the 2D unit square with $\partial u / \partial n = 0$ on $y = 0, 1$. The purpose is to verify the implementation of the line integrals associated with Neumann and Robin conditions (Chapter 3.5.2).

– Class `Poi2estimates` solves the special test problem in Chapter 2.10.6.

Details of a Subclass Solver. Let us take a closer look at class `Poi2sinesum`, which implements the test problem from class `Poisson1`, using the numerics of class `Poisson2`, the reporter facilities from class `ReportPoisson2`, and the generic error computations in class `Poisson2anal`. In addition to the inherited functionality from its base classes, class `Poi2sinesum` needs the data member `Ptv(real) A_parameters` for representing $A_1, \ldots, A_d$, it needs to read $A_1, \ldots, A_d$ from the menu, it needs to implement the analytical solution (u and ∇u), and it needs to compute the various measures of the error. The latter task is automatically performed and reported in the base class `Poisson2anal`. Our `Poi2sinesum` therefore becomes quite compact:

```
class Poi2sinesum : public Poisson2anal
{
protected:
  Ptv(real) A;       // parameters in the f function
public:
```

```
  Poi2sinesum() {}
 ~Poi2sinesum();
  virtual void define (MenuSystem& menu, int level = MAIN);
  virtual void scan ();
  virtual real valuePt (const Ptv(real)& x, real t = DUMMY);
  virtual Ptv(real) derivativePt (const Ptv(real)& x, real);
  virtual real f (const FiniteElement& fe,  real t = DUMMY);
};
```

The `define` function should define everything that the base classes define on the menu, but in addition include a menu item for $A_1, \ldots, A_d$. This is compactly accomplished by writing

```
void Poi2sinesum:: define (MenuSystem& menu, int level)
{
  Poisson2anal::define (menu, level);
  // add special menu item related to this test case:
  menu.addItem (level,          // menu level (submenu: level+1)
                "A parameters", // menu item command/name
                "A_1 A_2 ... in f expression", // help/description
                "3 2 1");       // default answer
}
```

The `scan` function follows relies on class `Poisson2anal` in the same way:

```
void Poi2sinesum:: scan ()
{
  Poisson2anal::scan ();
  A.redim (grid->getNoSpaceDim());
  MenuSystem& menu = SimCase::getMenuSystem();
  A.scan (menu.get ("A parameters"));
```

However, we should in the `scan` function also test that the data given to the `Poisson2` solver are correct. The present test problem can handle any domain or grid, since we set u equal to the analytical solution at the boundary, but boundary indicator 3 (user-specified variable Dirichlet condition) must be marked at all boundary nodes. We can force this setting of required boundary indicators by calling `GridFE::redefineBoInds` in the `scan` function; we simply redefine boundary indicator 3 as the sum of all previous (default) boundary indicators. The complete source code of `Poi2sinesum::scan` explains in detail how this is done.

The analytical solution of the test problem in class `Poi2sinesum` is implemented in terms of the `valuePt` function in the same spirit as we sketched for class `Poisson1` on page 248. Observe in Figure 3.10 that `Poi2sinesum` inherits from `FieldFunc` and thus can be treated as a functor[27]. Our simulator class can then be used in function calls where a `FieldFunc` functor is expected to represent the analytical solution.

---

[27] It is in fact the base class `Poisson2anal` of `Poi2sinesum` that is derived from `FieldFunc` and that encourages the subclasses to be their own functors for implementing the analytical solution.

```
real Poi2sinesum:: valuePt (const Ptv(real)& x, real)
{
  const int nsd = x.size();   real sum = 0;
  for (int j = 1; j <= nsd; j++)
    sum += A(j)*sin(j*M_PI*x(j)); // M_PI=3.14159...see math.h
  return sum;
}

Ptv(real) Poi2sinesum:: derivativePt (const Ptv(real)& x, real)
{
  const int nsd = x.size();   Ptv(real) g(nsd);
  for (int j = 1; j <= nsd; j++)
    g(j) = j*M_PI*A(j)*cos(j*M_PI*x(j));
  return g;
}
```

The `f` function is identical to `Poisson1::f`, whereas `g(x)` simply returns the analytical solution, i.e. `valuePt(x)` (this is the default implementation in class `Poisson2anal`).

Object-Oriented Programming. The reader who has studied the complete source code of class `Poi2sinesum` might wonder *how* the error computation actually takes place. Looking at the base class `Poisson2anal`, we can see a generic call to the `Lnorm` function in `resultReport`, using the solver class (`*this`) as the functor that represents the analytical solution:

> `ErrorNorms::Lnorm (*this, *u, ...);`

How such a generic call can work properly for all subclasses is now explained.

The `resultReport` function is virtual and called from the `multipleLoop` function in the menu system, which is usually the administering routine of all Diffpack simulators. The `multipleLoop` function takes a `SimCase&` argument that represents the simulator. This means that the `multipleLoop` function can only see our `Poi2sinesum` solver through the base class `SimCase`. All subclasses of `SimCase` objects must implement or inherit a `resultReport` function. C++ keeps track of *which* version of `resultReport` in the subclasses that is to be called. In the present example, this is `Poi2sinesum::resultReport`, and this function is actually inherited from the base class `Poisson2anal`. To summarize, `multipleLoop` calls a general virtual function `resultReport`, which is at run time translated by C++ to a specific call to `Poisson2anal::resultReport`.

The `Poisson2anal::resultReport` code performs the error estimation based on a functor (i.e. `*this`) for implementing the analytical solution. Notice that `*this` refers to the present object of type `Poi2sinesum` in this example. The `*this` argument is seen as a `FieldFunc&` argument by the `Lnorm` function in class `ErrorNorms` (this works well since our solver has `FieldFunc` as base class). All such `FieldFunc` objects have a `valuePt` function that evaluates the field. When `Lnorm` calls this `valuePt` function, C++ knows that the `FieldFunc&` argument is actually of type `Poi2sinesum` and that `Poi2sinesum::valuePt` is the function to be called. As soon as one gets used to this way of object-oriented thinking, the set-up is safe to use and gives a high degree of flexibility at very little programming cost. The larger your simulator project becomes, the more one appreciates object-oriented coding techniques.

Finally, we mention a remaining weak part of the `Poisson2` hierarchy: Altering the f, k, or g function usually requires derivation of a new subclass solver and recompilation. It would be advantageous to have more flexibility in the choice of the coefficients of the PDE at run time. For example, we should be able at run time to specify the type of the f function (constant, data on file, name of functor) and its values. This is indeed possible, using the `FieldFormat` class. Chapter 3.11.4 explains the details. The technique is also used in Chapter 6.1, where it is combined with virtual functions, giving an even higher degree of flexibility in handling variable coefficients.

3.5.7   Overriding Menu Answers in the Program

In all the subclass solvers mentioned in the preceding section, many of the menu items in class `Poisson2` are restricted to special choices. For example, class `Poi2flux` requires $D_1 = 0$ and $D_2 = 0$, and the boundary indicators must be redefined. This means that `Poi2flux` likes to be in charge of setting D_1 and D_2 as well as performing the mapping of default boundary indicators from the `PreproBox` preprocessor to the boundary indicator numbering required by the test case. In other words, we want the program to set appropriate menu answers that override any answers given by the user at run time. This functionality is accomplished by the `forceAnswer` function in class `MenuSystem`. Let us consider the menu item `Dirichlet value 1` defined by `Poisson2::define`,

```
menu.addItem(level,"Dirichlet value 1","const u value (ind. 1)","0.0");
```

used for assigning a value to D_1. The user might set the value on the menu by the command

```
set Dirichlet value 1 = 1.8
```

If we want to prescribe the menu answer inside the program, the appropriate statement is

```
menu.forceAnswer ("Dirichlet value 1 = 0.0");
```

When we issue the call `menu.get` for extracting the value of D_1 from the menu system, the commands given to `forceAnswer` take priority over the user-given answers or the default answers in the menu tree.

Class `Poi2flux` implements the test problem with a discontinuous k coefficient as documented in Chapter 2.8. This test problem is restricted to the unit interval, square, or cube, with Dirichlet conditions on $x = 0, 1$ and homogeneous Neumann conditions on all other boundaries. Regardless of default boundary indicators or what the user gives as boundary information, the `Poi2flux` class should ensure that correct boundary indicators are set. Using `forceAnswer`, we can adapt the too flexible menu system in class `Poisson2` to the special requirements of `Poi2flux`:

```
void Poi2flux:: scan ()
{
  MenuSystem& menu = SimCase::getMenuSystem();

  // override the menu system:
  menu.forceAnswer ("Dirichlet value 1 = 0.0");
  menu.forceAnswer ("Dirichlet value 2 = 1.0");
  menu.forceAnswer ("add boundary nodes = NONE");
  menu.forceAnswer ("beta = 0.0");

  Poisson2anal::scan ();

  // map default PreproBox indicators to the Poisson2 convention:
  String redef_boinds;
  const int nsd = grid->getNoSpaceDim();
  if (nsd == 1)
    redef_boinds = "n=2 names= u=0 u=1 1=(1) 2=(2)";
  else if (nsd == 2)
    redef_boinds =
    "n=4 names= u=0 u=1 u=g Neumann 1=(1) 2=(3) 3=() 4=(2 4)";
  else if (nsd == 3)
    redef_boinds =
    "n=4 names= u=0 u=1 u=g Neumann 1=(1) 2=(4) 3=() 4=(2 3 5 6)";
  grid->redefineBoInds (redef_boinds);

  // Note that redefineBoInds must be done *after* the grid is
  // generated - otherwise we do not know the number of space dim!
```

Notice that we must override the menu answers *before* Poisson2anal::scan reads the answers and initializes local variables like dirichlet_val1. The mapping of boundary indicators is performed *after* Poisson2anal::scan, because the grid must be initialized (in Poisson2::scan[28]) before we can call grid->redefineBoInds. Alternatively, we could used the redef_boinds string to form a menu.forceAnswer call. In that case, Poisson2anal::scan would perform the redefinition of boundary indicators. Other examples on customizing menu answers in a particular test case are found in the classes Poi2disk and Poi2Robin.

Although the forceAnswer function helps us to avoid incompatible input to specialized solvers, a more user-friendly program would offer just the menu items that are legal to change. A complete redefinition of the define and scan functions in Poi2flux is in this case required. Alternatively, one can make a scripting interface (see page 35) to a standard Diffpack solver and let the script take care of the input data handling. The script should then construct appropriate input files or command-line options for Diffpack's menu system. An example of such customization is presented in Chapter 3.10.5.

An important remark is that the forceAnswer function gives a possibility to initialize Diffpack's library classes via the menu system without forcing the user of the program to interact with the menu system. This is convenient if the solver implements its own user interface.

---

[28] Poi2flux::scan calls Poisson2anal::scan, which calls the scan function in class ReportPoisson2, which again calls the Poisson2::scan function.

Exercise 3.3. Consider the equation $\nabla \cdot (k\nabla u) = 0$ in a domain with boundary conditions as indicated in Figure 3.11. Explain how to create a subclass solver of `ReportPoisson2` that is adapted to the problem in Figure 3.11. The parameters C, D, and e should be adjusted on the menu. Regarding the k function, one can either just define it mathematically in terms of x and y coordinates, or one can apply the material concept from page 259 and test on `fe.grid().getMaterialType(fe.getElmNo())` inside the k function. The grid should be constructed such that the boundaries of the material in the interior of the domain coincides with the element boundaries. Also suggest how to reduce the size of the domain from symmetry arguments and what the boundary condition at the symmetry line(s) should be. $\diamond$

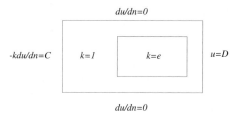

Fig. 3.11. Two-dimensional domain with boundary conditions for the Laplace equation $\nabla \cdot (k\nabla u) = 0$.

3.5.8   Estimating Convergence Rates

Tools for Estimating Convergence Rates. There is a class `ErrorRate` which makes it very easy to estimate the convergence rate of a numerical method, according to the technique described in Chapter 1.6.1. In a finite element context, the discretization measure h is typically a characteristic element length, and the error measure e is a norm of the difference between the numerical solution and an exact solution of the problem. Given a series of data points (h_i, e_i), $i = 1, 2, \ldots$, class `ErrorRate` computes the convergence rate r from (1.74) on page 93, based on two consecutive data points (h_i, e_i) and (h_{i+1}, e_{i+1}).

The usage of class `ErrorRate` is straightforward. First, one has to include the header file `ErrorRate.h`. Second, one declares an `ErrorRate` object `error_rate` as a member of the simulator class. Third, the `ErrorRate` object must be initialized by the name of the quantity whose error we work with:

```
error_rate.setNameOfUnknown ("u");
```

This call is normally placed in the `scan` function along with other initialization statements. After the measures of the error e are computed, one needs to add the new data point of $e(h)$ by calling the `add` function in the `ErrorRate` object. For example, `ErrorNorms::Lnorm` computes the L^1, L^2, and L^∞ norms of the error and each of these norms can be associated with an r parameter. Just call

```
error_rate.add (h, L1_error, L2_error, Linf_error);
```

The `ErrorRate` object requires multiple runs of the simulator. The easiest and most flexible way of accomplishing multiple runs is to use the multiple loop facility in Diffpack. After the runs are finished, the results of the estimation of r values are written to the output by the `print` or the `writeExtendedResults` functions in the `ErrorRate` object. The former writes to standard output, whereas the latter writes to a `MultipleReporter` object. Normally, one uses `writeExtendedResults` and calls this function in `closeReport`, since that function is automatically called at the end of a multiple loop, after all the experiments are finished.

The specialized subclass `Poisson2anal` of `Poisson2` is tailored to problems where the analytical solution is known, and this class employs the classes `ErrorRate`, `ErrorNorms`, and `MultipleReporter` in combination for estimating and reporting convergence rates. From the source code one can also see how to compute a suitable characteristic element length h from `GridFE` utilities.

Detecting Superconvergence at the Nodal Points. We shall now demonstrate an application of class `ErrorRate`, which also will highlight some programming conventions related to the use of enum types in Diffpack. The example concerns the fact that in some problems the accuracy of the finite element solution is higher at the nodal points that at arbitrary points in the domain. Occasionally, the numerical solution can also be exact at the nodes. To investigate such features, one can base the computation of error norms on nodal-point integration. That is, we sample the solution at the nodes only, when integrating the error, instead of using the more common Gauss-Legendre points. The type of integration points is a parameter to the `ErrorNorms::Lnorm` function. The parameter can have two self-explanatory enum values, `NODAL_POINTS` or `GAUSS_POINTS`. It is convenient to let the type of integration point be set on the menu. We now outline the necessary extensions of an `Poisson1`-like code to allow a new variable, the integration point type, to be included in the solver, with full menu support.

```
// declare new data member in class:
   NumItgPoints error_itg_pt_tp;   // this enum is defined in enum.h

// in the define function:
   menu.addItem (level, // determines main menu, submenu etc
                 "error integration point type", // menu command
                 "err_itg",  // command-line option: --err_itg
                 "type of itg rule used in ErrorNorms::Lnorm",
                 "GAUSS_POINTS", // default answer
                 "S/GAUSS_POINTS/NODAL_POINTS/"); // legal answers
```

```
// in the scan function:
   assignEnum (error_itg_pt_tp,
               menu.get("error integration point type"));

// in call to ErrorNorms::Lnorm, replace GAUSS_POINTS
// by error_itg_pt_tp

// in e.g. resultReport one can write the itg. pt. type like this:
   s_o << "Error integration over " << getEnumValue(error_itg_pt_tp);

// in input file
   set error integration point type = { NODAL_POINTS & GAUSS_POINTS }
```

Some of these statement demonstrate important features in Diffpack and deserve a few comments. The `addItem` call has more parameters than explained on page 225. The overloaded version of `addItem` used here allows explicit specification of the command-line option corresponding to the menu item, which is chosen as `--err_itg` in this case. Recall that the command-line option for the simpler version of `addItem` is automatically set equal to the menu command, with spaces replaced by underscores. Of more importance in the extended `addItem` function is the possibility to specify legal answers. Here, this is a string (S) that can take two values, `GAUSS_POINTS` or `NODAL_POINTS`. When one applies the graphical menu interface, enabled by the command-line option `--GUI` to the solver, the legal answers are listed in a pull-down menu.

The integration point type is an enum, and we need to map the menu answer, which is a string, to the appropriate enum value. The `assignEnum` functions accomplish this task, using the syntax demonstrated above. The opposite conversion, from an enum to a string, is enabled by the `getEnumValue` function, also demonstrated in the preceding code example. Most of the enums defined in `enum.h` have `assignEnum` and `getEnumValue` functions (see the header file `enumfunc.h` for a complete list of these functions).

The statements above are incorporated in the `Poisson2` solver, or more precisely, in the subclass `Poisson2anal`. The reader should now try to establish the convergence rate in L^∞ norm for the model problem from Chapter 3.2.1. The experiment can be accomplished by the following command:

```
./app --class Poi2sinesum --iscl --casename c1 --Default Verify/test1.i
      --err_itg '{ NODAL_POINTS & GAUSS_POINTS }' --casedir c1
```

The simulation program `app` is here run with the menu in pure command-line mode (`--iscl`), where menu items are given by their command-line option, like `--err_itg`, and in a file (`--Default Verify/test1.i`). All generated files appear in the subdirectory `c1` because of the `--casedir` option. Looking at the end of a report file in the `c1` directory, i.e., `c1-report.html`, `c1-report.tex`, or `c1-report.txt`, we see a table like the one shown in Table 3.1. Although we use only bilinear elements, we achieve fourth order convergence at the nodal points. The reader is encouraged to carry out this experiment for biquadratic elements as well as for linear and quadratic triangular elements.

Table 3.1. Error norms, computed by numerical integration, and corresponding rates for the unknown u in the model problem from Chapter 3.2.1. In the first five runs only nodal points are used in the integration, whereas a standard 2×2 Gauss-Legendre rule is applied in the next five runs. The grids consist of $2^k \times 2^k$ bilinear elements in the domain $(-0.25, 1.25) \times (0, 1.2)$, for $k = 2, 3, 4, 5, 6$. One can clearly see that the solution at the nodes converges to the exact solution as h^4, while the convergence at the Gauss-Legendre points (and other points) goes like h^2, where h is the length of the sides in the elements in this case.

run	h	L^1 error	L^1 rate	L^2 error	L^2 rate	L^∞ error	L^∞ rate
1	3.7500e-01	8.7934e-03		9.2068e-03		1.2923e-02	
2	1.8750e-01	5.0459e-04	4.12	4.7561e-04	4.27	7.0224e-04	4.20
3	9.3750e-02	3.0409e-05	4.05	2.8277e-05	4.07	4.2352e-05	4.05
4	4.6875e-02	1.8988e-06	4.00	1.7451e-06	4.02	2.6859e-06	3.98
5	2.3438e-02	1.1854e-07	4.00	1.0872e-07	4.00	1.6748e-07	4.00
6	3.7500e-01	4.8930e-01		4.5544e-01		6.3488e-01	
7	1.8750e-01	1.3246e-01	1.89	1.2054e-01	1.92	1.8320e-01	1.79
8	9.3750e-02	3.3819e-02	1.97	3.0552e-02	1.98	4.7011e-02	1.96
9	4.6875e-02	8.4991e-03	1.99	7.6641e-03	2.00	1.1829e-02	1.99
10	2.3438e-02	2.1271e-03	2.00	1.9177e-03	2.00	2.9622e-03	2.00

Exercise 3.4. Set up a family of convergence rate experiments for the model problem in class `Poi2disk`. This problem involves the Laplace equation in a hollow disk, i.e., the computations are performed on a curved mesh. Is the convergence rate higher at the nodal points? How much faster is the convergence for quadratic elements? How does the convergence rate vary with the type of norm? Is there any difference in the performance of triangular[29] versus quadrilateral elements? The answers are readily obtained within five minutes of human work and a bit of CPU time.   ◇

3.5.9   Axisymmetric Formulations and Cartesian 2D Code

It will often be advantageous to develop simulators that can handle axisymmetric 3D problems along with standard 2D problems in Cartesian coordinates. Considering the Poisson equation $-\nabla \cdot (\lambda \nabla u)$, the axisymmetric formulation in cylindrical coordinates (r, θ, z) reads

$$-\frac{1}{r}\frac{\partial}{\partial r}\left(\lambda(r, z)r\frac{\partial u}{\partial r}\right) - \frac{\partial}{\partial z}\left(\lambda(r, z)\frac{\partial u}{\partial z}\right) = f(r, z). \qquad (3.18)$$

Can this equation be easily incorporated in a code that solves

$$-\frac{\partial}{\partial x_1}\left(\lambda(x_1, x_2)\frac{\partial u}{\partial x_1}\right) - \frac{\partial}{\partial x_2}\left(\lambda(x_1, x_2)\frac{\partial u}{\partial x_2}\right) = f(x_1, x_2)$$

---

[29] We remark that nodal-point integration over the `ElmT6n2D` is not available.

by the finite element method? The answer to this question is positive.

Equation (3.18) is to be solved in a 2D (r, z) domain Ω. However, when we formulate the weighted residual method, we integrate the residual over the whole 3D space, i.e., we require

$$\int_{\Omega} \int_0^{2\pi} R W_i r \, dr \, dz \, d\theta = 0, \quad i = 1, 2, \ldots, n,$$

where R is the residual in the PDE and W_i is the weighting function in weighted residual method. Let us introduce $x_1 \equiv z$ and $x_2 \equiv r$. Multiplying (3.18) by W_i and integrating over a 3D cylindrical domain results in

$$2\pi \int_{\Omega} \lambda(x_2, x_1) \left(\frac{\partial W_i}{\partial x_1} \frac{\partial u}{\partial x_1} + \frac{\partial W_i}{\partial x_2} \frac{\partial u}{\partial x_2} \right) x_2 dx_1 dx_2 = 2\pi \int_{\Omega} f(x_2, x_1) x_2 dx_1 dx_2$$

for $i = 1, \ldots, n$. Observe that the factor 2π cancels. The only difference between the axisymmetric finite element equations and the corresponding equations in 2D Cartesian coordinates is the extra factor x_2. From an implementational point of view we can therefore treat axisymmetric problems in a Cartesian 2D code by replacing $\det J$ by $x_2 \det J$. This could be done as follows in `integrands`:

```
if (axisymmetric)
    detJxW *= fe.getGlobalEvalPt()(2);   // detJxW = detJxW*x(2)
```

3.5.10   Summary

Let us summarize how a typical finite element solver in Diffpack is constructed. As usual, the simulator is realized as a class. Standard finite element algorithms can be inherited by having `FEM` as base class. It is then necessary for the simulator class to implement the virtual function `integrands` for defining the integrands in the integrals of the finite element equations. Class `FEM` is derived from `SimCase`, which allows multiple loops and handling of numerical experiments through the menu system.

The data structures in a finite element solver are typically a grid (`GridFE`), a finite element field over the grid (`FieldFE`), an interface to linear systems and linear solvers (`LinEqAdmFE`), a mapping between the field representation and the linear system representation of the primary unknown(s) (`DegFreeFE`), and perhaps some problem-dependent constants and functions. The relations between a solver, its base classes, and its internal data structures are shown in Figure 3.12.

The standard member functions in a simulator are

- `adm` for administering the menu system,
- `define` for defining the items on the menu,
- `scan` for reading input data from the menu system and initializing the internal objects (grid, fields etc.),

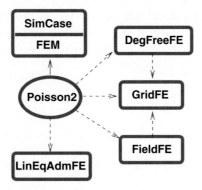

Fig. 3.12. A sketch of a simulator class (`Poisson2`), its base classes and internal objects. A solid line indicates class derivation ("is-a" relationship), whereas dashed lines represent a pointer/handle ("has-a" relationship).

- `integrands (ElmMatVec&,FiniteElement&)` for defining the integrands of the element-level finite element equations,
- `fillEssBC` for marking essential boundary conditions (in the `DegFreeFE` object),
- `solveProblem` for the main program flow of the solver,
- `resultReport` for writing results.

Optional routines are

- `calcElmMatVec(int,ElmMatVec&,FiniteElement&)` for computing the element matrix and vector (normally, this function will call the corresponding default version in class `FEM` and just do some additional task, like debug output or surface integrals),
- `integrands4side` for defining the integrands corresponding to boundary integrals in the weak formulation,
- `saveResults` for storing results on file.

In later chapters we shall introduce two additional key functions in finite element simulators:

- `solveAtThisTimeStep` for advancing the solution one time step in time-dependent problems (see Chapter 3.9.2),
- `makeAndSolveLinearSystem` for defining and solving the linear system in iteration methods for nonlinear algebraic equations (see Chapter 4.2).

Automatic report generation is conveniently coded by first letting the script `MkReport` generate a default code. This code can then be edited. It is a matter of taste whether one wants to separate the report generation statements in a subclass or not.

3.6   Adaptive Grids

The finite element method offers the possibility of having a dense mesh in regions where the solution changes rapidly, and a coarser mesh where the variations in the solution are smaller, thus allowing the grid to be *adapted* to the solution. For this purpose one needs grid software that enables *local mesh refinements*. Moreover, one needs a *refinement indicator* for selecting the elements to be refined. Diffpack has two grid classes that support local mesh refinements: class `GridFEAdT` and class `GridFEAdB`. These are subclasses of `GridFE` such that the finite element libraries work with objects of these grid classes in the same way as they work with `GridFE` objects. The refinement indicators are administered by a class `GridRefAdm`. Extending an existing solver with local mesh refinement capabilities is a matter of replacing the `GridFE` object by a `GridFEAdT` or `GridFEAdB` object, declaring a `GridRefAdm` object, and adding about 10 lines of code involving these new classes[30].

3.6.1   How to Extend an Existing Simulator

The `GridFEAdB` class works only with rectangle- and box-shaped elements of Lagrange type (`ElmB2n1D`, `ElmB3n1D`, `ElmB4n2D`, `ElmB9n2D`, `ElmB8n3D`, and `ElmB27n3D`, or `ElmTensorProd1` and `ElmTensorProd2`) and subdivides each element, being marked for refinement, into $q \times q$ new elements of the same type, where the integer q can be freely chosen by the user. Such a subdivision generally lead to irregular nodes (also called hanging nodes, improper nodes, or slave nodes), meaning that a node in one element does not coincide with a node in another element (and the node is not on the boundary), see Figures 2.8 and 3.14d for examples. The value of a field at an irregular node is constrained by the values at neighboring regular nodes. This gives rise to auxiliary constraint equations in the linear system [51]. The Diffpack libraries automatically take care of the additional constraint equations in the presence of a `GridFEAdB` grid.

   Class `GridFEAdT` works only with triangular or tetrahedral elements of the types `ElmT3n2D`, `ElmT6n2D`, `ElmT4n3D`, `ElmT10n3D` (in fact also `ElmB2n1D` and `ElmB3n1D`). Irregular nodes are not allowed in the grid, which means that refining an element implies refining also the neighbors to obtain a smooth transition from a refined region to a coarser region. Since irregular nodes are avoided, the refined grids can be used in combination with any other software tool in Diffpack or other finite element packages. Various strategies for dividing a triangle or tetrahedron into new elements are available. Both `GridFEAdT` and `GridFEAdB` can perform repeated refinements, yielding a hierarchy of refinement levels. A nice feature of class `GridFEAdT` is that the grids associated with the refinement levels are *nested*. The corresponding function spaces (V_h in Chapter 2.10 terminology) are also nested. This is advantageous for computational efficiency, especially when implementing multigrid methods based

---

[30] This requires that you have purchased the adaptive grid module in Diffpack.

on the hierarchy of grids in class `GridFEAdT`. Forthcoming documentation will deal with multigrid and adaptive mesh refinement in Diffpack.

We now describe the modifications of an existing simulator that are required for performing local mesh refinements. In the code segments below, the symbol `GridA` is used to represent either `GridFEAdT` or `GridFEAdB`.

1. Add
   ```
   #include <GridA.h>
   #include <GridRefAdm.h>
   ```
   in the header file of the simulator.
2. Add a refinement criterion in the simulator, i.e., include
   ```
   Handle(GridRefAdm) refcrit1;
   ```
   in the definition of the simulator class.
3. Add the refinement criteria menu at the end of the `define` function:
   ```
   GridRefAdm::defineStatic (menu, level+1);
   ```
4. Replace the creation of `GridFE` by
   ```
   grid.rebind (new GridA());
   ```
5. Create a `GridRefAdm` object:
   ```
   refcrit1.rebind (new GridRefAdm());
   ```
6. After `grid` is made, e.g. by calling `readOrMakeGrid`[31], the `GridRefAdm` object must be initialized:
   ```
   refcrit1->scan (menu);
   ```
7. All fields defined over the grid depend on the size of the grid. Since adaptive finite element methods imply regeneration of the grid many times, the fields must be also be redimensioned correspondingly. This means that we must create a function `redim`, where we update the `DegFreeFE` object, the `linsol` array, and all fields, typically by calling the `redim` function of these objects.

One can now perform the refinement by calling `refcrit1->refine(grid)`. Because the grid has changed, the simulator's `redim` function must be called immediately so that the fields match the new grid.

The use of `GridFEAdB` or `GridFEAdT` is transparent as we only operate with `GridFE` objects in the simulator. This makes it easy to allow for both grid types. In `scan` we can enable this flexibility by reading the grid type from the menu:

```
String grid_tp = menu.get ("grid type");
if (grid_tp == "GridFEAdB") {
  grid.rebind (new GridFEAdB());
  readOrMakeGrid (*grid, gridfile);
} else {
  grid.rebind (new GridFEAdT());
  readOrMakeGrid (*grid, gridfile);
  // box-shaped elements must be transformed to triangles/tetrahedra
  if (grid->getElmType(1).contains('B') ||
    grid->getElmType(1).contains("Tensor"))
    PreproBox::box2triangle (*grid);
}
```

---

[31] Note that `GridA` is a subclass of `GridFE` such that any preprocessor in Diffpack can make a `GridA` mesh.

The refinement will normally be carried out inside some loop, where one first computes with the original mesh, then refines the mesh according to the chosen criterion, and then repeats the process if the quality of the solution or the mesh has not met certain requirements. The most comprehensive way of accomplishing this is to apply an error estimator and classify the grid as satisfactory when the estimated error is below a critical level. Of course, one needs to prescribe a maximum number of elements such that too restrictive error criteria do not lead to unacceptable memory demands. The programmer of the simulator is completely in charge of the grid refinement loop and can employ several ready-made refinement criteria or make his own, e.g., an error estimator tailored to the PDE at hand. Programming of the error estimator is usually a fairly simple task using the Diffpack finite element tools, once the theory is known. We refer to Chapter 2.10.7 for examples on error estimators for the Poisson equation.

3.6.2   Organization of Refinement Criteria

Let m be number of elements in the current finite element mesh, and let $I = (v_1, \ldots, v_m)$, $v_i \in \mathbb{R}$, represent indicator values for each element. We refer to I as a *refinement indicator*. For example, v_i may be the estimated error of the numerical solution in element no. i. Intuitively, we would refine this element if v_i is larger than a critical value. We therefore introduce a *a refinement criterion* for evaluating the indicator I and deciding which elements that actually should be refined. The refinement criterion $\mathcal{C}$ can be expressed as $B = \mathcal{C}(I)$, $B = (b_1, \ldots, b_m)$, where b_i is a boolean value indicating whether element no. i is to be refined (true) or not (false). When a subset of the elements are marked for refinement, one must, at least in the case of a `GridFEAdT` grid, extend the set with additional elements, due to geometric restrictions on the refinement procedure. For example, if a triangle is to be refined, it often means that it is to be subdivided into four new triangles, with new nodes at the mid point of each side. The neighboring elements must then also be refined to avoid irregular nodes, although the indicator values in these elements may not qualify for direct refinement. In other cases one can impose certain smoothness properties of the element size distribution, which lead to increased refinement beyond the set of elements marked by $\mathcal{C}(I)$. We represent these additional refinement requirements by an operator $\mathcal{G}$ and write $B := \mathcal{G}(B)$.

Refinement of Diffpack grids is based on a four-step procedure:

1. evaluate the indicator I,
2. apply criterion: $B = \mathcal{C}(I)$,
3. add extra refinements: $B := \mathcal{G}(B)$,
4. call a refinement procedure in a grid object.

Class `GridRefAdm` administers these four steps. The class has a pointer to an object in the class `RefinementInd` hierarchy, representing the refinement indicator I. A parameter object of type `RefinementInd_prm` is used to select a desired indicator and associated parameters on the menu. Class `GridRefAdm` contains a set of criteria $C(I)$ that can also be selected through the menu system. Having I and $C(I)$, the `GridRefAdm` object calls an adaptive grid class, `GridFEAdB` or `GridFEAdT`, for determining the additional set of required refinements, $G(I)$, and carrying out the modification of the mesh. Several refinement criteria, e.g., for different fields when solving systems of PDEs, is trivially enabled by having several `GridRefAdm` objects in the simulator class. Figure 3.13 shows the relations between adaptive grid tools and a solver class `MySim`.

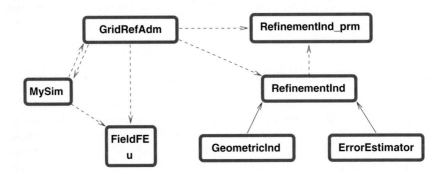

Fig. 3.13. Sketch of the relation between a simulator class `MySim` and the `GridRefAdm` utility for administering adaptive grid refinement. Dashed arrows represent pointers, whereas solid arrows denote inheritance. The subclasses `GeometricInd` and `ErrorEstimator` implement various refinement indicators based on simple geometric criteria or user-controlled error estimation, respectively.

The currently available refinement criteria $C(I)$ in class `GridRefAdm` are of two types:

1. b_i is true if $\ell_{lo} \leq I_i \leq \ell_{hi}$ (absolute criterion), or
2. b_i is true if I_i is among the largest p percent values in I (percentage criterion).

A common choice is to let $\ell_{hi} \to \infty$, which means that we refine all elements whose indicator values are larger than some critical value ℓ_{lo}.

Some examples on available refinement indicators I are listed next.

GeometricRegions: Set $I_i = 1$ if the centroid of element no. i is inside a prescribed geometric region, for example, a set of disks or hypercubes. Otherwise, $I_i \leq 0$. With $\ell_{lo} \leq 1$ and $\ell_{hi} \geq 1$, all elements inside the geometric regions will be refined. On the menu, one can specify these types of indicators:

```
set refinement criterion = GeometricRegions
set hypercube regions = Hypercubes d=2 n=1 [-0.2,1.0]x[0.85,1.0]
set disk regions = Disks d=2 n=2 (0,0)->0.2 (0.8,0.7)->0.1
```

This will lead to refinement of all elements whose centroids are inside a rectangle $[-0.2, 1] \times [0.85, 1]$ and two disks with radius 0.2 and 0.1, centered at $(0, 0)$ and $(0.8, 0.7)$, respectively.

OwnIndicator: This indicator calls a virtual function evalOwnIndicator in the simulator class and leaves it up to the user to provide a suitable indicator, for example, based on heuristic considerations, on a posteriori error estimation, or on other procedures that depend on the particular PDE being solved. This indicator requires access to the simulator, obtained by the statement

```
refcrit1->getRefinementInd().attach (*this);
```

Gradient: Given a field u over the grid, set $I_i = ||\nabla u||$, where the gradient is evaluated at the centroid of element no. i. This indicator is often used to refine the elements where u has large gradients. The field u must be attached to the refinement indicator class:

```
refcrit1->getRefinementInd().attach (*u);
```

if u is of type Handle(FieldFE). Such a statement is conveniently placed in the scan (or redim) function. It is not necessary that the attached field is the solution – it can be any field on the grid.

Value: Simply set I_i equal to the average of the nodal values of a field u over element no. i. The simulator must attach the u field to the refinement indicator tools as we have shown for the Gradient indicator.

ZZ: This is the indicator based on the ZZ error estimator, see Chapter 2.10.7 and [132]. The simulator must be attached to the refinement indicator as demonstrated for the OwnIndicator criterion.

The makegrid utility offers a menu with the GeometricRegions indicator and associated local mesh refinements. Hence, one can generate a grid, either by some external grid generation software and filter the grid file to Diffpack's format, or by a Diffpack preprocessor class, and then load this grid into makegrid for further manipulation in terms of local mesh refinements, redefining boundary indicators, adding material domains etc. Examples on such a process are provided in the Verify directory of the elasticity solver in src/app/Elasticity1 (see the self-explanatory *-makegrid.i files; no knowledge about the elasticity application is required to understand this use of makegrid).

Exercise 3.5. Introduce adaptive grids in a copy of the Poisson2 solver. Focus on the problem in Chapter 2.8.2, concerning the Poi2flux subclass and a large jump in the coefficient of a PDE. Try the GeometricRegions refinement indicator, using an interval in the vicinity of the jump as refinement area. Investigate possible advantages of using local mesh refinements.   ⋄

3.6.3   Example: Corner-Flow Singularity

A classical illustration of the benefits of local mesh refinements is the case of flow around a corner, as depicted in Figure 3.14a. In a fluid without vorticity, the stream function fulfills the Laplace equation. For the flow geometry in Figure 3.14, one can find the analytical solution of the stream function to be $u = r^{\frac{2}{3}} \sin\left(\frac{2}{3}(\theta + \frac{\pi}{2})\right)$, where r and θ are polar coordinates. The geometric features of this example enable us to a priori predict that large gradients in u will occur close to the corner. From Figure 3.14b we see that the magnitude of the velocity field, here equal to the norm $\|\nabla u\|$, grows fast towards the corner (it approaches infinity in the continuous case). Therefore, it may be appropriate to apply a simple refinement indicator, for example `GeometricRegion`, to specify where elements are to be refined. Such an indicator may lead to grids as we show in Figure 3.14. (Normally it is more optimal to use a criterion based on the solution, even if the criterion is heuristic.)

A solver `PoissonA` for the present case has been developed from a `Poisson2`-like class. The initial grid was generated by the `PreproStdGeom` preprocessor, using the `BOX_WITH_HOLE` option. The adaptive grids were incorporated adding the statements listed in Chapter 3.6.1. The source code can be found in `src/fem/adaptive/PoissonA`, with an appropriate input file for the present case in `Verify/corner.i`. Run this case and study various refinement strategies and corresponding error fields in the automatically generated HTML report. With the `PoissonA` solver one can at run time choose between the `GridFEAdT` and `GridFEAdB` type of grids.

A subclass `PoissonAee` of `PoissonA` implements a user-defined refinement indicator, which is a residual-based error estimator for the Poisson equation, more specifically (2.196) in Chapter 2.10.7. The input file `Verify/ee1.i` is a possible starting point for experimenting with this solver.

3.6.4   Transient Problems

The previous description of adaptive grids is also valid for time-dependent problems, but in such cases, the grid will be repeatedly refined at certain time levels. Since the PDE involves quantities at different time levels, it will be necessary to interpolate quantities at a previous time level onto the new refined grid. Our refinements will be based on an original coarse grid instead of, e.g., coarsening and refining a refined grid from a previous time step. The corresponding refinement step is implemented according to

```
// start refinements with the original grid:
grid.rebind (grid->getStartGrid());
refcrit->refine (grid);
```

Suppose we have a typical heat equation solver (Chapter 3.9) with the fields u and u_prev, representing the primary unknown at the present and the previous time level. Since it is necessary to interpolate u_prev onto the current grid, we need a copy of u_prev, e.g. called u_prev_save, in the simulator. Each time u_prev is updated from u, we also update u_prev_save. After the refinement call above, we must redimension the fields and interpolate u_prev_save, which is defined on the previous grid, onto u_prev on the new grid:

```
u_prev->redim (*grid, "u_prev");
u_prev->interpolate (*u_prev_save);
u_prev_save->redim (*grid, "u_prev_save");
// after u is computed: *u_prev_save=*u; *u_prev=*u;
```

Other fields might also need to be interpolated and copy fields like u_prev_save are then needed for these fields as well.

The code in src/fem/adaptive/AdvecA exemplifies the use of GridFEAdT and GridFEAdB for an advection equation $\partial u/\partial t + \boldsymbol{v} \cdot \nabla u = 0$.

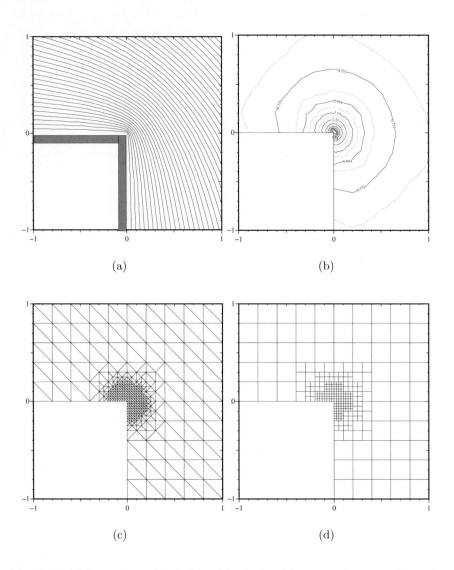

(a)   (b)

(c)   (d)

Fig. 3.14. (a) Streamlines of inviscid and irrotational flow around a corner (contour lines of u); (b) magnitude of the velocity (norm of the gradient of u). Three levels of refinement, using the `GeometricRegions` criterion and refinement of elements inside a disk of radius 0.2 at the corner $(0,0)$, are used for creating the refined grids. (c) Mesh made by class `GridFEAdT`; (d) mesh made by class `GridFEAdB`.

3.7   Projects

3.7.1   Flow in an Open Inclined Channel

Mathematical Problem. The boundary-value problem to be solved in this project reads

$$-\nabla^2 u(\boldsymbol{x}) = 1, \quad \boldsymbol{x} \in \Omega, \tag{3.19}$$

$$u = 0, \quad \boldsymbol{x} \in \partial\Omega_E, \tag{3.20}$$

$$\frac{\partial u}{\partial n} = \tau, \quad \boldsymbol{x} \in \partial\Omega_N. \tag{3.21}$$

The quantity τ is a dimensionless constant.

Physical Model. This project concerns incompressible viscous fluid flow in a straight open channel, as depicted in Figure 3.15a. The axis along the channel makes the angle α with the horizontal plane, and the fluid flow is driven by gravity and a wind stress, here represented in scaled form through τ, at the fluid surface $\partial\Omega_N$. Both positive and negative values of τ are relevant. The cross section of the channel can be arbitrary, but it is a basic assumption that the fluid surface is flat, that is, there are no waves. The boundary $\partial\Omega_E$ is the channel wall. We can introduce $\boldsymbol{x} = (x_1, x_2)$ as the coordinates in Ω. The x_3 coordinate is then directed along the axis of the channel. The primary unknown $u(\boldsymbol{x})$ is the velocity in x_3 direction.

Derive the boundary-value problem from the incompressible Navier-Stokes equations with the proper boundary conditions. (Hint: Make a basic assumption that the flow is directed along the x_3 axis and omit the pressure gradient term $\partial p/\partial x_3$.) Introduce a scaling to arrive at (3.19)–(3.21) and find the dimensionless parameter τ expressed by physical parameters in the problem.

Numerical Method. The problem (3.19)–(3.21) is to be solved by a Galerkin finite element method.

Analysis. If the "width" of Ω is much larger than the "height", the problem can be approximated by a thin film on an inclined plate of infinite extent (i.e., Ω is a rectangle whose width tends to infinity). Neglecting end-effects, we have hence reduced the problem to one space dimension, with the effective coordinate, here called x_1, running in the normal direction to the plate. (Calling the effective coordinate x_1 makes it easy to use the same code for a 1D test problem as well as for the original 2D problem.) The 1D test problem reads

$$-u''(x_1) = 1, \; x_1 \in (0,1), \quad u(0) = 0, \; u'(1) = \tau. \tag{3.22}$$

Show that this problem can be solved exactly by a finite element method with linear elements, regardless of the element size, thus making the test problem ideal for program verification.

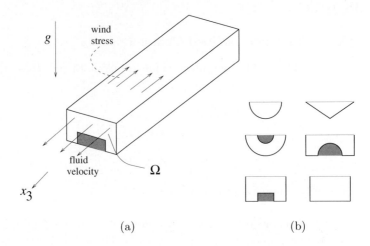

(a)   (b)

Fig. 3.15. (a) Viscous fluid flow in an open channel; (b) examples on various cross sections (Project 3.7.1).

Implementation. Create a standard Diffpack finite element solver for the problem, where the number of space dimensions is parameterized as usual (only 1D and 2D cases make sense in the current physical context). Provide debug output of the numerically computed and the hand-calculated element matrices and vectors in the 1D problem (linear elements), see page 270 for how to turn the debug statements on or off at compile time. Demonstrate in particular that the program works as expected with three elements. Visualize the solution u (the fluid velocity) and $||\nabla u||$ (a measure of the strain rate[32]). Furthermore, compute the effective fluid volume that passes through Ω per time unit, i.e., $Q = \int_\Omega u \, dx_1 dx_2 / \int_\Omega dx_1 dx_2$ (normalized volume flux). The material in Appendix B.5.1, especially on page 566, is useful for implementing the calculation of Q.

Computer Experiments. Run some examples with nontrivial geometries as depicted in Figure 3.15b (all the shown cross sections can be generated by the `PreproStdGeom` utility from Chapter 3.5.3). In all the cases, make sure to utilize symmetry. For example, the semi-disk can be reduced to a quarter of a disk, with $\partial u / \partial n = 0$ on the symmetry line. Determine which of the cross sections that has the maximum normalized volume flux Q.

---

[32] `FieldsFE::magnitude` can be used to compute $||\nabla u||$ when ∇u is available, see page 253.

3.7.2   Stress Concentration due to Geometric Imperfections

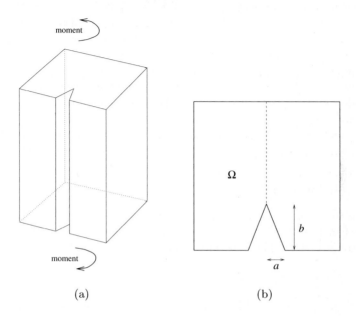

Fig. 3.16. (a) Torsion of a long cylinder, subject to moments at the ends; (b) cross section Ω of the cylinder (Project 3.7.2).

Mathematical and Physical Problem. We consider torsion of a long elastic cylinder with arbitrary cross section Ω as depicted in Figure 3.16. An important physical quantity is the absolute value of the shear stress in the cross section, which can be expressed as $||\nabla u||$, where u solves $\nabla^2 u = -2$ in Ω, with $u = 0$ on the boundary $\partial\Omega$. The torsion problem is derived in most textbooks on elasticity, see e.g. [50,56,103,118].

Numerical Method and Implementation. The current boundary-value problem is to be solved by a standard Galerkin finite element method. Implementation of the torsion problem is in fact described in Chapter 3.4.6, but one can also develop a tailored solver by straightforward editing of class `Poisson1` or `Poisson2`. Adaptive grids, as described in Chapter 3.6, are useful in the present context, especially since we know a priori that the area around the crack tip in Figure 3.16 should be refined. It is important to calculate and visualize the shear stress $||\nabla u||$ and write out the maximum value of $||\nabla u||$ and where it occurs.

Computer Experiments. The purpose of this project is to study the impact of cracks or geometrical imperfections, at the boundary, on the maximum shear stress. Figure 3.16b depicts a cross section containing an exaggerated crack with dimensions a and b. Due to symmetry it is only necessary to compute u in the area to the right of the dashed line. Along the dashed line we apply the symmetry condition $\partial u/\partial n = 0$. Generation of a grid over this domain can be accomplished by, e.g., transfinite mappings as explained in the report [64]. We suggest to read the crack-size parameters a and b from the menu and let the simulator transform a grid on the unit square to the actual form demanded by this particular problem. For each value of the pair (a, b), run through a series of grids with increasing refinement and find some criterion for picking a suffiently refined grid. One will find that even small imperfections of the geometry have a significant impact on the maximum shear stress.

3.7.3   Lifting Airfoil

Mathematical Problem. This project concerns flow around a 2D airfoil as depicted in Figure 3.17. We assume that the flow is inviscid, irrotational, and incompressible. The governing equation is then the Laplace equation

$$\nabla^2 \psi = 0 \qquad\qquad (3.23)$$

in the fluid domain Ω. The primary unknown ψ is the *stream function*. The flow velocity $\boldsymbol{v} = (u, v)^T$ is tangent to the isolines of ψ, and we have $u = \partial\psi/\partial y$ and $v = -\partial\psi/\partial x$. There is no velocity through the airfoil. This means that $\psi = c$, where c is a constant, on the airfoil boundary. On the outer boundary we set $\psi = \boldsymbol{U}_\infty \cdot (y, -x)$, where $\boldsymbol{U}_\infty$ is the free stream velocity.

At the trailing edge E we need a special condition called the *Kutta-Joukowski condition*. This means that the unknown constant c is determined so that $\partial\psi/\partial n$ is continuous at the point E.

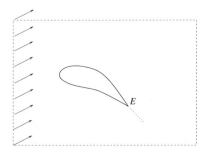

Fig. 3.17. Flow around a 2D airfoil (Project 3.7.3).

Analysis. Explain why $\nabla^2 \psi = 0$ reflects that the flow is irrotational ($\nabla \times \boldsymbol{v} = \boldsymbol{0}$). Also explain that any flow described by a stream function ψ must be incompressible ($\nabla \cdot \boldsymbol{v} = 0$), provided ψ is sufficiently smooth. Show that the pressure behaves like const $- \frac{1}{2}|\nabla\psi|^2$ (hint: use the Bernoulli equation). Finally, explain why the boundary condition $\psi = $ const ensures that the normal velocity $\boldsymbol{v} \cdot \boldsymbol{n}$ vanishes ($\boldsymbol{n}$ being the outward unit normal to the boundary).

Numerical Method. The challenge in this problem is to devise a procedure for incorporating the Kutta-Joukowski condition. One idea, taken from [74, Ch. 1], consists in setting

$$\psi = \psi_0 + c\psi_1,$$

where ψ_0 solves the problem with $c = 0$ and ψ_1 solves the problem with $c = 1$. The unknown constant c can then be computed by requiring a continuous normal derivative of ψ at the trailing edge E. The normal direction $\boldsymbol{n}$ is perpendicular to the dotted line starting at E in Figure 3.17. A numerical approximation to the normal derivative of ψ at E becomes

$$\frac{\partial\psi}{\partial n} \approx \frac{1}{\delta}(\psi_0(E + \delta\boldsymbol{n}) + c\psi_1(E + \delta\boldsymbol{n}) - (\psi_0(E) + c\psi_1(E)))$$

$$= \frac{1}{\delta}(\psi_0(E + \delta\boldsymbol{n}) + c\psi_1(E + \delta\boldsymbol{n}) - c).$$

The approximation is exact when the constant $\delta > 0$ approaches zero. The *jump* in the normal derivative can then be written

$$\left[\frac{\partial\psi}{\partial n}\right] \approx \frac{1}{\delta}(\psi_0(E + \delta\boldsymbol{n}) + c\psi_1(E + \delta\boldsymbol{n}) - c + \psi_0(E - \delta\boldsymbol{n}) + c\psi_1(E - \delta\boldsymbol{n}) - c).$$

Requiring this jump to be zero gives

$$c = \frac{\psi_0(E + \delta\boldsymbol{n}) + \psi_0(E - \delta\boldsymbol{n})}{\psi_1(E + \delta\boldsymbol{n}) + \psi_1(E - \delta\boldsymbol{n}) - 2}.$$

Implementation. We shall consider a symmetric airfoil, but with $\boldsymbol{U}_\infty$ making an angle α with the line of symmetry (i.e. α is the angle of attack and the flow becomes non-symmetric, resulting in a lift on the airfoil). The NACA0012 airfoil geometry is given by [74]:

$$y = 0.17735\sqrt{x} - 0.075597x - 0.212836x^2 + 0.17363x^3 - 0.06254x^4,$$

for $x \in [0, 1]$. Use a preprocessor, e.g. Geompack [64] or Triangle [104], to generate a suitable mesh around the complete airfoil.

Extend a typical linear stationary solver, like `Poisson1` or `Poisson2`, to this problem, by stripping unnecessary parts of the code, introducing three fields `psi0`, `psi1`, and `psi` instead of `u`, and adding a flag in the class that indicates whether we are solving the problem for ψ_0 or ψ_1. Test on this flag inside relevant routines (`fillEssBC`, `integrands`). The `solveProblem` must call up two assembly and solve processes, one for `psi0` and one for `psi1`. Implement the formula for c with a finite δ (one can use the `valuePt` function in the `FieldFE` class to evaluate $\psi_0(E + \delta n) = \psi_0(1, \delta)$ etc.). Visualize the final ψ function, the magnitude of the velocity field $||\nabla\psi||$ and the pressure $-\frac{1}{2}\varrho||\nabla\psi||^2$, where ϱ is the constant density of air. (Find a smooth $\nabla\psi$ from `makeFlux` and use the `FieldsFE::magnitude` to compute $||\nabla\psi||$, then take a copy of this field and apply the `FieldFE` functions `apply(sqr)` and thereafter `mult(-0.5*rho)` to calculate $-\frac{1}{2}\varrho||\nabla\psi||^2$). The streamlines are trivially visualized by plotting the isolines of ψ.

The total lift on the airfoil is obtained by integrating the pressure around the airfoil. Perform this integration numerically. (Hint: Use information from Appendix B.5.1 and construct a loop over all elements, check if current element contains the airfoil boundary, and if so, invoke a loop over the integration point on the sides, interpolate the pressure, and add the contribution to the lift.)

3.8   A Convection-Diffusion Solver

The Poisson equation solvers in Chapters 3.1 and 3.2 can almost trivially be extended to convection diffusion problems, involving equations like

$$\boldsymbol{v} \cdot \nabla u = \nabla \cdot (\lambda \nabla u), \tag{3.24}$$

with appropriate boundary conditions. We shall in this section outline a solver for a 2D version of (3.24), with λ constant and $\boldsymbol{v} = (v\cos\theta, v\sin\theta)^T$, being also constant, and $\theta \in [0, \pi/2]$. The boundary conditions are depicted in Figure 3.18. When $v \to \infty$ we get $u = 1$ for $y > x\tan\theta + 1/4$ and $u = 0$ for $y < x\tan\theta + 1/4$. There will hence be a steep front aligned with $y = x\tan\theta + 1/4$ for large velocities. If $u = 0$ at $y = 1$ there will be also a thin boundary layer close to $y = 1$. These rapid changes in u make strong demands on the discretization methods.

We employ the standard weighted residual method, with $\hat{u} = \sum_j u_j N_j$ and weighting functions W_i. The weighted residual statement for this problem takes the form

$$\sum_{j=1}^{n} \int_{\Omega} \left[W_i v \cos\theta \frac{\partial N_j}{\partial x} + W_i v \sin\theta \frac{\partial N_j}{\partial y} + \right.$$
$$\left. \lambda \frac{\partial W_i}{\partial x} \frac{\partial N_j}{\partial x} + \lambda \frac{\partial W_i}{\partial y} \frac{\partial N_j}{\partial y} \right] dx dy \, u_j = 0, \tag{3.25}$$

with the constraint that some u_i are required to fulfill the Dirichlet conditions.

As explained in Chapter 2.9, Galerkin methods $(W_i = N_i)$ for equations like (3.24) may lead to nonphysical oscillations if the convective term is much larger than the diffusion term. The local mesh Peclet number, here $\mathrm{Pe}_{\Delta} =$

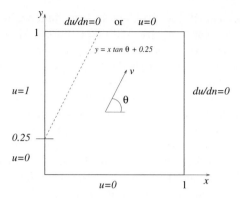

Fig. 3.18. Two-dimensional domain with boundary conditions for the convection-diffusion equation $\boldsymbol{v} \cdot \nabla u = \nabla^2 u$.

vh/λ, h being the characteristic element length, indicates whether a Galerkin approach is appropriate or not; a value of 2 is the critical size for a stable Galerkin solution. For large velocities v relative to λ and coarse grids one must in general apply Petrov-Galerkin methods where $W_i \neq N_i$ to stabilize the spatial discretization. Diffpack offers some common choices of weighting functions in class UpwindFE. To incorporate this tool in a solver, include the UpwindFE.h header file, declare an object UpwindFE PG as class member, put PG on the menu: UpwindFE::defineStatic or PG.define, initialize PG from the menu: PG.scan, and use PG as follows in the integrands routine:

```
void ConvDiff1::integrands (ElmMatVec& elmat,const FiniteElement& fe)
{
  const int nbf = fe.getNoBasisFunc(); // no of nodes in this element
  const real detJxW = fe.detJxW();     // Jacobian * integr. weight

  // compute Petrov-Galerkin weighting functions:
  PG.calcWeightingFunction (fe, velocity, diffusion, 0.0, true);

  int i,j;
  for (i = 1; i <= nbf; i++) {
    for (j = 1; j <= nbf; j++)
      elmat.A(i,j) +=
      (PG.W(i)*(velocity(1)*fe.dN(j,1) + velocity(2)*fe.dN(j,2)) +
      lambda*(PG.dW(i,1)*fe.dN(j,1) + PG.dW(i,2)*fe.dN(j,2)) )*detJxW;
    // not necessary: elmat.b(i) += 0;
  }
}
```

The parameters to the UpwindFE::calcWeightingFunction are the finite element information fe, the velocity vector at the current integration point, the diagonal of the diffusion tensor at the current integration point (represented as a Ptv vector, where each entry equals λ in the present problem), the time step length (0 in a time-independent problem) and a boolean parameter that is true if the problem is time independent. Some of the methods in class `UpwindFE` require additional information to be passed to `calcWeightingFunction`. We refer to the man page for further details.

A complete solver for our particular problem is found in the directory `src/fem/ConvDiff1`. This solver is tailored to the geometry and the boundary indicators in our problem, using tools like `GridFE::redefineBoInds` and `GridFE::addBoIndNodes`, which are described in Chapters 3.5.1 and 3.5.1. Using a multiple loop, we can investigate how nonphysical oscillations are suppressed by "upwind finite elements". Let us compare the Galerkin method with the classical weighting functions from Brooks and Hughes [18], denoted as methods 0 and 1, respectively, in class `UpwindFE`. Here are a series of experiments with the `ConvDiff1` solver that the reader should try:

```
./app --iscl --nel_x 18 --upwind '{0 & 1}' --u0y1 '{ true & false }'
       --velocity '{10 & 1000000}' --theta '{0.1 & 0.7}' --casename CD
plotmtv -nodate -color CD_m*.mtv &
```

We emphasize that the solver dumps the u field directly in Plotmtv format, that is, the more general `SaveSimRes` utility is not utilized. You will see that the solver adds a comment about v, θ, the number of nodes, and the upwind method, in the upper right corner of the plot. This makes it easy to identify a plot. More flexible visualization is of course enabled by adding a `SaveSimRes` object in the solver class.

From these experiments, one can easily see the improvements of the Petrov-Galerkin formulation when the velocity vector is large and skew to the grid. Especially when we force $u = 0$ at the outflow boundary $y = 1$ (the `--u0y1 true` option), the Galerkin method performs poorly. Method 1 in class `Upwind` is successful at stabilizing the results for θ not too small. Figure 3.19 demonstrates the significant improvement of the Petrov-Galerkin formulation over the traditional Galerkin method.

The current quite primitive convection-diffusion equation solver is in Chapter 6.1 combined with the more sophisticated elements of the `Poisson2` code.

3.9   A Heat Equation Solver

Chapters 3.1–3.6 all deal with stationary PDE solvers. In the rest of Chapter 3 we shall address transient PDEs. Our first task in this respect is to make a Diffpack program that solves the standard time-dependent heat (or diffusion) equation. The initial-boundary value problem reads

$$\frac{\partial u}{\partial t} = \nabla \cdot (k \nabla u) + f, \quad \boldsymbol{x} \in \Omega \subset \mathbb{R}^d, \ t > 0, \tag{3.26}$$

$$u(\boldsymbol{x}, t) = g(\boldsymbol{x}, t), \quad \boldsymbol{x} \in \partial\Omega_E, \ t > 0, \tag{3.27}$$

$$u(\boldsymbol{x}, 0) = I(\boldsymbol{x}), \quad \boldsymbol{x} \in \Omega, \ t = 0. \tag{3.28}$$

Here $u(\boldsymbol{x}, t)$ is the unknown temperature, $I(\boldsymbol{x})$ is a given initial temperature distribution, $k(\boldsymbol{x})$ is a coefficient related the conductive properties of the

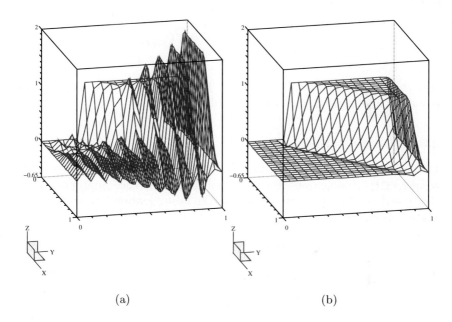

(a)   (b)

Fig. 3.19. Plot of $u(x, y)$, computed by the `ConvDiff1` solver, with $v = 1000$, $\theta = 0.7$, and 20×20 bilinear elements. (a) Pure Galerkin formulation; (b) Petrov-Galerkin formulation corresponding to upwind method 1 [18] in class `UpwindFE`.

medium[33], $f(\boldsymbol{x})$ represents a heat source, $g(\boldsymbol{x})$ is the Dirichlet-type boundary values for u, and $\partial\Omega_E$ is the whole boundary of Ω. As usual, the subscript E indicates that $\partial\Omega_E$ has essential boundary conditions.

Application of finite elements to time-dependent problems is frequently accomplished by first discretizing in time by finite differences and then solving the resulting recursive spatial problems using the finite element method. This strategy was explained in Chapter 2.2 and is frequently used throughout the book. It will become evident that a time-dependent finite element solver in Diffpack consists mainly of the objects and corresponding code we had in stationary problems, with only some small software extensions.

To verify the implementation, it is convenient to compare the numerical solution with an analytical solution. Let

$$u(x_1,\ldots,x_d,t) = e^{-d\pi^2 kt}\prod_{j=1}^{d}\sin(\pi x_j)$$

and $\Omega = (0,1)^d$. When $f = 0$, $g = u$, $I = u(\boldsymbol{x},0)$, and k is constant, this specific u fulfills the initial-boundary value problem.

3.9.1   Discretization

Let Δt be the time step length, and let $u^\ell(\boldsymbol{x})$, $\ell \in \mathbb{N}$, denote an approximation to $u(\boldsymbol{x},\ell\Delta t)$. Application of a backward finite difference approximation to the time derivative in (3.26) transforms the heat equation into a sequence of spatial problems,

$$u^\ell - \Delta t\nabla\cdot(k\nabla u^\ell) = \Delta t f + u^{\ell-1}. \tag{3.29}$$

A standard Galerkin finite element method can be used to convert (3.29) to a sequence of linear systems,

$$\boldsymbol{A}\boldsymbol{u}^\ell = \boldsymbol{b}(\boldsymbol{u}^{\ell-1}), \tag{3.30}$$

where $\boldsymbol{u}^\ell$ is the vector of nodal values of $u^\ell(\boldsymbol{x})$. The contribution to the element matrix and vector from a numerical integration point $\boldsymbol{\xi}_\ell$ becomes

$$[(N_iN_j + k\Delta t\nabla N_i\cdot\nabla N_j)\det J]_{\boldsymbol{\xi}=\boldsymbol{\xi}_\ell}\, w_\ell \tag{3.31}$$

for the matrix and

$$[N_i(f\Delta t + u^{\ell-1})\det J]_{\boldsymbol{\xi}=\boldsymbol{\xi}_\ell}\, w_\ell \tag{3.32}$$

for the vector. The notation was explained in Chapter 3.1.

---

[33] More precisely, k is the heat conduction coefficient (in Fourier's law) divided by ϱC_p (density times the heat capacity) and f is the heat source divided by ϱC_p.

Remarks. We can alternatively write the discrete equations in matrix form as

$$(\boldsymbol{M} + \Delta t \boldsymbol{K})\boldsymbol{u}^\ell = \Delta t f_T(t)\boldsymbol{c} + \boldsymbol{M}\boldsymbol{u}^{\ell-1},$$

assuming that the time dependence in f is separable, that is, $f(x,t) = f_T(t)\tilde{f}(x)$, and $\boldsymbol{c}$ is a vector arising from integrating the term $\tilde{f}(x)N_i$. The coefficient matrix in the linear system is $\boldsymbol{A} = \boldsymbol{M} + \Delta t \boldsymbol{K}$. In this case, the most efficient method for solving the problem is to compute $\boldsymbol{M}$, $\boldsymbol{K}$, and $\boldsymbol{c}$ initially and avoid the complete computation and assembly of these matrices at each time level. Moreover, if Gaussian elimination is used to solve the linear system, the decomposition of $\boldsymbol{A}$ should only be performed initially. The implementation of this procedure is given in Appendix B.6.2, and the associated efficiency increase can be substantial. However, in this introductory example on solving time-dependent PDEs in Diffpack, we will first use the general class FEM functionality and recompute the linear system at every time level. This approach is simple and completely general and is therefore the preferred method for rapid prototyping of a solver. We also know that in nonlinear problems the linear systems must be recomputed for each nonlinear iteration at each time step.

3.9.2   Implementation

Let Heat1 be the name of the solver. Class Heat1 must be derived from FEM, just as in the stationary case. Because time-dependent finite element problems are solved as a sequence of spatial problems, the reader should expect that a time-dependent finite element solver in Diffpack can reuse much of the code from the stationary solvers in Chapter 3. This is indeed the case. Our first time-dependent solver can be viewed as an extension of class Poisson1, and the reader should be familiar with that code and the material in Chapters 3.1 and 3.2 before proceeding with the present section.

In addition to the steps in the stationary solvers, a time-dependent solver needs to set the initial conditions for u in a function setIC and it needs to run a time loop in a function timeLoop. The extra data needed in a time-dependent solver is of course u at the previous time step $(u^{\ell-1})$, as well as information about Δt and the time interval in which our initial-boundary value problem is to be solved. These latter parameters, which are needed in all time-dependent problems, are collected in a class TimePrm in Diffpack, explained in Chapter 1.6.3 (see page 95).

Before discussing the contents of the various new functions in class Heat1, we present the class definition.

```
class Heat1 : public FEM
{
public:
  Handle(GridFE)      grid;
  Handle(DegFreeFE)   dof;
  Handle(FieldFE)     u;          // solution u^q
```

```
Handle(FieldFE)      u_prev;      // u^{q-1} (previous time level)
FieldSummary         u_summary;   // time history of extremes of u
Vec(real)            linsol;      // solution of linear system
Handle(LinEqAdmFE)   lineq;       // linear system and solvers
Handle(TimePrm)      tip;         // time discretization parameters

Handle(FieldFE)      error;
NumItgPoints         error_itg_pt_tp;
real L1_error,L2_error, Linf_error;
Handle(SaveSimRes)   database;

Handle(FieldsFE)     flux;        // flux = smooth -k*grad(u)
real                 diffusion_coeff;  // constant k
Handle(FieldFunc)    uanal;       // functor for analytical solution

// in case of surface integrals or debugging:
virtual void calcElmMatVec
   (int e, ElmMatVec& elmat, FiniteElement& fe);

virtual void integrands (ElmMatVec& elmat, const FiniteElement& fe);
virtual void timeLoop();
virtual void solveAtThisTimeStep();
virtual void fillEssBC ();              // set boundary conditions
virtual void setIC();                   // set u_prev equal to u(x,0)
virtual void define (MenuSystem & menu,int level=MAIN);
virtual void scan ();

Heat1 ();
~Heat1 ();
virtual void adm      (MenuSystem & menu);
virtual void solveProblem();
virtual void resultReport();

// f = source, k = diffusion, I = initial cond., g = Dirichlet cond.
virtual real f(const FiniteElement& fe, real t = DUMMY);
virtual real k(const FiniteElement& fe, real t = DUMMY);
virtual real I(const Ptv(real)& p);
virtual real g(const Ptv(real)& p, real t=DUMMY);
};
```

We have let all members of the class be public, just for convenient future access to various data from functors or other classes. The danger of accidentally manipulating internal data in class Heat1 is considered to be small.

Because most of the Heat1 class is similar to class Poisson1[34], we discuss only the new functions in the time-dependent case in the following. Assignment of initial conditions is performed in the setIC function:

```
void Heat1:: setIC ()
{
   u_prev->fill (*uanal, 0.0); // call uanal for t=0 at all nodes
   *u = *u_prev;               // make u ready for dump also at t=0
}
```

---

[34] This observation points in the direction of letting Heat1 be a subclass of Poisson1. Exercise 3.7 explores the idea.

We use a functor uanal in class Heat1 to hold the analytical solution of the problem. The field handle u_prev, representing $u^{\ell-1}(x)$ in our formulas, is filled with values from the analytical solution, evaluated at time zero, using

```
FieldFE::fill (FieldFunc& functor, real t)
```

The fill function calls the functor's valuePt(x,t) function at all grid points and sets the nodal values accordingly. As an alternative to FieldFE::fill, we could have coded a standard loop over the nodal points in setIC:

```
/* alternative syntax in setIC: user loop, node by node:
const int nno = grid->getNoNodes();
Ptv(real) x;                       // spatial point
for (int i = 1; i <= nno; i++) {
  x = grid->getCoor(i);            // get the coord. x of node i
  u_prev->values()(i) = I(x);      // assign value to node i
}
*/
```

The time loop is coded in a function timeLoop, which typically takes the form

```
void Heat1:: timeLoop()
{
  tip->initTimeLoop();
  setIC();
  database->dump (*u, tip.getPtr(), "initial condition");

  while(!tip->finished())
    {
      tip->increaseTime();
      solveAtThisTimeStep();
      *u_prev = *u;
    }
}
```

Notice that the call to the dump function takes two optional arguments in time-dependent problems. The second argument is a pointer to a TimePrm object and the third one is a string containing an optional comment. This extra information is stored along with u in the simres database.

The spatial problem to be solved at the current time level is coded in solveAtThisTimeStep. The reader should compare the contents of this function with a typical solveProblem function in a stationary solver.

```
void Heat1:: solveAtThisTimeStep ()
{
  fillEssBC ();                     // incorporate time-dep. ess. b.c.

  makeSystem (*dof,*lineq);         // FEM's assembly algorithm

  dof->field2vec (*u, linsol);      // use most recent u as start vector
  lineq->solve ();                  // solve linear system
  s_o << "t=" << tip->time();
  int niterations; bool c;          // for iterative solver statistics
  if (lineq->getStatistics(niterations,c))  // iterative solver?
```

```
    s_o << oform("  solver%sconverged in %3d iterations\n",
              c ? " " : " not ",niterations);
    s_o << '\n'; s_o.flush();    // flush() forces output _now_

    dof->vec2field (linsol, *u); // load linsol into the field u
    database->dump (*u, tip.getPtr(), "some comment if desired...");
    u_summary.update (tip->time()); // keep track of max/min u etc

    // compute smooth flux -k*grad(u);
    FEM::makeFlux (*flux, *u);
    database->dump(*flux, tip.getPtr(), "smooth flux -k*grad(u)");
}
```

Iterative solution methods for linear systems require a start vector for the iteration, and the default choice in Diffpack is to use the current value of the vector that is attached as solution vector in the LinEqAdmFE object. This means that linsol is used as start vector. For direct methods, like Gaussian elimination, the start vector has no meaning, but for iterative methods it is important to use as good start vector as possible. The situation is quite fortunate in linear time-dependent problems, since the solution at the previous time level provides a good guess for the solution at the current time level. We therefore copy the u field (or u_prev – they are equal after the update *u_prev=*u at the previous pass in the time loop) into linsol prior[35] to calling lineq->solve. When the solution is computed, the linsol vector must be copied into the FieldFE object representation of u. Copying between field representations and vector representations is provided by the DegFreeFE object.

The data item u_summary is of type FieldSummary and keeps track of the extreme values of u and where/when they occur. Such information can be convenient to get a quick summary of the properties of u in a simulation. We refer to the man page for class FieldSummary for information on the usage of the class. The computation of the flux $-k\nabla u$, represented in terms of a FieldsFE handle flux, is computed using the function makeFlux in class FEM.

The solveProblem function is only a call to timeLoop. Computations of errors can be performed at each time step, but to avoid the associated overhead, we do this only at the end of the simulation, in the function resultReport. An appropriate call reads

```
    ErrorNorms::Lnorm (*uanal, *u, tip->time(),
                L1_error, L2_error, Linf_error, error_itg_pt_tp);
```

The third argument, which was DUMMY when explaining the Lnorm function in Chapter 3.4.3, equals now the point of time for evaluating the uanal functor. When the time argument is DUMMY, it means that the simulation is stationary and the time parameter is to be ignored.

Finally, we show how the PDE itself enters the Heat1 simulator.

---

[35] This is not strictly required – linsol contains the solution from the last call to lineq->solve and therefore has the right contents. However, the explicit copy performed in terms of the field2vec call is a safe habit.

```
void Heat1:: integrands (ElmMatVec& elmat, const FiniteElement& fe)
{
  const real detJxW = fe.detJxW();
  const int  nsd = fe.getNoSpaceDim();
  const int  nbf = fe.getNoBasisFunc();
  const real dt  = tip->Delta();
  const real t   = tip->time();
  real gradNi_gradNj;
  // u at the previous time level at the current integration point:
  real up_pt = u_prev->valueFEM(fe);
  const real k_value = k (fe, t);
  const real f_value = f (fe, t);
  int i,j,s;
  for(i = 1; i <= nbf; i++) {
    for(j = 1; j <= nbf; j++) {
      gradNi_gradNj =0;
      for(s = 1; s <= nsd; s++)
        gradNi_gradNj += fe.dN(i,s)*fe.dN(j,s);

      elmat.A(i,j) +=(fe.N(i)*fe.N(j) +
                      dt*k_value*gradNi_gradNj)*detJxW;
    }
    elmat.b(i)+= fe.N(i)*(f_value*dt + up_pt)*detJxW;
  }
}
```

The other functions, like fillEssBC, define, scan, adm, f, g, and k are similar to those in the stationary solvers and should need no specific explanation. The reader should print out and study the complete source code of class Heat1. The code is found in the directory src/fem/Heat1.

Exercise 3.6. Run the Heat1 application with the Verify/test1.i input file. To be confident that the code is correct, we need to apply finer grids and observe that the error goes to zero. Using bilinear (or linear) finite elements in combination with the backward Euler scheme is expected to lead to errors in u^ℓ of order Δt and h^2, where h is a typical element length. Therefore, if we reduce the time step by a factor of four and the element length by a factor of two, the error should be reduced by a factor of four. We encourage the reader to make an input file with this finer grid in space and time and confirm that the errors are reduced by a factor of four. We can then expect that the error goes to zero as fast as the theory predicts and that our simulator implements the numerical method correctly. Such convergence tests should always be performed as partial verification of a simulation code.   ◇

Most of the functionality explained for the stationary solvers are also available for the present heat equation simulator. The reader is encouraged to play around with the menu system, multiple loops etc. Better design of the Heat1 solver, by e.g. extracting the code specific to the current test problem in a subclass, follows the same lines as we explained for the Poisson1 and Poisson2 solvers and is realized in Chapter 3.10. One can of course also create a time-dependent heat equation solver as a subclass of a stationary solver.

This is the subject of an exercise below. Visualization of time-dependent data, including making animations, is described in the next section.

Exercise 3.7. Because a time-dependent solver for the heat equation is nothing but a stationary solver with some additional data and functions, it could be convenient to implement time-dependent solvers as subclass extensions to stationary simulators[36]. Explore this idea by starting with class Poisson2, or ReportPoisson2, and making a subclass Heat1b in a separate directory (use AddMakeSrc to indicate that Heat1b files are to be compiled and linked with the Poisson2 files). ◇

A Comment on Termination Criteria for Linear Solvers. Finite element solution of PDE problems normally requires iterative solution of linear systems, such as (3.30), which we here write in general as $\boldsymbol{Ax} = \boldsymbol{b}$. All iterative solvers start with some guess $\boldsymbol{x}^0$ of the solution vector $\boldsymbol{x}$ and generate hopefully improved approximations $\boldsymbol{x}^1, \boldsymbol{x}^2, \ldots$ The iterations are stopped when a certain termination criterion is reached, for example

$$||\boldsymbol{r}^k|| \leq \epsilon_r ||\boldsymbol{r}^0||, \quad \boldsymbol{r}^j = \boldsymbol{b} - \boldsymbol{Ax}^j, \tag{3.33}$$

or

$$||\boldsymbol{r}^k|| \leq \epsilon_r . \tag{3.34}$$

A good start vector $\boldsymbol{x}^0$ can speed up the iterative process significantly. In linear time-dependent problems, the solution at the previous time step usually provides a very good start vector. Using the default criterion (3.33), which requires reduction of the *relative* residual, can be too restrictive if the start vector is close to the exact solution, because then the initial norm of the residual is small. As a consequence, (3.34) is often preferred in transient problems.

The convergence criterion (3.33) is represented by the name CMRelResidual, whereas (3.34) is called CMAbsResidual. A typical set of menu commands adjusting these and the convergence tolerance ϵ_r read

```
sub ConvMonitorList_prm
sub Define ConvMonitor #1
!set #1: convergence monitor name = CMRelResidual
! use absolute residual in linear time-dependent problems:
set #1: convergence monitor name = CMAbsResidual
set #1: residual type = ORIGINAL_RES
set #1: convergence tolerance = 1.0e-6
```

This submenu is invoked from the LinEqAdmFE submenu. More information on termination criteria for linear solvers is provided in Appendix D.6.

---

[36] In Chapter 3.10.4 we explain how a time-dependent solver can be applied to a stationary problem. The present exercise demonstrates the opposite possibility, i.e., how a stationary solver can be reused in a transient problem.

Exercise 3.8. The purpose of this exercise is to determine the relative efficiency of the two convergence criteria (3.33) and (3.34) in the test problem
solved by class `Heat1`, keeping ϵ_r fixed. First, run the `Heat1` simulator with the
`Verify/test1.i` file as input. You can now run the `InputManualInLaTeX` script
to produce a description of all menu items. Based on this information, set up
an appropriate input file where the matrix storage format is `MatSparse` (a general sparse matrix), the solver is `ConjGrad` (the Conjugate Gradient method),
and the preconditioner is `PrecRILU` with relaxation parameter 1.0 (corresponding to MILU preconditioning). Finally, specify a multiple answer, containing
`CMAbsResidual` and `CMRelResidual`, to the convergence monitor items shown
previously. Run the experiment with a 100×100 grid consisting of bilinear
elements on the unit square, and let the time step be 0.02 for $t \in [0, 0.4]$.
You will see that the `CMAbsResidual` criterion roughly halves the efficiency of
the solver compared with `CMRelResidual`, while the final errors in u are of
approximately the same size. (Of course, the efficiency of `CMRelResidual` can
be improved by increasing ϵ_r when using that criterion.)   ◇

3.10 A More Flexible Heat Equation Solver

Class `Heat1` offers the same degree of flexibility as class `Poisson1`. In Chapter 3.5 we motivated for more flexibility and presented the `Poisson2` solver,
which had a design and employed tools that encourage extensive reuse of the
solver for other stationary scalar PDEs. The design and tools can almost trivially be transferred to a time-dependent problem, resulting in the simulator
`Heat2`.

3.10.1 About the Model Problem and the Simulator

The Model Problem. The `Heat2` simulator solves

$$\beta \frac{\partial u}{\partial t} = \nabla \cdot [k \nabla u] + f, \quad \boldsymbol{x} \in \Omega, \ t > 0, \tag{3.35}$$

$$u(\boldsymbol{x}, 0) = I(\boldsymbol{x}), \quad \boldsymbol{x} \in \Omega, \tag{3.36}$$

$$u = D_1, \quad \boldsymbol{x} \in \partial \Omega_{E_1}, \ t > 0, \tag{3.37}$$

$$u = D_2, \quad \boldsymbol{x} \in \partial \Omega_{E_2}, \ t > 0, \tag{3.38}$$

$$u = g(\boldsymbol{x}), \quad \boldsymbol{x} \in \partial \Omega_{E_3}, \ t > 0, \tag{3.39}$$

$$-k \frac{\partial u}{\partial n} = 0, \quad \boldsymbol{x} \in \partial \Omega_N, \ t > 0, \tag{3.40}$$

$$-k \frac{\partial u}{\partial n} = \alpha u - U_0, \quad \boldsymbol{x} \in \partial \Omega_R. \tag{3.41}$$

The parameter β is a prescribed coefficient, and the other symbols are as
explained in Chapter 3.5.

We shall apply the θ-rule in time, combined with a Galerkin finite element method in space, see Chapter 2.2.2. The discrete initial-boundary value problem is implemented in the files in `src/fem/Heat2`, using the same conventions for the boundary indicators as in class `Poisson2`.

Automatic Report Generation. Automatic report generation in the simulator is a valuable tool when performing series of numerical experiments, as we explained in Chapter 3.5.5 for the `Poisson2` solver. Report generation functions can be generated by the `MkReport` script also for the `Heat2` solver. Now we type `MkReport -t Heat2`, where the `-t` option indicates that `Heat2` solves a time-dependent problem. Two new files are made by the `MkReport` script: `ReportHeat1.h` and `ReportHeat1.cpp`. These are slightly different from the files generated in the stationary case, simply due to the presence of additional information to be reported in a transient problem.

The file `testplot2D.i` in `Heat2`'s `Verify` directory represents a suitable test problem for report generation facility. Run the case and invoke the summary report in HTML format using your favorite browser (see page 273 for more information).

Overview of the Simulator Classes. The special test case used in the `Heat1` solver is located in the subclass `Heat2analsol` of `ReportHeat2`. Other subclasses, representing other test problems, can be added as explained for the `Poisson2` application. The reader is recommended to print out the source code of the files that make up the `Heat2` solver and compare these with class `Heat1` and the `Poisson2` solver.

Exercise 3.9. Derive the mathematical expressions to be coded in the function `integrands` when (3.35)–(3.41) is discretized by a θ-rule in time and a Galerkin finite element method in space.   ◇

Example 3.1. Thermal Conditions During Welding. To link (3.35)–(3.41) to a physical application, we consider *welding*, which is a widespread technique in industrial manufacturing of metal products. The thermal conditions associated with welding are important, because they influence the stress and deformation state of the material (this coupling is explored in Project 7.3.1). To compute the temperature measure T, we can solve the energy equation

$$\varrho C_p \frac{\partial T}{\partial t} = \nabla \cdot (\lambda \nabla T) + Q,$$

where ϱ is the density of the metal, C_p is the heat capacity, λ is the heat conduction coefficient, and Q incorporates the effects of welding as a heat source. Phase changes should also be included in the model, but this topic is omitted here for simplicity (see Chapter 6.1 for how to include this effect). A conventional welding source is

$$Q(x_1, \ldots, x_d, t) = Q_I \exp\left(-\sum_{i=1}^{d}(x_i - v_i t - y_i)^2 / (2E_i^2) \right).$$

The intensity Q_I of the source depends, among other things, on the electric voltage and the current from the heat source. The current position of the source is $(x_1, \ldots, x_d)$, whereas $(y_1, \ldots, y_d)$ is its position at $t = 0$, and the velocity of the heat source in the x_i direction is v_i. The quantity E_i reflects the extent of the source in the x_i direction.

The initial condition in the welding problem can be set as $T(\boldsymbol{x}, 0) = 0$, assuming that $T = T_R - T_0$, where T_R is the real (physical) temperature and T_0 is the uniform initial value of T_R. At the boundaries we apply Newton's cooling law, which can be expressed in terms of the Robin condition (3.41). The welding case described here is implemented in class `Heat2weld`, which is a subclass of `ReportHeat2`. As explained in Chapter 3.5.6, we add the problem dependent parameters Q_I, v_i, y_i, and E_i as data members in class `Heat2weld` and read their values from the menu. Moreover, the source Q is realized as a redefined virtual function that computes f in (3.35).

We shall use the `Heat2weld` solver actively in the demonstration of various visualization tools in the following section. For this purpose, it is not necessary to understand the source code – one only needs to run the application with ready-made input files.   ◇

3.10.2 Visualization and Animation of Time-Dependent Data

Our heat equation simulator applies the `SaveSimRes` tool for storing the computed fields on file. Filtering the resulting simres file format to a format required by a special plotting program is explained in Chapter 3.3. In time-dependent problems, some of the `SaveSimRes` and filter tools have additional features that are outlined next.

Filtering Time-Dependent Simres Data. Assume that you want to plot a field with name u at the time points 1.4, 2.6, and 3.9. The `simres2xxx` command must then be equipped with the -t option, e.g.,

```
simres2xxx -f SIMULATION -n u -t '1.4 2.6 3.9;' -a
```

This results in three files, one for each time point. The time value is a part of the filename. To plot *all* the u fields in the simres database, use the -A option instead of -t,

```
simres2xxx -f SIMULATION -n u -A -a
```

There is also an -E option that extracts *every* field in the simres database.

Storing Fields at Selected Time Points. By default, all fields that are subject to a `database->dump` call will be stored on file. In time-dependent problems this can easily fill up the disk space so class `SaveSimRes` defines a menu item that one can use to control the amount of data storage. Giving a menu command like

```
set time points for plot = [0:10,2]
```

on the SaveSimRes submenu, ensures that database->dump will only have effect when the time t *approximates an entry in the set* $(0, 2, 4, 6, 8, 10)$. The syntax [0:10,2] generates this set of time points (see page 70). The command

```
set time points for plot = ALL
```

makes all calls to dump active, while

```
set time points for plot = NONE
```

avoids dumping fields to the simres database, thus saving execution time and disk space. The effect of the latter menu answer is also obtained by the --nodump command-line option to app (cf. Appendix B.3.2).

Time Series at Selected Spatial Points. It can be of interest to plot the development of a field in time at fixed spatial points. Class SaveSimRes supports such curves, much in the same way as it supports plotting of fields along lines through the domain (see page 240). To enable time series curves $u(\boldsymbol{x}_p, t)$ at a number of specified points $\boldsymbol{x}_p$, we simply invoke the SaveSimRes submenu and write

```
set time series points = d=2 n=2 (0.5,0.5) (0.9,0.9)
```

This command specifies time series at two (n=2) spatial points in 2D (d=2): $(0.5, 0.5)$ and $(0.9, 0.9)$. Run the Heat2 solver with Verify/testplot2D.i as input. The time series of u at the specified spatial points can thereafter be plotted by the graphical curveplotgui tool or by the script curveplot (see Chapter 1.3.4 and Appendix B.4.1), or by the curve plot utility in the GUI on Windows platforms. For immediate testing we give the appropriate options to the curveplot script (mycase is the current casename):

```
curveplot gnuplot -f .mycase.curve.map -r 'Time' -r 'u' -r '.'
```

The code that activates plotting of time series at spatial points consists of three calls to the SaveSimRes object database:

1. an initializing call to initTimeSeriesPlot before the time loop is invoked,
2. a call to the function add2TimeSeriesPlot at every time step (located in Heat2::saveResults),
3. a final call to finishTimeSeriesPlot after the time loop has finished.

The reader can find these calls in the Heat2 source code.

Making Movies. Animation is an effective technique for visualizing time-dependent data, and Diffpack comes with some tools to support the production of animations. These tools are based on making a series of pictures, in PostScript or another format, and then using an mpeg encoder to make a movie in the mpeg format.

Generating Data for Animation. The usage of the animation tools is best illustrated through an example. The following command generates some simulation results, based on Example 3.1 on page 312, which we will make use of in the forthcoming text. To run the command, you need to be in the `Heat2` directory.

```
./app --class Heat2weld --iscl --Default Verify/embed.i
       --casename weld1 --source_center '0.2 0.3'
       --source_velocity '1 0.8' --source_intensity 4.0
       --plot_times ALL
```

The `--source*` options override default answers or commands in `embed.i`. The `--plot_times ALL` option is the counterpart to the `set time points for plot` command on the `SaveSimRes` submenu and ensures that all the computed fields are stored in the simres database. This is important for making a smooth movie. Alternatively, the command-line options to `app` can be added to `embed.i` (you are encouraged to do this as an exercise to help you gain experience with the menu system).

Animation Using Plotmtv. We can employ the public domain plotting program Plotmtv for producing the frames in a movie of u. A script `simres2mpeg` first runs `simres2mtv` to produce .mtv files, it then calls the script `plotmtvps` to produce PostScript files. These files are thereafter filtered by the `ps2mpeg` script to ppm format and transformed to a movie by an mpeg encoder. The result is a file `movie.mpeg` that can be shown using any mpeg player[37]. The script `ps2mpeg` employed by `simres2mpeg` is useful in other situations where one has a series of, e.g., PostScript files and wants to make a movie out of them.

The basic syntax of the `simres2mpeg` script is

```
simres2mpeg -s "simres2mtv flags" -p "plotmtv flags"
```

Each frame in the movie must have the same color scale and therefore we need to prescribe the global maximum and minimum u values, from the whole simulation, each time we generate a Plotmtv plot. This is done by the `cmin` and `cmax` directives in the Plotmtv file. The extreme values of u are in the `Heat2` solver computed by a `FieldSummary` object and written in the report. Alternatively, you can always compute the extreme values from a simres dataset using the `simres2summary` filter:

```
simres2summary -f weld1 -s -n u -A
```

This filter reads all (`-A`) the u fields (`-n u`) and applies a `FieldSummary` object to analyze the variations in the field. For the present simulation we find $u_{min} = 0$ and $u_{max} \approx 0.22$. The appropriate `simres2mpeg` command then becomes

---

[37] For example, `mpeg_play movie.mpeg`

```
simres2mpeg -s "-f weld1 -s -n u -A -o '%cmin=0 cmax=0.22' -b"
            -p "-scale 1.0"
```

The -s option contains the relevant simres2mtv commands, here reflecting that we should extract all (-A) scalar (-s) fields with name u (-n u) and write them to files in Plotmtv format, inserting the command %cmin=0 cmax=0.22 in the files (-o) and storing the numbers in binary format (-b). The next option, -p "-scale 1.0", just transfers the string -scale 1.0 to the Plotmtv program. If we want a 3D perspective plot instead, we could add the -3d option. However, in this latter case we must ensure that also the scale on the z axis is fixed throughout the simulation. That is, we must prescribe zmin and zmax in the Plotmtv files according to the minimum and maximum u values. The proper simres2mpeg command then becomes

```
simres2mpeg -s "-f weld1 -s -n u -A
            -o '%cmin=0 cmax=0.22 zmin=0 zmax=0.22' -b"
            -p "-scale 1.0 -3d"
```

Unfortunately, the production time of such a movie might be substantial on small computers in busy networks.

Animation Using Vtk. The Vtk system and the vtkviz interface, as introduced on page 238, offer a user-friendly way of making movies. Having data from the weld1 simulation generated by the app command in the preceding text, we first filter all u fields to Vtk format,

```
simres2vtk -f weld1 -s -n u -A -b
```

and then we start vtkviz, choose the Animation/Create Movie option on the File menu, and fill in the filenames.

Animation Using Matlab. When multiple fields are extracted from a simres dataset using the simres2matlab filter, the resulting Matlab script (with the default name dpd.m) automatically collects the plots of each field in a Matlab movie. With the previously generated weld1 dataset we can simply try

```
simres2matlab -f weld1 -s -n u -A -b
```

and thereafter issue the commands dpgui and dpd inside Matlab. Plots of the various fields are shown successively and then a movie is displayed. You can replay the movie X times with Y frames per second by typing movie(M,X,Y) at the Matlab prompt. Tuning of the visualization is enabled by editing the script file and repeating the execution of the script.

A plot of a single frame is easily accomplished since the various fields are available as Matlab variables. In the present example, one can see from the script file that the field u at $t = 0.075$ is stored in the variables x6_1, x6_2, and v6, where x6_1 and x6_2 are the coordinates of the nodal points and v6 contains the nodal field values. Typing surf(v6) produces an alternative visualization, which can be equipped with text or a different viewing angle using the GUI as explained in Chapter 3.3.4.

Occasionally you want the Matlab movie to be converted to a standard format like mpeg. This can be accomplished by dumping each frame in PostScript format and using the Diffpack script `ps2mpeg` to produce an mpeg movie. The Matlab script can dump each frame if the variable `movie2ps` is set equal to one prior to running the script. Try

```
>> movie2ps=1
>> dpd
>> !ps2mpeg *.ps
```

Be aware that the `ps2mpeg` command can be a slow process.

Real-Time Visualization in Matlab. Visualization in Matlab *during program execution* is enabled by using an object of class `MatlabEngine`, which opens a run-time connection to Matlab that we can use for computations as well. A `MatlabEngine` object is included in class `Heat2` for demonstrating real-time visualization. Comments in the file `Heat2.cpp` explains the `MatlabEngine` syntax needed to produce the plots. In short, we can place Diffpack arrays into Matlab's workspace and assign Matlab variables to them. Afterwards we can issue Matlab commands operating on the Diffpack data. The `MatlabEngine` tools require of course that Diffpack is compiled and linked with the Matlab libraries. The report [1] contains the necessary steps. You can try to compile the `Heat2` solver with the run-time Matlab-Diffpack coupling. Since we have placed statements that cause Diffpack to be linked to Matlab inside `#ifdef MATLAB`, you must provide the option `CXXUF=-DMATLAB` to `Make` in order to activate the `MatlabEngine` statements[38]. The menu system command `real-time visualization` must be set to `ON` and perhaps you also want to adjust the z scale in the Matlab plot through the `Matlab z axis` menu item. Run for example

```
./app --class Heat2weld --noreport < Verify/testmatlab.i
```

to see a demonstration of the real-time plotting capabilities. We refer to the source code in `Heat2.cpp` and to the report [1] for more information about the Matlab-Diffpack coupling.

3.10.3   Variable Time Step Size

The input string `"dt=0.1 t in [0,1]"` for the time integration parameters indicates a constant time step 0.1 throughout the time interval [0,1]. A slightly more complicated input string allows the time step to vary, for example,

---

[38] The `.cmake2` file in the `Heat2` directory contains the statement `SYSLIBS+=-lmat`, which causes the Matlab libraries to be linked with the `Heat2` application.

```
dt=[0.025 0.05 0.1]   t in [0 0.2 0.5 1]
```

indicates that in the interval [0,0.2] the time step size is 0.025, in [0.2,0.5] it is 0.05, while in [0.5,1] it is 0.1. Try this input string in the menu input file or on the command line.

It is also worth mentioning that one can easily alter the time step size inside the code, using the statement `tip->setTimeStep(r)`, where `r` is a `real` containing the new time step value.

A plot of the evolution of the time step size can be generated after the time loop is finished using the `tip->plotTimeSteps` function which takes a `CurvePlotFile` object as argument and generates a curve plot in Diffpack format. We can typically write

```
tip->plotTimeSteps(database->cplotfile);
```

at the end of the `timeLoop` function. The plot is recognized in the database by its title `Time step evolution`, its curve name `dt`, and its comment `tracked time steps`. To see the resulting curve, you can use curve plotting tools (`curveplotgui` or `curveplot`), as described in Chapter 1.3.4 and Appendix B.4.1, or you can include the plot in an automatically generated report. The script `Verify/test2.sh` exemplifies all the mentioned steps in a simulation with variable Δt.

3.10.4 Applying a Transient Solver to a Stationary PDE

Stationary solvers can be viewed as special cases of time-dependent solvers, where the time derivative term is set to zero and where the initial condition has no meaning. Mathematically, the stationary problem corresponds to the limit $t \rightarrow \infty$ in the time-dependent problem. We can easily modify class `Heat1` or `Heat2` such that the Poisson equation problem from Chapter 3.5 can be solved. The key to the easy implementation is that class `TimePrm` has a stationary mode, enabled by setting the time step to zero. This can for instance be done by giving the answer `dt=0` to the menu item `time parameters`. The function `bool TimePrm::stationary()` can then be tested in the `integrands` function for a true value, which implies that we should simply omit the time derivative term, i.e. set $\beta = 0$, and enforce $\Delta t = 1$. Before the `i,j`-loop in `integrands` we insert the code

```
if (tip->stationary()) { dt = 1.0;  t = DUMMY;  beta = 0.0; }
else                    { dt = tip->Delta();  t = tip->time(); }
```

The `Heat2` solver is now capable of solving the same basic boundary-value problems as class `Poisson2`.

3.10.5   Handling Simulation and Visualization from a Script

The purpose of the present section is to demonstrate how flexible computational engines, like Diffpack, can be combined with scripting languages, like Perl, for creating a user-friendly and efficient tool adapted to a special problem. The material here only describes a productivity-enhancing technique and is not a prerequisite for other parts of the book.

The menu system in Diffpack offers great flexibility, but sometimes the flexibility is too great, because the values of many of the input data are inter-dependent. For instance, if we alter the value of β in (3.35), i.e. ϱC_p in Example 3.1, we can view this as a change of the time scale:

$$\beta_{\text{new}} \frac{\partial u}{\partial t} = \beta_{\text{old}} \frac{\partial u}{\partial \bar{t}}, \quad \bar{t} = \frac{\beta_{\text{new}}}{\beta_{\text{old}}} t .$$

In other words, there is a relationship between the stopping time t_{stop} for the solver, Δt, and β. Moreover, if we desire to visualize the temperature in the report at four equally spaced points in the total time interval $[0, t_{\text{stop}}]$, these time points also depend on β. The problem with setting up a correct input file is then to make sure that all parameters are compatible.

Because a change in β affects many of the other parameters, it would be advantageous to constrain the flexibility of the input file. This could be done by deriving subclasses and implementing scan functions that adjust the menu answers according to some principles. Nevertheless, a more optimal solution would be to have *variables* and *formulas* for some of the menu answers *in the input file*. This functionality can be achieved by embedding the input file in a *script*. The scripting language Perl [123] has a convenient feature[39] for achieving this goal. Move to the Verify directory of the Heat2 solver, where you can find the input file embed.i:

```
set time parameters = dt =0.01875 t in [0,0.45]
set gridfile=P=PreproBox|d=2 [0,1]x[0,1]|d=2 e=ElmB4n2D [16,16] [1,1]
set time derivative coefficient = 0.3
sub SaveSimRes
set time points for plot = [0:0.45,0.0375]
ok
 set time points for report field plot = [0:0.45,0.1125]
ok
```

The command

```
    PerlifyInputFile embed.i
```

results in a Perl script embed.pl that writes a new input file embed.i. The generated embed.pl script looks like this:

---

[39] The elegant functionality to be used is often referred to as a "here document" in Unix shells and Perl.

```
#!/usr/bin/perl
# menu system input file embedded in a perl script

# possible processing of command line options to embed.pl:
while (@ARGV) {
  $option = shift;  # shift eats one command line argument
     if ($option =~ /-option1/) { } # action
  elsif ($option =~ /-option2/) { } # action
  elsif ($option =~ /-option3/) { } # action
}
rename("embed.i","embed.i.bak");  # take a copy
open(INPUTFILE,">embed.i");  # construct input file
print INPUTFILE <<EOF;  # print until next EOF mark
! input file - automatically constructed by embed.pl

set time parameters = dt =0.1 t in [0,1]
set gridfile=P=PreproBox|d=2 [0,1]x[0,1]|d=2 e=ElmB4n2D [10,10] [1,1]
set time derivative coefficient = 1.0
sub Store4Plotting
 set time points for plot = [0:1,0.2]
ok
set time points for report field plot = NONE
ok

EOF
close(INPUTFILE);
```

In this Perl script we might introduce variables and arithmetic expressions. Furthermore, we have the possibility of processing command-line arguments as indicated in the `while` loop in the script.

Suppose we want to give the β parameter and the number of nodal intervals n_x in each space direction as input variables to embed.pl. From this information we assume that it is appropriate to set $t_{\text{stop}} = 1.5\beta$ and $\Delta t = \beta/n_x$. We dump the solution at $t = 2\Delta t, 4\Delta t, \ldots$ and and plot u in the report at four equally spaced time points. To this end, we introduce variables `$beta`, `$nx`, `$tstop`, `$dt`, `$dumpincr`, and `$reportplotincr` in the Perl script. The `$beta` and `$nx` values are read from the command line, the other variables are thereafter computed, and finally the variables are *directly inserted in the output statements* after the line containing `print INPUTFILE` in embed.pl. Our edited embed.pl script then takes the following form:

```
#!/usr/bin/perl
# menu system input file embedded in a Perl script

# default values:
$beta = 0.3; $nx = 16;

# possible processing of command-line options to embed.pl:
while (@ARGV) {
  $option = shift;  # shift eats one command-line argument
     if ($option =~ /-beta/) { $beta = shift; } # beta = next param
  elsif ($option =~ /-nx/)   { $nx = shift; }   # nx   = next param
  else { die "Illegal option to $0\n"; }        # exit
}
```

```
$tstop = 1.5*$beta;
$dt = $beta*1.0/$nx;
$dumpincr = 2*$dt;
$reportplotincr = $tstop/4;

rename("embed.i","embed.i.bak");  # take a copy of the original file
open(INPUTFILE,">embed.i");       # write a new input file
print INPUTFILE <<EOF;            # print until next EOF mark
! input file - automatically constructed by embed.pl

set time parameters = dt =$dt t in [0,$tstop]
set gridfile=P=PreproBox|d=2 [0,1]x[0,1]|d=2 e=ElmB4n2D [$nx,$nx] [1,1]
set time derivative coefficient = $beta
sub Store4Plotting
 set time points for plot = [0:$tstop,$dumpincr]
ok
set time points for report field plot = [0:$tstop,$reportplotincr]
ok

EOF
# end of multi-line print statement
close(INPUTFILE);
```

As an example, run `embed.pl -beta 0.1 -nx 12` and look at the generated `embed.i` input file.

The `embed.pl` script can trivially be extended to run the simulator by adding these lines at the end:

```
$casename = "embed";
system "../app --class Heat2weld --casename $casename < $casename.i";
```

Visualization software can of course also be launched from the `embed.pl` script. An edited version of `embed.pl`, called `embed2.pl` in the `Heat2/Verify` directory, contains the appropriate Perl statements for generating an input file with compatible parameters, running the code, making an mpeg movie, and displaying the movie on the screen.

The Perl script `embed.pl` can easily be equipped with a graphical user interface [109, Ch. 14-16] if desired, using the Perl support for the popular Tk widgets. An extension of the `embed2.pl` script, named `embed3.pl`, offers such a graphical user interface.

3.10.6   Some Computer Exercises Involving Heat Transfer

Exercise 3.10. Consider the heat conduction problem in Figure 3.20. Make an input file for the `Heat2` solver with menu settings corresponding to this test case. Produce a movie of the evolution of u. One should observe the significant smoothing of the initial discontinuity at the boundary.   ◇

Exercise 3.11. Figure 3.21 shows an isolated circular tube. The goal is to compute the temperature distribution in the isolating material, bounded by

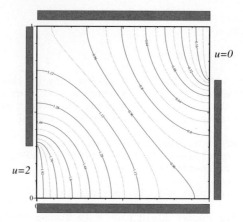

Fig. 3.20. Sketch of a heat conduction problem to be solved in Exercise 3.10. There are no internal heat sources ($f = 0$). Initially, $u = 1$. At two parts of the boundary, the temperature u is held at $u = 2$ and $u = 0$. The rest of the boundary is perfectly insulated, such that $-k\partial u/\partial n = 0$. The contour lines of u correspond to the steady-state solution as $t \to \infty$.

the tube and a square-shaped outer boundary. Air at constant temperature T_s is flowing through the inner tube. Outside the square-shaped boundary the temperature is oscillating, e.g. due to day and night variations, here modeled as $T_o = A\sin\omega t$. At both boundaries of this material we employ the cooling law $-k\partial T/\partial n = h_T(T - B)$, where B is either T_s or T_o, k is the heat conduction coefficient, and T is the unknown temperature. Make a subclass of `ReportHeat2` for simulating the steady-state oscillating temperature response (the initial condition is in this case immaterial so $T(\boldsymbol{x}, 0) = 0$ is a possible choice). Work with as small computational domain as possible by exploiting the symmetry in the problem (identify the symmetry lines and their associated boundary conditions). Mesh generation can conveniently be performed by the `PreproStdGeom` utility (see Figure 3.9 on page 268). Make animations of the resulting temperature field. ◇

Exercise 3.12. Figure 3.22 shows a two-material structure, where the heat conduction coefficient is k_1 (low) in one of the materials and k_2 (high) in the surrounding material. At $t = 0$ we assume a constant temperature T_0 in the whole domain. Then the temperature is suddenly increased to $T_1 > T_0$ at the black segment on the left boundary. At the right boundary, the temperature is fixed at T_0. The other boundaries are insulated (no-flux condition). Reduce the size of the domain by taking symmetry into account and specify the boundary condition to be used in the reduced domain. Make a subclass of `ReportHeat2` or `Heat2eff` (see Appendix B.6.2) to solve the present problem. Hardcode the **k** function for a specific geometric extent of the low-conductive

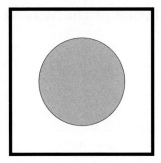

Fig. 3.21. Sketch of the geometry of the heat conduction problem to be solved in Exercise 3.11.

material. Construct two sequences of uniformly refined grids: one where the discontinuity in k always appear on the element boundaries and one where the material boundary is located in the interior of the elements. Compare the accuracy of the two mesh types. To increase your productivity in this numerical investigation, you should take full advantage of the multiple report and automatic report generation capabilities of the `Heat2` solver. (Chapter 2.8.2 is relevant background material on discontinuous coefficients in PDEs.)   ◇

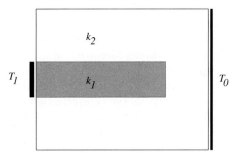

Fig. 3.22. Sketch of a heat conduction problem to be solved in Exercise 3.12. The domain consists of two materials with differing heat conduction properties. The temperature is kept at T_1 and T_0 at certain parts of the boundaries.

Exercise 3.13. Modify the `Heat2` solver such that it can solve a time-dependent convection-diffusion equation

$$\beta \frac{\partial u}{\partial t} + \boldsymbol{v} \cdot \nabla u = \nabla \cdot (k \nabla u) + f,$$

with appropriate initial and boundary conditions. The material in Chapter 3.8 can be useful.   ◇

3.11 Efficient Solution of the Wave Equation

This section describes how the multi-dimensional wave equation can be solved efficiently using finite element tools in Diffpack. The reader is probably able to modify, e.g., the Heat1 solver from Chapter 3.9 such that it solves the wave equation instead of the heat equation. Nevertheless, if the coefficients in the PDE are independent of time, the repeated assembly procedures at every time level, as performed in Chapter 3.9, are inefficient. One should rather compute the matrices that enter the discrete finite element equations only once and thereafter update the linear system at each time level by efficient matrix-vector operations. For the wave equation we can devise an explicit scheme such that also the solution of the linear system is computed with optimal efficiency. The result is a fast solver for the wave equation in arbitrary 1D, 2D, and 3D geometries.

Our model problem is the linear wave equation with some simple initial and boundary conditions:

$$\frac{\partial^2 u}{\partial t^2} = c^2 \nabla^2 u, \quad \boldsymbol{x} \in \Omega \subset \mathbb{R}^d, \; t > 0, \tag{3.42}$$

$$u(\boldsymbol{x}, 0) = I(\boldsymbol{x}), \quad \boldsymbol{x} \in \Omega, \tag{3.43}$$

$$\frac{\partial}{\partial t} u(\boldsymbol{x}, 0) = 0, \quad \boldsymbol{x} \in \Omega, \tag{3.44}$$

$$\frac{\partial u}{\partial n} = 0, \quad \boldsymbol{x} \in \partial\Omega_N, \; t > 0. \tag{3.45}$$

Here, $u(\boldsymbol{x}, t)$ is the primary unknown, c is the (constant) wave velocity, and d is the number of space dimensions. Moreover, $\boldsymbol{x}$ is a global point, t denotes time, and ∇^2 is the standard Laplacian operator in Cartesian coordinates. For simplicity, we assume that all the boundary conditions are of the homogeneous Neumann type, i.e., $\partial\Omega_N$ is the complete boundary of Ω. Extensions to a variable wave velocity $c(\boldsymbol{x})$ and other initial and boundary conditions are outlined in Chapter 3.11.3.

3.11.1 Discretization

We introduce a standard Galerkin finite element method in space and use the standard finite difference approximation to the second-order time derivative. See Chapter 2.2 for the basic ideas of the numerics. These procedures lead to a system of linear equations at each time step:

$$\boldsymbol{M}\boldsymbol{u}^{\ell+1} = 2\boldsymbol{M}\boldsymbol{u}^{\ell} - \boldsymbol{M}\boldsymbol{u}^{\ell-1} - \Delta t^2 \boldsymbol{K}\boldsymbol{u}^{\ell}. \tag{3.46}$$

In this expression, $\boldsymbol{M}$ and $\boldsymbol{K}$ are matrices, the superscript ℓ denotes the time level of the unknown discrete field values $\boldsymbol{u}$ of u, and Δt is the constant time step size. With a finite difference spatial discretization, the mass matrix $\boldsymbol{M}$ would be diagonal and the scheme becomes fully explicit. Using the finite element method, M is not diagonal, but it will be convenient to use a lumped (diagonal) mass matrix to increase the efficiency.

Equation (3.46) can be written on the computationally more attractive form

$$u^{\ell+1} = 2u^\ell - u^{\ell-1} - \Delta t^2 M^{-1} K u^\ell. \tag{3.47}$$

With a lumped M, the term $M^{-1} K u^\ell$ is efficiently computed as simple matrix-vector updates. We shall now give the explicit formulas for the entries in these matrices. Let

$$u(\boldsymbol{x}, t) \approx u^\ell(\boldsymbol{x}) = \sum_{j=1}^{\ell} u_j^\ell N_j(\boldsymbol{x})$$

and $\boldsymbol{u}^\ell = (u_1^\ell, \ldots u_n^\ell)^T$. The contributions to the various matrices and vectors at the element level can be given as

$$\tilde{M}_{i,j}^{(e)} = \int_{\tilde{\Omega}} N_i N_j \det J \, d\Omega \tag{3.48}$$

$$\approx \begin{cases} \sum_j \int_{\tilde{\Omega}} N_i N_j \det J \, d\Omega, & i = j \\ 0, & i \neq j, \end{cases} \tag{3.49}$$

$$\tilde{K}_{i,j}^{(e)} = \int_{\tilde{\Omega}} c^2 \nabla N_i \cdot \nabla N_j \det J d\Omega. \tag{3.50}$$

The matrix $\left\{ \tilde{M}_{i,j}^{(e)} \right\}$ is the contribution from element number e to the global matrix M. Furthermore, $\left\{ \tilde{K}_{i,j}^{(e)} \right\}$ is the similar contribution to K. Here, $\tilde{\Omega}$ denotes a reference element in local coordinates.

3.11.2 Implementation

Data Structures for the Linear System. By constructing M and K initially and lumping M, the solution at each time level can be computed by simple and highly efficient matrix-vector operations. Instead of a LinEqAdmFE object to hold the linear system and associated solvers, we now use two matrix objects: a MatDiag diagonal matrix to hold M and a MatSparse sparse matrix to hold K[40]. The mass matrix M is computed by a special FEM function makeMassMatrix, while K can be calculated in the ordinary way using an overloaded FEM::makeSystem function.

---

[40] Instead of having K explicitly as a sparse matrix object MatSparse, one can let K be a general Matrix handle, such that the user at run time can choose the matrix format freely. However, we have tried to keep the class as simple as possible.

The Computational Algorithm. The recursive scheme (3.47) is trivial to code for $\ell > 0$. For $\ell = 0$ we need to take the initial condition $\partial u/\partial t = 0$ into account. The discrete form of $\partial u/\partial t = 0$ results in $\boldsymbol{u}^{-1} = \boldsymbol{u}^1$. Elimination of $\boldsymbol{u}^{-1}$ in (3.47) for $\ell = 0$ then gives a special formula for $\boldsymbol{u}^1$ that must be used for the first time step. However, in Chapters 1.3.2 we suggested to apply (3.47) directly also for $\ell = 0$, but with a special definition of $\boldsymbol{u}^{-1}$ to obtain the right $\boldsymbol{u}^1$. Here we find that this fictitious $\boldsymbol{u}^{-1}$ should be defined as

$$\boldsymbol{u}^{-1} = \boldsymbol{u}^0 - \frac{1}{2}\Delta t^2 \boldsymbol{M}^{-1}\boldsymbol{K}\boldsymbol{u}^0 . \tag{3.51}$$

The computation of $\boldsymbol{u}^0$ stems from a Galerkin method applied to the equation $u = I$:

$$\sum_{j=1}^{n} \int_{\Omega} N_i N_j d\Omega u_j^0 = \int_{\Omega} I N_i d\Omega .$$

If we represent I as a finite element field, $\sum_{j=1}^{n} I_j N_j$, the mass matrix also appear on the right-hand side ($\sum_j \int_\Omega N_i N_j d\Omega I_j$) and multiplying by $\boldsymbol{M}^{-1}$ in the matrix equations gives $u_j^0 = I_j$, that is, the nodal values of the discrete u at $t = 0$ is simply set equal to the nodal values of $I(\boldsymbol{x})$. This is the technique that we shall apply for the finite element wave equation solver.

The algorithm now becomes:

1. Set $\boldsymbol{u}^0$ equal to the nodal values of $I(\boldsymbol{x})$.
2. Evaluate $\boldsymbol{u}^{-1}$ from (3.51).
3. Apply (3.47) for $\ell = 0, 1, 2, \ldots$ to compute $\boldsymbol{u}^1, \boldsymbol{u}^2, \boldsymbol{u}^3, \ldots$

The particular form of initial u values, $I(\boldsymbol{x})$, to be used in our test solvers reads

$$I(\boldsymbol{x}) = \frac{1}{2} - \arctan\left(\frac{\sigma}{\pi}\left(||\boldsymbol{x}|| - 2\right)\right), \tag{3.52}$$

where σ is an adjustable parameter that governs the steepness of the initial u profile. The impact of σ on the quality of the numerical approximation of the wave equation is discussed in Appendix A.4.8.

In d space dimensions, the stability criterion for the suggested method reads

$$\Delta t \leq c^{-1}\left(\sum_{i=1}^{d} \frac{1}{\Delta x_i^2}\right)^{-1/2} .$$

Stability of finite element methods for wave equations is treated in Chapter 2.4.3.

The Source Code of the Simulator. As usual, we represent the wave equation simulator as a class. We can follow the recipe from Chapter 3.9, with the main difference that the `LinEqAdmFE` object is replaced by matrix objects for $\boldsymbol{M}$ and $\boldsymbol{K}$ and a scratch vector. Only the key data members are shown; the rest of the class declares only the standard functions that should be familiar from the previous header file examples in this chapter.

```
class Wave0 : public FEM
{
protected:
  Handle(GridFE)      grid;
  Handle(FieldFE)     u;         // u at current time level
  Handle(FieldFE)     u_prev;    // u at previous time level, t-dt
  Handle(FieldFE)     u_prev2;   // u at time level t-2*dt
  Handle(DegFreeFE)   dof;       // mapping: field <-> equation system

  real                c2;        // square of wave velocity
  Handle(TimePrm)     tip;       // time interval, time step length etc
  MatDiag(real)       M;         // mass matrix
  MatSparse(real)     K;         // global matrix (the Laplacian term)
  Vec(real)           scratch;   // used in the updating formula for u
```

The implementational details of the functions that deviate significantly from the corresponding ones in the Heat1 and Heat2 solvers are listed next. The non-standard parts of the scan function consist of initializing the M and K matrices:

```
// special initialization of M and K:
const int nno = grid->getNoNodes();
scratch.redim (nno);
M.redim (nno);
// make sparsity pattern for the sparse matrix K:
Handle(SparseDS) sparse_ds (new SparseDS());
makeSparsityPattern (*sparse_ds, *dof);
K.redim (*sparse_ds);
```

The initialization of M, K, and u_prev2 (u^{-1}) is coded like this:

```
void Wave0:: solveProblem ()
{
  setIC ();                              // initial conditions

  makeMassMatrix (*grid, M, true);  // diagonal mass matrix M
  M.factLU();                       // factorize M (trivial!)
  makeSystem (*dof, (Matrix(NUMT)&) K);  // make matrix K

  // special u_prev2 (see the Wave1D solvers for explanation)
  prod (scratch, K, u_prev->values());   // scratch = K*u^0
  M.forwBack (scratch, scratch);          // scratch = M^{-1}*Ku^0
  add (u_prev2->values(), u_prev->values(),
       -0.5*sqr(tip->Delta()), scratch); // artificial u_prev2

  timeLoop ();
}
```

The numerical operations at a specific time step are as usual collected in a function solveAtThisTimeStep:

```
void Wave0:: solveAtThisTimeStep ()
{
  fillEssBC ();
  prod (scratch, K, u_prev->values());   // scratch = K*u^n
```

```
M.forwBack (scratch, scratch);          // scratch = M^{-1}*K*u^n
// scratch = u^{n-1} + dt^2*M^{-1}*K*u^n
add (scratch, u_prev2->values(), sqr(tip->Delta()), scratch);
// u^{n+1} = 2u^n - u^{n-1} - dt^2*M^{-1}*K*u^n
add (u->values(), 2, u_prev->values(), -1, scratch);
}
```

The integrands function is used for computing K so it should simply evaluate $c^2 \nabla N_i \cdot \nabla N_j \det J$ times the weight at the current integration point. The complete source code of the Wave0 solver is located in src/fem/Wave0.

A One-Dimensional Test Problem. The input file test1.i in the Verify directory defines a 1D test case with 40 linear elements in the domain $\Omega = (-10, 10)$. Run this case and make an animation of the solution using the animation tools for curves as described in Chapter 1.3.4 (see also Appendix B.4.1). Here are two equivalent examples, with cn as the casename of the run, just to get you started:

```
curveplot gnuplot -f .cn.curve.map -fps 1 -r . u .
                -o 'set yrange [-1.1:1.1];' -animate
curveplot matlab  -f .cn.curve.map -fps 1 -r . u .
                -o 'axis([min(x1) max(x1) -1.1 1.1]);' -animate
```

The -fps option controls the number of frames per second in the movie (if the value is greater than unity when using Gnuplot, the movie is shown in maximum speed). By substituting -animate by -psanimate in the plotting commands, we can automatically make mpeg movies of the wave motion.

The nice thing about this test case is that the Courant number is unity everywhere such that the numerical solution is exact at all the nodal points. By looking at an animation, it is fairly easy to convince oneself that the results are correct: The initial pulse is split into two waves of the same box shape, but with half the amplitude of the initial wave. These two waves are moving in opposite directions and reflected by the walls. There is no damping in the system. A similar solution, obtained by finite difference discretization, was demonstrated in Chapter 1.3.4.

A Two-Dimensional Test Problem. Since the Wave0 solver takes the number of space dimensions as an input parameter, we can easily solve a 2D wave equation on, for instance, the spatial domain $\Omega = (-10, 10) \times (0, 10)$. The initial condition implemented in class Wave0 is $u = 1$ for $r < 2$ and $u = 0$ otherwise, where $r = \sqrt{x^2 + y^2}$. This circular plug will propagate in both space directions and $y = 0$ is a symmetry line of the problem. With a partition [400, 200], the largest possible Δt is $0.005/\sqrt{2}$. Run the application with the input file Verify/test2.i, export all the stored fields to Plotmtv format[41] and take a look at the plots.

---

[41] See Verify/test2.sh for the correct statement on Unix systems.

The initial step function in this example can be represented by a two-dimensional Fourier series, with significant contribution from high frequencies. When the problem is solved numerically, the various Fourier components will move with different velocity. Therefore, after a while we can see waves with very short wave lengths as ripples on the plot. Choosing a smoother initial profile, reduces the contribution from high-frequency Fourier components and the numerical solution will look smoother. The reader is encouraged to experiment with this. A similar 1D problem is analyzed on page 518. The numerical noise is not damped when solving the wave equation. The heat equation, on the contrary, damps high-frequency waves very efficiently, see Figure A.1 on page 509.

Exercise 3.14. In the 2D test case described above (corresponding to the file `test2.i`), $x = 0$ is also a symmetry line. Modify the input file `test2.i` such that we can reduce the number of elements by a factor of two, utilizing that $x = 0$ and $y = 0$ are symmetry lines. (Hint: If $u = 0$ is symmetric about a curve, $\partial u / \partial n = 0$ at the curve, where n denotes the direction perpendicular to the curve.)   ◇

3.11.3   Extensions of the Model Problem

We consider the slightly generalized model problem

$$\frac{\partial^2 u}{\partial t^2} = \nabla \cdot (c^2(\boldsymbol{x})\nabla u), \quad \boldsymbol{x} \in \Omega \subset \mathbb{R}^d, \ t > 0, \tag{3.53}$$

$$u = g(\boldsymbol{x}, t), \quad \boldsymbol{x} \in \partial\Omega_E, \ t > 0, \tag{3.54}$$

$$\frac{\partial u}{\partial n} = 0, \quad \boldsymbol{x} \in \partial\Omega_N, \ t > 0, \tag{3.55}$$

$$u(\boldsymbol{x}, 0) = I(\boldsymbol{x}), \quad \boldsymbol{x} \in \Omega, \tag{3.56}$$

$$\frac{\partial}{\partial t} u(\boldsymbol{x}, 0) = \hat{I}(\boldsymbol{x}), \quad \boldsymbol{x} \in \Omega. \tag{3.57}$$

Here the boundary $\partial\Omega$ of Ω is split into two non-overlapping parts, $\partial\Omega = \partial\Omega_E \cup \partial\Omega_N$. In addition, the wave velocity $c(\boldsymbol{x})$ is space dependent. The nonzero "velocity" at the initial time level, i.e., the condition (3.57), will only affect the formula (3.51) for our fictitious $\boldsymbol{u}^{-1}$ value.

The Dirichlet boundary condition is straightforwardly implemented, but the technique is different from the method used in Chapter 3, mainly because the situation is much simpler in the present case. From (3.47) we see that the equation for $\boldsymbol{u}^{\ell+1}$ is $\boldsymbol{u}^{\ell+1} = \boldsymbol{b}$, where the contents of $\boldsymbol{b}$ should be evident from (3.47). We can compute $\boldsymbol{u}^{\ell+1} = \boldsymbol{b}$ without paying attention to the essential boundary conditions. Afterwards, we only need to adjust $\boldsymbol{u}^{\ell+1}$ such that it is in accordance with the Dirichlet conditions. This is possible since none of the new nodal values that disobey the boundary conditions can affect the computation of the others (a property of all explicit schemes). The implementation is realized as a loop over all nodal points:

```
const int nno = grid->getNoNodes();
for (int i = 1; i <= nno; i++) {
  if (grid->boNode (i,1))   // assume Dirichlet cond. 1 at indicator 1
    u->values()(i) = ....some formula or function....
  else if (grid->boNode (i,2))   // another Dirichlet cond at ind. 2
    u->values()(i) = ....some formula or function....
}
```

Exercise 3.15. Modify class `Wave0` such that it solves the initial-boundary value problem (3.53)–(3.57). Test the simulator on the same problem as in `Verify/test2.i`, but with $u = 0$ on the boundary and a variable phase velocity: $c = 1$ for $x <= 0$ and $c = c_0$ for $x \geq 0$. Is $x = 0$ now a symmetry line (cf. Exercise 3.14)?   ◇

3.11.4   Flexible Representation of Variable Coefficients

A User-Friendly Interface for Prescribing Variable Coefficients. The extended model problem in Chapter 3.11.3 with a spatially varying wave velocity needs a representation of the variable coefficient. In Chapter 3 we mainly used virtual functions for the coefficients and hardcoded particular formulas in subclass solvers. As an alternative, it is convenient at run time to decide among the following particular representations of c^2:

1. a constant,
2. an explicit formula, realized as a functor,
3. a piecewise constant field over material subdomains,
4. a finite element or finite difference field on file.

On the menu we can indicate these representations as follows.

1. `set c2 format = CONSTANT=3.2`
   The c^2 field is here prescribed to be the constant value 3.2.
2. `set c2 format = FUNCTOR=myc2`
   In this case the wave velocity field should be computed from a user-programmed functor[42] with the name `myc2`. The functor must have class `FieldFunc` as base class, and the member function `valuePt` evaluates the field at a space(-time) point.
3. `set c2 format = MATERIAL_CONSTANTS= 2.4 6.2 1 ;`
   This menu answer assigns the constant values 2.4, 6.2, and 1 to material[43] number 1, 2 and 3, respectively. The number of values listed in the answer must correspond to the number of materials.
4. `set c2 format = FIELD_ON_FILE=mycase("c2")`
   The c^2 field is now to be read from a simres[44] file with casename `mycase`. The name of the field on file is `c2`. One can also load a field from a time-dependent series of fields on file, using the syntax `mycase("c2",t=4.2)` to indicate the field with name `c2` at time 4.2.

---

[42] See Chapter 3.4.4 for basic information about the functor concept.
[43] See page 259 for basic information about the material concept.
[44] See Chapter 3.3 and Appendix B.4.2 for basic information about simres files.

Diffpack's Handling of Various Types of Fields. The implementation of this high degree of flexibility in specifying the c^2 field in the simulator class is straightforward. We represent the c^2 field by a `Handle(Field)` object. Class `Field` is the base class for all scalar fields in Diffpack. Special field formats are realized as subclasses: `FieldConst` for constant fields, `FieldFunc` for functors (i.e. explicit formulas), `FieldLattice` for fields over regular `GridLattice` grids, `FieldFE` for finite element fields, and `FieldPiWisConst` for piecewise constant fields over material subdomains. Similarly, class `Fields` is a base class for vector fields. Its subclasses have the same names as in the scalar field subclasses, except that the word `Field` is replaced by `Fields`. To hold the user's choice of field format, we use class `FieldFormat`. This class also offers functionality for allocating the proper subclass object and binding it to a `Field` (or `Fields`) handle as explained next.

Implementation. A revised version of class `Wave0`, simply called `Wave1`, implements a more flexible wave equation solver, using the `Handle(Field)` and `FieldFormat` tools. In the definition of class `Wave1` we include

```
Handle(Field) c2;
FieldFormat    c2_format;
```

To put the field format object on the menu, we first prescribe the field name and then call its **define** function with a suitable default value[45]:

```
c2_format.setFieldname("c2").define(menu,"CONSTANT=2.0",level);
```

This default answer leads to a representation of c^2 in terms of a `FieldConst` object, such that c^2 equals 2.0 throughout the domain. To read the user's field format answer, we simply call

```
c2_format.scan (menu);
```

Having the field format available, we can allocate the proper field object by calling

```
c2_format.allocateAndInit (c2, grid.getPtr());
```

Notice that we transfer the *handle* c2 to this function and that we also must provide a grid pointer[46]. If c2 is specified as a functor on the menu, the `allocateAndInit` function cannot create any of the functors provided by the programmer and returns a false value. In this case, the programmer must take care of the allocation. Here is a suitable set of statements illustrating these points.

---

[45] The `setFieldname` function returns a `FieldFormat&` reference, and that is why we can merge two calls into one statement.

[46] This pointer is used only in case c2 is to become a handle to a `FieldPiWisConst` object.

```
if ( !c2_format.allocateAndInit (c2, grid.getPtr()) )
  {
    // FieldFormat::allocateAndInit could not initialize,
    // perhaps a FUNCTOR format?

    if (c2_format.format == FieldFormat::FUNCTOR) {
      if (c2_format.function_name == "StepAtx0")
        // declare functor c=1, x<0 and c=10, x>0
        c2.rebind (new StepAtx0 (10.0));
      // else if (c2_format.function_name == "SomethingElse")
      //    c2.rebind (new ...);
      else
        errorFP("Wave1::scan",
                "c2_format.function_name=\"%s\" not implemented",
                c2_format.function_name.c_str());
    }
  }
```

Finally, we need to evaluate the c^2 field inside the integrands routine. This is efficiently done by calling the virtual valueFEM function in the Field hierarchy:

```
const real c2_pt = c2->valueFEM(fe);  // fe is a FiniteElement obj.
```

The subclasses of Field have optimal implementations of the valueFEM function. For example, FieldConst::valueFEM just returns a constant, whereas the implementation in FieldFE::valueFEM runs through a standard finite interpolation procedure in an element ($\sum_j u_j N_j$).

The reader should notice that the valuePt function, taking a space-time point as argument, is also a virtual function in the Field hierarchy that could be used for evaluating the c^2 field at a point. However, the valuePt function is *extremely inefficient* if c^2 is represented as a FieldFE object, since one has to find the element and the local coordinates of the spatial point prior to interpolating the finite element field. On the contrary, the valuePt function exhibits comparable efficiency with valueFEM in many other field classes, like FieldConst, FieldFunc, and FieldLattice, but the all-round choice for evaluating a Field, having a FiniteElement object at disposal, should always be the valueFEM function.

Prescribing Initial Conditions. The software tools we used for flexible representation of the variable coefficient $c^2(\boldsymbol{x})$ are also useful for representing, for instance, the initial condition $I(\boldsymbol{x})$. In this case we know that the initial field must be represented as a FieldFE object, but we can use a FieldFormat object for flexible assignment of the initial values; I can be constant, I can be given in terms of a functor, or the I values can be read from file. We typically insert a FieldFormat u0_format statement in the class definition, put the field format object on the menu and read the answer. The initial condition is set in the function setIC. We can follow the same basic ideas as when we initialized the c^2 field, but now we have already forced the field (u_prev) to be of type FieldFE. An overloaded allocateAndInit in class FieldFormat takes a `Handle(FieldFE)` argument and initializes its nodal values. Again, if the user prescribes a functor for I, `allocateAndInit` does not know about the user's functor names so the routine returns a false value, and the programmer must explicitly deal with various types of functor names and corresponding actions.

```
void Wave1:: setIC ()
{
  // fill u_prev from the prescribed initial condition for u
  if (!u0_format.allocateAndInit (u_prev))
    {
      // u0_format.allocateAndInit did not initialize u_prev,
      // perhaps u0_format indicates a FUNCTOR format?

      if (u0_format.format == FieldFormat::FUNCTOR) {
        if (u0_format.function_name == "plug") {
          // declare Plug0 functor and use FieldFE::fill
          // (u_prev is already allocated in scan - ready for use)
          Plug0 plug (sigma);
          u_prev->fill (plug, 0.0);  // values = plug for t=0.0
        }
        else
          errorFP("Wave1::setIC",
                  "u0_format.function_name=\"%s\" not implemented",
                  u0_format.function_name.c_str());
      }
    }
  // else: u0_format has allocated and initialized u_prev,
  *u = *u_prev;   // make u ready for plotting at t=0
}
```

The `Plug0` functor implements the formula (3.52). Notice that we now fill u_prev with functor values instead of letting a general `Field` handle point to a functor object as we did for the `c2` field, because u_prev is forced to be a field of known type. The `FieldFE::fill` function is explained on page 307.

Continuing Previous Simulations. The reader should observe that the flexible assignment of initial conditions automatically allows for restarting a previous simulation. Suppose you have run a case with casename `mycase1` and that the simulation was terminated by some reason, with the last dump of the solution to the simres database at time 6.2. To continue this simulation, we just specify the u field at time 6.2 in the `mycase1` simres database as initial condition:

```
set u0 format = FIELD_ON_FILE("mycase1",t=6.2)
```

The `setIC` function will load this field into u_prev, or more precisely, the call `u0_format.allocateAndInit` will do the job.

Remarks. In Chapter 3.5 we advocated a representation of coefficients in a PDE in terms of virtual functions, which could be specified in compact subclasses. The representation of variable coefficients in terms of field objects in the `Field` hierarchy is another approach to flexible handling of coefficients in the equation. An advantage of the latter approach is that a range of formats and assignment possibilities are already available, such that deriving a subclass, defining a function, and then compiling the application, is not necessary. The ultimate flexibility is of course to apply the field representation in the base class, but perform the evaluation of the field in a virtual function. Subclasses can then reimplement the virtual function and, e.g., not make use of the field representation. The simulator in Chapter 6.1 applies such a flexible implementation.

We should here remark that the combination of a virtual function calling a `Field`'s virtual `valueFEM` function decreases the computational efficiency. If the formula to be evaluated is comprehensive, the overhead of virtual function calls may not be important. Also, if the `integrands` routine involves complicated expressions inside the double loop over the nodes, the extra work for calling virtual functions outside the loop may not be significant. From a profile of the code (see page 542), one can quickly determine if flexible representation of coefficients is a critical efficiency factor or not. In the present application we call the evaluation of coefficients only at the first time step, so avoiding the overhead of virtual functions only contribute to decreasing the flexibility of the simulator, with negligible effect on the efficiency in long simulations.

Chapter 4

Nonlinear Problems

This chapter extends the numerical methods and software tools from Chapters 1–3 to nonlinear partial differential equations. Chapter 4.1 deals with discretization techniques for nonlinear terms in PDEs and algorithms for solving systems of nonlinear algebraic equations, whereas software tools supporting the implementation of the methods in Chapter 4.1 are described in Chapter 4.2. The application areas are limited to scalar PDEs, but the methodology in this chapter is applied to systems of PDEs and more challenging applications in Chapters 5.2, 6.3, and 7.

4.1 Discretization and Solution of Nonlinear PDEs

The present section concerns numerical methods for nonlinear PDEs. A brief introduction to this topic was given in Chapter 1.2.7. We start with demonstrating general finite difference and finite element discretization techniques through some specific examples. The discretization processes lead to systems of *nonlinear* algebraic equations. Such systems can be solved iteratively as a sequence of linear systems. The present chapter deals with two popular iteration techniques: Successive Substitutions, also known as Picard iteration, and the Newton-Raphson method. These methods can also be used at the PDE level to transform a nonlinear PDE into a sequence of linear PDEs. In addition, we outline how to embed the Newton-Raphson and Successive Substitution iterations in continuation methods for solving PDEs with severe nonlinearities.

4.1.1 Finite Difference Discretization

Let us start with two simple model problems on $\Omega = (0, 1)$,

$$\frac{d^2 u}{dx^2} + f(u) = 0, \quad u(0) = u_L, \ u(1) = u_R, \tag{4.1}$$

$$\frac{d}{dx}\left(\lambda(u)\frac{du}{dx}\right) = 0, \quad u(0) = u_L, \ u(1) = u_R. \tag{4.2}$$

The functions $f(u)$ and $\lambda(u)$ are assumed to be nonlinear in u. We shall discretize these two boundary-value problems by standard finite difference and finite element methods.

The standard finite difference method, as outlined in Chapters 1.2 and 1.2.6, gives rise to schemes on the form

$$\frac{1}{h^2}(u_{i-1} - 2u_i + u_{i+1}) + f(u_i) = 0, \tag{4.3}$$

$$\frac{1}{h^2}\left(\lambda_{i+\frac{1}{2}}(u_{i+1} - u_i) - \lambda_{i-\frac{1}{2}}(u_i - u_{i-1})\right) = 0. \tag{4.4}$$

Here, and in other equations, u_i is the numerical approximation to u at node i. A basic problem is that $\lambda_{i+\frac{1}{2}} \equiv \lambda(u_{i+\frac{1}{2}})$, but $u_{i+\frac{1}{2}}$ is not a primary unknown. We therefore approximate $\lambda_{i+\frac{1}{2}}$ by, for example, an arithmetic average,

$$\lambda_{i+\frac{1}{2}} = \frac{1}{2}[\lambda(u_{i+1}) + \lambda(u_i)]. \tag{4.5}$$

Due to the appearance of $f(u_i)$ and $\lambda(u_i)$, the algebraic equations become *nonlinear*.

4.1.2 Finite Element Discretization

We can apply the standard Galerkin finite element method from Chapter 2 to the model problems (4.1) and (4.2). The solution is approximated by a sum as usual,

$$u(x) \approx \hat{u}(x) = \sum_{j=1}^{n} u_j N_j(x).$$

Multiplying the PDEs by weighting functions N_i, $i = 1, \ldots, n$, and integrating over the domain, utilizing integration by parts of second-order derivatives, result in a system of algebraic equations. However, the equations are now nonlinear due to the nonlinear functions $\lambda(u)$ and $f(u)$. The model problem (4.1) results in

$$\sum_{j=1}^{n}\left(\int_0^1 N_i'(x)N_j'(x)dx\right)u_j = \int_0^1 f(\sum_{s=1}^{n} u_s N_s(x))N_i(x)dx \quad i = 1, \ldots, n. \tag{4.6}$$

The essential boundary conditions become $u_1 = u_L$ and $u_n = u_R$. The term on the left-hand side is similar to numerous examples from Chapter 2, but the integral

$$\int_0^1 f(\sum_s u_s N_s)N_i dx,$$

arising from the nonlinear term $f(u)$ in the PDE, is difficult to treat by analytical means. Normally, the integral must be computed by numerical integration methods. Let us employ linear finite elements and the trapezoidal rule (two-point Gauss-Lobatto rule). The contribution from the local nodes $r = 1, 2$ in element e becomes

$$\int_{-1}^{1} f(\sum_{s=1}^{2} \tilde{u}_s \tilde{N}_s(\xi)) \tilde{N}_r(\xi) \frac{h}{2} d\xi$$

which is approximated by

$$\frac{h}{2} \left(f(\sum_s \tilde{u}_s \tilde{N}_s(-1)) \tilde{N}_r(-1) + f(\sum_s \tilde{u}_s \tilde{N}_s(1)) \tilde{N}_r(1) \right) = \frac{h}{2} \begin{cases} f(\tilde{u}_1), r = 1 \\ f(\tilde{u}_2), r = 2 \end{cases}$$

Notice that $\tilde{u}_r$ here refers to the value of $\hat{u}$ at *local* node r. As usual, h is the element length, which is assumed constant here for simplicity. Assembling the element contributions gives the following set of global discrete equations:

$$u_1 = u_L, \tag{4.7}$$

$$\frac{1}{h}(u_{i-1} - 2u_{i+1} + u_{i+1}) + hf(u_i) = 0, \quad i = 2, \ldots, n-1, \tag{4.8}$$

$$u_n = u_R. \tag{4.9}$$

This is the same result as we achieved by the finite difference method. If we use a more accurate integration rule than the trapezoidal rule, the term $hf(u_i)$ becomes more complicated.

Example 4.1. As an illustrative example on the complexity arising from the nonlinear term $f(u)$ in a finite element context, we choose $f(u) = u^2$ and evaluate the integral analytically, using linear elements. A useful formula is

$$\int_{\Omega_e} N_i^p N_j^q dx = \frac{p! q!}{(p+q+1)!} h_e, \quad p, q \in \mathbb{R}, \tag{4.10}$$

with Ω_e being a one-dimensional element (cf. (2.125) and (2.131)). On the element, $\hat{u} = \sum_{s=1}^{2} \tilde{u}_s \tilde{N}_s$.

$$\int_{\Omega_e} f(\hat{u}) N_r dx = \frac{h}{2} \int_{-1}^{1} \left(\tilde{u}_1^2 \tilde{N}_1^2 + 2\tilde{u}_1 \tilde{u}_2 \tilde{N}_1 \tilde{N}_2 + \tilde{u}_2^2 \tilde{N}_2^2 \right) \tilde{N}_i d\xi$$

$$= h \begin{cases} \tilde{u}_1^2 \frac{3!0!}{(3+0+1)!} + 2\tilde{u}_1\tilde{u}_2 \frac{2!1!}{(2+1+1)!} + \tilde{u}_2^2 \frac{1!2!}{(1+2+1)!}, r = 1 \\ \tilde{u}_1^2 \frac{2!1!}{(2+1+1)!} + 2\tilde{u}_1\tilde{u}_2 \frac{1!2!}{(1+2+1)!} + \tilde{u}_2^2 \frac{0!3!}{(0+3+1)!}, r = 2 \end{cases}$$

$$= \frac{h}{12} \begin{cases} 3\tilde{u}_1^2 + 2\tilde{u}_1\tilde{u}_2 + \tilde{u}_2^2, r = 1 \\ \tilde{u}_1^2 + 2\tilde{u}_1\tilde{u}_2 + 3\tilde{u}_2^2, r = 2 \end{cases}$$

Assembling the contributions gives

$$\frac{h}{12} \left(3u_{i-1}^2 + 2u_i(u_{i-1} + u_i + u_{i+1}) + 3u_{i+1}^2 \right),$$

which can be interpreted as a smoothing of $(u^2)_i$ according to

$$\frac{h}{4}\left(u_{i-1}^2 + 2u_i\bar{u}_i + u_{i+1}^2\right), \quad \bar{u}_i = \frac{1}{3}(u_{i-1} + u_i + u_{i+1}).$$

$\diamond$

Turning the attention to the model problem (4.2), the appropriate weighted residual formulation using a Galerkin approach now reads

$$\sum_{j=1}^{n}\left(\int_0^1 \lambda(\hat{u})N_i'N_j'dx\right)u_j = 0, \quad i = 1,\dots,n, \tag{4.11}$$

with essential conditions $u_1 = u_L$ and $u_n = u_R$. The integral involving $\lambda(\hat{u})$ requires in general numerical integration. Employing linear elements and the trapezoidal rule, the following analytical formula for the element matrix can be derived:

$$\frac{\lambda(\tilde{u}_1) + \lambda(\tilde{u}_2)}{2h}\begin{pmatrix} 1 & -1 \\ -1 & 1 \end{pmatrix}. \tag{4.12}$$

Assembling these contributions, we see that the finite element method and the finite difference method give identical global equations.

Exercise 4.1. Derive the expression (4.12) for the element matrix corresponding to finite element discretization with linear elements and nodal-point integration in model problem (4.2). Show that the global discrete equations take the form (4.4), with $\lambda_{i+\frac{1}{2}} = (\lambda(u_i) + \lambda(u_{i+1}))/2$. The various steps in the calculations should be evident from the material on finite element discretization of (4.1).

$\diamond$

4.1.3 The Group Finite Element Method

We have just seen that when the finite element method is applied to nonlinear PDEs, the trapezoidal rule is a convenient tool for obtaining analytically tractable expressions and discrete equations that are similar to those of the finite difference method. Alternatively, one can apply the *group finite element method*, also called the *product approximation method*. A nonlinear function f is then represented like this:

$$f(\hat{u}) \approx \sum_{j=1}^{n} f(u_j)N_j.$$

Such expressions simplify the hand calculation of integrals significantly. For instance,

$$\int_\Omega \left(\sum_j f(u_j)N_j\right)N_i\,dx = \sum_j f(u_j)\int_\Omega N_iN_j\,dx.$$

At the element level, we recall that the integral $\int_\Omega N_i N_j dx$ results in the mass matrix in (2.90) on page 142. The element vector corresponding to $\int_\Omega f(\hat{u}) N_i dx$ therefore becomes

$$\frac{h}{6} \begin{pmatrix} 2 & 1 \\ 1 & 2 \end{pmatrix} \begin{pmatrix} f(\tilde{u}_1) \\ f(\tilde{u}_2) \end{pmatrix} = \frac{h}{6} \begin{pmatrix} 2f(\tilde{u}_1) + f(\tilde{u}_2) \\ f(\tilde{u}_1) + 2f(\tilde{u}_2) \end{pmatrix}.$$

Using the trapezoidal rule, or other means for lumping the mass matrix, yields an element vector $(f(\tilde{u}_1), f(\tilde{u}_2))^T h/2$.

Exercise 4.2. Use the group finite element representation for λ in (4.11) and derive the corresponding element matrix.   ◇

By assembling element contributions, we find that the group finite element method leads to the following system of nonlinear equations in our two model problems:

$$\frac{1}{h}(u_{i-1} - 2u_i + u_{i+1}) + \frac{h}{6}(f(u_{i-1}) + 4f(u_i) + f(u_{i+1})) = 0, \quad (4.13)$$

$$\frac{1}{h}(\lambda_{i+\frac{1}{2}}(u_{i+1} - u_i) - \lambda_{i-\frac{1}{2}}(u_i - u_{i-1})) = 0. \quad (4.14)$$

In the last equation, $\lambda_{i+\frac{1}{2}}$ means exactly $\frac{1}{2}(\lambda(u_i) + \lambda(u_{i+1}))$. The group finite element method and nodal-point integration are two features in the finite element framework that enable us to reproduce certain finite difference expressions, or alternatively, simplify the "default" finite element equations.

Summary. From the previous examples on finite difference and finite element discretization of nonlinear PDEs, we observe that model problem (4.2) leads to a nonlinear system on the special form

$$\boldsymbol{A}(\boldsymbol{u})\boldsymbol{u} = \boldsymbol{b}, \quad (4.15)$$

whereas model problem (4.1) results in a system

$$\boldsymbol{A}\boldsymbol{u} = \boldsymbol{b}(\boldsymbol{u}). \quad (4.16)$$

For a general nonlinear PDE, the nonlinear algebraic equations can be written in the generic form

$$\boldsymbol{F}(\boldsymbol{u}) = \boldsymbol{0}, \quad (4.17)$$

with $\boldsymbol{F}(\boldsymbol{u}) = \boldsymbol{A}(\boldsymbol{u})\boldsymbol{u} - \boldsymbol{b}$ and $\boldsymbol{F}(\boldsymbol{u}) = \boldsymbol{A}\boldsymbol{u} - \boldsymbol{b}(\boldsymbol{u})$ in our two model problems. In all these expressions, $\boldsymbol{u} = (u_1, u_2, \ldots, u_n)^T$, with a similar definition of $\boldsymbol{F}$ and $\boldsymbol{b}$. Vectors without argument are considered as constant, i.e., independent of $\boldsymbol{u}$. The next sections are devoted to solution algorithms for systems of nonlinear algebraic equations.

4.1.4   Successive Substitutions

Nonlinear algebraic systems on the form

$$A(u)u = b$$

can be solved by a simple iteration technique:

$$A(u^k)u^{k+1} = b, \quad k = 0, 1, 2, \ldots$$

until $||u^{k+1} - u^k||$ is sufficiently small. The iteration requires a start vector u^0, and a frequent choice is $u^0 = 0$. The direct application of this scheme to (4.4) is trivial. The present iteration strategy is often referred to as *Picard iteration* or the method of *Successive Substitutions*. Another name is *simple iterations* [40]. We will stick to the term Successive Substitutions in the following.

Similar equations, like $Au = b(u)$, are of course also good candidates for the Successive Substitution technique:

$$Au^{k+1} = b(u^k), \quad k = 0, 1, 2, \ldots.$$

The convergence of such iterations can be studied in terms of contraction mappings [57,32], but the direct practical applicability of the results to specific PDEs seems somewhat limited.

If the Successive Substitution method has problems with convergence, one can try a relaxation technique. This consists in first computing a tentative new approximation u^* from $A(u^k)u^* = b$ or $Au^* = b(u^k)$ and then set u^{k+1} as a weighted mean of the previous value u^k and the new u^*:

$$u^{k+1} = \omega u^* + (1 - \omega)u^k,$$

where $\omega \in (0, 1]$ is a prescribed *relaxation parameter*.

4.1.5   Newton-Raphson's Method

One Nonlinear Equation with One Unknown. Let us describe the famous and widely used Newton-Raphson method first for a single nonlinear equation $F(u) = 0$ in one scalar variable u. We assume that an approximation u^k to u is available, and our aim is to compute an improved approximation. The idea consists of approximating $F(u)$ in the vicinity of u^k, $F(u) \approx M(u; u^k)$, such that $M(u; u^k) = 0$ is an equation that is simple to solve. Its solution is taken as an improved approximation u^{k+1} to the root u of $F(u) = 0$. We may take $M(u; u^k)$ as the linear part of a Taylor-series approximation to F at the point $u = u^k$:

$$M(u; u^k) = F(u^k) + \frac{dF}{du}(u^k)(u - u^k).$$

We then let u^{k+1} be the solution of $M(u; u^k) = 0$, that is, we solve the equation $M(u^{k+1}; u^k) = 0$ with respect to u^{k+1} and get

$$u^{k+1} = u^k - \frac{F(u^k)}{\frac{dF}{du}(u^k)}.$$

This is the Newton-Raphson iteration scheme for solving $F(u) = 0$. The convergence rate is quadratic, i.e.,

$$|u - u^{k+1}| \leq C|u - u^k|^2$$

for a constant C, if u^0 is sufficiently close to the solution u [32].

Systems of Nonlinear Equations. The Newton-Raphson procedure can easily be extended to multi-dimensional problems $\boldsymbol{F}(\boldsymbol{u}) = \boldsymbol{0}$. We approximate $\boldsymbol{F}(\boldsymbol{u})$ by a linear function $\boldsymbol{M}(\boldsymbol{u}; \boldsymbol{u}^k)$ in the vicinity of an existing approximation $\boldsymbol{u}^k$ to $\boldsymbol{u}$:

$$\boldsymbol{M}(\boldsymbol{u}; \boldsymbol{u}^k) = \boldsymbol{F}(\boldsymbol{u}^k) + \boldsymbol{J}(\boldsymbol{u}^k) \cdot (\boldsymbol{u} - \boldsymbol{u}^k),$$

where $\boldsymbol{J} \equiv \nabla \boldsymbol{F}$ is the Jacobian of $\boldsymbol{F}$. If $\boldsymbol{F} = (F_1, \ldots, F_n)^T$ and $\boldsymbol{u} = (u_1, \ldots, u_n)^T$, entry (i, j) in $\boldsymbol{J}$ equals $\partial F_i / \partial u_j$. In order to find the next approximation $\boldsymbol{u}^{k+1}$ from $\boldsymbol{M}(\boldsymbol{u}^{k+1}; \boldsymbol{u}^k) = \boldsymbol{0}$, we have to solve a linear system with $\boldsymbol{J}$ as coefficient matrix. The idea of relaxation as we explained for the Successive Substitution method can be applied to Newton-Raphson iteration as well; the new solution $\boldsymbol{u}^{k+1}$ is set equal to weighted mean of the previous solution $\boldsymbol{u}^k$ and the solution $\boldsymbol{u}^*$ found from the equation $\boldsymbol{M}(\boldsymbol{u}^*; \boldsymbol{u}^k) = \boldsymbol{0}$. The computational steps are summarized in Algorithm 4.1.

Algorithm 4.1.

Newton-Raphson's method with relaxation.

given a guess $\boldsymbol{u}^0$ for the solution of $\boldsymbol{F}(\boldsymbol{u}) = \boldsymbol{0}$,
for $k = 0, 1, 2, \ldots$, until termination criterion is fulfilled
    solve the linear system $\boldsymbol{J}(\boldsymbol{u}^k)\delta\boldsymbol{u}^{k+1} = -\boldsymbol{F}(\boldsymbol{u}^k)$ wrt. $\delta\boldsymbol{u}^{k+1}$
    set $\boldsymbol{u}^{k+1} = \boldsymbol{u}^k + \omega\delta\boldsymbol{u}^{k+1}$

In the following examples we will mostly apply the Newton-Raphson method with $\omega = 1$. Relevant termination criteria for the method are

$$||\boldsymbol{u}^{k+1} - \boldsymbol{u}^k|| \leq \epsilon_u \quad \text{or} \quad ||\boldsymbol{F}(\boldsymbol{u}^{k+1})|| \leq \epsilon_r,$$

or variants where ratios are used instead, for example,

$$\frac{||\boldsymbol{u}^{k+1} - \boldsymbol{u}^k||}{||\boldsymbol{u}^k||} \leq \epsilon_u \quad \text{or} \quad \frac{||\boldsymbol{F}(\boldsymbol{u}^{k+1})||}{||\boldsymbol{F}(\boldsymbol{u}^0)||} \leq \epsilon_r.$$

Example 4.2. The finite difference approximation to $-u'' = f(u)$ leads to the nonlinear algebraic equations (4.3), where

$$F_i \equiv \frac{1}{h^2}\left(u_{i-1} - 2u_i + u_{i+1}\right) + f(u_i) = 0\,.$$

Calculation of $\boldsymbol{J}$ is normally done by analytic differentiation of the discrete nonlinear equations. The only nontrivial term to differentiate is the term $f(u_i)$. The contribution to entry (i,j) in $\boldsymbol{J}$ from $f(u_i)$ then becomes

$$\frac{\partial}{\partial u_j} f(u_i) = \begin{cases} f'(u_i), & i = j \\ 0, & i \neq j \end{cases} \tag{4.18}$$

We can now set up the coefficient matrix of the linear system in each Newton-Raphson iteration:

$$J_{i,j} \equiv \frac{\partial F_i}{\partial u_j} = \frac{1}{h^2}\begin{cases} 1, & j = i - 1 \\ -2 + h^2 f'(u_i), & j = i \\ 1, & j = i + 1 \\ 0, & j < i - 1 \text{ or } j > i + 1 \end{cases}$$

The equations to be solved in each iteration, $\sum_{j=1}^{n} J_{i,j}\delta u_j^{k+1} = -F_i$, for the indices $i = 2, \ldots, n-1$, then reduce to

$$\frac{1}{h^2}\left(\delta u_{i-1}^{k+1} + (-2 + h^2 f'(u_i^k))\delta u_i^{k+1} + \delta u_{i+1}^{k+1}\right) =$$

$$-\frac{1}{h^2}(u_{i-1}^k - 2u_i^k + u_{i+1}^k) - f(u_i^k), \tag{4.19}$$

where δu_i is component i in the solution vector $\delta\boldsymbol{u}$ in the system $\boldsymbol{J}\delta\boldsymbol{u} = -\boldsymbol{F}$. For $i = 1, n$ we have the equation $\delta u_i^{k+1} = 0$ if $u_1^0 = u_L$ and $u_n^0 = u_R$.   ◇

Example 4.3. The standard finite element representation of the nonlinear term $f(u)$ appears as

$$S_i \equiv \int_\Omega f(\sum_s u_s N_s)N_i dx\,.$$

For calculating the (i,j) entry in $\boldsymbol{J}$ we need

$$\frac{\partial}{\partial u_j} f(\sum_s u_s N_s) = \frac{df}{d\hat{u}}\frac{\partial\hat{u}}{\partial u_j} = f'(\hat{u})\frac{\partial}{\partial u_j}\sum_s u_s N_s = f'(\sum_s u_s N_s)N_j\,.$$

Thus,

$$\frac{\partial S_i}{\partial u_j} = \int_\Omega f'(\sum_s u_s N_s)N_j N_i dx\,. \tag{4.20}$$

Alternatively, we can apply the group finite element approximation to the nonlinear term $f(u)$. The associated finite element form reads

$$\bar{S}_i \equiv \int_\Omega \sum_s N_s f(u_s) N_i dx\,.$$

The (i, j) entry in $\boldsymbol{J}$ now becomes

$$\frac{\partial \bar{S}_i}{\partial u_j} = \int_\Omega f'(u_j) N_j N_i dx = f'(u_j) \int_\Omega N_i N_j dx\,.$$

Employing nodal-point integration gives the same results as in the finite difference case (the details are left for Exercise 4.3), otherwise we get a mass matrix-like contribution from the $f(u)$ term like we showed in the derivation of (4.13).   ◇

Example 4.4. Let us consider a multi-dimensional nonlinear problem,

$$-\nabla \cdot (\lambda(u)\nabla u) = f(\boldsymbol{x}),$$

with homogeneous Neumann conditions on the boundary of a domain $\Omega \subset \mathbb{R}^d$ and $\int_\Omega f d\Omega = 0$ for consistency. The Galerkin formulation leads to the nonlinear equations

$$F_i \equiv \int_\Omega (\lambda(\hat{u})\nabla N_i \cdot \nabla \hat{u} - f N_i)\, d\Omega = 0\,.$$

The associated (i, j) entry in $\boldsymbol{J}$ becomes

$$J_{i,j} \equiv \frac{\partial F_i}{\partial u_j} = \int_\Omega (\lambda'(\hat{u})N_j \nabla N_i \cdot \nabla \hat{u} + \lambda(\hat{u})\nabla N_i \cdot \nabla N_j)\, d\Omega\,.$$

◇

More generally, we can think of nonlinear terms on the form

$$\lambda = \lambda(\hat{u}, \hat{u}_{,1}, \hat{u}_{,2}, \hat{u}_{,3}),$$

where $\hat{u}_{,r}$ is a short form for differentiation with respect to x_r, that is, $\hat{u}_{,r} \equiv \partial \hat{u}/\partial x_r$ (cf. Appendix A.2). When $\hat{u} = \sum_j u_j N_j$ we now get

$$\frac{\partial \lambda}{\partial u_j} = \frac{\partial \lambda}{\partial \hat{u}} N_j + \sum_{r=1}^{d} \frac{\partial \lambda}{\partial \hat{u}_{,r}} \frac{\partial \hat{u}_{,r}}{\partial u_j}$$

$$= \frac{\partial \lambda}{\partial \hat{u}} N_j + \sum_{r=1}^{d} \frac{\partial \lambda}{\partial \hat{u}_{,r}} \frac{\partial N_j}{\partial x_r}\,.$$

More examples on calculating entries in $\boldsymbol{J}$ are found in Chapters 4.1.6, 6.3, and 7.2.

Example 4.5. As shown in Example 4.4, the concepts for solving 1D problems are readily extended to 2D and 3D in the finite element framework. The finite difference method, however, involves some additional technical details when extending the 1D techniques to higher dimensions. Consider, for instance,

$-\nabla^2 u = f(u)$ with $u = 0$ on the boundary. In 2D we can use the scheme $[\delta_x\delta_x u + \delta_y\delta_y u = -f(u)]_{i,j}$. The equations and unknowns are now naturally numbered by double indices, like (i,j). Equation no. (i,j) is written as $F_{i,j} = 0$, where

$$F_{i,j} \equiv \frac{1}{h^2}\left(u_{i,j-1} + u_{i-1,j} + u_{i+1,j} + u_{i,j+1} - 4u_{i,j}\right) + f(u_{i,j}), \qquad (4.21)$$

when $\Delta x = \Delta y = h$. The equation system in each Newton-Raphson iteration can be written

$$\sum_\ell \sum_m \frac{\partial F_{i,j}}{\partial u_{\ell,m}} \delta u_{\ell,m} = -F_{i,j} .$$

We realize that $\partial F_{i,j}/\partial u_{\ell,m}$ vanishes if not (ℓ,m) are "close to" (i,j). More precisely, from (4.21) we see that

$$\frac{\partial F_{i,j}}{\partial u_{\ell,m}} = 0 \quad \text{if } \ell \neq i-1, i, i+1 \text{ or } m \neq j-1, j, j+1 .$$

We can calculate the individual nonvanishing derivatives:

$$\frac{\partial F_{i,j}}{\partial u_{i-1,j}} = \frac{1}{h^2},$$

with an identical result for the derivatives with respect to $u_{i+1,j}$, $u_{i,j-1}$, and $u_{i,j+1}$. Furthermore,

$$\frac{\partial F_{i,j}}{\partial u_{i,j}} = -\frac{4}{h^2} + f'(u_{i,j}) .$$

The linear equations to be solved in an iteration can be written as follows:

$$\frac{1}{h^2}\left(\delta u_{i,j-1}^{k+1} + \delta u_{i-1,j}^{k+1} + \delta u_{i+1,j}^{k+1} + \delta u_{i,j+1}^{k+1} - 4\delta u_{i,j}^{k+1}\right) + f'(u_{i,j})\delta u_{i,j} =$$
$$-\frac{1}{h^2}\left(u_{i,j-1}^k + u_{i-1,j}^k + u_{i+1,j}^k + u_{i,j+1}^k - 4u_{i,j}^k\right) - f(u_{i,j}^k) \quad (4.22)$$

If we want to apply Neumann conditions at the boundary, we must modify the equations where the indices (i,j) correspond to a boundary point. Since $\delta u_{i,j}$ is a *correction* vector, its Neumann conditions should also vanish if u has prescribed (possibly nonhomogeneous) Neumann conditions and these are incorporated in the initial guess for the iteration[1].

In a certain sense the finite element method in Example 4.4 involves fewer technical details for calculating $\boldsymbol{J}$ in the problem corresponding to (4.22). The main reason for this is that the finite element method applies a single index for numbering the equations and the unknowns, regardless of the number of space dimensions.

The resulting coupled system of equations for $\delta u_{i,j}$ can be solved by the same type of methods that are relevant for the corresponding linear equation $-\nabla^2 u = f(\boldsymbol{x})$, see Appendix C.   ◇

---

[1] See Example C.1 on page 597 and the associated software for implementational details.

Exercise 4.3. Express the integral in (4.20) over one element in the reference coordinate system. Assume linear $\tilde{N}_i(\xi)$ and evaluate the integral by the trapezoidal rule. Interpret thereafter (4.20) as a general integral over a global domain $\Omega \in \mathbb{R}^d$ and employ a trapezoidal rule directly: $\int_\Omega g(\boldsymbol{x})\Omega \approx \sum_{q=1}^n q(\boldsymbol{x}^{[q]})w_q$, where w_q are the weights in the trapezoidal rule. Compare the results for $\partial S_i/\partial u_j$ with the expression (4.18) arising in a finite difference context.   ◇

Exercise 4.4. Identify $\boldsymbol{F}$ in (4.4) and the $\boldsymbol{J}$ matrix. Specialize the formulas for the entries in $\boldsymbol{J}$ to the case $\lambda(u) = 1 + u^2$.   ◇

Exercise 4.5. Explain why discretization of nonlinear PDEs by finite difference and finite element methods normally lead to a $\boldsymbol{J}$ matrix with the same sparsity pattern as one would encounter in an associated linear problem.   ◇

Exercise 4.6. Show that if $\boldsymbol{F}(\boldsymbol{u}) = \boldsymbol{0}$ is a *linear system* of equations, Newton-Raphson's method (with $\omega = 1$) finds the correct solution in the first iteration.   ◇

Exercise 4.7. The operator $\nabla \cdot (\lambda \nabla u)$, with $\lambda = ||\nabla u||^q$, $q \in \mathbb{R}$, and $||\cdot||$ being the Eucledian norm, appears in several physical problems, especially flow of non-Newtonian fluids (see Chapter 7.2). The quantity $\partial \lambda/\partial u_j$ is central when formulating a Newton-Raphson method, where u_j is the coefficient in the finite element approximation $u \approx \hat{u} = \sum_j u_j N_j$. Show that

$$\frac{\partial}{\partial u_j}||\nabla \hat{u}||^q = q||\nabla \hat{u}||^{q-2}\nabla \hat{u} \cdot \nabla N_j \,.$$

◇

Exercise 4.8. Consider the 3D equation $-\nabla \cdot (\lambda(u)\nabla u) = f(u)$, with u known on the boundary. Formulate a second-order accurate finite difference method for discretizing the equation, with Newton-Raphson's method for solving systems of nonlinear algebraic equations. Apply for simplicity Gauss-Seidel or SOR iteration, described in Appendix C.1, for solving the linear system in each Newton-Raphson iteration. Write the complete algorithm on implementational form. (The algorithm can be implemented as a small extension of the program associated with Example C.1 on page 597.)   ◇

4.1.6 A Transient Nonlinear Heat Conduction Problem

We shall now work through a complete nonlinear and time-dependent PDE case. The model problem of current interest is

$$\varrho C_p \frac{\partial u}{\partial t} = \nabla \cdot [\kappa(u)\nabla u], \quad \boldsymbol{x} \in \Omega \subset \mathbb{R}^d, \ t > 0, \tag{4.23}$$

$$u(\boldsymbol{x},t) = g(\boldsymbol{x},t), \quad \boldsymbol{x} \in \partial\Omega_E, \ t > 0, \tag{4.24}$$

$$u(\boldsymbol{x},0) = I(\boldsymbol{x}), \quad \boldsymbol{x} \in \Omega, \ t = 0 \,. \tag{4.25}$$

This mathematical model describes heat conduction in a body Ω, with $\kappa(u)$ being the heat conduction coefficient, which depends on the temperature level, ϱ is the density of the body, and C_p is the heat capacity. The primary unknown $u(\boldsymbol{x}, t)$ is the temperature in the body. To simplify the PDE, we can divide by ϱC_p and introduce $\lambda(u) = \kappa(u)/(\varrho C_p)$. At the complete boundary $\partial \Omega_E$ of Ω, the temperature is prescribed, according to the function g, while for $t = 0$ the temperature distribution is described by $I(\boldsymbol{x})$.

Equation (4.23) is discretized using the weighted residual method in space and the finite difference method in time. The time-discrete function $u^\ell(\boldsymbol{x})$ will be used to denote the approximation to $u(\boldsymbol{x}, t_\ell)$, where $t_\ell ll$ is the time level.

A fully implicit backward Euler scheme is used for the time discretization and results in a sequence of nonlinear spatial problems:

$$\frac{u^\ell - u^{\ell-1}}{\Delta t} = \nabla \cdot \left[\lambda(u^\ell) \nabla u^\ell \right], \quad \boldsymbol{x} \in \Omega \subset \mathbb{R}^d, \ \ell = 1, 2, \ldots, \quad (4.26)$$

$$u^\ell(\boldsymbol{x}) = g(\boldsymbol{x}), \quad \boldsymbol{x} \in \partial \Omega_E, \quad (4.27)$$

$$u^0(\boldsymbol{x}) = I(\boldsymbol{x}), \quad \boldsymbol{x} \in \Omega. \quad (4.28)$$

We then seek

$$u^\ell(\boldsymbol{x}) \approx \hat{u}^\ell(\boldsymbol{x}) = \sum_{j=1}^n u_j^\ell N_j(\boldsymbol{x}).$$

Inserting $\hat{u}$ in the time-discrete governing equations (4.26)–(4.28), requiring the residual to be orthogonal to n weighting functions W_i, $i = 1, \ldots, n$, and integrating the $\int_\Omega \nabla \cdot [\lambda(u) \nabla u] W_i d\Omega$ term by parts, lead to a system of nonlinear algebraic equations at each time level:

$$F_i(u_1^\ell, \ldots, u_n^\ell) = 0, \quad i = 1, \ldots, n,$$

where

$$F_i \equiv \int_\Omega \left[\left(\hat{u}^\ell - \hat{u}^{\ell-1} \right) W_i + \Delta t \lambda(\hat{u}^\ell) \nabla \hat{u}^\ell \cdot \nabla W_i \right] d\Omega. \quad (4.29)$$

The quantity $\hat{u}^{\ell-1}$ is considered as known and $\hat{u}^\ell$ is to be determined by solving the nonlinear equations $F_i = 0$ with respect to $u_1^\ell, \ldots, u_n^\ell$.

Let us first use Newton-Raphson's method to solve the system of nonlinear algebraic equations at each time level. Let $u_j^{\ell,k}$ denote the approximation to u_j^ℓ in the kth iteration of the Newton-Raphson method, and let $\hat{u}^{\ell,k}$ be the corresponding approximation to $\hat{u}^\ell$, that is, $\hat{u}^{\ell,k} = \sum_j u_j^{\ell,k} N_j$. In each iteration of the Newton-Raphson method, a linear system must be solved for the correction $\delta u_1^{\ell,k+1}, \ldots, \delta u_n^{\ell,k+1}$. Observe that when evaluating $J_{i,j}$ and $-F_i$, we use u values from the previous iteration ($u_j^{\ell,k}$). The entries in the Jacobian take the form

$$J_{i,j} \equiv \frac{\partial F_i}{\partial u_j^\ell} = \int_\Omega \left[W_i N_j + \Delta t \frac{d\lambda}{d\hat{u}}(\hat{u}^{\ell,k}) N_j \nabla W_i \cdot \nabla \hat{u}^{\ell,k} \right.$$

$$\left. + \Delta t \, \lambda(\hat{u}^{\ell,k}) \nabla W_i \cdot \nabla N_j \right] d\Omega. \quad (4.30)$$

The formula for the right-hand side is simply $-F_i$ with $\hat{u}^\ell$ replaced by $\hat{u}^{\ell,k}$ in (4.29).

Exercise 4.9. When solving $F_i = 0$ by Successive Substitutions, we end up with a linear system in each iteration that can be written on the form

$$\sum_{j=1}^{n} K_{i,j}(u_1^{\ell,k}, \ldots, u_n^{\ell,k}) u_j^{\ell,k+1} = b_i .$$

Derive formulas for $K_{i,j}$ and b_i.   ◇

Exercise 4.10. Replace (4.24) by the nonlinear cooling law

$$\kappa(u)\frac{\partial u}{\partial n} = \alpha(u)(u - U_S),$$

where $\alpha(u)$ is a heat transfer coefficient and U_S is the surrounding (constant) temperature. How are (4.29) and (4.30) modified by this boundary condition? ◇

Exercise 4.11. Many heat conduction applications involve large temperature variations such that the density ϱ and heat capacity C_p, which in principle depend on the temperature, can no longer be considered as constant. The governing PDE for the temperature u then takes the form

$$\varrho(u)C_p(u)\frac{\partial u}{\partial t} = \nabla \cdot (\kappa(u)\nabla u), \quad x \in \Omega \subset \mathbb{R}^d . \tag{4.31}$$

Formulate a finite element method for this PDE (choose appropriate boundary conditions). Derive formulas for the element matrix and vector in the linear system to be solved in each iteration of Newton-Raphson method for this problem.   ◇

Exercise 4.12. As a continuation of Exercise 4.11, we shall use a group finite element method for the nonlinear coefficients:

$$\varrho(u)C_p(u) \approx \sum_{q=1}^{n} \varrho(u_q)C_p(u_q)N_q,$$

with a similar expansion of $\kappa(u)$. Restrict the problem to one space dimension and linear elements. Develop the expressions for the element matrix and vector using analytical integration (hint: (4.10) is useful).   ◇

Example 4.6. Transient nonlinear PDEs, like (4.31), can be solved by *explicit* finite differences in time, thus avoiding solution of large systems of nonlinear algebraic equations at each time level. Discretizing (4.31) by a forward Euler method leads to

$$u^\ell = u^{\ell-1} + \Delta t \left(\varrho(u^{\ell-1})C_p(u^{\ell-1})\right)^{-1} \nabla \cdot (\kappa(u^{\ell-1})\nabla u^{\ell-1}).$$

This discretization has a truncation error of order Δt, while the stability criterion, according to Example A.19 on page 523, is roughly $\Delta t \leq h^2(\varrho C_p)_{min}/(2\kappa_{max})$, where h is a characteristic spatial mesh size. More precisely, h^2 is the harmonic mean of Δx^2 and Δy^2 for a finite difference discretization on a uniform grid, see (A.36). Finite element discretization with a consistent mass matrix reduces the critical Δt for stability by a factor $1/\sqrt{3}$, cf. Chapter 2.4.3.

To improve the accuracy of the time discretization, we could apply a *Leap-Frog* scheme (see also page 108) where we approximate (4.31) at time level ℓ by a second-order accurate centered time difference, involving the levels $\ell - 1$ and $\ell + 1$. That is,

$$\varrho(u^\ell)C_p(u^\ell)\frac{u^{\ell+1} - u^{\ell-1}}{2\Delta t} = \nabla \cdot (\kappa(u^\ell)\nabla u^\ell),$$

which is an explicit scheme for $u^{\ell+1}$. Unfortunately, stability analysis of the linear version of the discrete equation shows that the scheme is unstable for all choices of $\Delta t > 0$, see [111].

$\diamond$

Another popular technique for solving transient nonlinear PDE is *operator splitting*. The original equation is then split into two or more simpler equations. This strategy is exemplified in Project 4.3.1. Operator splitting is particularly convenient when solving systems of PDEs. Examples are provided in Chapters 6.2.2, 6.2.4, 6.2.5, 6.4.1, 6.5.1, 7.1.2, and 7.2.2.

4.1.7   Iteration Methods at the PDE Level

The ideas of the Newton-Raphson method and Successive Substitutions can be used directly at the PDE level, prior to any discretization. This means that we formulate an iteration method that replaces a nonlinear PDE by a sequence of linear PDEs. For each linear PDE we can apply standard discretization methods, and these will of course lead to linear systems[2]. Thus, there is no need for solving systems of nonlinear *algebraic* equations.

Consider the model problem (4.1). The Successive Substitution strategy can be applied at the PDE level by solving for u^{k+1} using the previously computed u^k in the nonlinear term:

$$-\frac{d^2 u^{k+1}}{dx^2} = f(u^k), \quad k = 0, 1, 2, \ldots \tag{4.32}$$

---

[2] The more mathematically-oriented reader will of course be conserned about the existence, uniqueness, and stability of the solutions of the linearized PDEs, or the convergence of the sequence of solutions of the linear problems towards the solution of the original nonlinear PDE.

Similarly, model problem (4.2) can be handled by the iteration

$$\frac{d}{dx}\left(\lambda(u^k)\frac{du}{dx}^{k+1}\right) = 0, \quad k = 0, 1, 2, \ldots \quad (4.33)$$

In the Successive Substitution method the unknown function $u^{k+1}(x)$ must obey the same boundary conditions as the exact solution $u(x)$.

Exercise 4.13. Show that the resulting sequence of linear systems in model problem (4.1) and (4.2) is independent of whether we first use Successive Substitution and then discretize the linear PDEs or we first discretize the PDEs and then apply Successive Substitutions to the resulting nonlinear algebraic equation systems. Consider both finite difference and finite element discretization.   ◇

   To formulate a Newton-Raphson inspired method, we might proceed as follows. We start with writing a new approximation u^{k+1} to u as

$$u^{k+1} = u^k + \delta u^k.$$

If the function $u^{k+1}(x)$ is close to the exact solution $u(x)$ of the PDE, the correction function $\delta u^k(x)$ is "small" in some sense, and products of δu^k quantities can be neglected. Inserting $u^{k+1} = u^k + \delta u^k$ in the nonlinear term $f(u)$ gives, upon Taylor-series expansion and omitting higher-order terms in δu^k:

$$f(u^{k+1}) = f(u^k) + f'(u^k)\delta u^k + \frac{1}{2}f''(u^k)\delta u^k \delta u^k + \cdots \approx f(u^k) + f'(u^k)\delta u^k.$$

Hence, we obtain the following PDE for the correction δu^k:

$$-\frac{d^2 u}{dx^2}^k - \frac{d^2}{dx^2}\delta u^k = f(u^k) + f'(u^k)\delta u^k,$$

which is more conveniently reordered as

$$\frac{d^2}{dx^2}\delta u^k + f'(u^k)\delta u^k = -\left(\frac{d^2 u}{dx^2}^k + f(u^k)\right). \quad (4.34)$$

If u^k obeys the prescribed essential boundary conditions, we realize that δu^k must vanish on the boundary. Notice that the right-hand side of (4.34) is the residual in the PDE in interation k, in the same way as we in the discrete Newton-Raphson method have the residual $\boldsymbol{F}(\boldsymbol{u}^k)$ on the right-hand side of the equation to be solved in each iteration. The computational procedure for (4.1) is summarized in Algorithm 4.2.

Algorithm 4.2.

Newton-Raphson's method at the PDE level.

given a guess u^0 for the PDE (4.1): $-u'' = f(u)$,
    such that $u^0(0) = u_L$, $u^0(1) = u_L$
for $k = 0, 1, 2, \ldots$ until termination criterion is fulfilled
    solve (4.34), with $\delta u^k(0) = \delta u^k(1) = 0$, with respect to δu^k
    set $u^{k+1} = u^k + \delta u^{k+1}$

Let us compute the resulting discrete equations that arise from Algorithm 4.2. For notational simplicity we drop the superscript k. Using the finite difference method, a typical linear system takes the form

$$\frac{1}{h^2}(\delta u_{i-1} - 2\delta u_i + \delta u_{i+1}) + f'(u_i)\delta u_i = -\frac{1}{h^2}(u_{i-1} - 2u_i + u_{i+1}) - f(u_i).$$

Comparing this equation with (4.19) shows that the Newton-Raphson method at the PDE level is mathematically equivalent to the Newton-Raphson method at the algebraic equation level in this particular example. When it comes to the finite element method, the only difference is the term $f'(u^k)\delta u^k$, or $f'(u)\delta u$ without the superscript k, which results in an integral

$$\int_\Omega f'(\sum_s u_s N_s)\delta u N_i dx.$$

This is indeed equivalent to (note that $\delta u = \sum_j N_j \delta u_j$)

$$\sum_j \int_\Omega f'(\sum_s u_s N_s) N_j N_i dx \, \delta u_j = \sum_j \frac{\partial S_i}{\partial u_j}\delta u_j,$$

where $\partial S_i/\partial u_j$ is the term that arose from the traditional Newton-Raphson method in (4.20). This means that the two versions of the Newton-Raphson method result in identical equations also in the finite element case.

Exercise 4.14. Use the Newton-Raphson method at the PDE level to solve the nonlinear problem (4.2). Compare the equations of the resulting linear systems with those generated by the traditional Newton-Raphson method. Apply both finite element and finite difference discretization.   ◇

Exercise 4.15. We consider the 2D extension of Exercise 4.14, that is, the same PDE as addressed in Examples 4.4 and 4.5. Apply a Newton-Raphson method at the PDE level and see if this approach is more straightforward than the techniques in Examples 4.4 and 4.5.   ◇

4.1.8   Continuation Methods

When solving PDEs with severe nonlinearities, the Newton-Raphson or the Successive Substitution method may diverge. Relaxation with $\omega < 1$ may help, but in highly nonlinear problems it can be necessary to introduce a *continuation parameter* Λ in the problem. For $\Lambda = 0$ the problem is easy to solve, and when $\Lambda = 1$ we recover the original problem. The idea is then to increase Λ in steps, from zero to unity, and use the solution corresponding to the most recent Λ value as start vector for a new nonlinear iteration. This approach is often referred to as a *continuation method*.

Suppose we intend to solve the equation from Exercise 4.7,

$$\nabla \cdot (||\nabla u||^q \nabla u) = 0$$

with some appropriate boundary conditions. We see that when $q = 0$, the problem is easy to solve. Introducing $\Lambda \in [0, 1]$ as the continuation parameter, we can solve the sequence of systems for the functions u^0, u^1, u^2, ... :

$$\nabla \cdot \left(||\nabla u^r||^{\Lambda q} \nabla u^r\right) = 0, \quad \Lambda = \Lambda_r, \ r = 0, 1, 2, ..., m,$$

where $0 = \Lambda_0 < \Lambda_1 < \cdots < \Lambda_m = 1$. The nonlinear problem for u^r, $r > 0$, is solved using u^{r-1} as an initial guess for the nonlinear iteration. Notice that u^0 is found from a linear PDE.

The sequence Λ_r, $r = 1, \ldots, m-1$, depends on physical insight into the mathematical model and should hence be given by the user. If convergence problems are faced with a particular Λ_r value, one can apply a bisection-like algorithm and try to solve the problem corresponding to $\Lambda^* = (\Lambda_r + \Lambda_{r-1})/2$. If this Λ^* also leads to divergence of nonlinear solvers, a new Λ value can obtained from halving the interval $[\Lambda_{r-1}, \Lambda^*]$. Algorithm 4.3 summarizes the steps in the continuation method.

Algorithm 4.3.

Continuation method for nonlinear problems.

$r = 0$
while $r \le m$
    $r \leftarrow r + 1$
    $\Lambda = \Lambda_r$
    solve PDE problem with this Λ
    if nonlinear solver diverges and $r > 0$
        $\bar{\Lambda} = \Lambda_{r-1}$
        do
            $\Lambda \leftarrow (\Lambda + \bar{\Lambda})/2$
            solve PDE problem with this Λ
        while nonlinear solver diverges
        $r \leftarrow r - 1$

Although most of Diffpack's tools for nonlinear iteration methods are described in the next section, the tools supporting continuation methods are briefly described on page 481, where the severe nonlinearities in the application demand a continuation method.

4.2   Software Tools for Nonlinear Finite Element Problems

We shall demonstrate the software tools in Diffpack for solving nonlinear PDEs by considering the specific model from Chapter 4.1.6. The reader should have a thorough understanding of the numerics in Chapter 4.1.6 before proceeding with the following text. Our purpose here is to create a simulator class NlHeat1 for the nonlinear model problem. It will become evident that class NlHeat1 is only a slight extension of class Heat1 from Chapter 3.9.

4.2.1   A Solver for a Nonlinear Heat Equation

A typical Diffpack simulator for nonlinear PDEs takes the following form.

```
class NlHeat1 : public FEM, public NonLinEqSolverUDC
{
protected:
  // general data:
  Handle(GridFE)      grid;    // finite element grid
  Handle(DegFreeFE)   dof;     // field <-> linear system mapping
  Handle(FieldFE)     u;       // primary unknown
  Handle(FieldFE)     u_prev;  // u at previous time step
  Handle(TimePrm)     tip;     // time step etc
  Vec(real)           linear_solution;
  Handle(LinEqAdmFE)  lineq;   // linear system & solution
  Handle(SaveSimRes)  database;

  virtual void setIC();
  virtual void timeLoop();
  virtual void solveAtThisTimeStep();
  virtual void fillEssBC();
  virtual void integrands (ElmMatVec& elmat, const FiniteElement& fe);

  // extensions for nonlinear problems:
  Vec(real)                   nonlin_solution;
  Handle(NonLinEqSolver_prm)  nlsolver_prm;
  Handle(NonLinEqSolver)      nlsolver;

  virtual void makeAndSolveLinearSystem ();
  virtual real lambda  (real u);  // nonlinear coefficient
  virtual real dlambda (real u);  // derivative of lambda

public:
  NlHeat1 ();
  ~NlHeat1 () {}

  virtual void adm     (MenuSystem& menu);
```

```
virtual void define (MenuSystem& menu, int level = MAIN);
virtual void scan    ();

virtual void solveProblem ();
virtual void resultReport ();
};
```

The `nonlin_solution` vector contains the solution of the system of nonlinear algebraic equations, whereas the `linear_solution` vector holds the solution of the linear system that must be solved in each nonlinear iteration. Notice that in the Successive Substitution method (Picard iteration) these two vectors are equal, whereas they differ in the Newton-Raphson method. The nonlinear solver has handles to `nonlin_solution` and `linear_solution`, such that these vectors can be updated by the solution algorithm.

In Diffpack there is a class hierarchy for solving systems of nonlinear algebraic equations. The base class of this hierarchy is `NonLinEqSolver`. It defines the virtual function `solve`, which solves the system of equations. Various subclasses implement specific nonlinear solvers. For instance, class `NewtonRaphson` offers the Newton-Raphson method. To initialize a nonlinear solver, we use a parameter object of class `NonLinEqSolver_prm` whose contents can be filled from the menu.

The initialization of the new data structures related to nonlinear systems and solvers is performed in `NlHeat1::scan`. Compared with, e.g., `Heat1::scan`, we need the following additional statements:

```
// create parameter objects this way:
nlsolver_prm.rebind (NonLinEqSolver_prm::construct());
nlsolver_prm->scan (menu);
// create iteration method (subclass of NonLinEqSolver):
nlsolver.rebind (nlsolver_prm->create());
nonlin_solution.redim (dof->getTotalNoDof());
linear_solution.redim (dof->getTotalNoDof());
lineq->attach (linear_solution);
nlsolver->attachUserCode (*this);
nlsolver->attachNonLinSol (nonlin_solution);
nlsolver->attachLinSol (linear_solution);
```

The `dof->getTotalNoDof()` call returns the number of unknowns in the algebraic systems. As an alternative, we could have used `u->getNoValues()` or `grid->getNoNodes()`[3]. The last three statements bind handles in the nonlinear solver class to our simulator and the two central vectors in the solution algorithm.

The solution of a system of nonlinear algebraic equations is accomplished by a call `nlsolver->solve()`.

Iterative methods for nonlinear systems usually require a linear system to be solved in each iteration. The simulator class is responsible for calculating this linear system and solving it by appropriate methods in the function

---

[3] This latter call cannot be used for determining the size of the linear and nonlinear solution vectors if there is more than one unknown per node.

```
void makeAndSolveLinearSystem ();
```

This is a virtual function declared by the base class `NonLinEqSolverUDC`. The nonlinear solver just sees the simulator class through a `NonLinEqSolverUDC` interface and assumes that there exists a function `makeAndSolveLinearSystem` in a subclass of `NonLinEqSolverUDC`. This explains why our simulator is derived from class `NonLinEqSolverUDC`.

The `makeAndSolveLinearSystem` function must set up and solve the relevant linear system and store the solution in the vector that was attached by `nlsolver->attachLinSol`, i.e., `linear_solution`. This design gives the programmer of the simulator class total control of the critical and problem dependent part of nonlinear solution algorithms, namely the specification and the solution of the linear systems. Notice that the definition of the nonlinear equations to be solved is implicitly contained in the definition of the linear system. Below is a suitable `makeAndSolveLinearSystem` function for our model problem.

```
void NlHeat1:: makeAndSolveLinearSystem ()
{
  dof->vec2field (nonlin_solution, *u); // copy most recent guess to u

  if (nlsolver->getCurrentState().method == NEWTON_RAPHSON)
    // essential boundary conditions must be set to zero because
    // the unknown vector in the linear system is a correction
    // vector (assume that nonlin_solution has correct ess. bc.)
    dof->fillEssBC2zero();
  else
    // normal (default) treatment of essential boundary conditions
    dof->unfillEssBC2zero();

  makeSystem (*dof, *lineq);

  // init start vector (linear_solution) for iterative linear solver:
  if (nlsolver->getCurrentState().method == NEWTON_RAPHSON)
    // start for a correction vector (expected to be close to 0):
    linear_solution.fill (0.0);
  else
    // use the most recent nonlinear solution:
    linear_solution = nonlin_solution;

  lineq->solve();   // invoke a linear system solver
  // the solution of the linear system is now available
  // in the vector linear_solution
}
```

The most recent nonlinear solution is contained in `nonlin_solution`. In the `integrands` function it is convenient to have the most recent values of u available in the field u. We therefore set u equal to `nonlin_solution`. Of course, `integrands` could access `nonlin_solution`, but the code is more readable if we use u when evaluating nonlinear coefficients in the PDE. Problems involving systems of PDEs, solved simultaneously in a common large system of nonlinear algebraic equations, normally require loading of `nonlin_solution` into various fields, because the indexing in `nonlin_solution` is then quite complicated. Recall that the `DegFreeFE` class keeps track of the transformation between degrees of freedom in a field (`FieldFE` or `FieldsFE`) and the ordering of the unknowns in a linear system (where the field values are collected in a single long vector). The switching between the field representation and the vector representation is therefore performed by `DegFreeFE` and its member functions `field2vec` and `vec2field`.

If the Newton-Raphson method is used to solve the nonlinear equations, the solution vector of each linear subsystem is a *correction* vector. This means that the essential boundary conditions to be enforced in these linear subsystems must be zero since it is assumed that the start vector for the nonlinear iteration contains the correct essential boundary condition values. The call to `DegFreeFE::fillEssBC2zero` ensures that nonzero essential condition values will be treated as zero values when inserted in a linear system. To turn this option off (default), there is a corresponding `unfillEssBC2zero` function.

We should emphasize a very important aspect of the `linear_solution` vector. If iterative solvers are used to solve the linear subsystem in each nonlinear iteration, the contents of `linear_solution` will affect the behavior of the iterative solvers[4]. Usually, one should fill the vector with a "good guess" of the solution. Since it is a correction vector, the expected entries will be small and the 0-vector may be a satisfactory guess. If one forgets to initialize the `linear_solution` vector, it may contain undefined values which may lead to serious run-time errors when running iterative solvers for linear systems. Always remember to explicitly initialize vectors that are used as solution vectors in iterative methods! If strange errors arise from iterative methods, check both the menu choice of the start vector and the contents of the vector.

As we see from the code segment in `makeAndSolveLinearSystem`,

```
nlsolver->getCurrentState().method
```

gives access to the current solution method for nonlinear equations. Since the linear subsystem in each nonlinear iteration will depend on the type of nonlinear solver, we need to impose appropriate tests both in the function `makeAndSolveLinearSystem` and in `integrands`.

Inside the time loop, the spatial problem at the current time level is solved by a call to `solveAtThisTimeStep`:

```
void NlHeat1:: solveAtThisTimeStep ()
{
  fillEssBC ();   // set essential boundary condition

  // initialize nonlin_solution with the nonlinear solution at
  // the previous time step:
```

---

[4] To be more precise, the user can choose between several types of start vectors for iterative solvers on the menu, see page 645. The default option is to apply the solution vector as it is when the iteration starts.

```
dof->field2vec (*u_prev, nonlin_solution);
// if there are new BC at this time step, update the present guess:
dof->insertEssBC (nonlin_solution);

// call nonlinear solver:
s_o << "t=" << tip->time() << '\n';
bool converged = nlsolver->solve ();

// load nonlinear solution found by the solver into the u field:
dof->vec2field (nonlin_solution, *u);
```

To obtain a start vector for the nonlinear iteration, the u field from the previous time step is loaded into nonlin_solution. This is performed by the DegFreeFE::field2vec function. Thereafter, essential boundary conditions are inserted in the start vector (nonlin_solution) to improve the quality of the initial guess. Notice that the essential boundary conditions may in general change with time such that the filling of essential conditions in nonlin_solution must be carried out at each time step.

The integrands function is as usual the most central function in a Diffpack finite element simulator. A possible implementation of integrands in the present model problem is listed next. We use a Galerkin method so $W_i = N_i$ in the formulas from Chapter 4.1.6.

```
void NlHeat1::integrands(ElmMatVec& elmat,const FiniteElement& fe)
{
    const real dt   = tip->Delta();          // current time step
    const int  nsd  = fe.getNoSpaceDim();    // no of space dims
    const real u_pt = u->valueFEM (fe);      // interpolate u
    const real up_pt= u_prev->valueFEM (fe); // interpolate u_prev
    Ptv(real) gradu_pt (nsd);                // grad u at present pt.
    u->derivativeFEM (gradu_pt, fe);         // compute gradu_pt
    Ptv(real) gradup_pt (nsd);               // grad u_prev --"--
    u_prev->derivativeFEM (gradup_pt, fe);   // compute gradup_pt

    const int nbf = fe.getNoBasisFunc();     // no of local nodes
    const real detJxW = fe.detJxW();
    real gradNi_gradNj, gradNi_gradu, h;
    int i,j,s;

    if (nlsolver->getCurrentState().method == NEWTON_RAPHSON)
      {
        for (i = 1; i <= nbf; i++) {
          gradNi_gradu = 0;
          for (s = 1; s <= nsd; s++)
            gradNi_gradu  += fe.dN(i,s)*gradu_pt(s);

          for (j = 1; j <= nbf; j++) {
            gradNi_gradNj = 0;
            for (s = 1; s <= nsd; s++)
              gradNi_gradNj += fe.dN(i,s)*fe.dN(j,s);
            h = fe.N(i)*fe.N(j) + dt*( lambda(u_pt)*gradNi_gradNj
                          + dlambda(u_pt)*fe.N(j)*gradNi_gradu );
            elmat.A(i,j)  += h*detJxW;
          }
```

```
        h = fe.N(i)*(u_pt - up_pt) + dt*u_pt*gradNi_gradu;
        elmat.b(i) += -h*detJxW;
      }
    }
  else
    errorFP("NlHeat1::integrands",
      "Linear subsystem for the nonlinear method %s is not impl.",
      getEnumValue(nlsolver->getCurrentState().method).c_str());
      // getEnumValue: returns a string of the enum and .c_str()
      // transforms the string to a const char* that can be fed
      // into errorFP (which applies an aform or C printf syntax)
}
```

The rest of the code in class `NlHeat1` is similar to the statements found in the classes `Poisson1` or `Heat1`. The reader is strongly encouraged to study the source code found in the `src/fem/NlHeat1` directory.

As a specific test case for verifying the implementation of the general parts of class `NlHeat1`, we consider $\lambda(u) = u$ and a boundary function

$$g(\boldsymbol{x}, t) = d \cdot t + \sum_{i=1}^{d} x_i, \quad \boldsymbol{x} = (x_1, \dots, x_d).$$

It is straightforward to realize that $u = g$ is also the analytical solution in the interior of the domain. This solution, which is linear in x_i and t, can be exactly reproduced by linear and higher-order elements in combination with any first- or higher-order scheme in time. Hence, we have a test problem where the numerical solution should be exact (modulo round-off errors) at the nodes, regardless of the mesh size. A possible input file to the program can look like this:

```
set gridfile = P=PreproBox | d=2 [0,1]x[0,1] |
               d=2  e_tp=ElmB4n2D partition=[5,5] grading=[1,1]
set time integration parameters = dt=0.1  t in [0,1]
sub NonLinEqSolver_prm
 set nonlinear iteration method = NewtonRaphson
 set max nonlinear iterations = 20
 set max estimated nonlinear error = 1.0e-5 ! eps in termination crit.
 set nonlinear relaxation prm = 0.8  ! omega, 1.0 is best here
ok
sub LinEqAdm
 sub Matrix_prm
  set matrix type = MatBand
 ok
 sub LinEqSolver_prm
  set basic method = GaussElim
 ok
ok
ok
```

Look for concrete input files and sample results in the subdirectory `Verify`.

The present model problem and its implementation are valid in any number of space dimensions. Try a 4D problem – all you have to do is to change the `gridfile` answer in the input file to

```
d=4 e=ElmTensorProd1 partition=[4,4,4,4] grading=[1,1,1,1]
```

The `ElmTensorProd1` element is a d-dimensional multi-linear element that co-incides with `ElmB2n1D` for $d = 1$, `ElmB4n2D` for $d = 2$, and `ElmB8n3D` when $d = 3$. Similar second-order and even higher-order elements are provided by `ElmTensorProd2`, `ElmTensorProd3` etc. We emphasize that the `ElmTensorProd` family is implemented as tensor products of one-dimensional elements, and that there is a significant performance penalty compared with specialized element classes, such as `ElmB4n2D`.

Suppose we want to compute the heat flux $-\kappa(u)\nabla u$, where $\kappa = \varrho C_p \lambda$. For simplicity we set $\varrho C_p = 1$. Flux computations are conveniently implemented by a call to the `FEM::makeFlux` function as explained in Chapter 3.4.5. How-ever, the `makeFlux` call shown in Chapter 3.4.5 requires the k function to be implemented as a virtual function[5] `k` with specified arguments. In the current context we could provide

```
real NlHeat1:: k (const FiniteElement& fe, real /*t*/)
{
  const real u_pt = u->valueFEM(fe);
  return lambda(u_pt);
}
```

A standard call `makeFlux`, as demonstrated in the `Poisson1` solver, will now compute a smooth flux $-\lambda(u)\nabla u$.

4.2.2   Extending the Solver

A natural improvement of the `NlHeat1` simulator is to extend the solution methods to cover the general θ-rule for time integration and enable the user to choose between Newton-Raphson's method and Successive Substitutions (Picard iteration) as the nonlinear solution methods. The θ-rule is defined in Chapter 2.2.2. Recall that $\theta = 0$ gives the explicit forward Euler scheme, a unit value of θ gives the fully implicit backward scheme, and $\theta = 1/2$ results in the mid-point rule or the Crank-Nicolson scheme.

For our diffusion equation these steps result in a system of nonlinear algebraic equations:

$$F_i(u_1^\ell, \ldots, u_n^\ell) = 0, \quad i = 1, \ldots, n.$$

where

$$F_i \equiv \int_\Omega \left[\left(\hat{u}^\ell - \hat{u}^{\ell-1}\right) W_i + \theta \Delta t \lambda(\hat{u}^\ell)\nabla \hat{u}^\ell \cdot \nabla W_i \right.$$

$$\left. + (1-\theta)\Delta t \lambda(\hat{u}^{\ell-1})\nabla \hat{u}^{\ell-1} \cdot \nabla W_i \right] d\Omega. \tag{4.35}$$

---

[5] You are not forced to follow this convention. An overloaded `makeFlux` function takes the coefficient k as a functor argument and gives full flexibility in defining k, see page 249.

Each linear problem in the Newton-Raphson method has a coefficient matrix

$$J_{i,j} \equiv \frac{\partial F_i}{\partial u_j^{\ell}} = \int\limits_{\Omega} [W_i N_j + \theta \Delta t \frac{d\lambda}{du}(\hat{u}^{\ell,k}) N_j \nabla W_i \cdot \nabla \hat{u}^{\ell,k}$$

$$+ \theta \Delta t \lambda(\hat{u}^{\ell,k}) \nabla W_i \cdot \nabla N_j)] \, d\Omega \qquad (4.36)$$

while the right-hand side is simply $-F_i$, with $\hat{u}^{\ell}$ replaced by $\hat{u}^{\ell,k}$ in (4.35).

The Successive Substitution method, which uses "old" values of u for the diffusivity $\lambda(u)$, is more straightforwardly implemented than the Newton-Raphson method, because we avoid differentiation of the nonlinear finite element equations:

$$F_i \equiv \int\limits_{\Omega} [(\hat{u}^{\ell} - \hat{u}^{\ell-1}) W_i + \theta \Delta t \lambda (\hat{u}^{\ell,k-1}) \nabla \hat{u}^{\ell,k} \cdot \nabla W_i$$

$$+ (1 - \theta) \Delta t \lambda (\hat{u}^{\ell-1}) \nabla \hat{u}^{\ell-1} \cdot \nabla W_i] d\Omega = 0. \qquad (4.37)$$

From this equation it should be straightforward to identify the matrix, the solution, and the right-hand side of a linear system for $u_j^{\ell,k}$, $j = 1, \ldots, n$.

What are the modifications to our previous class NlHeat1? We need θ as a parameter in the class, and its value should be read from the menu. Furthermore, the **integrands** routine must be extended. There are two types of linear subproblems, one for the Newton-Raphson method and one for the Successive Substitution method. How should we implement this? The most direct approach is to take a copy of class NlHeat1 and modify it. However, when one needs more than one version of a simulator, object-oriented programming becomes very convenient as explained in Chapters 3.4.6 and 3.5.6. We therefore derive a new extended class NlHeat1e from NlHeat1 and redefine the virtual functions that are not suitable.

```
class NlHeat1e : public NlHeat1
{
protected:
  real    theta;
  virtual void integrands (ElmMatVec& elmat, const FiniteElement& fe);
public:
  NlHeat1e ();
 ~NlHeat1e () {}
  void define (MenuSystem& menu, int level =MAIN); // add theta item
  void scan   ();                                  // read theta
};
```

In **define** and **scan** new statements are to be added, and the implementation conveniently becomes a call to the similar function in NlHeat1 followed by the additional statements.

```
void NlHeat1e:: define (MenuSystem& menu, int level)
{
  NlHeat1:: define (menu, level);
  menu.addItem (level, "theta", "parameter in theta-rule", "1.0");
}

void NlHeat1e:: scan ()
{
  NlHeat1:: scan ();
  theta = SimCase::getMenuSystem().get ("theta").getReal();
}
```

The `integrands` function from class `NlHeat1` must be significantly extended
(we still use $W_i = N_i$ in the expressions from Chapter 4.1.6).

```
void NlHeat1e::integrands(ElmMatVec& elmat,const FiniteElement& fe)
{
  const real dt   = tip->Delta();        // time step size
  const int  nsd  = fe.getNoSpaceDim(); // no of space dim.
  const real u_pt = u->valueFEM (fe);   // u at present itg.pt.
  const real up_pt = u_prev->valueFEM (fe);
  Ptv(real) gradu_pt (nsd);                    // grad u at present itg.pt.
  u->derivativeFEM (gradu_pt, fe);
  Ptv(real) gradup_pt (nsd);                   // grad u_prev --- " ----
  u_prev->derivativeFEM (gradup_pt, fe);
  const int nbf = fe.getNoBasisFunc();  // no of nodes in this elm.
  const real detJxW = fe.detJxW();
  real gradNi_gradNj, gradNi_gradu, gradNi_gradup, h;
  int i,j,s;
  if (nlsolver->getCurrentState().method == NEWTON_RAPHSON)
    {
      for (i = 1; i <= nbf; i++) {
        gradNi_gradu = gradNi_gradup = 0;
        for (s = 1; s <= nsd; s++) {
          gradNi_gradu  += fe.dN(i,s)*gradu_pt(s);
          gradNi_gradup += fe.dN(i,s)*gradup_pt(s);
        }
        for (j = 1; j <= nbf; j++) {
          gradNi_gradNj = 0;
          for (s = 1; s <= nsd; s++)
            gradNi_gradNj += fe.dN(i,s)*fe.dN(j,s);

          h = fe.N(i)*fe.N(j) + theta*dt*(
              lambda (u_pt)*gradNi_gradNj +
              dlambda(u_pt)*fe.N(j)*gradNi_gradu);

          elmat.A(i,j) += h*detJxW;
        }
        h = fe.N(i)*(u_pt - up_pt)
            + dt*theta*    lambda(u_pt )*gradNi_gradu
            + dt*(1-theta)*lambda(up_pt)*gradNi_gradup;
        elmat.b(i) += -h*detJxW;
      }
    }
  else if (nlsolver->getCurrentState().method == SUCCESSIVE_SUBST)
    {
      for (i = 1; i <= nbf; i++) {
```

```
        gradNi_gradup = 0;
        for (s = 1; s <= nsd; s++) {
          gradNi_gradup += fe.dN(i,s)*gradup_pt(s);
        }
        for (j = 1; j <= nbf; j++) {
          gradNi_gradNj = 0;
          for (s = 1; s <= nsd; s++)
            gradNi_gradNj += fe.dN(i,s)*fe.dN(j,s);
          h = fe.N(i)*fe.N(j) + theta*dt*(lambda(u_pt)*gradNi_gradNj);
          elmat.A(i,j) += h*detJxW;
        }
        h = fe.N(i)*up_pt - dt*(1-theta)*lambda(up_pt)*gradNi_gradup;
        elmat.b(i) += h*detJxW;
      }
    }
  else
    errorFP("NlHeat1e::integrands",
    "Linear subsystem for the nonlinear method %s is not implemented",
    getEnumValue(nlsolver->getCurrentState().method).c_str());
}
```

The reader can find the program in the directory `src/fem/NlHeat1/NlHeat1e`.

Exercise 4.16. Equip class `NlHeat1` with automatic report generation functionality (see page 312). The `NonLinEqSolver` class has report writing functionality (functions writing to a `MultipleReporter` object), but the output is too comprehensive and limited to the last call to the `solve` function. In time-dependent problems it is better to introduce a `NonLinEqSummary` object for collecting averages and extremes of the the nonlinear solver performance through the whole time simulation. The `NonLinEqSummary` object contains functions for writing to a `MultipleReporter` object.   ◇

Exercise 4.17. Adapt class `NlHeat1` to a stationary nonlinear PDE using the approach described in Chapter 3.10.4.   ◇

4.3 Projects

4.3.1 Operator Splitting for a Reaction-Diffusion Model

Mathematical Problem. This project concerns numerical methods for a nonlinear *reaction-diffusion equation*:

$$\frac{\partial u}{\partial t} = \alpha \nabla^2 u + f(u), \quad \boldsymbol{x} \in \Omega \subset \mathbb{R}^d, \ t > 0, \tag{4.38}$$

$$u(\boldsymbol{x}, 0) = I(\boldsymbol{x}), \quad \boldsymbol{x} \in \Omega, \tag{4.39}$$

$$\frac{\partial u}{\partial n} = 0, \quad \boldsymbol{x} \in \partial \Omega_N, \ t > 0. \tag{4.40}$$

Here, $u(\boldsymbol{x}, t)$ is the primary unknown, α is a dimensionless number, $\partial \Omega_N$ is the complete boundary of Ω, and $f(u)$ is nonlinear function of u that can be taken as $f(u) = -\beta u^m$, where β is a given dimensionless quantity.

Physical Model. The initial-boundary value problem (4.38)–(4.40) models diffusion of a substance that undergoes chemical reactions. Equation (4.38) reflects mass balance of the substance, (4.39) gives the initial distribution of the substance, and (4.40) ensures that there is no transport of the substance through the boundaries. The chemical reactions produce or extract mass and can hence be modeled as a source term $f(u)$ in the mass balance equation. Equation (4.38) results from neglecting temperature variations in (7.56) in the more complicated reaction-diffusion model treated in Project 7.3.3.

Numerical Method. Apply the Newton-Raphson method to (4.38) at the PDE level. Discretize the resulting sequence of linear equations by a θ-rule in time and a Galerkin finite element method in space. Develop precise formulas for the integrand expressions to be implemented in a Diffpack solver.

An alternative to solving (4.38) as a nonlinear PDE directly is to apply an *operator-splitting* technique. Starting with an approximation u^ℓ to u at time level ℓ, the aim is to construct $u^{\ell+1}$ through two or more intermediate steps, where we in each step solve a PDE involving only some of the terms in (4.38). A possible splitting is [93, Ch. 5.7]:

$$\frac{u^{\ell+\frac{1}{2}} - u^\ell}{\tau} = \alpha \nabla^2 u^{\ell+\frac{1}{2}} + f(u^\ell), \tag{4.41}$$

$$\frac{u^{\ell+1} - u^{\ell+\frac{1}{2}}}{\tau} = \rho \frac{u^{\ell+\frac{1}{2}} - u^\ell}{\tau} + f(u^{\ell+1}) - f(u^\ell), \tag{4.42}$$

where $\tau = \Delta t/(1+\rho)$ and $\rho \in (-1, 1]$ is an adjustable parameter. For $\rho = 0$ and $\rho = 1$ we recover the so-called Douglas-Rachford and Peaceman-Rachford splittings, respectively. For $\rho = 1$ (4.42) simplifies to

$$\frac{u^{\ell+1} - u^{\ell+\frac{1}{2}}}{\Delta t/2} = \alpha \nabla^2 u^{\ell+\frac{1}{2}} + f(u^{\ell+1}).$$

By eliminating $u^{\ell+\frac{1}{2}}$, one can show that the splitting is second-order accurate for $\rho = 1$ and first-order accurate for all other ρ values. The splitting results in a *linear* problem (4.41) for $u^{\ell+1/2}$, which can be solved by a standard Galerkin finite element method. Although (4.42) is a nonlinear equation, it does not involve spatial operators acting on the unknown $u^{\ell+1}$. Applying a group finite element approximation for the $f(u)$ term then results in a nonlinear equation at each node that is completely decoupled from the equations at the other nodes (carry out the details in this derivation). Hence, one only needs to solve a nonlinear equation in one variable at each node.

Operator-splitting methods are often also referred to as *fractional-step* methods in the literature[6]. More sophisticated operator-splitting methods for handling nonlinear reaction terms can be found in [31,80].

---

[6] See [93, p. 155] for a discussion of the terminology.

Implementation. Implement the full Newton-Raphson method for this problem as a slight modification of class `NlHeat1`. The solver based on operator splitting involves a heat equation a la class `Heat2`, with an additional loop over the nodes where we for each node must solve a nonlinear equation in one unknown.

Optimization of the Code. Assume that we apply a group finite element method for the nonlinear term $f(u)$ in all equations involving f. The discrete version of the nonlinear term can then be written as Mf, where $f = (f(u_1), \ldots, f(u_n))^T$. All the matrices that are involved in the discrete equations can now be considered as constant (time independent). This allows for a major optimization of the `Heat2` code, because the finite element assembly process can be carried out only once, as explained in Appendix B.6.2. The optimized algorithm is implemented in the subclass `Heat2eff` of `Heat2`. Incorporate these optimizations in the solver for (4.41). The same type of optimization also applies to the Newton-Raphson method, if we express the Jacobian of the Mf term as Mf', where f' is a diagonal matrix with entries $f'(u_1), \ldots, f'(u_n)$. Work with a lumped mass matrix when implementing the optimized versions of the solvers.

Computer Experiments. The operator-splitting technique is easier to implement and verify than the full Newton-Raphson method and involves less computational work at each time step. However, the Newton-Raphson method is expected to be more robust and allows for larger time steps. The relative performance of the methods is therefore an open question and dependent on the values of the parameters in the PDE. Set up a series of computational experiments and try to determine which of the two techniques that has the best overall efficiency (with respect to total CPU time). Use as large time step as possible, $\Delta t_{\max}$, for stability in the operator-splitting approach and convergence of the Newton-Raphson method. (Determination of $\Delta t_{\max}$ for each method must be performed through experiments. A corresponding linear diffusion equation discretized by the backward Euler scheme in time is stable for any value of Δt, but this does not hold when nonlinear iterations or operator splittings are introduced.) Check if halving $\Delta t_{\max}$ has a noticeable influence on the numerical results, i.e., if $\Delta t_{\max}$ leads to inaccurate results. (In strongly nonlinear problems the size of Δt might be dictated by stability rather than accuracy.)

4.3.2   Compressible Potential Flow

Mathematical Problem. A frequently used mathematical model for the flow of a compressible inviscid fluid involves the PDE

$$\nabla \cdot (\varrho \nabla \phi) = 0, \tag{4.43}$$

where ϱ is a nonlinear function of the primary unknown ϕ:

$$\varrho = \varrho_0 \left(1 - \frac{\gamma - 1}{\gamma + 1} \frac{1}{C_*^2} |\nabla \phi|^2 \right)^{\frac{1}{\gamma - 1}}. \tag{4.44}$$

The boundary conditions for (4.43) are $\varrho \partial \phi / \partial n = 0$ on a fixed body in the flow, a symmetry line, or a wall, and $\varrho \partial \phi / \partial n = \varrho \boldsymbol{U}_\infty \cdot \boldsymbol{n}$ far from the body, where $\boldsymbol{U}_\infty$ is the prescribed fluid velocity far from the body and $\boldsymbol{n}$ is the outward unit normal to this boundary. We assume that for flow around a body the velocity field is symmetric, otherwise we need additional conditions (the Kutta-Joukowsky condition, see Project 3.7.3).

Physical Model. In the model (4.43), $\nabla \phi$ is the fluid velocity, i.e., ϕ is the velocity potential, ϱ is the density of the fluid, γ is the ratio of specific heats ($\gamma = 1.4$ in air), ϱ_0 is the value of ϱ when $\nabla \phi = 0$, and C_* is the critical velocity. We shall require *subsonic flow*, $|\nabla \phi| < C_*$ everywhere in the domain $\Omega \subset \mathbb{R}^d$, such that the factor inside the parenthesis in (4.44) is always positive[7]

Equation (4.43) is the continuity equation for stationary compressible flow when the velocity is derived from a potential. Neglecting viscous effects, the counterpart to Newton's second law is the Bernoulli equation (here without body forces):

$$\int \frac{dp}{\varrho} + \frac{1}{2} |\nabla \phi|^2 = C_1,$$

where p is the pressure and C_1 is a constant. To close the model, we apply the equation of state $p = C_2 \varrho^\gamma$, C_2 being another constant. The constants C_1 and C_2 can be expressed by the conditions at one point, e.g., one may set $p = p_0$ and $\varrho = \varrho_0$ at the point where $\nabla \phi = 0$. Use this information as a starting point for deriving (4.43) and (4.44).

Show mathematically that the total integral of $\varrho \partial \phi / \partial n$ along the outer boundary must vanish. (Hint: Integrate the PDE over the domain and use the divergence theorem.) Interpret this result physically. Note that the result constrains the choice of $\boldsymbol{U}_\infty$.

Analysis. To learn about the model and find suitable cases for testing the implementation, we shall analyze two very simple problems. First, consider flow in a rectangular domain $\Omega = [0, a] \times [0, b]$, with $x = 0$ as an inflow boundary and $x = a$ as an outflow boundary. The remaining boundaries ($y = 0, b$) are either symmetry lines or walls. Choosing $\boldsymbol{U}_\infty = (U_0, 0)^T$ at $x = 0, a$ satisfies the compatibility condition that the integral of $\varrho \partial \phi / \partial n$ along the whole boundary vanishes. Demonstrate that $\phi(x, y) \sim x$ is a possible solution of this problem.

---

[7] Equation (4.43) can also be applied for *transonic* or *supersonic* flow, where $|\nabla \phi| \geq C_*$, but this requires more complicated solution procedures as the nature of the equation then changes completely.

A second test problem involves flow in a channel with varying width:

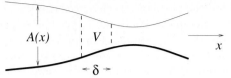

Integrating the PDE (4.43) over the volume V and using Gauss' divergence theorem gives $\int_{\partial V} \varrho \partial\phi / \partial n \, d\Gamma = 0$. The boundary integral vanishes on the solid walls due to the boundary condition. Hence, the only non-vanishing contributions come from the the two dashed lines:

$$-\int_{x_0} \varrho \frac{\partial\phi}{\partial x} dy + \int_{x_0+\delta} \varrho \frac{\partial\phi}{\partial x} dy = 0 \, .$$

Assuming that the variations of $\varrho \partial\phi / \partial x$ over the cross section are small, we might approximate the integral

$$-\int_{x_0} \varrho \frac{\partial\phi}{\partial x} dy \approx -A(x_0) \varrho \left. \frac{\partial\phi}{\partial x} \right|_{x_0} \, .$$

Show that this leads to the approximate continuity equation

$$\frac{\partial}{\partial x} \left(A(x) \varrho \frac{\partial\phi}{\partial x} \right) = 0 \, . \tag{4.45}$$

Introduce $u = \partial\phi / \partial x$, set $\phi = \phi(x)$ and obtain an algebraic equation for $u(x)$. Devise an iteration method for the numerical solution of this nonlinear equation.

Numerical Method. For the spatial discretization we shall use the finite element method. Since (4.43) is nonlinear, we need an outer iteration strategy. Formulate a standard Successive Substitution scheme and a Newton-Raphson method. In addition, we can expand $\nabla \cdot (\varrho \nabla \phi) = \varrho \nabla^2 \phi + \nabla \varrho \cdot \nabla \phi$ and apply the iteration

$$\nabla^2 \phi^{k+1} = -\frac{1}{\varrho^k} \nabla \varrho^k \cdot \nabla \phi^k, \tag{4.46}$$

where k is an iteration index. This approach is known as a *Poisson iteration* [48].

Implementation. Implement a Diffpack finite element simulator for (4.43) that offers the three iteration procedures Successive Subsition, Newton-Raphson iteration, and Poisson iteration. The simulator should also compute $\nabla \phi$ and the pressure $p = C_2 \varrho^\gamma$, where we for simplicity impose the particular scale $C_2 = 1$. The two test problems sketched previously should be implemented as subclasses (cf. Chapters 3.5.6 and 4.2.2). Partially verify the solver using the very simple analytical solution on a rectangular grid.

Computer Experiments. As main test case, we take flow in a channel with varying width, where the sides of the channel have the shape of a Gaussian bell function. Due to symmetry, only the upper half of the channel needs to be discretized. Apply the `PreproStdGeom` preprocessor and the `BOX_WITH_BELL` option (see the man page for class `PreproStdGeom`) to generate a suitable grid. The inflow and outflow velocities must be the same. On the channel wall and the symmetry line, the normal velocity $\partial\phi/\partial n$ vanishes. Let the program compute the average velocity $\bar{u}(x)$ over the cross section of the channel and make plots of $\bar{u}(x)$ and $u(x)$ computed from the approximate analytical solution (arising from solving (4.45)). Investigate through numerical experiments (i) the convergence behavior of the three iteration schemes and (ii) the quality of the approximate solution based on (4.45) for various bell shapes. Visualize the ϕ field, the velocity field, and the pressure field for two channel geometries, one that yields good agreement with the approximate analytical solution and one where the deviation is significant.

Chapter 5

Solid Mechanics Applications

The deformation of solid materials is a subject of importance in many fields of science and engineering. For example, the models and software in this chapter have applications in structural engineering, material science, seismology, geology, sensor technology, and bioengineering. The present chapter covers two mathematical models: (i) elastic deformations with thermal expansion effects and (ii) combined elastic and plastic deformations.

The mathematical modeling of elastic deformations is based on Newton's second law and a constitutive relation between the stresses and the deformations. Solution of the equations by the finite element method is a well established procedure that is available in many user-friendly and flexible software packages. The enormous success of finite element solution of elastic deformation problems has played a key role in promoting mathematical modeling and scientific computing in practical engineering.

We shall see that finite element methods for the elasticity equations, and the associated Diffpack implementations, are straightforward extensions of the concepts and techniques from scalar stationary boundary-value problems in Chapters 2 and 3. To see the strong links between the present chapter and Chapter 3, we use an *indicial notation* that is widespread in theoretical solid mechanics. An explanation of the notation is presented in Appendix A.2. However, the indicial notation is somewhat different from the mainstream *engineering notation* frequently used in many expositions on finite elements in solid mechanics, see for instance [15,16,30,37,131]. The present text emphasizes the finite element method as a general numerical approach for solving PDEs. For this purpose, the indicial notation represents a suitable tool for demonstrating the similarity between the general foundations of the finite element method in Chapter 2, the scalar PDE examples in Chapter 3, and the elasticity problem. Nevertheless, the engineering finite element notation is dominating in the literature and it is definitely very suitable for more complicated problems in solid mechanics, such as plasticity and viscoplasticity. We therefore introduce the engineering notation in Chapter 5.1.3 such that the reader can compare the two different languages of expressing the finite element equations in elasticity. The engineering notation simplifies the exposition of the numerics in Chapter 5.2 significantly.

Many applications in solid mechanics involve *permanent plastic deformations* beyond the elastic range. This calls for the simulation of combined elastic and plastic effects, using so-called *elasto-plastic* models. Herein we focus on a generalized model, called *elasto-viscoplasticity*, which contains the classical elasto-plastic model as a special case when time $t \to \infty$. The mathematical and numerical description of elasto-viscoplasticity fit well into our framework with initial-boundary value problems and PDEs. This model may therefore be easier to understand than the equations of classical elasto-plasticity. Chapter 5.2 is devoted to the elasto-viscoplastic model, solution algorithms, and a simple implementation. The corresponding simulator class is basically an elasticity solver equipped with a time loop.

5.1   Linear Thermo-Elasticity

The present section outlines a method and its Diffpack implementation for solving the time-independent Navier equations modeling isotropic linear elasticity with thermal effects. A quick introduction to elasticity is provided by the first two chapters of [118]. Readers who want a more comprehensive treatment of the physics and mathematical modeling of elastic deformations can consult a text on continuum mechanics, see for instance [50,56,75,77,103].

We start with presenting the equations of linear thermo-elasticity. The details of the finite element formulation are thereafter explained. Finally, we describe class `Elasticity1`, which is our Diffpack solver for stationary 2D and 3D thermo-elastic problems.

5.1.1   Physical and Mathematical Problem

We consider the deformation of an elastic continuum, where the fundamental equation is based on Newton's second law:

$$\varrho u_{r,tt} = \sigma_{rs,s} + \varrho b_r . \tag{5.1}$$

The indicial notation[1] used in this equation is explained in Appendix A.2. The first term represents the acceleration of the medium and is important for elastic waves, but will be neglected here. The quantity u_r is the displacement vector, and ϱ is the density. The first term on the right-hand side reflects the internal forces in the medium due to stresses. The final term ϱb_r represents body forces, e.g. gravity. Our purpose in the following is to solve the vector PDE

$$\sigma_{rs,s} = -\varrho b_r . \tag{5.2}$$

In elastic deformation problems, the main interest is often to study the stresses σ_{rs}. If the stresses are too large, the material may fracture, or the elastic model is no longer valid and must be replaced by a more relevant mathematical model.

---

[1] We shall here assume that r and s, and similar indices, run over $1, \ldots, d$ (the number of space dimensions), where d is 3 in the derivations, but $d = 2$ is allowed in the final equation.

Looking at the governing equation (5.2) we see that there are only three scalar components, but six unknowns[2] σ_{rs} (ϱ and b_r are prescribed quantities). To close the system, we therefore need to introduce additional information, reflecting the properties of elastic materials.

The stresses in elastic materials are related to deformation gradients $u_{r,s}$, represented through the strain tensor

$$\varepsilon_{rs} = \frac{1}{2}\left(u_{r,s} + u_{s,r}\right). \tag{5.3}$$

This expression and the mathematical model for elasticity presented here require that $|u_{r,s}|$ is small, such that products like $|u_{r,s}|^2$ can be neglected. Assuming isotropic elastic properties, the relation between the stress and strain tensors is given by Hooke's generalized law:

$$\sigma_{rs} = \lambda\varepsilon_{qq}\delta_{rs} + 2\mu\varepsilon_{rs}. \tag{5.4}$$

Here, λ and μ are *Lamé's elasticity constants*, which actually vary in space throughout heterogeneous materials, but are pure constants if the elastic material is homogeneous.

Heating or cooling a material leads to isotropic expansion or contraction. The strains associated with this deformation are referred to as *thermal strains* and denoted by ε_{rs}^T. An empirical model for ε_{rs}^T is

$$\varepsilon_{rs}^T = \alpha(T - T_0)\delta_{rs}, \tag{5.5}$$

where α is a thermal expansion coefficient, T is the temperature, and T_0 is a reference temperature where thermal deformations vanish. It is common to divide the total strain ε_{rs} into a thermal component ε_{rs}^T and an elastic component ε_{rs}^E,

$$\varepsilon_{rs} = \varepsilon_{rs}^T + \varepsilon_{rs}^E, \tag{5.6}$$

where the latter is related to the stresses via Hooke's law (5.4). Combining equations (5.3)–(5.6) then yields Hooke's generalized thermo-elastic law, which expresses the relation between stresses, temperature, and total deformation (u_r):

$$\sigma_{rs} = \lambda u_{q,q}\delta_{rs} + \mu(u_{r,s} + u_{s,r}) - \alpha(3\lambda + 2\mu)(T - T_0)\delta_{rs}. \tag{5.7}$$

This expression can now be inserted in Newton's second law (5.2), yielding a vector equation with d (the number of space dimensions) components for the d unknown components of the vector displacement field u_i:

$$((\lambda + \mu)u_{q,q})_{,r} + (\mu u_{r,q})_{,q} = (\alpha(3\lambda + 2\mu)(T - T_0))_{,r} - \varrho b_r. \tag{5.8}$$

In traditional vector notation, (5.8) takes the form

$$\nabla\left[(\lambda + \mu)\nabla \cdot \mathbf{u}\right] + \nabla \cdot \left[\mu\nabla\mathbf{u}\right] = \nabla\left[\alpha(3\lambda + 2\mu)(T - T_0)\right] - \varrho\mathbf{b}. \tag{5.9}$$

---

[2] The stress tensor is symmetric: $\sigma_{rs} = \sigma_{sr}$; hence only six entries can be distinct.

If the elasticity parameters are constant throughout the elastic medium we get the governing equation

$$(\lambda + \mu)u_{q,qr} + \mu u_{r,qq} = \alpha(3\lambda + 2\mu)T_{,r} - \varrho b_r, \tag{5.10}$$

or using vector notation,

$$(\lambda + \mu)\nabla(\nabla \cdot \boldsymbol{u}) + \mu\nabla^2\boldsymbol{u} = \alpha(3\lambda + 2\mu)\nabla T - \varrho\boldsymbol{b}. \tag{5.11}$$

Restricting d to 2, implies that $u_3 = 0$, $b_3 = 0$, and $\partial/\partial x_3 = 0$. This type of constrained deformation is called *plane strain* and is used as an approximate model for bodies having the shape of a long prismatic cylinder, where the loads are uniformly distributed in the third dimension and act perpendicular to the cylinder. Thin plate-like structures, with forces acting in the plane of the plate, can be described by the *plane stress* model. The corresponding governing equations are in fact (5.8) with $d = 2$, but then λ must be replaced by $\lambda' = 2\lambda\mu/(\lambda + 2\mu)$. A precise definition of plane stress and plane strain is given on page 374.

The equations (5.8) for the displacement field are commonly known as the Navier equations. These equations are to be solved in a domain Ω, representing the elastic body. The boundary conditions are of two types; the displacement u_r or the stress vector $\sigma_{rs}n_s$ (n_s is the outward normal vector to the boundary) must be prescribed at each boundary point. There must be d boundary conditions at each point of the boundary $\partial\Omega$. For example, at a point one may specify no shear stress ($d-1$ components of the stress vector are known) and vanishing normal displacement (one displacement component is known), giving a total of d conditions.

The present mathematical model is central in the field of structural analysis, but there are other important applications in soil mechanics as well. If we interpret $T - T_0$ to be the fluid pressure in a porous elastic medium, the present model governs the displacement and stresses of the porous medium caused by fluid flow. The quantity T will generally be found by solving a Laplace- or Poisson-type equation for the temperature or pressure field. In many cases, this equation for T is independent of the elastic deformations. We can then solve for T first and thereafter invoke the elasticity model with T as a prescribed quantity.

5.1.2   A Finite Element Method

We shall apply a straightforward Galerkin method to the Navier equations, as this will be an optimal method for the present problem[3]. To reduce the size of the expressions, we drop the body force term ϱb_r, which is trivial to

---

[3] The theory and main results from Chapter 2.10 can be extended to cover the current elasticity problem [17].

include in the forthcoming numerical formulation. The displacement field u_r is approximated by

$$\hat{u}_r = \sum_{j=1}^{n} u_j^r N_j(x_1, \ldots, x_d),$$

where N_j are prescribed finite element basis functions and u_j^r are $n \cdot d$ coefficients to be found by the method. Inserting the expression for $\hat{u}_r$ in each of the components in the Navier equations results in a residual. This residual multiplied by a weighting function is required to vanish for n linearly independent weighting functions. Employing the Galerkin method, where the weighting functions are identical to the trial functions N_j, we obtain after integration by parts of $\int_\Omega \sigma_{rs,s} N_i d\Omega$,

$$\int_\Omega \sigma_{rs} N_{i,s} d\Omega = \int_{\partial\Omega} N_i \sigma_{rs} n_s d\Gamma, \tag{5.12}$$

where a sum over the repeated index s is implied. The generalized Hooke's law relates the stresses to the deformation and temperature:

$$\sigma_{rs} = \lambda u_{q,q} \delta_{rs} + \mu(u_{r,s} + u_{s,r}) - \alpha(3\lambda + 2\mu)(T - T_0)\delta_{rs}. \tag{5.13}$$

Here δ_{rs} denotes the Kronecker delta. As usual, we get a linear system for the unknowns u_j^r, $j = 1, \ldots, n$ and $r = 1, \ldots, d$. These unknowns can be collected in a vector

$$\boldsymbol{u} = (u_1^1, \ldots, u_1^d, u_2^1, \ldots, u_2^d, \ldots, u_n^1, \ldots, u_n^d)^T. \tag{5.14}$$

The linear system of the form $\boldsymbol{Ku} = \boldsymbol{c}$ can be interpreted either at the element or the global level; n is then the number of nodes in the element or in the grid.

Let us derive the expressions for the matrix and right-hand side vector of the linear system. First, we insert (5.13) in (5.12). The indices i and r indicate the equation number, while j and s are used in the summation over the unknowns u_j^s, such that the system can be written as

$$\sum_{j=1}^{n} \sum_{s=1}^{d} A_{i,j}^{rs} u_j^s = c_i^r, \quad r = 1, \ldots, d, \; i = 1, \ldots, n. \tag{5.15}$$

In the discrete finite element equations we do not apply the summation convention and insert instead explicit summation symbols. For fixed i and j (node numbers), $A_{i,j}^{rs}$ is a $d \times d$ matrix, and $\boldsymbol{K}$ can be viewed as a $dn \times dn$ matrix with $n \times n$ blocks $A_{i,j}^{rs}$. To obtain the expression for $A_{i,j}^{rs}$, we recognize that (5.13) inserted in (5.12) gives rise to four basic terms, three of which comprise $A_{i,j}^{rs}$.

The integrand of the first term reads

$$\sum_j \sum_s \sum_k \lambda N_{i,s} N_{j,k} \delta_{rs} u_j^k = \sum_j \sum_k \lambda N_{i,r} N_{j,k} u_j^k$$

$$= \sum_j \sum_s \lambda N_{i,r} N_{j,s} u_j^s$$

by using the fact that $\delta_{rs} v_s = v_r$ for any vector v_s and replacing the dummy summation variable k by s. The second term has the form

$$\sum_j \sum_s \mu N_{i,s} N_{j,s} u_j^r = \sum_j \sum_s \sum_k \mu N_{i,s} N_{j,s} \delta_{rk} u_j^k$$

$$= \sum_j \sum_s \left(\sum_k \mu N_{i,k} N_{j,k} \right) \delta_{rs} u_j^s \, .$$

The third term is more straightforward as it takes the form of (5.15) directly:

$$\sum_j \sum_s \mu N_{i,s} N_{j,r} u_j^s \, .$$

The right-hand side contribution is also simple,

$$\sum_s \alpha(3\lambda + 2\mu)(T - T_0)\delta_{rs} N_{i,s} = \alpha(3\lambda + 2\mu)(T - T_0) N_{i,r} \, .$$

The general formula for block $\boldsymbol{K}_{i,j}$ in $\boldsymbol{K}$, representing the coupling between node i and j, can be written

$$\boldsymbol{K}_{i,j} = \begin{pmatrix} A_{i,j}^{11} & \cdots & A_{i,j}^{1d} \\ \vdots & \ddots & \vdots \\ A_{i,j}^{d1} & \cdots & A_{i,j}^{dd} \end{pmatrix}, \tag{5.16}$$

where

$$A_{i,j}^{rs} = \int_\Omega \left[\mu \left(\sum_k N_{i,k} N_{j,k} \right) \delta_{rs} + \mu N_{i,s} N_{j,r} + \lambda N_{i,r} N_{j,s} \right] d\Omega \tag{5.17}$$

at the global level. The corresponding element contribution is obtained by replacing the whole domain Ω by element e, Ω_e, restricting i and j to be nodes in element e, and perhaps transforming the integral to the reference coordinate system by including the determinant of the Jacobian of the isoparametric mapping. See Chapter 2.3 for more information about local and global coordinates.

The partitioning of $\boldsymbol{u}$ in n blocks $\boldsymbol{u}_j$ has $\boldsymbol{u}_j = (u_j^1, \ldots, u_j^d)^T$. Similarly, the corresponding partitioning of $\boldsymbol{c}$ has blocks $\boldsymbol{c}_i = (c_i^1, \ldots, c_i^d)^T$, where

$$c_i^r = \int_\Omega \left[(2\mu + 3\lambda) \, \alpha(T - T_0) N_{i,r} \right] d\Omega + \int_{\partial\Omega_N} N_i t_r d\Gamma \, . \tag{5.18}$$

Here, t_r is the stress vector $\sigma_{rs}n_s$ at the boundary, often referred to as the *traction*, and $\partial\Omega_N$ is the part of the boundary where t_r is prescribed. Again, the expressions can trivially be transformed to the element level and to integrals over the reference element.

Notice that traction-free boundaries appear as natural boundary conditions in the present problem. Essential boundary conditions (u_i known at a node) are incorporated in the matrix system by substituting the equation corresponding to the actual degree of freedom by the boundary condition. Since $\boldsymbol{K}$ is a symmetric matrix, the incorporation of essential conditions must preserve the symmetry. See Chapter 2.3 for information on the implementation of essential boundary conditions.

The stresses are usually of more interest than the displacement field. Having computed the latter, the stress tensor field can be computed using Hooke's law. Since the stresses are derivatives of the displacement components, the computed stress tensor field will be discontinuous over the element boundaries. For the purposes of plotting and analysis it is desirable to have a continuous stress field. It is also common to introduce a scalar stress measure which is easier to interpret than a tensor field. In class `Elasticity1`, we employ the von Mises stress, also called the *equivalent stress*, as a measure of the stress level. In terms of the stress tensor, the equivalent stress m reads

$$m = \sqrt{\frac{3}{2}\sigma'_{rs}\sigma'_{rs}}, \tag{5.19}$$

where σ'_{rs} is the stress deviator, $\sigma'_{rs} \equiv \sigma_{rs} - \frac{1}{3}\sigma_{kk}\delta_{rs}$.

5.1.3 Engineering Finite Element Notation

As mentioned in the introduction to this chapter, the indicial notation in the previous section is different from what is usually found in textbooks on the finite element method for elasticity problems. The more common engineering notation is explained below. As we shall see, the engineering notation makes the discrete equations very compact, and this has clear advantages in more complicated problems from solid mechanics, for example, elasto-viscoplastic deformations, which are treated in Chapter 5.2. The engineering notation is based on matrix and vector symbols. The stress tensor is not written as a tensor, but as a vector

$$\boldsymbol{\sigma} = \left(\sigma_{xx}, \sigma_{yy}, \sigma_{zz}, \sigma_{xy}, \sigma_{yz}, \sigma_{zx}\right)^T.$$

It is common in this notation to use σ_{xy} rather than σ_{12}. The strain tensor is written as a vector as well,

$$\boldsymbol{\varepsilon} = \left(\varepsilon_{xx}, \varepsilon_{yy}, \varepsilon_{zz}, \gamma_{xy}, \gamma_{yz}, \gamma_{zx}\right)^T,$$

where one applies the "engineering shear strain" $\gamma_{xy} = 2\varepsilon_{xy}$. Hooke's law can now be expressed as

$$\boldsymbol{\sigma} = \boldsymbol{D}\boldsymbol{\varepsilon}, \tag{5.20}$$

with

$$\boldsymbol{D} = \frac{E(1-\nu)}{(1+\nu)(1-2\nu)} \begin{pmatrix} 1 & \frac{\nu}{1-\nu} & \frac{\nu}{1-\nu} & 0 & 0 & 0 \\ & 1 & \frac{\nu}{1-\nu} & 0 & 0 & 0 \\ & & 1 & 0 & 0 & 0 \\ & & & \frac{1-2\nu}{2(1-\nu)} & 0 & 0 \\ & \text{symmetric} & & & \frac{1-2\nu}{2(1-\nu)} & 0 \\ & & & & & \frac{1-2\nu}{2(1-\nu)} \end{pmatrix} \tag{5.21}$$

for three-dimensional elasticity. The parameters E and ν are known as Young's modulus and Poisson's ratio, respectively.

The common cases of plane strain

$$\varepsilon_{xz} = \varepsilon_{yz} = \varepsilon_{zz} = 0, \quad \sigma_{zz} = \nu(\sigma_{xx} + \sigma_{yy})$$

and plane stress

$$\sigma_{xz} = \sigma_{yz} = \sigma_{zz} = 0, \quad \varepsilon_{zz} = -\frac{\nu}{E}(\sigma_{zz} + \sigma_{yy})$$

can be handled by a constitutive law on the form (5.20), but now with

$$\boldsymbol{\sigma} = (\sigma_{xx}, \sigma_{yy}, \sigma_{xy})^T, \quad \boldsymbol{\varepsilon} = (\varepsilon_{xx}, \varepsilon_{yy}, \gamma_{xy})^T,$$

and

$$\boldsymbol{D} = \frac{E}{1-\nu^2} \begin{pmatrix} 1 & \nu & 0 \\ \nu & 1 & 0 \\ 0 & 0 & \frac{1-\nu}{2} \end{pmatrix} \tag{5.22}$$

for plane stress and

$$\boldsymbol{D} = \frac{E(1-\nu)}{(1+\nu)(1-2\nu)} \begin{pmatrix} 1 & \frac{\nu}{1-\nu} & 0 \\ \frac{\nu}{1-\nu} & 1 & 0 \\ 0 & 0 & \frac{1-2\nu}{2(1-\nu)} \end{pmatrix} \tag{5.23}$$

for plane strain.

Let us approximate the displacement field, i.e. the primary unknown in the elasticity problem, with

$$\hat{\boldsymbol{u}} = \sum_{j=1}^{n} \boldsymbol{u}_j N_j(\boldsymbol{x}), \tag{5.24}$$

where $\boldsymbol{u}_j$ is the displacement vector at node no. j.

The discrete strain vector can now be expressed by

$$\boldsymbol{\varepsilon} = \sum_{j=1}^{n} \boldsymbol{B}_j \boldsymbol{u}_j, \tag{5.25}$$

with

$$\boldsymbol{B}_i = \begin{pmatrix} N_{i,x} & 0 & 0 \\ 0 & N_{i,y} & 0 \\ 0 & 0 & N_{i,z} \\ N_{i,y} & N_{i,x} & 0 \\ 0 & N_{i,z} & N_{i,y} \\ N_{i,z} & 0 & N_{i,x} \end{pmatrix}. \tag{5.26}$$

In plane stress and strain, where only the x and y components of $\boldsymbol{u}_i$ enter the equations, the $\boldsymbol{B}_i$ matrix takes the form

$$\boldsymbol{B}_i = \begin{pmatrix} N_{i,x} & 0 \\ 0 & N_{i,y} \\ N_{i,y} & N_{i,x} \end{pmatrix}. \tag{5.27}$$

Axisymmetric elasticity problems in (z,r) coordinates fits into the framework above, using the following definitions of $\boldsymbol{\varepsilon}$, $\boldsymbol{\sigma}$, $\boldsymbol{B}_i$, and $\boldsymbol{D}$.

$$\boldsymbol{\varepsilon} = \left(\varepsilon_{zz}, \varepsilon_{rr}, \varepsilon_{\theta\theta}, \gamma_{rz} \right)^T, \tag{5.28}$$

$$\boldsymbol{\sigma} = \left(\sigma_{zz}, \sigma_{rr}, \sigma_{\theta\theta}, \sigma_{rz} \right)^T, \tag{5.29}$$

$$\boldsymbol{B}_i = \begin{pmatrix} 0 & N_{i,z} \\ N_{i,r} & 0 \\ \frac{1}{r}N_i & 0 \\ N_{i,z} & N_{i,r} \end{pmatrix}, \tag{5.30}$$

$$\boldsymbol{D} = \frac{E(1-\nu)}{(1+\nu)(1-2\nu)} \begin{pmatrix} 1 & \frac{\nu}{1-\nu} & \frac{\nu}{1-\nu} & 0 \\ \frac{\nu}{1-\nu} & 1 & \frac{\nu}{1-\nu} & 0 \\ \frac{\nu}{1-\nu} & \frac{\nu}{1-\nu} & 1 & 0 \\ 0 & 0 & 0 & \frac{1-2\nu}{2(1-\nu)} \end{pmatrix}. \tag{5.31}$$

All integrals $\int_\Omega () d\Omega$ are in the axisymmetric case transformed to $2\pi \int () r dr dz$, see [131, Ch. 4].

The equilibrium equation (5.2) has the weak formulation (5.12), which in the engineering notation can be expressed as

$$\int_\Omega \boldsymbol{B}_i^T \boldsymbol{\sigma} d\Omega = \int_{\partial\Omega} N_i t d\Gamma. \tag{5.32}$$

The symbol $\boldsymbol{t}$ represents the traction $t_r = \sigma_{rs} n_s$. Hooke's law with temperature strains can now be written

$$\boldsymbol{\sigma} = \boldsymbol{D}(\boldsymbol{\varepsilon} - \boldsymbol{\tau}), \tag{5.33}$$

with $\boldsymbol{\tau} = \alpha(T - T_0)(1,1,1,0,0,0)^T$ in 3D, $\boldsymbol{\tau} = \alpha(T - T_0)(1,1,0)^T$ in plane stress, and $\boldsymbol{\tau} = (1+\nu)\alpha(T - T_0)(1,1,0)^T$ in plane strain. Inserting (5.25) in (5.33) and then the resulting (5.33) in (5.32) yields

$$\sum_{j=1}^n \int_\Omega \boldsymbol{B}_i^T \boldsymbol{D} \boldsymbol{B}_j d\Omega \, \boldsymbol{u}_j = \int_\Omega \boldsymbol{B}_i^T \boldsymbol{D} \boldsymbol{\tau} d\Omega + \int_{\partial\Omega} N_i t d\Gamma, \tag{5.34}$$

for $i = 1, \ldots, n$. For each i in (5.34) we have a vector equation with three components (or two in plane stress/strain or axisymmetry). The element matrix can be viewed as a block matrix with $n_e \times n_e$ blocks $\int_{\Omega_e} \boldsymbol{B}_i^T \boldsymbol{D} \boldsymbol{B}_j d\Omega$, $i, j = 1, \ldots, n_e$, where n_e is the number of nodes in the element. The element vector can also be written on block form, with block no. i taking the form of the right-hand side in (5.34). A nice feature of this formulation is that plane stress, plane strain, axisymmetric, and general three-dimensional problems can straightforwardly be treated in a unified notation and implementation.

Exercise 5.1. Show that the element matrix contribution from the coupling of local nodes no. i and j, in a finite element formulation of $-\nabla \cdot [\lambda \nabla u] = f$ in $\mathbb{R}^d$, can be expressed as $\int_{\Omega_e} \boldsymbol{B}_i^T \boldsymbol{D} \boldsymbol{B}_i d\Omega$ by proper definitions of $\boldsymbol{B}_i$ as a $d \times 1$ matrix and $\boldsymbol{D}$ as a 1×1 matrix. Much of the finite element literature works with $\boldsymbol{B}_i^T \boldsymbol{D} \boldsymbol{B}_i$ as a generic form of symmetric element matrices.   ⋄

5.1.4   Implementation

A simulator for thermo-elastic problems based on the formulation in Chapter 5.1.2 has been implemented in a class with name `Elasticity1`. The corresponding source code is located in `src/app/Elasticity1` and its subdirectories. In principle, class `Elasticity1` is very similar to class `Poisson2` in Chapter 3.5, but the elasticity model involves a stationary *vector* equation and some of the details of the solver are therefore slightly different from the scalar case. The primary unknown in class `Elasticity1` is a vector field of type `FieldsFE`. Roughly speaking, a `FieldsFE` object is just an array of handles to `FieldFE` objects for each component in the vector field. All the `FieldFE` functionality is therefore immediately available for `FieldsFE` objects.

```
class Elasticity1 : public FEM
{
public:
  Handle(GridFE)          grid;
  Handle(DegFreeFE)       dof;
  Handle(FieldsFE)        u;                 // displacement field
  Handle(FieldsFEatItgPt) stress_measures;   // von Mises equiv. stress
  Handle(FieldsFE)        smooth_stress_measures;
  Handle(SaveSimRes)      database;
  Vec(real)               solution;
  Handle(LinEqAdmFE)      lineq;

  Handle(Field)   T;              // temperature field
  FieldFormat     T_format;       // info about the type of T field
  Handle(Field)   E, nu;          // Young's modulus, Poisson's ratio
  FieldFormat     nu_format, E_format;
  Handle(Field)   rho;            // density
  FieldFormat     rho_format;
  Handle(Field)   alpha;          // thermal expansion coeff.
  FieldFormat     alpha_format;
  Ptv(real)       g_dir;          // direction of gravity: x(nsd)-dir
  real            pressure1;      // for boundary indicator 1
```

```
real              pressure2;      // for boundary indicator 2

enum Elasticity_type
    { PLANE_STRESS, PLANE_STRAIN, THREE_DIM, AXISYMMETRY };
Elasticity_type  elasticity_tp;

real              magnification;  // factor for exaggerated displ.
Handle(GridFE)    deformed_grid;  // grid + magnification*u
Handle(FieldFE)   equiv_stress2;  // equiv. stress over deformed_grid
Handle(FieldFE)   u_magnitude;    // magnitude of displacement vector

// internal structures for avoiding time consuming reallocation:
Mat(real)         matdxd;         // used in integrands
Ptv(real)         normal_vec;     // used in integrands4side
VecSimple(Ptv(real)) gradu_pt;    // used in derivedQuantitiesAtItgPt

// convert nu and E to Lame's elasticity constants:
static void nuE2Lame (real nu, real E, real& lambda, real& mu,
                      Elasticity_type el_tp = THREE_DIM);
```

We have here only shown the most important data members of the class for easy reference in the forthcoming discussion. One should notice that we use the flexible `FieldFormat` and `Handle(Field)` tools from Chapter 3.11.4 for representing variable coefficients like T, E, ν, and so on.

More than One Unknown at a Node. The fundamental difference between a scalar and vector PDE solver is that the latter involves more than one unknown per node. More specifically, our thermo-elastic model leads to d unknowns per node. We must therefore treat the interaction between the displacement vector field (`FieldsFE`) and the unknowns in the linear system carefully. As usual, we have a vector in class `Elasticity1` containing the unknowns in the linear system. After this vector has been calculated by the linear solver, we load it into the displacement field (`FieldsFE`) by calling the `vec2field` functionality in the `DegFreeFE` object. Recall that class `DegFreeFE` takes care of the relation between degrees of freedom in a field representation (here a `FieldsFE` vector field) and the ordering of the equations and unknowns in the linear system. The relation is quite simple in the present case: Degree of freedom no. j in displacement component field no. i has degree of freedom number $d(j-1)+i$ in the linear system, according to (5.14). This ordering of the unknowns and the algebraic equations is used at the element level as well and is therefore fundamental for the statements in the `integrands` function.

Looking at the basic expressions (5.17) and (5.18) in the `integrands` function, it can be wise to compute the element matrix and vector in terms of block contributions from the nodes. In the code below, we let the integers i and j run over the nodes, while r and s run over the d local degrees of freedom in each block. We compute with λ and μ in the discrete equations, but the elastic parameters on the menu are Young's modulus E and Poisson's ratio ν. There is a simple formula for computing λ and μ given E and ν.

```
void Elasticity1::integrands(ElmMatVec& elmat,const FiniteElement& fe)
{
  const int    d     = fe.getNoSpaceDim();
  const int    nbf   = fe.getNoBasisFunc();
  const real detJxW = fe.detJxW();

  // Handle(Field) T, E, etc must be interpolated at current point:
  const real T_pt     = T->valueFEM (fe);      // temperature
  const real E_pt     = E->valueFEM (fe);      // Young's modulus
  const real nu_pt    = nu->valueFEM (fe);     // Poisson's ratio
  const real alpha_pt = alpha->valueFEM (fe);  // expansion coeff.
  const real rho_pt   = rho->valueFEM (fe);    // density
  // convert to Lame's elasticity constants:
  real lambda, mu; nuE2Lame (nu_pt, E_pt, lambda, mu);

  int i,j;    // basis function counters
  int k,r,s;  // 1,..,nsd (space dimension) counters
  int ig,jg;  // element dof, based on i,j,r,s
  real gradNi_gradNj, shear_term, volume_term, body_force_term;
  // matdxd is a class member to avoid repeated local allocation

  for (i = 1; i <= nbf; i++) {
    for (j = 1; j <= nbf; j++) {
      gradNi_gradNj = 0;
      for (k = 1; k <= d; k++)
        gradNi_gradNj += fe.dN(i,k)*fe.dN(j,k);

      for (r = 1; r <= d; r++)
        for (s = 1; s <= d; s++)
          matdxd (r,s) = mu*fe.dN(i,s)*fe.dN(j,r);

      for (r = 1; r <= d; r++)
        matdxd (r,r) += mu*gradNi_gradNj;

      // add block matrix (i,j) to elmat.A:
      for (r = 1; r <= d; r++)
        for (s = 1; s <= d; s++) {
          shear_term  = matdxd(r,s);
          volume_term = lambda*fe.dN(i,r)*fe.dN(j,s);

          ig = d*(i-1)+r;
          jg = d*(j-1)+s;
          elmat.A(ig,jg) += (shear_term + volume_term)*detJxW;
        }
    }
    // add block matrix i to elmat.b:
    for (r = 1; r <= d; r++) {
      shear_term  = 2*mu*alpha_pt*T_pt*fe.dN(i,r);
      volume_term = 3*alpha_pt*lambda*T_pt*fe.dN(i,r);
      body_force_term = rho_pt*9.81*g_dir(r)*fe.N(i);

      ig = d*(i-1)+r;
      elmat.b(ig) += (shear_term+volume_term+body_force_term)*detJxW;
    }
  }
}
```

The surface integral over $\partial\Omega_N$ is implemented for normal tractions only. Moreover, the implementation restricts the essential conditions to be homogeneous: $u_i = 0$. Hence, the code can only be applied to problems where a point on the boundary of the body is either prevented from being displaced or subject to a pressure force. It is fairly straightforward to extend the code to treat general traction vectors. Prescribed nonvanishing boundary displacements are of course trivially implemented.

Boundary Indicators. Any Diffpack finite element simulator needs a convention for setting boundary conditions based on boundary indicators. In class Elasticity1 we introduce $d+2$ boundary indicators. The first two indicators are used for boundaries with normal stresses (pressure). The corresponding boundary conditions enter the finite element formulation through a surface integral term in the weak formulation and are hence implemented in the integrands4side function. Indicators $2+i$, $i = 1, \ldots, d$, mark boundary segments where $u_i = 0$. These indicators are fundamental to setting essential boundary conditions in the fillEssBC function:

```
dof->initEssBC ();
int nno = grid->getNoNodes();
int d   = grid->getNoSpaceDim();
int i,k;
for (i = 1; i <= nno; i++)
  for (k = 1; k <= d; k++) {
    if (grid->boNode (i, 2+k))
      dof->fillEssBC (i,k, 0.0);
  }
```

The implementation of the surface integral terms in the weak formulation follows the recipe from Chapter 3.5.2. The calcElmMatVec function is essentially the same as in class Poisson2, but the book-keeping of degrees of freedom in the integrands4side is slightly more demanding due to more than one unknown at each node.

```
void Elasticity1:: integrands4side
    (int /*side*/, int boind, ElmMatVec& elmat, const FiniteElement& fe)
{
  real pressure = DUMMY;
  if (boind == 1)        pressure = pressure1;
  else if (boind == 2)   pressure = pressure2;

  const int d   = fe.getNoSpaceDim();
  const int nbf = fe.getNoBasisFunc();
  const real JxW = fe.detSideJxW();
  fe.getNormalVectorOnSide (normal_vec /*class member*/);
  int i,r;
  for (i = 1; i <= nbf; i++)
    for (r = 1; r <= d; r++)
      elmat.b (d*(i-1)+r) += fe.N(i)*pressure*normal_vec(r)*JxW;
}
```

Modifications of Initializing Statements. A few adjustments of the initializing statements of a typical scalar PDE solver are necessary in the present case to deal with d unknowns per node. When constructing vector fields, say a `Handle(FieldFE) u`, the statement

```
u.rebind (new FieldsFE (*grid, "u"));
```

gives as many components in the field as there are space dimensions in the model[4].

Diffpack has a clear distinction between the `GridFE` class, which only contains *geometry* information about the grid and the elements, classes `FieldFE` and `FieldsFE`, which contain a grid with nodal *field values* and built-in evaluation at arbitrary spatial points, class `LinEqAdmFE`, which deals with linear system information only, and class `DegFreeFE`, which is the link between grid/field objects and linear system objects.

The number of unknowns per node is of course essential to the `DegFreeFE` object and must by set as an argument at construction time:

```
nsd = grid->getNoSpaceDim();
dof.rebind (new DegFreeFE (*grid, nsd);  // nsd unknowns per node
```

Scalar PDE solvers from Chapter 3 could use `grid->getNoNodes()` for extracting the number of unknowns in the linear system, but now we need more general tools,

```
u->getNoValues();       // u is Handle(FieldsFE)
dof->getTotalNoDof();   // alternative
```

The coefficients in the governing PDEs, like T, α, and the elasticity parameters, can be constants, explicit formulas, or discrete fields. The flexibility of the elasticity solver is enhanced by using general `Handle(Field)` representations of the variable coefficients and allowing the user to determine the format of each field at run time. The reader should consult Chapter 3.11.4 for an introduction to the usage and functionality of the classes `Handle(Field)` and `FieldFormat` for flexible field representations.

Stress Computation. From (5.7) we see that the stresses are linear combinations of the *derivatives* of the displacement field. Recall from Chapter 2.8 that the derivatives of finite element fields are in general discontinuous at the element boundaries. For plotting or analysis purposes it is often useful to work with a smooth stress measure. However, one should notice here that if (λ, μ), or alternatively (E, ν), are discontinuous, the exact stresses on the surface perpendicular to the surface of the discontinuity is in fact discontinuous. Smoothing can hence be physically incorrect. The optimal goal would be to smooth the discontinuities arising from the finite element interpolation

---

[4] There is of course another `FieldsFE` constructor that can also take the number of components in a vector or tensor field as argument.

functions, but keep the discontinuities due to layered media. This is a difficult problem and will not be addressed here.

Computation of a smooth scalar stress measure, e.g. m from (5.19), can be based on the outline in Chapter 3.4.5 regarding evaluation and smoothing of derivatives of finite element fields. The derived quantity m is represented by a FieldsFEatItgPt object, holding the values of m at the points in a reduced Gauss-Legendre integration rule on each element (notice that the derivatives of finite element fields have optimal accuracy at the points in a reduced Gauss-Legendre rule, cf. page 170). The FieldsFEatItgPt object has functionality for running through all elements and for each sampling point calling a virtual function derivedQuantitiesAtItgPt in the solver class. This virtual function computes the problem-dependent formulas for the derived quantities. Here, this function is supposed to evaluate m from (5.19). The FieldsFEatItgPt field can be reported as a set of scattered m values, or we can smooth the field. The smoothing is carried out by a call to the utility FEM::smoothFields.

```
void Elasticity1:: calcDerivedQuantities ()
{
  stress_measures->derivedQuantitiesAtItgPt
    (*this, *grid,  1  /* 1 derived quantity */,
     GAUSS_POINTS, -1  /* reduced Gauss-Legendre sampling points */);
  FEM::smoothFields (*smooth_stress_measures, *stress_measures);
}

void Elasticity1:: derivedQuantitiesAtItgPt
  (VecSimple(NUMT)& quantities, const FiniteElement& fe)
{
  const real T_pt     = T->valueFEM (fe);    // T at current point
  const real alpha_pt = alpha->valueFEM (fe);
  const real E_pt     = E->valueFEM (fe);
  const real nu_pt    = nu->valueFEM (fe);
  // convert to Lame's elasticity constants:
  real lambda, mu; nuE2Lame (nu_pt, E_pt, lambda, mu);

  // use the scratch matrix matdxd (class member) as stress tensor:
  Mat(real)& s = matdxd;
  s.redim (3,3);        // always 3x3, also in 2D problems
  s.fill (0);
  const int d = fe.getNoSpaceDim();
  gradu_pt.redim (d);   // (class member)

  real div_u = 0;  // divergence of displacement field
  for (int k = 1; k <= d; k++) {
    u()(k).derivativeFEM (gradu_pt(k), fe);
    div_u += gradu_pt(k)(k);
  }
  int i,j; const real temp_term = (3*lambda+2*mu)*alpha_pt*T_pt;

  for (i = 1; i <= d; i++) {
    // off-diagonal terms in the stress tensor:
    for (j = 1; j < i; j++) {
      s(i,j) = mu * (gradu_pt(i)(j) + gradu_pt(j)(i));
```

```
   s(j,i) = s(i,j);
   }
   // diagonal terms in the stress tensor:
   s(i,i) = lambda*div_u + 2*mu*gradu_pt(i)(i) - temp_term;
   }
   // augment s_zz for plane strain:
   if (elasticity_tp == PLANE_STRAIN)
     s(3,3) = lambda*div_u - temp_term;

   real e2 = 0.5*( 6*(sqr(s(1,2)) + sqr(s(2,3)) + sqr(s(1,3))) +
        sqr(s(1,1)-s(2,2)) + sqr(s(2,2)-s(3,3)) + sqr(s(3,3)-s(1,1)) );
   quantities(1) = sqrt(e2);
}
```

The code in class `Elasticity1` can handle any number of stress measures, e.g. all the stress tensor components in addition to several yield functions, but the current implementation computes only one measure, namely the von Mises equivalent stress (5.19).

Specializing Solvers in Subclasses. To verify the implementation of class `Elasticity1`, we need to compare the analytical and numerical solution in some standard test examples, such as a pressurized cylinder in plane strain and 3D elongation of a rod. The implementation of these examples are conveniently done in subclasses, like we explain in Chapters 3.4.6 and 3.5.6. Class `PressurizedCyl` implements a plane strained cylinder with pressure forces on the inner and outer boundaries. Comparison with the analytical solution [118, p. 69] and estimation of convergence rates follow the ideas of class `Poisson2anal`. Another subclass solver `Rod` simulates the elongation of a rod, with square-shaped cross section, due to normal stresses at the ends. The displacement field is linear in the spatial coordinates and the analytical solution should therefore be obtained within machine precision, regardless of the number of elements used in the structure.

Visualizing the Deformed Body. The elasticity computations are performed in the initial configuration of the elastic body. The primary unknown u_i describes how this body deforms under the action of loads, and the final shape of the deformed body might be of interest to the analyst. To this end, we can move the grid according to the displacement field and define computed fields over the deformed grid. The `scan` function makes an extra grid, `deformed_grid`, and a special field, `equiv_stress2`, to hold the smoothed equivalent stress m over `deformed_grid`.

In the `saveResults` function we easily move the original grid according to u_i times a magnification factor given by the user on the menu:

```
*deformed_grid = *grid;           // original configuration
deformed_grid->move (*u, magnification);
database->dump (*equiv_stress2); // equiv_stress2 uses deformed_grid
```

If the `magnification` factor is less than zero the program will compute a suitable value of `magnification` such that the deformed grid is displaced at most a distance $L/10$, where L is the characteristic size of the elastic body.

5.1.5 Examples

Some computational examples involving class `Elasticity1` are presented in the following. A Perl script `src/app/Elasticity1/plotEL.pl` runs the filter `simres2mtv` and other tools to provide color plots of m in the undeformed and deformed configurations, contour line plot of m in the deformed configuration, and color plot of the displacement magnitude $\sqrt{u_r u_r}$ in the undeformed domain. The `plotEL.pl` script is invoked with the casename as parameter. The simulation results must be computed prior to executing `plotEL.pl`.

L-Shaped Beam. An elasticity problem involving an L-shaped domain is depicted in Figure 5.1a. A suitable input file is found in `Verify/deformedL.i`. The reader is encouraged to run the simulator with this input and try the `plotEL.i` script:

```
./app --casename Lbody < Verify/deformedL.i
plotEL.pl Lbody
```

Load on an Arch. The next example concerns the deformation of an arch as depicted in Figure 5.2. In this problem we need to assign a pressure load, i.e. boundary indicator 1, over a portion of a side. The grid is first generated by the `DISK_WITH_HOLE` feature in the `PreproStdGeom` preprocessor. Thereafter we apply redefinition of boundary indicators and add boundary nodes to model the load and symmetry constraints. The file `Verify/arch1.i` contains explanations of our usage of basic Diffpack functionality. This example shows that the preprocessing capabilities covered in the text have some flexibility, but for real engineering applications, one should use professional preprocessor software and import the grid in the elasticity solver.

Run the simulator with the `Verify/arch1.i` input file and try the `plotEL.pl` script. Figure 5.2b shows the deformations and the stress state (m). The amount of deformation is exaggerated for visualization purposes. A plot of the `u_magnitude` field reveals the correct scales for the displacements.

Exercise 5.2. Figure 5.3 displays two thin plates in tension, where one of the plates has a crack and the other has an elliptic hole. First reduce the size of the computational domains by utilizing symmetry and formulate the boundary conditions to be applied on the symmetry boundaries. Then construct appropriate input files for the two geometries. The elliptic hole can be modeled as in Figure 3.9 on page 268. The grid for the crack problem is obtained by `PreproBox` and manipulating conditions at boundary nodes. However, the displacement gradients will be very large close to the crack tip so one should

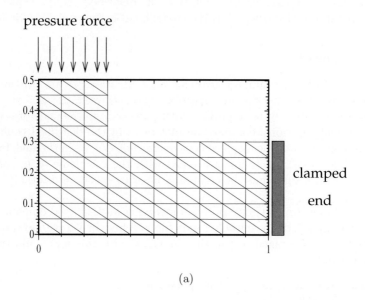

(a)

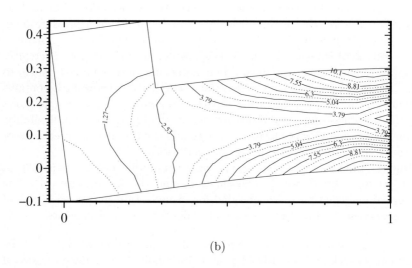

(b)

Fig. 5.1. L-shaped clamped beam under a pressure load; (a) sketch of the problem; (b) contour lines of the equivalent stress m from (5.19). When plotting the deformation in (b) we have scaled the displacement field such that the features of the displacements are more clearly visualized.

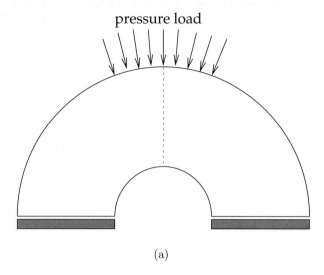

(a)

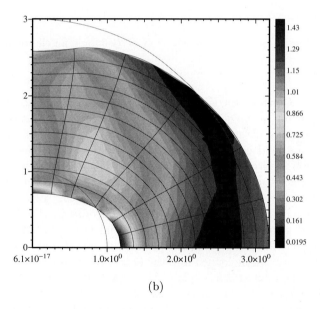

(b)

Fig. 5.2. Deformation of an arch with a load; (a) sketch of the problem; (b) gray-tone plot of the equivalent stress m from (5.19) over the magnified deformed grid (the boundary of the initial configuration is included in the plot). Biquadratic nine-node elements were used in the simulation. When plotting the deformation in (b) we have scaled the displacement field such that the features of the displacements are more clearly visualized.

use local mesh refinements (see e.g. Figure 5.6 on page 401). Running the `makegrid` utility with `crackedPlate-makegrid.i` as input (see the `Verify` directory) generates a possible mesh with refinements around the crack tip. Study the ratio of the maximum value of the stress measure m, divided by the stress as if there were no crack or hole in the plate, when the shape of the elliptic hole varies from a circle to an approximation of the crack[5]. The experiments are conveniently carried out by modifying a copy of the `plotEl.pl` script in the `Elasticity` directory, where you supply the width of the elliptic hole as input, run the simulator, visualize the results, and run `simres2summary -f casename -n equiv_stress -s -A` to grab the maximum value of m from the output.   ◇

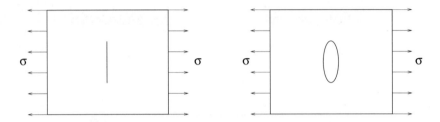

Fig. 5.3. Tension of a thin plate with a crack (left) and an elliptic hole (right).

Exercise 5.3. Consider the geometry in Figure 3.21 on page 323. Assume that the white portion of the domain is an elastic material and that the inner hole is filled with a gas at temperature T_g. Outside the elastic material the scaled temperature is held constant at T_0 (i.e. the temperature that corresponds to no thermal deformations in the material). There are no applied stresses at the boundaries, but temperature variations will lead to internal stresses. Compute first the temperature field with the `Poisson2` solver. Utilize symmetry to reduce the size of the computational domain (the grid can then be generated using the `PreproStdGeom` tool and the `BOX_WITH_ELLIPTIC_HOLE` option, see page 267). The temperature field is now stored in simres format on a datafile and can be loaded into the `Elasticity1` solver by specifying the field format `FIELD_ON_FILE` as explained on page 330. Neglect pressures on the physical boundaries in the elasticity problem, i.e., the only load stems from the spatial temperature variations. Extend the `plotEL.pl` script such that it visualizes the temperature field as well. Study the maximum stresses as the radius of the inner hole is varied.   ◇

---

[5] It can be interesting to know that for an infinite plate, the maximum tension stress σ_{xx} at the point of the ellipse that corresponds to the crack tip equals [50] $\sigma(1 + 2b/a)$, where x is in the direction of the external stress field σ, a is the half-axis in the x-direction, and b is the other half-axis of the ellipse.

Exercise 5.4. Modify the `Elasticity1` solver such that it computes with the B_i, D, t, σ, and τ quantities from Chapter 5.1.3. The resulting simulator should be capable of handling plane strain, plane stress, axisymmetry, and general three-dimensional elasticity.   ◇

5.2   Elasto-Viscoplasticity

The deformation of solid materials is usually purely elastic when the stresses are below a certain critical level, called the *yield stress*. When the stresses are above this threshold, a combination of elastic and *plastic* deformation occur, where the latter type of deformation is recognized by being permanent. Simulation of elasto-plastic deformation is fundamental in many engineering disciplines, and the finite element method has been successfully applied to this problem area during three decades [131]. From a numerical point of view, it might be easier to deal with a time-dependent extension of the classical elasto-plastic model, namely the elasto-viscoplasticity model, which recovers the elasto-plastic solution as time approaches infinity.

5.2.1   Basic Physical Features of Elasto-Viscoplasticity

The basic physical features of the elasto-viscoplastic model are best introduced by means of a one-dimensional mechanical system as depicted in Figure 5.4. The system consists of a spring with elastic properties, serially coupled to a friction slider and a dashpot, which comprise the viscoplastic part. The friction slider is inactive if the stress is below the yield point, resulting in no viscoplastic deformations. The dashpot reflects the viscous properties of viscoplasticity when the slider is active. The total deformation then consists of the elastic displacement in the spring plus the displacement of the viscous dashpot.

Equilibrium of any part of the structure implies that the sum of the normal stresses is σ everywhere. The onset of viscoplastic deformation occurs when $\sigma = C_Y$, where C_Y is the *yield stress*. For $\sigma < C_Y$ the deformation is purely elastic, whereas for $\sigma \geq C_Y$ the material is in a combined elasto-viscoplastic state. The friction slider develops an internal stress

$$\sigma^P = \begin{cases} C_Y, \sigma \geq C_Y \\ \sigma, \quad \sigma < C_Y \end{cases} \tag{5.35}$$

When the friction slider is active ($\sigma > C_Y$) the dashpot experiences a stress

$$\sigma^D = \sigma - \sigma^P, \tag{5.36}$$

which is related to the viscoplastic strain rate $\dot{\varepsilon}^P$ by the constitutive relation

$$\sigma^D = \mu \dot{\varepsilon}^P . \tag{5.37}$$

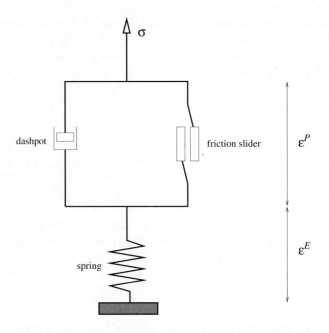

Fig. 5.4. Sketch of a mechanical model that reflects the basics physics of elasto-viscoplastic deformation.

In analogy with viscous fluid flow, μ is a viscosity coefficient.

The viscoplastic strain vanishes if the friction slider is inactive. The notation

$$\langle \Phi(F) \rangle = \begin{cases} \Phi(F), & F > 0 \\ 0, & F \le 0 \end{cases}$$

is then convenient, since it enables us to express the constitutive law for the viscoplastic part of the structure as

$$\dot{\varepsilon}^P = \langle \Phi(F) \rangle \frac{1}{\mu} \sigma^D . \tag{5.38}$$

Here, $F = \sigma - C_Y$. When discussing the model in Figure 5.4, we simply set $\Phi(F) = F$.

Normally, the material exhibits *strain hardening*. That is, after unloading from a viscoplastic state and then reloading again, it appears that the yield point has increased to

$$C_Y = C_{Y_0} + f(\varepsilon^P), \tag{5.39}$$

where C_{Y_0} is the yield point for an originally unstrained material (provided $f(0) = 0$). We should hence treat C_Y as a function of the *total* viscoplastic strain ε^P. The precise form of C_Y must be determined from physical experiments.

The elastic spring is characterized by a linear relation between stress and strain:

$$\sigma = E\varepsilon^E, \tag{5.40}$$

where the coefficient E is Young's modulus. Let ε be the total strain in the system. Obviously, ε is the sum of the strain in the spring, ε^E, and the strain in the viscoplastic part of the system, ε^P,

$$\varepsilon = \varepsilon^E + \varepsilon^P. \tag{5.41}$$

Our aim now is to develop a relationship between the total stress and strain in the two cases $\sigma < C_Y$ and $\sigma \geq C_Y$. In the purely elastic case, $\sigma < C_Y$ and $\varepsilon^P = 0$, we get

$$\sigma = E\varepsilon, \tag{5.42}$$

whereas

$$\dot{\varepsilon} = \frac{1}{E}\dot{\sigma} + \gamma(\sigma - C_Y(\varepsilon^P)) \tag{5.43}$$

in the combined elasto-viscoplastic state. We have introduced the *fluidity parameter* $\gamma = 1/\mu$.

Exercise 5.5. Derive (5.43). $\diamond$

Let us study the simple case where C_Y is a linear function of ε^P, with slope c: $C_Y = C_{Y_0} + c\varepsilon^P = C_{Y_0} + c(\varepsilon - E^{-1}\sigma)$. Equation (5.43) can then be expressed as

$$\dot{\varepsilon} + \gamma c\varepsilon = \frac{1}{E}\dot{\sigma} + \gamma\left(\frac{c}{E} + 1\right)\sigma - \gamma C_{Y_0}. \tag{5.44}$$

The solution becomes

$$\varepsilon(t) = e^{-\gamma ct} \int_{-\infty}^{t} e^{\gamma c\tau}\left(\frac{1}{E}\dot{\sigma}(\tau) + \gamma\left(\frac{c}{E} + 1\right)\sigma(\tau) - \gamma C_{Y_0}\right) d\tau. \tag{5.45}$$

Consider now a suddenly applied load q at time $t = 0$: $\sigma(t) = qH(t)$, where $H(t)$ is the Heaviside function, defined as $H(t) = 1$ for $t > 0$ and $H(t) = 0$ for $t < 0$. The delta function is the derivative of $H(t)$: $H'(t) = \delta(t)$. Using the general properties that

$$\int_{-\infty}^{t} f(x)\delta(x - t_0)dx = f(t_0)H(t - t_0)$$

and

$$\int_{\infty}^{t} f(x)H(x - t_0)dx = H(t - t_0)\int_{t_0}^{t} f(x)dx,$$

one finds that

$$\varepsilon(t) = \frac{q}{E} + \frac{q - C_{Y_0}}{c}\left[1 - e^{-\gamma ct}\right], \quad t > 0. \tag{5.46}$$

We see that the immediate response is purely elastic (q/E). Thereafter, a viscous deformation develops. The total deformation approaches a finite value as $t \to 0$. In the case of an *ideally plastic* material, C_Y is a constant, and letting $c \to 0$ reveals that ε increases linearly in time.

5.2.2   A Three-Dimensional Elasto-Viscoplastic Model

We shall write the equations of 2D and 3D elasto-viscoplasticity using the engineering notation from Chapter 5.1.3, because this notation simplifies the manipulations of the expressions considerably. The brief text below is an combination of the expositions in Zienkiewicz and Taylor [131, Ch. 7.11-7.12] and Owen and Hinton [87, Ch. 8], tailored to the numerical background from Chapters 2, 4.1, and 5.1. Another useful reference, also with emphasis on implementational aspects, is Smith and Griffiths [106]. We refer to these books and the references therein for more information about the broad topic of plasticity and viscoplasticity.

Basic Equations. The fundamental assumption of elastic-viscoplasticity is that the total deformation can be separated into elastic and viscoplastic parts. More specifically, one can express the total *rate of strain*, $\dot{\varepsilon}$, according to

$$\dot{\varepsilon} = \dot{\varepsilon}^E + \dot{\varepsilon}^P, \tag{5.47}$$

where $\dot{\varepsilon}^P$ is the viscoplastic strain rate, and $\dot{\varepsilon}^E$ is the elastic strain rate. The dot denotes partial differentiation in time.

The strain rate is related to the stresses through the time-differentiated version of Hooke's generalized law,

$$\dot{\sigma} = D\dot{\varepsilon}^E . \tag{5.48}$$

The viscoplastic strain rate vanishes if the stress intensity is below a critical level. In three-dimensional theory, this is more precisely expressed in terms of the *yield function* $f(\sigma_{ij})$. The deformation is purely elastic ($\dot{\varepsilon}^P = 0$) when $f(\sigma_{ij}) \leq C_Y$, where C_Y is generally given as in (5.39). No special numerical attention is paid to hardening in the following. Instead we refer to [131, Ch. 7.8, vol II] for more information. There are numerous choices of the yield function f, some of which are briefly described in the next section.

One can write the constitutive law for the viscoplastic strains as

$$\dot{\varepsilon}^P = \gamma \langle \Phi(F) \rangle \frac{\partial Q}{\partial \sigma}, \tag{5.49}$$

where $F = f(\sigma_{ij}) - C_Y$, γ is the fluidity parameter (inverse "viscosity"), which influences the time scale only, and Q is a *plastic potential*. The common case of *associated plasticity* corresponds to taking $Q = F$. A widespread choice of the $\Phi(F)$ function is $\Phi(F) = (F/C_Y)^N$, for some prescribed constant N.

The equation of motion, in terms of stresses and accelerations, is the same for all continua. Therefore, (5.1) is also valid here. Neglecting accelerations and employing Galerkin's method results in the same weak form (5.32) as in the elastic case:

$$\int_\Omega B_i^T \sigma \, d\Omega + f_i = 0, \quad i = 1, \ldots, n, \tag{5.50}$$

where $\boldsymbol{f}_i$ contains body forces and surface integrals of the tractions. Recall from Chapter 5.1.2 that the latter quantities appear as natural boundary conditions. Finally, we have the relation between the total strain and the displacement field $\boldsymbol{u} \approx \hat{\boldsymbol{u}} = \sum_j N_j \boldsymbol{u}_j$, which we need in the following time-differentiated and spatially discrete form:

$$\dot{\varepsilon} = \sum_{j=1}^{n} \boldsymbol{B}_j \dot{\boldsymbol{u}}_j \,. \tag{5.51}$$

Recall that quantities like $\boldsymbol{B}_j$, $\boldsymbol{u}_j$, and $\boldsymbol{D}$ are defined in Chapter 5.1.3.

In the elasticity problem, we can easily eliminate the stress and strain quantities and obtain a vector equation for the displacement field. This is more complicated in the present elasto-viscoplastic problem, because of the nonlinearities introduced by the $\dot{\varepsilon}^P$ term. By proper discretizations we can, however, derive an iterative procedure where we in each iteration solve a linear system with respect to the displacement field. In its simplest form, the linear system appearing in this numerical method has a coefficient matrix identical to that of the elasticity problem.

Combining (5.47), (5.48), and (5.51) yields

$$\dot{\boldsymbol{\sigma}} = \boldsymbol{D}(\sum_j \boldsymbol{B}_j \dot{\boldsymbol{u}}_j - \dot{\varepsilon}^P) \,. \tag{5.52}$$

According to the derivation of the elasticity equations, the next step would now be to eliminate $\boldsymbol{\sigma}$, by inserting the constitutive law, here (5.52), in (5.50) and thereby achieve a vector equation for the displacement. However, $\dot{\varepsilon}^P$ depends nonlinearly on $\boldsymbol{\sigma}$, making (5.52) a nonlinear ordinary differential equation for $\boldsymbol{\sigma}$. Hence, we cannot eliminate $\boldsymbol{\sigma}$ and must work with essentially two types of spatially discrete equations governing $\boldsymbol{\sigma}$ and $\boldsymbol{u}_j$:

$$\int_\Omega \boldsymbol{B}_i^T \boldsymbol{\sigma} d\Omega + \boldsymbol{f}_i = 0, \quad i = 1, \ldots, n, \tag{5.53}$$

$$\dot{\boldsymbol{\sigma}} - \boldsymbol{D} \sum_{j=1}^{n} \boldsymbol{B}_j \dot{\boldsymbol{u}}_j + \boldsymbol{D}\dot{\varepsilon}^P = 0 \,. \tag{5.54}$$

The time derivatives can be approximated by a θ-rule (cf. Chapter 2.2.2):

$$\frac{\Delta\boldsymbol{\sigma}}{\Delta t} - \boldsymbol{D} \sum_{j=1}^{n} \boldsymbol{B}_j \frac{\Delta\boldsymbol{u}_j}{\Delta t} + \theta\dot{\varepsilon}^{P,\ell} + (1 - \theta)\dot{\varepsilon}^{P,\ell-1} = 0 \,.$$

We have here made use of the notation

$$\Delta\boldsymbol{\sigma} \equiv \boldsymbol{\sigma}^\ell - \boldsymbol{\sigma}^{\ell-1}, \quad \Delta\boldsymbol{u}_j \equiv \boldsymbol{u}_j^\ell - \boldsymbol{u}_j^{\ell-1} \,.$$

Utilizing these approximations in (5.53)–(5.54), gives the following nonlinear discrete problem:

$$\int_\Omega \boldsymbol{B}_i^T \boldsymbol{\sigma}^\ell d\Omega + \boldsymbol{f}_i^\ell = 0, \quad i = 1, \ldots, n,$$

$$\Delta\boldsymbol{\sigma} - \boldsymbol{D}\sum_j \boldsymbol{B}_j \Delta\boldsymbol{u}_j + \Delta t \boldsymbol{D}\theta \dot{\boldsymbol{\varepsilon}}^{P,\ell} + \Delta t \boldsymbol{D}(1-\theta)\dot{\boldsymbol{\varepsilon}}^{P,\ell-1} = 0.$$

A Newton-Raphson-Based Iteration Method. The current nonlinear problem can be treated by a Newton-Raphson-like procedure, and the nature of the resulting approximations enables us to eliminate $\Delta\boldsymbol{\sigma}$ and derive a linear system for $\Delta\boldsymbol{u}_j$ $(j = 1, \ldots, n)$.

Define

$$\boldsymbol{\Psi}_i^\ell \equiv \int_\Omega \boldsymbol{B}_i^T \boldsymbol{\sigma}^\ell d\Omega + \boldsymbol{f}_i^\ell, \tag{5.55}$$

$$\boldsymbol{R}^\ell \equiv \Delta\boldsymbol{\sigma} - \boldsymbol{D}\sum_j \boldsymbol{B}_j \Delta\boldsymbol{u}_j + \Delta t \boldsymbol{D}\theta \dot{\boldsymbol{\varepsilon}}^{P,\ell} + \Delta t \boldsymbol{D}(1-\theta)\dot{\boldsymbol{\varepsilon}}^{P,\ell-1}. \tag{5.56}$$

We now consider $\boldsymbol{\Psi}_i^\ell = 0$ and $\boldsymbol{R}^\ell = 0$ as a simultaneous nonlinear system of algebraic equations for $\boldsymbol{\sigma}^\ell$ and $\boldsymbol{u}^\ell$, $i = 1, \ldots, n$. A Newton-Raphson approach consists in making first-order Taylor-series expansions of $\boldsymbol{\Psi}_i^\ell$ and $\boldsymbol{R}^\ell$ around an approximation $\boldsymbol{\sigma}^{\ell,k}$ and $\boldsymbol{u}^{\ell,k}$ in iteration k. Enforcing the linear Taylor-series expansion to vanish, results in linear equations for the increments $\delta\boldsymbol{\sigma}^{\ell,k+1}$ and $\delta\boldsymbol{u}_j^{\ell,k+1}$:

$$\boldsymbol{\Psi}_i^{\ell,k+1} \approx \boldsymbol{\Psi}_i^{\ell,k} + \int_\Omega \boldsymbol{B}_i^T \delta\boldsymbol{\sigma}^{\ell,k+1} d\Omega = 0, \tag{5.57}$$

$$\boldsymbol{R}^{\ell,k+1} \approx \boldsymbol{R}^{\ell,k} + \left(\frac{\partial\boldsymbol{R}}{\partial\boldsymbol{\sigma}^\ell}\right)^{\ell,k} \delta\boldsymbol{\sigma}^{\ell,k+1} + \sum_{j=1}^n \left(\frac{\partial\boldsymbol{R}}{\partial\boldsymbol{u}_j^\ell}\right)^{\ell,k} \delta\boldsymbol{u}_j^{\ell,k+1}$$

$$= \boldsymbol{R}^{\ell,k} + \delta\boldsymbol{\sigma}^{\ell,k+1} - \boldsymbol{D}\sum_{j=1}^n \boldsymbol{B}_j \delta\boldsymbol{u}_j^{\ell,k+1}$$

$$+ \Delta t \boldsymbol{D}\theta \boldsymbol{C} \delta\boldsymbol{\sigma}^{\ell,k+1} = 0, \tag{5.58}$$

with

$$\boldsymbol{C} = \left(\frac{\partial\dot{\boldsymbol{\varepsilon}}^P}{\partial\boldsymbol{\sigma}}\right)^{\ell,k}. \tag{5.59}$$

We can solve (5.58) with respect to the stress increment, resulting in

$$\delta\boldsymbol{\sigma}^{\ell,k+1} = \hat{\boldsymbol{D}}\sum_j \boldsymbol{B}_j \delta\boldsymbol{u}_j^{\ell,k+1} - \boldsymbol{Q}\boldsymbol{R}^{\ell,k}, \tag{5.60}$$

where

$$\hat{D} = (D^{-1} + \theta \Delta t C)^{-1}, \tag{5.61}$$
$$Q = (I + \theta \Delta t DC)^{-1}. \tag{5.62}$$

Inserting this $\delta \sigma^{\ell,k+1}$ in (5.57) yields a linear system for $\delta u_j^{\ell,k+1}$:

$$\sum_j \left(\int_\Omega B_i^T \hat{D} B_j d\Omega \right) \delta u_j^{\ell,k+1} = \int_\Omega B_i^T Q R^{\ell,k} d\Omega - \Psi_i^{\ell,k}. \tag{5.63}$$

The Computational Procedure. Let us summarize the equations from the previous section in a computational algorithm.

Initial Conditions. At time $t = 0$, $\dot{\varepsilon}^{P,0} = 0$, and the stresses correspond to a purely elastic state. That is, u^0 is determined from a standard elasticity problem:

$$\sum_j \left(\int_\Omega B_i^T D B_j d\Omega \right) u_j^0 = -f_i^0, \quad i = 1, \ldots, n,$$

with the associated stresses $\sigma^0 = D \sum_j B_j u_j^0$.

The Equations at an Arbitrary Time Level. Suppose $u^{\ell-1}$, $\sigma^{\ell-1}$, and $\dot{\varepsilon}^{P,\ell-1}$ are known. New displacements u^ℓ are generally computed by an iteration procedure. As an initial guess for the iterations, we set $u_j^{\ell,0} = u_j^{\ell-1}$, $\sigma^{\ell,0} = \sigma^{\ell-1}$, and $\dot{\varepsilon}^{P,\ell,0} = \dot{\varepsilon}^{P,\ell-1}$. For $k = 0, 1, \ldots$ until convergence of the Newton-Raphson method, we perform the following steps.

1. Compute C, $\hat{D}$, and $R^{\ell,k}$ from (5.59), (5.61), and (5.56) at an integration point in an element.
2. Compute Q from (5.62).
3. Compute the contribution $B_i^T \hat{D} B_j$ to the coefficient matrix.
4. Compute the contributions $B_i^T \sigma^{\ell,k}$ (needed in $\Psi_i^{\ell,k}$) and $B_i^T Q R^{\ell,k}$ to the right-hand side of (5.63).
5. Assemble and solve the linear system (5.63) for the correction $\delta u^{\ell,k+1}$ of the displacement field.
6. Calculate the displacements and stresses according to

$$\sigma^{\ell,k+1} = \sigma^{\ell,k} + \delta \sigma^{\ell,k+1}, \tag{5.64}$$
$$u^{\ell,k+1} = u^{\ell,k} + \delta u^{\ell,k+1}. \tag{5.65}$$

where $\delta \sigma^{\ell,k+1}$ is found from (5.60).
7. Calculate the new viscoplastic strain rate from

$$\dot{\varepsilon}^{P,\ell,k+1} = \gamma \langle \Phi(F(\sigma_{ij}^{\ell,k+1})) \rangle \left(\frac{\partial F}{\partial \sigma} \right)_i^{\ell,k+1}. \tag{5.66}$$

8. Proceed with the next iteration.

If the iterative procedure has converged in m iterations, we define $\boldsymbol{u}^{\ell,m}$ as the converged solution $\boldsymbol{u}^{\ell}$ at this time level.

When the effect of hardening is included in the viscoplastic model, the parameter C_Y depends on an accumulated quantity κ, which can be the total viscoplastic strain ε^P. This quantity is naturally updated according to

$$\varepsilon^{P,\ell} = \varepsilon^{P,\ell-1} + \theta \Delta t \dot{\varepsilon}^{P,\ell} + (1-\theta)\Delta t \dot{\varepsilon}^{P,\ell-1}. \tag{5.67}$$

We remark that some of the numerical formulas above differ from the seemingly corresponding ones in [87, Ch. 8].

5.2.3   Simplification in Case of a Forward Scheme in Time

The computational algorithm simplifies considerably if $\theta = 0$, which corresponds to a forward finite difference scheme in time. In this case, $\hat{\boldsymbol{D}} = \boldsymbol{D}$ and the coefficient matrix in (5.63) becomes identical to the one in elasticity. Moreover, the $\boldsymbol{C}$ matrix does not enter the algorithm, and $\boldsymbol{Q}$ is the identity matrix. Of course, we can incorporate these simplifications directly in (5.63), but it may be more instructive to go back to the original set of equations $\boldsymbol{\Psi}_i^{\ell} = 0$ and $\boldsymbol{R}^{\ell} = 0$. When $\theta = 0$ we can solve for $\boldsymbol{\sigma}^{\ell}$ from $\boldsymbol{R}^{\ell} = 0$, see (5.56), and insert $\boldsymbol{\sigma}^{\ell}$ in (5.55). Assuming $\boldsymbol{f}$ constant in time and that $\boldsymbol{\sigma}^{\ell-1}$ fulfills $\boldsymbol{\Psi}_i^{\ell-1} = 0$, we then get the simpler form of (5.63):

$$\sum_j \left(\int_\Omega \boldsymbol{B}_i^T \boldsymbol{D} \boldsymbol{B}_j d\Omega \right) \Delta \boldsymbol{u}_j = \Delta t \int_\Omega \boldsymbol{B}_i^T \boldsymbol{D} \dot{\varepsilon}^{P,\ell-1} d\Omega. \tag{5.68}$$

At initial time $t = 0$ we compute and store the coefficient matrix of a standard linear elasticity problem. Then we solve the elasticity problem for the initial elastic displacement, incorporating the prescribed loads on the right-hand side: $\boldsymbol{K}\boldsymbol{u}^0 = -\boldsymbol{f}$. When $\boldsymbol{f}$ is constant in time, the prescribed loads will disappear from the equations as is seen from (5.68). At an arbitrary time level ℓ, the following algorithm is executed.

1. Compute the contribution $\Delta t \boldsymbol{B}_i^T \boldsymbol{D} \dot{\varepsilon}^{P,\ell-1}$ to the right-hand side.
2. Assemble the right-hand side contributions and solve the linear system for $\Delta \boldsymbol{u}_j$, $j = 1, \ldots, n$.
3. Calculate the displacements and stresses according to

$$\boldsymbol{\sigma}^{\ell} = \boldsymbol{\sigma}^{\ell-1} + \boldsymbol{D} \sum_j \boldsymbol{B}_j \Delta \boldsymbol{u}_j - \Delta t \boldsymbol{D} \dot{\varepsilon}^{P,\ell-1} \tag{5.69}$$

$$\boldsymbol{u}_j^{\ell} = \boldsymbol{u}_j^{\ell-1} + \Delta \boldsymbol{u}_j. \tag{5.70}$$

4. Calculate the new viscoplastic strain rate from (5.66).
5. Proceed with the next time step.

The algorithm corresponding to $\theta = 0$ is significantly simpler to implement than the implicit one $(0 < \theta \leq 1)$, but has an expected restriction on the time step length,

$$\Delta t \leq \Delta t_{\text{crit}}, \tag{5.71}$$

where Δt_{crit} depends on the yield criterion [87]. Provided $Q = F$ and $\Phi(F) = F$, we have for the Tresca, von Mises, and Mohr-Coulomb yield criteria (see Chapter 5.2.4) that

$$\Delta t_{\text{crit}} = \frac{(1 + \nu)C_Y}{\gamma E} \qquad \text{Tresca,}$$

$$\Delta t_{\text{crit}} = \frac{4}{3}\frac{(1 + \nu)C_Y}{\gamma E} \qquad \text{von Mises,}$$

$$\Delta t_{\text{crit}} = \frac{4(1 - 2\nu)}{1 - 2\nu + \sin^2 \phi}\frac{(1 + \nu)c\cos\phi}{\gamma E} \qquad \text{Mohr-Coulomb.}$$

No simple expression for Δt_{crit} exists when the Drucker-Prager criterion is applied. Notice that the critical time step *does not depend on spatial discretization parameters*.

The general implicit algorithm is unconditionally stable for $\frac{1}{2} \leq \theta \leq 1$. However, the accuracy of the time discretization and the convergence properties of the Newton-Raphson method normally limits the choice of Δt.

5.2.4   Numerical Handling of Yield Criteria

Computation of $\partial F/\partial\boldsymbol{\sigma}$ and also $\partial\dot{\boldsymbol{\varepsilon}}^P/\partial\boldsymbol{\sigma}$ are required in the algorithm from the previous sections. Such computations can be conveniently handled in a unified numerical framework for yield criteria, see [131, Ch. 7.8, Vol II] and [87, Ch. 7.4]. This framework differs from the standard exposition of classical plasticity theory found in most textbooks on solid mechanics or material science.

We shall make frequent use of the following quantities:

$$J_1 = \frac{1}{3}\sigma_m = \sigma_{kk}, \tag{5.72}$$

$$\bar{\sigma} = \sqrt{\frac{1}{2}\sigma'_{ij}\sigma'_{ij}}, \qquad \sigma'_{ij} = \sigma_{ij} - \frac{1}{3}J_1\delta_{ij}, \tag{5.73}$$

$$J'_3 = \det\{\sigma'_{ij}\}, \tag{5.74}$$

where J_1 is the first invariant of σ_{ij}, whereas $\bar{\sigma}^2$ and J'_3 are the second and third invariants of the deviatoric stress tensor σ'_{ij}. We also define

$$\theta = \frac{1}{3}\sin^{-1}\left(-\frac{3\sqrt{3}}{2}\frac{J'_3}{\bar{\sigma}}\right), \qquad -\frac{\pi}{6} < \theta < \frac{\pi}{6}. \tag{5.75}$$

This θ must not be confused with the θ in the θ-rule for approximating time derivatives! We keep the θ symbol for both these quantities since this is so well established in the literature.

Several common yield criteria can be written in terms of J_1, $\bar{\sigma}$, and θ. To incorporate strain hardening, we assume that the critical yield stress C_Y in a uni-axial tensile test is a function of a hardening parameter κ [87, Ch. 7.2.2]. The special case $\kappa = \varepsilon^P$ was used in (5.39).

- Tresca's yield criterion:

$$F = 2\bar{\sigma}\cos\theta - C_Y(\kappa). \tag{5.76}$$

- von Mises' yield criterion:

$$F = \sqrt{3}\bar{\sigma} - C_Y(\kappa). \tag{5.77}$$

- Mohr-Coulomb's yield criterion:

$$F = \frac{1}{3}J_1\sin\phi + \bar{\sigma}\left(\cos\theta - \frac{1}{\sqrt{3}}\sin\theta\sin\phi\right) - c\cos\phi, \tag{5.78}$$

where $\phi(\kappa)$ and $c(\kappa)$ are the cohesion and angle of friction, respectively, which can depend on the strain hardening parameter κ.
- Drucker-Prager's yield criterion:

$$F = \alpha J_1 + \bar{\sigma} - k', \tag{5.79}$$

with

$$\alpha = \frac{2\sin\phi}{\sqrt{3}(3 - \sin\phi)}, \qquad k' = \frac{6c\cos\phi}{\sqrt{3}(3 - \sin\phi)},$$

where c and ϕ can depend on κ.

We now write $F = F(J_1, \bar{\sigma}, \theta)$ and introduce $\boldsymbol{a}^T \equiv \partial F/\partial\boldsymbol{\sigma}$. The chain rule gives

$$\boldsymbol{a}^T = \frac{\partial F}{\partial J_1}\frac{\partial J_1}{\partial\boldsymbol{\sigma}} + \frac{\partial F}{\partial\bar{\sigma}}\frac{\partial\bar{\sigma}}{\partial\boldsymbol{\sigma}} + \frac{\partial F}{\partial\theta}\frac{\partial\theta}{\partial\boldsymbol{\sigma}}. \tag{5.80}$$

A convenient form for computations is [87, Ch. 7.4]

$$\boldsymbol{a}^T = C_1\boldsymbol{a}_1^T + C_2\boldsymbol{a}_2^T + C_3\boldsymbol{a}_3^T, \tag{5.81}$$

where

$$\boldsymbol{a}_1^T \equiv \frac{\partial J_1}{\partial\boldsymbol{\sigma}} = (1, 1, 1, 0, 0, 0), \tag{5.82}$$

$$\boldsymbol{a}_2^T \equiv \frac{\partial\bar{\sigma}}{\partial\boldsymbol{\sigma}} = \frac{1}{2\bar{\sigma}}(\sigma'_{xx}, \sigma'_{yy}, \sigma'_{zz}, 2\sigma_{xy}, 2\sigma_{yz}, 2\sigma_{zx}), \tag{5.83}$$

$$\boldsymbol{a}_3^T \equiv \frac{\partial J_3}{\partial\boldsymbol{\sigma}} = (\sigma'_{yy}\sigma'_{zz} - \sigma_{yz}^2 + \frac{1}{3}\bar{\sigma}^2, \sigma'_{xx}\sigma'_{zz} - \sigma_{xz}^2 + \frac{1}{3}\bar{\sigma}^2,$$

$$\sigma'_{yy}\sigma'_{xx} - \sigma^2_{xy} + \frac{1}{3}\bar{\sigma}^2, 2(\sigma_{yz}\sigma_{zx} - \sigma'_{zz}\sigma_{xy}),$$
$$2(\sigma_{xz}\sigma_{xy} - \sigma'_{xx}\sigma_{yz}), 2(\sigma_{xy}\sigma_{yz} - \sigma'_{yy}\sigma_{xz})) \tag{5.84}$$

$$C_1 = \frac{\partial F}{\partial J_1}, \tag{5.85}$$

$$C_2 = \frac{\partial F}{\partial \bar{\sigma}} - \frac{\tan 3\theta}{\bar{\sigma}}\frac{\partial F}{\partial \theta}, \tag{5.86}$$

$$C_3 = -\frac{\sqrt{3}}{2\cos 3\theta}\frac{1}{\bar{\sigma}^3}\frac{\partial F}{\partial \theta}. \tag{5.87}$$

Different yield criteria are now reflected in different values of C_1, C_2, and C_3 only. Table 5.1 lists the expressions for these constants in the case of the four previously defined yield criteria.

Table 5.1. Values of the constants in (5.80) for various yield criteria.

yield criterion	C_1	C_2	C_3
Tresca, $\theta \neq \pi/6$	0	$2\cos\theta(1 + \tan\theta\tan 3\theta)$	$\frac{\sqrt{3}}{\bar{\sigma}}\frac{\sin\theta}{\cos 3\theta}$
Tresca, $\theta = \pm\pi/6$	0	$\sqrt{3}$	0
von Mises	0	$\sqrt{3}$	0
Mohr-Coulomb, $\theta \neq \pi/6$	$\frac{1}{3}\sin\phi$	$\cos\theta[(1 + \tan\theta\tan 3\theta)$ $+ \sin\phi(\tan 3\theta - \tan\theta)/\sqrt{3}]$	$\frac{\sqrt{3}\sin\theta + \cos\theta\sin\phi}{2\bar{\sigma}^2\cos 3\theta}$
Mohr-Coulomb, $\theta = \pm\pi/6$	$\frac{1}{3}\sin\phi$	$\frac{1}{2}\sqrt{3} \mp \frac{\sin\phi}{\sqrt{3}}$	0
Drucker-Prager	α	1	0

The case $\theta > 0$ in the time-discretization scheme requires evaluation of the matrix

$$\boldsymbol{C} = \frac{\partial\dot{\boldsymbol{\varepsilon}}^P}{\partial\boldsymbol{\sigma}} = \gamma\left(\Phi\frac{\partial\boldsymbol{a}^T}{\partial\boldsymbol{\sigma}} + \frac{d\Phi}{dF}\boldsymbol{a}\boldsymbol{a}^T\right).$$

Using the von Mises criterion, one can readily evaluate

$$\frac{\partial\boldsymbol{a}^T}{\partial\boldsymbol{\sigma}} = \frac{\sqrt{3}}{2\bar{\sigma}}\boldsymbol{M}^I - \frac{\sqrt{3}}{3\bar{\sigma}}\boldsymbol{a}\boldsymbol{a}^T,$$

where

$$\boldsymbol{M}^I = \frac{1}{9\sigma_m}\begin{pmatrix} \frac{2}{3} & -\frac{1}{3} & -\frac{1}{3} & 0 & 0 & 0 \\ & \frac{2}{3} & -\frac{1}{3} & 0 & 0 & 0 \\ & & \frac{2}{3} & 0 & 0 & 0 \\ & & & 2 & 0 & 0 \\ & \text{symm.} & & & 2 & 0 \\ & & & & & 2 \end{pmatrix}.$$

5.2.5   Implementation

Looking at the computational algorithm for the elasto-viscoplastic problem in the case $\theta = 0$, we realize that the simulation code can be made similar to class `Elasticity1`, except that we need

1. an outer time loop,
2. representation of the stresses $\boldsymbol{\sigma}$, the viscoplastic strain rates $\dot{\boldsymbol{\varepsilon}}^P$, and the total viscoplastic strain $\boldsymbol{\varepsilon}^P$, and
3. a hierarchy of yield criteria and associated functionality for computing quantities like $\boldsymbol{a}$.

The specialized algorithm corresponding to the time-discretization parameter $\theta = 0$ involves only repeated solutions of elasticity problems and are hence quite easy to implement. Here we shall demonstrate how this implementation can be achieved by extending class `Elasticity1`. The resulting solver is called `ElastoVP1` and its source code is located in `src/app/ElastoVP1`. The principal extensions of class `ElastoVP1`, in comparison with class `Elasticity1`, are listed next.

1. We need to supply functions for filling the matrices $\boldsymbol{D}$ and $\boldsymbol{B}_i$ in the case of 3D, plane stress, plane strain, and axisymmetry.
2. The computational algorithm requires extensive computing with stresses and viscoplastic strains. For this purpose, it is convenient to include a `FieldsFEatItgPt` structure for storing $\boldsymbol{\sigma}$, $\dot{\boldsymbol{\varepsilon}}^P$, and $\boldsymbol{\varepsilon}^P$ at the integration points in all elements. Built-in calculation procedures in this structure call up the solver's `derivedQuantitiesAtItgPt` routine for sampling the expressions of the components of all the fields in the `FieldsFEatItgPt` structure. The formulas for stresses and viscoplastic strain rates hence appear in the `derivedQuantitiesAtItgPt` function in the solver.

   Notice that we now need to sample the derived quantities at the standard integration points, rather than the reduced points, since the values are to be looked up in the `integrands` routine at the next time level (the use of a `FieldsFEatItgPt` object for the stresses in class `Elasticity1` used the reduced Gauss points for optimal accuracy of the derivatives).
3. At $t = 0$ we set stresses and strain rates to zero. The `integrands` routine must then solve the corresponding elastic problem.
4. At $t > 0$, we turn off body forces and surface tractions (we only solve for *corrections* in the displacement field). Note that `integrands` must handle the equations both at $t = 0$ (pure elasticity) and at $t > 0$ (combined elasto-viscoplasticity).
5. The implementation of the time loop follows the standards from Chapter 3, but we now also allow for termination if there are no plastic strains or if a steady state has been reached.
6. The coefficient matrix is constant during the simulation, while the right-hand side needs to be updated through a standard assembly procedure. Optimizations for this kind of problem are suggested in Appendix B.6.1 and incorporated in class `ElastoVP1`.

In the following, we present two computational examples. Figure 5.5 depicts a building on a two-material foundation. Under the action of gravity, the building will deform the foundation. Because the yield stress is greater in material 1 than in material 2, the building will be displaced to the right. All the input data, including values of the material parameters, are provided in the `building.i` file in the `Verify` directory.

The next example is more challenging. We consider a clamped beam, modeled by a 2D grid and plane strain conditions. Close to the clamped end we impose a crack. This is a central problem setting in fracture mechanics. Numerically, a crack here means that the nodes along the crack line are duplicated; one set belongs to the material to the left of the crack and the other set belongs to the material to the right. The boundary condition along the crack line is, of course, vanishing stress. The `makegrid` utility offers the possibility of defining such a crack, and the necessary steps are explained in the input file to `makegrid`, named `crackedBeam-makegrid.i`, in the `Verify` directory. Because severe stress concentrations are expected in the vicinity of the crack tip, smaller elements are needed in this area. This calls for local mesh refinements and the tools described in Chapter 3.6. We do not need to equip the solver with the adaptive grids tools, as the `makegrid` program offers access to local mesh refinements, provided that the refinement indicator can be based on geometric considerations alone. In the present example we mark a disk around the crack tip for local mesh refinements, see Figure 5.6a. From Figure 5.6b one can see the high stress concentrations at the crack tip and at the lower boundary. The plastic flow is limited to these localized areas.

Finally, we mention that Diffpack has been used for much more advanced solid mechanics problems than what is covered in this chapter. Nick Zabaras and co-workers have implemented fully implicit finite element models for large plastic deformations in Lagrangian coordinates [127,108,128]. These models have also been combined with sensitivity equations for optimization and control [101] of manufacturing processes.

Exercise 5.6. Repeat Exercise 5.2 on page 383, but apply an elasto-viscoplastic material model.   ◇

Exercise 5.7. Extend the `ElastoVP1` solver with software tools for nonlinear PDEs and implement the general elasto-viscoplastic algorithm from Chapter 5.2.2. To simplify the problem, you can restrict the implementation to the von Mises yield criterion. Run the examples in the `ElastoVP1/Verify` directory and establish the relative efficiency of the $\theta = 0$ algorithm and the more complicated, but also more stable, version with $\theta = 1/2$ and $\theta = 1$, when the simulation is run until the stationary elasto-plastic state is reached.   ◇

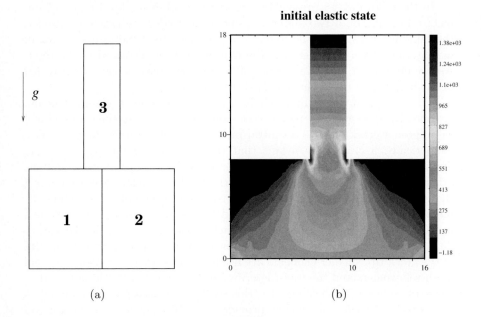

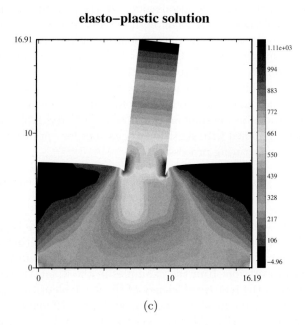

Fig. 5.5. Elasto-viscoplastic deformation of a building on a two-material foundation. (a) sketch of the three material domains; (b) elastic stress state (yield function F) at $t = 0$; (c) final converged state of the elasto-viscoplastic solution (the scalar F field is shown).

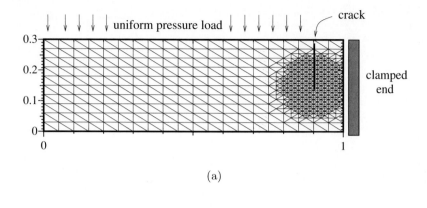

(a)

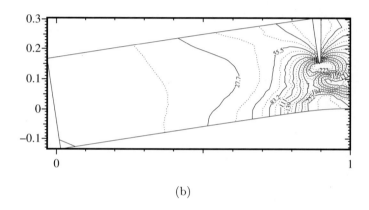

(b)

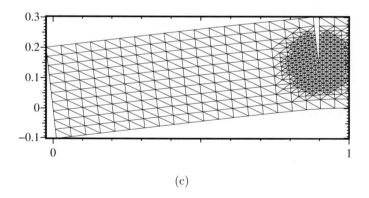

(c)

Fig. 5.6. Elasto-viscoplastic simulation of a beam in plane strain with a crack. (a) Sketch of the problem; (b) the equivalent stresses F in the final converged elasto-plastic state; (c) the grid corresponding to (a).

Chapter 6

Fluid Mechanics Applications

This chapter brings together numerical and implementational topics from the previous chapters in three application areas taken from fluid mechanics. First we present a solver for a general time-dependent and possibly nonlinear convection-diffusion equation, where the implementation constitutes a synthesis of most of the Diffpack tools mentioned in Chapters 3 and 4.2. The next application concerns waves in shallow water. We first treat finite difference methods for the system of PDEs on staggered grids in space and time. Thereafter we describe suitable finite element methods for weakly nonlinear and dispersive shallow water waves. The rest of the chapter is devoted to incompressible viscous flow governed by the Navier-Stokes equations. A classical finite difference method on staggered grid in 3D extends the ideas of the finite difference-based numerical model for shallow water waves. A penalty method for the Navier-Stokes equations, in combination with finite element discretization, demonstrates how numerical and implementational tools from the `Poisson2`, `NlHeat1`, and `Elasticity1` solvers in previous chapters can be combined to solve a time-dependent nonlinear vector PDE. Another finite element method for the Navier-Stokes equations, based on operator splitting, is also discussed, with special emphasis on efficient implementation.

6.1 Convection-Diffusion Equations

6.1.1 The Physical and Mathematical Model

The Governing Equations. Convection-diffusion equations appear in a wide range of mathematical models. The particular initial-boundary value problem to be addressed here reads

$$b\left(\alpha\frac{\partial u}{\partial t} + \boldsymbol{v}\cdot\nabla u\right) = \nabla\cdot(k\nabla u) - au + f, \quad \boldsymbol{x}\in\Omega\in\mathbb{R}^d,\ t>0, \quad (6.1)$$

$$u(\boldsymbol{x},0) = I(\boldsymbol{x}), \quad \boldsymbol{x}\in\Omega, \quad (6.2)$$

$$u(\boldsymbol{x},t) = D_1, \quad \boldsymbol{x}\in\partial\Omega_{E_1},\ t>0, \quad (6.3)$$

$$u(\boldsymbol{x},t) = D_2, \quad \boldsymbol{x}\in\partial\Omega_{E_2},\ t>0, \quad (6.4)$$

$$-k\frac{\partial u}{\partial n}(\boldsymbol{x},t) = \nu, \quad \boldsymbol{x}\in\partial\Omega_N,\ t>0, \quad (6.5)$$

$$-k\frac{\partial u}{\partial n}(\boldsymbol{x},t) = h_T(u-U_0), \quad \boldsymbol{x}\in\partial\Omega_R,\ t>0, \quad (6.6)$$

where k, v, a, and f are functions of x and possibly t, and b, D_1, D_2, ν, h_T, and U_0 are constants. Moreover, α is an indicator (1 or 0) that turns the time dependence on or off. The complete boundary $\partial \Omega$ consists of the four parts $\partial \Omega_{E_1}$, $\partial \Omega_{E_2}$, $\partial \Omega_N$, and $\partial \Omega_R$, having Dirichlet, Neumann, and Robin conditions.

Convection-diffusion equations are of particular importance in heat transfer and specie transport problems. Moreover, such equations also arise in the intermediate steps of numerical methods for the Navier-Stokes equations and multi-phase porous media flow.

Physical Interpretations. In the field of heat transfer, equation (6.1) stems from the first law of thermodynamics and expresses energy balance of a continuous medium. The primary unknown $u(x, t)$ represents the temperature, b is the product of the density of the medium times the heat capacity, v is the velocity of the medium, k is the medium's heat conduction coefficient, and $f - au$ represents external heat sources. The time-derivative term is the accumulation of internal energy at a fixed point in space, the *convective* term $v \cdot \nabla u$ models transport of internal energy with the flow, $\nabla \cdot (k \nabla u)$ reflects transport of thermal energy by molecular vibrations (i.e. heat conduction), and the source term $f - au$ might represent heat generation or extraction due to, for example, internal friction in the fluid, chemical reactions, or radioactivity. The boundary condition (6.6) is explained in Project 2.6.1.

One can also interpret equation (6.1) as a mass conservation equation that governs specie transport in a fluid. In this case, u is the concentration of the specie, b is the density of the specie, v is the velocity field of the fluid, k is a diffusion coefficient, which is normally constant, and $f - au$ represents specie production or destruction. The time-derivative term expresses accumulation of mass at a point in space, while the convection $(v \cdot \nabla u)$ and the diffusion $(k \nabla^2 u)$ terms reflects transport of mass with the flow and due to molecular diffusion, respectively. The source or sink term $f - au$ might model, for instance, injection or extraction of the specie or mass loss/gain due to chemical reactions. If the heat transfer or the specie transport takes place in a porous medium, the governing PDE is still the same, but the interpretation of the coefficients must be slightly adjusted.

It must also be mentioned that special cases of equation (6.1) appear in many other branches of engineering and science. For example, simple model equations like the Laplace, Poisson, and Helmholtz equations are contained in (6.1). We can also make (6.1) nonlinear, e.g., by letting $k = k(u)$ and replacing $f - au$ by $f(u)$. Such nonlinearities arise both in simple model problems as well as in the heat transfer (cf. Chapter 1.2.7) and specie transport problems.

6.1.2   A Finite Element Method

By means of a θ-rule in time and the weighted residual method in space we can derive the following discrete equations:

$$
\int_\Omega \Big[b\alpha \left(\hat{u}^\ell - \hat{u}^{\ell-1} \right) W_i + \theta \Delta t b W_i \boldsymbol{v}^\ell \cdot \nabla \hat{u}^\ell + (1-\theta) \Delta t b W_i \boldsymbol{v}^{\ell-1} \cdot \nabla \hat{u}^{\ell-1} +
$$

$$
\theta \Delta t k^\ell \nabla W_i \cdot \nabla \hat{u}^\ell + (1-\theta) \Delta t k^{\ell-1} \nabla W_i \cdot \nabla \hat{u}^{\ell-1} +
$$

$$
\theta \Delta t W_i a^\ell \hat{u}^\ell + (1-\theta) \Delta t W_i a^{\ell-1} \hat{u}^{\ell-1} - \theta \Delta t W_i f^\ell - (1-\theta) \Delta t W_i f^{\ell-1} \Big] d\Omega
$$

$$
+ \int_{\partial \Omega_N} W_i \Delta t \nu d\Gamma
$$

$$
+ \int_{\partial \Omega_R} W_i \Delta t h_T (\theta (u^\ell - U_0) + (1-\theta)(u^{\ell-1} - U_0)) d\Gamma = 0 . \tag{6.7}
$$

Superscript ℓ denotes the time level, $\hat{u}^\ell(\boldsymbol{x}) = \sum_{j=1}^n u_j^\ell N_j(\boldsymbol{x})$ is an approximation to $u(\boldsymbol{x}, t)$ at time level ℓ, and (6.7) is supposed to hold for n linearly independent weighting functions W_i, $i = 1, \ldots, n$. If some of the details in the derivation of (6.7) are unclear, we refer to similar examples in Chapter 2.

   The formula for the element matrix follows from restricting the domain of integration to an element, replacing $\hat{u}^\ell$ by $\sum_j u_j^\ell N_j$ and collecting the terms at level ℓ containing the indices i and j. The remaining terms belong to the element vector.

Exercise 6.1. Write down the precise expressions for the integrands of the element matrix and vector associated with (6.7). $\diamond$

6.1.3   Incorporation of Nonlinearities

A flexible convection-diffusion solver must handle nonlinear coefficients. Here we suppose that b, k, and f can possibly depend on u. To solve the resulting system of nonlinear algebraic equations, we introduce an iteration with q as iteration index, and where $\hat{u}^{\ell,q}$ is the approximation to $\hat{u}^\ell$ in iteration q. The Successive Substitution method (Picard iteration) implies that one simply evaluates the expressions b^ℓ, k^ℓ, and f^ℓ as $b(\hat{u}^{\ell,q-1})$, $k(\hat{u}^{\ell,q-1})$, and $f(\hat{u}^{\ell,q-1})$. The corresponding modifications of the expressions in the element matrix and vector are trivial to incorporate.

   As usual, the Newton-Raphson method involves more book-keeping. A term like $W_i b(\hat{u}^\ell) \hat{u}^\ell$ now gives the contribution

$$
\frac{\partial}{\partial u_j^\ell} \left(W_i b(\hat{u}^\ell) \hat{u}^\ell \right) = W_i b(\hat{u}^{\ell,q-1}) N_j + W_i \frac{db}{du} (\hat{u}^{\ell,q-1}) N_j \hat{u}^{\ell,q-1}
$$

to the integrands in the expression for the element matrix. Notice that only the first term is used in the Successive Substitution method.

Exercise 6.2. Derive the precise expressions for the integrands of the element matrix and vector when b, k, and f can depend on u. Try to write the expressions in a form that is valid both in the Successive Substitution and Newton-Raphson methods (introduce for example an on-off indicator as coefficient in the Newton-Raphson-specific terms related to derivatives of b, k, and f). ◇

6.1.4   Software Tools

Ideally, we would like to have a flexible solver for the linear model problem (6.1)–(6.6), with a fast specialized version in the case the coefficients are not time dependent and the matrix assembly process can be avoided at each time level, *and* another version that treats the computationally more demanding problem when b, k, and f depend on u. The different program modules should share as much common code as possible. This is straightforwardly realized using the ideas of Chapter 3.5.6 and the optimization technique from Appendix B.6.2.

We create a base class CdBase that implements the linear version of (6.1)–(6.6) in a flexible way, with the coefficients b, k, and f as virtual functions that can be customized in specialized subclasses written by a user. The default implementation of b, k, and f in class CdBase applies the general field representation Handle(Field) for the coefficients, like we explained in Chapter 3.11.4. This means that we typically implement f as

```
virtual real f (const FiniteElement& fe, real t = DUMMY)
  { return f_field->valueFEM (fe, t); }
```

where f_field is of type Handle(Field). A FieldFormat object is used to allocate and initialize f_field based on menu information at run time. In integrands we call f before the loop over the element matrix and vector entries,

```
const real f_value = f(fe,t);
```

A subclass CdEff specializes class CdBase in the case where the coefficients and boundary conditions in the PDE are time independent. Two matrices are then assembled initially, and the actual coefficient matrix and right-hand side in the linear system at each time level are obtained by efficient matrix-vector operations. The algorithms and software tools are explained in Appendix B.6.2.

Another subclass CdNonlin of CdBase implements the nonlinear version of (6.1)–(6.6). New virtual functions for db/du, dk/du, and df/du are introduced. The typical representation of, for example, the function k in the code becomes

```
virtual real ku   (real u, const FiniteElement& fe, real t = DUMMY)
  { return u*u/2; }
virtual real dkdu (real u, const FiniteElement& fe, real t = DUMMY)
  { return u; }
```

Here $k(u, \boldsymbol{x}, t) = u^2/2$ is just an example. In `integrands` we evaluate k and dk/du by statements like

```
const real u_pt       = u->valueFEM(fe);  // u at current point
const real k_value    = ku  (u_pt, fe, t);
const real dkdu_value = dkdu (u_pt, fe, t);
```

Notice that we actually do not use the function for k as defined in the base class `CdBase`, i.e. a `k(const FiniteElement&)` function as we had in the solvers in Chapter 3, because we find it more convenient to have u as an explicit argument. However, the `k` function is convenient when computing the flux by the `FEM::makeFlux` function, and its proper form for this purpose is

```
real CdNonlin:: k (const FiniteElement& fe, real t)
{
   const real u_pt = u->valueFEM(fe); return ku (u_pt, fe, t); }
}
```

Particular expressions for the coefficients b, k, and f, as well as their derivatives, must be hardcoded in subclasses of `CdNonlin`. Class `CdNonlin` must also implement a generalized edition of `integrands` and `integrands4side`.

In the case (6.1) is convection dominated, the numerical solution can develop non-physical oscillations at high mesh Peclet numbers $\mathrm{Pe}_\Delta = b||\boldsymbol{v}||h/k$ (h reflects the element size). Chapters 2.9 and 3.8 outline algorithms and software tools for handling numerical problems associated with convection-dominated phenomena. Class `UpwindFE` is a convenient tool for representing different choices of W_i in the `CdBase` class.

Figure 6.1 depicts the class hierarchy for the convection-diffusion solver. The source code is located in the directory `src/app/Cd`. Some demo scripts, with filename extension `.pl`, are found in the `Verify` subdirectory. The reader is encouraged to study the source and play around with the demo scripts.

Exercise 6.3. Suppose you want to apply the suggested framework as basis for your own software development, but that you need to make an efficient solver for the Poisson equation on grids with linear triangular elements. Although class `CdBase` will work in this problem, the implementation of `integrands` can be made much more efficient (see Appendix B.6.3). Suggest how to derive a subclass `CdTriPoisson` where you rely on data structures in class `CdBase`, but avoid the `integrands` function and fill analytically integrated expressions for the element matrix and vector directly in the `ElmMatVec` object in the `calcElmMatVec` function. The material in Chapter 2.7.3 is useful for developing the relevant analytical expressions.   ◇

Remark. The framework for the convection-diffusion solver as sketched in this chapter is quite flexible, but some users may find it too flexible and not very easy to use for a novice C++ programmer. In such cases it is advantageous to build an interface to the convection-diffusion solver and only apply

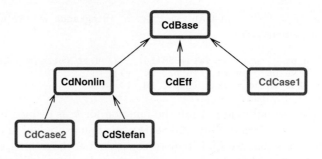

Fig. 6.1. A sketch of the convection-diffusion solver, with the base class `CdBase`, the efficient implementation `CdEff`, the more general nonlinear convection-diffusion solver `CdNonlin`, and the specialization `CdStefan` of `CdNonlin` to solve problems with freezing or melting. The classes `CdCase1` and `CdCase2` just indicate possible user-defined small classes that customize the b, k, and f functions in a particular problem. The arrows indicate class derivation.

the class hierarchy as a hidden computational engine. The ideas from Chapter 3.10.5 can be used as a starting point for building an easy-to-use, perhaps graphical, interface in Perl, Tcl, Python, or Java. The interface should allow the user to change only some of the input data to the solver, while others are kept at suitable default values. The script must process input data from the user, run the simulator, and visualize the results. Further development of such scripting interfaces might lead to truly easy-to-use flexible simulation environments, where the user can assign even mathematical expressions to b, k, etc. in the interface and interactively specify the grid and boundary conditions. Diffpack was designed for being a flexible computational engine in such problem solving environments.

6.1.5   Melting and Solidification

Many heat transfer applications also involve solidification or melting, i.e., phase changes. An example is heat conduction in a fluid, where parts of the fluid are frozen, while other parts are in a liquid state. The interface between the frozen and melted region is an unknown moving internal boundary. Let us consider a basic one-dimensional mathematical model for such a problem. There are two substances, denoted by the subscripts s (solid) and l (liquid). At the temperature $T = T_b$, a phase change between solid and liquid takes place. At time t we assume that the liquid part of the substance occupies the region $0 \leq x \leq b(t)$, whereas the solid part is located for $b(t) < x \leq a$. In each of these domains, a heat conduction equation is valid:

$$\varrho_l C_l \frac{\partial T_l}{\partial t} = \kappa_l \frac{\partial^2 T_l}{\partial x^2}, \quad 0 < x < b(t), \ t > 0, \tag{6.8}$$

$$\varrho_s C_s \frac{\partial T_s}{\partial t} = \kappa_s \frac{\partial^2 T_s}{\partial x^2}, \quad b(t) < x < a, \ t > 0, \tag{6.9}$$

$$T_l = \tau_0, \quad x = 0, \tag{6.10}$$

$$T_s = \tau_a, \quad x = a. \tag{6.11}$$

Here, ϱ is the density, C is the heat capacity, T is the temperature, κ is the heat conduction coefficient, and τ_0 and τ_a are prescribed temperatures at the end points of the domain. At the interface $x = b$ we have continuity in the temperature and a jump in the heat flux:

$$\kappa_s \frac{\partial T_s}{\partial x} - \kappa_l \frac{\partial T_l}{\partial x} = L \frac{db}{dt}, \quad x = b(t), \ t > 0, \tag{6.12}$$

$$T_l = T_s = T_b, \quad x = b(t), \ t > 0, \tag{6.13}$$

with L being the latent heat of phase transformation per unit volume. One difficulty with such a moving boundary problem is that different PDEs must be solved in different parts of the domain $(0, a)$. However, in the present problem it is possible to formulate a unified PDE that can be solved over the whole domain $(0, a)$, with the interface condition (6.12) being automatically satisfied without explicitly tracking the internal boundary. The key to this simplification is to employ an *enthalpy* formulation.

We introduce the enthalpy $H(T)$ according to

$$H(T) = \begin{cases} \varrho_s C_s T, & T < T_b \\ \varrho_l C_l T + L, & T > T_b \end{cases}$$

with

$$\varrho_s C_s T \leq H(T) \leq \varrho_l C_l T + L, \quad T = T_b.$$

The PDEs and the interface conditions in (6.8)–(6.13) can now be recast in the unified form

$$\frac{\partial H}{\partial t} = \frac{\partial}{\partial x} \left(\kappa(T) \frac{\partial T}{\partial x} \right), \quad 0 \leq x \leq a, \ t > 0, \tag{6.14}$$

where

$$\kappa(T) = \begin{cases} \kappa_s, & T < T_b \\ \kappa_l, & T > T_b \end{cases}$$

Usually, one solves (6.14) with respect to H. We then need the function $T(H)$:

$$T(H) = \begin{cases} H/(\varrho_s C_s), & H < \varrho_s C_s T_b \\ T_b, & \varrho_s C_s T_b \leq H \leq \varrho_l C_l T_b + L \\ (H - L)/(\varrho_l C_l), & H > \varrho_l C_l T_b + L \end{cases} \tag{6.15}$$

In the general heat transfer case, we have an initial-boundary value problem for $H(\boldsymbol{x}, t)$ on the form

$$\frac{\partial H}{\partial t} + \boldsymbol{v} \cdot \nabla H = \nabla \cdot (\kappa \nabla T(H)) + f, \quad \boldsymbol{x} \in \Omega \subset \mathbb{R}^d, \ t > 0, \tag{6.16}$$

$$H(\boldsymbol{x}, 0) = H(T_I(\boldsymbol{x})), \quad \boldsymbol{x} \in \Omega, \tag{6.17}$$

$$H(\boldsymbol{x}, t) = H(T_1), \quad \boldsymbol{x} \in \partial\Omega_{E_1}, \ t > 0, \tag{6.18}$$

$$H(\boldsymbol{x}, t) = H(T_2), \quad \boldsymbol{x} \in \partial\Omega_{E_2}, \ t > 0, \tag{6.19}$$

$$-\kappa\frac{\partial T}{\partial n}(\boldsymbol{x}, t) = \nu, \quad \boldsymbol{x} \in \partial\Omega_N, \ t > 0, \tag{6.20}$$

$$-\kappa\frac{\partial T}{\partial n}(\boldsymbol{x}, t) = h_T(T - T_0), \quad \boldsymbol{x} \in \partial\Omega_R, \ t > 0, \tag{6.21}$$

We see that we can use class CdNonlin for solving (6.16)–(6.21), but we need to apply the H function to the initial temperature condition $T_I(\boldsymbol{x})$ and the prescribed boundary values T_1 and T_2. Moreover, we have a special nonlinearity in the term $\nabla\cdot(\kappa\nabla T(H)) = \nabla\cdot(\kappa T'(H)\nabla H)$. The κ function is implemented as a separate function in the code, whereas the k function, intended for the effective heat conduction coefficient in Fourier's law, must evaluate $\kappa T'(H)$, such that $-k\nabla H = -\kappa\nabla T$.

Our finite element method is based on a standard weighted residual formulation, where we integrate by parts the term

$$\int_{\Omega} \nabla\cdot(\kappa\nabla T(H))W_i d\Omega = -\int_{\Omega} \kappa\Psi\nabla W_i \cdot \nabla H \, d\Omega -$$

$$\int_{\partial\Omega_R} W_i h_T(T(H) - T_0)d\Gamma - \int_{\partial\Omega_N} W_i \nu d\Gamma .$$

The factor Ψ equals $T'(H)$, but from a numerical point of view a more stable evaluation formula for Ψ is

$$\Psi = \begin{cases} \frac{\nabla T\cdot\nabla H}{||\nabla H||^2} & \text{if } ||\nabla H|| \neq 0 \\ T'(H) & \text{if } ||\nabla H|| = 0 \end{cases}$$

With a class CdStefan, derived from class CdNonlin, we simply implement a new integrands routine according to the formulas above. Successive Substitution is used as solution method for the nonlinear systems. In addition, we must provide scan and setIC functions that transform the T_1, T_2, and T_I values through the function $H(T)$. To track the solid and liquid region, it can be convenient to introduce an indicator field, which equals zero at the nodes where $T < T_b$ and unity at the nodes where $T \geq T_b$. The 0.5 contour of this field will then visualize the movement of the frozen and melted regions.

6.2   Shallow Water Equations

The purpose of this section is to present a widely used model in geophysics for large-scale water wave phenomena. This model involves a coupled system of transient PDEs, referred to as the shallow water equations. A linearized version of the system is first presented and discretized by finite differences on a staggered in time and space. Thereafter, a Diffpack implementation is explained. Finite element methods are also described, and the basic mathematical model is extended to include weak dispersion and nonlinearities.

The system of PDEs for modeling water waves in the present section has clear similarities to the Navier-Stokes equations. Much of the discretization reasoning for the shallow water equations will therefore be reused when formulating a finite difference method for the Navier-Stokes equations in Chapter 6.4.

6.2.1   The Physical and Mathematical Model

Water waves are usually described by an inviscid fluid model involving the Laplace equation $\nabla^2 \varphi = 0$, where φ is the velocity potential (such that $\nabla \varphi$ is the fluid velocity). The equation $\nabla^2 \phi = 0$ must be solved in the time-varying 3D water volume, with nonlinear boundary conditions at the surface. If the wave length is large compared with the depth ("shallow water"), it is possible to simplify the full water wave model and arrive at a system of equations to be solved in a two-dimensional *fixed* domain. This domain corresponds to the still-water surface. Although water waves are chosen as the principal physical interpretation of the PDEs in this section, similar equations also arise in acoustics and meteorology.

Small-amplitude three-dimensional shallow water waves, where the wave length is much larger than the still-water depth $H(x, y)$, and nonlinear effects are neglected, can be described by the two-dimensional mathematical model:

$$\frac{\partial \eta}{\partial t} = -\frac{\partial}{\partial x} (uH) - \frac{\partial}{\partial y} (vH), \quad \boldsymbol{x} \in \Omega, \ t > 0, \tag{6.22}$$

$$\frac{\partial u}{\partial t} = -\frac{\partial \eta}{\partial x}, \quad \boldsymbol{x} \in \Omega, \ t > 0, \tag{6.23}$$

$$\frac{\partial v}{\partial t} = -\frac{\partial \eta}{\partial y}, \quad \boldsymbol{x} \in \Omega, \ t > 0. \tag{6.24}$$

Figure 6.2 shows a sketch of the situation. We remark that the equations have been scaled; for example, gravity – which is the driving force of the waves – does not appear explicitly in the scaled model[1]. The primary unknowns are the depth-averaged velocity components $u(x, y, t)$ and $v(x, y, t)$ in addition to the surface elevation $\eta(x, y, t)$. Typical boundary conditions are (i) η prescribed or (ii) zero normal velocity: $\boldsymbol{v} \cdot \boldsymbol{n} = un_x + vn_y = 0$, where $\boldsymbol{n} = (n_x, n_y)^T$ is the normal vector to the boundary and $\boldsymbol{v} = (u, v)^T$ is the horizontal velocity vector. Condition (i) applies when the wave motion is known outside the domain, whereas condition (ii) applies to coastlines. A third type of condition, allowing waves to pass undisturbed out of the domain (so-called radiation condition), is also frequent, but is not treated here (see

---

[1] The scaling is the same as the one used for the 1D wave equation in Appendix A.1 on page 495.

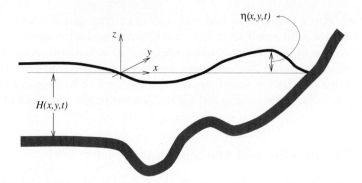

Fig. 6.2. A sketch of the shallow water wave problem.

Example A.21 on page 526). Initially, η, u, and v must be given for the model to be well posed.

Equation (6.22) is a continuity equation which ensures that the mass of water is conserved. The other two equations arise from Newton's second law and express balance between acceleration on the left-hand side and pressure forces on the right-hand side. The assumption of long waves implies that the pressure is hydrostatic, i.e., the pressure at a point equals the weight of the water column above the point: $p \sim \eta - z$.

To derive (6.22)–(6.24), one starts with the general Navier-Stokes equations for an incompressible fluid, which are listed in Chapter 6.3. Nonlinear and viscous terms are neglected, and the equations are integrated in the vertical direction, see [68, Ch. 8] or the more comprehensive references [79] and [89]. Alternatively, one may start out with an integral form of the incompressible Navier-Stokes equations and apply it to a volume that is infinitesimal in the horizontal directions, but of extent $[-H, \eta]$ in the vertical direction. This latter modeling technique is demonstrated in Chapter 7.1 in another physical setting.

For the discussion of finite difference methods, we assume that the domain is a rectangle $[0, \alpha] \times [0, \beta]$ in dimensionless coordinates. This domain may model a basin, e.g. a harbor, with walls where the normal velocity is zero, i.e., condition (ii) above applies to the whole boundary. The boundary conditions are then $u(0, y, t) = u(\alpha, y, t) = 0$ and $v(x, 0, t) = v(x, \beta, t) = 0$. The initial conditions are taken for simplicity as

$$u(x, y, 0) = 0, \quad v(x, y, 0) = 0, \quad \eta(x, y, 0) = \eta^0(x, y),$$

where $\eta^0(x, y)$ is a prescribed initial surface displacement.

6.2.2 Finite Difference Methods on Staggered Grids

For the discretization of (6.22)–(6.24), it is common to use a grid where the unknowns η, u, and v are located at different points in space and time. Such grids are denoted as *staggered*, or *alternating*, and play an important role in computational fluid dynamics, especially when solving Navier-Stokes-like equations. Figure 6.3 depicts a typical staggered grid, where the discrete η values are located at the center of the cell, and the discrete velocities u and v are located at the cell edges. A typical cell in Figure 6.3a, illustrated by the inner dashed square, is enlarged in Figure 6.4 and equipped with specification of the spatial and temporal indices in the primary discrete unknowns. We notice from Figure 6.4 that η is unknown at integer time levels ℓ, whereas the velocities are unknown at $\ell + 1/2$.

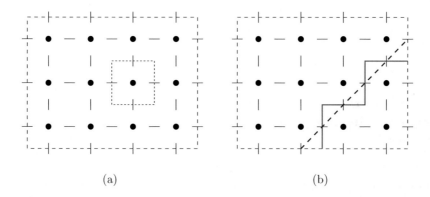

(a)   (b)

Fig. 6.3. Example on a staggered grid for the shallow water equations. The ● denotes η points, − denotes u points, whereas | denotes v points. This type of grid is often referred to as an Arakawa C-grid. The surrounding dashed line in (a) illustrates an impermeable (reflecting) boundary. Figure (b) exemplifies how a skew boundary can be approximated by a "stair-case" boundary.

Our model problem, involving waves in a rectangular basin $\Omega = [0, \alpha] \times [0, \beta]$, requires the points where the velocities are known, i.e. the cell edges, to lie on the boundaries, like we have indicated in Figure 6.3a. Of book-keeping reasons it is therefore natural to use integer indices for the cell edges (u and v points), cf. Figure 6.4. In other problems it may be convenient to have the integer indices at the cell center instead.

The discrete primary unknowns are

$$\eta^\ell_{i+\frac{1}{2},j+\frac{1}{2}}, \quad u^{\ell+\frac{1}{2}}_{i,j+\frac{1}{2}}, \quad v^{\ell+\frac{1}{2}}_{i+\frac{1}{2},j},$$

where i, j, and ℓ are integers. The spatial and temporal grid increments are denoted by Δx, Δy, and Δt. The basic idea of the discretization of equations (6.22)–(6.24) is to apply centered differences in time and space. Equation (6.22) is approximated at time level $\ell - \frac{1}{2}$ at the spatial point $(i + \frac{1}{2}, j + \frac{1}{2})$.

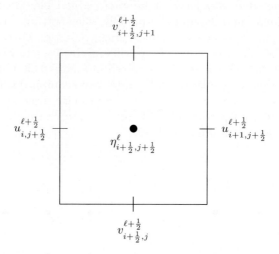

Fig. 6.4. A typical cell in a staggered grid, used for solving the shallow water equations.

In other words, we form finite differences for (6.22) at the space-time point $(i+\frac{1}{2}, j+\frac{1}{2}, \ell-\frac{1}{2})$. Equation (6.23) is approximated at $(i, j+\frac{1}{2}, \ell)$, and finally equation (6.24) is approximated at $(i+\frac{1}{2}, j, \ell)$. Writing out standard two-point central difference approximations to the derivatives at the indicated space-time points, gives the following discrete counterpart to the system (6.22)–(6.24):

$$\frac{1}{\Delta t}\left(\eta^{\ell}_{i+\frac{1}{2},j+\frac{1}{2}} - \eta^{\ell-1}_{i+\frac{1}{2},j+\frac{1}{2}}\right) = -\frac{1}{\Delta x}\left((Hu)^{\ell-\frac{1}{2}}_{i+1,j+\frac{1}{2}} - (Hu)^{\ell-\frac{1}{2}}_{i,j+\frac{1}{2}}\right)$$
$$-\frac{1}{\Delta y}\left((Hv)^{\ell-\frac{1}{2}}_{i+\frac{1}{2},j+1} - (Hv)^{\ell-\frac{1}{2}}_{i+\frac{1}{2},j}\right), \quad (6.25)$$

$$\frac{1}{\Delta t}\left(u^{\ell+\frac{1}{2}}_{i,j+\frac{1}{2}} - u^{\ell-\frac{1}{2}}_{i,j+\frac{1}{2}}\right) = -\frac{1}{\Delta x}\left(\eta^{\ell}_{i+\frac{1}{2},j+\frac{1}{2}} - \eta^{\ell}_{i-\frac{1}{2},j+\frac{1}{2}}\right), \quad (6.26)$$

$$\frac{1}{\Delta t}\left(v^{\ell+\frac{1}{2}}_{i+\frac{1}{2},j} - v^{\ell-\frac{1}{2}}_{i+\frac{1}{2},j}\right) = -\frac{1}{\Delta y}\left(\eta^{\ell}_{i+\frac{1}{2},j+\frac{1}{2}} - \eta^{\ell}_{i+\frac{1}{2},j-\frac{1}{2}}\right). \quad (6.27)$$

Looking at these equations, we observe a striking feature: The scheme is fully explicit due to the staggered grid in space and time. Assume that the quantities

$$\eta^{\ell-1}_{i+\frac{1}{2},j+\frac{1}{2}}, \quad u^{\ell-\frac{1}{2}}_{i,j+\frac{1}{2}}, \quad v^{\ell-\frac{1}{2}}_{i+\frac{1}{2},j}$$

are known for all i and j in the spatial grid. Equation (6.25) represents an explicit formula for computing $\eta^{\ell}_{i+\frac{1}{2},j+\frac{1}{2}}$ at all spatial grid points. Equation (6.26) can thereafter be used to update all the $u^{\ell+\frac{1}{2}}_{i,j+\frac{1}{2}}$ quantities for all i and j indices. Finally, equation (6.27) represents an explicit formula for computing new values of $v^{\ell+\frac{1}{2}}_{i+\frac{1}{2},j}$. Since all the finite difference approximations are based on two-point centered differences, we expect that the scheme is of order $\mathcal{O}(\Delta x^2, \Delta y^2, \Delta t^2)$. This can be shown by a standard truncation error analysis or by studying numerical dispersion relations (see Project 6.6.1). Moreover, since the scheme is fully explicit, we expect a stability criterion that limits the size of Δt. One can show (see Project 6.6.1) that this criterion reads, for constant H,

$$\Delta t \leq H^{-\frac{1}{2}} \left(\frac{1}{\Delta x^2} + \frac{1}{\Delta y^2} \right)^{-\frac{1}{2}}. \tag{6.28}$$

When H varies with x and y, we apply the "worst case" value of H, that is, $\sup\{H(x,y)\,|\,(x,y) \in \Omega\}$, in (6.28).

After having understood the details of the discretization procedure on staggered grids, it is convenient to condense the finite difference expressions and introduce the operator notation from Appendix A.3. This enables us to precisely define the numerical scheme in the following compact form:

$$[\delta_t \eta = -\delta_x(Hu) - \delta_y(Hv)]^{\ell-\frac{1}{2}}_{i+\frac{1}{2},j+\frac{1}{2}}, \tag{6.29}$$

$$[\delta_t u = -\delta_x \eta]^{\ell}_{i,j+\frac{1}{2}}, \tag{6.30}$$

$$[\delta_t v = -\delta_y \eta]^{\ell}_{i+\frac{1}{2},j}. \tag{6.31}$$

The original coupled system of PDEs has been split into a sequence of scalar PDEs by the particular time discretization. Such splitting of the original compound differential (vector) operator is a common strategy, often referred to as *operator splitting*, and is also encountered in Project 4.3.1 and Chapters 6.4, 6.5, 7.1, and 7.2.

Exercise 6.4. Eliminate u and v from the system (6.22)–(6.24) and show that η fulfills a wave equation

$$\frac{\partial^2 \eta}{\partial t^2} = \nabla \cdot [H(x,y)\nabla \eta]. \tag{6.32}$$

Also show that the boundary condition $\boldsymbol{v} \cdot \boldsymbol{n} = 0$ translates to $\partial \eta/\partial n = 0$ and that the initial condition $u = v = 0$ is equivalent to $\partial \eta/\partial t = 0$ at $t = 0$. ◇

Exercise 6.5. Eliminate the discrete u and v values from the system (6.25)–(6.27) and show that $\eta^{\ell}_{i+\frac{1}{2},j+\frac{1}{2}}$ fulfills a standard finite difference discretization of the wave equation (6.32). (For book-keeping simplicity it can be easier to work with integer spatial indices in η, that is, $\eta^{\ell}_{i,j}$, $u^{\ell+\frac{1}{2}}_{i+\frac{1}{2},j}$, and $v^{\ell+\frac{1}{2}}_{i,j+\frac{1}{2}}$.) ◇

Exercises 6.4 and 6.5 demonstrate that our coupled system of shallow water equations is mathematically and numerically equivalent to solving a standard wave equation for η (or u). We could therefore equally well stick to the simpler (6.32) and avoid systems of PDEs and staggered grids. Nevertheless, the system (6.22)–(6.24) is only a simplified model, which for relevant applications in geophysics must be extended with, for example, nonlinearities, dispersion, bottom friction, and surface drag from wind. These extensions are much more readily incorporated in the system (6.22)–(6.24) than in the scalar wave equation (6.32).

Exercise 6.6. Underwater slides can be represented as a time-varying bottom $H(x, y, t)$. The movement of the slide causes surface wave generation, modeled through an extra term $\partial H/\partial t$ on the right-hand side of (6.22). Show that the corresponding extra term in (6.32) reads $\partial^2 H/\partial t^2$.   ◇

6.2.3 Implementation

When coding the scheme (6.25)–(6.27) it would be convenient to use an array abstraction that allows half-integer indices. Moreover, the number of spatial points for u, v, and η differ, and a mesh abstraction for staggered grids that automatically calculates the proper length of arrays and loops, makes the programming easy and safe. Such convenient software abstractions are supported by Diffpack, but we shall first code the scheme (6.25)–(6.27) in terms of integer indices and standard array structures, because this implementation can easily be ported to any computer language and because this is a good exercise in nontrivial index book-keeping.

Data Structures and Index Handling. The following basic field abstractions can be useful for representing spatial values of η, u, and v at the current and the previous time level:

```
Handle(FieldLattice) eta, u, v, eta_prev, u_prev, v_prev;
```

Recall from Chapter 1.5.5 that every `FieldLattice` object has an array of field-point values, represented as an `ArrayGenSel(real)` object. We can access this array through the member function `values`. Let us store the quantity $\eta^{\ell}_{i+\frac{1}{2},j+\frac{1}{2}}$ in the array entry `eta->values()(i,j)`. The previous value $\eta^{\ell-1}_{i+\frac{1}{2},j+\frac{1}{2}}$ is stored in `eta_prev->values()(i,j)`. A similar idea is applied to the discrete u and v values. For instance, $u^{\ell+\frac{1}{2}}_{i,j+\frac{1}{2}}$ is stored in the array entry `u->values()(i,j)`. In general, we replace $i + \frac{1}{2}$ by the array index `i`, $j + \frac{1}{2}$ by the index `j`, $i - \frac{1}{2}$ by the index `i-1`, and $j - \frac{1}{2}$ by the index `j-1`. The `i` and `j` counters start at zero, and the spatial coordinates corresponding to the index `(i,j)` is the lower left corner of the cell containing $\eta_{i+\frac{1}{2},j+\frac{1}{2}}$.

The construction of a `FieldLattice` object implies allocation of grid-point values at all the corner of the cells. Since there are fewer center or edge points than corners, the arrays in the field objects will be too long. The extra space is padded with zeroes and never addressed by the program. Unfortunately, the extra space can cause strange results when arrays in the field objects are communicated diretly to visualization programs.

Computer Code Extracts. The source code for the program that solves the constant-depth shallow water equations ($H = 1$) is given below. The corresponding files are located in src/fdm/LongWave1. As mentioned, to access the value $u_{i,j+\frac{1}{2}}^{\ell+\frac{1}{2}}$ we can perform the array-look up u->values()(i,j), but this statement involves three inline function calls: one call to operator-> in class Handle(FieldLattice), one call to values in class FieldLattice, and one call to operator() in class ArrayGenSel. Not all compilers are good at optimizing such nested inline function calls. We might therefore help the compiler a bit and index the ArrayGenSel object directly:

```
ArrayGenSel(real)& Un = u->values();
```

Using Un(i,j) rather than u->values()(i,j) may give give a significant speed-up of the code on some machines. The numerical scheme can now be coded like this:

```
void LongWave1:: solveAtThisTimeStep ()
{
  ArrayGenSel(real)& Un    = u      ->values();
  ArrayGenSel(real)& Un_p  = u_prev->values();
  ArrayGenSel(real)& Vn    = v      ->values();
  ArrayGenSel(real)& Vn_p  = v_prev->values();
  ArrayGenSel(real)& ETAn   = eta     ->values();
  ArrayGenSel(real)& ETAn_p = eta_prev->values();

  const real mu = tip->Delta()/u->grid().Delta(1);   // dt/dx
  const real nu = tip->Delta()/v->grid().Delta(2);   // dt/dy
  const real m = grid->getMaxI(1);   // last grid point i
  const real k = grid->getMaxI(2);   // last grid point j
  int i,j;

  for (j = 0; j <= k-1; j++)
    for (i = 0; i <= m-1; i++)
      ETAn(i,j) = ETAn_p(i,j)
         - mu*(Un_p(i+1,j) - Un_p(i,j)) - nu*(Vn_p(i,j+1) - Vn_p(i,j));

  for (j = 0; j <= k-1; j++)
    for (i = 1; i <= m-1; i++)
      Un(i,j) = Un_p(i,j) - mu*(ETAn(i,j) - ETAn(i-1,j));

  for (j = 1; j <= k-1; j++)
    for (i = 0; i <= m-1; i++)
      Vn(i,j) = Vn_p(i,j) - nu*(ETAn(i,j) - ETAn(i,j-1));

  *u_prev   = *u;
  *v_prev   = *v;
  *eta_prev = *eta;
}
```

When studying the source code, one should have in mind that the purpose of the program is to make a compact, but not very flexible or computationally efficient implementation of the finite difference scheme for shallow water equations with constant depth.

Exercise 6.7. Explain in detail how to extend the `LongWave1` class such that it can handle a variable depth $H(x, y)$. Discuss three formats of H: $H(x, y)$ available as an explicit function, H available at the same points as η, and H available at the mid point of each cell side. (The relation between array and grid indices and the corresponding indices in the numerical scheme must be carefully considered when implementing a variable depth.)   ◇

Test problems. A possible choice of the initial surface disturbance is the plug

$$\eta^0_{i+\frac{1}{2}, j+\frac{1}{2}} = \begin{cases} 0, & i < p_1 \text{ or } i > p_2 \\ A/2, & i = p_1 \text{ or } i = p_2 \\ A, & p_1 < i < p_2 \end{cases}$$

where p_1 and p_2 are two prescribed i indices in the grid and A is the given amplitude. A particularly attractive property of this model is that the numerical scheme reproduces the analytical solution if the Courant number equals unity, which here means that $\Delta t = \Delta x$. This is very advantageous when developing software since we then know exactly what the output of the program in a particular example should be.

Another implemented test example involves $\eta(x, y, 0)$ as a Gaussian bell function.

Abstractions for Staggered Grid. Experience with programming of the shallow water equations or the Navier-Stokes equations shows that it is easy to make mistakes with the loop limits and array indices. For rapid prototyping it is thus convenient to program with grid and array abstractions that can handle staggered grids in a natural way. Diffpack offers class `GridLatticeC` in combination with class `FieldLattice` for this purpose. The `GridLatticeC` class contains actually three grids, one for u, one for v, and one for η. A `FieldLattice` object must be linked to one of these grids. From the `FieldLattice` object for, say η, we can only view the η grid. To set up a loop over the η points, one simply calls ordinary `GridLattice` functions for determining the loop lengths:

```
const int eta_i0 = eta->grid().getBase(1);   // first grid pt in i-dir
const int eta_j0 = eta->grid().getBase(2);   // first grid pt in j-dir
const int eta_in = eta->grid().getMaxI(1);   // last  grid pt in i-dir
const int eta_jn = eta->grid().getMaxI(2);   // last  grid pt in j-dir
for (j = eta_j0; i <= eta_jn; j++)
  for (i = eta_i0; i <= eta_in; i++)
    eta->valuesIndex(i+0.5,i+0.5) = ....
```

As we see, the `FieldLattice` object allows using half integer indices in the `valueIndex` function. This results in a code that is very close to the numerical specification of the scheme and hence easy to debug, especially since the staggered grid object also controls the lengths of the loops. Furthermore, there is no extra unused memory space in the fields, thus making interaction with visualization software simple, efficient, and safe. We refer to the report [4] for detailed documentation of Diffpack's staggered grid classes in general.

Computational Efficiency. As could be expected, the attractive index syntax `eta->valueIndex(i+0.5,j+0.5)` is associated with unacceptable low efficiency. This is therefore a tool for program development. As soon as the implementation is verified, alternative loops over η, u, and v, with a more efficient indexing technique should also be included in the `solveAtThisTimeStep` function. We emphasize the importance of first having a code that works before optimizing the loops, and that the old and new implementations must exist side by side and be activated by some run-time chosen indicator. Even the pure integer indices in class `LongWave1` have problems competing with a similar Fortran 77 implementation, and this Diffpack program can also benefit from improved indexing strategies in the `ArrayGenSel` class. Appendix B.6.4 provides an example on efficient indexing of `ArrayGen` (and hence `ArrayGenSel`) objects in a typical finite difference setting. The report [2] and the source code in the `opt` subdirectory demonstrate the implementation of efficient indexing techniques, also in combination with staggered grid classes.

Another candidate statement for optimization is the updating `*u_prev = *u`. It would here be more efficient to just switch pointers to the underlying arrays of grid point values. Calling `u_prev->exchangeValues(*u)` performs this task in a safe manner.

Parallel Computing. The updating finite difference formula for η at a point is independent of the similar formulas for all the other points in the grid. The updating statements can hence be performed concurrently. The same reasoning applies to the explicit finite difference formulas for the new u and v values as well. This opens up for a very efficient *parallel* algorithm for implementation on parallel computers. In theory, the present explicit finite difference method should be optimal from a parallel computing point of view; that is, on an n-processor machine the CPU time should be $1/n$ times the CPU time on a single-processor machine. Diffpack offers support for easy migration of class `LongWave1` to a parallel computing environment. The report [3] explains the details of the algorithms and implementations.

Animation of Waves. Since our `eta` field contains redundant values, visualization must be done with care, i.e., the `eta->values()` array cannot be used directly, but we must write out the physical significant entries in some suitable format[2]. In the `LongWave1` solver, we demonstrate how to dump the

---

[2] Using staggered grid classes in the `eta` field results in a `FieldLattice` object without redundant values. The `FieldLattice` object can then be managed by a `SaveSimRes` object or other tools operating directly on large arrays.

η values in a Matlab script (`casename.m`). After the simulation one can then invoke Matlab and see a movie of the wave motion. Try the program with a Gaussian bell-shaped initial surface, having the center of the bell located a corner such that we take advantage of symmetry and compute in only 1/4 of the physical domain (cf. page 44). An appropriate execution command is

```
./app --case 2 --casename bell -t 'dt=1 t in [0,2]'
      -g 'd=2 [0,1]x[0,1] [0:50]x[0:50]' -xc 0 -yc 1
```

Invoke Matlab and type `bell` (this will execute the `bell.m` script). Each frame of η will be first be plotted and thereafter the smooth movie is shown. You can play the movie again by typing `movie(M)` at the Matlab prompt.

6.2.4   Nonlinear and Dispersive Terms

The discretization of (6.22)–(6.24) appeared very natural in the staggered grid. If we incorporate nonlinear and dispersive wave effects, some new terms in the equations do not fit trivially with the previous discretization set-up. As an illustration, we consider the following extension of a one-dimensional version of the equations (6.22)–(6.24):

$$\frac{\partial \eta}{\partial t} = -\frac{\partial}{\partial x}\left\{(1+\alpha\eta)u\right\}, \tag{6.33}$$

$$\frac{\partial u}{\partial t} + \alpha \frac{\partial}{\partial x}\left(\frac{1}{2}u^2\right) = -\frac{\partial \eta}{\partial x} - \frac{\epsilon}{3}\frac{\partial^2}{\partial x^2}\left(\frac{\partial u}{\partial t}\right). \tag{6.34}$$

The dimensionless depth H is now assumed constant and has thus disappeared from the equations ($H = 1$) for simplicity. We have two new terms related to nonlinearity, marked with α, and one new term related to dispersion, marked with ϵ, where α and ϵ are dimensionless parameters measuring nonlinear and dispersive effects, respectively. For example, $\sqrt{\epsilon}$ is the ratio of the depth and the wave length and reflects how "shallow the water is" and thereby the importance of dispersion.

Let us introduce a staggered space-time grid with $\eta_{i+\frac{1}{2}}^{\ell}$ and $u_i^{\ell+\frac{1}{2}}$ as primary discrete unknowns. Applying central differences to all derivatives, we can write down the following discrete counterpart to (6.33)–(6.34):

$$[\delta_t \eta = -\delta_x((1+\alpha\eta)u)]_{i+\frac{1}{2}}^{\ell-\frac{1}{2}},$$

$$[\delta_t u + \frac{1}{2}\alpha\delta_x u^2 = -\delta_x \eta - \frac{1}{3}\epsilon\delta_x\delta_x\delta_t u]_i^{\ell}.$$

For convenience we also write these nonlinear difference equations in complete detail:

$$\frac{\eta_{i+\frac{1}{2}}^{\ell} - \eta_{i+\frac{1}{2}}^{\ell-1}}{\Delta t} = -\frac{1}{\Delta x}\left(\left\{(1+\alpha\eta_{i+1}^{\ell-\frac{1}{2}})u_{i+1}^{\ell+\frac{1}{2}}\right\} - \left\{(1+\alpha\eta_i^{\ell-\frac{1}{2}})u_i^{\ell+\frac{1}{2}}\right\}\right)$$

$$\frac{u_i^{\ell+\frac{1}{2}} - u_i^{\ell-\frac{1}{2}}}{\Delta t} = -\frac{\alpha}{2\Delta x}\left((u^2)^\ell_{i+\frac{1}{2}} - (u^2)^\ell_{i-\frac{1}{2}}\right) - \frac{\eta^\ell_{i+\frac{1}{2}} - \eta^\ell_{i-\frac{1}{2}}}{\Delta x}$$
$$- \frac{\epsilon}{3\Delta x^2}\left(\frac{u_{i-1}^{\ell+\frac{1}{2}} - u_{i-1}^{\ell-\frac{1}{2}}}{\Delta t} - 2\frac{u_i^{\ell+\frac{1}{2}} - u_i^{\ell-\frac{1}{2}}}{\Delta t} + \frac{u_{i+1}^{\ell+\frac{1}{2}} - u_{i+1}^{\ell-\frac{1}{2}}}{\Delta t}\right)$$

Quantities like $\eta_i^{\ell+\frac{1}{2}}$ and $u_{i+\frac{1}{2}}^\ell$ must be expressed in terms of our primary unknowns in the staggered grid. The standard technique is to introduce some kind of average. For example, $\eta_i^{\ell+\frac{1}{2}}$ can be averaged in space and time as

$$\eta_i^{\ell+\frac{1}{2}} \approx \frac{1}{4}\left(\eta^\ell_{i-\frac{1}{2}} + \eta^{\ell+1}_{i-\frac{1}{2}} + \eta^\ell_{i+\frac{1}{2}} + \eta^{\ell+1}_{i+\frac{1}{2}}\right).$$

A similar arithmetic average could also be used for $(u^2)^\ell_{i+\frac{1}{2}}$:

$$(u^2)^\ell_{i+\frac{1}{2}} \approx \frac{1}{4}\left((u^2)^{\ell-\frac{1}{2}}_i + (u^2)^{\ell-\frac{1}{2}}_{i+1} + (u^2)^{\ell+\frac{1}{2}}_i + (u^2)^{\ell+\frac{1}{2}}_{i+1}\right).$$

The disadvantage with this arithmetic average is that we get a nonlinear system of discrete equations due to the nonlinearity of u^2. If we apply a *geometric mean* in time and an arithmetic mean in space, that is,

$$(u^2)^\ell_{i+\frac{1}{2}} \approx \frac{1}{2}\left(u_i^{\ell-\frac{1}{2}}u_i^{\ell+\frac{1}{2}} + u_{i+1}^{\ell-\frac{1}{2}}u_{i+1}^{\ell+\frac{1}{2}}\right),$$

we obtain a linear system for the new u values at each time level. The reader should identify a tridiagonal linear system for $\eta^\ell_{i-\frac{1}{2}}$, $\eta^\ell_{i+\frac{1}{2}}$, and $\eta^\ell_{i+\frac{3}{2}}$ from the first difference equation and another tridiagonal system for $u_{i-1}^{\ell+\frac{1}{2}}$, $u_i^{\ell+\frac{1}{2}}$, and $u_{i+1}^{\ell+\frac{1}{2}}$ from the other equation. Since we have to solve linear systems, the scheme is implicit and one would expect better stability properties than (6.28). This is true; one can perform a stability analysis when $\alpha = 0$ (cf. Exercise 6.11) and arrive at

$$\Delta t \leq \sqrt{\Delta x^2 + \frac{4}{3}\epsilon}. \tag{6.35}$$

Exercise 6.8. Formulate a complete numerical algorithm for solving the system (6.33)–(6.34) with $u = 0$ at the boundaries. Initially, η is known and $u = 0$.   ◇

Finally, we mention that real applications of the shallow water equations involve complicated coastline geometries and often additional physical effects in the equations, such as the Coriolis force, bottom friction, and surface drag from wind. All these extensions are, at least in principle, straightforwardly included in the presented models. Complicated coastline geometries can be approximated by "stair-case" boundaries as indicated in Figure 6.3b. One can

hope that the successive setting of $u = 0$ and $v = 0$ yields a good approxima-
tion to $\boldsymbol{v} \cdot \boldsymbol{n} = 0$ at the real boundary. Unfortunately, "stair-case" boundaries
can generate non-physical wave motion (noise). When the geometry is com-
plicated, the finite element method is therefore an attractive technique for
spatial discretization.

6.2.5   Finite Element Methods

A finite element method for the shallow water equations (6.22)–(6.24) can
be established through a standard Galerkin procedure. However, there is
no immediate technique available to mimic the staggered grid[3]. Boundary
conditions along curved coastlines, $\boldsymbol{v} \cdot \boldsymbol{n} = 0$, constitute essential boundary
conditions in form of extra constraints, which demands a highly nontrivial
implementation. It turns out that a reformulation of (6.22)–(6.24) might be
advantageous for the application of finite element procedures. This reformulation consists in introducing the depth-averaged *velocity potential* $\phi(x, y, t)$
as a primary unknown instead of the velocities u and v. The latter quantities
are related to ϕ by $u = \partial\phi/\partial x$ and $v = \partial\phi/\partial y$, that is, $\boldsymbol{v} = \nabla\phi$. The number
of primary unknowns and PDEs is reduced from three to two. In this case
one also applies an integrated form of (6.23)–(6.24). The governing equations
for η and ϕ then take the form

$$\frac{\partial\eta}{\partial t} = -\nabla \cdot (H\nabla\phi), \tag{6.36}$$

$$\frac{\partial\phi}{\partial t} = -\eta. \tag{6.37}$$

The first of these is still a continuity equation derived from the principle
of conservation of mass, whereas the latter is an integrated Newton's second law (Bernoulli equation expressing balance of mechanical energy). The
boundary condition on the shoreline is now $\partial\phi/\partial n = 0$, which arises as a
natural boundary condition in the finite element formulation. Furthermore,
using the same basis functions for η and ϕ, combined with nodal-point integration, gives discrete equations of the same nature as those from the finite
difference method on staggered grids. With this alternative formulation, the
finite element method is a very handy tool for discretizing shallow water
equations in the complicated geometries frequently encountered in coastal
engineering applications.

The effects of weak dispersion and weak nonlinearity, as outlined in the
equations in Chapter 6.2.4, can also be incorporated in the η-ϕ formulation. A

---

[3] *Mixed finite element methods*, where different basis functions are used for different unknowns, constitute the counterpart to staggered grids in the finite element
world. However, mixed methods are beyond the scope of this text. Diffpack offers
a generalized version of the finite element toolbox that makes programming with
mixed methods quite easy.

possible set of relevant, commonly called the *Boussinesq equations* for water waves, might be written

$$\frac{\partial \eta}{\partial t} + \nabla \cdot \boldsymbol{q} = 0, \quad (6.38)$$

$$\frac{\partial \phi}{\partial t} + \frac{\alpha}{2} \nabla \phi \cdot \nabla \phi + \eta - \frac{1}{2} \epsilon H \nabla \cdot \left(H \nabla \frac{\partial \phi}{\partial t} \right) + \frac{1}{6} \epsilon H^2 \nabla^2 \frac{\partial \phi}{\partial t} = 0. \quad (6.39)$$

Also here, $H(x, y)$ denotes the still-water depth. The flux $\boldsymbol{q}$ in equation (6.38) is given by

$$\boldsymbol{q} = (H + \alpha \eta) \nabla \phi + \epsilon H \left(\frac{1}{6} \frac{\partial \eta}{\partial t} - \frac{1}{3} \nabla H \cdot \nabla \phi \right) \nabla H. \quad (6.40)$$

Exercise 6.9. Let $\alpha, \epsilon \to 0$ in (6.38)–(6.39), eliminate ϕ from the equations, and show that η fulfills a standard wave equation:

$$\frac{\partial^2 \eta}{\partial t^2} = \nabla \cdot (H \nabla \eta).$$

Suppose that the boundary condition is $\partial \phi / \partial n = 0$ and that the initial condition for ϕ is $\phi(x, y, 0) = 0$. Show that the corresponding conditions for η are $\partial \eta / \partial n = 0$ at the boundary and $\partial \eta / \partial t = 0$ for $t = 0$.   ◇

As usual, we shall first discretize the equations in time. To simplify the mathematical expressions and increase the focus on the principal ideas, we assume that H is constant. Equations (6.38)–(6.39) then become

$$\frac{\partial \eta}{\partial t} + \nabla \cdot [(H + \alpha \eta) \nabla \phi] = 0, \quad (6.41)$$

$$\frac{\partial \phi}{\partial t} + \frac{\alpha}{2} \nabla \phi \cdot \nabla \phi + \eta - \frac{\epsilon}{3} H^2 \nabla^2 \frac{\partial \phi}{\partial t} = 0. \quad (6.42)$$

It is convenient to think of a staggered grid in time (but not in space). This means that we have η^ℓ as the time discrete approximation to $\eta(x, y, \ell \Delta t)$ and $\phi^{\ell + \frac{1}{2}}$ as the similar approximation to $\phi(x, y, (\ell + \frac{1}{2}) \Delta t)$.

The ideas from Chapter 6.2.4 can be reused in the present context. We approximate the first (continuity) equation at time level $\ell - \frac{1}{2}$ and use a centered difference in time. The second (momentum) equation is more naturally approximated at time level ℓ. Quantities like $\eta^{\ell - \frac{1}{2}}$ and $[\nabla \phi \cdot \nabla \phi]^\ell$ arise from the time discretization, and these must be approximated by suitable averages. Using an arithmetic average, $\eta^{\ell - \frac{1}{2}} = \frac{1}{2}(\eta^{\ell-1} + \eta^\ell)$, and a geometric mean, $[\nabla \phi \cdot \nabla \phi]^\ell = \nabla \phi^{\ell - \frac{1}{2}} \nabla^{\ell + \frac{1}{2}}$ (see also Chapter 6.2.4) enables us to *decouple* the original system of two coupled nonlinear PDEs. The result is that we at each time level first can solve a linear PDE (6.43) for η^ℓ and then solve a linear PDE (6.44) for $\phi^{\ell + \frac{1}{2}}$.

$$\frac{\eta^\ell - \eta^{\ell-1}}{\Delta t} + \nabla \cdot (\frac{\alpha}{2} \eta^\ell \nabla \phi^{\ell - \frac{1}{2}}) + \nabla \cdot \left[(H + \frac{\alpha}{2} \eta^{\ell-1}) \nabla \phi^{\ell - \frac{1}{2}} \right] = 0, \quad (6.43)$$

$$\frac{\phi^{\ell+\frac{1}{2}} - \phi^{\ell-\frac{1}{2}}}{\Delta t} + \frac{\alpha}{2} \nabla \phi^{\ell-\frac{1}{2}} \cdot \nabla \phi^{\ell+\frac{1}{2}} -$$
$$\frac{\epsilon}{3} H^2 \nabla^2 \left(\frac{\phi^{\ell+\frac{1}{2}} - \phi^{\ell-\frac{1}{2}}}{\Delta t} \right) = -\eta^\ell . \quad (6.44)$$

Each of these equations are linear convection-diffusion equations that can be straightforwardly discretized by the finite element method in space, as explained in previous chapters.

The stability of the proposed scheme can be calculated for constant H and $\alpha = 0$. Applying the resulting criterion locally in each element and requiring that Δt must be less than the "worst case" of all elements, results in a generalization of (6.35):

$$\Delta t^2 \leq \min_e \left[H_e^{-1} \left(\frac{1}{\Delta x_e^2} + \frac{1}{\Delta y_e^2} \right)^{-1} + \frac{4}{3} H_e \epsilon \right], \quad (6.45)$$

where the subscript e means evaluation of the quantity in element no. e. We remark that the result (6.45) demands a lumped mass matrix (with a consistent mass matrix, we get the factor $1/\sqrt{3}$ as in Chapter 2.4.3).

Regarding implementation, one can start with a solver like Heat2 and include fields for η, ϕ, and H. The key idea is to introduce an enum variable for indicating the current equation to be solved. The enum variable is set prior to calling makeSystem, and integrands must test on the enum variable to determine the right expressions in the element matrix and vector. The principal steps regarding the enum variable is presented next.

```
class Boussinesq : public FEM
{
  Handle(FieldFE) eta, phi, H;
  enum Equation_type { CONTINUITY, MOMENTUM };
  Equation_tp eq_tp;
  Handle(LinEqAdmFE) lineq;
  Handle(DegFreeFE)  dof;
  ...
};

void Boussinesq::solveAtThisTimeStep()
{
  eq_tp = CONTINUITY;
  makeSystem (*dof, *lineq);
  lineq->solve(); // unknowns are in linsol
  dof->vec2field (linsol, *eta);

  eq_tp = MOMENTUM;
  makeSystem (*dof, *lineq);
  lineq->solve(); // unknowns are in linsol
  dof->vec2field (linsol, *phi);
  ...
}
```

```
void Boussinesq::integrands(ElmMatVec& elmat,const FiniteElement& fe)
{
  if (eq_tp == CONTINUITY)
    integrandsContinuity (elmat, fe);
  else if (eq_tp == MOMENTUM)
    integrandsMomentum (elmat, fe);
}
```

Complete finite element-based Diffpack simulators for the Boussinesq equations (6.38)–(6.39) have been developed [65]. The nonlinear shallow water equations in moving (Lagrangian) coordinates, aimed at modeling run-up of waves on beaches, have been solved by mixed finite element methods and moving-grid tools in Diffpack [66].

Exercise 6.10. Restrict (6.43) and (6.44) to one space dimension and assume that H has a finite element representation similar to that of η and ϕ. Derive the element matrices and vectors corresponding to each term in (6.43) and (6.44), using linear elements and nodal point integration. Try to express the resulting equations in the compact finite difference notation from Appendix A.3. Implement the difference schemes in a program (a suitable starting point might be the `Parabolic1D1` code in `src/fdm`, see Chapter A.5). The discrete solution to be calculated in Exercise 6.11 is valuable for verification and debugging of the program. Let $H(x)$ and $\eta(x, 0)$ be bell-shaped functions of the type outlined on page 44. Make a scripting interface to the simulator such that you can give the parameters of the initial state and the depth and get an mpeg movie with three simultaneous animations of $\eta(x, t)$ in return, one with $\alpha = \epsilon = 0$, one with $\alpha = 0$ and $\epsilon = 1$ and one with $\alpha = \epsilon = 1$. Play around with this script and try to demonstrate the effects of (weak) dispersion and nonlinearity on water waves.   ⬦

Exercise 6.11. Restrict (6.43) and (6.44) to the one-dimensional case with H constant and $\alpha = 0$. Use linear elements and nodal-point integration for establishing that the resulting difference equations can be written on the form

$$[\delta_t \eta + H \delta_x \delta_x \phi = 0]_i^{\ell - \frac{1}{2}}$$

and

$$[\delta_t \phi - \frac{\epsilon}{3} H^2 \delta_t \delta_x \delta_x \phi = -\eta]_i^\ell,$$

expect for the boundary points. Identifying $[\delta_x \phi]_{i - \frac{1}{2}}^{\ell + \frac{1}{2}}$ as $u_{i + \frac{1}{2}}^{\ell - \frac{1}{2}}$ shows that this finite element method is similar to a finite difference method on staggered grids with η_i^ℓ and $u_{i + \frac{1}{2}}^{\ell - \frac{1}{2}}$ as primary unknowns. Analyze the stability and accuracy of this scheme by employing the techniques from Appendix A.4. Also find a complete solution of the discrete problem in the case $\eta(x, y, 0) = \cos \pi x$, $\Omega = (0, 1)$, and $\partial \phi / \partial x = 0$ at $x = 0, 1$ (choose some appropriate initial values for ϕ).   ⬦

6.3   An Implicit Finite Element Navier-Stokes Solver

6.3.1   The Physical and Mathematical Model

The equations for viscous incompressible fluid flow are founded on the general mass and momentum balance equations for continuous media:

$$\varrho_{,t} + (\varrho v_s)_{,s} = 0, \tag{6.46}$$

$$\varrho(v_{r,t} + \varrho v_s v_{r,s}) = \sigma_{rs,s} + \varrho b_r. \tag{6.47}$$

These equations have been written using the indicial notation explained in Appendix A.2, where comma denotes differentiation and repeated indices imply a sum from 1 to the number of space dimensions d. The various symbols have the following meaning: ϱ is the fluid density, v_r is the fluid velocity, σ_{rs} is the stress tensor, and b_r denotes external body forces. Equation (6.46) reflects mass conservation, while (6.47) is a consequence of Newton's second law. The left-hand side of (6.47) is the "mass" (here represented through the density) times the acceleration, and the right-hand side is the sum of internal friction forces in the fluid and external forces such as gravity.

Incompressible flow is recognized by the fact that the mass conservation equation reduces to the constraint $v_{s,s} = 0$. This is compatible with constant ϱ, which we will assume in the following. To close the system of PDEs, the stress tensor must be related to the velocity field by a constitutive law for the fluid. Many common fluids, such water, air, and oil, exhibit a linear relation between shear stresses and shear deformation rates. Such fluids are classified as Newtonian and the corresponding constitutive law reads

$$\sigma_{rs} = -p\delta_{rs} + \mu(v_{r,s} + v_{s,r}), \tag{6.48}$$

where p is the fluid pressure and μ is a constant coefficient of viscosity. Inserting the constitutive law (6.48) in (6.47) and reducing (6.46) to the incompressibility condition $v_{s,s} = 0$, yields (see Exercise A.10)

$$\varrho(v_{r,t} + v_s v_{r,s}) = -p_{,r} + \mu v_{r,ss} + \varrho b_r, \tag{6.49}$$

$$v_{s,s} = 0. \tag{6.50}$$

This is the *Navier-Stokes equations* for incompressible flow. Some readers might appreciate to see the equations written in traditional vector notation as well:

$$\varrho\left(\frac{\partial v}{\partial t} + v \cdot \nabla v\right) = -\nabla p + \mu \nabla^2 v + \varrho b, \tag{6.51}$$

$$\nabla \cdot v = 0. \tag{6.52}$$

Our interest in indicial notation is primarily motivated by the fact that expressions like $\nabla \cdot v$ need to be written out in detail prior to implementation, whereas the indicial counterpart $v_{s,s}$ is a computational algorithm by itself;

we translate $v_{s,s}$ trivially to $\sum_{s=1}^{d} \partial v_s / \partial x_s$ and then to a corresponding for-loop in the code.

The primary unknowns in the mathematical model (6.49)–(6.50) are the velocity vector field v_r and the scalar pressure field p. If thermal effects cannot be neglected, the system (6.49)–(6.50) must augmented by a heat transport equation. The coupling of heat conduction and fluid flow is the subject of Chapter 7.2.

The boundary conditions associated with the system (6.49)–(6.50) are frequently of two types: either v_r is prescribed or the normal derivative of v_r vanishes. The latter condition can be replaced by a stress vector condition like we explained in the elasticity problem in Chapter 5.1, but this variant will not be considered herein. Uniqueness of the pressure requires p to be specified at a point. Furthermore, transient flow requires a prescribed velocity field v_r at initial time.

Equations for stationary flow arise by omitting the time derivative term $v_{r,t}$ in (6.49). It can therefore be convenient to insert an indicator α, which is zero in the stationary case and unity otherwise, in front of the $v_{,r}$ term, such that we can easily turn the time dependency on or off. This is done in the forthcoming equations.

6.3.2   A Finite Element Method

The two main problems with numerical methods for the incompressible Navier-Stokes equations (6.49)–(6.50) are the handling of the pressure term $p_{,r}$ and the incompressibility constraint $v_{s,s} = 0$. If there had been a time derivative term $p_{,t}$ in (6.50), the governing equations would exhibit the same structure as the shallow water equations in Chapter 6.2, modulo the terms $v_s v_{r,s}$ and $v_{r,ss}$. Introducing a slight (artificial) compressibility in the mass conservation equation may result in such a $p_{,t}$ term, and the system of PDEs can be solved by explicit time integration after the same lines as we followed in Chapter 6.2, see [40, Ch. 17]. However, the stability restriction on the time step is much more severe than in the wave equation application[4].

Omitting the pressure term from (6.49) gives an equation that has the same nature as a (nonlinear) convection-diffusion equation. The perhaps most dominating technique for solving the Navier-Stokes equations applies this observation and splits the differential operators such that one ends up with solving a series of Poisson and convection-diffusion equations for which very efficient numerical methods exist. This approach is the subject of Chapters 6.4 and 6.5.

An implementationally attractive method for dealing with the pressure term and the incompressibility constraint is the *penalty function* approach, where the equation of continuity is approximated by the equation $v_{s,s} =$

---

[4] This is natural because sound waves in a slightly compressible fluid propagate much faster than water surface waves.

$-\lambda^{-1}p$, where $\lambda \to \infty$. For numerical purposes, λ is chosen as a large number. The pressure can now be *eliminated* from (6.49), using the relation

$$p = -\lambda v_{s,s} \quad (= -\lambda \nabla \cdot \boldsymbol{v}), \quad \lambda \to \infty. \tag{6.53}$$

The new form of (6.49) has the same nature as the equations of elasticity, thus allowing us to apply solution concepts from Chapter 5.1. The penalty function method has a thorough mathematical justification in linear problems, and we refer the reader to Reddy's paper [95] for an overview of the theoretical background. Other sources regarding the numerics of the penalty method are [44], [55], and [96, Ch. 4]. For broad information about solution methods for the Navier-Stokes equations we refer to textbooks, e.g. [40,44,93,96], and the references therein.

Inserting (6.53) in (6.49) results in the governing equation

$$\varrho(\alpha v_{r,t} + v_s v_{r,s}) = \lambda v_{s,sr} + \mu v_{r,ss} + \varrho b_r, \tag{6.54}$$

which should be compared with the equation of elasticity (5.10). The velocity is first found from (6.54) and thereafter the pressure can be recovered from $p = -\lambda v_{s,s}$. Recall that the α parameter turns the time dependency on $(\alpha = 1)$ and off $(\alpha = 0)$.

Solving (6.54) by the finite element method starts with discretizing in time by finite differences, for instance the θ-rule from Chapter 2.2.2:

$$\varrho\alpha\frac{v_r^\ell - v_r^{\ell-1}}{\Delta t} + \theta\varrho v_s^\ell v_{r,s}^\ell + (1 - \theta)\varrho v_s^{\ell-1}v_{r,s}^{\ell-1} =$$
$$\theta\left(\lambda v_{s,sr}^\ell + \mu v_{r,ss}^\ell\right) + (1 - \theta)\left(\lambda v_{s,sr}^{\ell-1} + \mu v_{r,ss}^{\ell-1}\right). \tag{6.55}$$

Superscript ℓ denotes a quantity at time level ℓ. We have also dropped the source term b_r to save space in the formulas. We seek an approximation to v_r^ℓ on the form

$$\hat{v}_r^\ell(x_1, \ldots, x_d, t) = \sum_{j=1}^{n} v_j^{r,\ell} N_j(x_1, \ldots, x_d),$$

where N_j are prescribed finite element functions and $v_j^{r,\ell}$ are to be found by the method[5]. The parameter n represents the number of nodes in the finite element mesh. Inserting this approximation in (6.55) and applying a weighted residual method with weighting functions W_i, $u = 1, \ldots, n$, yields a system of nonlinear algebraic equations at time level ℓ:

$$F_i^r\left(v_1^1, \ldots, v_1^d, v_2^1, \ldots, v_2^d, \ldots, v_n^1, \ldots, v_n^d\right) = 0, \tag{6.56}$$

[5] We remark that comma in superscripts just separates indices; it does not indicate any differentiation.

with $i = 1, \ldots, n$ and $r = 1, \ldots, d$. The detailed expression for F_i^r reads

$$F_i^r \equiv \int_\Omega \left(\alpha(c_{ij} v_j^{r,\ell} - c_{ij} v_j^{r,\ell-1}) + \theta \Delta t d_{ij}^\ell v_j^{r,\ell} + (1-\theta) \Delta t d_{ij}^{\ell-1} v_j^{r,\ell-1} \right) d\Omega +$$

$$\int_\Omega \left(\theta \lambda \Delta t q_i^{r,\ell} (1-\theta) \lambda \Delta t q_i^{r,\ell-1} \right) d\Omega -$$

$$\int_{\partial\Omega_N^r} \left(\theta \Delta t f_i^{r,\ell} + (1-\theta) \Delta t f_i^{r,\ell-1} \right) d\Gamma . \tag{6.57}$$

The following symbols have been introduced to make the expressions above more compact:

$$d_{ij}^\ell = \mu W_{i,k} N_{j,k} + \varrho W_i \hat{v}_k^\ell N_{j,k} \tag{6.58}$$

$$c_{ij} = \varrho W_i N_j, \tag{6.59}$$

$$q_i^{r,\ell} = W_{i,r} \hat{v}_{k,k}^\ell \tag{6.60}$$

$$f_i^{r,\ell} = W_i \left(\frac{\partial \hat{v}_r^\ell}{\partial n} + p^\ell n_r \right). \tag{6.61}$$

Here, n_r denotes an outward unit normal to the boundary $\partial\Omega$. We see that $f_i^{r,\ell}$ appears in a boundary term over a part $\partial\Omega_N^r$ of $\partial\Omega$. On $\partial\Omega_N^r$ we shall assume that the following condition holds:

$$\frac{\partial v_r}{\partial n} + p n_r = 0 . \tag{6.62}$$

This is a natural "outflow" condition in many flow cases. As a consequence, the boundary integral vanishes.

The explicit version of the θ-rule, i.e. $\theta = 0$, is not suitable in combination with the penalty method because of the unfavorable stability condition: $\Delta t < \mathcal{O}(h^2/\lambda)$, where h is a characteristic element size.

The choice of finite elements is limited when using the penalty method. Best results are obtained with multi-linear elements, although multi-quadratic elements can also be used. Triangles or tetrahedra often lead to unacceptable results.

6.3.3   Solution of the Nonlinear Systems

The system of nonlinear equations $F_i^r = 0$ can be solved by, for example, Newton-Raphson iteration. We recall from Chapter 4 that the Newton-Raphson method leads to a linear system that must be assembled and solved in each iteration. The right-hand side of this linear system equals $-F_i^r$, whereas the coefficient matrix is the Jacobian of F_i^r. By differentiating F_i^r

with respect to the unknown values $v_j^{s,\ell}$, we can compute an entry in the Jacobian to be[6]

$$A_{i,j}^{rs} \equiv \frac{\partial F_i^r}{\partial v_j^{s,\ell}} = \int\limits_{\Omega} \left((\alpha c_{ij} + \theta \Delta t d_{ij}^\ell)\delta_{rs} + \theta \Delta t c_{ij} \hat{v}_{r,s}^\ell \right) d\Omega +$$

$$\int\limits_{\Omega} \lambda\theta\Delta t W_{i,r} N_{j,s} d\Omega . \qquad (6.63)$$

We recognize that $A_{i,j}^{rs}$ is similar to corresponding symbol in Chapter 5.1, representing the coupling of the degree of freedom r at node i with the degree of freedom s at node j. As in Chapter 5.1, we can view $A_{i,j}^{rs}$ as a block matrix, where i and j run over the blocks and r and s are local indices in a block. The coefficient matrix $\boldsymbol{J}$ of the linear system in each Newton-Raphson iteration can hence be partitioned into $n \times n$ blocks $\boldsymbol{J}_{i,j}$, each with dimension $d \times d$,

$$\boldsymbol{J}_{i,j} = \begin{pmatrix} A_{i,j}^{11} & \cdots & A_{i,j}^{1d} \\ \vdots & \ddots & \vdots \\ A_{i,j}^{d1} & \cdots & A_{i,j}^{dd} \end{pmatrix} .$$

The right-hand side of the equation system, $-F_i^r$, is similarly a vector partitioned into blocks, where i runs over the nodes (blocks) and r runs inside a block (degrees of freedom at a node). Recall that when evaluating the coefficient matrix and the right-hand side, we should use the most recent approximation to $v_i^{r,\ell}$. The correction vector in each Newton-Raphson iteration is of course also partitioned into block form, exactly as in (6.56) or the unknown displacement vector (5.14) in Chapter 5.1. The linear system to be solved then takes the form

$$\sum_{j=1}^{n} \sum_{s=1}^{d} A_{i,j}^{rs} \delta v_j^{r,\ell} = -F_i^r, \quad i = 1,\ldots,n, \ r = 1,\ldots,n .$$

This expression has a structure similar to the linear system in elasticity, cf. (5.15). We remark that the iteration index (k) in the nonlinear solution method has been skipped for notational simplicity.

With experience from the implementation of the elasticity model, the present solution algorithm can be regarded as an elasticity-like problem embedded in a nonlinear solver loop, which again is embedded in a time loop.

Discretizing the penalty version of the Navier-Stokes equations in space by the Galerkin method, results in a spatially discrete problem that can be written on matrix form

$$\varrho\boldsymbol{M}\frac{d\boldsymbol{v}}{dt} + \varrho\boldsymbol{N}(\boldsymbol{v})\boldsymbol{v} = \lambda\boldsymbol{G}\boldsymbol{v} + \mu\boldsymbol{K}\boldsymbol{v},$$

---

[6] $A_{i,j}^{rs}$ is a block matrix, and the comma between i and j separates the row and column blocks; the comma is not related to partial differentiation.

where M is the mass matrix, N stems from the nonlinear convection term $v_s v_{r,s}$, $\lambda G v$ is the penalty term, K is the discrete Laplace operator, and v is the vector of all $v_j^r(t)$ values. If we now let $\lambda \to \infty$, we must require $G v = 0$, which implies the undesired solution $v = 0$ if G is nonsingular. However, with a singular matrix G, the product $G v$ may approach zero for a nonzero v as $\lambda \to \infty$. A singular G can be obtained by a technique known as *reduced integration* [55,96], which means that we integrate the λ terms with a Gauss-Legendre rule of one order lower than what we normally would have applied. For example, a 2^d-point rule is sufficiently accurate for the involved integrals over multi-linear elements, and the reduced rule is therefore the $(2-1)^d = 1$ point rule. With multi-quadratic elements, a 3^d-rule is sufficient, and the reduced rule has $(3-1)^d = 2^d$ points. The computation of the element matrix and vector must hence deal with different numerical integration rules for different terms. The term *selective reduced integration* is often used for this type of integration strategy.

Another serious aspect of the penalty method is that the condition number of the coefficient matrix in the linear systems is proportional to λ. Choosing a large λ to approximate the incompressibility constraint $v_{s,s} = 0$ well, is thus not compatible with efficient *iterative solution* of linear systems, because the convergence speed of iterative solvers decreases with increasing condition number (see Appendix C.3.1). It turns out that standard preconditioning methods, such as the RILU technique, are not sufficiently effective at reducing the condition number. The consequence is that iterative linear solvers are too slow, and the linear system in each Newton-Raphson iteration must be solved by direct methods, such as banded Gaussian elimination. Computations with a solver based on the penalty method are therefore restricted to flow cases with up to a few thousand nodes. For such problem sizes the solution technique is reliable and quite efficient. In general we must admit that the penalty method is far from being the state-of-the-art finite element approach to viscous fluid flow simulations. Our motivation for discussing this method is mainly to demonstrate that the numerics and the implementational aspects represent a combination of building blocks from Chapters 3, 4, and 5.

Exercise 6.12. Formulate a Successive Substitution method for the equations $F_i^r = 0$ and derive detailed expressions for the element matrix and vector in the linear system that must be solved in each nonlinear iteration.   ◇

6.3.4   Implementation

Overview of the Simulator. The penalty-based Navier-Stokes solver is implemented in a class `NsPenalty1`. The basic data structures of this solver consists of the usual grid (`GridFE`) and degree of freedom handler (`DegFreeFE`), plus a vector field for the velocities v_r^ℓ and $v_r^{\ell-1}$, as well as a scalar field for the pressure p:

```
Handle(GridFE)         grid;    // finite element mesh
Handle(DegFreeFE)      dof;     // matrix dof <-> u dof
Handle(FieldsFE)       u;       // velocity field
Handle(FieldsFE)       u_prev;  // u at the previous time level
Handle(FieldFE)        p;       // pressure field
```

In addition we need tools from Chapters 3.9 and 4.2, namely a `TimePrm` object and nonlinear solvers. The latter tools are the same here as in class `NlHeat1`:

```
Vec(real)                    nonlin_solution; // sol. of nonlinear system
Vec(real)                    linear_solution; // sol. of linear subsystem
Handle(NonLinEqSolver_prm)   nlsolver_prm;    // init prm for nlsolver
Handle(NonLinEqSolver)       nlsolver;        // nonlinear solvers
Handle(LinEqAdmFE)           lineq;           // linear solvers & data
```

Furthermore, we need `real` variables for representing the constants λ, μ, ϱ, and θ.

The overall program flow remains the same as in class `NlHeat1`, with a `timeLoop` function that calls a nonlinear solver, which again jumps back to the simulator class in the function `makeAndSolveLinearSystem`. The contents of the two mentioned functions are the same as in class `NlHeat1`. The only major difference from `NlHeat1` is the `calcElmMatVec` and `integrands` functions, because we now have to deal with (i) a vector PDE and (ii) different integration rules for different terms in the finite element integrals.

Implementation of Reduced Integration. In `calcElmMatVec` we split the integration into two parts, one over the λ terms and one over the remaining terms. A class member flag `integrands_tp` of type

```
    enum Integrand_type { LAMBDA_TERMS, ORDINARY_TERMS }
```

is used to indicate the type of integrands to be evaluated in the `integrands` function:

```
void NsPenalty1:: calcElmMatVec
  (int e, ElmMatVec& elmat, FiniteElement& fe)
{
  // itg_rules is inherited from base class FEM
  itg_rules.setRelativeOrder (0);   // request ordinary rule
  fe.refill (e, itg_rules);
  integrands_tp = ORDINARY_TERMS;
  numItgOverElm (elmat, fe);        // ordinary integration

  itg_rules.setRelativeOrder (-1);  // request reduced rule
  fe.refill (e, itg_rules);
  integrands_tp = LAMBDA_TERMS;
  numItgOverElm (elmat, fe);        // reduced integration
}
```

In the `integrands` function, we simply check the `integrands_tp` flag and call either a `integrandsReduced` function for the λ terms or `integrandsNonReduced` for the ordinary terms. We refer to the computer code in `src/app/NsPenalty1` for details of the two integrands functions. The source code variables and expressions should follow closely the numerical formulas given here in the text. Going through the details of the integrands functions is a straightforward and highly recommended exercise[7].

As any Diffpack solver, the present Navier-Stokes code must have a boundary indicator convention. We have found it convenient to implement seven indicators: no. 1 marks the inflow boundary, 2 the outflow boundary, 3-5 mark boundaries where v_1, v_2, or v_3 are zero, respectively, whereas indicators 6 and 7 are left for the stream function (see below) and not used in class NsPenalty1.

Pressure Computation. After the velocity has been computed at a time level, the pressure $p = -\lambda v_{s,s}$ can be found by differentiation of the velocity field. Because $v_{s,s}$ is given by the derivatives of the primary unknown finite element field, the discrete pressure becomes discontinuous at the element boundaries, with the highest accuracy at the reduced Gauss points in an element (see Chapter 2.8). Normally, we want a smooth finite element representation of the pressure. This can be obtained by a computational procedure of the same nature as smoothing derivatives of finite element fields, and we can apply the tools presented in Chapter 3.4.5. The Galerkin or least-squares formulation of approximating $-\lambda v_{s,s}$ by $\hat{p} = \sum_{j=1}^{n} p_j N_j$ results as usual in a linear system, with the mass matrix as coefficient matrix and a right-hand side $c_i = -\int_{\Omega} \lambda v_{s,s} N_i d\Omega$.

The implementation of the pressure calculation in class NsPenalty1 makes use of an *integrand functor* for evaluating c_i. Obviously, we need a finite element assembly routine and a specification of the integrand for computing c_i, but the integrands function in the simulator class NsPenalty1 is already used for the penalty-modified Navier-Stokes equation. Integrand functors allow us to define additional integrands functions, such that we can assemble several finite element systems[8]. More information on integrand functors are presented in Appendix B.5.2, but now we focus directly on an application. In the present case the integrand functor is a class PressureIntg containing a virtual function integrands for evaluating the contribution from $-\lambda v_{s,s}$ at an integration point within an element. To perform this evaluation, the PressureIntg needs access to λ and the computed velocity field. This is enabled by a pointer to the NsPenalty1 class. Class FEM has special makeSystem functions that can use this PressureIntg class for specifying the integrands in finite element equations. More precisely, we can specify the integrands function as an argument to FEM::makeSystem (notice that such a function argument is in C++ most naturally represented as a class with a virtual evaluation function).

---

[7] It is definitely advantageous to be familiar with the details of the somewhat simpler Elasticity1::integrands function from page 378.

[8] Using an indicator such as integrands_tp on page 432 is another possible implementation of multiple integrands functions.

The `PressureIntg` class is defined by

```
class PressureIntg: public IntegrandCalc
{
  NsPenalty1* data; // access to input data and computational results
public:
  PressureIntg (NsPenalty1* data_ ) { data = data_; }
  virtual void integrands(ElmMatVec& elmat,const FiniteElement& fe);
};
```

The `integrands` function takes the form

```
void PressureIntg:: integrands
  (ElmMatVec& elmat, const FiniteElement& fe)
{
  const int nsd = fe.getNoSpaceDim();
  real div_v=0;
  for (int k = 1; k <= nsd; k++) {
    data->u()(k).derivativeFEM (data->gradu_pt(k), fe);
    div_v += data->gradu_pt(k)(k);
    // (gradu_pt is a VecSimple(Ptv(real)) structure)
  }
  const real pressure = - data->lambda*div_v;

  const int nbf = fe.getNoBasisFunc();
  const real detJxW = fe.detJxW();
  for (int i = 1; i <= nbf; i++)
    elmat.b(i) += pressure*fe.N(i)*detJxW;
}
```

A scratch variable like `gradu_pt`, containing $v_{r,s}$, is most naturally declared locally in the functions where it is needed, but since the type is a vector in dynamic memory, more precisely `VecSimple(Ptv(real))`, the allocation time of such variables may be significant. Therefore we let `gradu_pt` be a member of the `NsPenalty1` class and allocate it only once. This is a useful programming principle that applies to all variables which allocate free memory and which are used in `integrands` or other functions being called a large number of times.

To compute a smooth pressure field, we simply call `FEM::smoothField` with our functor specification of c_i:

```
PressureIntg penalty_integrand (this);
FEM::smoothField (*p, penalty_integrand);
```

Visualizing Streamlines. Fluid flow simulations are often visualized in terms of streamlines, i.e., curves that have the velocity field as tangent vector. In 2D flow, the most convenient way of visualizing the streamlines consists of first computing the stream function ψ and then using the property that the isolines of ψ are the streamlines. The stream function fulfills the equation

$$\nabla^2 \psi(x_1, x_2) = \frac{\partial v_2}{\partial x_1} - \frac{\partial v_1}{\partial x_2}.$$

The right-hand side is the *vorticity* (curl) of the 2D velocity field (v_1, v_2). The boundary conditions for ψ are of two types: $\partial \psi / \partial n = 0$ at boundaries where the tangential velocity vanishes (typical inflow and outflow sections), while ψ is prescribed at solid walls or symmetry lines (i.e. walls or symmetry lines are streamlines). Considering flow in a channel-like geometry, we can set $\psi = 0$ at the lower wall (or symmetry line). The value at the upper boundary must equal the total volume flux Q in the area between the lower and upper streamlines, obtained by, e.g., integrating the inlet profile.

Having computed the velocity field, the stream function can be calculated as a post process. Since the stream function is limited to 2D and mainly used for stationary flow cases, it can be convenient to make a separate stream-function solver `Streamfunc1` that reads the velocity from a simres database and stores ψ in a new simres database for later visualization. The implementation of `Streamfunc1` is a matter of a slight modification of class `Poisson1` and can be found in the directory `src/fem/NsPenalty1/Streamfunc1`. The boundary conditions associated with the stream function are conveniently stored as part of the grid used for the velocity and pressure calculations. Hence, we reserve the boundary indicators 6 and 7 for the lower and upper streamline boundary, respectively. The ψ value at the lower boundary can always be set to zero. The user is then responsible for calculating the volume flux Q and assign this (on the menu) as the known ψ value at the upper streamline boundary.

The `Streamfunc1` solver takes the simres database name for the velocity field as input and applies class `SimResFile` for loading the velocity field into a `FieldsFE` object. In Chapter 3.11.4 we explain how to read fields from simres files using the `FieldFormat` tools, but in class `Streamfunc1` we do not need the great flexibility of class `FieldFormat` and prefer instead to have a code performing the read operations directly (our code is merely a copy of the example on reading simres databases from the man page for class `SimResFile`). A perl script `plotStreamlines.pl` in the `Streamfunc1` directory automates the execution of the streamline computation and visualization of the streamlines as the isolines of ψ.

Channel Flow with Constriction. We end the description of the `NsPenalty1` solver by a computational example involving flow in a channel with constriction. The grid is depicted in Figure 6.7a. The inlet profile is plug formed, and the Reynolds number in the plots equals 60. At the outflow boundary we apply the condition (6.62). A relevant input file is `constriction.i` in the `Verify` directory. The simulation required 10 Newton-Raphson iterations, starting with zero velocities. Figures 6.7b and 6.8 show the velocities, the pressure, and the streamlines.

Exercise 6.13. Extend class `NsPenalty1` with flexible assignment of the initial velocity field, by adopting ideas from Chapter 3.11.4. This new feature makes it easy to continue a previous transient simulation. However, it also makes it easy to solve a high Reynolds number stationary flow problem using the solution corresponding to a lower Reynolds number as start vector for the nonlinear iteration. This is nothing but a manual continuation method (see Chapter 4.1.8) with the Reynolds number as continuation parameter. One should scale the low Reynolds number velocity field such that the initial guess for the higher Reynolds number simulation has the correct characteristic velocity and hence the correct (new) Reynolds number. The Reynolds number is a key parameter in viscous fluid flow and is derived in Exercise A.1 in Appendix A.1.   ◇

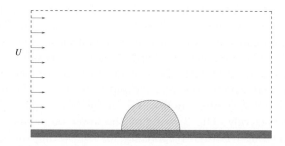

Fig. 6.5. Flow over a semi-cylinder (Exercise 6.14).

Exercise 6.14. A viscous fluid is flowing over a semi-cylinder attached to a plane wall as depicted in Figure 6.5. The boundary conditions are as follows: $v_1 = U$ and $v_2 = 0$ at the inlet, $v_1 = U$ and $v_{2,n} = 0$ at the top boundary, $v_{1,n} = v_{2,n} = 0$ at the outflow boundary, and $v_1 = v_2 = 0$ at the solid boundaries. Generate a corresponding input file for the NsPenalty1 solver (the geometry can be generated by the BOX_WITH_ELLIPTIC_HOLE feature in the PreproStdGeom utility[9], see Figure 3.9 on page 268). The size of the domain must be large enough for the condition $v_1 = U$ to hold at the top boundary and for $v_{1,n} = v_{2,n} = 0$ to hold at the outlet (the recirculating region behind the cylinder should not intersect the outflow boundary).   ◇

Exercise 6.15. Figure 6.6 shows a channel with an abrupt change in width, resulting in what is referred to as flow over a backward-facing step, which is a common benchmark test for numerical methods for the Navier-Stokes equations. Calculate the stationary analytical solution for flow in an infinite channel with plane walls and use this solution as inflow condition in the geometry in Figure 6.6. Reduce the size of the domain by taking symmetry

---

[9] For high Reynolds number flow the downstream area needs to be significantly larger than the upstream part of the geometry, and this will require somewhat more flexible grid generation utilities [64]).

Fig. 6.6. Flow in a channel with a sudden expansion (Exercise 6.15).

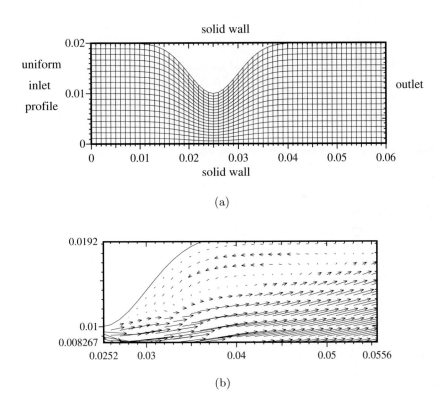

Fig. 6.7. Flow in a channel with constriction (Re = 60). (a) Sketch of the problem, with the computational grid; (b) a close-up plot of the velocities in the recirculating region. See also Figure 6.8.

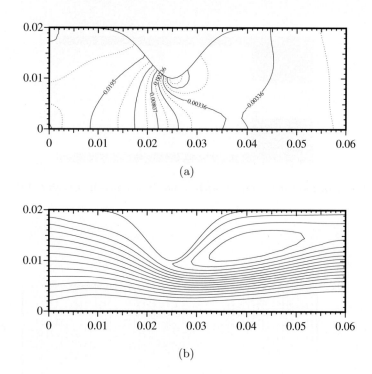

(a)

(b)

Fig. 6.8. Flow in a channel with constriction (Re = 60), see also Figure 6.7. (a) The pressure field; (b) streamlines.

into account. Assign appropriate conditions at the boundaries. Set up a corresponding input file for the NsPenalty1 solver (the grid can be generated by the BOX_WITH_BOX_HOLE feature in the PreproStdGeom utility, see Figure 3.8 on page 267). Run different values of the maximum velocity in the inlet profile and observe the extent of the recirculating region.   ◇

Exercise 6.16. Instead of having Streamfunc1 as a stand-alone application, one could think of calling Streamfunc1 from the NsPenalty1 solver. Implement this idea by letting an object of type Streamfunc1 be a member of class NsPenalty1 (or perhaps a member in a subclass of NsPenalty1 to separate the new code from the old). The v field in Streamfunc1 can be bound directly to the v field in the Navier-Stokes solver, such that one avoids loading the velocity from file.   ◇

6.4   A Classical Finite Difference Navier-Stokes Solver

In this section we shall present one of the most fundamental and widely used techniques for simulating incompressible viscous fluid flow, namely an operator-splitting method in combination with finite differences in time and space. The specific scheme to be presented here dates back to the late 1960s and work by Chorin [27] and Temam [117]. Since then numerous variations over the basic classical theme have been proposed, see [40] and [90] for an overview. Recently, similar schemes have also become popular among finite element practitioners [96,97], and we address this topic in Chapter 6.5.

The spatial discretization relies on the use of staggered grids. Before proceeding, the reader should therefore study Chapter 6.2.2, where we explain discretization on staggered grids for a set of PDEs (6.22)–(6.24) that is simpler than the three-dimensional Navier-Stokes equations (6.49)–(6.50) we aim to solve in the following. The forthcoming description of the numerics is quite brief, and the reader can consult the comprehensive and computationally-oriented book by Griebel et al. [45] for more information on numerical details, model extensions, and a large collection of fluid flow applications. Other comprehensive books on finite difference techniques for the Navier-Stokes equations are Anderson [8] and Ferziger and Perić [38]. We refer to Chapter 6.3.1 for a quick introduction to viscous fluid flow and the Navier-Stokes equations.

6.4.1   Operator Splitting

As mentioned in Chapter 6.5, the main difficulties with numerical solution of the incompressible Navier-Stokes equations are the handling of the pressure term $p_{,r}$ and incompressibility constraint $v_{s,s} = 0$. One way of dealing with these difficulties is to split the full Navier-Stokes equations into more tractable equations. Most operator splitting techniques for the Navier-Stokes equations split the equations into a vector convection-diffusion equation and a Poisson equation. These simpler type of equations can then be solved efficiently by numerous methods.

We first rewrite the convective term $v_s v_{r,s}$ in (6.49) by the aid of $v_{s,s} = 0$:

$$v_s v_{r,s} = (v_r v_s)_{,s} - v_r v_{s,s} = (v_r v_s)_{,s} .$$

We then drop the body force term (b_r) for simplicity, divide (6.49) by ϱ, and introduce the kinematic viscosity $\nu = \mu/\varrho$. The system of PDEs to be solved is now

$$v_{r,t} + (v_r v_s)_{,s} = -\frac{1}{\varrho} p_{,r} + \nu v_{r,ss} \tag{6.64}$$

$$v_{s,s} = 0 . \tag{6.65}$$

The basic idea of the splitting technique to be applied here consists of discretizing the momentum equation (6.64) by an explicit forward difference in time:

$$\frac{v_r^{\ell+1} - v_r^\ell}{\Delta t} = -(v_r v_s)^\ell_{,s} - \frac{1}{\varrho} p^\ell_{,r} + \nu v^\ell_{r,ss} \,. \tag{6.66}$$

The superscript ℓ denotes as usual the time level. The fundamental obstacle is that (6.66) yields velocities $v_r^{\ell+1}$ not satisfying the continuity equation. We therefore have to change the pressure field in such a way that the incompressibility constraint $v_{s,s}^{\ell+1} = 0$ is fulfilled. To this end, we seek a solution $(v_r^{\ell+1}, p^{\ell+1})$ of the time discrete system

$$\frac{v_r^{\ell+1} - v_r^\ell}{\Delta t} = -(v_r v_s)^\ell_{,s} - \frac{1}{\varrho} p^{\ell+1}_{,r} + \nu v^\ell_{r,ss} \tag{6.67}$$

$$v_{s,s}^{\ell+1} = 0 \,. \tag{6.68}$$

To accomplish the solution of this *implicit system* for $v_r^{\ell+1}$ and $p^{\ell+1}$, we first split the velocity into two components,

$$v_r^{\ell+1} = v_r^* + \delta v_r,$$

where v_r^* is found by stepping (6.67) forward in time, using old pressure values, and δv_r is found by imposing the constraint (6.68). The expression for v_r^* becomes

$$v_r^* = v_r^\ell + \Delta t \left(-(v_r v_s)^\ell_{,s} - \frac{\beta}{\varrho} p^\ell_{,r} + \nu v^\ell_{r,ss} \right) . \tag{6.69}$$

The parameter $\beta \in [0, 1]$ is used to adjust the amount of pressure information that carries over to the tentative velocity field v_r^*. The incompressibility condition $v_{s,s}^{\ell+1} = 0$ is equivalent to

$$(\delta v_r)_{,r} = -v^*_{s,s} \,. \tag{6.70}$$

Subtracting (6.69) from (6.67) yields

$$\delta v_r = v_r^{\ell+1} - v_r^* = -\frac{\Delta t}{\varrho} \Phi_{,r},$$

where $\Phi = p^{\ell+1} - \beta p^\ell$. Inserting this expression for δv_r in (6.70) yields

$$\Phi_{,rr} = \frac{\varrho}{\Delta t} v^*_{s,s} \,. \tag{6.71}$$

We recognize $\Phi_{,rr}$ as the Laplacian of Φ: $\Phi_{,rr} = \nabla^2 \Phi$. The solution of the Navier-Stokes equations is then reduced to a forward step in the convection-diffusion equation (6.69), solving a Poisson equation for the "pressure" Φ, and finally updating the velocity and the pressure. The steps are summarized in Algorithm 6.1.

Algorithm 6.1.

Operator-splitting approach for the Navier-Stokes equations.

at each time level:
   compute tentative velocity v_r^* from (6.69)
   solve the Poisson equation (6.71) for Φ
   correct tentative field: $v_r^{\ell+1} = v_r^* - \Phi_{,r}\Delta t/\varrho$
   update $p^{\ell+1} = \Phi + \beta p^\ell$

The forward step in (6.69) leads of course to a stability restriction on Δt. Better stability properties can be obtained by using implicit schemes for the convection-diffusion equation for the tentative velocity [40]. Here we keep things simple and stick to the explicit forward scheme.

6.4.2 Finite Differences on 3D Staggered Grids

Algorithm 6.1 is essentially a special time discretization technique used to split the differential operators in the incompressible Navier-Stokes equations. The technique can in principle be combined with any type of spatial discretization, but here we focus on the finite difference method on a three-dimensional staggered grid. This results in the most widespread numerical method for the Navier-Stokes equations.

The staggered grid is recognized by having the pressure as unknown in the center of a cell and the velocity components[10] $u \equiv v_1$, $v \equiv v_2$, and $w \equiv v_3$ as unknowns at the sides of the cell. Figure 6.9 depicts the staggered grid and the primary unknowns in the scheme:

$$p^\ell_{i+\frac{1}{2},j+\frac{1}{2},k+\frac{1}{2}}, \quad u^\ell_{i,j+\frac{1}{2},k+\frac{1}{2}}, \quad v^\ell_{i+\frac{1}{2},j,k+\frac{1}{2}}, \quad w^\ell_{i+\frac{1}{2},j+\frac{1}{2},k}.$$

We notice that, contrary to Chapter 6.2.2, the temporal grid is *not* staggered.

The reason for using a staggered grid is that standard grids, where all the unknowns are located at the corners of each cell, often lead to non-physical oscillations in the pressure[11].

Implementation of the finite difference discretization first requires that we write the vector equation (6.69) on component form. The x-component reads

$$u^* = u^\ell + \Delta t\left(-(uu)^\ell_{,x} - (uv)^\ell_{,y} - (uw)^\ell_{,z} - \frac{\beta}{\varrho}p^\ell_{,x} + \nu\left(u^\ell_{,xx} + u^\ell_{,yy} + u^\ell_{,zz}\right)\right).$$

---

[10] From now on it is convenient to work with u, v, and w rather than the indicial notation v_r.

[11] Instead of applying staggered grids, one can add stabilizing terms to the discrete equations. This has become popular among finite volume and finite element practitioners in recent years, see [47] for these and other advances in the field of incompressible computational fluid dynamics.

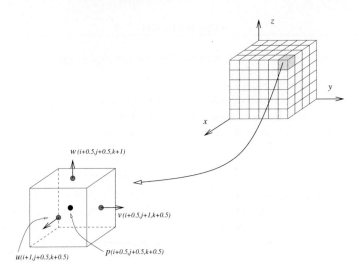

Fig. 6.9. Example on a three-dimensional staggered grid for solving the incompressible Navier-Stokes equations. The indices i, j, and k refer to the x, y, and z directions, respectively.

This equation is to be discretized by centered differences at the spatial point $(i, j + \frac{1}{2}, k + \frac{1}{2})$. Looking at a staggered grid sketch, for example the 2D grid in Figure 6.3 on page 413, or the 3D cell in Figure 6.9, it is straightforward to discretize the term $u^\ell_{,xx}$:

$$[u_{,xx}]^\ell_{i,j+\frac{1}{2},k+\frac{1}{2}} \approx \frac{1}{\Delta x^2} \left(u^\ell_{i+1,j+\frac{1}{2},k+\frac{1}{2}} - 2u^\ell_{i,j+\frac{1}{2},k+\frac{1}{2}} + u^\ell_{i-1,j+\frac{1}{2},k+\frac{1}{2}} \right).$$

Applying the operator notation from Appendix A.3, we can make the last expression compact and precise:

$$[\delta_x \delta_x u]^\ell_{i,j+\frac{1}{2},k+\frac{1}{2}}.$$

In a similar way we can discretize other terms:

$$[u_{,yy}]^\ell_{i,j+\frac{1}{2},k+\frac{1}{2}} \approx \frac{1}{\Delta y^2} \left(u^\ell_{i,j+\frac{3}{2},k+\frac{1}{2}} - 2u^\ell_{i,j+\frac{1}{2},k+\frac{1}{2}} + u^\ell_{i,j-\frac{1}{2},k+\frac{1}{2}} \right)$$
$$= [\delta_y \delta_y u]^\ell_{i,j+\frac{1}{2},k+\frac{1}{2}},$$

as well as

$$[u_{,zz}]^\ell_{i,j+\frac{1}{2},k+\frac{1}{2}} \approx [\delta_z \delta_z u]^\ell_{i,j+\frac{1}{2},k+\frac{1}{2}}.$$

The $(uu)_{,z}$ term requires somewhat more treatment. We start with writing a centered difference approximation,

$$[(uu)_{,x}]^\ell_{i,j+\frac{1}{2},k+\frac{1}{2}} \approx \frac{1}{\Delta x} \left((uu)^\ell_{i+\frac{1}{2},j+\frac{1}{2},k+\frac{1}{2}} - (uu)^\ell_{i-\frac{1}{2},j+\frac{1}{2},k+\frac{1}{2}} \right)$$

$$= [\delta_x uu]^{\ell}_{i,j+\frac{1}{2},k+\frac{1}{2}} \cdot$$

The quantity $u^{\ell}_{i+\frac{1}{2},j+\frac{1}{2},k+\frac{1}{2}}$ is not a primary unknown point value in the grid and must therefore be represented by an arithmetic average:

$$u^{\ell}_{i+\frac{1}{2},j+\frac{1}{2},k+\frac{1}{2}} \approx \frac{1}{2}(u^{\ell}_{i+1,j+\frac{1}{2},k+\frac{1}{2}} - u^{\ell}_{i,j+\frac{1}{2},k+\frac{1}{2}}) = [\overline{u}^x]^{\ell}_{i+\frac{1}{2},j+\frac{1}{2},k+\frac{1}{2}} \cdot$$

We can then write

$$[(uu)_{,x}]^{\ell}_{i,j+\frac{1}{2},k+\frac{1}{2}} \approx [\delta_x \overline{u}^x \overline{u}^x]^{\ell}_{i,j+\frac{1}{2},k+\frac{1}{2}} \cdot$$

With the average operator at hand we can continue the approximation of the convective terms:

$$[(uv)_{,y}]^{\ell}_{i,j+\frac{1}{2},k+\frac{1}{2}} \approx \frac{1}{\Delta y}\left((uv)^{\ell}_{i,j+1,k+\frac{1}{2}} - (uv)^{\ell}_{i,j,k+\frac{1}{2}}\right)$$

$$= \frac{1}{\Delta y}\left(u^{\ell}_{i,j+1,k+\frac{1}{2}} v^{\ell}_{i,j+1,k+\frac{1}{2}} - u^{\ell}_{i,j,k+\frac{1}{2}} v^{\ell}_{i,j,k+\frac{1}{2}}\right)$$

$$\approx [\delta_y \overline{u}^y \overline{v}^x]^{\ell}_{i,j+\frac{1}{2},k+\frac{1}{2}}$$

and

$$[(uw)_{,z}]^{\ell}_{i,j+\frac{1}{2},k+\frac{1}{2}} \approx [\delta_z \overline{u}^z \overline{w}^x]^{\ell}_{i,j+\frac{1}{2},k+\frac{1}{2}} \cdot$$

Finally, the pressure term is discretized by a natural centered difference,

$$[p_{,x}]^{\ell}_{i,j+\frac{1}{2},k+\frac{1}{2}} \approx [\delta_x p]^{\ell}_{i,j+\frac{1}{2},k+\frac{1}{2}} \cdot$$

The reader should write out the other components of the equation for the tentative velocity field and discretize them in a similar way.

The Poisson equation for the "pressure" Φ is discretized by centered differences at the spatial point $(i + \frac{1}{2}, j + \frac{1}{2}, k + \frac{1}{2})$:

$$[\delta_x \delta_x \Phi + \delta_y \delta_y \Phi + \delta_z \delta_z \Phi = \frac{\varrho}{\Delta t}(\delta_x u^* + \delta_y v^* + \delta_z w^*)]_{i+\frac{1}{2},j+\frac{1}{2},k+\frac{1}{2}} \cdot$$

Notice that the differences $[\delta_x u^*]_{i+\frac{1}{2},j+\frac{1}{2},k+\frac{1}{2}}$ etc. fit naturally into the staggered grid.

Boundary conditions related to the scheme above require careful consideration. What type of conditions should be placed on the tentative velocity field v_r^*? It turns out that $v_r^{\ell+1}$ is independent of v_r^* on the boundary [90]. We can therefore apply the physical conditions for $v_r^{\ell+1}$ to v_r^* as well. Doing so implies the use of homogeneous Neumann boundary conditions for Φ: $\partial\Phi/\partial n = 0$. This follows directly from taking the dot product of $v_r^{\ell+1} - v_r^* = -\Delta t \Phi_{,r}/\varrho$ and the normal vector at the boundary. If we apply $\partial\Phi/\partial n = 0$ on the complete boundary, Φ is only determined up to an additive constant, thus requiring Φ to be specified at a point.

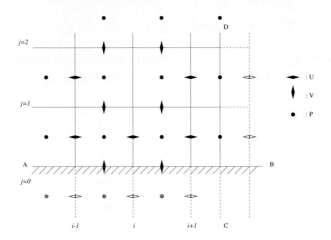

Fig. 6.10. Sketch of two 3D cells and associated velocity components.

The staggered grid should be constructed such that the boundary passes through velocity nodes at the parts of the boundary where the velocity is prescribed. Similarly, if the pressure is given on a boundary, the boundary should be located at the pressure nodes. Figure 6.10 shows a solid wall A-B and an outflow boundary C-D. At the solid wall A-B we demand $u = v = w = 0$ and $\partial \Phi / \partial y = 0$. For the velocity component v located at the wall face, implementation of $v = 0$ is trivial. On the other hand, the condition $u = 0$ must be implemented in an average sense by demanding

$$u^\ell_{i,0,k+\frac{1}{2}} \approx \frac{1}{2}(u^\ell_{i,\frac{1}{2},k+\frac{1}{2}} + u^\ell_{i,-\frac{1}{2},k+\frac{1}{2}}) = 0 \,.$$

This condition involves the fictitious value $u^\ell_{i,-\frac{1}{2},k+\frac{1}{2}}$. The same type of fictitious values appear in the condition $\partial \Phi / \partial y = 0$:

$$\Phi_{i,\frac{1}{2},k+\frac{1}{2}} - \Phi_{i,-\frac{1}{2},k+\frac{1}{2}} = 0 \,.$$

The fictitious values can be eliminated by applying the difference equation at the boundary, as explained in Chapter 1.3.6, or by using a ghost boundary. In the present Navier-Stokes solver the implementation makes use of field objects FieldLattice with ghost boundaries. This means that the fields can also access the fictitious values at the grid points outside the physical domain. We can hence sample the difference equations also at the boundary. After the solution is computed, the fictitious values on the ghost boundary must be updated according to the boundary conditions. In introduction to programming with ghost boundaries and Diffpack finite difference fields is provided by the source code related to Example C.1 on page 597.

At the outflow boundary C-D pressure values can be set directly, whilst averaging of the normal component of the velocity may be applied. At an inflow boundary one can easily implement a prescribed pressure gradient $(\partial \Phi / \partial n)$.

We refer to the source files for complete documentation of the details regarding the implementation of boundary conditions.

6.4.3   A Multigrid Solver for the Pressure Equation

The most time-consuming part of Algorithm 6.1 is the solution of the Poisson equation $\nabla^2 \Phi = v^*_{s,s} \varrho / \Delta t$. In a finite difference context it is tempting to apply a simple iterative solution method for the Poisson equation, for example, the Gauss-Seidel or SOR method. However, as explained in Appendix C.1, the classical iterative methods are not efficient for fine-grid problems, and our 3D Navier-Stokes solver is typically aimed at grids with millions of unknowns. A sufficiently fast method is multigrid, which is described in Appendix C.4.2. The multigrid algorithm has been implemented to solve the equation for Φ, assuming that the grids are of box-shape with uniform partition. The cell size is doubled in each space direction when going from one grid to the coarser neighboring grid. Since V-cycle multigrid has proved to be very efficient for solving Poisson equations, we have restricted the solver to this strategy.

6.4.4   Implementation

The present solution method for the Navier-Stokes equations is implemented in a class NsFD. This class has two basic components: a geometry handler and a Poisson-equation solver. Both these components are represented as class hierarchies, and the NsFD class contains bass class pointers (handles) to the base classes Geometry and Poisson in these two hierarchies. The geometry object is responsible for setting correct boundary conditions in the finite difference schemes, and the various subclasses of Geometry implement specific flow cases, such as 2D/3D cavity flow and 2D/3D channel flow (for verification). The Poisson class hierarchy allows to switch between different Poisson-equation solvers, e.g., SOR (PoissonSOR) and multigrid (PoissonMG) solvers.

Programming of finite difference schemes on staggered grids is most conveniently done using the grid and field objects for staggered finite difference grids, see page 418. This safe approach is followed in class NsFD. However, the efficiency of half-integer indexing of the type p(i+0.5,j+0.5,k+0.5) is unacceptable in 3D viscous flow simulations, so after the NsFD class is debugged, we derive a subclass NsFDFast, where the index operations are optimized as outlined in Appendix B.6.4. The same implementation philosophy is applied to the Poisson-equation solvers. The multigrid method is implemented in the most obvious way on a uniform grid and does not make use of the multigrid toolbox in Diffpack [76] (which is geared towards finite element methods).

An overview of the classes in this finite difference-based Navier-Stokes solver is provided in Figure 6.11.

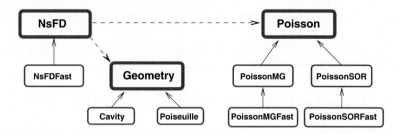

Fig. 6.11. Sketch of the classes in the NsFD solver. Solid lines indicate inheritance ("is-a" relationship), while dashed lines indicate pointers ("has-a" relationship).

Three-dimensional cavity flow is a typical example of nontrivial viscous flow in box-shaped grid. The fluid is confined in a box, where the top face moves with constant velocity. Figure 6.12 displays a subset of the velocity vectors for a simulation on a grid with $96^3 = 884,736$ cells and a Reynolds number of 1000. Stationary flow phenomena, such as cavity flow, must with the present numerical approach be simulated as a time-dependent problem, usually starting from rest. Most of the work in such simulations is spent in the multigrid method. Naturally, finding the optimal combination of parameters and numerical strategies in multigrid might reduce the simulation time substantially. Contrary to common practice in smaller 2D examples, we do not use the same number of pre- and post-smoothing steps at the various levels. Improved efficiency in the present problem is obtained by increasing the number of pre-smoothing iterations for each coarser grid, keeping the number of post-smoothing iterations constant. The optimal number of grid levels in a V-cycle is another basic issue. Figure 6.13 shows the residual in the linear system arising from the pressure-correction equation as a function of work units (WU). A work unit is here defined as the cost of performing one smoothing iteration on the finest grid. The optimal convergence rate was obtained with four grid levels. Figure 6.14 displays the effect of varying the number of pre- and post-smoothing iterations on the finest grid level (using a total of four levels). One post-iteration and one or two pre-iterations had the best performance.

Utilizing Diffpack's menu system and the multiple loop facility, which is available in the NsFD solver, one can easily set up a number of experiments and find a good combination of parameters for a given flow case.

The current finite difference-based solution method is restricted to a uniformly partitioned box, but more complicated geometries can be treated by blocking cells in the box. Such seemingly crude approaches have been widely used and prove to be effective [45], although the finite element method might be a more accurate way of dealing with nontrivial geometries.

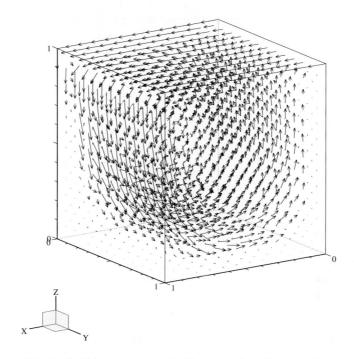

Fig. 6.12. Velocity vectors in 3D cavity flow (Re = 1000).

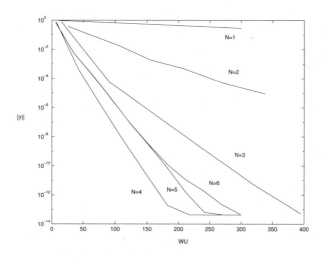

Fig. 6.13. Experimentation with the effect of different number of grid levels in the multigrid solver for the Poisson equation for Φ. The residual is plotted against the computational work (measured in work units).

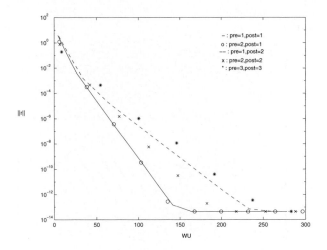

Fig. 6.14. Experimentation with the effect of different pre- and post-smoothing iterations in the multigrid solver for the Poisson equation for Φ. The residual is plotted against the computational work (measured in work units).

6.5   A Fast Finite Element Navier-Stokes Solver

The goal of this section is to outline a fast finite element method for the incompressible Navier-Stokes equations (6.49)–(6.50) based on the ideas from the operator-splitting approach in Chapter 6.4. The resulting algorithm involves only vector updates and solution of a Poisson equation, which can be efficiently performed also in 3D problems with geometrically complicated grids. Since the time discretization is explicit, there is a restriction on the time step length, and stationary problems must be computed as the limiting stationary state of an appropriate transient problem.

The operator-splitting technique in Chapter 6.4 is normally combined with spatial finite difference discretization on a staggered grid. The corresponding finite element discretization should then apply different basis functions for the velocity and pressure (mixed finite elements). This is quite straightforward in Diffpack, but requires concepts that are beyond the scope of the present text. By using a particular time discretization, it appears that we can use identical basis functions for the velocity and pressure.

With regard to the implementation, we shall devise an optimized algorithm where we avoid invoking full `makeSystem` calls at each time level, utilizing software constructions from Chapter 3.11 and Appendix B.6.2. The basic ideas will be outlined, but the actual implementation is left as a project for the reader.

For further reading about finite element methods based on operator splitting of the transient Navier-Stokes equations, we refer to [44], [47], and [93, Ch. 13] and the references therein.

6.5.1 Operator Splitting and Finite Element Discretization

A Second-Order Operating-Splitting Technique. The basic idea of the current operator-splitting technique to be applied to the Navier-Stokes equations is most clearly explained for a simple equation on the form $u_{,t} = F(u)$, where F is some nonlinear spatial operator. A centered time scheme for this equation can be expressed as

$$\frac{u^{\ell+1} - u^{\ell}}{\Delta t} = [F(u)]^{\ell+\frac{1}{2}} \approx \frac{1}{2}\left(F(u^{\ell}) + F(u^{\ell+1})\right).$$

The value $F(u^{\ell+1})$ is of course unknown, leading to an implicit nonlinear equation for $u^{\ell+1}$. However, we can *predict* $F(u^{\ell+1})$ by an explicit forward Euler step: $F(u^{\ell+1}) \approx F(\hat{u})$, where $\hat{u} = u^{\ell} + \Delta t F(u^{\ell})$. This leads to a centered explicit scheme of general applicability to nonlinear transient PDEs:

$$k^{(1)} = \Delta t F(u^{\ell}), \tag{6.72}$$

$$\hat{u} = u^{\ell} + k^{(1)}, \tag{6.73}$$

$$k^{(2)} = \Delta t F(\hat{u}), \tag{6.74}$$

$$u^{\ell+1} = u^{\ell} + \frac{1}{2}\left(k^{(1)} + k^{(2)}\right). \tag{6.75}$$

Operator Splitting Applied to the Navier-Stokes Equations. We can combine the scheme (6.72)–(6.75) with the basic algorithm in Chapter 6.4. When performing the steps (6.72) and (6.74) with the Navier-Stokes equations, we omit the pressure and the body force terms (these will be included in the corrector step). The resulting combination of techniques was originally proposed by Ren and Utnes [97] and consists of four steps.

1. Calculation of coefficients for an intermediate velocity field:

$$k_r^{(1)} = -\Delta t(v_s^{\ell}v_{r,s}^{\ell} - \nu v_{r,ss}^{\ell}) \tag{6.76}$$

$$\hat{v}_r = v_r^{\ell} + k_r^{(1)}, \tag{6.77}$$

$$k_r^{(2)} = -\Delta t(\hat{v}_s^{\ell}\hat{v}_{r,s}^{\ell} - \nu\hat{v}_{r,ss}^{\ell}) \tag{6.78}$$

2. Calculation of an intermediate velocity field v_r^*:

$$v_r^* = v_r^{\ell} + \frac{1}{2}\left(k_r^{(1)} + k_r^{(2)}\right). \tag{6.79}$$

3. Solution of a Poisson equation for the new pressure (arising from the incompressibility constraint $v_{s,s}^{\ell+1} = 0$):

$$\nabla^2 p^{\ell+1} = \frac{\varrho}{\Delta t}v_{s,s}^*. \tag{6.80}$$

4. Correction of the intermediate velocity field:

$$v_r^{\ell+1} = v_r^* - (p_{,r}^{\ell+1} - \varrho b_r)\Delta t/\varrho. \tag{6.81}$$

The velocity and pressure are represented using identical basis functions:

$$v_r^\ell = \sum_{j=1}^n v_j^{r,\ell} N_j, \quad p^\ell = \sum_{j=1}^n p_j^\ell N_j .$$

The fields $k_r^{(1)}$, $k_r^{(2)}$, $\hat{v}_r$, and v_r^* have representations similar to v_r^ℓ.

Equation (6.76) can be considered as d decoupled equations for the d fields $k_1^{(1)}, \ldots, k_d^{(1)}$. Introducing

$$\boldsymbol{v}_r^\ell = (v_1^{r,\ell}, \ldots, v_1^{r,\ell})^T,$$

with similar definitions of $\boldsymbol{k}_r^{(1)}$, $\boldsymbol{k}_r^{(2)}$, $\boldsymbol{v}_r^*$, and $\hat{\boldsymbol{v}}_r$, we can write a Galerkin finite element formulation of (6.76) on the form

$$\boldsymbol{M}\boldsymbol{k}_r^{(1)} = -\Delta t \boldsymbol{a}_r(\boldsymbol{v}_1, \ldots, \boldsymbol{v}_d) - \nu \Delta t \boldsymbol{K}\boldsymbol{v}_r, \tag{6.82}$$

for $r = 1, \ldots, d$. Notice that for each value of r, (6.82) is a linear system for $\boldsymbol{k}_r^{(1)}$. We can hence solve d linear systems for $\boldsymbol{k}_1^{(1)}, \ldots, \boldsymbol{k}_d^{(1)}$ instead of solving one compound vector system like we focused at in Chapters 5.1.2 and 6.3.2.

The entries of the mass matrix $\boldsymbol{M}$ have the usual form $M_{i,j} = \int_\Omega N_i N_j d\Omega$, where Ω is the fluid domain and $\boldsymbol{M}$ may be lumped to simplify the solution of the linear systems. The matrix $\boldsymbol{K}$ arises from the Laplace operator (recall that $v_{r,ss} = \nabla^2 v_r$), with entries $K_{i,j} = \int_\Omega N_{i,s} N_{j,s} d\Omega$. The nonlinear term $\boldsymbol{a}_r(\boldsymbol{v}_1, \ldots, \boldsymbol{v}_d)$ is somewhat more complicated. Entry no. i, here denoted by a_i^r, can be expressed as an integral over a triple sum:

$$a_i^r = \int_\Omega N_i N_k v_k^s N_{j,s} v_j^r \, d\Omega . \tag{6.83}$$

The discrete form of (6.78) should be obvious from (6.82). Equation (6.77) achieves the form

$$\boldsymbol{M}\hat{\boldsymbol{v}}_r = \boldsymbol{M}\boldsymbol{v}_r^\ell + \boldsymbol{M}\boldsymbol{k}_r^{(1)} \quad \Rightarrow \quad \hat{\boldsymbol{v}}_r = \boldsymbol{v}_r^\ell + \boldsymbol{k}_r^{(1)} . \tag{6.84}$$

In other words, the computation of $\hat{\boldsymbol{v}}_r$ involves no matrices, just nodal values vectors in a node-by-node formula. The same reasoning applies to (6.79), with the result

$$\boldsymbol{v}_r^* = \boldsymbol{v}_r^\ell + \frac{1}{2}\left(\boldsymbol{k}_r^{(1)} + \boldsymbol{k}_r^{(1)}\right) . \tag{6.85}$$

Equation (6.80) is a standard Poisson equation, yielding the discrete form

$$\boldsymbol{K}\boldsymbol{p}^{\ell+1} = \frac{\varrho}{\Delta t}\boldsymbol{B}_s \boldsymbol{v}_s^*, \tag{6.86}$$

where $\boldsymbol{K}$ has the same form as in (6.82) and the entries of $\boldsymbol{B}_s$ are given by

$$B_{i,j}^s = \int_\Omega N_i N_{j,s} d\Omega . \tag{6.87}$$

Finally, we have the discrete form of the update (6.81):

$$\boldsymbol{M}\boldsymbol{v}_r^{\ell+1} = \boldsymbol{M}\boldsymbol{v}_r^* - (\boldsymbol{B}_r\boldsymbol{p}^{\ell+1} - \varrho\boldsymbol{c}_r)\,. \tag{6.88}$$

The vector $\boldsymbol{c}_r$ contains the body forces b_r: $c_i^r = \int_\Omega N_i b_r \, d\Omega$.

The original coupled system, i.e. the vector equation (6.49) and the scalar PDE (6.50), has now been transformed into a sequence of linear systems, where each linear system involves only n unknowns. Put in another way, we have approximated the incompressible Navier-Stokes equations by a sequence of standard scalar PDE problems of the type covered in Chapter 3.

The time step of the suggested scheme is limited by the explicit treatment of the convection-diffusion equation (6.76). The stability criterion takes the form

$$\Delta t \leq \min_e \frac{h_e^2}{2\nu + U_e h_e},$$

where h_e is the characteristic length of element no. e and U_e is the characteristic velocity of the element.

6.5.2 An Optimized Implementation

We observe that the matrices $\boldsymbol{M}$, $\boldsymbol{K}$, and $\boldsymbol{B}_s$ are all constant in time. A linear system like (6.86) can then be updated by performing d matrix-vector products for the right-hand side, which is much more efficient than going through all the computational details of the assembly process at each time level. The corresponding implementation techniques in Diffpack are explained in Chapter 3.11 and Appendix B.6.2. The reader should be familiar with these techniques before working through the forthcoming material in detail.

The term $\boldsymbol{a}_r(\boldsymbol{v}_1, \ldots, \boldsymbol{v}_d)$ changes in time and seemingly demands a full assembly procedure at each time level. From the formula (6.83) we can write

$$a_i^r = C_{isjk} v_k^s v_j^r, \quad C_{isjk} = \int_\Omega N_i N_k N_{j,s} \, d\Omega\,. \tag{6.89}$$

If we interpret this formula at the element level, i.e. n is n_e (the number of nodes in the element) and Ω is an element, we see that we can generate the element vector a_i^r from precomputed C_{isjk} values (at the element) through a triple sum over s, j, and k. Hence, we cannot (easily) avoid assembly of element contributions, but we can avoid computing basis functions, their derivatives, and the expressions (6.83) at each time level.

When it comes to suitable storage structures, the $\boldsymbol{M}$ and $\boldsymbol{K}$ matrices can be represented as explained in Chapter 3.11 and Appendix B.6.2[12]. The matrices $\boldsymbol{B}_s$, $s = 1, \ldots, d$, can be represented as a

---

[12] If essential pressure boundary conditions are to be prescribed, $\boldsymbol{K}$ in (6.86) must be modified, and the same $\boldsymbol{K}$ can no longer be used for (6.84) and (6.86). Alternatively, we can integrate by parts in (6.88) to obtain a surface integral involving pressure boundary conditions (the transpose of the $\boldsymbol{B}_r$ matrix then enters that equation).

```
VecSimplest(Handle(Matrix(real))) B
```

object, i.e., a vector of d matrices. The computation of the lumped mass matrix M is carried out by FEM::makeMassMatrix. The K matrix is conveniently stored in a LinEqAdmFE object, since we need to solve linear systems later with K, and LinEqAdmFE gives immediate access to Diffpack's full range of storage formats and linear solvers. An appropriate implementation consits of making an integrands functor for K (see Appendix B.5.2) and calling a FEM::makeSystem function that applies this functor for filling a LinEqAdmFE object.

The B_s matrices can be allocated from the K matrix[13] and then be computed by the FEM::makeSystem function that takes a Matrix object instead of LinEqAdmFE. Again we recommend to implement the integrands function for the B_s matrices in terms of a functor with s as parameter. One could also think of letting all makeSystem calls end up in the simulator's integrands function, but apply an enum variable in the simulator for indicating which integrand formula that is to be used.

The C_{isjk} quantity on each element requires somewhat more consideration. An obvious data structure is a five-dimensional array C(e,i,s,j,k). However, to have full control of the efficiency when accessing such large arrays, we propose the application of just a one-dimensional vector Vec(real) C. Let nsd be the number of space dimensions, nne the number of nodes in an element (assumed constant), and nel the number of elements. The length of the C array is then nel*nne*nne*nne*nsd. Since we will compute and use C in an element-by-element fashion, the entries associated with one element should be collected in a continuous memory block. This is enabled by letting e be the "most slowly-varying" index, while the summation indices s, j, and k should have the fastest variation. The conceptual index C(e,i,s,j,k) is obtained by

```
C( (e-1)*nne*nsd*nne*nne + (i-1)*nsd*nne*nne + (s-1)*nne*nne +
   (j-1)*nne + k )
```

An equivalent, but more optimized index is

```
// n1=nne*nne n2=n1*nsd n3=n2*nne n4=n1+n2+n3+nne
C( e*n3 + i*n2 + s*n1 + j*nne + k - n4 )
```

However, one should avoid this type of indexing and instead try to access C entry-by-entry in the way this vector is stored in memory. To this end, a triple sum $C_{isjk} v_k^s v_j^r$ in element e can be implemented as follows.

```
// given VecSimple(Vec(real)) v_e, where v_e(s)(i) is vector field
// component s at local node i, this v_e is updated like this:
for (s = 1; s <= nsd; s++)
```

---

[13] The appropriate statement is lineq->A().makeItSimilar(B(s)), cf. the code example in Appendix B.6.2.

```
    v()(s).localValues (v_e(s), e);  // uses a Handle(FieldsFE) v
// triple sum:, store result in a(i), i=1,...,nne
real sum = 0, v_j_r, v_k_s;
int idx = (e-1)*n3 + (i-1)*n2;    // +1 gives start index in C
Vec(real)* v_s;                   // efficient pointer to v_e(s)
Vec(real)* v_r = &(v_e(r));
for (s = 1; s <= nsd; s++) {
  v_s = &(v(s));
  for (j = 1; j <= nne; j++) {
    v_j_r = (*v_r)(j);
    for (k = 1; k <= nne; k++) {
      idx++; v_k_s = (*v_s)(k);
      sum += C(idx)*v_k_s*v_j_r;
    }
  }
}
a(i) = sum;  // local a_i^r in the formulas
// --- assemble a into global system: ---
// given VecSimple(int) loc2glob and Vec(real) global_vec:
dof->loc2glob (loc2glob, e);      // uses a Handle(DegFreeFE) dof
global_vec.assemble (a, loc2glob, e);
```

The computation of the C array is slightly more complicated as we need to write a tailored makeSystem routine, where we do not make use of any LinEqAdmFE or ElmMatVec objects. Fortunately, this is a matter of combining the information in Appendix B.5.1: We merge the code from the functions FEM::makeSystem, FEM::calcElmMatVec, FEM::numItgOverElm, and integrands into one makeSystem function, removing statements regarding LinEqAdmFE and ElmMatVec. The core part of the new makeSystem function in the simulator can be sketched as follows.

```
// inherits ElmItgRules itg_rules and FiniteElement finite_elm
// from base class FEM, takes DegFreeFE dof and Vec(real) C as
// arguments
finite_elm.attach (*dof);
C.redim (n3*nel);
int idx=0;
real N_i, N_js;
const Vec(real)& N = fe. N();  // faster array access
const Mat(real)& dN = fe.dN();
for (e = 1; e <= nel; e++) {
  fe.refill (e, this /*attach solver to fe*/);
  fe.setLocalEvalPts (rules);    // tell fe about intgr. points
  fe.initNumItg();
  while (fe.moreItgPoints()) {   // integration loop in an elm.
    fe.update4nextItgPt();
    for (i = 1; i <= nne; i++) {
      N_i = N(i);
      for (s = 1; s <= nsd; s++) {
        for (j = 1; j <= nne; j++) {
          N_js = dN(j,s);
          for (k = 1; k <= nne; k++) {
            idx++;
            C(idx) = N_i*N_js*N(k);
          }
```

```
                }
            }
        }
    }
}
```

One could also think of restricting the element type to triangles or tetrehedra and compute the C_{isjk} quantity by analytical integration.

The rest of the solver code consists of statements that generate the proper right-hand sides from matrix-vector multiplications and perform the four solution phases at each time level. Pressure-driven flow between two flat plates (see Project 1.7.2) can be a suitable test problem when debugging the code.

6.6   Projects

6.6.1   Analysis of Discrete Shallow Water Waves

Mathematical Problem. We consider shallow water waves in a rectangular basin with constant depth. The mathematical model is then equations (6.25)– (6.27), with $H = $ const and zero normal velocity on the boundaries. The initial condition reads $u = v = 0$ (at $t = \Delta t/2$), and

$$\eta(x, y, 0) = A \sin kx, \tag{6.90}$$

where A is a constant (the initial wave amplitude). The problem can be reduced to one space dimension such that $v = 0$, $u = u(x, t)$, and $\eta = \eta(x, t)$.

Numerical Method. Set up an explicit finite difference method for the one-dimensional equations, using a staggered grid in space and time. This will be referred to as method 1. Method 2 is an explicit finite difference scheme on a non-staggered grid:

$$\frac{\eta_j^{\ell+1} - \eta_j^{\ell-1}}{2\Delta t} = -\frac{u_{j+1}^\ell - u_{j-1}^\ell}{2\Delta x}$$

$$\frac{u_j^{\ell+1} - u_j^{\ell-1}}{2\Delta t} = -\frac{\eta_{j+1}^\ell - \eta_{j-1}^\ell}{2\Delta x}$$

Such centered time differences over two time intervals are frequently known as *Leap-Frog schemes*.

Analysis. Find the analytical solution of the problem by assuming that $\eta(x, t) = A_\eta \exp(i(kx - \omega t + \phi_\eta))$ and $u(x, t) = A_u \exp(i(kx - \omega t + \phi_u))$. Adjust the phase ϕ_η (and ϕ_u) such that the real part of these complex functions gives the physically relevant solution (see Appendix A.4). Identify the analytical dispersion relation $\omega = \omega(k)$.

In the numerical scheme we also search for exponential solutions:

$$\eta_j^\ell = \tilde{A}_\eta \exp\left(i(kjh - \tilde{\omega}t)\right), \quad u_j^\ell = \tilde{A}_u \exp\left(i(kjh - \tilde{\omega}t)\right).$$

Insert these expressions in the numerical schemes for method 1 and 2 and eliminate $\tilde{A}_u$ to find the numerical dispersion relation. Control the answer in the method 1 case by first eliminating the discrete velocities from the scheme, then identifying the resulting equation as a standard finite difference discretization of the wave equation $\partial^2\eta/\partial t^2 = H\partial^2\eta/\partial x^2$, and then looking up the numerical dispersion relation for this equation in Appendix A.4.5. Discuss the accuracy and stability issues of method 1 and 2.

Implementation. Extend the `Wave1D1` class from Chapter 1.6.3 to handle a system of two equations. Implement both method 1 and method 2 (in method 2 one can mimic the boundary condition $u_0^\ell = 0$ by setting the mean value $u_{-1}^\ell + u_1^\ell$ to zero). Compute the error as a field and dump curve plots of this field for animation. Verify the implementation.

6.6.2 Approximating the Navier-Stokes Equations by a Laplace Equation

Mathematical Problem. This project deals with flow between two channel walls, where the upper wall is deformed according to a Gaussian bell function, as depicted in Figure 6.15. The viscous fluid flow is governed by the

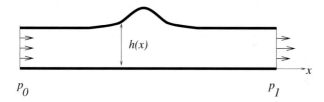

Fig. 6.15. Viscous fluid flow in a narrow gap between two surfaces, where the upper surface has the shape of a bell function.

incompressible Navier-Stokes equations (6.49)–(6.50). We assume that the flow is stationary. At the inlet we prescribe a plug profile and at the outlet we demand constant pressure and zero normal derivative of the velocity (the natural boundary condition in the penalty method in Chapter 6.3). On the walls the velocity vector must vanish. The shape of the upper wall is given according to a Gaussian bell function $h(x) \sim A\exp\left(-(x - x_0)^2/(2\sigma^2)\right)$, where x_0 is the location of the top of the bell, A is the amplitude, and σ is a measure of the width of the bell.

Flow in "thin geometries" can be approximated by the so-called Reynolds lubrication equation. Instead of solving the incompressible Navier-Stokes equations in a deformed geometry, one solves the Laplace equation with the gap between the walls, h, as a variable coefficient:

$$\nabla \cdot \left(\frac{h^3}{12\mu} \nabla p \right) = 0, \tag{6.91}$$

p being the fluid pressure and μ the viscosity of the fluid. In our application, $p = p(x)$ and $h = h(x)$. (Equation (6.91) is derived in Chapter 7.1.1.) The purpose of the this project is to evaluate, through computer experiments, the validity of the lubrication approximation.

Numerical Method. The incompressible Navier-Stokes equations can be solved by the method explained in Chapter 6.3. The implementation is available as class NsPenalty. Adapt a subclass of NsPenalty to the present problem.

Equation (6.91) is straightforward to solve using finite difference methods and programs from Chapter 1. A fundamental problem is how to assign pressure values at the inlet and outlet. We suggest for the present investigation of models to first run the Navier-Stokes solver to compute the pressure, and then apply these pressure values as boundary conditions in the lubrication equation (6.91).

The velocity

$$u = \frac{h^2}{12\mu} \frac{\partial p}{\partial x}$$

is constant in the continuous model (6.91), but not necessarily so when solving (6.91) numerically and differentiating p, whereas the Navier-Stokes solver computes a spatially varying velocity field. One must thus implement some kind of average of the latter field in order to compare the two models. Suggest some suitable averaging procedure.

Meshing the geometry in Figure 6.15 is easy if we apply the BOX_WITH_BELL feature of the PreproStdGeom preprocessor. Details are given in the man page for class PreproStdGeom.

Analysis. Define the error measure $e = U_N - U_L$, where U_L is the velocity from the lubrication model, and U_N is the average velocity from the full Navier-Stokes model. Organize your program such that you can give A and σ as input and get e as output. For all runs you must use a fine enough grid. This is checked by doubling the grid size and controlling that the difference in the solution is satisfactorily small.

The lubrication approximation is assumed to be good when A is small and σ is large, such that the gap $h(x)$ is smoothly varying. Our interest now is to see how severe the deformations can be before the lubrication approximation becomes inaccurate. Present plots or tables of the variation of e with A and σ. Visualize some severely distorted geometries and include the corresponding e value in the plots.

Chapter 7

Coupled Problems

This chapter deals with two specific examples on systems of PDEs concerning fluid-structure interaction and coupled heat and fluid flow. The exposition includes derivation of the PDEs, precise description of the numerical solution algorithms, and software design principles based on object-oriented programming and Diffpack tools. Contrary to Chapters 5 and 6, where *vector* PDEs were in main focus, we now address systems of PDEs where the different equations reflect different fundamental physical principles. Each scalar equation in the PDE system then has a life on its own. For example, the system treated in Chapter 7.2 consists of a momentum equation and an energy equation. From an implementational point of view, it would be advantagous to realize the compound solver for the system of PDEs as a simple combination of well-tested stand-alone solvers for the various scalar PDEs in the system. We shall pursue this idea in the present chapter.

7.1 Fluid-Structure Interaction; Squeeze-Film Damping

Fluid-structure interaction is a basic problem in many engineering disciplines. The flow of a fluid around a structure gives rise to forces and associated motion of the structure, which again influences the fluid flow. The motion of the fluid is often described by the incompressible Navier-Stokes equations, whereas the structural deformations might be governed by the equations of linear elasticity. We could therefore, at least in principle, solve a class of fluid-structure interaction problems by combining solvers from Chapters 5 and 6. However, the implementational details of such a coupling of classes are better explained in a simplified problem. Of this reason, we address a fluid-structure problem where certain simplifications can be made, such that we end up with a PDE for the fluid flow and an ODE for the motion of the structure. This reduces the mathematical and numerical complexity and helps to expose the details of the software design. Furthermore, the simplifications demonstrate important mathematical modeling techniques for deriving PDEs.

7.1.1 The Physical and Mathematical Model

Sensors and actuators frequently contain small vibrating plates whose motion can be considerably influenced by induced viscous air flow in the surroundings. We shall focus on the situation where there are two vibrating plates separated by a narrow gap.

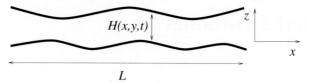

If the characteristic length of the plates is L and $H(x, y, t)$ is the width of the gap, we make the assumption that $L \gg H$ and that H is slowly varying. This particular physical problem is often referred to as *squeeze films*. The viscous fluid motion in the film can have a significant damping effect on the vibrations of the plates. In the design of sensors subject to large accelerations, e.g., during impact, one can rely on squeeze film damping to avoid destructive displacement of plate-like structures in the sensor system.

The equations governing squeeze-film damping will now be derived. Readers whose primary interest is the special software implementation technique being used for this simulator, can safely move on to Chapter 7.1.2.

The mathematical model for the fluid motion consists of the well-known incompressible or compressible Navier-Stokes equations. Depending on the constitutive law of the fluid, an energy equation might also be required. The fluid equations are defined in a domain whose shape is coupled to the movement of the plates. This movement can be modeled by standard equations for vibrating plates, with the fluid pressure as a driving force.

Derivation of the Fluid Flow Model. Physical problems involving domains where one dimension is much smaller than the others, can be effectively modeled by introducing quantities that are averaged over the thickness of the small dimension. In the present problem, we can introduce averaged velocities and pressure in the z direction. To derive the resulting equations, it is convenient to work with the fundamental balance laws on integral form. The general equation of continuity on integral form for an arbitrary control volume V reads

$$\int_V \frac{\partial \varrho}{\partial t} d\Omega + \int_{\partial V} \varrho \boldsymbol{v} \cdot \boldsymbol{n} d\Gamma = 0,$$

where ϱ is the fluid density, $\boldsymbol{v} = (u, v, w)^T$ is the velocity field, $\boldsymbol{n}$ is the outward unit normal vector to the surface ∂V of V, and t denotes time. For the purpose of deriving an averaged differential form of the continuity equation, we let

$$V = \{(x, y, z) \mid x \in [x_0, x_0 + \Delta x], \ y \in [y_0, y_0 + \Delta y],$$
$$z \in [h_B(x, y, t), h_T(x, y, t)]\},$$

The position of the lower plate is $z = h_B$, while $z = h_T$ is the equation for the upper plate. The gap is hence $H(x, y, t) = h_T(x, y, t) - h_B(x, y, t)$.

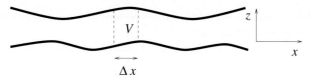

Evaluating the integrals for this particular choice of V gives

$$\Delta x \Delta y \int_{h_B}^{h_T} \frac{\partial \varrho}{\partial t} dz + \Delta y (U|_{x_0+\Delta x} - U|_{x_0})$$

$$+ \Delta x (V|_{y_0+\Delta y} - V|_{y_0}) + \varrho \Delta x \Delta y \left(\frac{\partial h_T}{\partial t} - \frac{\partial h_B}{\partial t} \right) = 0, \qquad (7.1)$$

Here,

$$U = \int_{h_B}^{h_T} \varrho u \, dz, \quad V = \int_{h_B}^{h_T} \varrho v \, dz$$

are the averaged horizontal mass fluxes. Moreover, we have used the boundary conditions on $z = h_B$ and $z = h_T$: $\boldsymbol{v} \cdot \boldsymbol{n} = \boldsymbol{v}_s \cdot \boldsymbol{n}$, where $\boldsymbol{v}_s$ is the velocity of the surface. The latter quantity equals $(\partial h_T / \partial t) \boldsymbol{k}$ for the surface $z = h_T$, and $(\partial h_B / \partial t) \boldsymbol{k}$ for the lower surface.

In the limit $\Delta x, \Delta y \to 0$, we obtain an equation of continuity that is averaged over the gap:

$$H \frac{\partial \varrho}{\partial t} + \frac{\partial U}{\partial x} + \frac{\partial V}{\partial y} + \varrho \frac{\partial H}{\partial t} = 0. \qquad (7.2)$$

Here we have assumed that $\partial \varrho / \partial t$ is independent of z. For an incompressible fluid, we take $\varrho = $ constant, whereas in the compressible case we simply assume that ϱ is constant[1] in z direction.

The U and V quantities couple to the equation of motion. If $H \ll L$ and the variations of H are small (more precisely, the typical wave length of the deformation of the plates is much larger than H), one can assume that the flow is locally similar to steady rectilinear flow between two flat plates. With this assumption we can apply the expressions for U and V found from the equation of fluid motion in a channel of width H (Poiseuille flow [68]):

$$U = -\varrho \frac{H^3}{12\mu} \frac{\partial p}{\partial x}, \quad V = -\varrho \frac{H^3}{12\mu} \frac{\partial p}{\partial y}, \qquad (7.3)$$

where μ is the fluid viscosity. Equation (7.3) is now our approximate form of the momentum equation in the squeeze-film problem. Note that the expressions for the mass fluxes stem from a simplified version of the *incompressible*

---

[1] This can be justified for barotropic fluids, since ϱ is then directly related to the pressure, which is expected to be approximately constant in z direction.

Navier-Stokes equations. We will, however, employ the same approximation in the compressible case as well.

Inserting U and V in the equation of continuity results in

$$H \frac{\partial \varrho}{\partial t} + \varrho \frac{\partial H}{\partial t} = \nabla \cdot \left[\frac{H^3}{12\mu} \varrho \nabla p \right]. \tag{7.4}$$

For incompressible flow this equation reduces to

$$\nabla \cdot \left[\frac{H^3}{12\mu} \nabla p \right] = \frac{\partial H}{\partial t}. \tag{7.5}$$

We employ the barotropic model $p/\varrho^\gamma = \text{const}$ in the case of compressible flow. From a numerical point of view, the equation for the fluid motion is easier to solve if we introduce $\tilde{u} = p^{1/\gamma}$. This results in

$$\frac{\partial}{\partial t}(H\tilde{u}) = \nabla \cdot \left[\frac{H^3}{12\mu} \gamma \tilde{u}^\gamma \nabla \tilde{u} \right]. \tag{7.6}$$

The fluid flow equations are to be solved in a domain Ω, covering the extent of the smallest plate. At the boundary $\partial \Omega$ of Ω, the pressure must match the atmospheric pressure, denoted here by p_0. The initial state is taken as $p = p_0$.

Exercise 7.1. Derive (7.1) and (7.3) in detail.   ◇

Motion of the Plates. For the motion of the plates we can apply the standard equation for small deflection of thin plates. The total pressure $p(x, y, t) - p_0$ and the force $\varrho_s f(t)$ in an accelerated coordinate system comprise the loads on the plate. Here, $f(t)$ is the external acceleration of the coordinate system, and ϱ_S is the density per unit area of the plate. The governing equation for a vibrating plate can then be written as [60]

$$\varrho_S \frac{\partial^2 d}{\partial t^2} + D\nabla^4 d = p - p_0 + \varrho_S f, \quad D = \frac{Eq^3}{12(1 - \nu^2)}. \tag{7.7}$$

The function $d(x, y, t)$ is the deflection of the plate, $\nabla^4 \equiv \nabla^2 \nabla^2$ is the biharmonic operator, ν is Poisson's ratio, E is Young's modulus, and q is the thickness of the plate. We shall be particularly concerned with impulsively started vibrations from rest, modeled as rapid variation of $f(t)$:

$$f(t) = \begin{cases} I \sin^2 \omega t, & t < \pi/\omega \\ 0, & t \geq \pi/\omega \end{cases} \tag{7.8}$$

The boundary conditions depend on the support of the plate. For example, on a clamped part of the plate, $d = \partial d/\partial n = 0$.

A finite element method for (7.7) requires quite complicated elements with twice differentiable basis functions, or we need to rewrite the plate equation as a system of two PDEs, involving ∇^2 operators instead of ∇^4. A much simpler approach is to apply a spectral method for the spatial discretization of (7.7). This reduces the PDE to an initial-value problem involving only ordinary differential equations (ODEs). The disadvantage is that the spectral method outlined below is restricted to plates of rectangular or circular shape. However, such shapes are highly relevant in squeeze-film applications.

Assuming that the plate is rectangular and simply supported at $x = 0, L$ ($d = 0$), with the two other ends free ($\partial^2 d/\partial n^2 = 0$), a possible spatial expansion of d can be written as

$$d(x, y, t) \approx \hat{d}(x, t) = \sum_{j=1}^{M} a_j(t) \sin\left(j\pi\frac{x}{L}\right), \qquad (7.9)$$

where $a_j(t)$ are amplitude functions to be computed. By this particular expansion we also assume that the fluid pressure does not vary with y, otherwise d would also depend on y. Since we know that damping effects may be significant in the squeeze-film problem, high-frequency basis functions are expected to have very small amplitude. We therefore attempt a one-term expansion,

$$\hat{d}(x, t) = a(t) \sin \pi\frac{x}{L}. \qquad (7.10)$$

The equation for $a(t)$ can be derived from a Galerkin method applied to the vibrating plate equation with $\sin \pi x/L$ as weighting function. The result becomes[2]

$$\varrho_s \ddot{a} + \left(\frac{\pi}{L}\right)^4 Da = \frac{4}{\pi}\varrho_s f + \frac{2}{L} \int_0^L (p - p_0) \sin\left(\pi\frac{x}{L}\right) dx. \qquad (7.11)$$

Exercise 7.2. Approximate the plate displacement d by a sum of M sinusoidal basis functions as in (7.9) and derive a decoupled system of ODEs for $a_j(t)$, $j = 1, \ldots, M$. Explain why the system becomes decoupled. $\diamond$

Exercise 7.3. Suggest a generalization of (7.9) in the case of a rectangular plate and a two-dimensional pressure field, $p = p(x, y, t)$. Assume that all sides of the plate are simply supported. Compute the equation for the time-dependent coefficient in a one-term expansion. $\diamond$

Summary of the Mathematical Model. For a sample application in micromechanical sensor technology, the lower plate is often stiff enough to remain plane ($h_B = \text{constant}$). The gap between the plates in the nondeformed state is h_0, and we place the z axis such that $z = h_B = 0$. The relation between H and d then becomes $H(x, y, t) = h_0 + d(x, y, t)$. Since h_0 is expected to be small, the coefficient H^3 can be very small and cause numerical problems.

---

[2] Useful formulas: $\int_0^L \sin(x\pi/L)dx = 2L/\pi$, $\int_0^L \sin^2(x\pi/L)dx = L/2$. Later we will also make use of $\int_0^L \sin^3(x\pi/L)dx = 4L/(3\pi)$.

A proper scaling of the equations would cure such problems. However, scaling of the present initial-boundary value problem quickly becomes a tedious procedure. We therefore apply the simpler approach of multiplying the flow equation by the factor $12\mu h_0^{-3}$, such that we avoid very small values in the coefficient in the Laplace term. Another problem is that $p_0 = 0$ implies $p = 0$ at all times. It is therefore advantageous to have the primary unknown u as a perturbation around unity. This is accomplished by introducing $u = (p/p_0)^{1/\gamma}$ as primary unknown.

The initial-boundary value problem for compressible flow can now be summarized.

$$\varrho s \ddot{a} + \left(\frac{\pi}{L}\right)^4 Da = \frac{4}{\pi}\varrho s f + \frac{2}{L}\int_0^L (p(\boldsymbol{x},t) - p_0)\sin\left(\pi\frac{x}{L}\right) dx, \quad (7.12)$$

$$a(0) = \dot{a}(0) = 0, \quad (7.13)$$

$$12\mu h_0^{-3}\frac{\partial}{\partial t}(Hu) = \nabla\cdot\left[\left(\frac{H}{h_0}\right)^3 p_0\gamma u^\gamma \nabla u\right], \quad \boldsymbol{x}\in\Omega,\ t>0, \quad (7.14)$$

$$u = \frac{p^{1/\gamma}}{p_0^{1/\gamma}}, \quad (7.15)$$

$$H = h_0 + a(t)\sin\left(\pi\frac{x}{L}\right), \quad (7.16)$$

$$u(\boldsymbol{x},t) = 1, \quad \boldsymbol{x}\in\partial\Omega_E, \quad (7.17)$$

$$u(\boldsymbol{x},0) = 1, \quad \boldsymbol{x}\in\Omega. \quad (7.18)$$

The domain Ω is either one- or two-dimensional, and $\partial\Omega_E$ denotes the complete boundary of Ω.

For incompressible flow, we simply replace (7.14) by

$$\nabla\cdot\left[\left(\frac{H}{h_0}\right)^3 p_0\nabla u\right] = 12\mu h_0^{-3}\frac{\partial H}{\partial t}, \quad \boldsymbol{x}\in\Omega,\ t>0. \quad (7.19)$$

In this case, $u = p/p_0$. To handle both the compressible and incompressible case within the numerical expressions and the same code lines, it is convenient to introduce a variable coefficient $k(u)$ in the Laplace term, where $k(u) = \gamma u^\gamma$ in compressible flow and $k(u) = 1$ when the flow is incompressible.

Analysis of a Simple Case. Valuable insight into the problem can be obtained by analyzing a special case where an analytical solution is straightforwardly derived. This is also fundamental for partial verification of a computer implementation. With $H(x,t) = h_0 + a(t)\sin(\pi x/L)$, the boundary-value problem of *an incompressible* fluid, in the one-dimensional case, becomes

$$\frac{\partial}{\partial x}\left(\left(1 + \frac{a}{h_0}\sin\pi\frac{x}{L}\right)^3\frac{\partial p}{\partial x}\right) = 12\mu h_0^{-3}\dot{a}\sin\pi\frac{x}{L}, \quad p(0) = p(L) = p_0. \quad (7.20)$$

We can integrate (7.20) and make a first-order approximation to the resulting right-hand side:

$$\frac{\partial p}{\partial x} = 12 \mu \dot{a} h_0^{-3} \left(\frac{L}{\pi} \cos \pi \frac{x}{L} + C_1 \right).$$

Here C_1 is an integration constant. Integrating once more and inserting the boundary values yields

$$p(x,t) = p_0 - \dot{a} \frac{12 \mu L^2}{h_0^3 \pi^2} \sin \pi \frac{x}{L}. \tag{7.21}$$

In the case of a *flat plate*, $d = a(t)$, we would get

$$p = p_0 - \dot{a} \frac{12 \mu}{h_0^3} \frac{1}{2} x(L - x),$$

which leads to the same qualitative behavior of $p - p_0$. The maximum pressure disturbance is, however, affected by a factor of $\pi^2/8 \approx 1.23$.

The $p - p_0$ function can now be inserted in our simplified vibrating plate equation. The result becomes

$$\varrho_S \ddot{a} + \kappa \dot{a} + \left(\frac{\pi}{L} \right)^4 Da = \frac{4}{\pi} \varrho_S f, \quad \kappa = \frac{12 \mu L^2}{h_0^3 \pi^2}. \tag{7.22}$$

Observe that the "driving force" $\partial H/\partial t$ in the fluid flow equation leads to $p \sim \dot{a}$, which then gives rise to a *damping* term $\kappa \dot{a}$ in the vibrating plate equation. This means that for incompressible fluids, the squeeze film will always damp the structural vibration.

Exercise 7.4. Consider compressible flow with a prescribed $d = d_0 \sin \omega t$. Scale the flow equation, using ω^{-1} as time scale, and show that only one dimensionless number, $\sigma = 12 \mu \omega L^2/(p_0 h_0^2 \gamma)$, appears in the scaled equation. One often refers to σ as the *squeeze film number*.   ◇

7.1.2   Numerical Methods

Our mathematical model for the squeeze-film problem consists of a linear or nonlinear heat-conduction-like PDE, i.e. (7.19) or (7.14), coupled with a linear second-order ODE (7.12). The simplest solution strategy is to solve the equations in sequence. At each time level, we first solve for the fluid motion and compute the pressure load on the plate. Thereafter, we carry out one time step in the discrete plate equation.

The fluid flow equation (7.19) or (7.14) can be discretized by a Galerkin finite element method and a θ-rule in time (see Chapter 2.2.2). This yields a system of linear or nonlinear algebraic equations to be solved at each time level. In the nonlinear case, the system can be solved by Newton-Raphson or Successive Substitution (Picard iteration) techniques. We refer to Chapter 4 for details regarding discretization and implementation of a PDE like (7.14). The coefficient matrix $J_{i,j}$ and the right-hand side vector $-F_i$ of the linear system to be solved in each Newton-Raphson iteration take the following form:

$$J_{i,j} = \int_{\Omega} \left[\tau 12\mu h_0^{-3} H N_i N_j + \right.$$

$$\left. \theta \left(\frac{H}{h_0} \right)^3 \Delta t \left(k(u) \nabla N_i \cdot \nabla N_j + k'(u) \nabla N_i \cdot \nabla u N_j \right) \right] d\Omega, \quad (7.23)$$

$$-F_i = -\int_{\Omega} \left[12\mu h_0^{-3} N_i \Delta \Psi + \right.$$

$$\left. \left(\frac{H}{h_0} \right)^3 \Delta t \left(\theta k(u) \nabla N_i \cdot \nabla u + (1-\theta) k(\bar{u}) \nabla N_i \cdot \nabla \bar{u} \right) \right] d\Omega. \quad (7.24)$$

The current time step size is denoted as Δt. Quantities with a bar denotes evaluation at the previous time level. For example, if Δt is constant, u is a short notation for the numerical approximation to $u(\boldsymbol{x}, \ell \Delta t)$, while $\bar{u}$ is the corresponding notation for the approximation to $u(\boldsymbol{x}, (\ell-1)\Delta t)$. To simplify the notation, we have dropped the iteration number as superscript, that is, the symbol u in the numerical formulas refers to the most recent approximation to the primary unknown function u. The parameter τ equals unity in compressible flow and vanishes for incompressible flow. Moreover,

$$\Delta \Psi = \begin{cases} Hu - \bar{H}\bar{u}, & \text{compressible flow} \\ H - \bar{H}, & \text{incompressible flow} \end{cases}$$

In the Successive Substitution method the linear system in each iteration has a coefficient matrix

$$A_{i,j} = \int_{\Omega} \left(\tau 12\mu h_0^{-3} H N_i N_j + \theta \left(\frac{H}{h_0} \right)^3 \Delta t k(u) \nabla N_i \cdot \nabla N_j \right) d\Omega \quad (7.25)$$

and a right-hand side vector

$$b_i = \int_{\Omega} \left(12\mu h_0^{-3} N_i \Phi - (1-\theta) \left(\frac{H}{h_0} \right)^3 \Delta t k(\bar{u}) \nabla N_i \cdot \nabla \bar{u} \right) d\Omega. \quad (7.26)$$

Here, $\Phi = \bar{H}\bar{u}$ in compressible flow and $\Phi = \bar{H} - H$ in incompressible flow.

Exercise 7.5. Derive the expressions (7.23)–(7.26).   ◇

The second-order ODE for the plate motion can be compactly written as

$$c_1 \ddot{a} + c_2 \dot{a} + c_3 a = c_4, \quad (7.27)$$

with suitable definitions of c_1, c_2, c_3, and c_4 according to (7.12). A widely used *Newmark scheme* for (7.27) can be formulated as [131, Ch. 10, Vol II]

$$\ddot{a} = \left(c_1 + \beta_1 c_2 \Delta t + c_3 \beta_2 \frac{\Delta t^2}{2} \right)^{-1} \times$$

$$\left(c_4 - c_2 \left(\bar{a} + (1 - \beta_1)\Delta t \dot{\bar{a}} \right) \right) - c_3 (\bar{a} + \Delta t \dot{\bar{a}} + (1 - \beta_2) \frac{\Delta t^2}{2} \ddot{\bar{a}} \right) \quad (7.28)$$

$$a = \bar{a} + \Delta t \dot{\bar{a}} + (1 - \beta_2) \frac{\Delta t^2}{2} \ddot{\bar{a}} + \beta_2 \frac{\Delta t^2}{2} \ddot{a} \quad (7.29)$$

$$\dot{a} = \dot{\bar{a}} + (1 - \beta_1)\Delta t \ddot{\bar{a}} + \beta_1 \Delta t \ddot{a} \quad (7.30)$$

Again, the bar indicates quantities at the previous time level. Initially, $\bar{a}$ and $\dot{\bar{a}}$ are prescribed. The value of $\ddot{\bar{a}}$ for $t = 0$ follows from (7.27).

Various choices of the parameters correspond to different well-known schemes. For example, $\beta_1 = \beta_2 = 1/2$ results in a Crank-Nicolson-like scheme. In the present scalar case, we obtain an explicit formula for the new a value, although the finite difference scheme is implicit (an implicit *linear* equation in one variable can always be converted to an explicit form).

7.1.3 Implementation

The coupled fluid-structure problem modeling squeeze films is rather complicated in its original form. Through some reasonable assumptions the model has been reduced to a coupled system of a generally nonlinear scalar PDE (7.14) and a linear ODE (7.12).

For simplicity, we decide to solve the PDE and the ODE in sequence at each time level. The implementation of coupled models can quickly become an error-prone process. To ease the coding and the associated debugging, and at the same time increase the extensibility of the mathematical model, we propose to make separate solvers for the fluid PDE and the plate ODE. The equations can then be tested separately before we couple the two classes in a compound solver for the fluid-structure interaction problem. At any time in the debugging of the compound solver, the component simulators can be pulled apart again and retested separately. This is an attractive implementational approach, but we need to clarify the details of the software design. It is advantageous to have studied Chapter 3.5.6 before proceeding. The complete source code of the squeeze film solver is located in the directory tree `src/app/SqueezeFilm`.

The PDE and ODE are coupled through the H and p quantities. Therefore, if we want to solve the fluid PDE on its own, H must be a prescribed quantity. Similarly, when solving the plate ODE separately from the fluid equation, the p field must be specified.

Assume that we make a class `PlateVib1` that solves the general version of (7.27). For testing purposes, we derive a subclass `PlateVibSin` where we

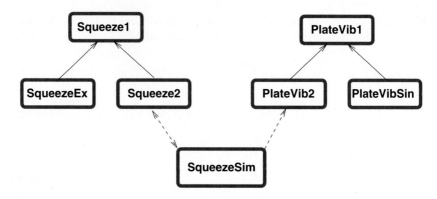

Fig. 7.1. Sketch of the squeeze film solver `SqueezeSim`, which consists of a fluid flow solver `Squeeze2` and a vibrating plate solver `PlateVib2`. The classes `SqueezeEx` and `PlateVibSin` are aimed at simplified test problems for the fluid flow and the structural vibration, respectively. Solid lines indicate inheritance ("is-a" relationship), while dashed lines indicate pointers ("has-a" relationship).

specialize $f(t)$ as $f(t) = q \sin^2 \omega t$, set $p - p_0 = 0$, and compare the numerical and analytical solution. In another subclass `PlateVib2` we implement the more application-relevant shock pulse (7.8) for $f(t)$. This version of the vibrating plate solver is to be coupled to the fluid flow equation. As usual, the code in subclass solvers is only a few lines.

The fluid PDE is implemented as a standard Diffpack finite element solver, like class `NlHeat1` from Chapter 4.2. The name of the class is `Squeeze1`, and it can handle both compressible and incompressible flow in one or two space dimensions. We leave the choice of H as a virtual function `hmodel`. In a subclass `SqueezeEx`, `hmodel` is implemented according to $H = h_0 + t$. The solution, assuming incompressible flow, is then $p = p_0 + 6x(x - 1)/(h_0 + t)^3$. We can hence use class `SqueezeEx` for partial verification of the implementation of the fluid flow solver.

Another subclass `Squeeze2` of `Squeeze1` implements the `hmodel` function with access to the plate deflection field: $H = h_0 + d$. To couple the `Squeeze2` and `PlateVib2` solvers, one can introduce a manager class `SqueezeSim` that holds a fluid flow solver `Squeeze2` and a vibrating plate solver `PlateVib2`. Figure 7.1 depicts the class relations in the squeeze film solver. As advocated in Chapter 3.5.6, the source code of the various subclasses in a solver hierarchy can conveniently be stored in different subdirectories. With the `AddMakeSrc` script one can tell the make program that the source is distributed across directories. The squeeze-film solver has a directory `coupling` for the compound simulator, `vib-test` for verifying the vibrating plate solver, and `film-test` for verifying the fluid flow simulator as a stand-alone solver.

For a more detailed explanation of the software design of the squeeze film simulator, it is convenient to start with the manager, class `SqueezeSim`:

```
class SqueezeSim : public SimCase
{
public:
  Handle(PlateVib2)  plate;
  Handle(Squeeze2)   film;

  Handle(GridFE)     grid;
  Handle(TimePrm)    tip;
  Handle(SaveSimRes) database;

  virtual void define (MenuSystem& menu, int level = MAIN);
  virtual void scan   ();
  virtual void timeLoop ();
  real computePressureLoad (FieldFE& pressure, real p0);
  SqueezeSim ();
 ~SqueezeSim () {}

  virtual void adm (MenuSystem& menu);
  virtual void solveProblem ();
  virtual void resultReport ();
};
```

The manager SqueezeSim is in charge of the grid, the time integration parameters, and the storage for later visualization. Other tasks are distributed to the fluid flow and vibrating plate solvers. The define function exemplifies how some menu items are specific to SqueezeSim and how others are completely handled by calling the define functions in the stand-alone solvers for each equation.

```
void SqueezeSim:: define (MenuSystem& menu, int level)
{
  menu.addItem (level, "gridfile", "readOrMakeGrid syntax",
                "P=PreproBox | d=1 [0,10] | d=1 elm_tp=ElmB2n1D "
                "div=[20], grading=[1]");
  menu.addItem (level, "time integration parameters",
                "TimePrm::scan(Is) syntax", "dt=0.1  t in [0,1]");
  SaveSimRes::defineStatic (menu, level+1);

  menu.setCommandPrefix ("plate");
  plate->define (menu, level, true);
  menu.setCommandPrefix ("film");
  film ->define (menu, level, true);
  menu.unsetCommandPrefix ();
}
```

The fluid flow and the vibrating plate simulators could in general happen to define menu items with the same name. This is the case if both of the solvers put a linear system and solver interface object (LinEqAdmFE) on the menu, a situation that occurs when coupling two or more finite element solvers (cf. the simulator in Chapter 7.2). However, the menu system offers the possibility to set a *command prefix* for all the proceeding command names. As we see from the define function above, one sets the command prefix "plate" before calling plate->scan. This means that the menu command density in the vibrating plate solver actually gets the name `plate density`. Similarly, all the fluid flow menu items are prefixed by "film". If the fluid flow solver had defined a menu item `density` as well, the name of this item would be `film density`. During scanning of menu items, one can activate or deactivate the prefix feature of the menu system. For example, with the prefix "film", a command `menu.get("density")` will actually search for `film density`. The `scan` function can look like this:

```
void SqueezeSim:: scan ()
{
  MenuSystem& menu = SimCase::getMenuSystem();
  String gridfile = menu.get ("gridfile");
  grid.rebind (new GridFE());              // create empty grid object
  readOrMakeGrid (*grid, gridfile);        // fill grid
  tip.rebind (new TimePrm());
  tip->scan (menu.get ("time integration parameters"));
  database.rebind (new SaveSimRes());
  database->scan (menu, grid->getNoSpaceDim());

  menu.setCommandPrefix ("plate");
  plate->scan (menu, database->cplotfile, tip.getPtr());
  menu.setCommandPrefix ("film");
  film ->scan (grid.getPtr(), tip.getPtr(), database.getPtr());
  menu.unsetCommandPrefix ();
```

The `Squeeze1` solver must create its own grid in order to work as a stand-alone solver. However, when the class is used in conjunction with `SqueezeSim`, it is natural for the manager class to be responsible for the grid[3]. The `scan` function of a solver could then take a grid pointer as argument. If the pointer is null, the grid is allocated internally in class `Squeeze1`, otherwise the grid handle in `Squeeze1` is rebound to an external grid. The same strategy applies to the `TimePrm` and `SaveSimRes` objects, which are either internal in the fluid flow simulator or managed by class `SqueezeSim`.

```
void Squeeze1:: define (MenuSystem& menu, int level, bool externals)
{
  if (!externals)
    menu.addItem (level, "gridfile", "readOrMakeGrid syntax",
                  "P=PreproBox | d=1 [0,10] | d=1 elm_tp=ElmB2n1D "
                  "div=[20], grading=[1]");


void Squeeze1:: scan (GridFE* grid_, TimePrm* tip_,
                      SaveSimRes* database_)
{
  MenuSystem& menu = SimCase::getMenuSystem();
  if (grid_ != NULL)
    grid.rebind (grid_); // bind to some external grid
  else {
```

---

[3] In the present case, only the fluid solver needs a grid, but in a more general problem setting, the manager class creates a common grid and distributes it to all the solvers.

```
    String gridfile = menu.get ("gridfile");
    grid.rebind (new GridFE());          // create empty grid object
    readOrMakeGrid (*grid, gridfile);    // fill grid
}
```

With the use of handles, the origin of an object is of no interest; all solver classes can access the object through the handle, as if it were created by that class.

The heart of the SqueezeSim class is the timeLoop routine. This function demonstrates how the fluid flow and the vibrating plate solvers are supposed to work together. The basic numerical steps consist of advancing the fluid flow solver one time level to compute a new pressure field p. Then $p - p_0$ is integrated over the plate, using the function SqueezeSim::computePressureLoad. The resulting pressure load on the plate is transferred to the vibrating plate solver before asking that solver to update the d field at the new time level.

```
void SqueezeSim:: timeLoop ()
{
  tip->initTimeLoop();
  film ->timeLoopSetUp ();
  plate->timeLoopSetUp ();

  while(!tip->finished())
    {
      tip->increaseTime();
      film ->solveAtThisTimeStep ();
      plate->effectivePressureLoad
        (this->computePressureLoad (*film->p, film->p0));
      plate->solveAtThisTimeStep ();
      film ->saveResults ();
      plate->saveResults ();

      if (tip->getTimeStepNo() % 100 == 0)
        s_o << "t=" << tip->time() << endl;
    }
  plate->timeLoopFinish();
  film ->timeLoopFinish();
}
```

The fluid flow and vibrating plate solvers have split the traditional contents of a time loop function into timeLoopSetUp for storing/plotting the initial fields, solveAtThisTimeStep for solving the equations in a solver at a given time level, saveResults for calling SaveSimRes or related functionality for storing fields for later visualization, and timeLoopFinish for closing curve plots or post processing time series results. The need for this high degree of modularity is not apparent for a single solver, but is very advantageous when combining stand-alone solvers into a simulator for a system of differential equations.

The basic fluid flow solver Squeeze1 is very similar to, e.g., class NlHeat1 so there is no need to explain the details here. The purpose of the Squeeze2 subclass simulator is to implement the virtual function hmodel for computing H by calling the vibrating plate solver's function computeDeflection. This latter function loads the deflection into a `FieldFE` object. The communication between the fluid flow solver and the vibrating plate solver is enabled by a two-way pointer between class `Squeeze2` and the manager, class `SqueezeSim`.

```
class Squeeze2 : public Squeeze1
{
public:
  SqueezeSim* manager;
  Squeeze2 (SqueezeSim* manager_) : manager(manager_) {}
  virtual void hmodel (FieldFE& H);
};
```

We can then simply compute $H = d + h_0$ like this:

```
void Squeeze2:: hmodel (FieldFE& H)
{
  manager->plate->computeDeflection (H);   // H  = d
  H.add (h0);                              // H += h0
}
```

Based on our brief review of the `SqueezeSim` solver, the reader is encouraged to study the source code in detail. Hopefully, one will realize that the classes `Squeeze1` and `PlateVib1` are similar to standard Diffpack simulators and that the extra glue for communication when solving the compound system is just very small subclasses. The ideas of coupling stand-alone simulators for solving a system of differential equations are developed further in Chapter 7.2.

Exercise 7.6. The squeeze-film simulator handles in principle 1D and 2D pressure fields. However, class `PlateVib1` does not support 2D functions for spectral discretization of the vibrating plate equation. Describe how class `PlateVib1` can be extended to handle rectangular plates (e.g. simply supported along all sides). ◇

Exercise 7.7. The efficiency of the simulator can be enhanced by introducing a spectral approximation for the pressure field p, similar to the representation of d. Formulate such a numerical method and discuss its impact on the design of the compound simulator `SqueezeSim`. Explain how we can choose between a finite element or a spectral solver at run time. ◇

Finally, we show a typical solution of the squeeze-film problem. Figure 7.2 displays the characteristic pressure and displacement response to an external acceleration pulse on the fluid-structure system. As expected, the vibration of the plate is damped due to the motion of the fluid film.

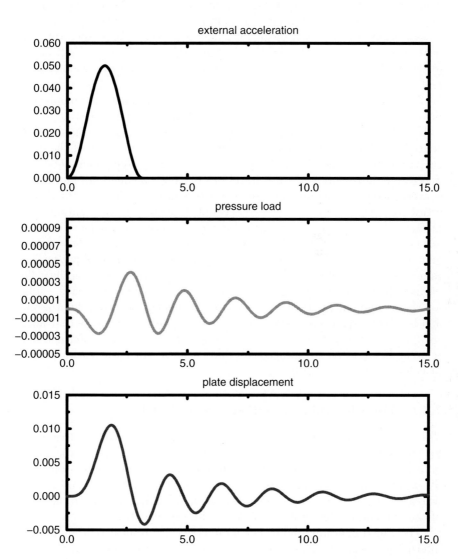

Fig. 7.2. Time series of the external acceleration $f(t)$ (top), the pressure load on the plate (middle), and the plate displacement $a(t)$ (bottom). The parameters were $E = 5000$, $\nu = 0.25$, $q = 0.01$, $L = 1$, $\beta_1 = \beta_2 = 0.5$, $\omega = 1$, $I = 0.05$, $\varrho_S = 0.5$, $\theta = 1$, $\gamma = 0$, $h_0 = 0.2$, $p_0 = 0$, and $\mu = 1.7 \cdot 10^{-5}$.

7.2   Fluid Flow and Heat Conduction in Pipes

This section addresses a coupled fluid-heat flow problem. We consider laminar flow of a non-Newtonian fluid in a straight pipe with a geometrically arbitrary cross section. The viscosity of the fluid may depend on the temperature, and heat generation by internal friction in the fluid is an effect we shall include in the model. Possible applications of such a model cover extrusion of metal and flow of highly viscous polymers.

The present flow problem posed in a general 3D geometry is complicated, but the restriction of the flow region to a straight pipe makes it possible to introduce substantial simplifications. The resulting model consists of two coupled nonlinear Poisson equations. For some relevant values of the physical parameters the nonlinearities are severe, and many common methods, like Successive Substitution (Picard iteration) or Newton's method, face serious convergence problems. In the case of Newtonian flow, where the fluid properties are constant, we achieve two decoupled linear Poisson equations. This means that the present mathematical model is a nice test problem for numerical solution of a system of PDEs where the coupling between the equations and the nonlinearities can be adjusted by varying the value of physical parameters in the problem. The main purpose of including this fluid-heat flow problem in the present book is for demonstrating how we can make a solver for the compound problem by combining independent solvers for the fluid and heat flow equations, respectively. This is a general technique that makes software development of simulators solving systems of PDEs faster and more reliable. It might be advantageous to have studied the basic software design principles in Chapter 7.1.3.

7.2.1   The Physical and Mathematical Model

The General Mathematical Model. Our derivation of the mathematical model for non-Newtonian pipeflow with thermal effects will make use of the indicial notation, including the summation convention and the comma notation for derivatives. Appendix A.2 explains this effective notation in detail. Readers who are not interested in the derivation of the model, can jump to the listing of the boundary-value problem (7.37)–(7.40) on page 474.

The Newtonian fluid model is characterized by a linear relation between internal forces in the fluid and the fluid motion. More precisely, there is a linear relation between the stress tensor σ_{rs} and the velocity gradients $v_{r,s}$:

$$\sigma_{rs} = -p\delta_{rs} + \mu(v_{r,s} + v_{s,r}) \,. \tag{7.31}$$

Here, p is the pressure and δ_{rs} is the Kronecker delta. This is a constitutive law, which reflects physical properties of a particular fluid. The other equations that enter the complete mathematical model are of general type and apply to "all" fluids. For constant μ the presented equations reduce to the well-known Navier-Stokes equations. If the relation between the stress and the velocity gradients is nonlinear, the fluid is classified as non-Newtonian.

Non-Newtonian fluid flow is a huge topic and may involve complicated large-deformation viscoelastic and viscoplastic models. Nevertheless, many fluids that deviate from the common Newtonian model are satisfactorily described by a conceptually simple extension of (7.31). Instead of treating μ as a constant for the fluid in question, we allow the apparent viscosity to depend on the motion. Introducing a measure $\dot{\gamma}$ of the "intensity" of the deformation in the flow, $\dot{\gamma} = \sqrt{2\dot{\varepsilon}_{rs}\dot{\varepsilon}_{rs}}$, we simply postulate that μ is a function of $\dot{\gamma}$. This gives rise to the family of the so-called *generalized Newtonian* fluid models, which will attract our attention in the rest of this section.

A complete mathematical model for coupled fluid and heat flow involves a mass conservation equation, an equilibrium equation, a constitutive law like (7.31), and an energy equation. Many common non-Newtonian fluids are highly viscous and incompressible. The relevant mass conservation equation then becomes

$$v_{r,r} = 0. \tag{7.32}$$

The equilibrium equation for a continuum takes the form

$$\varrho(v_{r,t} + v_s v_{r,s}) = \sigma_{rs,s} + \varrho b_r, \tag{7.33}$$

where the left-hand side is the density ϱ times the acceleration and the right-hand side reflects the sum of forces on the continuum: stresses and body forces (b_r). A suitable form of the constitutive law for a generalized Newtonian fluid is

$$\sigma_{rs} = -p\delta_{rs} + \mu(v_{r,s} + v_{s,r}), \tag{7.34}$$

where μ depends on

$$\dot{\gamma} = \sqrt{\frac{1}{2}(v_{r,s} + v_{s,r})(v_{r,s} + v_{s,r})} \tag{7.35}$$

and possibly the temperature T. The latter quantity is governed by an energy equation, e.g.,

$$\varrho C_p(T_{,t} + v_k T_{,k}) = \kappa T_{,kk} + \mu\dot{\gamma}^2. \tag{7.36}$$

Here, C_p is the heat capacity at constant pressure, and κ is the coefficient of thermal conduction. The left-hand side reflects the change of internal energy of a small fluid element, which is due to the right-hand side terms involving conduction ($\kappa T_{,kk} \equiv \kappa\nabla^2 T$) and dissipation ($\mu\dot{\gamma}^2$). The latter term represents internal heat generation from the work done by the stresses.

Combining the equations (7.32)–(7.36) yields a coupled system of a non-linear convection-diffusion equation for T and a generalized Navier-Stokes equation for v_i and p. With a suitable Navier-Stokes solver, we could devise a finite element method for this system of PDEs, embedded in an iteration technique for handling the nonlinearities. Nevertheless, our purpose is to derive a simpler mathematical model by restricting the flow geometry to a straight pipe.

The Simplified Mathematical Model. The fundamental simplification of the coupled heat-fluid flow model is that we consider laminar flow in a *straight* pipe. Let the z axis be directed along the pipe. The velocity field is then expected to be $\boldsymbol{v} = w\boldsymbol{k}$, where $\boldsymbol{k}$ is a unit vector in z direction. The equation of continuity, $v_{r,r} = 0$, now immediately gives $\partial w / \partial z = 0$, which implies that $w = w(x, y, t)$. The special form of the velocity field leads to a dramatically simplified equation of motion for the fluid; the Navier-Stokes equations with a nonlinear viscosity model are reduced to a nonlinear scalar PDE $\varrho w_{,t} = \nabla \cdot [\mu \nabla w] + \text{const}$. If the problem is considered as stationary, as we do in the following, w is governed by a Poisson equation $\nabla \cdot [\mu \nabla w] = -\beta$, where β is the constant pressure gradient that drives the flow. Another basic assumption is that T does not vary along the pipe. The simplifications in the heat flow model due to rectilinear flow are not substantial; the convection-diffusion equation in the general case is only reduced to a Poisson equation $\nabla^2 T = f(T, w)$ in the present stationary problem. The equations for w and T are to be solved in a domain Ω which represents the *cross section of the pipe*.

The Boundary-Value Problem. The complete set of differential equations can be written as

$$\nabla \cdot [\mu \nabla w] = -\beta, \tag{7.37}$$

$$\nabla^2 T = -\kappa^{-1} \mu \dot{\gamma}^2, \tag{7.38}$$

$$\mu = \mu_T(T) \mu_w(\dot{\gamma}), \tag{7.39}$$

$$\dot{\gamma} = \sqrt{(w_{,x})^2 + (w_{,y})^2}. \tag{7.40}$$

The simplifications due to flow in a straight pipe also apply if we consider flow in a straight channel, both of Poiseuille type (driven by a pressure gradient β) or Couette type (driven by a moving channel wall). The PDEs are the same, but the number of effective space dimensions and the boundary conditions are different. By allowing some flexibility in setting the boundary conditions, the mathematical model and its computer implementation will be applicable to both pipe and channel flow.

Let $\partial \Omega_1$ be the part of the boundary corresponding to a fixed wall where $w = 0$, let $\partial \Omega_2$ be a wall moving with velocity W_1 in z direction, let $\partial \Omega_3$ be a possible symmetry plane where $\partial w / \partial n = 0$ and $\partial T / \partial n = 0$, let $\partial \Omega_4$ be a wall with fixed temperature $T = 0$, let $\partial \Omega_6$ be another wall with fixed temperature $T = T_1$, and finally let $\partial \Omega_7$ be a wall where a cooling condition

$$-\kappa \frac{\partial T}{\partial n} = h_T(T - T_s) \tag{7.41}$$

applies. Here, h_T is a coefficient that reflects the heat transfer through the channel wall to the pipe's surroundings, which have a constant temperature T_s. When assigning boundary conditions in a flow case, one must recall that we need exactly one condition on w and one condition on T at every point on the boundary.

Constitutive Laws. Some common models for μ_T and μ_w are listed next.

- The *Sisko* model:

$$\mu = \mu_\infty + \mu_0 \dot{\gamma}^{n-1}, \tag{7.42}$$

  where μ_∞ is the viscosity at very high shear rates, μ_0 is a reference viscosity[4], and n is the "power-law exponent", which is usually in the interval $[0.15, 0.6]$. When $\mu_\infty = 0$, this model reduces to the standard power-law model. The choice $n = 1$ and $\mu_\infty = 0$ leads to Newtonian flow with viscosity μ_0.
- The *Cross* model takes the form

$$\mu = \frac{\mu_0}{1 + (\mu_0 \dot{\gamma}/\tau_0)^{1-n}}, \tag{7.43}$$

  where τ_0 is the shear stress level at which the flow undergoes a transition from the Newtonian nature to the power-law region.
- The *Herschel-Bulkley* model is more general:

$$\mu = \begin{cases} \mu_\infty \to \infty, & \tau \leq \tau_0 \\ \tau_0/\dot{\gamma} + \mu_0 \dot{\gamma}^{n-1} \end{cases} \tag{7.44}$$

  where $\tau = 2\mu\dot{\gamma}$ is the effective stress for plastic flow[5] and τ_0 is a critical value of τ for the transition between a rigid-body movement of the fluid ($\mu_\infty \to \infty$) and a modified power-law behavior. Notice that $\tau = 0$ recovers the standard power-law model, whereas $n = 1$ corresponds to a Bingham model.
- For the temperature dependence we may choose

$$\mu_T(T) = \exp\left(-\alpha(T - T_0)\right), \tag{7.45}$$

  which reflects that the viscosity is reduced when the temperature is increased, T_0 being a reference temperature.

Looking at the specific constitutive laws above, we see that the parameters n, α, μ_∞, and τ_0 are central. Setting $n = 1$, $\alpha = 0$, $\tau_0 = 0$, and $\mu_\infty = 0$ results in two decoupled linear PDEs for w and T, where we first can solve for w and then for T. Decreasing n towards zero and increasing α sharpen the coupling and the degree of nonlinearity. The present system of PDEs should therefore be everything from easy to very difficult to solve, depending on the values of n and α in particular.

---

[4] One should notice that μ_0 does not have the dimension of viscosity unless $n = 1$.
[5] τ is similar to the von Mises stress in solid mechanics.

7.2.2 Numerical Methods

Finite Element Formulation. The nonlinear Poisson equations for w and T are straightforwardly solved by a Galerkin finite element method. We set

$$w(x,y) \approx \sum_{j=1}^{m} w_j N_j(x,y), \quad T(x,y) \approx \sum_{j=1}^{m} T_j N_j(x,y),$$

where $N_j(x,y)$ are finite element basis functions in the grid over Ω. Notice that we use m as the number of nodes, and not n as in previous chapters, because n is a standard symbol for the "power-law exponent" in the literature on generalized Newtonian fluids. Multiplying the PDEs by N_i, integrating over Ω, and integrating the second order derivatives by parts, lead to a system of discrete equations on the form

$$F_i^{(w)}(w_1, \ldots, w_m, T_1, \ldots, T_m) = 0, \tag{7.46}$$

$$F_i^{(T)}(w_1, \ldots, w_m, T_1, \ldots, T_m) = 0, \tag{7.47}$$

for $i = 1, \ldots, m$. This is a system of $2m$ coupled nonlinear algebraic equations. The exact expressions for $F_i^{(w)}$ and $F_i^{(T)}$ are given below.

$$F_i^{(w)} \equiv \int_{\Omega} \left(\mu(T, \dot{\gamma}) \nabla w \cdot \nabla N_i - \beta N_i \right) d\Omega,$$

$$F_i^{(T)} \equiv \int_{\Omega} \left(\nabla T \cdot \nabla N_i - \kappa^{-1} \mu(T, \dot{\gamma}) \dot{\gamma}^2 \right) d\Omega + \int_{\partial \Omega_7} h_T (T - T_s) N_i d\Gamma.$$

Solution of Nonlinear Algebraic Equations. When μ depends on $\dot{\gamma}$ or on T, the algebraic equations $F_i^{(w)} = 0$ and $F_i^{(T)} = 0$ are nonlinear. There are two different basic strategies for solving these equations: either (i) solve the $F_i^{(w)} = 0$ and $F_i^{(T)} = 0$ equations in sequence with an outer iteration[6], or (ii) apply a standard nonlinear solution method, like the Newton-Raphson method or Successive Substitutions (Picard iteration) to the compound system $(F_i^{(w)} = 0, F_i^{(T)} = 0)$, and solve for w_i and T_i simultaneously. Strategy (ii) is often referred to as a *fully implicit approach*, whereas strategy (i) will be denoted as *Gauss-Seidel* or *Jacobi* iteration on the PDE level. To see why the names Gauss-Seidel and Jacobi are natural[7], we write the algorithm associated with strategy (i) in more detail. Let q be an iteration parameter. Quantities with superscript q denote approximations in the qth iteration. The Gauss-Seidel procedure can then be expressed as in Algorithm 7.1.

---

[6] This is also referred to as an operator-splitting technique.

[7] We refer to Appendix C.1 for an introduction to the ideas of Gauss-Seidel and Jacobi iteration for solving systems of (linear) equations.

Algorithm 7.1.

Gauss-Seidel-type method for systems of nonlinear PDEs.

for $q = 1, 2, \ldots$ until convergence
  solve $F_i^{(w)}(w_1^q, \ldots, w_m^q, T_1^{q-1}, \ldots, T_m^{q-1}) = 0$
    with respect to $(w_1^q, \ldots, w_m^q)$, $i = 1, \ldots, m$
  solve $F_i^{(T)}(w_1^q, \ldots, w_m^q, T_1^q, \ldots, T_m^q) = 0$
    with respect to $(T_1^q, \ldots, T_m^q)$, $i = 1, \ldots, m$

In other words, we first solve (7.37) with respect to w, using the most recently computed T_i values in the formulas for μ. Thereafter we solve (7.38) with respect to T, using the most recently computed w_i values in the nonlinear term on the right-hand side.

The equation for w is still nonlinear and can be solved by, e.g., a Newton-Raphson method or Successive Substitutions. Jacobi's method is similar to the Gauss-Seidel approach, except that we use the old w_i^{q-1} values when solving for T in iteration q, see Algorithm 7.2.

Algorithm 7.2.

Jacobi-type method for systems of nonlinear PDEs.

for $q = 1, 2, \ldots$ until convergence
  solve $F_i^{(w)}(w_1^q, \ldots, w_m^q, T_1^{q-1}, \ldots, T_m^{q-1}) = 0$
    with respect to $(w_1^q, \ldots, w_m^q)$, $i = 1, \ldots, m$
  solve $F_i^{(T)}(w_1^{q-1}, \ldots, w_m^{q-1}, T_1^q, \ldots, T_m^q) = 0$
    with respect to $(T_1^q, \ldots, T_m^q)$, $i = 1, \ldots, m$

The attractive feature of the Jacobi or Gauss-Seidel iteration approach to the nonlinear problem is that we only need to solve standard scalar PDEs. The fully implicit approach, on the contrary, requires us to consider a system of two PDEs with two unknowns per node, resulting in nonlinear algebraic equations in $2m$ unknowns.

Let us explain the details of the Newton-Raphson method applied to the fully implicit system of nonlinear algebraic equations. At each node i we have two equations, $F_i^{(w)} = 0$ and $F_i^{(T)} = 0$, and two unknowns w_i and T_i. The total system has $2m$ equations and $2m$ unknowns. We order the equations as follows.

$$F_1^{(w)} = 0, F_1^{(T)} = 0, \ F_2^{(w)} = 0, F_2^{(T)} = 0, \ldots, F_m^{(w)} = 0, F_m^{(T)} = 0. \quad (7.48)$$

The corresponding numbering of the unknowns is

$$(w_1, T_1, w_2, T_2, \ldots, w_m, T_m)^T. \quad (7.49)$$

This numbering gives smaller bandwidth compared with stacking together all the w equations and unknowns first, followed by all the T equations and unknowns, see the next exercise.

Exercise 7.8. As an alternative to the numbering of equations and unknowns in (7.48)–(7.49) we consider

$$F_1^{(w)} = 0, F_2^{(w)} = 0, \ldots, F_m^{(w)} = 0, \ F_1^{(T)} = 0, \ F_2^{(T)} = 0, \ldots, F_m^{(T)} = 0\,.$$
(7.50)

and

$$(w_1, w_2, \ldots, w_m, T_1, T_2, \ldots, T_m)^T\,.$$
(7.51)

Find the bandwidth of the associated coefficient matrix for each of the two numbering strategies (7.48)–(7.49) and (7.50)–(7.51) by considering a 2D lattice domain with bilinear elements and a line-by-line node numbering. You can assume for simplicity that the equations $F_i^{(w)} = 0$ and $F_i^{(T)} = 0$ are linear, although the same reasoning can easily be extended to the nonlinear case as well.   ◇

From each node i we get a 2×2 linear system that is to be included in the element matrix and vector and thereafter assembled into the global $2m \times 2m$ system. In the Newton-Raphson method the local 2×2 system reads

$$\begin{pmatrix} \frac{\partial F_i^{(w)}}{\partial w_j} & \frac{\partial F_i^{(w)}}{\partial T_j} \\ \frac{\partial F_i^{(T)}}{\partial w_j} & \frac{\partial F_i^{(T)}}{\partial T_j} \end{pmatrix} \begin{pmatrix} \delta w_j^q \\ \delta T_j^q \end{pmatrix} = \begin{pmatrix} -F_i^{(w)} \\ -F_i^{(T)} \end{pmatrix}\,.$$
(7.52)

In the matrix and the right-hand side we evaluate the expressions using old values, w^{q-1} and T^{q-1}:

$$F_i^{(w)} \equiv \int_\Omega \left(\mu_T(T^{q-1})\mu_w(\dot\gamma^{q-1})\nabla w^{q-1} \cdot \nabla N_i - \beta N_i \right) d\Omega,$$
(7.53)

$$F_i^{(T)} \equiv \int_\Omega \left(\nabla T^{q-1} \cdot \nabla N_i - \kappa^{-1}\mu_T(T^{q-1})\mu_w(\dot\gamma^{q-1})(\dot\gamma^{q-1})^2 \right) d\Omega$$

$$+ \int_{\partial\Omega_7} h_T(T^{q-1} - T_s)N_i d\Gamma\,.$$
(7.54)

In the case where $\mu_w(\dot\gamma) = \dot\gamma^{n-1}$, the derivatives become as follows.

$$\frac{\partial F_i^{(w)}}{\partial w_j} = \int_\Omega \left(\mu_T(T^{q-1})(n-1)\mu_w(\dot\gamma^{q-1})^{n-2} \frac{\partial \dot\gamma}{\partial w_j} \nabla w^{q-1} \cdot \nabla N_i + \right.$$

$$\left. \mu_T(T^{q-1})\mu_w(\dot\gamma^{q-1})^{n-1} \nabla N_j \cdot \nabla N_i \right) d\Omega,$$

$$\frac{\partial F_i^{(w)}}{\partial T_j} = \int_\Omega \left(\frac{\partial \mu_T}{\partial T_j} \mu_w (\dot{\gamma}^{q-1})^{n-1} \nabla N_i \cdot \nabla w^{q-1} \right) d\Omega,$$

$$\frac{\partial F_i^{(T)}}{\partial w_j} = -\int_\Omega \left(\kappa^{-1} \mu_T (T^{q-1})(n+1)\mu_w(\dot{\gamma}^{q-1})^n N_i \frac{\partial \dot{\gamma}}{\partial w_j} \right) d\Omega,$$

$$\frac{\partial F_i^{(T)}}{\partial T_j} = \int_\Omega \left(\nabla N_i \cdot \nabla N_j - \kappa^{-1} \frac{\partial \mu_T}{\partial T_j} \mu_w (\dot{\gamma}^{q-1})^{n+1} N_i \right) d\Omega$$

$$+ \int_{\partial\Omega_7} h_T N_i N_j d\Gamma,$$

$$\frac{\partial \dot{\gamma}}{\partial w_j} = \mu_w(\dot{\gamma}^{q-1})^{-1} \nabla N_j \cdot \nabla w^{q-1},$$

$$\frac{\partial \mu_T}{\partial T_j} = -\alpha \mu_T (T^{q-1}) N_j.$$

In the Successive Substitution (Picard iteration) method we also have a 2×2 system at each node:

$$\begin{pmatrix} A_{ww} & A_{wT} \\ A_{Tw} & A_{TT} \end{pmatrix} \begin{pmatrix} w_j^q \\ T_j^q \end{pmatrix} = \begin{pmatrix} b_w \\ b_T \end{pmatrix}, \tag{7.55}$$

where

$$A_{ww} \equiv \int_\Omega \mu_T(T^{q-1})\mu_w(\dot{\gamma}^{q-1}) \nabla N_i \cdot \nabla N_j \, d\Omega,$$

$$b_w \equiv \beta \int_\Omega N_i d\Omega,$$

$$A_{TT} \equiv \int_\Omega \nabla N_i \cdot \nabla N_j \, d\Omega + \int_{\partial\Omega_7} h_T N_i N_j \, d\Gamma,$$

$$b_T \equiv \int_\Omega \kappa^{-1} \mu_T(T^{q-1})\mu_w(\dot{\gamma}^{q-1})(\dot{\gamma}^{q-1})^2)d\Omega + \int_{\partial\Omega_7} h_T N_i T_s d\Gamma,$$

$$A_{wT} \equiv 0,$$
$$A_{Tw} \equiv 0.$$

(We remark that the result $A_{wT} = A_{Tw} = 0$ is not a general property of this method.)

A Continuation Method. We know that $n = 1$ is an easy problem to solve, whereas convergence problems are expected as n approaches zero or n is significantly greater than unity. This points in the direction of formulating a continuation method, see Chapter 4.1.8, using $\lambda = (1-n)/(1-n_p)$ as continuation parameter, with n_p being the target value of n for the computations. By defining a set of proper values $\lambda_0 = 0 < \lambda_1 < \cdots < \lambda_p = 1$ of λ, and using the solution obtained with λ_{i-1} as initial guess for the nonlinear solvers in the problem corresponding to λ_i, we might hope to establish convergence for small n values. The α parameter can be used as continuation parameter in a similar way.

Remark. From the theory and practice of iterative methods for linear systems it is known that Jacobi's method is generally slower than Gauss-Seidel iteration. This is intuitively expected in the present nonlinear situation as well, since the Gauss-Seidel algorithm incorporates new approximations as soon as they are available. Nevertheless, when applying these iterative strategies at the PDE level, Jacobi iteration sometimes have important advantages with respect to conservation properties of the PDEs. For example, a Jacobi method conserves mass in multi-phase reactive flow problems. This is occasionally a fundamental property of the numerical formulation, although it is not of that importance in the present relatively simple flow case.

7.2.3 Implementation

Looking back at the mathematical and numerical model, there are several open questions regarding the choice of constitutive laws and nonlinear iteration strategies. The influence of the element type and preconditioning strategies for the linear systems is also not known. This calls for *flexibility* in the implementation, like we have emphasized many other places in this book. More specifically, a flexible simulation tool must deal with the following aspects of the present problem.

- It must be easy to switch between the Gauss-Seidel, Jacobi, or fully implicit solution strategies.
- The nonlinear algebraic equations in the inner iterations of the Gauss-Seidel, Jacobi, or fully implicit methods must be solved by either Newton-Raphson iteration or Successive Substitutions.
- Several methods must be available for solving the linear systems arising in each nonlinear iteration.
- Any combination of solution strategies for nonlinear and linear equations must be easily available at run time.
- The implementation must work for any finite element grid with any isoparametric element.
- It should be easy to redefine boundary indicators such that the solver can also handle channel flow of Poiseuille or Couette type. Analytical solutions are known for these flow cases and will therefore constitute an important tool in the verification of the implementation.

The purpose of the present section is to introduce a software design for the coupled heat-fluid flow simulator that meets the flexibility requirements listed above. Despite the fact that the present problem is a quite simple coupled problem, the basic software design principles are general and applicable to much more complicated systems of PDEs. The material to be presented constitute a further development and improvement of the ideas from [21].

We assume that the reader is familiar with standard finite element-based PDE solvers in Diffpack. Our basic idea for the present simulator is to develop independent standard Diffpack solvers for the momentum and energy equations and then couple these solver classes. This will work when the PDE system is solved by Gauss-Seidel or Jacobi iteration strategies. We follow the design from Chapter 7.1 and emphasize the possibility to pull the classes apart at any time such that we can check that each PDE in the system is correctly solved when its coefficients are not coupled to other equations. Moreover, we present a way of programming constitutive laws and other common relations in a separate module that can be accessed by all PDE solvers in the system. Chapter 3.5.6 provides valuable background information for the design issues discussed below.

The source code of the coupled heat-fluid flow solver is located in the directory `src/app/Pipeflow`.

The Momentum Equation Solver. Equation (7.37) is implemented in a class `Momentum1`, but with a simple prescribed form of μ to assist the debugging and verification of the implementation. Introducing a virtual function `viscosity` for evaluating μ, `Momentum1` can let `viscosity` return a constant, whereas subclasses of `Momentum1` can implement `viscosity` with a call to physically relevant viscosity models. Class `Momentum1` is similar to class `NlHeat1` from Chapter 4.2, but with automatic report generation facilities built into the class. The nonlinear solver in `Momentum1` is a `NonLinEqSolvers` object, which has the same behavior as class `NonLinEqSolver` and its subclasses, but one can switch between different iteration methods within a call to `solve`. For example, one can apply Successive Substitutions for the first iterations and then switch to Newton-Raphson for hopefully faster convergence when the approximate solution is sufficiently close to the exact solution. A `NonLinEqSolvers` object typically contains an array of `NonLinEqSolver` handles; here we use two such handles, pointing to a `NewtonRaphson` and a `SuccessiveSubst` object.

The `NonLinEqSolver` class offers a function `continuationSolve` that implements the continuation method in Algorithm 4.3 on page 351. We make use of this functionality in the `Momentum1` class. The `NonLinEqSolver_prm` object can read a set of continuation parameters from the menu, and if there are more than one parameter, the `solve` function in `NonLinEqSolver` calls the `continuationSolve` function. Each time this latter function invokes an ordinary nonlinear solve phase, it first calls a virtual function

```
void beforeSolveInContinuationMethod (real lambda, int niter);
```

in the simulator class. The purpose of this function is to use the information about the current value of the continuation parameter, $\Lambda \in [0, 1]$ (`lambda`), for adjusting the corresponding physical continuation parameter in the simulator prior to computing the linear systems in each nonlinear iteration. Continuation methods are hence available from the `NonLinEqSolver` tool by just writing a small additional function for linking $\Lambda \in [0, 1]$ to a physical parameter (here $n = 1 - (1 - n_p)\Lambda$). The `niter` parameter reflects the current number of iterations in the continuation method itself.

Class `Momentum2`, which is a subclass of `Momentum1`, implements a new version of the `viscosity` function, where a real generalized Newtonian viscosity model is used: $\mu = \mu_w(\dot{\gamma})\mu_T(T)$. To this end, `Momentum2` needs to access the temperature field. We will come back to the details on how this is achieved technically. One could also think of several stepwise refinements of class `Momentum1`, for example, first making a subclass for a prescribed variable coefficient μ, then a subclass for a nonlinear μ, before the coupling to the real physical viscosity model is realized. The purpose of each step is to create test problems of increasing complexity as this will aid the debugging of the final momentum equation solver. We refer to Chapter 3.5.6 for a detailed explanation of the ideas of making a class hierarchy for solving various versions of $\nabla \cdot [\mu \nabla w] = -\beta$.

The energy equation solver is also realized as a class hierarchy, with class `Energy1` as a stand-alone solver for $\nabla^2 T = f$, $f = $ const, and subclass `Energy2` with the physically relevant $f = \dot{\gamma}^2 \mu_w(\dot{\gamma})\mu_T(T)$.

Software Components for the Coupled System of PDEs. The PDEs are coupled through the coefficients in the equations. We represent these coefficients by virtual functions taking a `FiniteElement` object as argument, e.g.,

```
void Momentum1::viscosity (const FiniteElement& fe);
```

With the `fe` variable at hand we can perform evaluation of explicit formulas as well as efficient interpolation in Diffpack's field objects (see pages 249 and 332 for more information).

The base class solvers have simple versions of the virtual coefficient functions, often corresponding to constant coefficients, such that the solver can be tested separately from the other solvers. Subclasses implement more complicated forms of the variable coefficients and couple the coefficients to other solvers. This coupling can be accomplished by calling a module that holds formulas for μ_w and μ_T.

The coefficients in the original PDEs involve in general a set of constitutive laws that are common to several PDEs. This is the case in the present problem, where both PDE solvers need the quantities $\dot{\gamma}$, $\mu_w(\dot{\gamma})$, and $\mu_T(T)$. All common relations for a system of PDEs can be collected in a separate class, here called `CommonRel`. This class contains physical parameters like μ_0, μ_∞, T_0, α, n, and τ_0, which are initialized using the menu system. In addition, class `CommonRel` needs to interpolate and store the values of w, $\dot{\gamma}$, T, $\nabla \dot{\gamma}$, and so on, to avoid unnecessary recomputation of mathematical expressions. A function `tabulate` performs this task and must be called prior to functions for calculating $\mu_w(\dot{\gamma})$ and $\mu_T(T)$. The latter functions are virtual in class `CommonRel` and can be redefined in subclasses that implement different physical viscosity models. Class `CommonRel` implements the widely used power-law viscosity model, in the slightly extended Sisko form (7.42). A subclass `Cross` implements the Cross model (7.43). Class `CommonRel` is sketched below.

```
class CommonRel : public HandleId
{
public:
    real mu_0;              // μ₀
    real mu_inf;           // μ∞
    real T_0;              // T₀
    real alpha;            // α
    real n;                // power law exponent n
    real tau;              // τ₀
    Handle(FieldFE) w, T;  // w(x,y) and T(x,y)

    // tabulated values at a point:
    real w_pt, T_pt;       // w and T
    Ptv(real) dw_pt;       // ∇ w
    real gamma_pt;         // γ̇
    real gamma_n;          // γ̇ⁿ
    real gamma_nm1;        // γ̇ⁿ⁻¹
    Vec(real) grad_gamma_dot_gradN;   // ∇γ̇ · ∇Nⱼ

    virtual void tabulate (const FiniteElement& fe);

    virtual real muw () const;   // fast evaluation of μ_w
    virtual real muT () const;   // fast evaluation of μ_T

    real viscosity () const    { return muw()*muT(); }
    real dissipation () const  { return viscosity()*sqr(gamma_pt); }
    ...
};
```

Class `CommonRel` is derived from `HandleId` because we want to access `CommonRel` objects through a handle (cf. page 84).

Newton-like methods require differentiation of the quantities in the constitutive relations with respect to the nodal unknowns w_j and T_j. In a general case with a nonlinear function $f(u, u_{,1}, u_{,2}, u_{,3})$, $u \approx \sum_{j=1}^{n} N_j u_j$, one needs

$$\frac{\partial}{\partial u_j} f(u, u_{,1}, u_{,2}, u_{,3}) = \frac{\partial f}{\partial u} N_j + \sum_{k=1}^{d} \frac{\partial f}{\partial u_{,k}} N_{j,k},$$

which is conveniently implemented in a function returning the vector $\partial f / \partial u_j$ at an evaluation point inside the element. Here, $j = 1, \ldots, n_e$, where n_e is the number degrees of freedom of u in an element. Class `CommonRel` hence offers the functions `dmuw_dwj` for $\partial \mu_w / \partial w_j$ and `dmuT_dTj` for $\partial \mu_T / \partial T_j$.

```
class CommonRel : public HandleId
{
public:
  ...

  virtual void dmuw_dwj (Vec(real)& ddwj, const FiniteElement& fe);
  virtual void dmuT_dTj (Vec(real)& ddTj, const FiniteElement& fe);

  void viscosity_dwj (Vec(real)& ddwj, const FiniteElement& fe)
    { dmuw_dwj(ddwj,fe); ddwj.mult(muT()); }

  void viscosity_dTj (Vec(real)& ddTj, const FiniteElement& fe)
    { dmuT_dTj(ddTj,fe); ddTj.mult(muw()); }
  ...
};
```

So far we have developed separate solvers for the momentum and energy equation, in addition to a common pool of relations that are needed in both PDEs. The final step is to make a class `Manager`, which holds the momentum and energy equation solvers, the common data structures (grid, linear system), and a `CommonRel` object. The main purpose of class Manager is to administer the whole solution process.

```
class Manager : public SimCase
{
public:
  Handle(Energy2)        energy_eq;
  Handle(Momentum2)      momentum_eq;
  Handle(CommonRel)      constrel;

  // common data structures for the momentum and energy eq solvers:
  Handle(GridFE)         grid;
  Handle(SaveSimRes)     database;
  Handle(LinEqAdmFE)     lineq;

  virtual void scan ();
  virtual void define (MenuSystem& menu, int level=MAIN);
  virtual void solveProblem ();
  virtual real solveThisIteration ();
  ...
};
```

The `define` and `scan` functions put the local data, like `grid`, `database`, `lineq`, and parameters for the outer Gauss-Seidel/Jacobi nonlinear iteration, on the menu and initializes these data structures. Thereafter the `define` and `scan` functions call the `define` and `scan` functions in the momentum and energy equation solvers. The `solveProblem` function implements the nonlinear outer iteration method. This is basically a loop with calls to `solveThisIteration`. The latter function calls up `solve` functions in the momentum and energy equation solvers (in sequence) to compute new w and T fields. The `solve` functions have the same purpose as `solveProblem` in the a stationary solver like `Poisson1` in Chapter 3.2, but the current solution should not be dumped to file, because we do not yet know if the solution has converged.

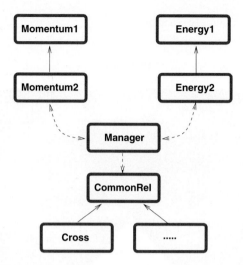

Fig. 7.3. Relation between solver classes and the pool of common relations in the simulator for coupled heat and fluid pipeflow. Solid arrows indicate inheritance ("is-a" relationship). Dashed arrows indicate pointers ("has-a" relationship).

The Momentum1 solver applies boundary indicators 1 and 2 for indicating the conditions $w = 0$ and $w = W_1$. Also in the Energy1 solver, indicators 1 and 2 are used to set Dirichlet conditions, here $T = 0$ and $T = T_1$. A major problem is now that these two classes are supposed to use the same grid, but the boundary indicators for w and T might differ. A solution is to use two Indicators objects in the Manager class, one for the w conditions and one for the T conditions. The manager then attaches the w indicators to the grid before calling the velocity solver and attaches the T indicators to the grid before computing the temperature. On the menu the manager offers two items for redefining the original grid boundary indicators to the proper indicators required by the Momentum1 and Energy1 solvers. The source code of class Manager provides further details.

An overview of the various classes in the heat-fluid flow simulator is presented in Figure 7.3. We are now able to explain how Momentum2 can override the viscosity function in order to compute a physically relevant expression for μ:

```
class Momentum2 : public Momentum1
{
  Manager* manager;
  virtual real viscosity (const FiniteElement& fe)
    { return manager->constrel->viscosity(); }
  ...
};
```

velocity

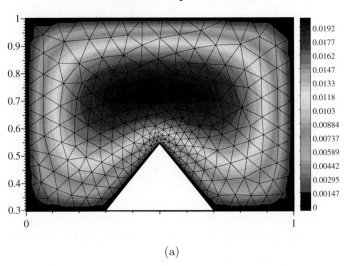

(a)

temperature

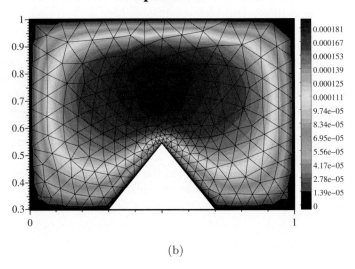

(b)

Fig. 7.4. Coupled heat and fluid flow in a straight pipe. (a) $w(x,y)$ in a cross section; (b) $T(x,y)$ in a cross section. A power-law viscosity model with $n = 0.8$ and $\alpha = 0$ was used.

The Fully Implicit Simulator. It would be advantageous to combine independent solvers for each PDE also in the fully implicit case, but this is technically more difficult. However, using the Diffpack module for generalized and mixed finite element methods, the previously described software design can be applied also to the fully implicit solver, but we will not present the details here. Instead, we realize the fully implicit solution method in class `FullyImplicit`, where the two PDEs are tightly coupled also in the implementation. Recall that the nonlinear algebraic equation system has two unknowns at node no. i (w_i and T_i). The related book-keeping is managed by the `DegFreeFE` object, much in the same way as we did in the elasticity solver in Chapter 5.1 and the Navier-Stokes solver in Chapter 6.3. The `integrands` function in class `FullyImplicit` builds the element matrix and vector in terms of blocks as explained in Chapters 5.1 and 6.3. The reader is encouraged to study the numerics of the fully coupled formulation and the corresponding implementation of the `integrands` function. The rest of class `FullyImplicit` is some kind of a sum of class `Momentum1-2` and `Energy1-2`. Class `CommonRel` is of course a valueable tool also in the fully implicit solver.

Without doubt, it is considerably more difficult to code and especially debug class `FullyImplicit` compared to the more modular approach of the classes `Momentum`, `Energy`, and `Manager`. An effective implementation strategy for a fully implicit solver is therefore to first establish a working sequential solver using the component-based software design described in this chapter and then utilize parts of the code and results from test problems when developing a separate implicit simulator.

The source code of the pipeflow simulator employs some constructs that are not completely described in this book. However, the material here gives an overview of the problem and a motivation for the basic features of the software design. Remaining details can be looked up in the man pages. We recommend readers who are interested in a modular approach for solving system of PDEs to study the source code of the pipeflow simulator in detail.

7.3   Projects

7.3.1   Transient Thermo-Elasticity

Mathematical Problem. The current project aims at solving the PDE system (3.35) and (5.10), where the u in (3.35) is identical to the temperature T in (5.10).

Physical Model. The elasticity solver in Chapter 5.1 is able to handle the effect of a general thermal "load" $T(\boldsymbol{x})$. A time-dependent temperature field $T(\boldsymbol{x}, t)$ is also allowed in the formulation of the governing equations, as long as the time rate of change of T is not so fast that accelerations are induced in the elastic medium. In general, the temperature field $T(\boldsymbol{x}, t)$ must be found from a heat equation. The aim of this project is therefore to couple the elasticity solver `Elasticity1` from Chapter 5.1 with the heat equation solver `Heat2` from Chapter 3.10, using the software techniques of Chapters 7.1 or 7.2. The result is a *transient* thermo-elastic solver.

We assume that the (rate of) elastic deformations do not modify the heat equation, such that (3.35) is a sufficient model for the temperature evolution. That is, (5.10) depends on (3.35), but (3.35) is independent of (5.10). A sequential solution approach is hence exact.

Implementation. The coupling of the classes `Elasticity1` and `Heat2` can follow the recipe for coupling in Chapter 7.1.3. Notice that the situation is simpler than in Chapter 7.2 as there is no need for a common pool of fields and constitutive relations; instead the `T` handle in class `Elasticity1` can just be bound to the `u` field in class `Heat2`. Create an administering class that has an `Elasticity1` and a `Heat2` solver as members. At each time level, we first compute the temperature field in the `Heat2` solver and then compute the displacement field and stresses in the `Elasticity1` solver. It is advantageous if the managing class distributes a common grid and a `SaveSimRes` object to the two solvers (i.e. the `define` and `scan` functions in classes `Elasticity1` and `Heat2` should be generalized as demonstrated in Chapters 7.1.3 and 7.2.3).

The elasticity solver needs to be equipped with a `Handle(TimePrm)` object, pointing to `Heat2`'s time data, for marking deformation and stress fields in the simres database with the proper time value. The fieldname of `u` in class `Heat2` should be altered to, e.g., `temperature` in `Heat2::scan` to distinguish it from `Elasticity1`'s `u` field in the database. Class `Heat2` also need to implement the functions `timeLoopSetUp` and `timeLoopFinish` as explained for the `Squeeze1` class in Chapter 7.1. The `Elasticity1` and `Heat2` solvers apply different conventions for the boundary indicators, and the ideas on page 484 can be applied to overcome this problem. The report generation facilities in classes `Elasticity1` and `Heat2` can be combined to a single report by following the source code of class `Manager` in the pipeflow simulator from Chapter 7.2.

A basic test problem for verification of the implementation is to consider a box or rod with no normal displacement at two ends and the other sides free of stress. Develop a formula for the stress in the body as a function of a uniform temperature. Arrange the temperature boundary conditions such that the temperature becomes constant in space, but linearly varying in time. (The solvers should then reproduce stress and temperature variations exactly.)

Another suitable test problem is transient thermal loading of a hollow cylinder. The elastic solution for a general temperature variation is listed in [118, Ch. 13]. The temperature field must be found from solving the transient heat equation in radial coordinates analytically[8], with some appropriate initial and boundary conditions.

---

[8] The solution involves Bessel functions, but fortunately the standard C math library offers some Bessel functions, see `math.h`.

Computer Experiments. Consider the welding model as explained in Example 3.1 on page 312, now coupled with elastic deformations. Suggest a suitable geometry, assign stress free conditions at all surfaces of the body, make a scripting interface for efficient handling of simulation and visualization, and demonstrate how the parameters in the welding model from Example 3.1 affect the stress picture in the elastic material.

7.3.2   Convective-Diffusive Transport in Viscous Flow

Mathematical Problem. This project concerns computation of convective-diffusive transport, governed by (6.1), in a fluid whose motion is computed from the Navier-Stokes equations (6.50). That is, we shall solve the system (6.50) and (6.1), where the former couples to the latter through the velocity field v.

Implementation. The coupling can be realized by combining the CdBase solver from Chapter 6.1 and the NsPenalty1 solver from Chapter 6.3, using the software methodology from Project 7.3.1. As the present coupling is almost identical to the one in Project 7.3.1, we recommend to study the text in that project and understand the software details before carrying out the current project. We assume that the velocity field is time-dependent such that both equations must be solved at each time level. (If the velocity field is stationary, it is easier to compute this by the NsPenalty1 solver and load the field into the CdBase solver.)

As test problem for program verification one can try channel flow in a 2D geometry $[0,1] \times [-1,1]$, where the concentration u of a specie in the flow is 0 initially, but with $u = 0.1$ at the inlet boundary for all times $t > 0$. Without diffusion, the specie will be passively transported along the streamlines, that is, $u(x, y, t) = 1 - H(x - U_0(1 - y^2)t)$, where $U_0(1 - y^2)$ is the inlet velocity profile and H is the Heaviside function. (We remark that it will be hard to approximate $u(x, y, t)$ well).

7.3.3   Chemically Reacting Fluid

Mathematical Problem. The current project considers the temperature distribution in a chemically reacting fluid. A simplified mathematical model, where a chemical specie with concentration a is transformed into another specie, consists of a system of two reaction-diffusion equations:

$$\frac{\partial a}{\partial t} = D\nabla^2 a - sa^m \exp\left(-\frac{E}{RT}\right), \tag{7.56}$$

$$\varrho C_p \frac{\partial T}{\partial t} = \kappa\nabla^2 T + Qsa^m \exp\left(-\frac{E}{RT}\right). \tag{7.57}$$

The functions $a(\boldsymbol{x}, t)$ and $T(\boldsymbol{x}, t)$ are primary unknowns, while s, m, E, R, D, ϱ, C_p, κ, and Q are viewed as prescribed constants.

Physical Model. Equation (7.56) reflects conservation of the specie mass, where $D\nabla^2 a$ models transport by diffusion and the last term models mass reduction due to chemical reactions. Equation (7.57) is a standard energy equation where the effect of heat generation from chemical reactions is taken into account. The interpretation of the constants in the model is as follows: m is the order of the reaction, E is the activation energy, R is the universal gas constant, s is an adjustable constant, D is the diffusion coefficient of the chemical specie (according to Fick's law), ϱ is the density of the fluid, C_p is the heat capacity of the fluid, κ is the heat conduction coefficient in the fluid (according to Fourier's law), and Q is the heat released by the chemical reactions. A brief description of the model can be found in [72, p. 238].

We remark that (7.56) and (7.57) can be extended to mass and energy transport in a flowing fluid by including appropriate convection terms. If the flow field is unknown, the equations for a and T must be coupled to the Navier-Stokes equations. Using the suggested software design and a sequential solution method, these extensions of the model are easily accomplished. One will often also include more than one substance. This gives rise to a series of equations like (7.56) for S substances $a_1, \ldots, a_S$. Such extensions are also quickly incorporated in our suggested design of the simulator.

Numerical Method. The system of PDEs can be discretized by a finite element method in space and a θ-rule in time. Develop complete expressions for the integrands in the element matrices and vectors, using both a squential and a fully implicit method (follow the ideas in Chapter 7.2.2). Use the Successive Substitution method for solving the nonlinear systems of equations.

Implementation. The implementation can make use of the solvers described in Chapter 7.2.3. The modifications consists in adding functionality for the time-dependency and changing the `integrands` functions. In the sequential solution approach, we make separate solvers for (7.56) and (7.57), and each of these can be tested separately as explained in Chapter 7.2.3. The source terms in (7.56) and (7.57) are conveniently implemented in a `CommonRel` class such that it is easy at a later stage to include additional models for the chemical reactions. Also make a fully implicit solver for the two PDEs.

Computer Experiments. Limit the computational study to a rectangular 2D domain with $\partial a/\partial n = \partial T/\partial n = 0$ on the boundary. The initial concentration a can be taken as a Gaussian bell function (exploit symmetry and place the center of the bell at the lower left corner of the domain). The initial temperature can be constant. Try to assess the impact of the parameters m, E, κ, and D. Investigate the relative efficiency and stability of the fully implicit versus the sequential solution strategy.

Appendix A

Mathematical Topics

A.1 Scaling and Dimensionless Variables

Initial-boundary value problems arising from physical problems frequently contain many parameters. By introducing dimensionless independent and dependent variables, the number of physical parameters can often be reduced because only certain combinations of the parameters appear in the resulting equations. Moreover, by finding the right scales when converting variables to dimensionless form, it is possible to precisely identify the relative size of the various terms in the equations. This is crucial for simplifying models by omitting terms or invoking special approximation techniques such as perturbation methods. Successful scaling relies on physical understanding of the problem in question, although the technical steps of the scaling procedure are simple. These steps are briefly explained in the following. More information about scaling can be found in [69, Ch. 6.3], [41, Ch. 2], or [71, Ch. 1.3].

Scaling a Two-Point Boundary-Value Problem. Our first example concerns pressure-driven steady viscous flow between two flat plates, $x = a$ and $x = b$. Let $u(x)$ be the velocity in the direction of the planes, let β be the magnitude of the pressure gradient, and let μ denote the viscosity of the fluid. From the Navier-Stokes equations one can derive the following model for $u(x)$:

$$\frac{d^2 u}{dx^2} = -\frac{\beta}{\mu}, \quad u(a) = u(b) = 0. \tag{A.1}$$

The scaling consists of introducing dimensionless independent and dependent variables. In general, a quantity q is made dimensionless by

$$\bar{q} = \frac{q - q_0}{q_c},$$

where q_0 is a characteristic reference value and q_c is a characteristic magnitude of $q - q_0$. The ultimate goal of the scaling is to obtain a unit magnitude of $\bar{q}$. Physical insight into the problem is usually required to find the right scale q_c. In the present problem we can introduce

$$\bar{x} = \frac{x - a}{b - a}, \quad \bar{u} = \frac{u}{u_c},$$

where u_c is the (unknown) maximum velocity or average velocity. Sometimes it can be convenient to just perform the scaling and postpone the precise estimation of some scales. Noting that

$$\frac{du}{dx} = \frac{d(u_c \bar{u})}{d\bar{x}} \frac{d\bar{x}}{dx} = \frac{u_c}{b-a} \frac{d\bar{u}}{d\bar{x}},$$

we can derive the dimensionless version of the boundary-value problem (A.1).

$$\frac{d^2 \bar{u}}{d\bar{x}^2} = -\alpha, \quad \alpha = \beta \frac{(b-a)^2}{\mu u_c}, \quad \bar{u}(0) = \bar{u}(1) = 0, \tag{A.2}$$

where α is a dimensionless parameter.

How can we estimate u_c? The easiest way in the present problem is to compute the exact solution of (A.1): $u(x) = \beta(x-a)(b-x)/(2\mu)$. The maximum value of $u(x)$ appears at the mid point $x = (a+b)/2$, which results in $u_c = \beta(b-a)^2/(8\mu)$, or $\alpha = 8$. In more complicated problems, we can hope to find a simple or approximate solution of the PDE. This solution does not necessarily need to fulfill all the initial or boundary conditions, but it must give enough information to reflect the proper scales. Such an approach is demonstrated in the scaling of the heat equation to be presented later. Another strategy for determining the scales relies on known maximum principles for certain PDEs. For example, the solution of the scaled problem (A.2) has the following property see [120, Ch. 1]: $\sup_{\bar{x} \in [0,1]} |\bar{u}(\bar{x})| \leq \alpha/8$. The inequality is sharp so we may conclude that a unit maximum value of $\bar{u}$ corresponds to taking $\alpha = 8$. With $\alpha = 8$ we have the velocity scale $u_c = \beta(b-a)^2/(8\mu)$.

What has actually been achieved by the scaling in (A.2)? The original problem involved four physical input parameters to the problem: a, b, β, and μ. The solution can therefore be written on the form $u(x; a, b, \beta, \mu)$. To obtain a complete description of this problem, e.g. by numerical experimentation, it is necessary to investigate the $u(x; a, b, \beta, \mu)$ function in a four-dimensional parameter space. The dimensionless version of the boundary-value problem does not involve any physical parameters. A single graph of $\bar{u}(\bar{x})$ contains all information about $u(x; a, b, \beta, \mu)$, because

$$u(x; a, b, \beta, \mu) = \frac{\beta(b-a)^2}{8\mu} \bar{u}\left(\frac{x-a}{b-a}\right).$$

Numerical investigation of the original function $u(x; a, b, \beta, \mu)$, by e.g. letting each of the parameters a, b, β, and μ vary on ten levels, requires 10,000 computer experiments. A single experiment with the scaled problem produces the same (and much more) information in a dramatically clearer way.

We remark that after the scaling is carried out, it is common to omit the bars (or other labels) in the dimensionless quantities. That is, one proceeds with x, u etc. as the scaled variables.

Our goal of finding a scale such that the dimensionless variable gets a unit magnitude is often hard to reach. In some problems one must be satisfied with a dimensionless variable of finite value. A PDE may describe different physical regimes, and in such problems a particular scaling is seldom useful for all the regimes.

Introducing a Variable Coefficient. The heat conduction problem from Chapter 1.2.1 is a natural extension of the problem (A.1):

$$-k\frac{d^2u}{dx^2} = s(x), \quad u(0) = T_s, \quad -ku'(b) = -Q.$$

Again, x is scaled according to $\bar{x} = x/b$ such that $\bar{x} \in [0, 1]$. We might choose $\bar{u} = (u - T_s)/u_c$, where the scale u_c must be determined from insight into the problem. A scaling of s is performed according to $\bar{s} = s/s_c$, where the characteristic value s_c is taken as $\sup_{x \in [0,b]} |s(x)|$, since $s(x)$ is a known function. With the special choice $s(x) = R \exp(-x/L_R)$, as in Chapter 1.2.1, we get that $s_c = R$. These steps result in

$$-\frac{d^2\bar{u}}{d\bar{x}^2} = \gamma\bar{s}(\bar{x}), \quad \gamma = \frac{b^2 s_c}{k u_c}, \quad \bar{u}(0) = 0, \ \bar{u}'(1) = \frac{bQ}{k u_c}. \tag{A.3}$$

How should we choose u_c? Of course, we could try to apply a known maximum principle for this problem, but this turns into messy algebra. Instead we shall perform some simple reasoning. Let T_b be the unknown temperature at $x = b$. If the heat generation is not a dominating effect, we expect that u will lie between the boundary values, such that u_c can be taken as $T_b - T_s$. A rough estimation of T_b can be based on the boundary condition (Fourier's law) $-Q = -ku'(b)$, and then approximating $u'(b)$ by the finite difference $(T_b - T_s)/b = u_c/b$. This gives $u_c = Qb/k$ and

$$-\frac{d^2\bar{u}}{d\bar{x}^2} = \gamma\bar{s}(\bar{x}), \quad \gamma = \frac{bs_c}{Q}, \quad \bar{u}(0) = 0, \ \bar{u}'(1) = 1.$$

The scaling leads to a problem with one dimensionless parameter γ. To explore the model, we would compute curves $\bar{u}(\bar{x}; \gamma)$ for various choices of γ and shapes of $\bar{s}(\bar{x})$. For the particular choice $s(x) = R \exp(-x/L_R)$, $\bar{s}(\bar{x}) = \exp(-\beta\bar{x})$, where $\beta = b/L_R$ is another dimensionless parameter. Instead of experimenting with different shapes of $\bar{s}(\bar{x})$, we can experiment with different values of β. The problem has been reduced to studying $\bar{u}(\bar{x}; \beta, \gamma)$.

Scaling the Heat Equation. Our next example concerns the two-dimensional heat equation

$$\varrho c_p \frac{\partial u}{\partial t} = k\left(\frac{\partial^2 u}{\partial x^2} + \frac{\partial^2 u}{\partial y^2}\right),$$

where ϱ is the density, c_p is the heat capacity, k is the heat condution coefficient, and $u(x, y, t)$ is the unknown function. The domain is taken as $(a, b) \times (c, d)$. We assign the boundary values $u = 0$ on $x = a, b$ and $y = d$, as well as $-k\partial u/\partial n = C$, where C is a constant, on $y = c$. The initial condition reads $u(x, y, 0) = f(x, y)$.

The obvious dimensionless form of the coordinates is

$$\bar{x} = \frac{x - a}{b - a}, \quad \bar{y} = \frac{y - c}{d - c}.$$

Nevertheless, this scaling give rise to anisotropic dimensionless diffusion. Using the same length scale for x and y, preserves the isotropic diffusion term. In the following we scale both x and y by $b - a$. The time coordinate is scaled by t_c, whose value must be estimated. Similarly, $u(x, y, t)$ is scaled by the unknown quantity u_c, whereas $f(x, y)$ is scaled by its maximum absolute value, here referred to as f_c. Inserting the new dimensionless variables in the initial-boundary value problem results in

$$\frac{\partial \bar{u}}{\partial \bar{t}} = \alpha \left(\frac{\partial^2 \bar{u}}{\partial \bar{x}^2} + \frac{\partial^2 \bar{u}}{\partial \bar{y}^2} \right),$$

$$-\frac{\partial \bar{u}}{\partial \bar{n}} = \beta, \quad \bar{y} = 0,$$

$$\bar{u} = 0, \quad \bar{x} = 0, 1, \quad \bar{y} = \delta,$$

$$\bar{u}(\bar{x}, \bar{y}, 0) = \gamma \bar{f}(\bar{x}, \bar{y}).$$

The dimensionless parameters α, β, γ, and δ are given as

$$\alpha = \frac{kt_c}{\varrho c_p (b - a)^2}, \quad \beta = \frac{C(b - a)}{ku_c}, \quad \gamma = \frac{f_c}{u_c}, \quad \delta = \frac{d - c}{b - a}.$$

It remains to determine the scales t_c and u_c. For this purpose we need to find a solution of the PDE that displays the principal characteristics of u. A possible guess is

$$u(x, y, t) = e^{-\nu t} \sin \pi \frac{x - a}{b - a} \sin \pi \frac{y - c}{d - c},$$

which upon insertion in the heat equation gives

$$\nu = \frac{k\pi^2}{\varrho c_p} \left((b - a)^{-2} + (d - c)^{-2} \right).$$

The characteristic time t_c can be chosen such that the solution is reduced by a factor of e at time t_c, i.e. $u(x, y, t_c) = e^{-1} u(x, y, 0)$, giving $t_c = 1/\nu$, which is often referred to as the *e-folding time*. We simplify the expression for t_c by replacing $d - c$ by the other length scale $b - a$. The result becomes

$$t_c = \frac{\varrho c_p (b - a)^2}{2\pi^2 k},$$

which implies $\alpha = 2\pi^2$. We could skip the $2\pi^2$ factor in t_c without any significant loss of important information in the time scale. Then α equals unity, which is more esthetic in the final dimensionless PDE.

Determination of u_c relies on a fundamental property of the current PDE: u is bounded above by its initial value (see [120, Ch. 4] for more precise information about this property in a 1D problem with Dirichlet boundary conditions). Hence, $u_c = f_c$ is a suitable choice. We can then summarize the scaled initial-boundary value problem:

$$\frac{\partial \bar{u}}{\partial \bar{t}} = \frac{\partial^2 \bar{u}}{\partial \bar{x}^2} + \frac{\partial^2 \bar{u}}{\partial \bar{y}^2},$$

$$-\frac{\partial \bar{u}}{\partial \bar{n}} = \beta, \quad \bar{y} = 0,$$

$$\bar{u} = 0, \quad \bar{x} = 0, 1, \quad \bar{y} = \delta,$$

$$\bar{u}(\bar{x}, \bar{y}, 0) = \bar{f}(\bar{x}, \bar{y}).$$

The original problem, involving $u(x, y, t; \varrho, c_p, k, a, b, c, d, C)$ and $\bar{f}(\bar{x}, \bar{y})$, is now reduced to a problem involving $\bar{u}(\bar{x}, \bar{y}, \bar{t}; \beta, \delta)$ and $\bar{f}$. This is a significant reduction in complexity when it comes to investigation of the problem through computer experiments.

Scaling the Wave Equation. Our next example is devoted to the one-dimensional wave equation with a variable coefficient $q(x)$:

$$\frac{\partial^2 u}{\partial t^2} = \frac{\partial}{\partial x}\left(q(x)\frac{\partial u}{\partial x}\right), \quad x \in (a, b).$$

The boundary conditions read $u(a) = u(b) = 0$, whereas the initial conditions are taken as $u(x, 0) = f(x)$ and $\partial u(x, 0)/\partial t = 0$. With $\bar{x} = (x - a)/(b - a)$, $\bar{t} = t/t_c$, $\bar{u} = u/u_c$, $\bar{q} = q/q_c$, $q_c = \sup_{x \in [a,b]} |q(x)|$, $\bar{f}(\bar{x}) = f(a + \bar{x}(b - a))/f_c$, $f_c = \sup_{x \in [a,b]} f(x)$, we obtain

$$\frac{\partial^2 \bar{u}}{\partial \bar{t}^2} = \gamma \frac{\partial}{\partial \bar{x}}\left(\bar{q}(\bar{x})\frac{\partial \bar{u}}{\partial \bar{x}}\right), \quad \bar{x} \in (0, 1),$$

$$\bar{u}(0) = 0,$$

$$\bar{u}(1) = 0,$$

$$\bar{u}(\bar{x}, 0) = \delta \bar{f}(\bar{x}),$$

$$\frac{\partial}{\partial \bar{t}}\bar{u}(\bar{x}, 0) = 0.$$

The dimensionless parameters γ and δ read

$$\gamma = \frac{t_c^2 q_c}{(b - a)^2}, \quad \delta = \frac{f_c}{u_c}.$$

It remains to determine u_c and t_c. Again, we do this by studying a prototype solution of the PDE. Let $q(x) = q_c$. The general solution of the wave equation is then

$$u(x, t) = F(x - \sqrt{q_c}t) + G(x + \sqrt{q_c}t),$$

which is easily justified by inserting this expression in the PDE. With $u = f$ and $\partial u/\partial t = 0$ at $t = 0$, we get $F = G = f/2$ such that $u(x, t) = (f(x - $

$\sqrt{q_c}t) + f(x + \sqrt{q_c}t))/2$. It follows that $u_c = f_c$ and consequently $\delta = 1$. The characteristic time scale t_c can be chosen equal to the traveling time of a wave across the domain: $t_c = (b - a)/\sqrt{q_c}$, realizing that the function f (i.e. the wave pulse) is propagated to the left and right with velocity $\sqrt{q_c}$. Then γ becomes equal to unity. Notice that when $q(x)$ is constant (equal to q_c), all physical parameters are "scaled away" from the initial-boundary value problem. It can be convenient, nevertheless, to keep γ in front of the $\partial^2 \bar{u}/\partial \bar{x}^2$ term just for labeling this term in hand calculations.

Scaling the Convection-Diffusion Equation. The convection-diffusion equation

$$\frac{\partial u}{\partial t} + \boldsymbol{v} \cdot \nabla u = k \nabla^2 u \qquad (A.4)$$

appears in many fluid flow contexts. Scaling of initial and boundary conditions for this equation will be similar to the previous heat equation example, so we just focus at scaling the PDE (A.4) in the following. It is assumed that $\boldsymbol{v}$ in (A.4) is a prescribed spatially varying vector (velocity) field, k is a known parameter, and $u(\boldsymbol{x}, t)$ is the primary unknown. Equation (A.4) is referred to as a *convection-diffusion* equation.

Let L be the characteristic length of the domain Ω in which the equation above is to be solved. Furthermore, let U be a characteristic measure of the velocity field $\boldsymbol{v}$. It is then natural to introduce the following dimensionless variables:

$$\bar{\boldsymbol{x}} = \frac{\boldsymbol{x}}{L}, \quad \bar{\boldsymbol{v}} = \frac{\boldsymbol{v}}{U}, \quad \bar{u} = \frac{u}{u_c},$$

where u_c is a characteristic size of the solution. Inserting these expressions in (A.4) results in

$$\bar{\boldsymbol{v}} \cdot \bar{\nabla} \bar{u} = \frac{1}{\mathrm{Pe}} \bar{\nabla}^2 \bar{u} .$$

The bar in $\bar{\nabla}$ indicates derivation with respect to scaled coordinates $\bar{\boldsymbol{x}}$. Contrary to the previous examples, the present one has a dimensionless parameter in the governing PDE. This parameter is the Peclet number $\mathrm{Pe} = UL/k$. We can interpret the Peclet number as the ratio between the $|\boldsymbol{v} \cdot \nabla u|$ and $k|\nabla^2 u|$ terms, which physically expresses the relative importance of convective and diffusive effects:

$$\frac{|\boldsymbol{v} \cdot \nabla u|}{|k\nabla^2 u|} \sim \frac{UL^{-1}u_c}{kL^{-2}u_c} = \frac{UL}{k} = \mathrm{Pe} .$$

If we extend (A.4) with a time derivative term, $\partial u/\partial t$ on the left-hand side, we also need to scale the time: $\bar{t} = t/t_c$. The natural time scale depends on whether diffusion or convection dominates. Assuming that convection is most important, the typical velocity in the problem is U, and t_c can be taken as the time it takes to propagate a signal through the medium, that is, $t_c = L/U$. The resulting equation becomes

$$\frac{\partial \bar{u}}{\partial \bar{t}} + \bar{\boldsymbol{v}} \cdot \bar{\nabla} \bar{u} = \frac{1}{\mathrm{Pe}} \bar{\nabla}^2 \bar{u} . \qquad (A.5)$$

However, if diffusion is dominant, we should choose a time scale like we did in the heat equation example. With our present symbols this results in $t_c = L^2/k$. The corresponding dimensionless PDE reads

$$\frac{\partial \bar{u}}{\partial \bar{t}} + \text{Pe}\, \bar{\boldsymbol{v}} \cdot \bar{\nabla} \bar{u} = \bar{\nabla}^2 \bar{u}\,. \tag{A.6}$$

In the limit $\text{Pe} \to \infty$, when convection dominates over diffusion, equation (A.5) tends to the expected form where the diffusion term is neglected. Conversely, when $\text{Pe} \to 0$ we can neglect the term $\boldsymbol{v} \cdot \nabla u$, which is clearly indicated by (A.6).

Exercise A.1. Consider the Navier-Stokes equations

$$\frac{\partial \boldsymbol{v}}{\partial t} + (\boldsymbol{v} \cdot \nabla)\boldsymbol{v} = -\frac{1}{\varrho}\nabla p + \nu \nabla^2 \boldsymbol{v}, \tag{A.7}$$

where ϱ and ν are known constants, representing the density and the viscosity of the fluid, while $\boldsymbol{v}$ is the fluid velocity, and p is the fluid pressure. Explain how we can derive the following dimensionless form of the Navier-Stokes equations,

$$\frac{\partial \bar{\boldsymbol{v}}}{\partial \bar{t}} + (\bar{\boldsymbol{v}} \cdot \bar{\nabla})\bar{\boldsymbol{v}} = -\text{Eu}\bar{\nabla}\bar{p} + \frac{1}{\text{Re}}\bar{\nabla}^2 \bar{\boldsymbol{v}}, \tag{A.8}$$

where $\text{Re} = UL/\nu$ and $\text{Eu} = p_c/(\varrho U^2)$ are dimensionless numbers, and the bar indicates dimensionless quantities. The parameters U, L, and p_c are the characteristic velocity, length, and pressure of the problem, respectively. In many flows, the motion depends on pressure *differences* and not on pressure levels like p_c. This implies that Eu can be taken as unity. Re is called the *Reynolds number* and play a fundamental role in viscous fluid flow. (An excellent treatment of the present exercise is found in [100, Ch. 5.2], while [124, Ch. 3.9], [68, Ch. 2], and [126, Ch. 2.9] represent alternative references that contain more advanced material on scaling the equations of fluid flow.)   ⋄

Exercise A.2. Extend the Navier-Stokes equations (A.7) with an additional term $-g\boldsymbol{k}$ on the right-hand side. This term models gravity forces, with g being the constant acceleration of gravity and $\boldsymbol{k}$ being an associated unit vector. Perform the scaling and identify an additional dimensionless number, the so-called Froude number $\text{Fr} = U/\sqrt{gL}$.   ⋄

Scaling of Models with Many Parameters. The examples in this section demonstrate the two main strengths of scaling: (i) the size of each term in a PDE is reflected by a dimensionless coefficient, and (ii) the number of parameters in the problem is reduced because only certain combinations of the parameters appear in the scaled equations. In more complicated mathematical models, involving systems of PDEs and a large number of parameters, the advantages of scaling might be more limited. The scaling is normally restricted to a particular physical regime, while advanced models typically exhibit several different physical regimes. Equations (A.5) and (A.6) illustrate that even for a simple convection-diffusion equation there are two possible time scales. Furthermore, the reduction in active parameters in the model is not as substantial as in simpler problems. The danger of introducing errors through tedious manipulations in scaling procedures is another negative aspect. Therefore, if the aim is to develop a flexible simulation code for exploring a complicated mathematical model, it is often convenient to use quantities with dimension, or to introduce only a partial scaling, if the magnitude of some variables is far from unity and thereby can cause numerical problems. These comments explain why the simple PDE examples in this text are usually written in dimensionless form, while the more complicated models in the application chapters appear in their original form with physical dimensions.

A.2   Indicial Notation

This appendix introduces an indicial notation that helps to condense large mathematical expressions, yet with a syntax that translates directly to program code. In this notation, v_i denotes a vector and σ_{ij} represents a tensor[1]. Whether the notation v_i means component no. i of the vector, or the whole vector, will be evident from the context. The same convention applies to tensors as well. The index i in v_i is just a dummy index; we could also have written v_j or v_r, or σ_{rs} for σ_{ij}. In this book we often use r and s for indices ranging from one to the number of space dimensions, while i and j are frequently used for indexing basis functions, nodes, grid points, or unknowns in linear systems.

Let d be the number of space dimensions. Using Einstein's summation convention, we can write $\sum_{k=1}^d a_k b_k$ simply as $a_k b_k$. That is, we sum over an index that is repeated *twice* in an expression.

Let x_i denote the spatial coordinates. We can then to introduce a convenient and compact notation for differentiation:

$$v_{r,s} \equiv \frac{\partial v_r}{\partial x_s}, \quad \sigma_{rs,s} \equiv \sum_{s=1}^d \frac{\partial \sigma_{rs}}{\partial x_s}.$$

In other words, a comma in the index expression denotes derivation. As we see, the comma notation and summation convention are useful tools for reducing the size of equations and thereby improving the readability. For example, the incompressible Navier-Stokes equations can easily fit within a line:

$$v_{r,t} + v_s v_{r,s} = -p_{,r} + \frac{1}{\text{Re}} v_{r,ss} + b_r \,.$$

---

[1] Readers who are unfamiliar with the tensor concept can roughly think of tensors as matrices when reading the present book. Explanation of the properties of tensors are given in most books on continuum mechanics [75,77,103,118]. These references also provide more comprehensive introductory material on the indicial notation.

The reader is encouraged to write out these equations for a 3D problem $(r, s = 1, 2, 3)$ and observe the amount of space that is saved.

The comma notation is also much in use when the subscripts are x, y, and z, rather than 1, 2, or 3. For example,

$$N_{i,x} \equiv \frac{\partial N_i}{\partial x}.$$

Repeated x, y, and z does not imply summation: $w_{,xx} \equiv \partial^2 w / \partial x^2$.

The identity tensor is denoted by δ_{rs}, also referred to as the Kronecker delta. We have that $\delta_{rs} = 0$ when $r \neq s$, while δ_{rr} (without sum) equals 1. Normally, δ_{rr} implies a sum, $\sum_{r=1}^d \delta_{rr} = \sum_{r=1}^d 1 = d$, so we must explicitly state if the summation convention is not to be applied. There are many important rules for contracting products of vectors (or tensors) with the Kronecker delta. For example,

$$v_r \delta_{rs} = v_s, \qquad \sigma_{rs} \delta_{rq} = \sigma_{qs}.$$

These results are easily shown by writing out the sums (over r), choosing some specific values of free indices (s and q), and using the values of δ_{rs}.

Remark. The indicial notation explained above is in its presented form restricted to Cartesian coordinate systems. Other tools, e.g. dyadic notation, are attractive for hand calculations with cylindrical or spherical coordinates. However, in Cartesian coordinates the indicial notation offers compact expressions and at the same time the details of the algorithm for computing the expressions.

Exercise A.3. Express the matrix-vector product using the indicial notation. Develop a similar formula for a vector times a matrix.   ◇

Exercise A.4. Explain that the Laplace operator, $\nabla^2 u$ can be written as $u_{,rr}$ using the indicial notation. (Notice that r is a dummy index such that $u_{,rr}$ and $u_{,jj}$ are equivalent forms.)   ◇

Exercise A.5. Explain that the variable-coefficient Laplace operator can be written as $(k u_{,r})_{,r}$. Generalize this result to the case where k is a tensor.   ◇

Exercise A.6. Explain that the divergence operator, $\nabla \cdot \boldsymbol{v}$, can be written as $v_{r,r}$ using the indicial notation.   ◇

Exercise A.7. Write explicitly out all terms in the vector equation $\sigma_{rs,s} = 0$, $r, s = 1, \ldots, 3$.   ◇

Exercise A.8. Write the heat equation

$$\varrho C_p \left(\frac{\partial T}{\partial t} + \boldsymbol{v} \cdot \nabla T \right) = \kappa \nabla^2 T,$$

using the indicial notation.   ◇

Exercise A.9. Explain why $\sigma_{ik}\delta_{kj} = \sigma_{ij}$ and $u_{i,j}\delta_{i,j} = u_{k,k}$. These results are useful when deriving numerical methods for the elasticity and the Navier-Stokes equations. $\diamond$

Exercise A.10. Given the relations $\sigma_{ij,j} = 0$, $\sigma_{ij} = -p\delta_{ij} + 2\mu\dot{\varepsilon}_{ij}$, $\dot{\varepsilon}_{ij} = (v_{i,j} + v_{j,i})/2$, and $v_{i,i} = 0$, show that these relations can be combined into

$$-p_{,i} + \mu v_{i,jj} = 0.$$

This is essentially the derivation of a simplified version of the Navier-Stokes equations, where acceleration and body force terms are neglected. Similar mathematical manipulation with index expressions appears in the derivation of the equations for linear elasticity. $\diamond$

A.3 Compact Notation for Difference Equations

The discrete equations arising from finite difference or finite element techniques become much more lengthy than the underlying PDEs. It can therefore be convenient to introduce a compact notation that aids to make difference equations short, clear, and intuitive. Furthermore, mathematical manipulation of difference schemes, which is required in accuracy and stability calculations (cf. Appendix A.4), is significantly simplified using the compact notation.

Let $u_{i,j,k}^{\ell}$ be the numerical approximation to the function $u(x, y, z, t)$. We then define the difference operator

$$[\delta_x u]_{i,j,k}^{\ell} \equiv \frac{u_{i+\frac{1}{2},j,k}^{\ell} - u_{i-\frac{1}{2},j,k}^{\ell}}{\Delta x},$$

with similar definitions for δ_y, δ_z, and δ_t. Sometimes we need a difference over two cells,

$$[\delta_{2x} u]_{i,j,k}^{\ell} \equiv \frac{u_{i+1,j,k}^{\ell} - u_{i-1,j,k}^{\ell}}{2\Delta x}.$$

Compound operators, like $\delta_x \delta_x$ can now be defined. To simplify the subscript expressions, we restrict the attention to functions $u(x, t)$ without loss of generality. We then have

$$[\delta_x \delta_x u]_i^{\ell} = [\delta_x \phi]_i^{\ell}, \quad \phi_i^{\ell} \equiv [\delta_x u]_i^{\ell} = \frac{u_{i+\frac{1}{2}}^{\ell} - u_{i-\frac{1}{2}}^{\ell}}{\Delta x},$$

$$= \frac{\phi_{i+\frac{1}{2}}^{\ell} - \phi_{i-\frac{1}{2}}^{\ell}}{\Delta x},$$

$$= \frac{1}{\Delta x}\left(\frac{u_{i+1}^{\ell} - u_i^{\ell}}{\Delta x} - \frac{u_i^{\ell} - u_{i-1}^{\ell}}{\Delta x}\right),$$

$$= \frac{1}{\Delta x^2}\left(u_{i-1}^{\ell} - 2u_i^{\ell} + u_{i+1}^{\ell}\right).$$

In equations with variable coefficients we need the arithmetic average operator

$$[\overline{u}^x]_i^\ell \equiv \frac{1}{2}\left(u_{i+\frac{1}{2}}^\ell + u_{i-\frac{1}{2}}^\ell\right).$$

Sometimes we also need one-sided differences, like the forward difference

$$[\delta_x^+ u]_i^\ell \equiv \frac{u_{i+1}^\ell - u_i^\ell}{\Delta x},$$

and the corresponding backward difference

$$[\delta_x^- u]_i^\ell \equiv \frac{u_i^\ell - u_{i-1}^\ell}{\Delta x}.$$

Example A.1. The equation $-u''(x) = f(x)$, with conditions $u(0) = u(1) = 0$ and grid points $(i-1)\Delta x$, $i = 1, \ldots, n$, can now be written

$$-[\delta_x\delta_x u]_i = [f(x)]_i, \quad i = 2, \ldots, n-1, \quad [u]_1 = [u]_n = 0.$$

It is convenient to place the whole discrete equations inside brackets:

$$[-\delta_x\delta_x u = f]_i.$$

With this notation we have a strong link between the original differential equation and the discretized version.   ◇

Example A.2. The wave equation

$$\frac{\partial^2 u}{\partial t^2} = \gamma^2 \frac{\partial^2 u}{\partial x^2}, \quad x \in (0,1), \quad t > 0,$$

for $u(x,t)$, with initial and boundary conditions $u(x,0) = I(x)$, $\frac{\partial}{\partial t}u(x,0) = 0$, $u(0,t) = 0$, and $u(1,t) = 0$, can be discretized by a standard finite difference method as in Chapter 1.3.2. Using the compact notation, the difference equation corresponding to the PDE can be written

$$[\delta_t\delta_t u = \gamma^2 \delta_x\delta_x u]_i^\ell, \quad i = 2, \ldots, n-1, \; \ell \geq 1, \tag{A.9}$$

whereas the discrete initial and boundary conditions can be expressed as $[u = f]_i^0$, $[\delta_t u = 0]_i^0$, $[u = 0]_1$, and $[u = 0]_n$. Observe the clear similarity in the notation of the continuous and discrete problem.   ◇

Example A.3. The variable-coefficient PDE $(\lambda u')' = 0$ is normally discretized according to

$$\frac{1}{\Delta x}\left(\frac{1}{2}(\lambda_i + \lambda_{i+1})\frac{u_{i+1} - u_i}{\Delta x} - \frac{1}{2}(\lambda_{i-1} + \lambda_i)\frac{u_i - u_{i-1}}{\Delta x}\right) = 0, \tag{A.10}$$

see Chapter 1.2.6. This can be compactly and more intuitively written as

$$[\delta_x\overline{\lambda}^x\delta_x u = 0]_i. \tag{A.11}$$

Observe again the close similarity between the continuous and discrete notation. The reader is encouraged to write out the left-hand side of (A.11) in detail and verify that the expression becomes identical to the more conventional form (A.10).   ◇

Example A.4. The 3D wave equation $\partial^2 u/\partial t^2 = \nabla \cdot (\lambda \nabla u)$ can be discretized using standard centered (second-order accurate) finite differences in time and space, combined with arithmetic averaging of λ. The specification of such a scheme in the compact notation reads

$$[\delta_t \delta_t u = (\delta_x \overline{\lambda}^x \delta_x u + \delta_y \overline{\lambda}^y \delta_y u + \delta_z \overline{\lambda}^z \delta_z u)]^\ell_{i,j,k} \, .$$

We see that the compact notation not only saves space, it also gives a more intuitive explanation of the reasoning behind the discretization.   ◇

Example A.5. Discretizing the 2D heat equation $\partial u/\partial t = \kappa \nabla^2 u$ with a forward difference in time and standard centered differences in space, yields a scheme that takes the following form in the compact notation:

$$[\delta_t^+ u = \kappa \, (\delta_x \delta_x u + \delta_y \delta_y u)]^\ell_{i,j} \, . \tag{A.12}$$

This scheme is often referred to as the forward Euler method for the heat equation.   ◇

Example A.6. The θ-rule for the heat equation $\partial u/\partial t = \kappa \partial^2 u/\partial x^2$ can be written

$$[\delta_t u]_i^{\ell - \frac{1}{2}} = \theta \kappa [\delta_x \delta_x u]_i^\ell + (1 - \theta) \kappa [\delta_x \delta_x u]_i^{\ell - 1} \, . \tag{A.13}$$

The choice $\theta = 0$ results in the explicit forward Euler scheme (A.12), $\theta = 1/2$ gives the Crank-Nicolson scheme, and with $\theta = 1$ we get the implicit backward Euler scheme. (See Project 1.7.1 for an introduction to these three well-known schemes for the heat equation.)   ◇

A.4   Stability and Accuracy of Difference Approximations

A fundamental concern of all numerical methods is the errors arising from the approximations. Another critical aspect is accumulation of round-off errors due to finite precision arithmetic, as such accumulation may destroy the solution. These topics bring us to measures of the accuracy of finite difference approximations and to the concept of stability. The forthcoming discussion of accuracy and stability is centered around exact solutions of the difference equations, which also enables easy construction of test problems for verifying computer implementations. This approach is somewhat different from the standard approach in many other textbooks. Nevertheless, we also present classical subjects like truncation error, consistency, and the von Neumann method for investigating stability. A comprehensive extension of the approach advocated in this appendix, using the convection-diffusion equation as example, appears in the recent text by Gresho and Sani [44, Ch. 2].

The methods of analysis presented in the following are traditionally applied to finite difference schemes only. However, the methods can equally well be used to analyze finite element approximations. This is demonstrated in Chapter 2.4. The reader should be familiar with the compact finite difference notation from Appendix A.3 before studying Appendices A.4.5–A.4.11.

A.4.1   Typical Solutions of Simple Prototype PDEs

Separation of Variables. Homogeneous linear PDEs with constant coefficients allow exponential or trigonometric functions as solutions. Separation of variables might be used to show this property. Consider, for instance, the damped wave equation,

$$\varrho\frac{\partial^2 u}{\partial t^2} + \beta\frac{\partial u}{\partial t} = \gamma^2\nabla^2 u, \tag{A.14}$$

which represents a mixture of the standard wave equation, the heat equation, and the Laplace equation. The coefficients ϱ, β, and γ are assumed to be constant in space and time. Separating the variables in the solution according to

$$u(x_1, \ldots, x_d, t) = X_1(x_1)\cdots X_d(x_d)T(t),$$

inserting this expression in the equation, and dividing by u gives

$$\varrho\frac{T''(t)}{T(t)} + \beta\frac{T'(t)}{T(t)} = \gamma^2\sum_{s=1}^{d}\frac{X_s''(x_s)}{X_s(x_s)}.$$

The left-hand side is a function of t only, whereas each term on the right-hand side depends on x_s only, $s = 1, \ldots, d$. If the equation is to be fulfilled, the left-hand side must be a constant, and each term on the right-hand side must also be constant. The constants must sum up to zero for the equation to hold. In other words, separation of variables lead to $d+1$ ordinary differential equations for T and X_s:

$$\varrho T''(t) + \beta T'(t) - \lambda T(t) = 0, \tag{A.15}$$

and

$$\gamma^2 X_s''(x_s) - \mu_s X_s = 0, \quad s = 1, \ldots, d. \tag{A.16}$$

The constants λ and μ_s fulfill $\lambda = \sum_s \mu_s$.

Solution of the Separated Equations. The solution of equations (A.15) and (A.16) takes the form $T = \exp(\omega t)$ and $X_s = \exp(k_s x_s)$, where ω and k_s are *complex numbers* to be determined. The total solution u is then $\exp(\omega t + \sum_s k_s x_s)$.

Inserting the exponential solution for T and X_s into (A.15) and (A.16) gives $\varrho\omega^2 + \beta\omega - \lambda = 0$ and $\gamma^2 k_s^2 - \mu_s = 0$ for $s = 1, \ldots, d$. The ω and

k_s parameters are hence solutions of quadratic algebraic equations, and in general we achieve two complex roots. Denoting the two ω roots as $\omega^{(1)}$ and $\omega^{(2)}$, the function $T(t)$ is the linear combination

$$T(t) = A \exp\left(\omega^{(1)} t\right) + B \exp\left(\omega^{(2)} t\right),$$

where A and B are unknown constants. The same reasoning applies equally well for the X_s functions. The constants in all these functions, as well as λ and μ_s, must be determined from the initial and boundary conditions.

Complex Notation. Any complex number ω can be written as $\omega = \omega_r + i\omega_i$, where ω_r is the real part of ω, ω_i is the imaginary part, and i is the imaginary unit: $i = \sqrt{-1}$. By elementary properties of complex numbers we have

$$e^{\omega t} = e^{\omega_r t + i\omega_i t} = e^{\omega_r t} \left(\cos \omega_i t + i \sin \omega_i t\right) . \tag{A.17}$$

A similar decomposition of k_s is also useful.

In many physical applications, the trigonometric behavior is often dominating in space, which means that k_s is often purely imaginary. Therefore, it is convenient to seek $X_s = \exp\left(i k_s x_s\right)$, such that k_s becomes real. For wave phenomena, the same comments apply to ω, and it is again convenient to write $T = \exp\left(i\omega t\right)$. Diffusion problems, on the other hand, have their typical time dependence as $\exp\left(-\omega t\right)$, with ω real and greater than zero.

We will mainly work with solutions $u \sim \exp\left(i(\sum_s k_s x_s - \omega t)\right)$ in the following. The physical solution is not complex, so we must take the real or imaginary part of $\exp\left(i(\sum_s k_s x_s - \omega t)\right)$ prior to physical interpretation. If we multiply by a factor $e^{i\phi}$, where ϕ is free, we can *always* take the real part: $\mathrm{Re} \exp\left(i(\sum_s k_s x_s - \omega t + \phi)\right)$; adding $\phi = \pi/2$ in this function argument leads to the same results as taking the imaginary part when $\phi = 0$.

The outlined example demonstrates the basic ideas behind the technique known as *separation of variables*. This is a general technique for calculating analytical solutions of linear constant-coefficient PDEs [71,110,120]. The purpose of separating variables in the current context is mainly to show that the solution of linear homogeneous PDEs with constant coefficients can be conveniently sought on the form $\exp\left(i(\sum_s^d k_s x_s - \omega t + \phi)\right)$, with complex ω. We shall make use of this generic form of the solution when analyzing properties of mathematical models and numerical schemes.

Working with complex functions might seem unnecessarily complicated, but the complex notation is very efficient for practical hand calculations, and we only make use of a few very basic properties of complex numbers, essentially formulas like (A.17) and $i^2 = -1$.

A One-Dimensional Example. Let us restrict the governing PDE (A.14) to one space dimension. We write $k \equiv k_1$, $x \equiv x_1$, and insert the candidate solution $u = A \exp\left(i(kx - \omega t + \phi)\right)$, with A being a constant amplitude, in the governing PDE. This yields

$$-\varrho\omega^2 - i\omega\beta + \gamma^2 k^2 = 0 .$$

With $\beta = 0$ (no damping) we have $\omega = \pm \gamma k/\sqrt{\varrho}$, otherwise we have two complex ω values as solution. In both cases we see that $\omega = \omega(k)$. The solution is hence parameterized by k, and we can write

$$u(x, t; k) = A(k)e^{i(kx - \omega(k)t + \phi)}.$$

In the following, it appears to be convenient to work with $u(x, t; k)$ as a complex quantity. It then goes without saying that only the real part is of physical significance.

Forming the General Solution. The general solution of the one-dimensional version of our PDE (A.14) can now be obtained by forming the linear combination of the different $u(x, t; k)$ over the set of legal k values. If the PDE is defined on an interval, some boundary conditions will usually restrict k to a set of discrete values. A typical example is the condition $u(0, t) = u(1, t) = 0$. These are fulfilled when $u \sim \sin q\pi x$, $q \in \mathbb{N}$. Hence, $k = q\pi$ and $\phi = \pi/2$, ensuring that $\mathrm{Re} \exp(iq\pi x + i\pi/2) = \sin q\pi x$. The general solution is now a linear combination of all $u(x, t; q)$:

$$u(x, t) = \sum_{q=-\infty}^{\infty} u(x, t; q) = \sum_{q=-\infty}^{\infty} A_q \exp(i(q\pi x - \omega(q\pi)t + \pi/2)). \quad \text{(A.18)}$$

The generally complex amplitudes A_q are unknown and can be determined from the initial conditions. Setting $t = 0$ in (A.18) gives an ordinary complex Fourier series, and the coefficients A_q are then determined by standard Fourier techniques.

When the PDE is defined on an infinite interval, there are usually no restrictions on the k values so the linear combination is an integral:

$$u(x, t) = \int_{-\infty}^{\infty} A(k) \exp(i(kx - \omega(k)t))dk. \quad \text{(A.19)}$$

Notice that the coefficient $A(k)$ is now a continuous function. Standard Fourier integral or transform techniques can be used to calculate $A(k)$. Nevertheless, in this book we will not need explicit expressions for the amplitudes $A(k)$ or A_q, and therefore Fourier transforms or Fourier series are not used further. It appears that the nature of the solution is sufficiently well reflected by the argument $kx - \omega(k)t$ of the exponential solution function.

A.4.2   Physical Significance of Parameters in the Solution

The parameters ω and k have important physical interpretations, especially in wave phenomena when ω and k are frequently real numbers. Consider the solution

$$u(x, t) \sim \mathrm{Re}\, e^{i(kx - \omega(k)t)} = \cos(kx - \omega t),$$

with real k and ω. The reader should verify that this is a solution of the one-dimensional wave equation

$$\frac{\partial^2 u}{\partial t^2} = \gamma^2 \frac{\partial^2 u}{\partial x^2} \tag{A.20}$$

provided $\omega^2 = \gamma^2 k^2$, i.e., $\omega = \pm \gamma k$. This means that two values of ω are allowed and that the complete solution is a linear combination on the form

$$u(x,t) = C_1 \cos k(x - \gamma t) + C_2 \cos k(x + \gamma t). \tag{A.21}$$

If the initial conditions are $\partial u/\partial t = 0$ and $u(x,0) = A \cos kx$ we simply get $C_1 = C_2 = A/2$.

The purpose now is to give a physical interpretation of such a solution. Since u is a solution of a wave equation, we expect u to reflect typical properties of "waves". Fixing t, say $t = 0$ for simplicity, we see that $u = A \cos kx$, which is a periodic wave-like function with amplitude A and period $2\pi/k$. The period is actually the spatial length between two peaks of the function, usually referred to as the *wave length* $\lambda = 2\pi/k$. Fixing x, say $x = 0$, gives $u = A \cos \omega t$, which means that we watch the up and down movement of a fixed x point on the u surface. The period of this up and down movement is $2\pi/\omega$, a quantity that is obviously referred to as the *period T* of the wave.

Another important physical quantity is the phase velocity of the wave, which can be defined as the velocity of a peak. The peaks of $\cos \theta$ are given by $\theta = 2n\pi$, $n \in \mathbb{Z}$. In the present example the peaks are characterized by $\theta = kx - \omega t = 2\pi n$, or by using λ and T: $x/\lambda - t/T = n$. A particular peak $x_p(t)$ then moves in time according to $x_p/\lambda - t/T = n$, that is, $x_p(t) = \lambda n + \lambda t/T$. The velocity of the peak is therefore $dx_p/dt = \lambda/T$. We refer to this velocity as the *phase velocity c*. If $\omega = \omega(k)$ it follows that $T = T(\lambda)$, which means that c depends in general on the wave length: $c = c(\lambda) = \lambda/T(\lambda) = \omega(k)/k$. The following sketch exemplifies the central parameters in the mathematical description of waves:

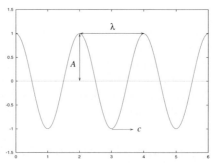

The relation $c = c(\lambda)$ or $\omega = \omega(k)$ is referred to as the *dispersion relation*. It connects the wave length and the phase velocity of a wave component. The general solutions (A.18) or (A.19) can be interpreted as a discrete or continuous weighted sum of wave components $u(x,t;k)$ with generally different phase velocities. Because the wave components propagate with different speed, the shape of u is modified as time increases. This effect is called *dispersion*. Nondispersive waves are recognized by c being a true constant or ω being a linear function of k. That is, all wave components move with the same velocity and the graph of $u(x, t)$ moves with preserved shape in time. Solutions of (A.20) have this property. Oral communication relies on the nondispersive nature of pressure waves in the air; the pressure signal resulting from our voice propagates with undisturbed shape through the air and can be recognized by other humans at different positions.

More information about basic description of waves can be found in most textbooks on general physics, see e.g. [7, Ch. 28] or [88, Ch. 15-18].

Also in problems where ω (or k) is complex, it makes sense to think of the solution as composed of waves, but in those cases the waves will be damped or, more seldom, amplified. Fundamental qualitative properties of the solution are reflected in an arbitrary wave component $u(x, t; k)$. We shall therefore focus on studying a single wave component in the following. The technique of analysis is often referred to as dispersion relation analysis and is widespread in geophysics and fluid mechanics. The strength of the method is its simplicity and that it gives a strong coupling between numerical properties of the scheme and the underlying properties of the physical phenomenon. Moreover, the method can be used to study accuracy and stability of numerical schemes.

A.4.3   Analytical Dispersion Relations

The analytical dispersion relation $\omega = \omega(k)$ is found by inserting the exponential solution

$$u(x_1, \ldots, x_d, t) = A \exp\left(i\left(\sum_s k_s x_s - \omega t + \phi\right)\right)$$

in the PDE. For the one-dimensional wave equation (A.20) we get the relation $\omega = \pm \gamma k$. Some examples and exercises concerning analytical dispersion relations are given next.

Example A.7. One important property of the dispersion relation $\omega = \omega(k)$ for the wave equation is that ω is real, that is, there is no damping or growth of the waves. Turning our attention to the heat equation

$$\frac{\partial u}{\partial t} = \kappa \frac{\partial^2 u}{\partial x^2}$$

and inserting a wave component $u = A \exp\left(i(kx - \omega t + \phi)\right)$, we easily find that $\omega = -i\kappa k^2$, which means that ω is imaginary. The typical wave solution of the heat equation is then

$$u = Ae^{-\kappa k^2 t} e^{i(kx+\phi)} . \tag{A.22}$$

This is a damped wave since $\kappa > 0$ is an important physical condition in the heat equation (in fact, $\kappa < 0$ implies that heat flows from cold to hot regions!). Notice that high frequency wave components (k large) are significantly damped, since the damping factor behaves like $\exp(-\kappa k^2 t)$. This property is demonstrated in detail in Example A.8.

We can add wave components, either as Fourier series or Fourier integrals, to obtain an analytical solution that fulfills the prescribed initial and boundary conditions. If the aim is just to construct an analytical solution, e.g. for verifying a computer implementation, we can look at (A.22) and *adjust* the initial and boundary conditions. For instance, working on a domain $(0, 1)$ with $u = 0$ at the boundary, requires the spatial part of the solution, $\exp(i(kx + \phi))$, to be $\sin q\pi x$, $q \in \mathbb{N}$. That is, $k = q\pi$. Moreover, using the real part of (A.22) as the physical significant part, implies $\phi = \pi/2$. It then remains to fit a suitable initial condition. The simplest choice is to use only one k value, say $k = \pi$ ($q = 1$). The analytical solution with physical significance is then $u(x, t) = \operatorname{Re} A \exp(-\kappa\pi^2 t + i\pi x + i\pi/2) = A \exp(-\kappa\pi^2 t) \sin \pi x$ for an arbitrary real constant A. $\diamond$

Example A.8. Suppose we have the rapidly varying initial condition $u(x, 0) = \sin \pi x + 0.6 \sin 100\pi x$ in the one-dimensional heat equation. The general solution for a wave component is given in (A.22), and in the present case the total solution can be obtained by adding two such components with wave numbers $k = \pi$ and $k = 100\pi$. Choosing $\kappa\pi^2 = 1$ to condense the expressions, we get

$$u(x, t) = e^{-t} \sin \pi x + e^{-10000t} 0.6 \sin 100\pi x.$$

The highly oscillatory component, which is significant at $t = 0$, is very quickly damped. Figure A.1 shows that after $1/1000$ second, all the oscillations have disappeared and u looks like an average of the initial shape. This is an illustration of the property that the solutions of the heat equation are smoothly varying, regardless of the initial function. $\diamond$

Example A.9. Consider the 3D wave equation $\partial^2 u/\partial t^2 = \gamma^2 \nabla^2 u$. A wave component can then be written as

$$u(x, y, z, t) = A \exp(i(k_x x + k_y y + k_z z - \omega t + \phi)).$$

Now $\mathbf{k} = (k_x, k_y, k_z)$ is the wave-number vector, indicating the spatial direction of the wave. The wave length is $\lambda = 2\pi/k$, with $k = \sqrt{k_x^2 + k_y^2 + k_z^2}$. The phase velocity becomes $\mathbf{c}(\lambda) = c(\lambda)\mathbf{k}/k$, c being the length of $\mathbf{c}$. Inserting the wave component in the PDE gives

$$-\omega^2 = \gamma^2 \left(-k_x^2 - k_y^2 - k_z^2\right),$$

that is, $\omega = \pm\gamma k$ as in the one-dimensional counterpart. $\diamond$

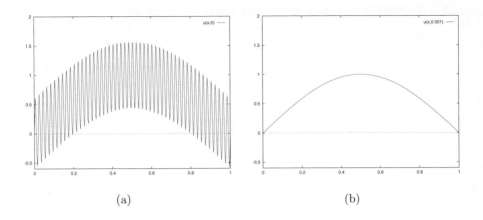

(a)   (b)

Fig. A.1. Illustration of the damping properties in the heat equation: (a) initial condition; (b) solution after $1/1000$ s.

Example A.10. We can extend the analysis in Example A.7 to a 2D heat equation, $\partial u/\partial t = \kappa \nabla^2 u$. A characteristic wave component can now be written

$$u(x, y, t) = A \exp\left(i(k_x x + k_y y - \omega t + \phi)\right).$$

Inserting this u in the equation gives the dispersion relation $\omega = -i\kappa(k_x^2 + k_y^2)$. Again we can use this information to construct a special analytical solution to the 2D heat equation to help us in program verification. Suppose the PDE is to be solved on $\Omega = (0, 1) \times (0, 1)$ with $\partial u/\partial n = 0$ on the boundary. To fulfill the boundary conditions, $u \sim \cos \pi q x \cos \pi r y$, $q, r \in \mathbb{N}$. This gives $k_x = q\pi$, $k_y = r\pi$, and $\phi = 0$. Let us pick $q = 2$ and $r = 1$. The solution then becomes

$$u(x, y, t) = \operatorname{Re} A e^{-\kappa 5\pi^2 t} e^{i(2\pi x + \pi y)} = A e^{-\kappa 5\pi^2 t} \cos 2\pi x \cos \pi y.$$

◇

Exercise A.11. Sound or light waves in three-dimensional space, propagating with perfect spherial symmetry from some source, can be described by the wave equation in spherical coordinates with radial symmetry:

$$\frac{\partial^2 u}{\partial t^2} = \gamma^2 \left(\frac{\partial^2 u}{\partial r^2} + \frac{2}{r}\frac{\partial u}{\partial r}\right),$$

where r represents the distance to the origin. Show that we can write this equation alternatively as

$$\frac{\partial^2 ru}{\partial t^2} = \gamma^2 \frac{\partial^2 ru}{\partial r^2}.$$

Find the dispersion relation and discuss whether the waves are damped or not.

◇

A.4.4   Solution of Discrete Equations

Consider a linear homogeneous difference equation, for example,

$$u_{j-1} - 2u_j + u_{j+1} = \alpha u_j . \tag{A.23}$$

The solution of this equation takes the form $u_j = Q^j$, where Q can be determined by inserting $u_j = Q^j$ in (A.23). This gives

$$Q = 1 + \frac{\alpha}{2} \pm \sqrt{\alpha(1 + \frac{\alpha}{4})} .$$

The general solution is a linear combination of the two roots:

$$u_j = A \left(1 + \frac{\alpha}{2} + \sqrt{\alpha(1 + \frac{\alpha}{4})} \right)^j + B \left(1 + \frac{\alpha}{2} - \sqrt{\alpha(1 + \frac{\alpha}{4})} \right)^j ,$$

where A and B are constants to be determined from the boundary conditions. We see that the solution Q^j can be written in exponential form if desired: $Q^j = e^{j \ln Q} = e^{j\tilde{Q}}$, now with $\tilde{Q} = \ln Q$ as the unknown quantity to be calculated.

When $\alpha = 0$ we have a double root $Q = 1$. Two linearly independent solutions are then $Q^j\ (= 1)$ and $jQ^j\ (= j)$. We can now write $u_j = AQ^j + BjQ^j = A + Bj$.

Separation of variables works in the case of multi-dimensional problems. For example, $u_{r,s} = Q^r P^s = \exp(r \ln Q + s \ln P)$, where the parameters Q and P (or $\ln Q$ and $\ln P$) are determined by inserting the assumed solution in the discrete equations. As in the continuous case, we find that linear homogeneous difference equations allow complex exponential solutions. In fact, if $u^{\ell}_{j_1,\dots,j_d}$ is the discrete solution at the point with spatial indices $(j_1,\dots,j_d)$ and time level ℓ, a generic form of $u^{\ell}_{j_1,\dots,j_d}$ is

$$u^{\ell}_{j_1,\dots,j_d} = \exp\left(i\left(\sum_{s}^{d} k_s(j_s - 1)\Delta x_s - \tilde{\omega}\ell\Delta t + \phi\right)\right). \tag{A.24}$$

Here, $\Delta x_1, \dots, \Delta x_d$ are the constant spatial grid spacings, and Δt is the time step. The exact appearance of the indices j_s depends on the numbering of the grid points; we see that $j_s = 1$ corresponds to $x_s = 0$. Changing the number of grid points is equivalent to multiplying the exponential function by a constant, which is of no significance in a linear homogeneous difference equation. We will therefore usually write $\sum_s k_s j_s \Delta x_s$ to keep the argument in the exponential function as simple as possible. As in the continuous problem, only the real part of the right-hand side in (A.24) has physical significance. The k_{ℓ} parameters are supposed to be the same as in the analytical case, but $\tilde{\omega}$ is in general different from ω (otherwise the discrete solution would be identical to the exact solution). We can determine $\tilde{\omega}$ by inserting (A.24) in the discrete equations. This results in a numerical dispersion relation taking the form

$$\tilde{\omega} = \tilde{\omega}(k_1, \ldots, k_d, \Delta x_1, \ldots, \Delta x_d, \Delta t).$$

That is, the dispersion properties depend on the discretization parameters.

In difference equations arising from PDEs with first-order time derivative, like the heat equation, the calculations are often more conveniently carried out by introducing $\xi = \exp(-\tilde{\omega}\Delta t)$, i.e., we seek solutions on the alternative form $\xi^\ell \exp(i \sum_s k_s j_s \Delta x_s + \phi)$.

We have seen that the analytical dispersion relation is calculated by straightforward differentiation. Calculation of numerical dispersion relations requires much more algebra. By using the discrete operator notation from Appendix A.3 and some convenient rules from Table A.1, the algebra is substantially reduced. Having these tools at hand, we look at explicit expressions for numerical dispersion relations in Appendix A.4.5.

Table A.1. Useful formulas involving difference operations on complex exponential functions.

operator	u_j	result
$[\delta_x \delta_x u]_j$	$\exp(ikx_j)$	$u_j \frac{2}{\Delta x^2}(\cos k\Delta x - 1) = -u_j \frac{4}{\Delta x^2}\sin^2 k\Delta x/2$
$[\delta_x^+ u]_j$	$\exp(ikx_j)$	$u_j \frac{1}{\Delta x}(\exp(ik\Delta x) - 1)$
$[\delta_x^- u]_j$	$\exp(ikx_j)$	$u_j \frac{1}{\Delta x}(1 - \exp(-ik\Delta x))$
$[\delta_x u]_j$	$\exp(ikx_j)$	$u_j \frac{2}{\Delta x}\sin k\Delta x/2$
$[\delta_{2x} u]_j$	$\exp(ikx_j)$	$u_j \frac{1}{\Delta x}\sin k\Delta x$

Example A.11. With the aid of Table A.1 we can study the accuracy of numerical derivatives of wave-like functions. Assume that $u = \exp(ikx)$. We then have

$$\left[\left|\frac{\delta_{2x} u}{du/dx}\right|\right]_j = \left|\frac{\exp(ikx_j)\frac{1}{h}\sin kh}{-ik\exp(ikx_j)}\right| = \frac{1}{kh}\sin kh$$

$$= 1 - \frac{1}{6}k^2 h^2 + \mathcal{O}(k^4 h^4) \qquad (A.25)$$

and

$$\left[\left|\frac{\delta_x \delta_x u}{d^2 u/dx^2}\right|\right]_j = \left(\frac{2}{kh}\sin\frac{kh}{2}\right)^2 = 1 - \frac{1}{12}k^2 h^2 + \mathcal{O}(k^4 h^4). \qquad (A.26)$$

As we see, the critical quantity is the dimensionless number $(kh)^2 = 4\pi^2(h/\lambda)^2$, or in other words, the square of the ratio of the wavelength and the grid spacing. Having 20 cells per wavelength, we obtain a relative error of 1.6% in the

Table A.2. Useful formulas involving difference operations on quadratic and linear polynomials. The result of applying any of the present difference operators to a constant equals zero.

operator	u_j	result
$[\delta_x \delta_x u]_j$	j^2	$2/\Delta x^2$
$[\delta_x^+ u]_j$	j^2	$(2j+1)/\Delta x$
$[\delta_x^- u]_j$	j^2	$(2j-1)/\Delta x$
$[\delta_{2x} u]_j$	j^2	$2j/\Delta x$
$[\delta_x \delta_x u]_j$	j	0
$[\delta_x^+ u]_j$	j	$1/\Delta x$
$[\delta_x^- u]_j$	j	$1/\Delta x$
$[\delta_{2x} u]_j$	j	$1/\Delta x$

first derivative and 0.8% in the second derivative, using the leading terms in the expressions above. With only 4 cells per wavelength, the relative errors are 41% and 21%, respectively. It is therefore crucial to resolve wave-like functions sufficiently in the grid.   ⋄

Exercise A.12. A possible higher-order finite difference approximation to u'' is

$$\frac{1}{12h^2}\left(-u_{i-2} + 16u_{i-1} - 30u_j + 16u_{i+1} - u_{i+2}\right).\qquad(A.27)$$

Calculate the relative error of the second-order derivative approximation as indicated in (A.26). Show that the accuracy of the five-point difference is superior to the accuracy of the three-point scheme for u'' on fine grids, but that neither scheme is accurate on coarse grids.   ⋄

A.4.5 Numerical Dispersion Relations

It was stated in Appendix A.4.4 that the numerical dispersion relation is found by inserting the discrete exponential solution in the discrete equations. Let us demonstrate the relevant calculations in an example concerning the 1D discrete wave equation, $[\delta_t \delta_t u = \gamma^2 \delta_x \delta_x u]_j^\ell$. A solution $u_j^\ell = A \exp\left(i(kjh - \tilde{\omega}\ell\Delta t)\right)$ is inserted into the discrete equations. Using the formulas from Table A.1, we easily get

$$-\frac{4}{\Delta t^2}\sin^2\frac{\tilde{\omega}\Delta t}{2} = -\gamma^2 \frac{4}{h^2}\sin^2\frac{kh}{2},$$

which is simplified to

$$\sin\frac{\tilde{\omega}\Delta t}{2} = \pm\frac{\gamma\Delta t}{h}\sin\frac{kh}{2}.\qquad(A.28)$$

Solving with respect to $\tilde{\omega}$ and then inserting this expression in the discrete wave component yield an *analytical solution of the finite difference equation*. If we define

$$\Omega = \frac{2}{\Delta t} \sin \frac{\tilde{\omega}\Delta t}{2}, \quad K = \frac{2}{h} \sin \frac{kh}{2},$$

we see that the numerical dispersion relation takes the form $\Omega = \pm \gamma K$. Moreover,

$$\lim_{\Delta t \to 0} \Omega = \tilde{\omega}, \quad \lim_{h \to 0} K = k.$$

In the limit $h, \Delta t \to 0$ we therefore recover the analytical dispersion relation.

Example A.12. The numerical dispersion relation of the scheme

$$[\delta_t \delta_t u = \gamma^2 (\delta_x \delta_x u + \delta_y \delta_y u + \delta_z \delta_z u)]_{p,q,r}^\ell$$

for the 3D wave equation can be obtained by inserting

$$u_{p,q,r}^\ell = A \exp\left(i(k_x p \Delta x + k_y q \Delta y + k_z r \Delta z - \tilde{\omega}\ell\Delta t)\right).$$

With the aid of Table A.1, we find that

$$\Omega^2 = \gamma^2 \left(K_x^2 + K_y^2 + K_z^2\right). \tag{A.29}$$

Here,

$$K_x = \frac{2}{\Delta x} \sin \frac{k_x \Delta x}{2},$$

with similar definitions of K_y and K_z. Again we see that the analytical dispersion relation is recovered as the grid parameters $\Delta x, \Delta y, \Delta z, \Delta t$ go to zero. To find $\tilde{\omega}$, we multiply (A.29) by $\Delta t^2/4$ and take the square root,

$$\sin \frac{\tilde{\omega}\Delta t}{2} = \pm \gamma \Delta t \left(\frac{1}{\Delta x^2} \sin^2 \frac{k_x \Delta x}{2} + \frac{1}{\Delta y^2} \sin^2 \frac{k_y \Delta y}{2} + \frac{1}{\Delta z^2} \sin^2 \frac{k_x \Delta x}{2}\right)^{\frac{1}{2}}. \tag{A.30}$$

This equation can be solved with respect to $\tilde{\omega}$, thus yielding an explicit expression for the numerical dispersion relation. ◇

Example A.13. It was explained in Appendix A.4.4 how the analytical solution of the discrete equations could in principle be calculated. Such solutions are of fundamental importance for verifying computer implementations, because these solutions should be exactly reproduced by the program (within machine precision), regardless of the uniform grid size.

Suppose we have found an analytical formula for the numerical frequency $\tilde{\omega}$. The physical solution of the discrete equations can be taken as

$$u_j^\ell = \text{Re } A e^{i(kx_j - \tilde{\omega}\ell\Delta t + \phi)}$$

in a one-dimensional problem. This solution can be adapted to a particular test problem. Assume that the aim of our simulator is to solve a wave equation problem like (1.44)–(1.48). To fulfill the boundary and initial conditions, we can have an exact discrete solution that behaves like $\sin kx_j \cos\tilde{\omega}\ell\Delta t$, which is obtained by letting $\phi = \pi/2$ and $k = q\pi$, $q \in \mathbb{N}$. Thus we can try

$$u_j^\ell = A\sin(\pi(j-1)h)\cos(\ell\tilde{\omega}\Delta t), \tag{A.31}$$

with

$$\tilde{\omega} = \pm\frac{2}{\Delta t}\sin^{-1}\left(\frac{\gamma\Delta t}{h}\sin\frac{kh}{2}\right) \tag{A.32}$$

from (A.28). It suffices to use the plus sign in (A.32) since $\cos-\tilde{\omega} = \cos\tilde{\omega}$. This discrete solution (A.31) is compatible with the boundary conditions $u(0,t) = u(1,t)$ and the initial conditions $u(x,0) = A\sin\pi x$ and $\partial u/\partial t = 0$.

The case $C \equiv \gamma\Delta t/h = 1$ was used for testing the implementation of the numerical method for (1.44)–(1.48) in Chapter 1.3. We see from (A.32) that $C = 1$ implies $\tilde{\omega} = \pm\gamma k = \omega$, that is, the numerical dispersion relation is exact, and the discrete solution coincides with the analytical solution at the grid points.

The solver `Wave1D3e` in a subdirectory of `src/fdm/Wave1D3` computes the difference between the analytical solution (A.31) of the discrete equations and the u_j^ℓ values computed from the numerical scheme. Run this solver with 4 cells, Courant number 0.7 and integrate to, e.g., $t = 294$: `./app -n 4 -C 0.7 -t 294`. The error should be as close to zero as the machine precision allows. (We remark that the solver employs programming techniques introduced in Chapters 1.6 and 3.4.6.)   ◇

A.4.6 Convergence

The numerical scheme is said to be convergent when the difference between the discrete and continuous problem approaches zero as the grid parameters $(h, \Delta t)$ go to zero. This is of course a fundamental requirement of any numerical method, but proving convergence of a scheme is normally a difficult task. Nevertheless, we managed in the previous analysis, based on exact representation of the numerical solution, to quite easily show that a numerical wave component converges to the corresponding analytical component as $h, \Delta t \to 0$. By means of a famous theorem by Lax, convergence can fortunately be established by simple arguments in a wide range of problems without constructing exact prototype solutions of the discrete equations.

Theorem A.14. The Lax Equivalence Theorem. *Given a well-posed mathematical problem and a* consistent *finite difference approximation to it,* stability *is a necessary and sufficient condition for* convergence.

Proof. See [99] or [111] for a full proof. LeVeque [67, Ch. 10] presents an intuitive justification of the theorem.   □

To establish convergence, we do not need to show that the *error* itself goes to zero; it is sufficient (i) to show that the scheme is consistent, which is normally quite trivial, and (ii) to find the conditions for stability. Consistency is treated in Appendix A.4.9, whereas stability is the topic of the next section.

A.4.7   Stability

The concept of stability can be approached in many different ways. Here, we say that the numerical scheme is stable if the numerical solution

$$u_j^\ell = A \exp\left(i(kx_j - \tilde{\omega} t_\ell)\right)$$

mirrors the qualitative properties of the corresponding solution

$$u(x, t) = A \exp\left(i(kx - \omega t)\right)$$

of the continuous problem. For example, we know that ω is real in our wave equation example. This means that a wave is neither damped nor amplified in time. A similar requirement of the numerical solution demands that $\tilde{\omega}$ is real, which is the case when the right-hand side of (A.28) is less than or equal to unity. Otherwise, the sine function on the left-hand side of (A.28) has a magnitude larger than unity and the argument of the sine function must then be complex (recall that $\sin i\psi = i \sinh \psi$). Requiring the right-hand side of (A.28) to have a magnitude less than or equal to unity leads to $C \equiv \gamma \Delta t / h \leq 1$.

We might tolerate a slight damping in the numerical scheme, that is, we can accept that $\tilde{\omega}$ is complex, with a small negative imaginary part. Complex solutions of (A.28) are possible when the right-hand side is greater than 1 or less than -1. However, in those cases the roots of (A.28) when $C > 1$ will appear in complex conjugate pairs $\tilde{\omega} = \tilde{\omega}_r \pm i\tilde{\omega}_i$. The negative imaginary part corresponds to damping, but the positive imaginary part leads to exponential growth, which will dominate the whole solution after sufficiently long time. This is not in accordance with the properties of the continuous problem and cannot be accepted. We are therefore left with real roots of (A.28) and the requirement $C \leq 1$. The stability criterion on Δt becomes $\Delta t \leq h/\gamma$. To show that this criterion is of great practical importance, the reader is encouraged to run the wave equation simulator from Chapter 1.3.3 with $C > 1$ and observe that instabilities grow in time and destroy the solution.

Example A.15. Let us investigate the stability of the numerical scheme for the 3D wave equation treated in Example A.12 on page 513. We see from (A.30) that $\tilde{\omega} \in \mathbb{R}$ demands the right-hand side to be equal to or less than unity. The squared sine functions can at most be unity in size, with a corresponding magnitude of the right-hand side

$$C = \gamma \Delta t \left(\frac{1}{\Delta x^2} + \frac{1}{\Delta y^2} + \frac{1}{\Delta z^2} \right)^{\frac{1}{2}}.$$

The stability criterion therefore becomes $C \leq 1$, and this C is the Courant number for the 3D problem. $\diamond$

We remark that stability is also an issue in stationary problems, see Project 1.4.2.

A.4.8   Accuracy

The optimal measure of numerical accuracy is of course the numerical error as a function of the grid spacing parameters as well as the space and time coordinates. Our wave component analysis provides tools for investigating the numerical error and will be demonstrated below.

Since we have expressions for the analytical solution of the continuous and discrete problems, it is natural to define the error at the point (x_j, t_ℓ) as

$$e(x_j, t_\ell; k, h, \Delta t) = Ae^{i(kx_j - \omega t_\ell)} - Ae^{i(kx_j - \tilde{\omega}t_\ell)}$$
$$= Ae^{i(kx_j - \omega t_\ell)} \left(1 - e^{i(\omega - \tilde{\omega})t_\ell}\right).$$

The critical quantity in $e(x_j, t_\ell; k, h, \Delta t)$ is then the error in frequency:

$$E_\omega(k, h, \Delta t) \equiv \omega(k) - \tilde{\omega}(k; h, \Delta t).$$

The expression for E_ω is normally quite complicated due to the functional form of $\tilde{\omega}$. Thus it is customary to make a Taylor-series expansion of $\tilde{\omega}$ in powers of h and Δt.

The multi-dimensional extension of e and E_ω follows straightforwardly; it is just a matter of additional independent variables in the Taylor series.

Example A.16. We shall now analyze the accuracy of the 1D wave equation problem from Appendix A.4.5. There we had

$$\tilde{\omega} = \frac{2}{\Delta t} \sin^{-1}\left(\frac{\gamma \Delta t}{h} \sin \frac{kh}{2}\right).$$

By means of, e.g., a few Maple commands, we can easily compute the Taylor series expansion of E_ω as

$$E_\omega(k, h, \Delta t) = -\frac{1}{24}\gamma k^3 (h^2 - \gamma^2 \Delta t^2) + \mathcal{O}(h^2 \Delta t^2, h^4, \Delta t^4).$$

It turns out that when $\Delta t = h/\gamma$, i.e. $C = 1$, all terms in E_ω cancel, and the solution of the continuous problem is obtained at all grid points (recall that we have evaluated e at the grid points), regardless of the size of h or Δt.

When $C < 1$ we see that $E_\omega = \mathcal{O}(h^2, \Delta t^2)$. We say that the scheme is of *second order* in h and Δt. Investigating the error in the numerical dispersion relation is only one way of determining the accuracy and the order of a scheme. The most common technique involves the truncation error, which is covered in Appendix A.4.9.

Let us examine the error E_c in the numerical phase velocity:

$$E_c = \frac{E_\omega}{k} = -\frac{1}{24}\gamma k^2 h^2 (1 - C^2) + \mathcal{O}(h^4, \Delta t^4).$$

Note that we have replaced Δt by Ch/γ. The shortest wave that can be represented on the grid has wave length $\lambda_{min} = 2h$, with values $\pm A$ at the grid points. The corresponding minimum value of k in a uniform grid is hence $k_{min} = 2\pi/\lambda_{min} = \pi/h$. It is of interest to plot the normalized error E_c/γ as a function of $p \equiv kh \in (0, \pi]$ for various Courant numbers, see Figure A.2. We see from the figure or the formula for E_c/γ that the relative error decreases with increasing C and decreasing $p = kh$. With $C = 0.9$ the relative error is less than five percent for $p < 2$. For practical purposes one needs at least four grid points per wave length, i.e. $\lambda = 4h$, which implies $p = \pi/2 \approx 1.57$. This is therefore the largest relevant value of p. The important message from a practical computational point of view is that reducing Δt increases the error unless h is reduced correspondingly. The optimal ratio of Δt and h is to have C as close to unity as possible. We remark that this information about the accuracy is not evident from the order of the scheme.

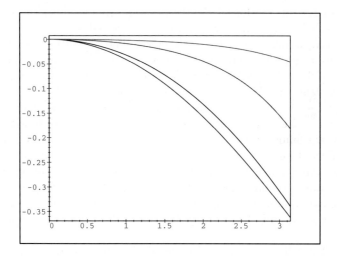

Fig. A.2. Error in numerical phase velocity, normalized by the exact phase velocity, as a function of $p = kh$. The different curves correspond to different Courant numbers; $C = 0.98$ (top curve), $C = 0.9$, $C = 0.5$, and $C = 0.1$ (bottom curve).

Figure A.2 can be used to explain numerical artifacts when solving the wave equation. Consider the initial function

$$f(x) = \begin{cases} 0.5 - \pi^{-1}\arctan(\sigma(x-2)), \ x > 0, \\ 0.5 + \pi^{-1}\arctan(\sigma(x+2)), \ x \le 0. \end{cases}$$

Here, σ is a parameter that controls the steepness of $f(x)$. The wave equation with the above $f(x)$ is solved in an application called `Wave1D3` found in the directory `src/fdm/Wave1D3`. The source code is designed according to ideas presented in Chapter 1.6, but for the present purpose it is not necessary to look into and understand the program. You can start the program either through the GUI `gui.pl` or through a command like `./app -C 0.9 -s 1000 -t 20 -n 60 --casename acc`. The `-s` option is used to set the σ value, `-C` is used for the Courant number, and the `-t` option assigns the time when the simulation is to be stopped. Animation of the string movie is enabled by the Visualize button on the GUI or by the a simple `curveplotmovie` command, `curveplotmovie gnuplot acc.map -0.5 1.2`. You should observe an undesired phenomenon: small, nonphysical waves are created. Choosing a smaller σ value, e.g. $\sigma = 10$, leads to a smoother initial profile $f(x)$ and the noise is much smaller. The Perl script `Verify/demo.pl` runs a series of examples illustrating the effect of numerical noise. An interesting problem is now to explain this generation of noise from the information that lies in the numerical dispersion relation.

The analytical solution of the discrete equations, taking the initial condition $u(x,0) = f(x)$ into account, can be viewed as a sum (or integral) of an infinite number of wave components. The amplitudes of the various wave components are determined from the initial condition by Fourier series methods. If $f(x)$ is a smooth function the wave components with small k (long wave length) will have the dominating amplitudes $A(k)$, whereas $A(k)$ becomes more significant for larger k values when $f(x)$ has steep gradients. We know that the numerical error in the phase velocity grows with kh for a fixed Courant number. Hence, when the initial condition requires short waves with significant amplitude to build up the steep initial profile, and we know that short waves move with wrong velocities, the inexact movement of high frequency wave components becomes visible. This is what we have demonstrated in our experiment. Lowering the value of σ decreases the amplitude of the shorter waves, and the wrong velocities of these waves are more difficult to observe.

Recall that all wave components move with the same velocity in the continuous problem such that a wave pulse keeps its original form. In the numerical case the phase velocity decreases with increasing kh, and the short waves move more slowly. The reader is encouraged to play around with various values of C, σ, and h to see how these parameters influence the visible quality of the simulation results.   ⬦

A.4.9   Truncation Error

Consider a PDE written on the form $\mathcal{L}(u) = f$, where $\mathcal{L}$ is a linear partial differential operator. The corresponding discrete problem reads $\hat{\mathcal{L}}(\hat{u}) = \hat{f}$, with $\hat{u}$ being the solution of the discrete problem. If we insert the solution u of the continuous problem in the discrete equations, the equations will of course normally not be fulfilled, and we get a residual τ:

$$\tau = \hat{\mathcal{L}}(u) - \hat{f}\,.$$

The $\hat{\mathcal{L}}$ and $\hat{f}$ quantities involve sampling functions at some neighboring points, which allows us to express $\hat{\mathcal{L}}(u)$ and $\hat{f}$ in terms of Taylor-series expansions of u and f in powers of the grid parameters. This enables us to investigate the size of the residual τ as a function of the grid parameters. Our primary interest is of course the error $u - \hat{u}$ as a function of the grid parameters, but this error may be difficult to estimate in many problems, whereas the residual can always be computed. We refer to τ as the *local truncation error* of the numerical scheme. It will be apparent that τ typically contains the error terms in the finite difference approximations to spatial and temporal derivatives. The computation of τ is demonstrated through two examples.

Example A.17. Our first example is a simple two-point boundary-value problem, $-u'' = f$, i.e., $\mathcal{L}(u) = -u''$, solved by the scheme $[\delta_x \delta_x \hat{u} = -f]_i$. We have $\mathcal{L}(\hat{u}) = [\delta_x \delta_x \hat{u}]_i$ and $\hat{f} = [f]_i$. In the following, it is important to distinguish clearly between the expressions $[\delta_x \delta_x u]_i$ and $[\delta_x \delta_x \hat{u}]_i$. The definition of τ applied to this example becomes

$$\tau = [\delta_x \delta_x u - f]_i\,.$$

The expression on the right-hand side involves quantities like u_{i-1}, u_i, and u_{i+1}, that is, the analytical solution of the continuous problem, u, sampled at x_{i-1}, x_i, and x_{i+1}. We make a Taylor series of these quantities around the principal point of the equation, x_i. For example,

$$u_{i+1} = u_i + \left[\frac{du}{dx}\right]_i h + \frac{1}{2}\left[\frac{d^2u}{dx^2}\right]_i h^2 + \cdots$$

We then achieve

$$[\delta_x \delta_x u]_i = \left[\frac{d^2u}{dx^2}\right]_i + \frac{h^2}{12}\left[\frac{d^4u}{dx^4}\right]_i + \mathcal{O}(h^4)\,.$$

The Taylor-series expansion of f_i is of course trivial. The truncation error hence reads

$$\tau = \left[\frac{d^2u}{dx^2} + f\right]_i + \frac{h^2}{12}\left[\frac{d^4u}{dx^4}\right]_i + \mathcal{O}(h^4)\,.$$

The first term vanishes since *the differential equation is supposed to be fulfilled by the solution u at all points in the domain.* Therefore,

$$\tau = \frac{h^2}{12}\left[\frac{d^4 u}{dx^4}\right]_i + \mathcal{O}(h^4).$$

We observe that residual tends to zero as $h \to 0$. Schemes with this property are said to be *consistent*. Inconsistent schemes are recognized by the fact that they do not necessarily solve the original PDE when the grid spacings tend to zero. The residual also reflects the quality of the approximation; here we see that $\tau = \mathcal{O}(h^2)$. One can hope that a small residual (truncation error) corresponds to a small error. For some model PDEs this property can be proved, see e.g. Chapter 2.10.5.   ◇

Example A.18. Let us consider the 1D wave equation with the numerical scheme $[\delta_t \delta_t \hat{\eta} = \gamma^2 \delta_x \delta_x \hat{\eta}]_i^\ell$. Inserting the solution of the continuous problem yields

$$\tau = [\delta_t \delta_t \eta - \gamma^2 \delta_x \delta_x \eta]_i^\ell.$$

Taylor-series expansion of η around the space-time point (x_i, t_ℓ) results in

$$\tau = \left[\frac{\partial^2 \eta}{\partial t^2}\right]_i^\ell + \frac{\Delta t^2}{12}\left[\frac{\partial^4 \eta}{\partial t^4}\right]_i^\ell + \mathcal{O}(\Delta t^4) - \gamma^2\left[\frac{\partial^2 \eta}{\partial x^2}\right]_i^\ell - \gamma^2 \frac{h^2}{12}\left[\frac{\partial^4 \eta}{\partial x^4}\right]_i^\ell + \mathcal{O}(h^4).$$

The second-order derivatives cancel since the solution of the continuous problem fulfills the PDE at all grid points. We then obtain

$$\tau = \mathcal{O}(h^2, \Delta t^2).$$

This result is consistent with our previous measures of the quality of the approximation, obtained from an expression for the real numerical error.   ◇

Table A.3 is useful for calculating the truncation error of finite difference schemes, since it gives the contribution to τ from many of the common finite difference operators. Equivalently, this table provides a list of the errors in finite difference approximations to derivatives.

Exercise A.13. Calculate the truncation error of the scheme (A.12) for the 2D heat equation.   ◇

Exercise A.14. Calculate the truncation error of the following scheme for the 2D wave equation: $[\delta_t \delta_t u = \gamma^2(\delta_x \delta_x + \delta_y \delta_y)u]_{i,j}^\ell$. Extend the analysis to the variable-coefficient case

$$[\delta_t \delta_t u = \gamma^2(\delta_x \overline{\lambda}^x \delta_x + \delta_y \overline{\lambda}^y \delta_y)u]_{i,j}^\ell.$$

◇

Exercise A.15. Set up a finite difference method for the Poisson equation $-\nabla^2 u = f$ in 3D and find the truncation error of the scheme.   ◇

Table A.3. Taylor-series expansions of some common finite difference operators applied to the solution u of a continuous problem. The table is useful for calculating the truncation error and for evaluating the error of finite difference approximations to derivatives.

operator	Taylor-series expansion
$[\delta_x\delta_x u]_i$	$\left[\frac{\partial^2 u}{\partial x^2}\right]_i + \frac{h^2}{12}\left[\frac{\partial^4 u}{\partial x^4}\right]_i + \mathcal{O}(h^4)$
$[\delta_x^- u]_i$	$\left[\frac{\partial u}{\partial x}\right]_i - \frac{1}{2}\left[\frac{\partial^2 u}{\partial x^2}\right]_i h + \mathcal{O}(h^2)$
$[\delta_x^+ u]_i$	$\left[\frac{\partial u}{\partial x}\right]_i + \frac{1}{2}\left[\frac{\partial^2 u}{\partial x^2}\right]_i h + \mathcal{O}(h^2)$
$[\delta_{2x} u]_i$	$\left[\frac{\partial u}{\partial x}\right]_i + \frac{1}{6}\left[\frac{\partial^3 u}{\partial x^3}\right]_i h^2 + \mathcal{O}(h^4)$
$[\delta_x u]_i$	$\left[\frac{\partial u}{\partial x}\right]_i + \frac{1}{6}\left[\frac{\partial^2 u}{\partial x^2}\right]_i \left(\frac{h}{2}\right)^2 + \mathcal{O}(h^4)$
$[\overline{\lambda}^x]_i$	$\lambda_i + \frac{1}{2}\left[\frac{\partial^2 \lambda}{\partial x^2}\right]_i \left(\frac{h}{2}\right)^2 + \mathcal{O}(h^4)$

A.4.10 Traditional von Neumann Stability Analysis

A popular definition of stability is described next. As model problem we choose the one-dimensional wave equation

$$\frac{\partial^2 u}{\partial t^2} = \gamma^2 \frac{\partial^2 u}{\partial x^2}$$

with initial conditions

$$u(x,0) = f(x), \quad \frac{\partial u}{\partial t}(x,0) = 0.$$

The corresponding scheme reads[2]

$$[\delta_t\delta_t u = \gamma^2\delta_x\delta_x u]_j^\ell, \quad u_j^0 = f_j, \quad u_j^{-1} = u_j^0 + \frac{1}{2}C^2(u_{j+1}^0 - 2u_j^0 + u_{j-1}^0), \quad \text{(A.33)}$$

for $1 \leq j \leq n$, $\ell \geq 0$, and $C = \gamma\Delta t/h$. Suppose we perturb the initial condition by a function $\epsilon(x)$, where ϵ can represent approximation and round-off errors in the representation of the initial function. The solution of this perturbed problem is denoted by v. Intuition tells that the error $e = u - v$ should be small if ϵ is small. Hence, we can define stability in terms of the property that the error $e(x,t)$ should be bounded in space and time. Then we require the scheme for e to preserve this property.

Let us first indicate that our requirement of bounded e in space and time is reasonable. By subtracting the equations for u and v, we see that the

---

[2] Contrary to Appendix A.4.9, we drop the special notation $\hat{u}$ for the numerical solution. It is only when we compute the truncation error that we really need a special notation to distinguish between u and $\hat{u}$.

function $e(x,t)$ also fulfills the wave equation, but with $e(x,0) = \epsilon(x)$ and $\partial e/\partial t = 0$ as initial conditions. The solution $e(x,t)$ has the form

$$e(x,t) = \frac{1}{2}\epsilon(x - \gamma t) + \frac{1}{2}\epsilon(x + \gamma t).$$

That is, the error is of the same order as the perturbation for all times.

The von Neumann method for stability analysis usually starts with the discrete problem for the error. The error e is then represented as a Fourier series of wave components. As explained before, it is sufficient to study one wave component in a linear problem, for instance,

$$e_j^\ell = \exp\left(i(kjh - \tilde{\omega}\ell\Delta t)\right) = \xi^\ell \exp\left(ikjh\right),$$

where $\xi = \exp\left(-i\tilde{\omega}\Delta t\right)$. Demanding that e_j^ℓ is bounded as $t \to \infty$ leads to $|\xi| \le 1$ as the primary stability requirement. Notice that this is equivalent with requiring $\tilde{\omega}$ to be real in our previous stability analysis. The numerical scheme for the error is identical to (A.33), with f_j replaced by ϵ_j. We insert the exponential form for e_j^ℓ in the scheme and find a condition on Δt from the requirement $|\xi| \le 1$. The calculations end up with $\Delta t \le h/\gamma$. For more details on this type of von Neumann stability analysis, or the alternative matrix method, see Fletcher [40, Ch. 4]. Tveito and Winther's book [120] also covers other methods for stability analysis, including energy arguments and maximum principles.

Alternative versions of the von Neumann stability analysis work with the original equation for u rather than the PDE for the error. The demand is then that the discrete u_j^ℓ remains bounded as $t \to \infty$, leading to $|\xi| \le 1$ when $u_j^\ell = \xi^\ell \exp\left(ikjh\right)$. This is a suitable definition for many physical problems, including the wave and heat equation models dealt with in this appendix. Note that if the underlying PDE is homogeneous and linear (otherwise the exponential solution will not work), the discrete equation for u and e are similar.

This latter von Neumann stability analysis approach is actually mathematically equivalent to our previous method on page 515. This will also be clear from the example in the next section.

Exercise A.16. Go through the details of applying the von Neumann method to the discrete 1D wave equation problem.   ◇

A.4.11   Example: Analysis of the Heat Equation

The ideas from the previous sections concerning the analysis of finite difference schemes via dispersion relations are now applied to the heat equation. An explicit finite difference scheme for the heat equation

$$\frac{\partial u}{\partial t} = \kappa \frac{\partial^2 u}{\partial x^2}$$

can be compactly written as $[\delta_t^+ u = \kappa \delta_x \delta_x u]_j^\ell$. We have already seen that the heat equation allows damped wave solutions according to (A.22). For calculations regarding stability and accuracy we can either work with the form $u_j^\ell = A \exp\left(i(kx_j - \tilde{\omega} t_\ell)\right)$ or we can utilize the analytical knowledge that ω is complex, which points us to $u_j^\ell = A\xi^\ell \exp\left(ikx_j\right)$ as a more appropriate form (note that $\xi = \exp\left(-i\tilde{\omega}\ell\Delta t\right)$). Inserting the latter expression for u_j^ℓ in the scheme yields

$$u_j^\ell = A\left(1 - \kappa \frac{4\Delta t}{h^2} \sin^2 \frac{kh}{2}\right)^\ell e^{ikx_j}.\tag{A.34}$$

The stability follows directly from the principle that $|\xi| \leq 1$ for the solution to be damped, and this gives

$$\Delta t \leq \frac{h^2}{2\kappa}.\tag{A.35}$$

On the other hand, working with $u_j^\ell = A \exp\left(i(kx_j - \tilde{\omega} t_\ell)\right)$ demands more algebra, but it can be instructive to go through these general calculations. After inserting the proposed u_j^ℓ in the scheme, we solve for $\tilde{\omega}$:

$$\tilde{\omega} = i\frac{1}{\Delta t} \ln\left(1 - \kappa \frac{4\Delta t}{h^2} \sin^2 \frac{kh}{2}\right).$$

The argument $\alpha \equiv 1 - \kappa \frac{4\Delta t}{h^2} \sin^2 \frac{kh}{2}$ in the ln function requires some considerations since we deal with complex variables. We know from the solution of the continuous problem that the waves are damped. Hence, we should find $\tilde{\omega} = \tilde{\omega}_r + i\tilde{\omega}_i$ with $\tilde{\omega}_r = 0$ and $\tilde{\omega}_i < 0$. For $0 < \alpha < 1$ these requirements are fulfilled, while $\alpha > 1$ implies $\tilde{\omega}_i > 0$, i.e. growth of waves, which is unacceptable. When $\alpha < 0$, the logarithmic function has complex values. In this case, $\ln \alpha = \ln |\alpha| + i\pi$. The imaginary term only gives rise to a factor -1 when inserted in the wave component so it is of no importance. To obtain a numerical solution that has the damping properties of the solution of the continuous problem, we must require $-1 \leq \alpha \leq 1$. This results in $\Delta t \leq h^2/(2\kappa)$.

Discussion of accuracy follows the same lines as in the example involving the wave equation. We can consider the difference $E_\omega = \omega - \tilde{\omega}_i$ between the continuous and discrete dispersion relations. Taylor-series expansion of $\ln \alpha$ helps us to simplify the expression for $\tilde{\omega}_i$, such that we get

$$E_\omega = -\frac{1}{2}k^4\kappa^2\Delta t + \frac{1}{12}k^4\kappa h^2 + \mathcal{O}(\Delta t^2, h^2\Delta t, h^4).$$

The scheme is hence of first order in Δt and of second order in h. This is of course expected from the approximation properties of the finite differences involved.

Example A.19. Let us demonstrate the von Neumann method for investigating the stability of the forward scheme (A.12) for the *two-dimensional* heat equation. The appropriate form of the wave component is

$$u^\ell_{p,q} = \xi^\ell \exp\left(i(k_x p\Delta x + k_y q\Delta y)\right).$$

By inserting this expression in the scheme, we simply get

$$\xi = 1 - 4\kappa\Delta t \left(\frac{1}{\Delta x^2} \sin^2 \frac{k_x \Delta x}{2} + \frac{1}{\Delta y^2} \sin^2 \frac{k_y \Delta y}{2}\right).$$

Requiring $-1 \le \xi \le 1$ leads to

$$\Delta t \le \frac{1}{2\kappa}\frac{1}{2}\left(\frac{1}{\Delta x^2} + \frac{1}{\Delta y^2}\right)^{-1} \tag{A.36}$$

as the stability criterion.   ◇

Example A.20. The convection-diffusion equation

$$\frac{\partial u}{\partial t} + v\frac{\partial u}{\partial x} = \kappa\frac{\partial^2 u}{\partial x^2},$$

with constant v and κ, can be discretized by a backward difference in time and centered differences in space: $[\delta^-_t u + v\delta_{2x}u = \kappa\delta_x\delta_x u]^\ell_j$. Inserting the wave component $u^\ell_j = \xi^\ell \exp(ikjh)$ in the scheme leads to

$$\xi = \left(1 + \frac{v\Delta t}{h}\sin kh + \frac{4\kappa\Delta t}{h^2}\sin^2\frac{kh}{2}\right)^{-1}.$$

Since the shortest possible wave in the grid has wave length $2h$, kh can at most equal π. Therefore, $\sin kh \ge 0$ and $\xi \le 1$ for all grid spacings, i.e., the scheme is unconditionally stable.

Changing the discretization of the convection term to an upwind difference (assuming $v \ge 0$), i.e. $[\delta^-_x u]^\ell_j$, leads to a complex ξ:

$$\xi = \left(1 + \frac{v\Delta t}{h}(1 - \cos kh + i\sin kh) + \frac{4\kappa\Delta t}{h^2}\sin^2\frac{kh}{2}\right)^{-1}.$$

Knowing that $|a^{-1}| = |a|^{-1}$ for a complex number a, we get

$$|\xi| = \left(\left(1 + \frac{v\Delta t}{h}(1 - \cos kh) + \frac{4\kappa\Delta t}{h^2}\sin^2\frac{kh}{2}\right)^2 + \sin^2 kh\right)^{-\frac{1}{2}}.$$

Since $1 - \cos kh \ge 0$ when $kh \le \pi$, the sum of the terms inside the square root is larger than or equal to unity, making $|\xi| \le 1$ for all h and Δt.   ◇

A.5   Exploring the Nature of Some PDEs

Linear PDEs with first- or second-order derivatives are frequently classified as *elliptic, parabolic,* or *hyperbolic.* The qualitative properties of the solution of a PDE depend on this classification, which in turn influence the numerical discretization strategies. We refer to the literature, for instance Fletcher [40], for precise mathematical approaches to the classification of PDEs. Here we shall focus on some specific prototype hyperbolic, parabolic, and elliptic PDEs and briefly demonstrate the most important characteristic properties of their solutions.

A Hyperbolic Equation. A typical hyperbolic equation is the wave equation

$$\frac{\partial^2 u}{\partial t^2} = \gamma^2 \frac{\partial^2 u}{\partial x^2},$$

whose physical significance is treated in Chapter 1.3.1. The general form of the solution reads

$$u(x,t) = f(x - \gamma t) + g(x + \gamma t),$$

where f and g are functions determined by the initial conditions involving $u(x,0)$ and $\partial u(x,0)/\partial t$. From the general form of the solution we see that the initial u profile, $u(x,0) = f(x) + g(x)$, is split into two parts, moving in positive and negative x direction with speed γ. An important property of the solution is that the shapes of f and g are preserved in time – the argument $x \pm \gamma t$ just leads to a translation of f and g. This means that if the initial profile is discontinuous, the discontinuity will propagate (with speed γ) through the domain[3]. This feature is best demonstrated through visualizing the results of a simulation, where we start with $u(x,0)$ as a plug: $u(x,0) = 1$ for $x \in [0.4, 0.6]$, and $u(x,0) = 0$ elsewhere on the unit interval $[0,1]$. The boundary conditions might be taken as $u(0,t) = u(1,t) = 0$. A suitable simulator for this problem is located in the directory `src/fdm/Wave1D5`. Running `./app -n 61 -C 1.0 --casename t1` solves the wave equation with 61 grid points and Courant number equal to 1.0. A movie can thereafter be produced by

```
curveplot gnuplot -f t1.map -r '.' 'u' '.' -animate -fps 1
              -o 'set yrange [-1.2:1.2]; set data style lines'
```

Alternatively, you can use the `matlab` option to `curveplot`. The Perl script `gui.pl` provides a graphical user interface for simulation and visualization (using Gnuplot).

We observe from the animation that the initial plug is split into two parts with identical shape, but traveling in opposite directions. At the boundary, the waves are reflected in an anti-symmetric fashion, i.e., u changes sign.

---

[3] From a mathematical point of view, discontinuous solutions demand a reformulation of the original PDE [72], such that differentiation of u is avoided.

An interesting question is how the solution is affected by changing the boundary condition at, e.g., $x = 1$. Let us apply the alternative condition[4] $\partial u/\partial x = 0$ at $x = 1$. This is enabled by the `-b 2` option to the `Wave1D5` solver (`-b 1` is default and implies $u(1, t) = 0$) or by activating the second choice on the pull-down menu for boundary conditions in the GUI (`gui.pl`). Making an animation out of the wave motion now shows that the left component of the initial profile is reflected from $x = 0$ as in the previous case, but the right component is reflected from $x = 1$ in a symmetric fashion. When the two components meet again, they cancel each other. Figure A.3 compares the two solutions at four points of time. One can clearly see that a qualitative difference in the wave to the right arises from changing the boundary condition.

Changing the initial conditions changes the shape of the wave for all times. In both simulation examples the value of u at a point x_p is not affected by the type of boundary condition before the reflected wave propagates into the medium again and hits the point x_p. The solution of the wave equation at the point x_p is therefore unaffected by the boundary conditions up to a certain time point. The wave velocity γ determines how fast information is exchanged in the model. This is a characteristic feature of hyperbolic equations. Another feature is that we can solve for new values (at a time level) of u at the grid points separately, i.e., the finite difference scheme is explicit.

Project 1.4.1 deals with another hyperbolic equation, $\frac{\partial u}{\partial t} + \gamma \frac{\partial u}{\partial x} = 0$, where the information is transported with speed γ in one direction only.

Example A.21. In many wave applications it is desirable to transmit waves out of the domain with no reflections. This is enabled by so-called radiation or open boundary conditions. For the constant-coefficient 1D wave equation, an exact radiation condition at $x = 1$ takes the form

$$\frac{\partial u}{\partial t} + \gamma \frac{\partial u}{\partial x} = 0. \tag{A.37}$$

The solution to this equation is a wave $u(x, t) = F(x - \gamma t)$, i.e., a wave traveling to the right, while a reflected wave traveling to the left, $u(x, t) = G(x - \gamma t)$ is not permitted. Equation (A.37) can be discretized by centered differences in space and time:

$$\frac{u_n^{\ell+1} - u_n^{\ell-1}}{2\Delta t} + \gamma \frac{u_{n+1}^\ell - u_{n-1}^\ell}{2h} = 0.$$

Solving this equation with respect to the fictitious value u_{n+1}^ℓ and inserting this expression in the scheme for the wave equation at the space-time point

---

[4] A physical interpretation of the problem can be sound waves in a clarinet, where u is the air pressure. The clarinet is modeled as a straight pipe with a closed end (vanishing air velocity, $\partial u/\partial x = 0$) at $x = 1$ where the blowing is done and an effective open end ($u = 0$) at the location $x = 0$ of an open side hole.

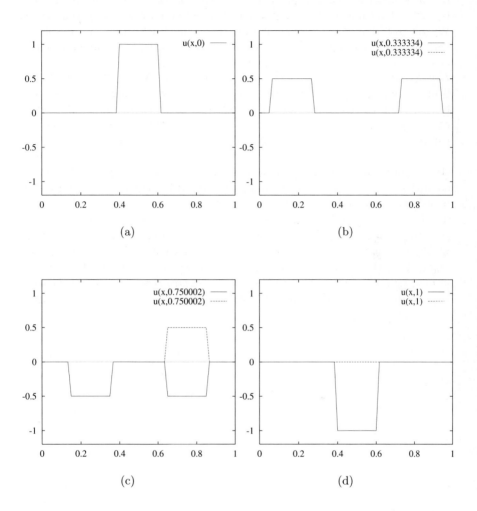

Fig. A.3. Solution of the wave equation with initial plug profile, $\partial u/\partial t = 0$ at $t = 0$, and $u(0,t) = 0$. The solid line corresponds to a solution with boundary conditions $u(1,t) = 0$, whereas the dashed line corresponds to a simulation with boundary condition $\partial u/\partial x = 0$ at $x = 1$. (a) Initial condition; (b) $t = 1/3$, the two disturbances are moving away from each other, but they have not yet reached the boundaries; (c) $t = 3/4$, the two pulses have been reflected from the boundaries and are approaching each other; (d) $t = 1$, the solution is a superposition of the two pulses.

(n, ℓ), results in a special difference formula at $x = 1$. This formula is activated in the Wave1D5 solver by giving the option -b 3. Run the case again, but with the new radiation condition at $x = 1$, and observe from the movie that the wave traveling to the right leaves the domain without any reflections. We remark that construction of radiation condititions that work well for variable-coefficient or higher-dimensional wave equations is a difficult task. ⋄

Exercise A.17. This is a continuation of Example A.21, but we now also want to impose a radiation condition at $x = 0$. Derive the correct form of this condition and the corresponding difference equation to be implemented at the space-time point $(1, \ell)$. Perform the implementation and demonstrate through an animation the effect of the two radiation conditions. ⋄

An Elliptic Equation. The simple equation $-u''(x) = f(x)$ with boundary conditions at $x = 0, 1$ is an example on an elliptic equation. In higher dimensions this equation generalizes to $-\nabla \cdot (\lambda \nabla u) = f$ or just $-\nabla^2 u = f$, which are two "famous" elliptic equations. Let us set $f(x) = 2$, $u(0) = 0$, and investigate the effect of $u(1) = 0$ or $u'(1) = 0$ as we did for the wave equation. Straightforward integration twice and determination of the integration constants from the boundary conditions give $u(x) = 1 - x^2$ when $u(1) = 0$ and $u(x) = 2x - x^2$ when $u'(1) = 0$. Now, *the boundary conditions affect the solution at all points in the domain,* see also Figure A.4. This is a typical feature

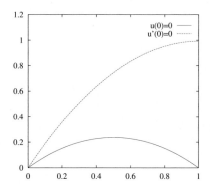

Fig. A.4. Solution of the elliptic equation $-u''(x) = 2$ with $u(0) = 0$ and (a) $u(1) = 0$ or (b) $u'(1) = 0$.

of elliptic equations. Numerically, we have to solve for all discrete values of u simultaneously, i.e., the discrete equations are coupled in a linear system. The resulting set of program statements in the computational algorithm are therefore completely different from a simulator for the wave equation.

A Parabolic Equation. A time-dependent version of our elliptic model problem $-u'' = f$ can read

$$\frac{\partial u}{\partial t} = \frac{\partial^2 u}{\partial x^2} + f(x).\tag{A.38}$$

This is an example on a parabolic equation. A corresponding multi-dimensional parabolic equation is

$$\frac{\partial u}{\partial t} = \nabla^2 u + f.\tag{A.39}$$

Now we focus on (A.38), with $f = 2$, in combination with the initial condition $u(x, 0) = 0$ and the boundary condition $u = 0$ at $x = 0$. At $x = 1$ we also need a boundary condition, and as in the other examples, we shall investigate the difference between $u = 0$ and $\partial u/\partial x = 0$ at $x = 1$. As $t \to \infty$, the solution $u(x, t)$ becomes independent of time $(\partial u/\partial t = 0)$ and (A.38) approaches the elliptic model problem $-u'' = 2$ and the solutions in Figure A.4. A suitable simulator for this test case can be found in `src/fdm/Parabolic1D1`. (Numerical methods and software for (A.38) is the topic of Project 1.7.1.) Running

```
./app -b 1 -n 41 -t 'dt=0.015 [0,3]' -theta 0.5
```

corresponds to using Crank-Nicolson-type time scheme (`-theta 0.5`), $u = 0$ at $x = 1$ (`-b 1`), 41 grid points (`-n 41`), $\Delta t = 0.015$, and a total time interval $[0, 3]$ (the `-t` option). The case $\partial u/\partial x = 0$ at $x = 1$ is simulated by giving the option `-b 2` and enlarging the time interval to, e.g., $[0, 2]$. An extra option `-i` allows to play with different initial conditions: `-i 1` gives $u = 0$, `-i 2` gives uncorrelated random u values between 0 and 0.1, and `-i 3` gives an initial step function: $u = 0.1$ for $x < 1/2$ and $u = 0$ for $x > 1/2$. There is also a graphical user interface `gui.pl` where the user can perhaps more easily adjust the input data.

Animation of the resulting curves reveals a smooth development of the solution from $u(x, 0) = 0$ towards the stationary profiles in Figure A.4. Again, we observe that changing the boundary condition affects the solution at all points and all times. The initial condition influences the solution at all finite times, but parabolic equations have fading memory such that the impact of the initial condition decreases with time. Different initial conditions end up with the same stationary solution. The reader should run experiments with the three initial conditions and observe how irregularities in $u(x, 0)$, or a discontinuity, are smoothed as time increases. From a numerical point of view," the impact of θ is also interesting. Running $\theta = 0.5$ and $\theta = 1$ with $n = 161$ and random initial values shows clearly different behavior. The rapid oscillations in u are quickly damped with the backward scheme ($\theta = 1$), while they tend to be only modestly damped in the Crank-Nicolson scheme ($\theta = 0.5$). From the analysis in Example A.8 on page 508 it is evident that high-frequency oscillations in the exact solution u are very quickly damped out. We can therefore conclude that the Crank-Nicolson scheme in this case leads to (finite amplitude) numerical noise.

Example A.22. Apply the analysis from Appendix A.4.11 to the general θ scheme and compute the damping factor ξ. Compare the damping properties of the exact and the numerical solutions and try to explain the peculiar behavior of the Crank-Nicolson scheme. $\diamond$

From a numerical point of view, parabolic equations like (A.38) can be solved either by explicit schemes of the nature we used for the wave equation or by implicit schemes where we need to solve linear systems of equations at each time level as in the elliptic model problem. Explicit schemes have quite restrict stability requirements[5] on Δt, while implicit schemes can be made stable for all values of Δt (cf. Appendix A.4.11).

Contrary to hyperbolic equations, parabolic equations quickly smooth discontinuities. This property is demonstrated in a two-dimensional problem in Example 3.10 on page 321, where we solve a parabolic equation like (A.39). An mpeg movie of the evolution of the discontinuities in this example is found in the file `discont3d.mpeg` in the directory `src/fem/Heat2`.

The Laplace Equation Solved by a Wave Simulator. The Laplace equation $\nabla^2 u = 0$ in 2D is an elliptic equation whose discretization leads to a linear system. Generating and solving this linear system is considerably more complicated in 2D and 3D than in 1D. (Details about algorithms and implementations are given in Appendices C and D.) However, looking at the equation,

$$\frac{\partial^2 u}{\partial x^2} + \frac{\partial^2 u}{\partial y^2} = 0$$

and rewriting this as

$$\frac{\partial^2 u}{\partial y^2} = -\frac{\partial^2 u}{\partial x^2}$$

indicates a possible interpretation of the Laplace equation as a wave equation with "wrong" sign. It could then be tempting to apply a simple explicit finite difference scheme for solving the Laplace equation. We shall here exploit the idea in a specific problem, $\nabla^2 u = 0$ on the unit square $(0,1) \times (0,1)$, with $u = 1 + x^2 - y^2$ on the boundary. The exact solution is then also given by $u = 1 + x^2 - y^2$.

One immediate problem is that the Laplace equation is associated with boundary conditions at all points in the 2D (x, y) domain, while the wave equation in this example requires a condition on u *and* $\partial u/\partial y$ at $y = 0$ and *no condition* at $y = 1$. Two conditions at $y = 0$ and none at $y = 1$ would also normally be unphysical in problems leading to the Laplace equation. However, since we have the exact solution, we can straightforwardly calculate the consistent condition on $\partial u/\partial y$ for $y = 0$. Changing the name of y to t,

---

[5] In the present problem we have $\Delta t \leq h^2/2$, to be compared with $\Delta t \leq h$ for the wave equation (when $\gamma = 1$).

leads to the following alternative mathematical formulation of the Laplace equation problem:

$$\frac{\partial^2 u}{\partial t^2} = -\frac{\partial^2 u}{\partial x^2}, \quad (x,t) \in (0,1) \times (0,1), \tag{A.40}$$

$$u(x,0) = 1 + x^2, \quad x \in [0,1], \tag{A.41}$$

$$\frac{\partial}{\partial t} u(x,0) = 0, \quad x \in [0,1], \tag{A.42}$$

$$u(0,t) = 1 - t^2, \quad t > 0, \tag{A.43}$$

$$u(1,t) = 2 - t^2, \quad t > 0. \tag{A.44}$$

A comparison with, e.g., (1.44)–(1.48) on page 25 reveals that we can apply a Wave1D1-like simulator for solving the present problem – the only modification is the minus sign in front of the second-order spatial derivative and different values in the initial and boundary conditions. An appropriate program is found in src/fdm/Wave1D4. The algorithm consists of solving for $u(x,t)$ along a line $t = t_\ell$, one point at a time, and then proceed with the next line $t = t_{\ell+1}$. We stop the simulation when $t = 1$ and compare the resulting profile at $t = 1$ with the analytical solution $u(x,1) = x^2$.

The simulation program takes two command-line options, -n for the number of grid points $(1 + 1/h)$ and -c for the Courant number $C = \Delta t/h$. Alternatively, you can use the graphical interface gui.pl with ready-made visualization commands. Let us first use 21 grid points and run varying Δt, corresponding to $C = 1, 0.8, 0.05$. Figure A.5a shows that the error in the first two solutions may be acceptable, while the $C = 0.05$ profile exhibits significant numerical noise in the form of oscillations. Running 11, 21, and 22 grid points, all with $C = 0.8$, see Figure A.5b, shows an alarming phenomenon: As we refine the grid, the oscillations grow and the problem is completely unstable as we go from 21 to 22 grid points; the solution corresponding to 22 grid points has values of magnitude 10^6.

The reason for the observed instability could be bugs in the implementation, a too large Δt in the numerical scheme (i.e. numerical instability), or that the underlying mathematical problem is unstable. That we run into problems, should not come as a big surprise; we said that all boundary values influence the solution of an elliptic PDE at any point in the domain, but here we applied an algorithm where no information about u at $y = 1$ had a chance to be communicated throughout the domain.

The Laplace equation problem, with boundary values on all parts of the boundary, is a stable problem in the sense that if we perturb the boundary values, the perturbation in the solution is bounded[6]. However, when we try to solve the Laplace equation problem as an initial-boundary value problem

---

[6] The precise form of this stability estimate follows from Theorem 2.5 on page 189 and Exercise 2.20 on page 194, if we transform $\nabla^2 u = 0$ with $u = g$ on the boundary to the form $-\nabla^2 v = f = \nabla^2 g$ with $v = 0$ on the boundary ($v = u - g$).

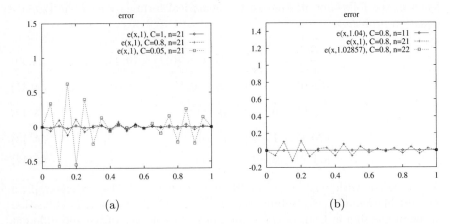

Fig. A.5. The error in the solution of the Laplace equation in 2D by a wave equation algorithm. (a) 21 grid points and varying $\Delta t = Ch$; (b) Fixed Courant number, i.e. $\Delta t = 0.8h$, but varying $n = 1 + 1/h$. The solution in (b) corresponding to $n = 22$ lead to values of order 10^6 and is hence not visible in the plot.

(A.40)–(A.44), small perturbations in the "initial" conditions at $y = 0$ $(t = 0)$ can give arbitrary large perturbations in the solution at $y = 1$ $(t = 1)$. To show this, we first set the right-hand sides in (A.41)–(A.44) to zero. This results in the trivial solution $u = 0$. Then we perturb the condition (A.42) sligthly,

$$\frac{\partial}{\partial t} u(x, 0) = \frac{\sin n\pi x}{n\pi}, \quad x \in [0, 1],$$

which is close to zero for large n. The function

$$u(x, t) = \frac{\sinh n\pi t \sin n\pi x}{n^2 \pi^2}$$

is easily verified to be a solution of the perturbed problem. Although the perturbation is small for $t = 0$ (if n is large), the perturbation is proportional to $\exp(n\pi)$ at $t = 1$, which is definitely large when n is large. The mathematical problem is therefore unstable. Small perturbations, e.g. due to finite precision arithmetic, are always present in a numerical simulation, and these perturbations will then destroy the solution. This means that it does not make sense to apply a modified wave equation solver to the Laplace equation. Zauderer [129, p. 139] presents an alternative analysis of the present problem in terms of the dispersion-relation tools from Appendix A.4.

Well-Posed Problems. An initial-boundary value problem is said to be well-posed mathematically if the following three conditions are met: (i) the solution exists, (ii) the solution is unique, and (iii) the solution depends continuously on the input data. The latter requirement means that small changes in the initial or boundary data, or in the coefficients in the PDEs, should only lead to small changes in the solution. Before we can start finding numerical solutions to PDEs, the underlying mathematical problems must obviously be well-posed. If the solution of the continuous problem does not exist, there is nothing to compute. Fundamental algorithmic problems arise if the solution is not unique. Finally, if the third condition is not met, small round-off errors due to finite precision in arithmetic operations can alter the solution dramatically.

When a well-posed mathematical problem is discretized by numerical methods, we must ensure that also the resulting discrete problem is well-posed. That is: (i) the discrete solution exists, (ii) the discrete solution is unique, and (iii) the discrete solution depends continuously on the approximate representation of initial and boundary data and prescribed functions in the PDEs. In addition, we want the discrete solution to be close, in some sense, to the solution of the continuous problem.

Unfortunately, well-posedness has only been established for rather simple model problems. With the lack of such fundamental theoretical results, one must approach computer experiments with care. One of the major challenges in scientific computing is therefore to interpret the computational results in light of knowledge from physics, numerical analysis, mathematical analysis, and previous experience in order to determine the quality of the results.

Appendix B

Diffpack Topics

B.1 Brief Overview of Important Diffpack Classes

Table B.1. List of the most important classes in Diffpack simulators. The table is continued on the next page.

classname	description
`VecSimplest`	C++ encapsulation of a primitive C vector
`VecSimple`	`VecSimplest` with `operator=`, `print`, and `scan`
`VecSort`	`VecSimple` with sort functionality
`Vec`	`VecSort` with numerics (inner product etc.)
`ArrayGenSimplest`	`VecSimplest` with multiple indices and free base index
`ArrayGenSimple`	`ArrayGenSimplest` with `VecSimple` functionality
`ArrayGen`	`Vec` with multiple indices and arbitrary index base
`ArrayGenSel`	`ArrayGen` with ghost (fictitious) boundary
`Vector`	base class for `Vec`, `ArrayGen`, `ArrayGenSel`
`MatSimplest`	C++ encapsulation of a primitive C matrix
`MatSimple`	`MatSimplest` with `operator=`, `print`,
`Mat`	`MatSimple` with numerics (matrix-vector product etc.)
`MatDense`	synonym for `Mat`
`MatTri`	tridiagonal matrix
`MatDiag`	diagonal matrix
`MatBand`	banded matrix
`MatStructSparse`	structured sparse matrix (nonzeroes on diagonals)
`MatSparse`	general sparse matrix
`Matrix`	base class for `Mat`, `MatTri`, `MatSparse` etc.
`GridFE`	ordinary finite element grid
`GridFEAdB`	adaptive finite element grid, box elements
`GridFEAdT`	adaptive finite element grid, triangles/tetrahedra
`GridDynFE`	dynamic finite element grid (lists of nodes/elements)
`BasisFuncGrid`	`GridFE` overlay for, e.g., mixed interpolation
`Field`	abstract base class for scalar fields
`Fields`	abstract base class for vector fields
`FieldFunc`	explicit function (formula) as scalar field
`FieldsFunc`	explicit function (formula) as vector field

`FieldFE`	scalar field over `GridFE`
`FieldsFE`	vector field over `GridFE`
`FieldsFEatItgPt`	fields at discrete (integration) points in elements
`FieldPiWisConst`	scalar field, constant over materials or elements
`FieldsPiWisConst`	vector field, constant over materials or elements
`GridLattice`	(finite difference) lattice grid with uniform partition
`FieldLattice`	scalar field over `GridLattice`
`FieldsLattice`	vector field over `GridLattice`
`DegFreeFE`	mapping: field d.o.f. $\leftrightarrow$ linear system d.o.f.
`LinEqAdmFE`	simple interface to linear systems and solvers
`LinEqSummary`	summary statistics of linear solver performance
`NonLinEqSolver`	solvers for systems of nonlinear algebraic eq.
`NonLinEqSolver_prm`	governing parameters for `NonLinEqSolver`
`NonLinEqSolverUDC`	interface to nonlinear solvers
`NonLinEqSolvers`	flexible switch between `NonLinEqSolver` objects
`NonLinEqSolvers_prm`	governing parameters for `NonLinEqSolvers`
`NonLinEqSummary`	summary statistics of nonlinear solver performance
`TimePrm`	Δt, time interval for simulation, etc.
`SaveSimRes`	administrator for storing fields on simres files
`SimRes2xxx`	filters simres data to visualization package `xxx`
`FEM`	base class for and interface to finite element solvers
`ErrorNorms`	tool for error computations
`ErrorRate`	estimation of convergence rates
`MenuSystem`	reading input data from a menu
`SimCase`	general interface to a simulator
`MultipleReporter`	tool for automatic report generation

Table B.2. Overview of the most important features in some central Diffpack classes. See the man page for each class for more detailed information. The table is continued on the next pages.

class `MenuSystem`	
`addItem`	defines a menu item
`addSubMenu`	starts a submenu
`get("length")`	returns menu answer for item `length` as `String`
`forceAnswer`	specifies menu answers inside the program
`multipleLoop`	controls program execution
`init`	initialization of a menu system object

class `String`	
`const char* c_str()`	returns string as a standard C `char` array

`contains("bit")`	returns true if string contains the word `bit`
`int getInt()`	extracts an integer
`real getReal()`	extracts a real
`bool getBool()`	extracts a boolean
`bool operator==`	true if two strings are equal

class `FEM`	
`makeSystem`	standard finite assembly algorithm
`calcElmMatVec`	computes element matrix and vector
`numItgOverElm`	numerical integration over element
`integrands`	samples the integrands at a point
`numItgOverSides`	numerical integration over sides of an element
`integrands4side`	"integrands" function for `numItgOverSides`
`makeFlux`	compute smooth flux $-k\nabla u$ from u
`makeMassMatrix`	makes a mass matrix

global functions	
`initDiffpack`	init function for all Diffpack codes
`initFromCommandLineArg`	defines a command-line option and reads its value
`warningFP`	issues a warning
`errorFP`	issues an error message
`fatalerrorFP`	issues an error message and abort execution
`readOrMakeGrid`	reads grid from file or run a preprocessor

class `OpSysUtil`	
`execute("cmd")`	executes the operating system command `cmd`
`removeFile("f1.p")`	removes the file `f1.p`
`renameFile("f1.p","f2.p")`	moves the file `f1.p` to `f2.p`
`makeDir("myplots")`	makes a subdirectory `myplots`
`fileExists("f3.p")`	returns true if file `f3.p` exists
`dirExists("myplots")`	returns true if subdirectory `myplots` exists
`appendFile("f4.p","fold.p")`	appends file `f4.p` to `fold.p`

class `FieldFE`	
`Vec(real)& values()`	returns nodal values vector
`fill(5.3)`	sets all nodal values to 5.3
`valueFEM`	evaluates the field at a point in an element
`valuePt`	evaluates the field at an arbitrary global point
`derivativeFEM`	as `valueFEM`, but for the gradient
`derivativePt`	as `valuePt`, but for the gradient
`GridFE& grid()`	returns access to the underlying grid

class `FieldLattice`	
`Vec(real)& values()`	returns vector of grid-point values
`fill(5.3)`	sets all grid-point values to 5.3

valueFEM	evaluates the field at a point in an element
valuePt	evaluates the field at an arbitrary global point
derivativeFEM	as valueFEM, but for the gradient
derivativePt	as valuePt, but for the gradient
GridLattice& grid()	returns access to the underlying grid

class GridLattice

real Delta(int i)	returns cell size in x_i direction
int getBase(int i)	start index in x_i direction
int getMaxI(int i)	maximum index in x_i direction
real xMin(int i)	minimum x_i value in grid
real xMax(int i)	maximum x_i value in grid
int getDivisions(int i)	number of cells in x_i direction
int getPt(int k, int i)	x_k coord. of grid-point i in k dir.
int getNoPoints()	returns number of grid points

class GridFE

int getNoSpaceDim	returns number of space dimensions
int getNoNodes	returns number of nodes
int getNoElms	returns number of elements
int getNoNodesInElm	returns the number of nodes in an element
String getElmType	returns the ElmDef class name of an element
bool boNode	checks if a node is subject to boundary indicators
getCoor	extracts the coordinates of a node
getMinMaxCoord	computes a hypercube surrounding the grid
addBoIndNodes	adds nodes to a boundary indicator
redefineBoInds	redefines boundary indicators
addMaterial	defines a patch of elements as a new material
getMaterialType	gets the material number of an element
setMaterialType	sets the material number of an element
scanLattice	inits GridFE object by a GridLattice init string
isLattice	true if the grid is actually a lattice
getLattice	extracts underlying GridLattice object (if lattice)
move	deforms the grid according to a displacement field
findElmAndLocPt	finds element and local coord. of a global point
int getNoPoints()	returns number of grid points

class FiniteElement

setLocalEvalPt	specifies evaluation point in local coordinates
getGlobalEvalPt	finds current eval. point in physical coordinates
getLocalEvalPt	gets current eval. point in local coordinates
real N(int i)	returns value of N_i at current eval. point
real dN(int i, int j)	returns value of $\partial N_i/\partial x_j$

`real detJxW()`	returns det J times numerical integration weight
`int getNoBasisFunc()`	returns number of basis func. in element
`bool boSide`	checks if a side is subject to a boundary indicator

class DegFreeFE

`initEssBC`	init object, destroy previous information
`fillEssBC`	assigns known values to degrees of freedom
`getEssBC`	extracts known values of degrees of freedom
`insertEssBC`	inserts known values of degrees of freedom in a vector
`vec2field`	transforms vector to field representation
`field2vec`	transforms field representation to vector
`loc2glob`	transforms local d.o.f. number to global d.o.f.
`GridFE& grid()`	gets access to underlying grid

class TimePrm

`real Delta()`	returns current Δt
`real time()`	returns current time
`setTimeStep`	sets new Δt value
`initTimeLoop`	inits object for monitoring a time loop
`getTimeStepNo`	returns the time step number
`increaseTime`	updates t, time step number, etc.
`bool finished()`	is the time loop finished? true if $t > t_{\text{stop}}$
`stopTimeLoop`	forces next `finished` call to return true
`stepBack`	$t \leftarrow t - \Delta t$, ready to increase time with new Δt

class Prepro

`smooth`	smooths an unstructured grid
`triangle2prism`	extends a 2D grid to a 3D volume

class FieldFunc

`valuePt`	evaluates field formula at an arbitrary global point
`derivativePt`	gradient at a point (formula or difference approx.)

B.2   Diffpack-Related Operating System Interaction

B.2.1   Unix

Customizing Your Unix Environment. To get access to Diffpack, you need to go through the following steps once. The specific actions to be made depends on whether your default Unix shell is bash, sh, csh, or tcsh (type echo $SHELL to see your current shell).

- bash:
  Load your $HOME/.bashrc file into a text editor and add the line
      source pathdp/NO/etc/setup/dpprofile
  at the end of the file.
- sh:
  Load your $HOME/.profile file into a text editor and add the line
      . pathdp/NO/etc/setup/dpprofile
  at the end of the file. The opening dot in this statement is important!
- csh, tcsh:
  Load your ~/.cshrc file into a text editor and add the line
      source pathdp/NO/etc/setup/dpcshrc
  at the end of the file.

In these instructions, the string pathdp is a textual representation of the path to the Diffpack installation on the system[1]. For example, pathdp could be /usr/local/diffpack. Never just type the characters "pathdp"! To activate the set up, either make a new window (i.e. start a new shell) or log out and in again. Then type echo $NOR and check that the output equals the proper Diffpack path (pathdp/NO). We remark that prior to the initialization command listed above, you might have to set the environment variable MACHINE_TYPE to a proper value. Consult the Diffpack installation instructions for your operating system type or ask your computer system manager.

Copying a Directory. Say you want to copy a directory from this book, for example src/fem/Poisson0, to your own directory tree such that you can edit and compile the files and run the application. You must then recall that all paths in this book are given relative to the root $NOR/doc/Book. The proper copy command is therefore

    cp -r $NOR/doc/Book/src/fem/Poisson0 .

The -r option copies the Poisson0 directory and all its subdirectories, and the final dot specifies that the files are to be copied to your current working directory.

---

[1] More precisely, the complete package is located in the directory tree pathdp/NO.

Cleaning Up the Disk. Diffpack applications tend to fill up the disk space, unless you have installed Diffpack with shared libraries. You should therefore remove your big app files regularly. Diffpack offers a script Clean for automatic removal of app files as well as lots of other redundant files[2]: *~, *.bak, core, *tmp, tmp*, to mention a few. Executing the Unix command Clean dir results in a clean up of files in the directory dir *and all its subdirectories.* For instance, Clean $HOME starts in your home directory and cleans all your directories. If you make temporary files that you need for a while, but that should be automatically removed at some later stage, you can use tmp in the filename, since a Clean command removes all files containing the string tmp.

Compiling and Linking Diffpack Applications. Compilation is performed using the Make script, which is nothing but an intelligent search for and execution of GNU's make program. If you know the path to GNU's make program on your computer system, you can always substitute Make by this path. The Make command in Diffpack allows several options:

- MODE=opt turns on compiler optimization. All safety checks of, e.g., array indices and null pointers in Diffpack are turned off.
- MODE=nopt (default) turns on internal safety checks in Diffpack and can thereby detect, for example, array indices out of bounds or null pointer handles[3]. Moreover, MODE=nopt also generates debugging information so that you can run the program inside a debugger and thereby set breaks, examine variables etc. Always use the MODE=nopt option during program development!
- CXXUF= specifies the user's flags to the C++ compiler, as in Make MODE=opt CXXUF="-DDP_DEBUG=2", where the compiler option -D is used to define the preprocessor variable (macro) DP_DEBUG. When you mix C++ code with C or Fortran codes, there are similar flags CCUF and FFUF for the C and Fortran compilers. The LDUF= option is used for specifying flags to the linker.
  The FAQ [63] contains information on how to change the default compiler flags permanently, how to work with different platforms and compiler versions, as well as how to link Diffpack with external software packages.

Compiling in Emacs. We strongly recommend compiling programs inside the emacs editor, since emacs automatically moves to the line in the source code files where a compiler error has occured. This makes it very easy to locate and correct errors. To compile in emacs, write M-x compile, where M-x is usually typed as ESC x, and edit the compilation command to the form you

---

[2] The wildcard * in a filename denotes "any sequence of characters".

[3] MODE=nopt defines the macro SAFETY_CHECKS that is used in preprocessor directives to enclose tests on array indices, null pointers, and other checks for consistency.

would have used on the Unix command line, e.g., `Make`. If the compilation results in errors, type `ctrl-x` ' (control-x and then a backquote) to move the cursor to the file and line where the error occurred. More information about compilation inside `emacs` can be found in [73, p. 77].

Profiling. Unix systems offer programs, called `prof` or `gprof`, for analyzing the CPU-time consumption of the various functions in an application. Such an analysis is referred to as *profiling* and constitutes an indispensable tool for optimizing numerical codes. Profiling can point out computational bottlenecks and thereby direct the programmer's attention to a few critical parts of the code. To allow for profiling, the code must be compiled and linked with a special option, normally `-p` or `-pg`. This can be done either by the command `Make MODE=prof` or more manually by

```
Make MODE=opt CXXUF=-p LDUF=-p  # some compilers require -pg
```

Run the program with input data that give a CPU time of at least 20 seconds. Then you can produce the profile by `gprof app` or `prof app`. The amount of output is quite comprehensive so printing only the first 30 lines can occasionally be convenient, e.g.,

```
gprof app | head -30 > profile.tmp
```

The file `src/fem/Heat2/Verify/testprof.sh` contains a test case involving the steps in profiling. We refer to Appendix B.6 for various ideas regarding optimization of Diffpack applications.

Customizing Makefiles. The Diffpack makefiles should never be edited, because they are under constant improvement, and new versions of Diffpack usually require all the old makefiles to be replaced by new ones (using the script `CpnewMakefiles`, which automatically copies the right makefiles from the repository `$NOR/etc/Makefiles` to the actual application directories). Customization of makefiles takes place in two separate files, named `.cmake2` and `.cmake1`, which are found in every Diffpack application directory (generated by `Mkdir`). In these latter files the user can set various makefile variables and in this way customize the compilation. This is seldom necessary for the average Diffpack user. More information about makefile variables can be found in [63] and in the `.cmake2` and `.cmake1` files.

B.2.2   Windows

Customizing the Visual C++ Environment. The description given here assumes that you are using the Microsoft Visual C++ compiler[4] 5.0 or newer. First, you have to set the paths used when searching for header files. Choose the menu item **Options** from the **Tools** menu. This pops up a new

---

[4] Diffpack also works with the C++ Builder 3.0 or newer from Borland.

window with several tabbed property pages. Choose the property page labeled Directories. To the right you have a list of categories, from which you should choose Include files. In the list of include directories enter:

- `pathdp\WinDP\src\bt\include`
- `pathdp\WinDP\src\la\include`
- `pathdp\WinDP\src\dp\include`
- `pathdp\WinDP\src\extern\include`

Here, `pathdp` should be replaced by the name of your installation's root directory, e.g. `C:\Program Files`.

Moreover, you have to provide paths leading to the library files. From the same window, choose the category Library files and enter:

- `pathdp\WinDP\src\bt\lib`
- `pathdp\WinDP\src\la\lib`
- `pathdp\WinDP\src\dp\lib`
- `pathdp\WinDP\src\extern\lib`

still replacing `pathdp` with your installation's root directory.

After these simple steps, your C++ compiler is all set for Diffpack development, and you can try the example applications located in the directory `pathdp\WinDP\src\extern\projects`.

For the rest of this description, we assume that you are working with the application located in a typical Diffpack finite element solver, e.g., the `Poisson2` application.

Load the Visual Studio workspace by double-clicking the file `DP.dsw` or using the Open Workspace option from the File menu. In the subdirectory `Sim` you will find the application code for this problem. Notice that this is only the Diffpack application code, which with a few exceptions is identical to its Unix counterpart. When compiling the application (using the Build menu) it is linked to the Diffpack GUI library, thus leading to an application with integrated visualization and a fully functional GUI. The compiled executables have the file extension `.exe` and are located in the subdirectories `Debug` and `Release`, respectively for the debug and release compilation modes.

Naturally, this procedure can be repeated for any of the supplied demo applications.

Creating a New Diffpack Application. When making an application of your own, you should start by copying the directory `GUI Application Template`. In the subdirectory `Sim` you will find a dummy application that can be replaced by your own code (or files). If you introduce new files, add them to the project description using the option Add To Project/Files from the Project menu, and repeat the compilation procedure sketched above.

If you want to lay out the workspace yourself, please make sure that the following preprocessor macros are defined:

 – Debug mode: _DEBUG,WIN32, WINDOWS,_AFXDLL,MFCDP_MENUS, NUMT=double,
   VTK_GRAPHICS, SAFETY_CHECKS
 – Release mode: NDEBUG,WIN32,_WINDOWS,_AFXDLL,MFCDP_MENUS,NUMT=double,
   VTK_GRAPHICS, POINTER_ARITHMETIC

Also, you have to link to MFC and use multi-threaded versions of the system libraries. This can be set using the option Settings from the Project menu. This menu choice results in a window in which you should select the property page named C/C++ and the category called Code Generation.

If you want to create a console application without the graphical user interface, omit the definition of the preprocessor macro MFCDP_MENUS and disable linkage to MFC. Further details are given below.

Moving a Diffpack Application from Unix to Windows. First, you should either make a copy of a template project or create a new project from the File menu in Visual Studio. Then, place your Unix application source code in the subdirectory Sim. Check that the file extension of all implementation files is *.cpp. All source code files can now be added to your project using the option Add To Project/Files from the Project menu.

Depending on the type of application you are going to make, you will have to impose a few minor adjustments to the application code. These adjustments are explained below.

Win32 Console Applications. Edit the file containing your application's main function and insert the include directive

```
#include <dpgui/LibsDP.h>
```

This include file contains special instructions on which libraries to link with. Make sure that your project defines the preprocessor macros listed above, except for MFCDP_MENUS which should be left undefined. In case you started out by creating a new Win32 console application project, you will only have to define NUMT=double, and if you also want internal Diffpack consistency checks in a non-optimized version, SAFETY_CHECKS must be defined.

Your project must link the multi-threaded version of the Windows run-time libraries, otherwise you will encounter duplicate definitions of run-time symbols during linkage.

MFC-Based Applications with a GUI. In order to utilize the graphical user interface, add the MFCDP_MENUS preprocessor macro to your project settings. You also have to:

 – Rename your main(int, char**) function to GUImain(int, char**).
 – Add the include directives

```
// MFC Specific includes
#include <dpgui/stdafx.h>    // For precompiled header
#include <dpgui/WinDPGlobals.h>
#include <dpgui/LibsGUI.h>
```

to the file containing the function `GUImain`. Note that you should now use the include file `LibsGUI.h` instead of `LibsDP.h` which is used for the console applications.
- For each of the `*.cpp` files, insert the following include directives at the beginning of the file:

```
// MFC Specific includes
#include <dpgui/stdafx.h>    // For precompiled header
#include <dpgui/WinDPGlobals.h>
```

- In the header file for your simulator's base class (typically derived from class `FEM`), add the include statement

```
// MFC specific includes
#if defined(WIN32) && defined(MFCDP_MENUS)
#include <dpgui/GUIGlue.h>
#endif
```

and add `GUIGlue` as a public base class to the simulator class.

Optionally you may add calls to the functions `synchronize` (needed to update the status of the simulation thread) and `setProgressCtrlData`. See one of the supplied examples for further details.

Once the listed modifications are done, you can compile your application. With a little practice you will be able to port a Unix application to a fully fledged Windows GUI application in the matter of minutes.

A Quick Tour of the Diffpack Application GUI. The Diffpack GUI has three main window areas. At the top there is a browser for Diffpack simulation results integrated with the visualization rendered in the bottom right window. To the left you will find property pages for control of the simulator and for different visualization options, see Figure B.1.

Most of the GUI widgets have associated tool tips. Let the mouse pointer rest for half a second over a button or list item to view a short description of its functionality.

The Toolbar. In the toolbar there are two special buttons to the right giving you online access to the web server with Diffpack information. The rightmost button will load the online reference manuals. This functionality requires the presence of Internet Explorer 3.01 or newer.

The Simres Browser Window. From the simres browser window you can load data for visualization in the rendering window. Notice that the scale factors for scalar and vector data manipulates the data set by multiplying the field values by the specified factor. Using the check box below the rendering button allows you to toggle between processing of vector fields or viewing a single component of the vector data as a scalar field.

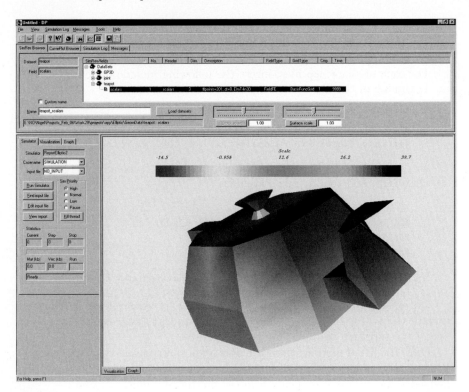

Fig. B.1. An example of a Diffpack application using the MFC-based graphical user interface.

The Simulation Control Window. From the simulation control window you can choose and edit input files to the menu system. Pressing the Run Simulator button invokes the menu system dialog in which you can give values to the individual menu items. Please notice that when you change a menu item you have to explicitly accept the change by clicking the button labeled Click to accept new value.

The Visualization Control Window. In the visualization control window you can choose between different tools:

General: From this page you have the possibility of adding a bounding box for your data set and axes to show the orientation of the scene. The origin of the axes is always placed in the corner of the bounding box corresponding to the minimum value in each spatial direction.
   This property page also provides information on the type of data set currently loaded in the rendering window.

Camera: You can choose between a number of predefined camera positions or move (with immediate rendering) the camera through different angles.

The three buttons at the bottom of this property page give you the possibility of zooming in or out step by step or resetting to normal view.

Light: You have the possibility of choosing between an infinite light model (default) or positional light. In each case you can vary the intensity of the light. For positional lighting you can also set the source point and the focal point. Any movement of the scene will switch back to infinite lighting.

Color preferences: Depending on whether you have chosen Light or Background you can set the color of the light source and the background, respectively. Each color channel (red, green, blue) can be adjusted individually. Checking the Lock channels option causes all channels to be simultaneously adjusted to identical values, resulting in different grayscale levels. There are also two buttons for shortcuts to black and white colors, respectively.

Colormap: Using the sliders you can set the minimum and maximum values of the colormap. By default, these values correspond to the range of the visualized data set. The bottom slider sets the range for the two other sliders by specifying an enlargement factor of the data range. If this factor is one, the sliders for minimum and maximum values of the colormap will be limited to the data range. The reset button reverts the extreme colormap values to the range of the chosen data set and sets the enlargement factor to 2.

Extract: From this page you can extract a subdomain of your data set by setting minimum and maximum values of the interesting volume either in x, y, z coordinates (unstructured data) or as I, J, K indices (structured data). The button labeled Apply must be switched on to have immediate rendering effect. To get an impression of the location of the subdomain, turn on the bounding box for the complete data set.

Surface: You can scale the surface depicting a scalar field by using the slider or supplying a numerical value as text input. The reset button reverts to scale value 1.

Contour/isosurface: This option gives you the possibility of providing a minimum and maximum value to be used for computation of contour curves (2D scalar data) or isosurfaces (3D scalar data). You can also vary the number of contours/isosurfaces to be computed. In order to render the result you have to click the Apply button. Leaving the Apply button on gives immediate rendering of the contour curves and isosurfaces.

Slice: For 3D scalar data sets you can render plane slices by specifying the origin (a point located in the slice plane) and the normal of the slice plane. You must click the Apply button to render the slice.

Vector: As for the surface rendering, you have the possibility of scaling the vector data by numerical text input or a slider. The Reset button reverts to scale value 1.

You are also allowed to switch between different vector visualization techniques like hedgehogs, stream ribbons, and stream tubes. For the latter case, there is also a slider setting the number of sides used for the tube profile. The effect of these visualization modes depends heavily on the current data set and the scaling used for its presentation.

Animation: By providing a root file name (using a relative path such as `..\Data\Anim\lat2D`) and the number of the first and last frame, a click on the Apply button shows an animation based on the data files. All frames must be stored as Vtk data files.

Export field(s): You can export one or more fields to different file formats including Plotmtv, UCD (used by IRIS Explorer and AVS), Vtk, and VRML. See `Projects\Data\VRML` for samples generated this way.

Save Image(s): Here, you can save one or more fields as PPM images. These images can be processed by several Unix utilities or the Windows shareware program LViewPro.

The Rendering Window. This is a standard Vtk rendering window. You can

- move the object when pressing the left mouse button,
- zoom when pressing the right mouse button,
- pan when holding the shift key and pressing the left mouse button,
- switch to wireframe rendering by pressing the W key,
- switch to surface rendering by pressing the S key,
- resetting the scene by pressing the R key.

B.3   Basic Diffpack Features

B.3.1   Diffpack man pages

Every Diffpack class has an associated manual page ("man page") where technical details of the class are documented[5]. The man page is automatically generated from information in the corresponding header file. To invoke the man page for class X on Unix systems, type `dpman X` (or `dpxman` and choose class X from the graphical display). The man page for parameterized classes, like `Vec(real)`, is shown by typing `dpman Vec`. Recall that class `Vec` inherits functionality from several base classes (`VecSort`, `VecSimple`, `VecSimplest`, `Vector`) and that you need to invoke the man pages for these classes as well to get the complete documentation of class `Vec`.

Fortunately, there is a web-based interface to the Diffpack man pages that allows easy overview of functions and data, including those inherited from base classes, see

> `http://www.nobjects.com/Diffpack/refmanuals/current`

---

[5] Man pages might be difficult to read for novice Diffpack programmers. However, extracting useful information from comprehensive documentation, far beyond one's own competence, is in general a very important ability to develop.

Windows NT/95 Remark B.1: When using the special graphical user interface on Win32 platforms, the web-version of the Diffpack man pages can be loaded by clicking on one of the small icons on the toolbar.   ◇

B.3.2 Standard Command-Line Options

Diffpack programs automatically process several command-line options:

- `--help` writes an updated list of the standard command-line options in Diffpack. Options related to `initFromCommandLineArg` calls or the menu system are not listed (run the script `InputManualInLaTeX` to obtain documentation of these simulator-dependent command-line options).
- `--casename` sets a casename for the run. The global variable `casename` is a `String` object that contains the current casename. The casename is automatically manipulated in multiple loops, being in the form `casename_mX` in run no. `X` (the global variable `casename_orig` then contains the original casename as given by the `--casename` command-line option).
- `--casedir mydir` changes the application's current working directory to `mydir`. That is, all files generated by the application are stored in the subdirectory `mydir`.
- `--tolerance` sets the global variable `comparison_tolerance` used for the `eq`, `lt`, `gt`, and also some `operator==` functions in Diffpack. For example, the statement `if (lt(a,b,comparison_tolerance))` actually performs the test `if (a<b-1.0e-4)`, if we have given the option `--tolerance 1.0e-4` to app. The man pages for `Ptv` and `genfc` list some useful `eq`, `lt`, and `le` functions. A suitable size of the tolerance is often governed by the spatial scales of a problem, e.g., the smallest element size divided by a factor. The function `GridFE::calcTolerance` returns a suggested tolerance based on a finite element grid[6].
- `--advice` turns on some internal analysis in Diffpack for checking if inappropriate numerical methods or Diffpack options are being used. For example, `--advice` can detect if banded Gaussian elimination is used on an unstructured grids with large bandwidth – one of the most common reasons for extreme CPU times and memory consumption.
- `--nowarnings` turns off all warning messages.
- `--nounix` avoids execution of operating system commands and saves them in a file `casename.unix`, which can be executed after the simulation, e.g., on another machine. This option works only if you run operating system commands through `OpSysUtil::execute`. Always use this function and

---

[6] Some grid generation algorithms test if two spatial points are identical, using `comparison_tolerance`. If the characteristic length of the grid is several orders of magnitude different from unity, one should adjust the global tolerance variable, either on the command line or in the program.

avoid the C library's `system` function. The global variable `nounix` is set to true if `--nounix` is present as command-line argument.

- `--noreport` avoids automatic report generation by turning off calls to the simulator class' `openReport`, `closeReport`, and `resultReport` functions from the `MenuSystem::multipleLoop` function. This can save substantial execution time, especially during debugging of small test problems, where the generation of reports take much longer time than computing the numerical results. For this option to work, the `main` program must run the code through a `multipleLoop` call. The global variable `noreport` is set to true if `--noreport` is present as command-line argument.

- `--nographics` turns off generation of graphics during executions. All the associated graphics commands are available in an executable script, which after the run has the name `casename.makegraphics`. The global variable `nographics` equals a true value if the `--nographics` option is present as command-line argument.

  The combination of `--nounix` and `--nographics` can be used to suppress execution of time-consuming operating system and visualization tasks during a simulation. The options are also useful for batch runs on high-performance computers where there is no appropriate visualization support. Once the simulation is finished, the scripts can be run (first the operating system commands, then the graphics commands) to complete plots etc.

- `--verbose N` triggers extra output from the Diffpack libraries, often aimed at debugging or detailed information about numerical actions. The amount of output increases with increasing value of the integer `N`. The option `--verbose 1` results in messages about major steps in numerical solution of PDEs, for example, LU factorization, (R)ILU factorization in preconditioners, linear system assembly, and linear system solution. This information, combined with output of memory usage from `--allocreport`, is useful when optimizing the code.

  The integer `verbose` is a global variable containing the value of `N`. By default, `N` is zero. You can use `verbose` in applications to turn debug output on or off at *run time*[7].

- `--allocreport N` reports allocation and deallocation of `int` and `real` matrices and vectors that are larger than `N` megabytes. This option is useful for tracking the memory usage of large-scale numerical applications. We recommend using the option together with `--verbose 1` as the total output is then more readable.

- `--GUI` turns on a graphical user interface in the menu system.

- `--iscl` turns on the command-line mode for reading menu answers.

- `--iss` turns on the terminal (standard input) prompt mode for reading menu commands and answers (this is the default menu system interface).

---

[7] This represents an alternative to the `DBP` macros and `#ifdef DP_DEBUG` preprocessor directive outlined on page 270, which turn debug output on or off at *compile time*.

- `--nodump` turns off all dumping of field and grid data structures from the `SaveSimRes` class. The feature is useful for saving disk space and CPU time when you run large-size numerical experiments and do not need the complete fields afterwards. There is an associated global variable `nodump`, which is true if the `--nodump` option appears on the command line.
- `--exceptions T` tells Diffpack how to handle error messages (exceptions). When `T` equals `LIBRARY` (default), the Diffpack libraries try and catch exceptions, i.e., error messages are reported directly in textual form from the libraries. With `T` as `USER`, the libraries throw exceptions and the user (application programmer) is responsible for the try and catch statements. See the FAQ [63] for more information about various types of exceptions thrown by your current Diffpack version. A corresponding global integer variable `exceptions` equals 0 if the libraries handle exceptions or 1 if it is the responsibility of the application programmer. You can use this variable to enforce a particular exception behavior in an application.

B.3.3 Generalized Input and Output

The Background for Generalized I/O. As Diffpack programmer you can deal with I/O in basically three different ways: (i) use basic `printf`-like C functions, (ii) use `iostream` classes in the standard C++ library, or (iii) use `Is` and `Os` classes in the Diffpack library. We recommend that you use the `Is` and `Os` classes since they are more flexible and general than the other alternatives.

The background for the `Is` and `Os` classes was that C++ I/O syntax is different for `iostream`, strings, and C files. Ideally, a programmer would like to write a single `print` function that can be applied to various output sinks, such as `ostream` (standard output), `ofstream` (files), strings (Diffpack's `String` class), and file pointers in C. Using only standard C++, we would need to write several overloaded `print` functions to support these different output media.

Class `Os` is an interface to various output sinks and offers a unified syntax to `ostream`, `ofstream`, strings, and C file pointers. Moreover, output in ASCII and binary format is transparent in class `Os`. In case of the C file pointer one can employ the xdr binary format[8].

Similar functionality for reading input sources like `istream`, `ifstream`, strings, and C file pointers is provided by class `Is`.

Many of the library classes in Diffpack provide general `print(Os)` and `scan(Is)` functions for writing and reading the contents of the class.

Example on Writing `scan` and `print` Functions. Let us assume that we want to write a class `MyClass` that holds the following data: an interval $[a, b]$, a real vector, and an integer vector. The class is to be equipped with read and write functionality based on the following format.

---

[8] The xdr format is hardware independent and makes it easy to exchange binary files between different platforms.

```
6 numbers in [0,2]
1.1 1.2 1.4 1.6 1.65 0.1
0 1 4 2 8 9
```

The two last rows contain the real and integer vectors and should be written
or read in binary format if the output or input medium is set in binary
mode. The first line should always be in ASCII form[9]. For full flexibility in
the choice of input and output media we use the Is and Os classes. As the
reader will notice, the interface to the Is and Os tools is much inspired by the
interface to the iostream library [91, Ch. 15]. The move from standard input
and output in C++ to our Is and Os classes should therefore be trivial. The
following code illustrates some of the most important functions in the Is and
Os classes.

```
class MyClass {
  real a,b;
  VecSimple(real) reals;
  VecSimple(int)  ints;
public:
  void init  (int n, real a=0, real b=1);
  void scan  (Is is);
  void print (Os os) const;
};

void MyClass:: scan (Is is)
{
  // first part is in ASCII (save format of is)
  Format_type orig_format = is->getFormat();
  is->setFormat(ASCII);  // force ASCII output format
  int n; is >> n;
  is->ignore('[');   // ignore all text up to [ (including [)
  is >> a;           // or is->get(a); read a from is
  is->ignore(',');   // ignore all text up to next comma
  is >> b;           // or is->get(b); read b from is
  is->ignore(']');
  reals.redim(n); ints.redim(n);
  is->setFormat(orig_format);       // set back to original format
  reals.scan(is); ints.scan(is);  // read vectors (binary or ASCII)
}

void MyClass:: print (Os os) const
{
  // first part is written in ASCII
  Format_type orig_format = os->getFormat();
  os->setFormat(ASCII);
  os << oform("%d numbers in [%g,%g]\n",reals.size(),a,b);
  os->setFormat(orig_format);
  if (os->getFormat() == ASCII)  os << '\n';
  reals.print(os);  // write in ASCII/BINARY format
  if (os->getFormat() == ASCII)  os << '\n';
  ints.print(os);
```

---

[9] Only sequence of numbers should be written in binary format, whereas strings
should always be in pure ASCII, since the string termination character '\0' can
only be properly treated in the ASCII format.

```
    if (os->getFormat() == ASCII)  os << '\n';
}
```

Example on Using Is *and* Os. Here is an example demonstrating the flexibility of the scan(Is) and print(Os) functions.

```
MyClass t;
t.scan (cin);                      // read from std input (istream)
t.scan (s_i);                      // read from std input (=cin)
String s = "2 [4,9] 4.1 4.2 8 9";
t.scan (s);                        // read from string
t.scan ("FILE=td.2");              // read from file "td.2"
t.scan (Is("td.2",INFILE));        // read from file "td.2"
ifstream ifile ("td.2",ios::in);
t.scan (ifile);                    // read from file "td.2"
Is ixdr1 ("td.2",BINARY,INFILE,true); // binary xdr file source
t.scan (ixdr1);                    // read from file using C file ptr
Is ixdr2 ("td.2",ASCII,INFILE,false); // ordinary C file (ASCII)
t.scan (ixdr2);                    // read from file using xdr format
Is ixdr3 ("td.2",BINARY,INFILE,false); // ordinary C file (BINARY)
t.scan (ixdr3);                    // read from file using C file ptr
t.print (cout);                    // print to std output (ostream)
t.print (s_o);                     // print to std output (=cout)
t.print (Os("td.2.Os",NEWFILE));   // print to file "td.2.Os"
t.print (Os("td.2.Os",APPEND));    // append to file "td.2.Os"
Os ofile ("td.2.Osb",NEWFILE);     // declare output file "td.2.Osb"
ofile->setFormat(BINARY);          // binary output format
t.print (ofile);                   // write t in binary format
Os oxdr1 ("td.2.Osxdr",BINARY,NEWFILE,true); // xdr file
t.print (oxdr1);                   // print to file in xdr format
Os oxdr2 ("td.2.OsC",BINARY,NEWFILE,false); // standard C file ptr
t.print (oxdr2);                   // print to file in BINARY format
```

When you write to strings, the appropriate syntax is as follows:

```
String s = "";                     // we intend to write to s
Os ostr (s);                       // output destination as a string
t.print (ostr);                    // append output to string s
```

The observant reader has probably noticed that there is automatic type conversion to Is and Os from the standard I/O objects such as istream, ifstream, ostream, ofstream, String, as well as const char* strings.

If you have a file tied to an Os object, e.g. Os ofile ("myfile",APPEND), the file will be closed when ofile goes out of scope, or you can close it manually by ofile->close().

The Input/Output Class Hierarchies. The classes Is and Os are basically pointers (handles) to class hierarchies for input and output. Figure B.2 depicts the relations between the classes for input. This explains why most of the Is and Os functionality must be accessed by an arrow ->; the class Is has a pointer to class Is_base, which serves as base class, defining the the generalized input interface. Various subclasses represent different input sources:

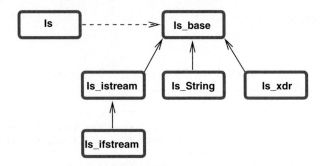

Fig. B.2. Sketch of the class hierarchy for generalized input.

Is_istream is an encapsulation of the standard istream class that comes with C++, Is_ifstream is the corresponding encapsulation of class ifstream for file handling in C++, Is_String offers reading of Diffpack String objects (through the Is_base interface), and Is_xdr offers reading of files, using C file pointers, possibly combined with the binary xdr format. The design of generalized output is similar: Class Os has a pointer to class Os_base, which acts as base class for Os_ostream, Os_ofstream, Os_String, and Os_xdr. The man page for class Is contains most of the information about generalized I/O in Diffpack.

The Is and Os classes do not give access to all the features of the iostream library. Suppose you work with an ifstream object hidden by an Is_ifstream object and accessed by a general Is object. If you want to use the seekg function, this is not supported by our generalized input interface, and you will need to extract the underlying ifstream object and call its seekg function. This is accomplished by casting the Is object's pointer to the Is_base class to the Is_ifstream subclass object. The latter object contains an ifstream member that can be accessed directly. Here is an appropriate code segment.

```
// we have Is inp, check if it is really an Is_ifstream:
if (TYPEID_REF(inp.getRef(),Is_ifstream)) {
  Is_ifstream& inps = CAST_REF(inp.getRef(),Is_ifstream);
  // inps.inputfile is now a reference to an ifstream object:
  ifstream& file = inps.inputfile;
  // call, e.g., file.seekg
```

Similar casts can be performed for other input and output media.

The tools TYPEID_REF and CAST_REF enable run-time checking of the class type and safe *downward* casting from a base class object (here Is_base) to a subclass object (here Is_ifstream). With the corresponding TYPEID_PTR and CAST_PTR one can also check and cast pointers instead of references. Other useful type identification tools[10] cover TYPEID_NAME(p), which returns a const

---

[10] The TYPEID* macros make use either of the C++ RTTI feature [114] in modern compilers or the corresponding functionality in Diffpack.

`char*` string containing the name of the class object being pointed to by `p`, and `TYPEID_STR(FieldFE)`, which converts the class name `FieldFE` to a `const char*` string[11].

B.3.4 Automatic Verification of a Code

The `Make` command has a convenient feature that allows automatic verification of a program in terms of regression tests. Suppose you have developed a program and made a test example with results that are thoroughly verified. You can then use this test case to check that future versions of the program produce results that are in accordance with the verified results. The following steps implement the automatic verification procedure.

- Make a subdirectory `Verify` of the application directory. Move to `Verify`.
- Choose a name for your test example, say `myex1`.
- If needed, make an input file (to the program) for the test example: `myex1.i`.
- Make a script `myex1.sh` that executes `../app` for this example and collects all the relevant results in a file `myex1.v`. Here is an example on such a script, written in Perl:

```
#!/usr/bin/perl
system "../app --casename myex1 < myex1.i > myex1.v";
# append key results:
system "simres2summary -f myex1 -n u -s -A >> myex1.v";
system "perl -pe '' myex1.dp >> myex1.v";  # append myex1.dp
```

As long as the filename ends in `.sh`, you can use any scripting language inside the script, for example, Bourne shell, C shell, Korn shell, Python, or Tcl (the first line specifies the language of the script).
- Run the script `myex1.sh`. If you believe the results are correct, copy `myex1.v` to `myex1.r`. This latter file will be the reference file for later verifications.

You can have many `*.i`, `*.r`, and `*.sh` files in the same `Verify` directory, or you can collect files in subdirectories of `Verify`.

Numerical results will normally change slightly when you run the program on different hardware. It may therefore be convenient to mark the result files with the machine type, that is, one can work with[12] `myex1.v-$MACHINE_TYPE` instead of `myex1.v`. This ensures that new results of the test are always compared to old results on the same hardware.

To perform the verification, issue the command `Make verify`. The make program will now look for `*.sh` files in the `Verify` directory, and all its subdirectories, execute each one of them and compare the recently generated `*.v*`

---

[11] Note that this works for `TYPEID_STR(MatBand(Type))`, while simply putting quotes around `MatBand(Type)` will not work!

[12] `$MACHINE_TYPE` is an environment variable set by the Diffpack start-up scripts.

files with the reference files `*.r*`. The differences are detected by the standard Unix command `diff`, and the output of `diff` is stored in the file `verify.log` in the application directory. Sometimes you have changed the output format or the amount of output from your program. This will result in differences reported in the `verify.log` file. If these differences are expected to be permanent, it is necessary to use the new results in `*.v*` as reference results. The command `Make newverify` copies the present `*.v*` files to their corresponding `*.r*` files.

Making `Verify` directories for all your applications enables easy checking of whether new versions of the program work satisfactorily or not. Furthermore, the perhaps most practical application of the verification procedure described above is for *documenting the usage of your code*; each `*.sh` file tells others how to run your program, how input data can look like, and where to find the expected results of the test case.

B.4   Visualization Support

Simulation software for numerical solution of PDEs tend to generate very large amounts of data. The user therefore needs tools for browsing large data sets and selecting a fraction of the data for visualization. Diffpack supports storage of data for later browsing and visualization. The tools aim at two different classes of simulation data:

– curves $y = f(x)$,
– stationary or time dependent scalar and vector fields.

This classification is motivated from a purely practical point of view, because the visualization tools for curves and fields are usually quite different. For curves one applies sophisticated curve-plotting programs, like Xmgr or Gnuplot, which offer axes, labels, titles, multiple curves in the same plot etc. Visualization of fields normally requires a real visualization system, like Vtk, IRIS Explorer, or AVS. These are big codes, offering advanced visualization algorithms for 2D and 3D fields. Simpler programs, like Plotmtv, can be used for fast 2D visualization. Some curve-plotting programs, e.g. Gnuplot, also offer primitive visualization of 2D functions $z = f(x, y)$, and some visualization systems support curve plotting (e.g. AVS).

The programmer of a Diffpack application can anywhere in the simulator simply dump a curve or a field to a special graphics database. Examples on relevant calls are found in the wave equation simulator from Chapter 1.3, where curves are dumped using the `CurvePlot` class directly, and in the `Poisson1` simulator from Chapter 3.2, where the `SaveSimRes` utility is used to dump scalar and vector fields. The dumped fields and curves are stored in a database, represented as a collection of files. Normally, the compact storage format for fields cannot be directly used in visualization programs. After having selected the fields to be plotted, one must run a filter to transform the

Diffpack-specific storage format to a format that the visualization system can interpret. For curves such format transformation is seldom required. Below we describe some details of the basic tools in Diffpack for storing curves and fields and browsing the stored data sets. The reader should be familiar with the corresponding tutorial material in Chapters 1.3 and 3.3.

B.4.1   Curves

The term *curves* is used for data on the form of (x, y) pairs, and the visualization consists of, e.g., drawing a line between the data points. Diffpack applies the class `CurvePlot` to represent a single curve. Since there might be hundreds of curves from a simulation, a managing class `CurvePlotFile` is used to administer the storage of the individual curves. We normally recommend to use *one* `CurvePlotFile` object in a simulator such that all curves appear to be in the same database. Any `CurvePlot` object needs to be linked to a managing `CurvePlotFile` object. Class `SaveSimRes` contains a data member `cplotfile` of type `CurvePlotFile` that can be used as managing object, especially in finite element simulators, which normally contains a `SaveSimRes` object (see class `Poisson1`). Simpler applications, such as those from Chapter 1.3, need to declare and store a `CurvePlotFile` object explicitly.

Dumping Curves. A curve database with name `casename` is declared and initialized according to

```
CurvePlotFile cplotfile; cplotfile.init(casename);
```

To make a specific curve, one first declares a `CurvePlot` object and binds it to a `CurvePlotFile` managing object:

```
CurvePlot cp; cp.bind(cplotfile);   // or: CurvePlot cp(cplotfile);
```

A curve is recognized by three items: the *title*, the *function name* (sometimes also referred to as curve name), and a *comment*. These data are set in the initialization call to `cp.initPairs`, e.g.,

```
cp.initPairs ("Displacement", "u", "t", aform("u(t=%g)",t));
```

The curve here is perhaps on the form $y = u(t)$, where u has the interpretation of being a displacement. Hence, the title is `Displacement`, the curve name is `u`, the name of the independent variable is `t`, and the final argument to `initPairs` is the comment. As soon as the curve is initialized, we can dump pairs to `cp` by calling `addPair`: `cp.addPair(t,u(t))`. Each `addPair` call adds a new discrete point to the curve. When all pairs are dumped, the curve is closed by the call `cp.finish()`. The `CurvePlot` object can then be reused for a new curve, starting with a `cp.initPairs` call.

CurvePlot objects can be used for spatial 1D functions, time series, convergence rates, time step evolution, and so on. Some of the curves are therefore declared and dumped locally in a simulator, while others are open through the whole simulation. Sometimes it is necessary to work with n related curves throughout the whole simulation. This is straightforwardly done by working with a vector of `CurvePlot` objects: `VecSimplest(CurvePlot)`.

When dumping a 1D field, there is a simplified interface to `CurvePlot`,

```
SimRes2gnuplot::makeCurvePlot
  (v,                       // 1D FieldLattice or FieldFE object
   cplotfile,               // CurvePlotFile manager
   "velocity",              // title of plot
   "v",                     // curve (function) name
   aform("v at t=%g",t));   // comment
```

which was used in the `Wave1D1` solver in Chapter 1.6.3. This function declares its own local `CurvePlot` object and performs the initialization, add, and finish steps. The pairs of $(x, v(x))$ are dumped, and v can be a *one-dimensional* `FieldLattice` or `FieldFE` object.

The `CurvePlot` object stores the data pairs in a file. Each curve has its own file, whose name is generated by the managing `CurvePlotFile` object. The stem of the filename is specified when initializing the `CurvePlotFile` object. Say the stem is S. The data pairs of curve no. 15 handled by this `CurvePlotFile` object are then stored in the file .S_15. The `CurvePlotFile` object keeps track of all its curves in a mapfile, whose name is S.map. The contents of this file is a list of all the dumped curves, with the data file name, title, function name, and comment written line by line. Class `SaveSimRes` applies the stem .casename.curve for its `CurvePlotFile` object cplotfile, so with a casename mc1, the map file has the name .mc1.curve.map and curves are stored in ..mc1.curve_X, X being the curve number.

Generation of hundreds of curves in a simulation causes a blow-up of data files in the directory. In practice it is difficult for a programmer to assign sensible filenames to all these curves, and a system like we have described above is then useful. Notice that the name of all data files starts with a dot such that the files are "invisible" in standard Unix directory listings. This is convenient when a large number of curves are generated.

Browsing and Selecting Curves. The reader is encouraged to compile and run an application that dumps curve plots, e.g., the simulator from Chapter 1.3, the `Wave1D1` simulator from Chapter 1.6.3, or one of the simulators from Chapter 3. Take a look at the mapfile (the mapfile syntax might be a bit odd since it is aimed at being interpreted by an `awk` or `perl` script; you can run the script `cmappr`, with the mapfile name as argument, to get a nice output of the contents of the mapfile).

There are three ways of plotting the curves: (i) using the graphical interface, `curveplotgui .mc1.curve.map`, or the similar functionality in the GUI on Windows platforms, (ii) using the Diffpack curve-plotting script `curveplot`, and (iii) plotting curve files (..mc1.curve_13 etc.) directly in a plotting program.

After having gained access to the list of all stored curves, the next step is to select a few curves for plotting. The selection can be performed by clicking on the desired curves in the graphical interface or by specifying *regular expressions* for the title, function name, and comment of a set of curves. Both procedures are briefly described in Chapter 1.3. Here, we dive a bit more into the specification of regular expressions, because this is a powerful tool and because this is the way we select curves in the Diffpack curve-plotting scripts.

Suppose you specify the following regular expressions for the title, the function name, and the comment:

```
'.'  'f[(]x=0,t=[3-6]\.0'  '.'
```

The first '.' matches everything, and the plot title can therefore be arbitrary. The next pattern specifies that the function name must start with `f(x=0,t=` (notice that the parenthesis () are special characters in regular expressions so we need a backslash or brackets to turn off their special interpretation, e.g. as in `f\(x` or `f[(]x`). The next characters in the function name must be among the numbers 3, 4, 5, or 6, followed by a dot (which has a special interpretation of matching everything, so when we mean the character '.' it must be preceded by a backslash) and 0. No more specification of the function name is given. Examples on names that match the given regular expression are

```
f(x=0,t=3.0)    f(x=0,t=3.01)    f(x=0,t=6.03253)
```

This is perhaps not what we intended. If the goal was to extract the curves with names

```
f(x=0,t=3.0)  f(x=0,t=4.0)  f(x=0,t=5.0)  f(x=0,t=6.0)
```

we should add a quoted) at the end of the regular expression. We refer to [123, Ch. 2] and [42] for a thorough explanation of regular expressions and their many applications in modern software development.

If you intend to produce plots in batch, in a program during execution or from a script, you will need to access the scripting interfaces to curve plotting in Diffpack. These interfaces are briefly demonstrated in Chapter 1.3.4. Diffpack offers a script `curveplot` for selecting and plotting curves, interactively or in batch. The important advantage of this script is that the plotting program is just a parameter, that is, the script provides a unified interface to several plotting programs. The script allows for single plots or animations (both with multiple curves). The visualization is shown on the screen or stored in PostScript or mpeg format. Here is a list of the most common options to the `curveplot` script.

`-f mapfile` specifies the mapfile that should be used when searching for regular expressions.

-r `title funcname comment` specifies three regular expressions, for the plot title, the function name, and the comment, of the curves to be selected for visualization.

-g `WxH+X+Y` specifies the geometry and position of the window (in standard X11 syntax).

-ps `psfile` indicates that the plot is not to be shown on the screen. Instead, a PostScript file `psfile`, containing the plot, is produced. Some plotting programs, such as Xmgr and Plotmtv, offer buttons in their plotting window for dumping the graph in PostScript format. However, the -ps option suppresses the graphical interface of the plotting program and is hence convenient for non-interactive plotting.

-c `commandfile` specifies that the plotting program-specific commands used to produce the plot are to be stored in a file `commandfile`. By default, the `curveplotgui` and `curveplot` store the actual plotting commands in a file `.prog.commands`, where `prog` means the name of the plotting program (Gnuplot, Xmgr, or Plotmtv). The command file can be used for examining or editing the plotting commands, which is particularly useful when you need to adjust the fine details of a plot. Studying this file also helps to learn the basics of the plotting programs. When running `curveplot matlab`, -c is used to set the name of the resulting Matlab script file (`dpc.m` by default, see below).

-o `'plotting program specific commands'` is an option that enables the user to give some plotting program-specific commands regarding the plot. For example, when using `curveplot gnuplot` one might say

    -o 'set title "F-curve"; set xrange [-1:1];'

-animate makes an animation (only available if first argument to `curveplot` is `gnuplot` or `matlab`).

-psanimate works as -animate, but instead of displaying the movie on the screen, all plots (frames) are stored consecutively in PostScript files with names `tmpdpcNNNN.ps`, where `NNNN` is a four digit number. Having generated the PostScript files, the script runs the command `ps2mpeg tmpdpc*.ps` to produce an mpeg movie `movie.mpeg`.

Sometimes you want to build the movie with a special opening page, e.g., with a sketch of the problem being solved, values of various parameters etc. You can then create a PostScript file with the front page in a drawing program and save the file as, e.g., `front.ps`. Rebuilding the movie with `front.ps` as the first frame is then accomplished by the command `ps2mpeg front.ps tmpdpc*.ps`. Several front pages are trivially added if desired.

-fps N results in an animation visualized with `N` frames per second. This option is useful for playing movies in slow motion. Notice that inside Matlab, you can issue the command `movie(M,N,FPS)` for playing the movie `M` `N` times at `FPS` frames per second.

-stop avoids starting up the plotting program. All the commands are available in a script file (`.prog.commands` or `dpc.m`, see the -c option above).

When using `curveplot` with `matlab` as first argument, the plotting session is available in a Matlab script, with default name `dpc.m`. The `dpc.m` file is copied to a `startup.m` file such that the plots are automatically produced when Matlab is started. Inside Matlab, one can rerun the plotting session by typing the name of the script file (`dpc`). All the curve data are stored in internal Matlab variables and can be manipulated and used in various ways. The script file provides documentation of the meaning of these variables. If you apply `curveplot matlab` for animation, the movie is available as the variable M, such that you can replay the movie by `movie(M)`, perhaps augmented with extra arguments for controlling the speed and the number of repetitions.

The curve-plotting tools in Diffpack are built in layers. The `curveplotgui` interface is a fairly short script that calls `curveplot`, which builds a command file for the actual plotting program, where the commands operate on two-column text files for each curve. A user has of course the possibility to invoke the most convenient layer for the problem at hand. For example, in an application where you know exactly the names of the curve files to be plotted, you can create sophisticated commands like we demonstrate in the Perl script `src/app/SqueezeFilm/coupling/xmgr.pl`. Such tailored plotting scripts offer a degree of user control that is of course not possible with tools like `curveplot` or `curveplotgui`. However, one can use the latter scripts to simply generate a basic command file and then develop this file manually to the desired level of sophistication.

Run-Time Plotting of Curves. Sometimes it is desirable to plot curves during the simulation. If the curve is already stored on file by a `CurvePlot` object, one can execute a standard `curveplot` command. For example,

```
OpSysUtil::execute
    (aform("curveplot gnuplot -f .%s.curve.map -r '.' 'u',"
           casename.c_str()));
```

Class `CurvePlot` can also store the data pairs in internal arrays, which can be sent directly to a curve-plotting program. Xmgr and Matlab allow data to be piped, i.e., the arrays can be visualized directly as soon as they are computed.

B.4.2   Scalar and Vector Fields

The support for visualization of scalar and vector fields in Diffpack enables fields to be dumped in a compact binary- or ASCII-based file format anywhere in the simulator. It is also easy and efficient to locate and retrieve a particular group of data from the file. The connection to visualization programs is constructed in such a way that one can select among several programs for visualizing the data. Interactive visualization during the simulation is also supported.

File formats for visualization programs can be classified in two categories: *geometric* formats containing lists of geometric primitives, like points, lines,

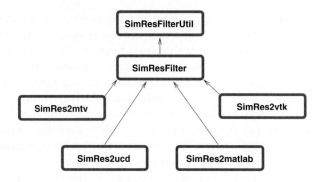

Fig. B.3. Sketch of the class hierarchy for simres filters.

and triangles, and *field* formats containing specification of field values (scalar or vector) over a grid. Data in a field format are usually piped through some sort of drawing algorithms, whose output is in a geometric format. The drawing algorithms and the associated transformation of field data to geometric data are performed inside the visualization system. The normal way of visualizing Diffpack fields is therefore to *filter* Diffpack's field format to a visualization program specific field format and then handle the whole visualization process over to the visualization system.

The simulation program can write field data to file in the Diffpack-specific simres format using class `SaveSimRes`. There exist a number of different programs for filtering the simres format to various visualization system-specific file formats. An introduction to the `SaveSimRes` tool and associated filters is given in Chapters 3.3 and 3.10.2.

Writing Fields Directly in a Specific Format. If the simulator is supposed to be tightly integrated with a particular visualization system, it might be convenient to produce data files with the right format at once instead of going through the intermediate simres format step. Various subclasses of `SimResExport`, see Figure B.3, offer functions that take a Diffpack scalar or vector field object and dumps the field to file in a plotting program specific format. These functions are normally static. For example, the `SimRes2ucd` class exports fields to the ucd format which can be read by AVS or IRIS Explorer. The usage is simple: One sends a vector field object to `SimRes2ucd::plotVector` or a scalar field object to `SimRes2ucd::plotScalar`. Alternatively, we can dump fields in Vtk format by calling the function `SimRes2vtk::plotVector` or its counterpart for scalar fields: `SimRes2vtk::plotScalar`.

The `SimRes2vtk` class also contains many functions for transforming Diffpack field objects to the corresponding objects in Vtk, e.g., `FieldLattice` can be converted to `vtkStructuredGrid`. This is convenient for interactive visualization by means of Vtk in a Diffpack simulator.

It is also possible to use drawing algorithms in class `SimRes2gb` to transform field data to a geometric format inside Diffpack. An example, provided by `src/fem/Poisson1/Verify/s2gb.demo`, involves contouring of finite element fields.

Writing Fields to a Simres Database. Class `SaveSimRes` is only a high-level interface to class `SimResFile`, which is again an interface to the low-level classes `FieldWriter` and `FieldReader`, which actually perform the reading of and writing to simres databases. This layered design enables both easy-to-use tools as well as detailed control of the storage and retrieval of fields.

Fig. B.4. Layers of interfaces to the simres functionality.

`SaveSimRes` offers dumping of a scalar or vector field u to the simres database by the call `database->dump(u)`, where `database` is a `SaveSimRes` object.

`SimResFile` gives slightly more control than class `SaveSimRes`. With this class one can read and write fields of unknown type from and to a simres database. If the type of the field is known, some simpler functions are offered as well. Suppose you want to load a `FieldFE` structure u, with name `temp`, at time 3.2 from a simres database. Having a `SimResFile` resfile connected to this simres dataset, one can write

```
real time = 3.2;
SimResFile::readField (u, resfile, "temp", time);
```

If there is no `temp` field stored at time 3.2, the `readField` function will find the field that is closest to time 3.2 among all the stored `temp` fields.

`FieldWriter` is the low-level class for writing fields to a simres database and `FieldReader` is its counterpart.

Browsing and Filtering Simres Data. Suppose you have generated a simres dataset with casename myc1. There are basically two ways of browsing the contents of this dataset, either examining the text file .myc1.simres or executing the graphical simres browser `simresgui myc1` (on Win32 platforms the general GUI contains a special simres browser). Both the file and the graphical tool should be self-explanatory. Having chosen some interesting fields to visualize, one can filter the data to the desired format by running a `simres2xxx` filter as explained in Chapter 3.3 or one can click on Filter in the graphical interface. During intensive experimentation with a simulation program one will normally write a tailored script that runs the simulator and produces the right kind of visualization automatically. Such scripts must of course rely on executing the non-interactive `simres2xxx` filters. An example on the approach is provided in Chapter 3.10.5.

The Storage Structure of the Simres Format. When storing a field using `SaveSimRes::dump`, the grid is stored on a file `.casename.grid`, while the field values are stored on `.casename.field`. If several fields are dumped to file, they can share the same grid information in the `.casename.grid` file. The reader is encouraged to run a simulator, say `Poisson1`, with a small grid and examine the field and grid files. During the test phase of a simulator, it is often convenient to look at the field file directly to examine the nodal values of the solution.

Writing New Filters. Writing a new filter for transforming the simres format to a new visualization format is most easily accomplished by copying and modifying an existing filter. The functions that may differ from problem to problem are declared as virtual in class `SimResFilter` and can be redefined in the subclass. Frequently, only a couple of the virtual functions need to be implemented. The class hierarchy of some filters appear in Figure B.3.

B.5   Details on Finite Element Programming

B.5.1   Basic Functions for Finite Element Assembly

Overview. Diffpack-based finite element simulators are normally implemented as subclasses of class `FEM`. Class `FEM` contains default versions of the most important finite element algorithms, such as assembly of the linear system (`makeSystem`), computation of the element matrix and vector (`calcElmMatVec`), as well as numerical integration over elements (`numItgOverElm`) and sides (`numItgOverSide`). The `numItgOverElm` routine calls the user-defined function `integrands` for sampling the integrands in the volume integrals in the weak formulation at an integration point in an element. Similarly, `numItgOverSide` calls the user-defined function `integrands4side` for evaluating the integrands in surface or line integrals entering the weak formulation.

The `makeSystem`, `calcElmMatVec`, `numItgOverElm`, and `numItgOverSide` functions are virtual. Class `FEM` provides general versions of these functions, but the programmer can customize the functions in the finite element simulator if desired. When line or surface integrals enter the weak formulation, as in the problem solved by class `Poisson2` in Chapter 3.5, it is necessary to extend the default version of the `calcElmMatVec` function in the finite element solver class and provide both `integrands` and `integrands4side`.

The Body of the Finite Element Assembly Functions. To increase the understanding of the finite element toolbox in Diffpack, it might be instructive to study the source code of some functions in class FEM, for example, makeSystem, calcElmMatVec, numItgOverElm, and numItgOverSide. The bodies of these functions are very short, mainly due to the rich functionality of class FiniteElement. This means that customization of the toolbox, according to special needs in an application, is straightforwardly accomplished. Before we present the bodies of the functions, we need to know about three basic data members in class FEM:

```
FiniteElement finite_elm;
ElmMatVec     elm_matvec;
ElmItgRules   itg_rules;
```

The ElmItgRules object handles various integration rules over the interior of elements and over the sides. The element matrix and vector are represented by the ElmMatVec object. Finally, the FiniteElement object represents basis functions, their derivatives, the Jacobian of the isoparametric mapping, and the integration rule. By the default, the FiniteElement object uses a pointer to the itg_rules object, but if the programmer implements a customized calcElmMatVec function, one can feed other integration-rule objects into the FiniteElement object if desired.

Let us start with the code of the makeSystem function.

```
void FEM::makeSystem (DegFreeFE& dof, LinEqAdmFE& lineq,
                      bool compute_A, bool compute_RHS, bool dummy)
{
  // if compute_A   is true: preserve coefficient matrix
  // if compute_RHS is true: preserve right-hand side

  lineq.initAssemble (dof, compute_A, compute_RHS);
  elm_matvec.attach (dof);
  finite_elm.attach (dof.grid());
  const int nel = dof.grid().getNoElms();
  for (int e = 1; e <= nel; e++)
    {
      elm_matvec.refill (e);
      calcElmMatVec (e, elm_matvec, finite_elm);
      elm_matvec.enforceEssBC ();
      lineq.assemble (elm_matvec);
    }
}
```

The algorithm starts with initializing the global coefficient matrix and right-hand side in the linear system to zero. The boolean arguments can mark the LinEqAdmFE object such that the existing matrix and/or right-hand side are preserved (cf. Appendix B.6.1). Thereafter we initialize the ElmMatVec and FiniteElement objects. The ElmMatVec object needs a pointer to the DegFreeFE object for direct access to the number of degrees of freedom in an element, i.e. the size of the element matrix and vector, as well as the relation between the local and global degrees of freedom numbering needed when assembling element contributions. The `FiniteElement` object is of course dependent on some information about the grid.

The loop over all elements starts with initializing the element matrix and vector, that is, loading the mapping between local and global degrees of freedom and setting all matrix and vector entries to zero. Thereafter, the `elm_matvec` object is computed by invoking either the default version of `calcElmMatVec` or the programmer's own version in the finite element solver. If some of the local degrees of freedom are subjected to essential boundary conditions, the element matrix and vector are modified accordingly. Finally, the element matrix and vector are assembled in the global linear system (depending on the state of `compute_A` and `compute_RHS`).

The default `calcElmMatVec` function is very short:

```
void FEM::calcElmMatVec(int elm_no,ElmMatVec& elmat,FiniteElement& fe)
{
  fe.refill (elm_no, this /*attach solver to fe*/);
  fe.setLocalEvalPts (itg_rules);    // tell fe about intgr. points
  numItgOverElm (elmat, fe);
}
```

The `numItgOverElm` function applies some functionality in class `FiniteElement` to implement the loop over the integration points.

```
void FEM::numItgOverElm (ElmMatVec& elmat, FiniteElement& fe)
{
  fe.initNumItg ();
  while (fe.moreItgPoints()) {
    fe.update4nextItgPt ();
    integrands (elmat, fe);
  }
}
```

In case of line or surface integrals, `calcElmMatVec` must be extended as demonstrated on page 261, with an additional loop over the sides and associated calls to `numItgOverSide`. The latter function has the form

```
void FEM::numItgOverSide
  (int /*side*/, int bound, ElmMatVec& elmat, FiniteElement& fe)
{
  fe.refill4side (side);
  fe.initNumItg ();
  while (fe.moreItgPoints()) {
    fe.update4nextItgPt ();
    integrands4side (side, bound, elmat, fe);
  }
}
```

Example: Integrating the Numerical Error. Suppose you have computed at finite element field `FieldFE` and want to compute the L^2 norm e of the error, provided that a suitable analytical solution functor (`FieldFunc`) is available.

To this end, we need to integrate the squared difference between the numerical and exact solution over all the elements. This is straightforwardly accomplished in the solver class by using some finite element tools and the material in the preceding text:

```
// given Handle(GridFE) grid, Handle(FieldFE) u, and
// Handle(FieldFunc) a (for the exact solution)
ElmItgRules rules (GAUSS_POINTS, 0 /*std order of the rule*/);
FiniteElement fe (*grid);
const int nel = grid->getNoElms();
real L2_norm = 0;
for (int e=1; e<=nel; e++) {
  fe.refill (e, this /*attach solver to fe*/);
  fe.setLocalEvalPts (rules);     // tell fe about intgr. points
  fe.initNumItg();
  while (fe.moreItgPoints()) {    // integration loop in an elm.
    fe.update4nextItgPt();
    L2_norm += sqr(u->valueFEM(fe) - a->valueFEM(fe))*fe.detJxW();
  }
}
L2_norm = sqrt(L2_norm);
}
```

B.5.2   Using Functors for the Integrands

The standard way of providing information about the weak formulation is to write an integrands (and perhaps an integrands4side) function. If we need to work with several weak formulations, we face a fundamental problem since there is only one virtual integrands function available in a simulator class. The most obvious solution is to declare a data member in the simulator class that keeps track of which weak formulation that is to be evaluated. With an if-else statement in the integrands function we can easily jump to an appropriate local routine for evaluating the relevant integrands. Chapters 6.2.5 and 6.3.4 briefly comment on these technical issues in concrete applications. However, class FEM offers another flexible possibility, namely representation of the integrands function as a *functor*.

The integrands functors must be derived from class IntegrandCalc, which defines two virtual functions, integrands and integrands4side, with the same signature as the corresponding virtual functions in class FEM. When using a functor for the integrands function, one must also use a functor for the calcElmMatVec function. Such functors are derived from class ElmMatVecCalc, which (not surprisingly) has a virtual function calcElmMatVec with the same signature as in class FEM. The ElmMatVecCalc functor also allows specification of a (virtual) integrands function. Since most problems can utilize a very simple calcElmMatVec function, like the default one in class FEM, there is a ready-made functor, class ElmMatVecCalcStd, that implements a standard three-line calcElmMatVec function. A programmer can therefore often use this functor and only write a problem-dependent IntegrandCalc functor. If a new

ElmMatVecCalc functor is written, it might be natural to place the integrands function in that functor and skip the special subclass of IntegrandCalc.

A typical IntegrandCalc functor, tied to a finite element solver MySim, can have a MySim* pointer as data member for accessing the solver's data in the integrands function of the functor. If boundary integrals appear in the weak formulation, the functor must also provide an integrands4side function. The body of the integrands (and integrands4side) function is the same as it would be in class MySim, except that access to physical data in class MySim is now enabled through the MySim* pointer. All functors in Diffpack follow the same basic philosophy, so understanding a functor for representing an analytical solution (see Chapter 3.4.4) is sufficient for understanding the IntegrandCalc and ElmMatVecCalc functors.

The assembly of a linear system, using functor representation of the calcElmMatVec and integrands functions, can be performed as follows.

```
HelpPDE1 integrand1 (*this);  // functor, derived from IntegrandCalc
ElmMatVecCalcStd emv; // standard functor provided by FEM.h
makeSystem (*dof, emv, &integrand1, *lineq);
```

A null pointer as IntegrandCalc functor indicates that we use our own ElmMatVecCalc functor, which contains the integrands and integrands4side functions. An example on using integrand functors in a simulator appears on page 433.

Class FEM contains numerous functions that allow assembly into Matrix and Vector structures (cf. Chapter 3.11.2 and Appendix B.6.2). With the aid of functors for representing the integrands in the weak formulation, one can therefore operate with several coefficient matrices, right-hand sides, and weak formulations.

B.5.3   Class Relations in the Finite Element Engine

Although Diffpack programmers normally see only the high-level classes of the finite element and linear algebra toolboxes, it is advantageous for the general understanding of the toolboxes to see an overview of the key classes at lower levels. These classes can also be used directly to build new toolboxes or extend the present ones.

A map over the most important classes and their relations in a typical Diffpack finite element simulator is outlined in Figure B.5. The principal code of the simulator is class MyPDE, as usual derived from class FEM, which again is derived from class SimCase. Specialization or generalization of MyPDE takes place in subclasses, of which MyPDE2 is an example. The MyPDE class contains three fundamental data structures: a grid (GridFE), a field (FieldFE) for the unknown function in the PDE, and a linear system with solvers (LinEqAdmFE). These structures are built of lower-layer classes in the Diffpack libraries.

Let us start with the field structure. The principal content is a vector Vec(real) of nodal values and a pointer to a grid. This grid is actually not an ordinary `GridFE` object, but rather an overlay grid (of type `BasisFuncGrid`) that adds information about basis functions and element types in an ordinary (`GridFE`) grid. The `GridFE` class is basically used as information about the grid *geometry* only. In case of isoparametric elements, the functions used for mapping reference elements are also used as basis functions in the finite element expansion. Sufficient grid information for finite element computations is in this case provided by a `GridFE` object. However, one can think of elements where the geometry is like the 4-node quadrilateral (`ElmB4n2D`), but where the basis function is constant, i.e., there is only one basis function. Such type of elements appear in mixed finite element methods. We can then use a `BasisFuncGrid` to add information about the basis functions over the grid geometry. Many fields may share the same grid geometry, but have different `BasisFuncGrid` overlays. For isoparametric elements, the `BasisFuncGrid` is merely a transparent overlay; all functionality in `BasisFuncGrid` is provided by the underlying `GridFE` object.

The field structure needs to perform interpolation inside the element, that is, it needs to evaluate the basis functions and their derivatives. This is enabled by an internal `FiniteElement` object. The `FiniteElement` object has information about integration rules (`ElmItgRules`), i.e., points where the basis functions are frequently evaluated, the collection of elements in Diffpack (the `ElmDef` class hierarchy), and arrays with the values of basis functions and their derivatives at a point (`BasisFuncAtPt`). For optimization purposes, the `FiniteElement` object can use a table of `BasisFuncAtPt` objects for efficient switch between basis functions at different points. This yields a significant speed-up in problems where all elements are equal, because the basis functions and associated quantities can be computed only once.

The linear system and linear solver tools are managed by class `LinEqAdmFE`. It contains a pointer to a linear solver class hierarchy (`LinEqSolver`) and a pointer to a linear system (`LinEqSystemPrec`). The latter structure consists naturally of the matrix (`LinEqMatrix`), the vector of unknowns and the right-hand side (2 `LinEqVector` objects), as well as a pointer to a class hierarchy of preconditioners (`Precond`). We refer to Appendix D for more detailed information about the linear algebra classes.

B.6   Optimizing Diffpack Codes

The example programs presented in this text mainly demonstrate *flexible* C++ and Diffpack constructions for making codes that are easy to read and debug. Moreover, the codes should be straightforward to extend to more complicated models and allow any type of grid to be used. There is of course an efficiency penalty due to this reliability and flexibility. Experience shows that the efficiency concern is usually limited to the `makeSystem` function (and the function it calls) in finite element simulators. Compared with solution of

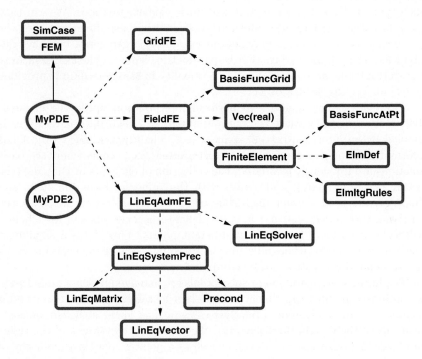

Fig. B.5. A sketch of a simulator class (MyPDE), its base classes, and some layers of the Diffpack library classes that are used by the solver. A solid line indicates class derivation ("is-a" relationship), whereas dashed lines represent a handle ("has-a" relationship).

linear systems, the assembly process[13] involves lots of initializations for each element, short loops, and many scattered function calls. The consequence is that the degree of code optimization depends strongly on the problem at hand and especially on the element type being employed. On the other hand, the linear algebra part of a Diffpack simulator traverses long arrays and is in this respect efficient and general at the same time.

The default generic finite element assembly in Diffpack is usually sufficiently efficient in nonlinear problems when complicated constitutive relations enter the coefficients of the PDE, because in those cases most of the CPU time is spent on evaluating constitutive relations anyway. For simple constant-coefficient PDEs, on the contrary, fine tuning of `makeSystem`-related operations can improve the performance significantly, depending on the particular PDE problem at hand and the willingness to tailor the code to a special class of simulation cases.

The optimization is usually a combination of improving the numerical solution method and making the C++ statements more efficient. The modifications to be done to achieve a performance close to that of a special-purpose program for the problem are normally limited to a couple of pages of extra code. Combination of safety, flexibility, and high performance can therefore be accomplished in Diffpack.

At this point we must emphasize the importance of a proper attitude towards optimization. Too many numerical programmers tend to perform *premature optimizations* [105, Ch. 1.5.1] at places in the code where the effect is hardly noticeable and where "efficient statements" only decrease the readability and extensibility. Most of the CPU time is spent in small parts of the code, which means that one should first develop a simulator that is safe and flexible, then thoroughly verify this implementation, and thereafter perform a profiling (see page 542). The profiling can uncover unexpected bottlenecks and point the programmer to (usually a few) CPU-critical functions in the code. Diffpack and C++ offer various tools to keep the old code intact along with several new optimized versions. At any time the old reliable results can be compared with new ones to quickly detect errors (which are indeed very easy to introduce in an optimization process!).

The relatively simple PDE simulators in Chapters 3, 4.2, 5.1, 6.1–6.3 can benefit greatly from the optimizations suggested below, at the cost of limited extension of the optimized versions to more demanding applications.

B.6.1 Avoiding Repeated Matrix Factorizations

Optimizing Heat Equation Solvers. The heat equation simulators `Heat1` and `Heat2` from Chapters 3.9 and 3.10, respectively, solve a problem where the coefficient matrix in the linear system at each time level remains the same,

---

[13] By *assembly process* we here mean all the tasks in Algorithm 2.1 on page 140, not just the addition of the element contribution into the global linear system.

but the implementation does not exploit this feature. If we choose a direct solver (GaussElim), the code performs in addition a costly LU factorization and a forward-backward solve at every time step. Since the coefficient matrix is constant for all time steps, it is more efficient to construct and factorize this matrix only at the first time step. Normally, one would apply an iterative solver to the linear system with e.g. the popular *incomplete* LU factorization (ILU) preconditioner (see Appendix C.3.3). Also in this case one should reuse an initial incomplete factorization. In addition to avoiding refactorization of matrices that are constant in time, the assembly process of the matrix should of course also be carried out only once, but the right-hand side depends, unfortunately, on u at the previous time level and must be assembled. With a certain adjustment of the algorithm, we can completely avoid the assembly process after the first time step. The details are given in Appendix B.6.2. Optimization of the assembly process itself, which is of primary concern when the matrix changes in time and the coefficient matrix *must* be recalculated, is a technically more comprehensive subject that is dealt with in Appendix B.6.3. Right now we shall limit the attention to optimizing the calls to FEM::makeSystem and LinEqAdmFE::solve such that repeated costly factorizations are avoided.

Implementation. The function LinEqAdmFE::solve can take a boolean argument, indicating whether the coefficient matrix has changed since the last call to solve. If this argument is false, the old factorization can still be used. In (R)ILU preconditioned iterative methods, solve will instead utilize the preconditioner computed in the last call. When we want to reuse a factorized coefficient matrix, we must make sure that makeSystem does not overwrite the matrix with new element matrices in the assembly process. This is accomplished by a third boolean argument to makeSystem. The following code segment exemplifies the modified calls to makeSystem and solve in the function solveAtThisTimeStep in class Heat1 or Heat2:

```
bool first_step = getbool(tip->getTimeStepNo ()==1); // NEW
makeSystem (*dof, *lineq, first_step);               // MODIFIED
lineq->solve (first_step);                           // MODIFIED
```

Notice that this modification is only correct if the coefficients in the PDE are independent of time such that the coefficient matrix is constant. A flexible heat equation solver should have the statements related to special cases located in subclass solvers like we explained in Chapter 3.5.6. The present optimizations are therefore performed in a subclass Heat1iLU of Heat1 in the subdirectory iLU. Since the source code is so short, it is put together with the main function in one file. The code is coupled to the Heat1 files by using the AddMakeSrc tool.

Computational Results. To investigate the effect of the proposed optimizations, we need to have a measure of the CPU time of various parts of the simulator. This is obtained by specifying the command-line option `--verbose 1`. The particular timings to be reported here were obtained on an Intel 200 MHz processor running Linux and the egcs C++ compiler.

Our first test concerns an unstructured 2D grid with the shape of a quarter of a disk and 2016 triangular elements (1084 nodes). The gridfile, found in `iLU/Verify/disk-orig.grid` was created with the `PreproStdGeom` preprocessor (which actually called the Geompack software in this case). Using the default linear solver, i.e. Gaussian elimination on a banded matrix, results in a CPU time of 87 s. For comparison, generation of the linear system takes only 0.4 s. Grid generation techniques that result in unstructured grids usually lead to large a bandwidth. This is exactly the case with our quarter of a disk; the half-bandwidth is 1083. In fact, using a dense matrix is more efficient than a banded matrix! By giving the command-line option `--advice` to `app`, Diffpack will notify you about this undesired phenomenon.

Running the Diffpack program `redband`[14] on the gridfile reduces the bandwidth from 1083 to 41. Now the storage of the coefficient matrix is dramatically reduced and the CPU time of the solver is only 0.16 s. Nevertheless, if we apply an iterative solver, more precisely the Conjugate Gradient method with MILU preconditioning on a sparse (`MatSparse`) matrix, which is a natural and good choice for the present problem with a symmetric and positive definite coefficient matrix, the CPU time of the solve phase is reduced to 0.03 s using the *original* grid with large bandwidth. The corresponding CPU time for the renumbered grid was in fact unchanged. This shows that iterative methods operating on sparse matrix storage structures exhibit minor sensitivity to the nodal numbering and the bandwidth[15]. Also the assembly process is much faster when using the bandwidth-reduced banded matrix or the sparse matrix, since it is faster to access a compact matrix structure than a very large banded matrix, especially if the large structure implies use of virtual memory. In the current example, the CPU time of the assembly phase was reduced by a factor of 4.

The next computational example concerns a box-shaped grid with $7 \times 7 \times 7$ triquadratic elements and 3375 nodes. The bandwidth produced by the `PreproBox` preprocessor is large, but optimal, in this special case. The assembly process took 4.5 s and the solve process, using Gaussian elimination on a banded matrix, required 99 s. Most of the time of the solve process is devoted to factorization of the matrix. This is evident by running `Heat1iLU` and observing that the solve process takes only 0.4 s when we avoid factorization at the second and later time levels. The efficiency gain of our improved `solveAtThisTimeStep` function is hence a factor of over 200. Switching to an iterative solver, i.e. the Conjugate Gradient method with MILU precondi-

---

[14] This program calls functionality in class `Puttonen` to renumber the unknowns in a grid with the purpose of reducing the bandwidth.

[15] This is not completely true; the MILU preconditioner will throw away less fill-in entries if the bandwidth is smaller, cf. Appendix C.3.3.

tioning, led to 1.8 s for solving the linear system at the first time level, and only 0.4 s at later time levels when the initial MILU factorization could be reused.

From these test examples we summarize five important (and widely known) observations: (i) the factorization part of the Gaussian elimination process is often very much more time consuming than the forward elimination and backward substitution process and the assembly the linear system, (ii) the speed of a solver can be dramatically improved by factorizing the matrix only once, (iii) renumbering the unknowns in unstructured grids, with the purpose of reducing the bandwidth, is an essential step before trying to apply Gaussian elimination, (iv) iterative solution of linear systems avoids most of the concerns in point (i)–(iii), and (v) avoiding repeated factorizations of ILU/MILU-type preconditioners can speed up the solve phase by a significant factor.

B.6.2   Avoiding Repeated Assembly of Linear Systems

In the `Heat1iLU` solver from the previous section, all element matrices and vectors were computed at each time level, but only the vectors were used in the assembly process. Below we will show how one can avoid the assembly process completely, after the first time step, and thereby obtain a significant speed-up of the code. This important optimization consists in adjusting the numerical algorithm, using some Diffpack tools that support such adjustments. We shall modify the `Heat2` solver and place the optimizations in a subclass `Heat2eff`.

Formulation of an Efficient Algorithm. The linear systems corresponding to (3.35)–(3.41), with a θ-rule in time and a Galerkin method in space, can be written

$$(\boldsymbol{M} + \theta \Delta t \boldsymbol{K})\boldsymbol{u}^{\ell} = (\boldsymbol{M} + (\theta - 1)\Delta t \boldsymbol{K})\boldsymbol{u}^{\ell-1} + \boldsymbol{c}, \tag{B.1}$$

where $\boldsymbol{M}$ is the mass matrix with entries $M_{i,j} = \int_{\Omega} \beta N_i N_j d\Omega$ and $\boldsymbol{K}$ is a matrix with entries

$$K_{i,j} = \int_{\Omega} k \nabla N_i \cdot \nabla N_j \, d\Omega + \int_{\partial \Omega_R} \alpha N_i N_j \, d\Omega \,.$$

The vector $\boldsymbol{c}$ contains contributions from f and the surface integral of prescribed quantities,

$$c_i = \theta \Delta t \int_{\Omega} f(\boldsymbol{x}, t_{\ell}) N_i \, d\Omega + (1 - \theta) \Delta t \int_{\Omega} f(\boldsymbol{x}, t_{\ell-1}) N_i \, d\Omega$$

$$+ \Delta t \int_{\partial \Omega_R} U_0 N_i \, d\Omega - \Delta t (1 - \theta) \int_{\partial \Omega_R} \alpha u_i^{\ell-1} N_i \, d\Omega \,.$$

Notice that we here assume U_0 to be time independent.

To avoid the finite element assembly process when computing the system (B.1), we must be able to reuse M, K, and c from an initial computation. This is straightforward if β, k, and f are independent of time, because then M, K, and c are also independent of time.

Exercise B.1. If the time dependency in β, k, and f is separable, that is, if we can write $\beta = \beta_T(t)\tilde{\beta}(x)$, $k = k_T(t)\tilde{k}(x)$, and $f = f_T(t)\tilde{f}(x)$, the linear system can be written as a sum of time-independent matrices and vectors, multiplied by appropriate factors consisting of Δt, θ, β_T, k_T, and f_T. For example, $\int_\Omega k\nabla N_i \cdot \nabla N_j d\Omega$ is factorized into $k_T(t)\tilde{K}_{i,j}$, where $\tilde{K}_{i,j} = \int_\Omega \tilde{k}(x)\nabla N_i \cdot \nabla N_j d\Omega$. Carry out the details in this generalization.   ◇

For the rest of this section we assume that β, k, and f are time independent and that Δt and θ are constant, as this gives a simpler algorithm than what is demanded in the more general case considered in Exercise B.1. The computational steps are summarized in Algorithm B.1.

Algorithm B.1.

Solving the heat equation with time-independent coefficients.

for $\ell = 1, 2, 3 \ldots$
    if $\ell = 1$
        compute $A = M + \theta\Delta t K$, $A_{\mathrm{rhs}} = M + (\theta - 1)\Delta t K$, and c.
    compute $b = A_{\mathrm{rhs}}u^{\ell-1} + c$
    solve $Au^\ell = b$ with respect to u^ℓ

Notice that the key to avoiding the assembly process at each time level is to write $b = A_{\mathrm{rhs}}u^{\ell-1} + c$ and update b from the precomputed A_{rhs} and c. The computation of b at every time level in Algorithm B.1 then involves very efficient matrix-vector operations. The result is that the simulation code will spend almost all its time on solving linear systems. When we recompute b by a complete assembly process for each ℓ, the formation of the linear system through a call to FEM::makeSystem is often a significantly time-consuming part of the simulation. One should notice, however, that the assembly algorithm is a process involving $C_1 n$ arithmetic operations, where C_1 is a constant. A method for solving linear systems typically requires $C_2 n^{1+\alpha}$, where C_2 is another constant, usually smaller than C_1, and $\alpha \geq 0$. For small n, generating linear systems can therefore be more time consuming than solving them, while as $n \to \infty$, the solution of linear systems will always dominate, unless one applies special multigrid or domain decomposition methods for which $\alpha = 0$.

When implementing Algorithm B.1, it would be convenient not to touch the original solver Heat2 to minimize the danger of introducing errors in a well-tested code. The new algorithm can be realized in a subclass of Heat2, where we simply provide a more efficient `makeSystem` function. Recall that `makeSystem` is a virtual function that can be redefined in any simulator class derived from FEM (cf. Appendix B.5.1). The new `makeSystem` must compute A, A_{rhs}, and c at the first time level. At all later times, `makeSystem` just computes $b = A_{rhs}u^{\ell-1} + c$ directly from a matrix-vector product and a vector addition.

Handling of Essential Boundary Conditions. Our standard procedure for incorporating essential boundary conditions affects the entries in A and b. Assuming that $\partial\Omega_{E_1}$ and $\partial\Omega_{E_2}$ are fixed in time, we can modify A initially to incorporate essential boundary conditions. The matrix A_{rhs} and the vector c are computed without any adjustments due to essential boundary conditions. The vector b, however, must be modified at each time level to reflect the prescribed u_i values. These modifications consist in the following steps. First, one computes a vector $q = -\sum_{i \in J} \alpha_i A^{(i)}$. Here, J is the set of degrees of freedom (node) numbers subject to essential boundary conditions. The constant α_i is the value of the primary unknown (u) at degree of freedom no. $i \in J$, and $A^{(i)}$ is the ith column of A. The reader can see the full algorithm for incorporation of a single essential boundary condition on page 135. The next step is to compute $b = A_{rhs}u^{\ell-1} + c + q$, and finally one overwrites entry no. i in b with α_i.

The standard behavior of `FEM::makeSystem` is to enforce essential boundary conditions by modifying both the element matrix and vector simultaneously. However, in the present context we want to modify A, but not A_{rhs}, while the modifications of b are to be performed at the global level as a postprocess. Diffpack offers this flexibility through the `DegFreeFE` class. It has two functions

```
void modifyMatDue2essBC (bool onoff);
void modifyVecDue2essBC (bool onoff);
```

used to indicate whether or not the element matrix and vector should be modified due to essential boundary conditions during the assembly process. If the modifications are *not* carried out, the `DegFreeFE` object assembles the element contributions to $-\sum_{i \in J} \alpha_i A^{(i)}$ in a local vector `b_mod`. This means that the `DegFreeFE` object computes our q vector.

Class `DegFreeFE` has another useful function for the present application:

```
void insertEssBC
     (VecSimple(NUMT)& rhs, bool account4multiplicity = false);
```

The `insertEssBC` function replaces entries in the vector `rhs` by known values of the unknowns in the linear systems. If the coefficient matrix has been modified at the element level, the value of a modified diagonal entry in the global matrix is not unity, but an integer that equals the number of element contributions to this degree of freedom. This integer number must also multiply the corresponding known values in the right-hand side vector (`rhs`), which is the case if the second argument to `insertEssBC` is true. Otherwise, `insertEssBC` inserts the known value in `rhs`, assuming that the coefficient matrix has a unit entry on the diagonal.

How to Maximize Reuse of the Original Solver Code. The optimized solution algorithm for the heat equation is to be included in the `Heat2` solver as an extension in a subclass. We would expect that the `Heat2::integrands` and `Heat2::integrands4side` functions are well tested. Perhaps other types of optimization, such as analytical integration, have also been implemented in class `Heat2`. Our new optimized algorithm should therefore utilize the existing, hopefully well-tested, implementations of the discrete equations to as large degree as possible. Ideally, we would compute A, A_{rhs}, and c by just calling `FEM::makeSystem`, which then can jump to the integrands functions of class `Heat2` for evaluating the finite element formulas. However, this requires that we can control the parts of the integrands of the weak formulation to be filled in the coefficient matrix and right-hand side and that we can control the handling of essential boundary conditions.

Looking at the discrete equations (B.1), we see that a standard `makeSystem` operation in class `Heat2` fills the coefficient matrix in the `LinEqAdmFE` object with a matrix $M + \theta \Delta t K$, whereas the right-hand side vector in the `LinEqAdmFE` object becomes equal to $(M + (\theta - 1)\Delta t K)u^{\ell-1} + c$. These observations lead to the following conclusions regarding reuse of class `Heat2` code.

- We can at the first time step compute A by a call to `FEM::makeSystem`, which then invokes existing `Heat2` functionality for making the matrix $A = M + \theta \Delta t K$ and the right-hand side $b = (M + (\theta - 1)\Delta t K)u^0 + c$. (The A and b arrays are after assembly accessible through the functions `LinEqAdmFE::A()` and `LinEqAdmFE::b()`.)
- We can compute A_{rhs} by a call to `FEM::makeSystem`, *provided that we have redefined θ as $1 - \theta$.*
- Having A_{rhs} and b, we can compute $c = b - A_{\mathrm{rhs}}u^0$.

These steps explain how we can carry out the special actions for $\ell = 1$ in Algorithm B.1. A precise computational plan is presented in Algorithm B.2.

Algorithm B.2.

Computation of A, A_{rhs}, *and* c *by reusing class* `Heat2` *code.*

Given `DegFreeFE dof` and `LinEqAdmFE lineq`
compute A_{rhs} without impl. essential boundary cond.:
$\theta_{\text{orig}} = \theta$; $\theta = \theta_{\text{orig}} - 1$
`dof.modifyMatDue2essBC (OFF)`
`dof.modifyVecDue2essBC (OFF)`
`FEM::makeSystem (dof, lineq, true, false)`[a]
store A_{rhs} in a matrix `A_rhs`
`A_rhs = lineq.A()`
compute A and b:
   $\theta = \theta_{\text{orig}}$
   `dof.modifyMatDue2essBC (ON)`
   `dof.modifyVecDue2essBC (OFF)`
   `FEM::makeSystem (dof, lineq, true, true)`
compute $c = b - A_{\text{rhs}}u^0$ with b in `lineq.b()`:
   `A_rhs().prod (u_prev().values(), scratch);`
   `c.add (lineq.b(), '-', scratch);`

---

[a] The `FEM::makeSystem` function takes two boolean arguments,
`compute_A` and `compute_RHS`, which are true by default and indicate whether we should assemble contributions from the element
matrix and vector or not (see also page 572). The present call
tells `makeSystem` to throw away all right-hand side contributions,
but compute the coefficient matrix in the normal way.

In the second call to `FEM::makeSystem` in this algorithm, b is not modified due to essential boundary conditions, but the q vector is computed and
available as the vector `dof.b_mod`. Moreover, the `LinEqAdmFE` object contains
the correct coefficient matrix $A = M + \theta \Delta t K$, but not the correct right-
hand side b in the linear systems to be solved at each time level. To compute
the correct b vector at a time level, we first evaluate $b = A_{\text{rhs}}u^{\ell-1} + c + q$.
Thereafter we call `dof.insertEssBC` to insert correct boundary values in b.
Finally, we load the computed b into `lineq.b()`. The `LinEqAdmFE` object now
contains the proper matrix and right-hand side, and we can proceed with the
call to `lineq`'s `solve` function.

The implementation of the optimized version of the `Heat2` solver is realized
as a subclass `Heat2eff` of `Heat2` (or rather `ReportHeat2` so we get the report
facilities as well). The definition of class `Heat2eff` is shown next.

```
class Heat2eff : public ReportHeat2
{
protected:
  Handle(Matrix(real)) A_rhs;    // right-hand side matrix
  Vec(real)            c;        // source terms, Neumann conditions
  Vec(real)            scratch; // used to compute the right hand side
  real                 c_norm;  // if zero: no need to calculate c

public:
  Heat2eff ();
  ~Heat2eff () {}
  // modify Heat2's solveAtThisTimeStep:
  virtual void solveAtThisTimeStep ();
  // extend makeSystem:
  virtual void makeSystem (DegFreeFE& dof_, LinEqAdmFE& lineq_,
                           bool compute_A, bool compute_RHS,
                           bool only_safe_opt = true);
};
```

The algorithm above can be coded directly in a new function makeSystem.

```
void Heat2eff:: makeSystem
  (DegFreeFE& dof_, LinEqAdmFE&  lineq_,
   bool compute_A, bool compute_RHS, bool only_safe_opt)
{
  // we use dof_ and lineq_ here to avoid confusion with the
  // inherited variables dof and lineq from class Heat2

  FEM::cpu.initTime();  // measure CPU time of the assembly process
  FEM::cpu.lock();       // no FEM functions can now alter cpu state

  if (compute_A)   // assemble matrices (and help vectors)?
    {
      DBP("assembling matrices and vectors in Heat2eff::makeSystem");
      // First we compute the coefficient matrix on the right-hand
      // side, i.e. A_rhs, which is simply the coefficient matrix
      // on the left hand side with theta replaced by theta-1, i.e.,
      // we can call FEM::makeSystem (and Heat2::integrands) to
      // compute lineq_.A and then load lineq_.A into A_rhs.
      // Do not modify the matrix due to essential boundary cond.
      // Notice that the right-hand side is never used.

      const real theta_orig = theta;
      theta = theta_orig - 1;
      dof_.modifyMatDue2essBC (OFF); // default ON
      dof_.modifyVecDue2essBC (OFF);
      FEM::makeSystem (dof_, lineq_, true, false, only_safe_opt);

      // make A_rhs of the same type and size as lineq_.A:
      lineq_.A().makeItSimilar (A_rhs);
      *A_rhs = lineq_.A();

      // Now, compute A, the real coefficient matrix in the linear
      // systems at each time level, and c. Modifications due to
      // essential boundary conditions should be performed for the
      // left-hand side, but not for the right hand side.
```

```
        theta = theta_orig;
        dof_.modifyMatDue2essBC (ON);
        dof_.modifyVecDue2essBC (OFF);

        // if theta = 0 we could use a lumped mass matrix and
        // explicitly demand lineq to use MatDiag and GaussElim
        // (global_menu.forceAnswer("basic method = GaussElim") etc
        // and lineq.scan(global_menu) ... otherwise theta=0 implies
        // consistent mass and full (but efficient) solution of Mu=...

        FEM::makeSystem (dof_, lineq_, true, true, only_safe_opt);

        const int n = lineq_.b().getNoEntries();  // no of equations
        scratch.redim(n); scratch.fill(0.0); c.redim(n); c.fill(0.0);
        // compute c = lineq_.b - M_rhs*u_prev:
        A_rhs().prod (u_prev().values(),scratch);
        c.add (lineq_.b(), '-', scratch);
        // c is often zero, which can save some operations...
        c_norm = c.norm();
    }
  if (compute_RHS)
    {
        // always compute right-hand side of the linear system:
        // lineq_.b = A_rhs*u_prev + c + dof_.b_mod
        // use the scratch vector for intermediate computations
        A_rhs->prod (u_prev->values(),scratch);
        if (!eq(c_norm,0.0,comparison_tolerance))  // c approx 0?
          scratch.add (scratch, c);
        scratch.add (scratch, dof_.b_mod);
        // overwrite degrees of freedom by the boundary values:
        dof_.insertEssBC (scratch, true);
        // store the computed result in lineq's right-hand side:
        lineq_.b() = scratch;
        // lineq_.A and lineq_.b are ready for lineq_.solve()
    }
  FEM::cpu.unlock();
  FEM::cpu_time_makeSystem = FEM::cpu.getInterval();
  if (compute_A)
    FEM::reportCPUtime("Heat2eff::makeSystem(full assembly, step 1)");
  else
    FEM::reportCPUtime("Heat2eff::makeSystem(optimized, rhs only)");
}
```

Remark. Instead of letting the first call to `FEM::makeSystem` fill a `LinEqAdmFE` object and then copying it over to a separate matrix `A_rhs`, we can use an overloaded `FEM::makeSystem` function that computes on `A_rhs` directly. We refer to the `Wave0::solveProblem` function in Chapter 3.11 for an example on this approach. We also mention that multiplying `A_rhs` by `u_prev->values()` is correct only when the vector of nodal values in u has the same numbering as the unknown vector in the linear system. In general one needs to apply `DegFreeFE::field2vec` for loading `u_prev` into a `Vec(real)` array (and back again). The latter vector object is then to be multiplied by the `A_rhs` matrix.

Looking at the boolean arguments in the new `makeSystem` function, we see that `compute_A` must be true at the first time step, when A, A_{rhs}, and c are to be assembled. At future time steps, `compute_A` must be false. The `compute_RHS` argument should always be true, since we always need to compute b. Moreover, the `LinEqAdmFE` object should never touch the coefficient matrix (or its derived preconditioners) after the first step. We recall from page 572 that `LinEqAdmFE::solve` has a boolean parameter that should match `compute_A`. The adjustments of the `makeSystem` and `lineq->solve` calls make it necessary to redefine the first part of the `solveAtThisTimeStep` function in class `Heat2eff`:

```
void Heat2eff:: solveAtThisTimeStep ()
{
  fillEssBC ();
  bool first_step = getbool(tip->getTimeStepNo()==1);        // NEW

  makeSystem (*dof, *lineq, first_step, true);               // MODIFIED

  dof->field2vec (*u, linsol); // most recent sol. as start vector
  lineq->solve (first_step);   // solve linear system       // MODIFIED
```

No other modifications of the original `Heat2` solver should be necessary. Notice that the `makeSystem` call invokes the `Heat2eff::makeSystem` which (at the first time level) calls `FEM::makeSystem` to compute subexpressions in the linear system.

Figure 3.20 on page 322 depicts a model problem in heat conduction. For a mesh with 1280 triangular elements the `Heat2eff` solver increases the efficiency by a factor of about 3.5. (The CPU time of the assembly phase can easily be examined by giving the option `--verbose 1` to `app` or by looking at the `casename.dp` file.) A 3D extension of the example in Figure 3.20 has also been tried. The efficiency factor then becomes about 8 for a grid with 82944 tetrahedra. One should observe that the relative efficiency of the `Heat2` and `Heat2eff` solvers depends on the element type. The effect of the `Heat2eff` optimizations become larger when the ratio of elements and nodes increases, i.e., the maximum gain occurs for tetrahedral elements.

More Flexible Approaches to Precomputed Matrices. A main goal of the customized `makeSystem` function in the preceding section was to reuse as much of the `Heat2` code as possible, with particular emphasis on utilizing the `integrands` (and possibly the `integrands4side`) function for computing A, A_{rhs}, and c. The technique was based on the observation that a routine for computing A can also compute A_{rhs} if θ is replaced by $\theta - 1$. Such simple strategies are not always possible. Also, if Δt or θ changes during the simulation, or if we have time-variable coefficients as explained in Exercise B.1, we need to compute and store separate matrices and vectors instead of A and A_{rhs}. We shall now show how this can be done, but we still limit the details to the case where the coefficients β, k, and f are independent of time.

One way to flexibly compute two matrices M and K in addition to the vector c is to introduce an indicator as class member:

```
enum Integrands_term { M_term, K_term, c_term };
Integrands_term       integrands_term;
```

Using this indicator, we can in `integrands` turn on or off the terms that relate to M, K, and c:

```
if (integrands_term == M_term) {
  for (i = 1; i <= nbf; i++) {
    for (j = 1; j <= nbf; j++) {
      elmat.A(i,j) += beta*fe.N(i)*fe.N(j)*detJxW;
    }
  }
}
if (integrands_term == K_term) {
  for (i = 1; i <= nbf; i++) {
    for (j = 1; j <= nbf; j++) {
      gradNi_gradNj = 0;
      for(s = 1; s <= nsd; s++)
        gradNi_gradNj += fe.dN(i,s)*fe.dN(j,s);

      elmat.A(i,j) += k_value*gradNi_gradNj*detJxW;
    }
  }
}
if (integrands_term == c_term) {
  for (i = 1; i <= nbf; i++) {
    elmat.b(i) += dt*fe.N(i)*f_value*fe.N(i)*detJxW;
  }
}
```

The `makeSystem` function follows the same structure as `Heat2eff::makeSystem`, but we now need to compute two versions of M, one without modifications for essential boundary conditions that is to be used for the right-hand side computations, and one with modifications for essential boundary conditions that is to be used as part of the coefficient matrix $M + \theta \Delta t K$:

```
// M matrix, with and without modifications due to ess. b.c.
lineq_.A().makeItSimilar (M);        // without ess. b.c.
lineq_.A().makeItSimilar (M_webc);   // with ess. b.c.
integrands_term = M_term;
dof_.modifyMatDue2essBC (OFF);
FEM::makeSystem (dof_, *M);
dof_.modifyMatDue2essBC (ON);
FEM::makeSystem (dof_, *M_webc);
```

To compute M, we call an overloaded `makeSystem` function in class `FEM` that fills a matrix directly instead of a `LinEqAdmFE` object.

The same procedure is repeated for the K matrix, except that we in this case must ensure that zeroes are placed on the main diagonal when modifying the matrix for essential boundary conditions (M is also modified). The function `unitMatDiagonalDue2essBC` in class `DegFreeFE` is used to switch between zero or unity on the diagonal in the element matrix.

The c vector needs to exist in only one version where essential boundary conditions are not taken into account. Finally, we compute the coefficient matrix and the right-hand side by efficient matrix-matrix and matrix-vector operations. We refer to `Heat2eff2::makeSystem` for the remaining details regarding the computation of the linear system.

Finally, we remark that instead of using the `integrands_term` indicator, we could apply integrand functors for M, K, and c as outlined in Appendix B.5.2.

B.6.3   Optimizing the Assembly Process

The optimizations in Appendix B.6.2 apply only to transient PDEs with time-independent coefficients. For other (e.g. nonlinear) PDEs, the assembly process must be repeated for each linear system. In such cases one can often gain a significant speed-up by optimizing the assembly process itself. This is the topic of the present section.

We still consider the heat equation solver, in this case class `Heat1` from Chapter 3.9 augmented with a θ-rule for the time discretization. Of course, the optimal finite element implementation for this particular numerical problem was devised in Appendix B.6.2, but the simplicity of the `Heat1` code makes it well suited for illustrating and comparing various optimization techniques, although these are suboptimal for the particular test case at hand. Our purpose here is to list a series of modifications of class `Heat1` that will increase the computational efficiency of the assembly process. All the listed items have been implemented in a subclass `Heat1opt`, see the directory `src/fem/Heat1/optimize`, thus allowing the reader to study the optimizations in detail and experiment with the code. We remark that most of the suggested optimizations are directly applicable to any Diffpack finite element solver for stationary or nonlinear PDEs as well.

1. Writing large fields to file, e.g. by using the `SaveSimRes` tool, is frequently a time-consuming process. All `dump` functions in `SaveSimRes` can be made inactive by the `--nodump` command-line option. Alternatively, one can introduce a user-given flag, which controls the calls to all features not strictly required for the numerical compuatations, including dump actions in the `SaveSimRes` object. The reader can study the usage of such a flag, `stripped_code`, in the `Heat1opt` class.

2. The `makeFlux` function computes *smooth* finite element representations of the flux $-k\nabla u$, but the default smoothing algorithm involves d assembly processes for a problem in $\mathbb{R}^d$, thus requiring `makeFlux` to be called with care. In class `Heat1opt` we call `makeFlux` only if `stripped_code` is false.

3. There is a command-line option `--FEM::optimize ON` that turns on some internal optimizations in Diffpack's finite element toolbox. These optimizations are not always safe so the option must be used with care and always after the program is thoroughly verified.

4. The `integrands` function in classes like `Poisson0`, `Poisson1`, `Poisson2`, `Heat1`, and `Heat2` are written for clarity and safety and can be optimized

by utilizing the fact that the element matrix is symmetric and by moving as many arithmetic operations as possible outside the i-j loops. Moreover, not all C++ compilers are able to inline and optimize calls like fe.dN(i,s). Instead of calling fe.dN(i,s), we can extract a reference to the underlying array and index the array structure directly. This will normally improve the performance.

The following extract from an optimized version of an integrands function from class Heat1opt should be compared to the original code on page 308.

```
// compute the contribution to the element matrix/vector:
const real dt_k_detJxW = dt*k_value*detJxW;
const real rhs = detJxW*up_pt + dt*f_value*detJxW;
int i,j,s;  real N_i;
const Vec(real)& N = fe.N();
const Mat(real)& dN = fe.dN();

for (i = 1; i <= nbf; i++) {
  N_i = N(i);
  for (j = 1; j <= i; j++) {
    gradNi_gradNj = 0;
    for (s = 1; s <= nsd; s++)
      gradNi_gradNj += dN(i,s)*dN(j,s);

    elmat.A(i,j) += detJxW*N_i*N(j) + dt_k_detJxW*gradNi_gradNj;
  }
  elmat.b(i)+= N_i * rhs;
}
for (i = 1; i < nbf; i++)
  for (j = i+1; j <= nbf; j++)
    elmat.A(i,j) = elmat.A(j,i);
```

5. Restricting the integrands routine to a specific element, e.g., the ElmT4n3D element, allows removal of all the short loops. A typical code segment might take this form:

```
// specialized code for ElmT4n3D:
const real dt_k_detJxW = dt*k_value*detJxW;
const real rhs = detJxW*up_pt + dt*f_value*detJxW;

gradNi_gradNj=dN(1,1)*dN(1,1)+dN(1,2)*dN(1,2)+dN(1,3)*dN(1,3);
elmat.A(1,1) += detJxW*N(1)*N(1) + dt_k_detJxW*gradNi_gradNj;

gradNi_gradNj=dN(2,1)*dN(1,1)+dN(2,2)*dN(1,2)+dN(2,3)*dN(1,3);
elmat.A(2,1) += detJxW*N(2)*N(1) + dt_k_detJxW*gradNi_gradNj;

gradNi_gradNj=dN(2,1)*dN(2,1)+dN(2,2)*dN(2,2)+dN(2,3)*dN(2,3);
elmat.A(2,2) += detJxW*N(2)*N(2) + dt_k_detJxW*gradNi_gradNj;
```

6. The numerical integration over elements is costly, and a significant performance improvement can be achieved by turning to analytical integration. This means that we restrict the element type to triangles or tetrahedra and use the formulas from Chapters 2.7.3 and 2.7.5. For example,

$\int_{\Omega_e} N_i N_j d\Omega$ and $\int_{\Omega_e} \nabla N_i \cdot \nabla N_j d\Omega$ are easily calculated from these formulas. The analytical expressions for the element matrix and vectors are then loaded into the `ElmMatVec` object in the function `calcElmMatVec`.

In nonlinear problems one can also take advantage of analytical integration. Consider the term $\int_{\Omega_e} \lambda(u)\nabla N_i \cdot \nabla N_j d\Omega$. Expanding $\lambda(u) \approx \sum_k \lambda(u_k)N_k$ leaves us with the integral $a_{ijk} = \nabla N_i \cdot \nabla N_j \int_{\Omega_e} N_k d\Omega$ on linear elements. The a_{ijk} expressions on each element can be stored in an array `a(k,i,j,e)`, where `e` runs over the elements and `i,j,k` run from 1 to 3 (2D) or 4 (3D). One can also merge the `i` and `j` indices into one index and take advantage of symmetry. The contribution to the element matrix can now be efficiently computed by a loop over the `k` index and multiplying by the most recently computed u values. This type of optimization technique is explained in detail in Chapter 6.5.2.

7. The default `makeSystem` function in class `FEM` performs a series of nested loops with calls to several virtual functions, including the `calcElmMatVec` and `integrands` routines. One can merge the whole assembly process into one function, with the `integrands` statements in the inner loop. An example on such a function is `Heat1opt::makeSystem2`.

   It appears that the overhead of having virtual functions `calcElmMatVec`, `numItgOverSide`, and `integrands` is negligible, because the number of arithmetic operations in the innermost function, i.e. `integrands`, is quite large.

8. Grids containing linear triangular and tetrahedral elements allow for special optimizations. There are two classes in Diffpack supporting such optimizations: `ToolsElmT3n2D` and `ToolsElmT4n3D`. With these tools, one can implement an efficient combination of analytical and numerical integration. Furthermore, Diffpack offers optimized versions of class `ElmMatVec` for representing element matrices and vectors, called `ElmMatT3n2D` for `ElmT3n2D` elements and `ElmMatT4n3D` for `ElmT4n3D` elements. Examples on using these techniques are found in `makeSystem3-6` in the `Heat1opt` class.

9. After the first time step it is not necessary to recompute the coefficient matrix. One can introduce a flag for this in the `makeSystem` and `LinEqAdmFE::solve` calls, like we demonstrated in Appendix B.6.1. (With this technique one also avoids unnecessary recalculation of incomplete factorizations in the RILU preconditioner.) Class `Heat1opt` has a user-given flag, `recompute_A`, that can be used to control if the coefficient matrix should be recomputed or not. All relevant functions in class `Heat1opt` can make use of `recompute_A`.

10. The functions `f`, `k`, and `g` are called quite heavily. These can be made inline instead of virtual. In class `Heat1opt` this is enabled by a conditional compilation command.

11. We mention that `Heat1opt::makeSystem1` implements the optimal algorithm where there is no assembly except at the first time level[16] (see Appendix B.6.2).

12. The computational work when computing element matrices and vectors is proportional to the number of integration points in the element. Reducing the number of integration points by changing the `relative quadrature order` menu item is therefore a quickly performed optimization trick, if it is numerically sufficient to apply a rule with fewer points. For example, the default rule for `ElmT3n2D` elements in Diffpack has three points, but integration of the term $\int_{\Omega_e} \lambda \nabla N_i \cdot \nabla N_j d\Omega$, $\lambda = \sum_k \lambda_k N_k$, only involves linear functions N_k in the integrand and a one-point rule is sufficient. With the `ElmB4n2D` and `ElmB8n3D` a one-point rule for the term $\int_{\Omega_e} \lambda \nabla N_i \cdot \nabla N_j d\Omega$ is also feasible, but this requires special stabilization techniques [70].

13. Finally, we can make a list of actions that reduce the efficiency: (i) using `valuePt` to evaluate fields when `valueFEM` could be called instead (this might be very inefficient in `integrands` routines), (ii) using a too strict termination criterion for linear or nonlinear solvers, (iii) using inefficient linear solvers or preconditioners, (iv) forgetting the ampersand `&` in the declaration of a potentially large data structure as argument in a function, and (v) using the generic `ElmTensorProd1` element when a tailored element like `ElmB4n2D` or `ElmB8n3D` would be more efficient (avoid all the `ElmTensorProd`-type elements if you are concerned about efficiency).

Let us exemplify the effect of the suggested optimizations in the numerical problem from Chapter 3.9. We run the `Heat1opt` code on a grid consisting of $10 \times 10 \times 10$ boxes, where each box is divided into six tetrahedra, giving a total of 6000 elements. The heat equation is integrated over 60 time steps. As equation solver we used the Conjugate Gradient method with SSOR preconditioning and an absolute residual (`CMAbsResidual`) less than 10^{-7} as stopping criterion. The fields are dumped via `SaveSimRes` at every time point in the simulation. Table B.3 shows some timing results.

The specification of the optimizations in the first column of Table B.3 corresponds to the preceding list. A striking result is that the suggested optimizations speed up the code by a factor of almost 15! The best result corresponds, not surprisingly, to the method in Appendix B.6.2, which we actually has defined as not relevant in the current section because we want to focus at problems where the assembly process must be repeated at each time level. Turning off file writing and just optimizing the `integrands` routine, keeping all generality of the code, gives almost a factor of 2.5 in speed-up. Analytical integration gives another factor of two. Specializing the code to the `ElmT4n3D` element also contributes with a factor of two, but the flexibility with respect

---

[16] Notice the difference between items 9 and 11: In 9 an assembly over the right-hand side contributions to the linear system is performed at every time level, whereas in 11 no assembly takes place after the first time step.

Table B.3. Effect of various optimizations in the solver `Heat1opt`. The optimization strategy numbers correspond to the list in the text. The tests were run on a dual Pentium 200 MHz Intel processor under the Linux operating system and the egcs C++ compiler with -O3 optimization.

optimization strategy	CPU time		
	total	assembly	solve
default	58.63	63%	4%
1+2	38.67	94%	6%
1+2+3	37.01	94%	6%
1+2+3+4	23.97	89%	10%
1+2+3+5	19.86	88%	11%
1+2+3+6	13.18	81%	18%
1+2+3+4+7	21.76	87%	11%
1+2+3+4+8	7.16	66%	31%
1+2+3+4+8+9	4.44	46%	50%
1+2+11	4.02	40%	55%

to extension and choice of elements is reduced and the implementational and debugging work becomes significant. A clear conclusion is that the overhead of having `calcElmMatVec` and `integrands` as virtual functions is negligible.

Reordering the Loops in the Assembly Procedure. Let us briefly mention some further possible optimizations of the assembly process. The assembly algorithm is essentially a series of nested `for` loops:

```
for e=1,2,...,no of elements
  for p=1,2,...,no of integration points
    for i=1,2,...,no of element degrees of freedom
      for j=1,2,...,no of element degrees of freedom
        assemble contribution from point p in element e to (i,j) entry
```

When the elements are of the same type and size, the efficiency of the nested loops increases if we avoid nested function calls, have the longest loop as the innermost loop, and operate directly on a large array in the inner statements. Hence, we could "reverse" the loops, which is an important optimization technique on vector computers.

```
for i=1,2,...,no of element degrees of freedom
  for j=1,2,...,no of element degrees of freedom
    for p=1,2,...,no of integration points
      for e=1,2,...,no of elements
        compute contribution from point p in element e to (i,j) entry
```

The inner details of `integrands` could be copied into this loop. An additional set of loops is needed for the line or surface integrals. To improve the efficiency, all the element matrices can be stored in a three-dimensional array, `em(e,j,i)`, the purpose being to avoid function calls and indirect array addressing in the innermost loop. A new loop over `i`, `j`, and `e` is thereafter invoked for assembling the element contributions into a specific Diffpack matrix format for the global matrix.

B.6.4 Optimizing Array Indexing

Implementation of finite difference schemes are conveniently done in terms of grids and fields abstractions, represented by the `GridLattice` and `FieldsLattice` classes. The computationally intensive part of finite difference schemes is the traversal of the grid points, involving indexing in the underlying `ArrayGenSel` array of point values in the field object.

Having a `Handle(FieldLattice)` object with the name `u`, we can index the field values by `u->values()(i,j,k)`. However this is actually a quite complicated set of nested function calls:

```
Handle(FieldLattice)::operator->
FieldLattice::values()          // returns an ArrayGen(real)
ArrayGenSel(real)::operator()   // ArrayGen(real)::operator()
ArrayGenSimplest(real)::operator()
ArrayGenSimplest(real)::multiple2single  // (i,j,k) -> single index
```

The final return statement makes an array look up on the form `A[a*k+b*j+i]`, where `A` is the basic one-dimensional C array containing the array entries and `a` and `b` are constants. Although all these functions are defined as inline in the source code, many C++ compilers have problems with really performing the inlining and optimizing the resulting expressions to the fullest extent. We have therefore in this book emphasized extracting references to lower level structures, e.g.,

```
ArrayGen(real)& Up = u->values();
```

and then indexing the `ArrayGen(real)` object[17] `Up(i,j,k)` since we then get rid of two nested function calls compared with `u->values()(i,j,k)`.

A typical code segment involving a finite difference scheme for the 3D heat equation, $[\delta_t^+ u = \delta_x \delta_x u + \delta_y \delta_y u + \delta_z \delta_z u]_{i,j,k}^\ell$ (in the notation of Appendix A.3), might look as follows:

```
ArrayGen(real)& Up = u_prev->values();
ArrayGen(real)& U  = u->values();
// arrays have dimension (1:imax,1:jmax,1:kmax)

for (k = 2 ;k <= kmax-1; k++) {
  for (j = 2; j <= jmax-1; j++) {
    for (i = 2; i <= imax-1; i++) {
      U(i,j,k) =   Up(i,j,k)+(
                   (Up(i+1,j,k) - 2*Up(i,j,k) + Up(i-1,j,k))
```

---

[17] `ArrayGen` is a base class for `ArrayGenSel`, see page 71, but we could also used an `ArrayGenSel` reference, or an `ArrayGenSimplest` reference for that sake.

```
        +  (Up(i,j+1,k)  -  2*Up(i,j,k)  +  Up(i,j-1,k))
        +  (Up(i,j,k+1)  -  2*Up(i,j,k)  +  Up(i,j,k-1)));
    }
  }
}
```

We emphasize that in a multi-dimensional `ArrayGen`-type array, the entries are stored in "column order", exactly as in Fortran, such that the first index has the fastest variation. It is therefore important to have the loops in the order `k`, `j`, and `i`. In this way, the index `(i,j,k)` traverses the array data as they are stored in memory. Performance will usually decrease substantially if the order of the loops is changed.

Class `ArrayGenSimple` offers special functions for fast indexing in finite difference contexts. The technique consists in setting a reference index `(i,j,k)` in the array object and then performing indexing relatively to the reference index:

```
ArrayGen(real)& U = u->values();
ArrayGen(real)& Up = u_prev->values();
U.getDim(xl,yl,zl);
real xyl=xl*yl;
for (k = 2; k <= kmax-1; k++ ) {
  for (j = 2; j <= jmax-1; j++ )
    // setLocalIndex converts (i,j,k) to single index:
    U. setLocalIndex(2,j,k);
    Up.setLocalIndex(2,j,k);
    for (i = 0; i <= imax-3; i++) {
      // local(i) adds i to the single index set by setLocalIndex:
      U.local(i) =  Up.local(i) + (
            (Up.local(i+1)   -2*Up.local(i) + Up.local(i-1))
          + (Up.local(i+xl)  -2*Up.local(i) + Up.local(i-xl))
          + (Up.local(i+xyl) -2*Up.local(i) + Up.local(i-xyl)));
    }
  }
}
```

This indexing gives a significant speed-up compared with `U(i,j,k)`. However, the relative efficiency between different indexing techniques depends considerably on compilers and hardware.

Complete control of the indexing is enabled by accessing the underlying one-dimensional C array directly. The statement

```
real* U = u->values().getPtr1();
```

gives direct pointer access to the C array. The entries are then `U[1]`, `U[2]`, `U[3]` and so on[18]. The programmer is now responsible for indexing the one-dimensional array correctly; the 3-tuple `(i,j,k)` must efficiently be turned into a single index. This can be done by the function

---

[18] Notice here that the `getPtr1` function returns a pointer value such that `U[1]` is the first entry. Standard C convention, i.e., `U[0]` as the first entry, is obtained by calling `getPtr0`.

```
ArrayGenSimplest::multiple2single(i,j,k)
```

which is available to all `ArrayGen` or `ArrayGenSel` objects. Alternatively, one can calculate this index manually as we do in the code segment below[19]. The finite difference code segment now takes the form

```
real* U = u->values().getPtr1();
real* Up = u_prev->values().getPtr1();
const int ijmax = imax*jmax;
// (i,j,k) -> (k-1)*imax*jmax + (j-1)*imax + i

for (k = 2; k <= kmax-1; k++ ) {
  for (j = 2; j <= jmax-1; j++ ) {
    // call U.multiple2single(0,j,k) or use explicit formula:
    localbase = (k-1)*ijmax + (j-1)*imax;
    for (i = 2; i <= imax-1; i++ ) {
      base = localbase + i;
      U[base] =   Up[base]+(
                  (Up[base+1]      -2*Up[base] + Up[base-1])
            + (Up[base+imax]   -2*Up[base] + Up[base-imax])
            + (Up[base+ijmax] -2*Up[base] + Up[base-ijmax]));
    }
  }
}
```

It goes without saying that this optimized version should not be considered before the safe version is thoroughly debugged. With a working code at hand, it is much easier to detect the bugs in a complicated set of optimized statements.

---

[19] Full overview of indexing formulas is important for debugging anyway. The particular formula we use requires a unit base index for i, j, and k, while the `multiple2single` function would handle arbitrary base indices.

Appendix C

Iterative Methods for Sparse Linear Systems

When discretizing PDEs by finite difference or finite element methods, we often end up solving systems of linear algebraic equations (frequently just called *linear systems*). For large classes of problems, the computational effort spent on solving linear systems dominate the overall computing time. Consequently, it is of vital importance to have access to efficient solvers for linear systems (often called *linear solvers*). Since the optimal choice of a linear solver is highly problem dependent, this calls for software that allows and even encourages you to test several alternative methods. The linear algebra tools available in Diffpack provide a flexible environment tailored for this purpose. The current appendix aims at introducing the basic concepts of common iterative methods for linear solvers and thereby equipping the reader with the required theoretical knowledge for efficient utilization of the corresponding software tools in Diffpack. The next appendix is devoted to the practical usage of the software tools.

Linear systems can be solved by *direct* or *iterative* methods. The term direct methods normally means some type of Gaussian elimination, where the solution is computed by an algorithm with precisely known complexity, independent of the underlying PDE and its discretization. On the contrary, the performance of an iterative method is often strongly problem dependent.

For most large-scale applications arising from PDEs, iterative methods are superior to direct solvers with respect to computational efficiency. The main reason for this superiority is that iterative methods can take great advantage of the sparse matrices that arise from finite difference and finite element discretizations of PDEs. The effect is especially pronounced in 3D problems, where the pattern of the corresponding coefficient matrices tend to be extremely sparse. To exemplify, only 0.004% of the coefficient matrix will be occupied by nonzeroes when the 3D Poisson problem is discretized on a $60 \times 60 \times 60$ grid by standard finite differences. This mesh spacing, leading to $n = 216,000$ unknowns, is moderate and further refinements would decrease the sparsity factor considerably. In general, the fraction of nonzeroes in the coefficient matrix is about $7q^{-6}$ on a $q \times q \times q$ grid. Even such rough estimates reveal the potential of iterative methods that compute with the nonzeroes only. In contrast to most direct methods, iterative solvers can usually be formulated in terms of a few basic operations on matrices and vectors, e.g. matrix-vector products, inner products, and vector additions. The way these operations are combined distinguish one iterative scheme from another. Naturally, the critical issue for any iterative solver is whether the iterations will converge sufficiently fast. For this reason, a major concern in this field is the construction of efficient *preconditioners* capable of improving the convergence rate.

Bruaset [19] presents some exact figures on the efficiency of preconditioned iterative methods relative to direct methods for the finite difference discretized 3D Poisson equation on the unit cube with homogeneous Dirichlet boundary conditions. The direct solver was banded Gaussian elimination, whereas a MILU preconditioned Conjugate Gradient method was used as iterative solver. The ratio C of the CPU time of the direct and iterative methods as well as the ratio M of the memory requirements were computed for various grid sizes[1]. For a small grid with 3 375 unknowns, $C = 72$ and $M = 20$, which means that the MILU preconditioned Conjugate Gradient method is 72 times faster than banded Gaussian elimination and reduces the memory requirements by a factor of 20. For 27,000 unknowns, the largest grid where the direct method could be run on the particular computer used in [19], $C = 924$ and $M = 95$. As the number of unknowns increases, the effect of using iterative methods become even more dramatic. With 8,000,000 unknowns, theoretical estimates give $C = 10^7$ and $M = 4871$. This means that the iterative method used about 2,600 seconds, wheras the direct method would use 832 years, if there had been a machine with 2,400 Gb RAM to store the banded matrix!

We begin the presentation of iterative solvers with a compact review of the simple classical iterative methods, such as Jacobi and Gauss-Seidel iteration. Then we describe the foundations for Conjugate Gradient-like methods and show that we can reuse much of the reasoning from Chapter 2 on finite element methods also for solving *algebraic equations* approximately. Thereafter we treat some fundamental preconditioning techniques. The fundamental ideas of domain decomposition and multigrid methods are described at the end of the appendix.

C.1   Classical Iterative Methods

Classical iterative methods, like Jacobi, Gauss-Seidel, SOR, and SSOR iteration, are seldom used as stand-alone solvers nowadays. Nevertheless, these methods are frequently used as preconditioners in conjugate gradient-like techniques and as smoothers in multigrid methods. Our treatment of classical iterative methods is here very brief as there is an extensive literature on the subject. An updated and comprehensive textbook is Hackbusch [49],

---

[1] On large grids, banded Gaussian elimination is too slow for practical computations, but the expected CPU time, neglecting the effect of swapping etc., can be estimated from the formula for the work involved in the Gaussian elimination algorithm.

while Bruaset [19] is well suited as a compact extension of the present exposition.

C.1.1   A General Framework

The linear system to be solved is written as

$$Ax = b, \quad A \in \mathbb{R}^{n,n}, \ x, b \in \mathbb{R}^n.$$

Let us split the matrix A into two parts M and N such that $A = M - N$, where M is invertible and linear systems with M as coefficient matrix are in some sense cheap to solve. The linear system then takes the form

$$Mx = Nx + b,$$

which suggests an iteration method

$$Mx^k = Nx^{k-1} + b, \quad k = 1, 2, \ldots \tag{C.1}$$

where x^k is a new approximation to x in the kth iteration. To initiate the iteration, we need a start vector x^0.

An alternative form of (C.1) is

$$x^k = x^{k-1} + M^{-1}r^{k-1}, \tag{C.2}$$

where $r^{k-1} = b - Ax^{k-1}$ is the *residual* after iteration $k - 1$. For further analysis, but not for the implementation, the following version of (C.1) is useful

$$x^k = Gx^{k-1} + c, \quad k = 1, 2, \ldots,$$

where $G = M^{-1}N$ and $c = M^{-1}b$. The matrix G plays a fundamental role in the *convergence* of this iterative method, because one can easily show that

$$x^k - x = G^k(x^0 - x).$$

That is, $\lim_{k\to\infty} \|G^k\| = 0$ is a necessary and sufficient condition for convergence of the sequence $\{x^k\}$ towards the exact solution x. This is equivalently expressed as

$$\varrho(G) < 1,$$

where $\varrho(G)$ is the *spectral radius* of G. The spectral radius is defined as $\varrho(G) = \max_{i=1,\ldots,n} |\lambda_i|$, where λ_i are the real or complex eigenvalues of G. Making $\varrho(G)$ smaller increases the speed of convergence. Of special interest is the *asymptotic rate of convergence* $R_\infty(G) = -\ln \varrho(G)$. To reduce the initial error by a factor ϵ, that is, $\|x - x^{k-1}\| \le \epsilon \|x - x^0\|$, one needs $-\ln \epsilon / R_\infty(G)$ iterations [19].

We shall first give examples on popular choices of splittings (i.e. the matrices M and N) that define classical iterative solution algorithms for linear systems. Thereafter, we will return to the convergence issue and list some characteristic results regarding the value of $R_\infty(G)$ for the suggested iterative methods in a model problem.

In the following, we need to write A as a sum of a diagonal part D, a lower triangular matrix L, and an upper triangular matrix U:

$$A = L + D + U.$$

The precise definition of these matrices are $L_{i,j} = A_{i,j}$ if $i > j$, else zero, $U_{i,j} = A_{i,j}$ if $i < j$, else zero, and $D_{i,j} = A_{i,j}$ if $i = j$, else zero.

As a PDE-related example on a common linear system, we shall make use of the five-point 2D finite difference discretization of the Poisson equation $-\nabla^2 u = f$ with $u = 0$ on the boundary. The domain is the unit square, with m grid points in each space direction. The unknown grid-point values $u_{i,j}$ are then coupled in a linear system

$$u_{i,j-1} + u_{i-1,j} + u_{i+1,j} + u_{i,j+1} - 4u_{i,j} = -h^2 f_{i,j}, \qquad \text{(C.3)}$$

for $i, j = 2, \ldots, m-1$. The parameter h is the grid increment: $h = 1/(m-1)$. In the start vector we impose a zero value at the boundary points $i = 1, m$, $j = 1, m$. Updating only the inner points, $2 \leq i, j \leq m-1$, then preserves the correct boundary values in each iteration.

C.1.2   Jacobi, Gauss-Seidel, SOR, and SSOR Iteration

Simple Iteration. The simplest choice is of course to let $M = I$ and $N = I - A$, resulting in the iteration

$$x^k = x^{k-1} + r^{k-1}. \qquad \text{(C.4)}$$

Jacobi Iteration. Another simple choice is to let M equal the diagonal of A, which means that $M = D$ and $N = -L - U$. From (C.2) we then get the iteration

$$x^k = x^{k-1} + D^{-1}r^{k-1}. \qquad \text{(C.5)}$$

We can write the iteration explicitly, component by component,

$$x_i^k = x_i^{k-1} + \frac{1}{A_{i,i}}\left(b_i - \sum_{j=1}^n A_{i,j}x_j^{k-1}\right), \quad i = 1, \ldots, n. \qquad \text{(C.6)}$$

An alternative (and perhaps more intuitive) formula arises from just solving equation no. i with respect to x_i, using old values for all the other unknowns:

$$x_i^k = \frac{1}{A_{i,i}}\left(b_i - \sum_{j=1}^{i-1} A_{i,j}x_j^{k-1} - \sum_{j=i+1}^n A_{i,j}x_j^{k-1}\right), \quad i = 1, \ldots, n.$$

Applying this iteration to our model problem (C.3) yields

$$u_{i,j}^k = \frac{1}{4}\left(u_{i,j-1}^{k-1} + u_{i-1,j}^{k-1} + u_{i+1,j}^{k-1} + u_{i,j+1}^{k-1} + h^2 f_{i,j}\right), \qquad \text{(C.7)}$$

for $i, j = 2, \ldots, m-1$.

Relaxed Jacobi Iteration. Let $\boldsymbol{x}^*$ denote the predicted value of $\boldsymbol{x}$ in iteration k with the Jacobi method. We could then compute $\boldsymbol{x}^k$ as a weighted combination of $\boldsymbol{x}^*$ and the previous value $\boldsymbol{x}^{k-1}$:

$$\boldsymbol{x}^k = \omega \boldsymbol{x}^* + (1-\omega)\boldsymbol{x}^{k-1}. \qquad \text{(C.8)}$$

For $\omega \in [0,1]$ this is a weighted mean of $\boldsymbol{x}^*$ and $\boldsymbol{x}^{k-1}$. The corresponding scheme is called *relaxed* Jacobi iteration. As an example, the relaxed Jacobi method applied to the model problem (C.3) reads

$$u_{i,j}^k = \frac{\omega}{4}\left(u_{i,j-1}^{k-1} + u_{i-1,j}^{k-1} + u_{i+1,j}^{k-1} + u_{i,j+1}^{k-1} + h^2 f_{i,j}\right) + (1-\omega)u_{i,j}^{k-1}, \quad \text{(C.9)}$$

for $i, j = 2, \ldots, m-1$.

On the Equivalence of Jacobi Iteration and Explicit Time Stepping. Consider the heat equation

$$\alpha\frac{\partial u}{\partial t} = \nabla^2 u + f \qquad \text{(C.10)}$$

on a 2D grid over the unit square. If we assume that a stationary state, defined as $\partial u/\partial t = 0$, is reached as $t \to \infty$, we can solve the associated stationary problem $-\nabla^2 u = f$ by solving the time-dependent version (C.10) until $\partial u/\partial t$ is sufficiently close to zero. This is often called *pseudo time stepping*. Let us discretize (C.10) using an explicit forward Euler scheme in time, combined with the standard 5-point finite difference discretization of the $\nabla^2 u$ term. With $u_{i,j}^k$ as the approximation to u at grid point (i, j) and time level k, the scheme reads

$$u_{i,j}^k = u_{i,j}^{k-1} + \frac{\Delta t}{\alpha h^2}\left(u_{i,j-1}^{k-1} + u_{i-1,j}^{k-1} + u_{i+1,j}^{k-1} + u_{i,j+1}^{k-1} - 4u_{i,j}^{k-1} + h^2 f_{i,j}\right),$$
$$\text{(C.11)}$$

with Δt being the time step length. This scheme can alternatively be expressed as

$$u_{i,j}^k = \left(1 - 4\frac{\Delta t}{\alpha h^2}\right)u_{i,j}^{k-1} + \frac{\Delta t}{\alpha h^2}\left(u_{i,j-1}^{k-1} + u_{i-1,j}^{k-1} + u_{i+1,j}^{k-1} + u_{i,j+1}^{k-1} + h^2 f_{i,j}\right).$$
$$\text{(C.12)}$$

Comparison with (C.9) reveals that (C.12) is nothing but a relaxed Jacobi iteration for solving a finite difference form of $-\nabla^2 u = f$. The relaxation parameter ω is seen to equal $4\Delta t/(\alpha h^2)$. The stability criterion of the explicit forward scheme for the heat equation in two dimensions is given by (A.36):

$\Delta t \leq \alpha h^2/4$, or equivalently, $\omega \leq 1$. The original Jacobi method is recovered by the choice $\omega = 1$, which in light of the transient problem indicates that this is the most efficient value of the relaxation parameter since it corresponds to the largest possible time step.

The analogy between explicit time integration of a heat equation and Jacobi iteration for the Poisson equation, together with the smoothing properties of the heat equation (see Example A.7 on page 507), suggest that Jacobi's method will efficiently damp any irregular behavior in $\boldsymbol{x}^0$. This property is utilized when relaxed Jacobi iteration is used as a smoother in multigrid methods, and $\omega = 1$ is in this context not the optimal value with respect to efficient damping of high frequencies in the solution (cf. Exercise C.4).

Gauss-Seidel Iteration. Linear systems where the coefficient matrix is upper or lower triangular are easy to solve. Choosing $\boldsymbol{M}$ to be the lower or upper triangular part of $\boldsymbol{A}$ is therefore attractive. With $\boldsymbol{M} = \boldsymbol{D} + \boldsymbol{L}$ and $\boldsymbol{N} = -\boldsymbol{U}$ we obtain *Gauss-Seidel's method*:

$$(\boldsymbol{D} + \boldsymbol{L})\boldsymbol{x}^k = -\boldsymbol{U}\boldsymbol{x}^{k-1} + \boldsymbol{b}\,. \tag{C.13}$$

Combining this formula on component form with the standard algorithm for solving a system with lower triangular coefficient matrix yields

$$x_i^k = \frac{1}{A_{i,i}} \left(b_i - \sum_{j=1}^{i-1} A_{i,j} x_j^k - \sum_{j=i+1}^{n} A_{i,j} x_j^{k-1} \right), \quad i = 1, \dots, n\,. \tag{C.14}$$

This equation has a simple interpretation; it is similar to the ith equation in Jacobi's method, but we utilize the new values $x_1^k, \dots, x_{i-1}^k$, which we have already computed in this iteration, on the right-hand side. In other words, the Gauss-Seidel always applies the most recently computed approximations in the equations.

For our Poisson equation scheme, the Gauss-Seidel iteration takes the form

$$u_{i,j}^k = \frac{1}{4} \left(u_{i,j-1}^k + u_{i-1,j}^k + u_{i+1,j}^{k-1} + u_{i,j+1}^{k-1} + h^2 f_{i,j} \right)\,. \tag{C.15}$$

We here assume that we run through the (i,j) indices in a double loop, with i and j going from 1 to m. However, if we for some reason should reverse the loops (decrementing i and j), the available "new" (k) values on the right-hand side change. The efficiency of the Gauss-Seidel method is strongly dependent on the ordering of the equations and unknowns in many applications. The ordering also affects the performance of implementations on parallel and vector computers. A quick overview of various orderings and versions of Gauss-Seidel iteration is provided in [125, Ch. 4.3].

From an implementational point of view, one should notice that the new values x_i^k or $u_{i,j}^k$ can overwrite the corresponding old values x_i^{k-1} and $u_{i,j}^{k-1}$. This is not possible in the Jacobi algorithm.

Example C.1. Consider the Poisson equation $\nabla^2 u = 2$ on the unit square, with $u = 0$ for $x = 0$, and $u = 1$ for $x = 1$. At the boundaries $y = 0, 1$ we impose homogeneous Neumann conditions: $\partial u / \partial y = 0$. The exact solution is then $u(x, y) = x^2$. We introduce a grid over $(0, 1) \times (0, 1)$ with points $i, j = 1, \ldots, m$ as shown in Figure C.1. The discrete equations are to be solved by the Gauss-Seidel method. At all inner points, $i, j = 2, \ldots, m - 1$, we can use the difference equation (C.15) directly. At $j = 1$ we have by the boundary condition $u_{i,2} = u_{i,0}$. Similarly, $u_{i,m-1} = u_{i,m+1}$. These relations can be used to eliminate the fictitious values $u_{i,m+1}$ and $u_{i,0}$ in the difference equations at $j = 1$ and $j = m$:

$$u_{i,1}^k = \frac{1}{4}\left(u_{i,2}^k + u_{i-1,1}^k + u_{i+1,1}^{k-1} + u_{i,2}^{k-1} + h^2 f_{i,1}\right),$$

$$u_{i,m}^k = \frac{1}{4}\left(u_{i,m-1}^k + u_{i-1,m}^k + u_{i+1,m}^{k-1} + u_{i,m-1}^{k-1} + h^2 f_{i,1}\right).$$

The start values $u_{i,j}^0$ can be arbitrary as long as $u_{1,j}^0 = 0$ and $u_{m,j}^0 = 1$ such that the preceding iteration formulas sample the correct values at $i = 1, m$.

The demo program `src/linalg/GaussSeidelDemo/main.cpp` contains a suggested implementation of the solution algorithm. In that program, the field values $u_{i,j}^k$ overwrite $u_{i,j}^{k-1}$. Moreover, the start values are $u_{i,j}^0 = 0$ or $u_{i,j}^0 = (-1)^i$, except for $u_{1,j}$ and $u_{m,j}$, which must equal the prescribed boundary conditions. Included in this directory is a little Perl script `play.pl` that takes m and the start-vector type as argument, runs the program, and thereafter displays an animation of the error along x as the iteration index k increases. There is also a GUI version of the script, called `play-GUI.pl`.

For small values of m the Gauss-Seidel method is seen to converge rapidly to the exact solution. For $m = 80$ we see that the error is initially steep, and during the first iterations it is effectively reduced and smoothed. However, the convergence slows down significantly. After 1000 iterations the maximum error is still as high as 20 percent of the initial maximum error.

The property that the first few Gauss-Seidel iterations are very effective in reducing and smoothing the error, is a key ingredient in multigrid methods.
◇

Successive Over-Relaxation. The idea of relaxation as introduced in the Jacobi method is also relevant to Gauss-Seidel iteration. If x^* is the new solution predicted by a Gauss-Seidel step, we may set

$$x^k = \omega x^* + (1 - \omega) x^{k-1}. \tag{C.16}$$

The resulting method is called *Successive Over-Relaxation*, abbreviated SOR. It turns out that $\omega > 1$ is a good choice when solving stationary PDEs of the Poisson-equation type. One can prove that $\omega \in (0, 2)$ is a necessary condition for convergence. In terms of M and N, the method can be expressed as

$$M = \frac{1}{\omega} D + L, \quad N = \frac{1 - \omega}{\omega} D - U.$$

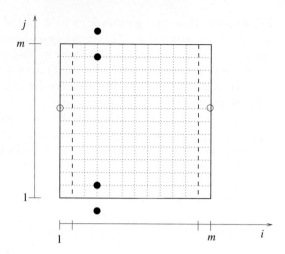

Fig. C.1. Sketch of a grid for finite difference approximation of $\nabla^2 u = 2$ on the unit square, with u prescribed for $i = 1, m$ and $\partial u/\partial y = 0$ for $j = 1, m$. The difference equations must be modified for points at $j = 1, m$ by using the boundary condition $\partial u/\partial y = 0$, involving the points marked with bullets. The boundary points marked with circles have prescribed u values and can remain untouched.

The algorithm is a trivial combination of (C.14) and (C.16):

$$x_i^k = \frac{\omega}{A_{i,i}} \left(b_i - \sum_{j=1}^{i-1} A_{i,j} x_j^k - \sum_{j=i+1}^{n} A_{i,j} x_j^{k-1} \right) + (1 - \omega) x_i^{k-1}, \qquad \text{(C.17)}$$

for $i = 1, \ldots, n$. The SOR method for the discrete Poisson equation can be written as

$$u_{i,j}^k = \frac{\omega}{4} \left(u_{i,j-1}^k + u_{i-1,j}^k + u_{i+1,j}^{k-1} + u_{i,j+1}^{k-1} + h^2 f_{i,j} \right) + (1 - \omega) u_{i,j}^{k-1}, \quad \text{(C.18)}$$

for $i, j = 2, \ldots, m - 1$. By a clever choice of ω, the convergence of SOR can be much faster than the convergence of Jacobi and Gauss-Seidel iteration. A guiding value is $\omega = 2 - \mathcal{O}(h)$, but the optimal choice of ω can be theoretically estimated in simple model problems involving the Laplace or Poisson equations. One should observe that SOR with $\omega = 1$ recovers the Gauss-Seidel method.

Symmetric Successive Over-Relaxation. The *Symmetric SOR* method, abbreviated SSOR, is a two-step SOR procedure where we in the first step apply a standard SOR sweep, whereas we in the second SOR step run through the unknowns in reversed order (backward sweep). The matrix M is now given by

$$M = \frac{1}{2 - \omega} \left(\frac{1}{\omega} D + L \right) \left(\frac{1}{\omega} D \right)^{-1} \left(\frac{1}{\omega} D + U \right). \qquad \text{(C.19)}$$

In the algorithm we need to solve a linear system with M as coefficient matrix. This is efficiently done since the above M is written on a factorized form. The solution can hence be performed in three steps, involving two triangular coefficient matrices and one diagonal matrix. As for SOR, $\omega \in (0, 2)$ is necessary for convergence, but SSOR is less sensitive than SOR to the value of ω. To formulate an SSOR method for our Poisson equation problem, we make explicit use of the ideas behind the algorithm, namely a forward and backward SOR sweep. We first execute

$$u_{i,j}^{k,*} = \frac{\omega}{4} \left(u_{i,j-1}^{k,*} + u_{i-1,j}^{k,*} + u_{i+1,j}^{k-1} + u_{i,j+1}^{k-1} + h^2 f_{i,j} \right) + (1 - \omega) u_{i,j}^{k-1}, \quad (C.20)$$

for $i, j = 2, \ldots, m - 1$. The second step consists in running through the grid points in reversed order,

$$u_{i,j}^{k} = \frac{\omega}{4} \left(u_{i,j-1}^{k,*} + u_{i-1,j}^{k,*} + u_{i+1,j}^{k} + u_{i,j+1}^{k} + h^2 f_{i,j} \right) + (1 - \omega) u_{i,j}^{k,*}, \quad (C.21)$$

for $i, j = m - 1, m - 2, \ldots, 2$. Notice that $u_{i,j}^{k,*}$ are the "old" values in step two.

Line SOR Iteration. The methods considered so far update only one entry at a time in the solution vector. The speed of such *pointwise* algorithms can often be improved by finding new values for a subset of the unknowns simultaneously. For example, in our Poisson equation problem we could find new values of $u_{i,j}^{k}$ simultaneously along a line. The SOR algorithm could then be formulated as

$$-u_{i-1,j}^{*} + 4u_{i,j}^{*} - u_{i+1,j}^{*} = u_{i,j-1}^{k} + u_{i,j+1}^{k-1} + h^2 f_{i,j}, \quad (C.22)$$

$$u_{i,j}^{k} = \omega u_{i,j}^{*} + (1 - \omega) u_{i,j}^{k-1}. \quad (C.23)$$

For a fixed j, (C.22) is a tridiagonal system coupling u values along a line $j = \text{const}$. We can solve for these values simultaneously and then continue with the line $j + 1$. The algorithm is naturally called *line SOR* iteration. Instead of working with lines, we could solve simultaneously for an arbitrary local block of unknowns. This general approach is referred to as *block SOR* iteration. The ideas of lines and blocks apply of course to Jacobi and Gauss-Seidel iteration as well.

The solution of a stationary PDE (like the Poisson equation) at a point is dependent on all the boundary data. Therefore, we cannot expect an iterative method to converge properly before the boundary conditions have influenced sufficiently large portions of the domain. The information in the pointwise versions of the classical iterative methods is transported one grid point per iteration, whereas line or block versions transmit the information along a line or throughout a block per iterative step. This gives a degree of implicitness in the iteration methods that we expect to be reflected in an increased convergence rate. The figures in the next section demonstrate this feature.

Exercise C.1. Implement the line SOR method in the program from Example C.1.   ◇

Convergence Rates for the Poisson Problem. Bruaset [19] presents some estimates of $R_\infty(G)$ when solving the 2D Poisson equation $-\nabla^2 u = f$ by a standard centered finite difference method on a uniform grid over the unit square, with $u = 0$ on the boundary. For the point Jacobi, Gauss-Seidel, SOR, and SSOR methods we have the asymptotic estimates:

rate	Jacobi	Gauss-Seidel	SOR	SSOR
$R_\infty(G)$	$\pi^2 h^2/2$	$\pi^2 h^2$	$2\pi h$	$> \pi h$

The estimates for the SOR and SSOR methods are based on using a theoretically known optimal value of the relaxation parameter ω for the present model problem.

Similar convergence results for the line versions of the classical iterative methods are given in the next table.

rate	Jacobi	Gauss-Seidel	SOR	SSOR
$R_\infty(G)$	$\pi^2 h^2$	$2\pi^2 h^2$	$2\sqrt{2}\pi h$	$\geq \sqrt{2}\pi h$

Recalling that $1/R_\infty(G)$ is proportional to the expected number of iterations needed to reach a prescribed reduction of the initial error, we see that Gauss-Seidel is twice as fast as Jacobi, and that SOR and SSOR are significantly superior to the former two on fine grids (h vs. h^2). The line versions are about twice as fast as the corresponding point versions.

From the results above it is clear that Jacobi iteration is a slow method. However, because of its explicit updating nature, this type of iterative approach, either used as a stand-alone solver, as a preconditioner, or as a smoother in multigrid, has been popular on vector and parallel machines, where more sophisticated methods may have difficulties in exploiting the hardware features.

C.2   Conjugate Gradient-Like Iterative Methods

A successful family of methods, usually referred to as Conjugate Gradient-like algorithms, can be viewed as Galerkin or least-squares methods applied to a linear system $Ax = b$. This view is different from the standard approaches to deriving the classical Conjugate Gradient method in the literature. Nevertheless, the Galerkin or least-squares framework makes it possible to reuse the principal numerical ideas from Chapter 2.1 and the analysis from Chapter 2.10 when solving linear systems approximately. We believe that this view may increase the general understanding of variational methods and their applicability.

Our exposition focuses on the basic reasoning behind the methods, and a natural continuation of the material here is provided by two review texts; Bruaset [19] gives an accessible theoretical overview of a wide range of Conjugate Gradient-like methods, whereas Barrett et al. [13] present a collection of computational algorithms and give valuable information about the practical use of the methods. Every Diffpack practitioner who needs to access library modules for iterative solution of linear systems should spend some time on these two references. For further study of iterative solvers we refer the reader to the comprehensive books by Hackbusch [49] and Axelsson [11].

C.2.1 Galerkin and Least-Squares Methods

Given a linear system

$$\boldsymbol{A}\boldsymbol{x} = \boldsymbol{b}, \quad \boldsymbol{x}, \boldsymbol{b} \in \mathbb{R}^n, \ \boldsymbol{A} \in \mathbb{R}^{n,n} \tag{C.24}$$

and a start vector $\boldsymbol{x}^0$, we want to construct an iterative solution method that produces approximations $\boldsymbol{x}^1, \boldsymbol{x}^2, \ldots$, which hopefully converge to the exact solution $\boldsymbol{x}$. In iteration no. k we seek an approximation

$$\boldsymbol{x}^k = \boldsymbol{x}^{k-1} + \sum_{j=1}^{k} \alpha_j \boldsymbol{q}_j, \tag{C.25}$$

where $\boldsymbol{q}_j \in \mathbb{R}^n$ are known vectors and α_j are constants to be determined by a Galerkin or least-squares method. The corresponding error in the equation $\boldsymbol{A}\boldsymbol{x} = \boldsymbol{b}$, the residual, becomes

$$\boldsymbol{r}^k = \boldsymbol{b} - \boldsymbol{A}\boldsymbol{x}^k = \boldsymbol{r}^{k-1} - \sum_{j=1}^{k} \alpha_j \boldsymbol{A}\boldsymbol{q}_j \,.$$

The Galerkin Method. Galerkin's method states that the inner product of the residual and k independent weights $\boldsymbol{q}_i$ vanish:

$$(\boldsymbol{r}^k, \boldsymbol{q}_i) = 0, \quad i = 1, \ldots, k \,. \tag{C.26}$$

Here $(\cdot, \cdot)$ is the standard Euclidean inner product on $\mathbb{R}^n$. Inserting the expression for $\boldsymbol{r}^k$ in (C.26) gives a linear system for α_j:

$$\sum_{j=1}^{k} (\boldsymbol{A}\boldsymbol{q}_i, \boldsymbol{q}_j)\alpha_j = (\boldsymbol{r}^{k-1}, \boldsymbol{q}_i), \quad i = 1, \ldots, k \,. \tag{C.27}$$

The Least-Squares Method. The idea of the least-squares method is to minimize the square of the norm of the residual with respect to the free parameters $\alpha_1, \ldots, \alpha_k$. That is, we minimize (r^k, r^k):

$$\frac{\partial}{\partial \alpha_i}(r^k, r^k) = 2(\frac{\partial r^k}{\partial \alpha_i}, r^k) = 0, \quad i = 1, \ldots, k.$$

Since $\partial r^k / \partial \alpha_i = -Aq_i$, this approach leads to the following linear system:

$$\sum_{j=1}^{k}(Aq_i, Aq_j)\alpha_j = (r^{k-1}, Aq_i), \quad i = 1, \ldots, k. \tag{C.28}$$

Equation (C.28) can be viewed as a weighted residual method with weights Aq_i, also called a Petrov-Galerkin method.

A More Abstract Formulation. The presentation of the Galerkin and least-squares methods above follow closely the reasoning in Chapter 2. We can also view these methods in a more abstract framework like we did in Chapter 2.10.1. Let

$$\mathcal{B} = \{q_1, \ldots, q_k\}$$

be a basis for the k-dimensional vector space $V_k \subset \mathbb{R}^n$. The methods can then be formulated as: *Find* $x^k - x^{k-1} \in V_k$ *such that*

$$(r^k, v) = 0, \quad \forall v \in V_k \text{ (Galerkin)}$$
$$(r^k, Av) = 0, \quad \forall v \in V_k \text{ (least-squares)}$$

Provided that the solutions of the resulting linear systems exist and are unique, we have found a new approximation x^k to x. When $k = n$, the only solution to the equations above is $r^n = 0$. This means that the exact solution is found in at most n iterations, neglecting effects of round-off errors.

The Galerkin condition can alternatively be written as

$$a(x^k, v) = L(v), \quad \forall v \in V_k,$$

with $a(u, v) = (Au, v)$ and $L(v) = (b, v)$. The analogy with the abstract formulation of the finite element method in Chapter 2.10.1 is hence clearly demonstrated.

Krylov Subspaces. To obtain a complete algorithm, we need to establish a rule to update the basis $\mathcal{B}$ for the next iteration. That is, we need to compute a new basis vector $q_{k+1} \in V_{k+1}$ such that

$$\mathcal{B} = \{q_1, \ldots, q_{k+1}\} \tag{C.29}$$

is a basis for the space V_{k+1} that is used in the next iteration. The present family of methods applies the *Krylov subspace*

$$V_k = \text{span}\left\{r^0, Ar^0, A^2 r^0, \ldots A^{k-1} r^0\right\}. \tag{C.30}$$

Some frequent names of the associated iterative methods are therefore Krylov subspace iteration, Krylov projection methods, or simply Krylov methods.

Computation of the Basis Vectors. Two possible formulas for updating $\boldsymbol{q}_{k+1}$, such that $\boldsymbol{q}_{k+1} \in V_{k+1}$, are

$$\boldsymbol{q}_{k+1} = \boldsymbol{r}^k + \sum_{j=1}^{k} \beta_j \boldsymbol{q}_k, \tag{C.31}$$

$$\boldsymbol{q}_{k+1} = \boldsymbol{A}\boldsymbol{r}^k + \sum_{j=1}^{k} \beta_j \boldsymbol{q}_k, \tag{C.32}$$

where the free parameters β_j can be used to enforce desirable orthogonality properties of $\boldsymbol{q}_1, \ldots, \boldsymbol{q}_{k+1}$. For example, it is convenient to require that the coefficient matrices in the linear systems for $\alpha_1, \ldots, \alpha_k$ are diagonal. Otherwise, we must solve a $k \times k$ linear system in each iteration. If k should approach n, the systems for the coefficients α_i are of the same size as our original system $\boldsymbol{A}\boldsymbol{x} = \boldsymbol{b}$! A diagonal matrix ensures an efficient closed form solution for $\alpha_1, \ldots, \alpha_k$. To obtain a diagonal coefficient matrix, we require in Galerkin's method that

$$(\boldsymbol{A}\boldsymbol{q}_i, \boldsymbol{q}_j) = 0 \quad \text{when } i \neq j,$$

whereas we in the least-squares method require

$$(\boldsymbol{A}\boldsymbol{q}_i, \boldsymbol{A}\boldsymbol{q}_j) = 0 \quad \text{when } i \neq j.$$

We can define the inner product

$$\langle \boldsymbol{u}, \boldsymbol{v} \rangle \equiv (\boldsymbol{A}\boldsymbol{u}, \boldsymbol{v}) = \boldsymbol{u}^T \boldsymbol{A}\boldsymbol{v}, \tag{C.33}$$

provided $\boldsymbol{A}$ is symmetric and positive definite. Another useful inner product is

$$[\boldsymbol{u}, \boldsymbol{v}] \equiv (\boldsymbol{A}\boldsymbol{u}, \boldsymbol{A}\boldsymbol{v}) = \boldsymbol{u}^T \boldsymbol{A}^T \boldsymbol{A}\boldsymbol{v}. \tag{C.34}$$

These inner products will be be referred to as the $\boldsymbol{A}$ product, with the associated $\boldsymbol{A}$ norm, and the $\boldsymbol{A}^T\boldsymbol{A}$ product, with the associated $\boldsymbol{A}^T\boldsymbol{A}$ norm.

The orthogonality condition on the $\boldsymbol{q}_i$ vectors are then $\langle \boldsymbol{q}_{k+1}, \boldsymbol{q}_i \rangle = 0$ in the Galerkin method and $[\boldsymbol{q}_{k+1}, \boldsymbol{q}_i] = 0$ in the least-squares method, where i runs from 1 to k. A standard Gram-Schmidt process can be used for constructing $\boldsymbol{q}_{k+1}$ orthogonal to $\boldsymbol{q}_1, \ldots, \boldsymbol{q}_k$. This leads to the determination of the $\beta_1, \ldots, \beta_k$ constants as

$$\beta_i = \frac{\langle \boldsymbol{r}^k, \boldsymbol{q}_i \rangle}{\langle \boldsymbol{q}_i, \boldsymbol{q}_i \rangle} \quad \text{(Galerkin)} \tag{C.35}$$

$$\beta_i = \frac{[\boldsymbol{r}^k, \boldsymbol{q}_i]}{[\boldsymbol{q}_i, \boldsymbol{q}_i]} \quad \text{(least squares)} \tag{C.36}$$

for $i = 1, \ldots, k$.

Computation of a New Solution Vector. The orthogonality condition on the basis vectors $\boldsymbol{q}_i$ leads to the following solution for $\alpha_1, \ldots, \alpha_k$:

$$\alpha_i = \frac{(\boldsymbol{r}^{k-1}, \boldsymbol{q}_i)}{\langle \boldsymbol{q}_i, \boldsymbol{q}_i \rangle} \quad \text{(Galerkin)} \tag{C.37}$$

$$\alpha_i = \frac{(\boldsymbol{r}^{k-1}, \boldsymbol{A}\boldsymbol{q}_i)}{[\boldsymbol{q}_i, \boldsymbol{q}_i]} \quad \text{(least squares)} \tag{C.38}$$

In iteration $k-1$, $(\boldsymbol{r}^{k-1}, \boldsymbol{q}_i) = 0$ and $(\boldsymbol{r}^{k-1}, \boldsymbol{A}\boldsymbol{q}_i) = 0$, for $i = 1, \ldots, k-1$, in the Galerkin and least-squares case, respectively. Hence, $\alpha_i = 0$, for $i = 1, \ldots, k-1$. In other words,

$$\boldsymbol{x}^k = \boldsymbol{x}^{k-1} + \alpha_k \boldsymbol{q}_k .$$

When $\boldsymbol{A}$ is symmetric and positive definite, one can show that also $\beta_i = 0$, for $i = 1, \ldots, k-1$, in both the Galerkin and least-squares methods [19]. This means that $\boldsymbol{x}^k$ and $\boldsymbol{q}_{k+1}$ can be updated using only $\boldsymbol{q}_k$ and not the previous $\boldsymbol{q}_1, \ldots, \boldsymbol{q}_{k-1}$ vectors. This property has of course dramatic effects on the storage requirements of the algorithms as the number of iterations increases.

For the suggested algorithms to work, we must require that the denominators in (C.37) and (C.38) do not vanish. This is always fulfilled for the least-squares method, while a (positive or negative) definite matrix $\boldsymbol{A}$ avoids break-down of the Galerkin-based iteration (provided $\boldsymbol{q}_i \neq \boldsymbol{0}$).

The Galerkin solution method for linear systems was originally devised as a *direct* method in the 1950s. After n iterations the exact solution is found in exact arithmetic, but at a higher cost compared with Gaussian elimination. Naturally, the method did not receive significant popularity before researchers discovered (in the beginning of the 1970s) that the method could produce a good approximation to $\boldsymbol{x}$ for $k \ll n$ iterations.

Finally, we mention how to terminate the iteration. The simplest criterion is $||\boldsymbol{r}^k|| \leq \epsilon_r$, where ϵ_r is a small prescribed quantity. Sometimes it is appropriate to use a relative residual, $||\boldsymbol{r}^k||/||\boldsymbol{r}^0|| \leq \epsilon_r$. Termination criteria for Conjugate Gradient-like methods is a subject on its own [19], and Diffpack offers a framework for monitoring convergence and combining termination criteria. Appendix D.6 deals with the details of this topic.

C.2.2   Summary of the Algorithms

Summary of the Least-Squares Method. In Algorithm C.1 we have summarized the computational steps in the least-squares method. Notice that we update the residual recursively instead of using $\boldsymbol{r}^k = \boldsymbol{b} - \boldsymbol{A}\boldsymbol{x}^k$ in each iteration since we then avoid a possibly expensive matrix-vector product.

Algorithm C.1.

Least-squares Krylov iteration.

given a start vector $\boldsymbol{x}^0$,
compute $\boldsymbol{r}_0 = \boldsymbol{b} - \boldsymbol{A}\boldsymbol{x}^0$ and set $\boldsymbol{q}_0 = \boldsymbol{r}^0$.
for $k = 1, 2, \ldots$ until termination criteria are fulfilled:
$\qquad \alpha_k = (\boldsymbol{r}^{k-1}, \boldsymbol{A}\boldsymbol{q}_k)/[\boldsymbol{q}_k, \boldsymbol{q}_k]$
$\qquad \boldsymbol{x}^k = \boldsymbol{x}^{k-1} + \alpha_k \boldsymbol{q}_k$
$\qquad \boldsymbol{r}^k = \boldsymbol{r}^{k-1} - \alpha_k \boldsymbol{A}\boldsymbol{q}_k$
$\qquad$ if $\boldsymbol{A}$ is symmetric then
$\qquad\qquad \beta_k = [\boldsymbol{r}^k, \boldsymbol{q}_k]/[\boldsymbol{q}_k, \boldsymbol{q}_k]$
$\qquad\qquad \boldsymbol{q}_{k+1} = \boldsymbol{r}^k - \beta_k \boldsymbol{q}_k$
$\qquad$ else
$\qquad\qquad \beta_j = [\boldsymbol{r}^k, \boldsymbol{q}_j]/[\boldsymbol{q}_j, \boldsymbol{q}_j], \quad j = 1, \ldots, k$
$\qquad\qquad \boldsymbol{q}_{k+1} = \boldsymbol{r}^k - \sum_{j=1}^{k} \beta_j \boldsymbol{q}_j$

Exercise C.2. Write the algorithm above on implementational form, ready for coding. Store the $\boldsymbol{A}\boldsymbol{q}_j$ vectors for computational efficiency, but otherwise try to minimize the storage requirements.   ◇

Remark. The algorithm above is just a summary of the steps in the derivation of the least-squares method and should not be directly used for practical computations without further developments.

Truncation and Restart. When $\boldsymbol{A}$ is nonsymmetric, the storage requirements of $\boldsymbol{q}_1, \ldots, \boldsymbol{q}_k$ may be prohibitively large. It has become a standard trick to either *truncate* or *restart* the algorithm. In the latter case one restarts the algorithm every Kth step, i.e., one aborts the iteration and starts the algorithm again with $\boldsymbol{x}^0 = \boldsymbol{x}^K$. The other alternative is to truncate the sum $\sum_{j=1}^{k} \beta_j \boldsymbol{q}_j$ and use only the last K vectors:

$$\boldsymbol{x}^k = \boldsymbol{x}^{k-1} + \sum_{j=k-K+1}^{k} \beta_j \boldsymbol{q}_j \,.$$

Both the restarted and truncated version of the algorithm require storage of only K basis vectors $\boldsymbol{q}_{k-K+1}, \ldots, \boldsymbol{q}_k$. The basis vectors are also often called *search direction vectors*, and this name is used in Diffpack. The truncated version of the least-squares method in Algorithm C.1 is widely known as Orthomin, often written as Orthomin(K) to explicitly indicate the number of search direction vectors. In the literature one encounters the name *Generalized Conjugate Residual method*, abbreviated CGR, for the restarted version of Orthomin. When $\boldsymbol{A}$ is symmetric, the method is known under the name *Conjugate Residuals*.

One can devise very efficient implementational forms of the truncated and restarted Orthomin algorithm. We refer to [62] for the details of such an algorithm.

Summary of the Galerkin Method. In case of Galerkin's method, we assume that A is symmetric and positive definite. The resulting computational procedure is the famous Conjugate Gradient method, listed in Algorithm C.2. Since A must be symmetric, the recursive update of q_{k+1} needs only one previous search direction vector q_k, that is, $\beta_j = 0$ for $j < k$.

Algorithm C.2.

Galerkin Krylov iteration (Conjugate Gradient method).

given a start vector x^0,
compute $r_0 = b - Ax^0$ and set $q_0 = r^0$.
for $k = 1, 2, \ldots$ until termination criteria are fulfilled:
$$\alpha_k = (r^{k-1}, q_k)/\langle q_k, q_k \rangle$$
$$x^k = x^{k-1} + \alpha_k q_k$$
$$r^k = r^{k-1} - \alpha_k A q_k$$
$$\beta_k = \langle r^k, q_k \rangle/\langle q_k, q_k \rangle$$
$$q_{k+1} = r^k - \beta_k q_k$$

The previous remark that the listed algorithm is just a summary of the steps in the solution procedure, and not an efficient algorithm that should be implemented in its present form, must be repeated here. An efficient Conjugate Gradient algorithm suitable for implementation is given in [13, Ch. 2.3].

Looking at Algorithms C.1 and C.2, one can notice that the matrix A is only used in matrix-vector products. This means that it is sufficient to store only the nonzero entries of A. The rest of the algorithms consists of vector operations of the type $y \leftarrow ax + y$, the slightly more general variant $q \leftarrow ax + y$, as well as inner products.

C.2.3   A Framework Based on the Error

Let us define the error $e^k = x - x^k$. Multiplying this equation by A leads to the well-known relation between the error and the residual for linear systems:

$$Ae^k = r^k. \tag{C.39}$$

Using $r^k = Ae^k$ we can reformulate the Galerkin and least-squares methods in terms of the error. The Galerkin method can then be written

$$(r^k, q_i) = (Ae^k, q_i) = \langle e^k, q_i \rangle = 0, \quad i = 1, \ldots, k. \tag{C.40}$$

For the least-squares method we obtain

$$(r^k, Aq_i) = [e^k, q_i] = 0, \quad i = 1, \ldots, k. \tag{C.41}$$

This means that

$$\langle e^k, v \rangle = 0 \quad \forall v \in V_k \text{ (Galerkin)}$$
$$[e^k, v] = 0 \quad \forall v \in V_k \text{ (least-squares)}$$

In other words, the error is A-orthogonal to the space V_k in the Galerkin method, whereas the error is $A^T A$-orthogonal to V_k in the least-squares method. This formulation of the Galerkin principle should be compared with similar statements in the finite element method, see the proof of Theorem 2.13 in Chapter 2.10.2.

We can unify these results by introducing the inner product $(u, v)_B \equiv (Bu, v)$, provided B is symmetric and positive definite. The associated norm reads $||v||_B = (v, v)_B^{\frac{1}{2}}$. Given a linear space V_k with basis (C.29), $x^k = x^{k-1} + \sum_j \alpha_j q_j$ can be determined such that

$$(e^k, v)_B = 0 \quad \forall v \in V_k. \tag{C.42}$$

When the error is orthogonal to a space V_k, the approximate solution x^k is then the best approximation to x among all vectors in V_k. A proof of this well-known result was given on page 195. In the present context, where that proof must be slightly modified for an $x^0 \neq 0$, we can state the best approximation principle more precisely as [19]

$$||x - x^k||_B \leq ||x - (x^0 + v)||_B \quad \forall v \in V_k. \tag{C.43}$$

One can also show that the error is nonincreasing: $||e^k||_B \leq ||e^{k-1}||_B$, which is an attractive property. The reader should notify the similarities between the results here and those for the finite element method in Chapter 2.10.

Exercise C.3. Let (C.42) be the principle for determining $\alpha_1, \ldots, \alpha_k$ in an expansion (C.25). The updating formula for q_{k+1}, like (C.31) and (C.32), can be written more generally as $q_{k+1} = z_k + \sum_{j=1}^{k} \beta_j q_k$, where different choices of z_k yield different methods. Derive the corresponding generalized algorithm and present it on the same form as Algorithms C.1 or C.2.   ◇

Choosing $B = A$ when A is symmetric and positive definite gives the Conjugate Gradient method, which then minimizes the error in the A norm. With $B = A^T A$ we recover the least-squares method. Many other choices of B are possible, also when $(\cdot, \cdot)_B$ is no longer a proper inner product. If $BA = A^T B$, the recurrence is short, and there is no need to store all the basis vectors q_i (cf. Algorithm C.1 in the case A is symmetric). We refer to Bruaset [19] for a framework covering numerous Conjugate Gradient-like methods based on (C.42).

Several Conjugate Gradient-like methods have been developed during the last two decades, and some of the most popular methods do not fit directly into the framework presented here. The theoretical similarities between the methods are covered in [19], whereas we refer to [13] for algorithms and practical comments related to widespread methods, such as the SYMMLQ method (for symmetric indefinite systems), the Generalized Minimal Residual (GMRES) method, the BiConjugate Gradient (BiCG) method, the Quasi-Minimal Residual (QMR) method, and the BiConjugate Gradient Stabilized (BiCGStab) method. When A is symmetric and positive definite, the Conjugate Gradient method is the optimal choice with respect to computational efficiency, but when A is nonsymmetric, the performance of the methods is strongly problem dependent. Diffpack offers all the aforementioned iterative procedures.

C.3   Preconditioning

C.3.1   Motivation and Basic Principles

The Conjugate Gradient method has been subject to extensive analysis, and its convergence properties are well understood. To reduce the initial error $e^0 = x - x^0$ with a factor $0 < \epsilon \ll 1$ after k iterations, or more precisely, $\|e^k\|_A \le \epsilon \|e^0\|_A$, it can be shown that k is bounded by

$$\frac{1}{2} \ln \frac{2}{\epsilon} \sqrt{\kappa},$$

where κ is the ratio of the largest and smallest eigenvalue of A. The quantity κ is commonly referred to as the spectral *condition number*[2] of A. Actually, the number of iterations for the Conjugate Gradient method to meet a certain termination criterion is influenced by the complete distribution of eigenvalues of A.

Common finite element and finite difference discretizations of Poisson-like PDEs lead to $\kappa \sim h^{-2}$, where h denotes the mesh size. This implies that the Conjugate Gradient method converges slowly in PDE problems with fine grids, as the number of iterations is proportional to h^{-1}. However, the performance is better than for the Jacobi and Gauss-Seidel methods, which in our example from page 600 required $\mathcal{O}(h^{-2})$ iterations. Although SOR and SSOR have the same asymptotic behavior as the Conjugate Gradient method, the latter does not need estimation of any parameters, such as ω in SOR and SSOR. The number of unknowns in a hypercube domain in $\mathbb{R}^d$ is approximately $n = (1/h)^d$ implying that $\sqrt{\kappa}$ and thereby number of iterations goes like $n^{1/d}$.

---

[2] The spectral condition number is defined as the ratio of the magnitudes of the largest and the smallest eigenvalue of A [93, Ch. 2].

To speed up the Conjugate Gradient method, we should manipulate the eigenvalue distribution. For instance, we could reduce the condition number κ. This can be achieved by so-called *preconditioning*. Instead of applying the iterative method to the system $Ax = b$, we multiply by a matrix M^{-1} and apply the iterative method to the mathematically equivalent system

$$M^{-1}Ax = M^{-1}b. \tag{C.44}$$

The aim now is to construct a nonsingular *preconditioning matrix* M such that $M^{-1}A$ has a more favorable condition number than A.

For increased flexibility we can write $M^{-1} = C_L C_R$ and transform the system according to

$$C_L A C_R y = C_L b, \quad y = C_R^{-1} x, \tag{C.45}$$

where C_L is the *left* and C_R is the *right* preconditioner. If the original coefficient matrix A is symmetric and positive definite, $C_L = C_R^T$ leads to preservation of these properties in the transformed system. This is important when applying the Conjugate Gradient method to the preconditioned linear system[3]. It appears that for practical purposes one can express the iterative algorithms such that it is sufficient to work with a single preconditioning matrix M only [13,19]. We shall therefore speak of preconditioning in terms of the left preconditioner M in the following.

Optimal convergence rate for the Conjugate Gradient method is achieved when the coefficient matrix $M^{-1}A$ equals the identity matrix I. In the algorithm we need to perform matrix-vector products $M^{-1}Au$ for an arbitrary $u \in \mathbb{R}^n$. This means that we have to solve a linear system with M as coefficient matrix in each iteration since we implement the product $y = M^{-1}Au$ in a two step fashion: First we compute $v = Au$ and then we solve the linear system $My = v$ for y. The optimal choice $M = A$ therefore involves the solution of $Ay = v$ in each iteration, which is a problem of the same complexity as our original system $Ax = b$. The strategy must hence be to compute an $M \approx A$ such that the algorithmic operations involving M are cheap.

The preceding discussion motivates the following demands on the preconditioning matrix M:

1. M should be a good approximation to A,
2. M should be inexpensive to compute,
3. M should be sparse in order to minimize storage requirements, and
4. linear systems with M as coefficient matrix must be efficiently solved.

Regarding the last property, such systems must be solved in $\mathcal{O}(n)$ operations, that is, a complexity of the same order as the vector updates in the Conjugate Gradient-like algorithms. These four properties are contradictory and some sort of compromise must be sought.

---

[3] Even if A and M are symmetric and positive definite, $M^{-1}A$ does not necessarily inherit these properties.

C.3.2   Classical Iterative Methods as Preconditioners

Consider the basic iterative method (C.4),

$$\boldsymbol{x}^k = \boldsymbol{x}^{k-1} + \boldsymbol{r}^{k-1} .$$

Applying this method to the preconditioned system $\boldsymbol{M}^{-1}\boldsymbol{A}\boldsymbol{x} = \boldsymbol{M}^{-1}\boldsymbol{b}$ results in the scheme

$$\boldsymbol{x}^k = \boldsymbol{x}^{k-1} + \boldsymbol{M}^{-1}\boldsymbol{r}^{k-1},$$

which is nothing but a classical iterative method, cf. (C.2). This motivates for choosing $\boldsymbol{M}$ from the matrix splittings in Appendix C.1 and thereby defining a class of preconditioners for Conjugate Gradient-like methods. To be specific, the appropriate choices of the preconditioning matrix $\boldsymbol{M}$ are as follows.

- Jacobi preconditioning: $\boldsymbol{M} = \boldsymbol{D}$.
- Gauss-Seidel preconditioning: $\boldsymbol{M} = \boldsymbol{D} + \boldsymbol{L}$.
- SOR preconditioning: $\boldsymbol{M} = \omega^{-1}\boldsymbol{D} + \boldsymbol{L}$.
- SSOR preconditioning:

$$\boldsymbol{M} = (2 - \omega)^{-1} \left(\omega^{-1}\boldsymbol{D} + \boldsymbol{L}\right) \left(\omega^{-1}\boldsymbol{D}\right)^{-1} \left(\omega^{-1}\boldsymbol{D} + \boldsymbol{U}\right) .$$

Line and block versions of the classical schemes can also be used as preconditioners.

Turning our attention to the four requirements of the preconditioning matrix, we realize that the suggested $\boldsymbol{M}$ matrices do not demand additional storage, linear systems with $\boldsymbol{M}$ as coefficient matrix are solved effectively in $\mathcal{O}(n)$ operations, and $\boldsymbol{M}$ needs no initial computation. The only questionable property is how well $\boldsymbol{M}$ approximates $\boldsymbol{A}$, and that is the weak point of using classical iterative methods as preconditioners.

The implementation of the given choices for $\boldsymbol{M}$ is very simple; solving linear systems $\boldsymbol{M}\boldsymbol{y} = \boldsymbol{v}$ is accomplished by performing exactly one iteration of a classical iterative method. The Conjugate Gradient method can only utilize the Jacobi and SSOR preconditioners among the classical iterative methods, because the $\boldsymbol{M}$ matrix in that case is on the form $\boldsymbol{M}^{-1} = \boldsymbol{C}_L\boldsymbol{C}_L^T$, which is necessary to ensure that the coefficient matrix of the preconditioned system is symmetric and positive. For certain PDEs, like the Poisson equation, it can be shown that the SSOR preconditioner reduces the condition number with an order of magnitude, i.e., from $\mathcal{O}(h^{-2})$ to $\mathcal{O}(h^{-1})$, provided we use the optimal choice of the relaxation parameter ω. The performance of the SSOR preconditioned Conjugate Gradient method is not very sensitive to the choice of ω, and for PDEs with second-order spatial derivatives a reasonably optimal choice is $\omega = 2/(1 + ch)$, where c is a positive constant.

We refer to [13,19] for more information about classical iterative methods as preconditioners.

C.3.3 Incomplete Factorization Preconditioners

Imagine that we choose $M = A$ and solve systems $My = v$ by a direct method. Such methods typically first compute the LU factorization $M = \bar{L}\bar{U}$ and thereafter perform two triangular solves. The lower and upper triangular factors $\bar{L}$ and $\bar{U}$ are computed from a Gaussian elimination procedure. Unfortunately, $\bar{L}$ and $\bar{U}$ contain nonzero values, so-called *fill-in*, in many locations where the original matrix A contains zeroes. This decreased sparsity of $\bar{L}$ and $\bar{U}$ increases both the storage requirements and the computational efforts related to solving systems $My = v$. An idea to improve the situation is to compute *sparse* versions of the factors $\bar{L}$ and $\bar{U}$. This is achieved by performing Gaussian elimination, but neglecting the fill-in. In this way we can compute approximate factors $\widehat{L}$ and $\widehat{U}$ that become as sparse as A. The storage requirements are hence only doubled by introducing a preconditioner, and the triangular solves become an $\mathcal{O}(n)$ operation since the number of nonzeroes in the $\widehat{L}$ and $\widehat{U}$ matrices (and A) is $\mathcal{O}(n)$ when the underlying PDE is discretized by finite difference or finite element methods. We call $M = \widehat{L}\widehat{U}$ an *Incomplete LU Factorization* preconditioner, often just referred to as the ILU preconditioner.

Instead of throwing away all fill-in entries, we can add them to the main diagonal. This yields the *Modified Incomplete LU Factorization* method, commonly known as the MILU preconditioner. If the fill-in to be added on the main diagonal is multiplied by a factor $\omega \in [0, 1]$, we get the *Relaxed Incomplete LU Factorization* preconditioner, with the acronym RILU. MILU and ILU preconditioning are recovered with $\omega = 1$ and $\omega = 0$, respectively.

For certain second-order PDEs with associated symmetric positive definite coefficient matrix, it can be proven that the MILU preconditioner reduces the condition number from $\mathcal{O}(h^{-2})$ to $\mathcal{O}(h^{-1})$. This property is also present in numerical experiments going beyond the limits of existing convergence theory. When using ILU or RILU factorization (with $\omega < 1$), the condition number remains of order $\mathcal{O}(h^{-2})$, but the convergence rate is far better than for the simple Jacobi preconditioner. Some work has been done on estimating the optimal relaxation parameter ω in model problems. For the 2D Poisson equation with $u = 0$ on the boundary, the optimal ω is $1 - \delta h$, where h is the mesh size and δ is independent of h. It appears that $\omega = 1$ can often give a dramatic increase in the number of iterations in the Conjugate Gradient method, compared with using an ω slightly smaller than unity. The value $\omega = 0.95$ could be regarded as a reasonable all-round choice. However, in a particular problem one should run some multiple loops in Diffpack to determine a suitable choice of ω and other parameters influencing the efficiency of iterative solvers.

The general algorithm for RILU preconditioning follows the steps of traditional exact Gaussian elimination, except that we restrict the computations to the nonzero entries in A. The factors $\widehat{L}$ and $\widehat{U}$ can be stored directly in the sparsity structure of A, that is, the algorithm overwrites a copy M of A

with its RILU factorization. The steps in the RILU factorizations are listed in Algorithm C.3.

Algorithm C.3.

Incomplete LU factorization.

given a sparsity pattern as an index set $\mathcal{I}$
copy $M_{i,j} \leftarrow A_{i,j}$, $i, j = 1, \ldots, n$
for $k = 1, 2, \ldots, n$
    for $i = k + 1, \ldots, n$
        if $(i, k) \in \mathcal{I}$ then
            $M_{i,k} \leftarrow M_{i,k}/M_{k,k}$
        else
            $M_{i,k} = 0$
        $r = M_{i,k}$
        for $j = k + 1, \ldots, n$
            if $j = i$ then
                $M_{i,j} \leftarrow M_{i,j} - rM_{k,j} + \omega \sum_{p=k+1}^{n} \left(M_{i,p} - rM_{k,p} \right)$
            else
            if $(i, j) \in \mathcal{I}$ then
                $M_{i,j} \leftarrow M_{i,j} - rM_{k,j}$
            else
                $M_{i,j} = 0$

We also remark here that the algorithm above needs careful refinement before it should be implemented in a code. For example, one will not run through a series of (i, j) indices and test for each of them if $(i, j) \in \mathcal{I}$. Instead one should run more directly through the sparsity structure of $\boldsymbol{A}$. See [61] for an ILU/MILU algorithm on implementational form.

The RILU methodology can be extended in various ways. For example, one can allow a certain level of fill-in in the sparse factors. This will improve the quality of $\boldsymbol{M}$, but also increase the storage and the work associated with solving systems $\boldsymbol{My} = \boldsymbol{v}$. Block-oriented versions of the pointwise RILU algorithm above have proven to be effective. We refer to [13,19] for an overview of various incomplete factorization techniques. A comprehensive treatment of incomplete factorization preconditioners is found in the text by Axelsson [11].

C.4   Multigrid and Domain Decomposition Methods.

The classical iterative methods and the Conjugate Gradient-like procedures we have sketched so far are general algorithms that have proven to be successful in a wide range of problems. A particularly attractive feature is the simplicity of these algorithms. However, the methods are not *optimal* in the sense that they can solve a linear system with n unknowns in $\mathcal{O}(n)$ operations. The MILU preconditioned Conjugate Gradient method typically demands $\mathcal{O}(n^{1+1/2d})$ operations, which means $\mathcal{O}(n^{1.17})$ in a 3D problem. RILU preconditioning in general leads to $\mathcal{O}(n^{1.33})$ operations in 3D. This unfavorable asymptotic behavior has motivated the research for optimal algorithms.

Two classes of optimal iterative strategies that can solve a system with n unknowns in $\mathcal{O}(n)$ operations, are the *multigrid* and *domain decomposition* methods. These methods are more complicated to formulate and analyze, and the associated algorithms are problem dependent, both with respect to the algorithmic details and the performance. On parallel computers, however, multigrid and domain decomposition algorithms are much easier to deal with than RILU-like preconditioned iterative methods. The recent book by Smith et al. [105] gives an excellent introduction to the algorithms and analysis of domain decomposition and multigrid methods and their applications to several types of stationary PDEs. Our very brief introduction to the subject is meant as an appetizer for studying [105] and applying, for instance, Diffpack's tools for multigrid and domain decomposition methods [25].

C.4.1   Domain Decomposition

We consider again our model problem $-\nabla^2 u = f$ in $\Omega \in \mathbb{R}^d$ with $u = g$ on the boundary $\partial\Omega$. As the name implies, domain decomposition methods consists in decomposing the domain Ω into subdomains Ω_s, $s = 1, \ldots, D$. The basic idea is then to solve the PDE in each subdomain rather than in the complete Ω. A fundamental problem is to assign appropriate boundary conditions on the inner non-physical boundaries. The classical *alternating Schwarz* method accomplishes this problem through an iterative procedure. The subdomains must in this case be *overlapping*. Let u_s^k be the solution on Ω_s after k iterations. Given an initial guess u_s^0, we solve for $s = 1, \ldots, D$ the PDE $-\nabla^2 u_s^k = f$ on Ω_s using $u_s^k = g$ on physical boundaries. On the inner boundaries of Ω_s we apply Dirichlet conditions with values taken from the most recently computed solutions in the overlapping neighboring subdomains. These values are of course not correct, but by repeating the procedure we can hope for convergence towards the solution u. This can indeed be proved for the present model problem.

The Schwarz method can be viewed as a kind of block Gauss-Seidel method, where each block corresponds to a subdomain. If we only use Dirichlet values on inner boundaries from the previous iteration step, a block Jacobi-like procedure is obtained. All the subproblems in an iteration step can now be solved in parallel. This is a very attractive feature of the domain decomposition approach and forms the background for many parallel PDE solvers. Parallel computing with Diffpack is also founded on such domain decomposition strategies [20]. The subdomain solvers can be iterative or direct and perhaps based on different discretization techniques. The approach can also be extended to letting the physical model, i.e. the PDEs, vary among the subdomains.

Although domain decomposition can be used as an efficient stand-alone solver, the Schwarz method is frequently applied as preconditioner for a Krylov subspace iteration method. There is only need for an approximate solve on each subdomain in this case. A popular class of domain decomposition-based preconditioners employs *non-overlapping* grids. We refer to [13] for a quick overview of domain decomposition preconditioners and to [105] for a thorough explanation of the ideas and the associated computational algorithms.

The primitive alternating Schwarz method as explained above exhibits rather slow convergence unless it is combined with a so-called *coarse grid correction*. This means that we also solve the PDE on the complete Ω in each iteration, using a coarse grid. The coarse grid solution over Ω is then combined with the fine grid subdomain solutions. Such combination of different levels of discretizations is the key point in *multigrid methods*.

C.4.2   Multigrid Methods

In Example C.1 we developed a demo program for the 2D Poisson equation, where the linear system arising from a finite difference discretization is solved by Gauss-Seidel iteration. The animation of the error as the iteration index k grows, shows that the error is efficiently smoothed and damped during the first iterations, but then further reduction of the error slows down. Considering a one-dimensional equation for simplicity, $-u'' = f$, we can easily write down the Gauss-Seidel iteration in terms of quantities u_j^ℓ, where $j = 1, \ldots, m$ is the spatial grid index and ℓ *represents the iteration number:*

$$2u_j^\ell = u_{j-1}^\ell + u_{j+1}^{\ell-1} + h^2 f_j \,.$$

The error $e_i^\ell = u_i^\ell - u_i^\infty$ satisfies the homogeneous version of the difference equation:

$$2e_j^\ell = e_{j-1}^\ell + e_{j+1}^{\ell-1} \,. \tag{C.46}$$

As explained in Appendix A.4.4, we may anticipate that the solution of (C.46) can be written as a sum of wave components:

$$e_j^\ell = \sum_k A_k \exp\left(i(kjh - \tilde{\omega}\ell\Delta t)\right) \,.$$

An alternative form, which simplifies the expressions a bit, reads

$$e_j^\ell = \sum_k A_k \xi^\ell \exp\left(ikjh\right), \quad \xi = \exp\left(-i\tilde{\omega}\Delta t\right) \,.$$

Inserting a wave component in the Gauss-Seidel scheme results in

$$\xi = \exp\left(-i\tilde{\omega}\Delta t\right) = \frac{\exp\left(ikh\right)}{2 - \exp\left(-ikh\right)} \,. \tag{C.47}$$

The reduction in amplitude due to one iteration is represented by the factor ξ, because $e_j^\ell = \xi e_j^{\ell-1}$. The absolute value of this damping factor, here named $F(kh)$, follows from (C.47): $F(kh) = (5 - 4\cos kh)^{-1/2}$. In the grid we can represent waves with wave length $\lambda \to \infty$ (a constant) down to waves with wave length $\lambda = 2h$. The corresponding range of k becomes $(0, \pi/h]$. Figure C.2 shows a graph of $F(p)$, $p = kh \in [0, \pi]$. As we see, only the shortest waves are significantly damped. This means that if the error is rough, i.e., wave components with short wave length have significant amplitude, the Gauss-Seidel iteration will quickly damp these components out, and the error becomes smooth relative to the grid. The convergence is then slow because the wave components with longer wave length are only slightly damped from iteration to iteration.

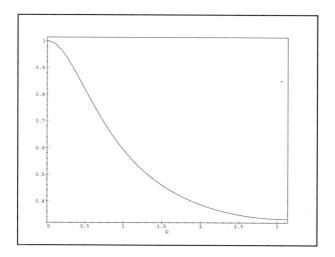

Fig. C.2. Damping factor $F(p)$ of the error in one Gauss-Seidel iteration for the equation $-u'' = f$, where $p = kh$ ($\lambda = 2\pi/k$ being the wave length and h being the grid cell size).

The basic observation in this analysis is that $F(p)$ decreases as $p = kh$ increases. Hence, if the error is dominated by wave components whose p values are too small for efficient damping, we can increase h. That is, transferring the error to a coarser grid turns low-frequency wave components into high-frequency wave components on the coarser grid. These high-frequency components are efficiently damped by a few Gauss-Seidel iterations. We can repeat this process, and when the grid is coarse enough, we can solve the error equation exactly (e.g. by a direct method). The error must then be transferred back to the original fine grid such that we can derive the solution of the fine-grid problem. An iterative method that is used to damp high-frequency components of the error on a grid is often called a *smoother*. Instead of Gauss-Seidel iteration, one can use a relaxed version of Jacobi's method or incomplete factorization techniques as smoothers [125].

Exercise C.4. Apply the relaxed Jacobi method from page 595 to the one-dimensional error equation $[\delta_x \delta_x e]_j = 0$ and analyze its damping properties. ◇

We now assume that we have a sequence of K grids, G^q, $q = 1, \ldots, K$, where the cell size decreases with increasing q. That is, $q = 1$ corresponds to the coarsest grid and $q = K$ is the fine grid on which our PDE problem of interest is posed. With each of these grids we associate a linear system $A^q u^q = b^q$, arising from discretization of the PDE on the grid G^q. Any vector v^q of grid point values can be transferred to grid G^{q-1} by the *restriction operator* R^q: $v^{q-1} = R^q v^q$. The opposite action, i.e., transferring a vector v^q on G^q to v^{q+1} on a finer grid G^{q+1}, is obtained by the *prolongation operator* P^q: $v^{q+1} = P^q v^q$. We mention briefly how the restriction and prolongation operators can be defined. Consider a one-dimensional problem and assume that if G^q has grid size h; then G^{q-1} has grid size $2h$. Going from the fine grid G^q to the coarse grid G^{q-1} can be done by a local weighted average as illustrated in Figure C.3a, whereas the prolongation step can make use of standard linear interpolation (Figure C.3b).

On every grid we can define a smoother and introduce the notation $S(\tilde{u}^q, v^q, f^q, \nu_q, q)$ for running ν_q iterations of the smoother on the system $A^q x^q = f^q$, yielding the approximation v^q to x^q, with $\tilde{u}^q$ as start vector for the iterations.

In the multigrid method, we start with some guess $\tilde{u}^q$ for the exact solution on some grid G^q and proceed with smoothing operations on successively coarser grids until we reach the coarsest grid $q = 1$, where we solve the associated linear system by a direct method. One can of course also use an iterative method at the coarsest level; the point is to solve the linear system with sufficient accuracy. The method is commonly expressed as a recursive procedure, as the one in Algorithm C.4. The LMG($\tilde{u}^q, v^q, f^q, q$) function in that algorithm has the arguments $\tilde{u}^q$ for the start vector for the current iteration, v^q for the returned (improved) approximation, f^q for the right-hand side in the linear system, and q is the grid-level number.

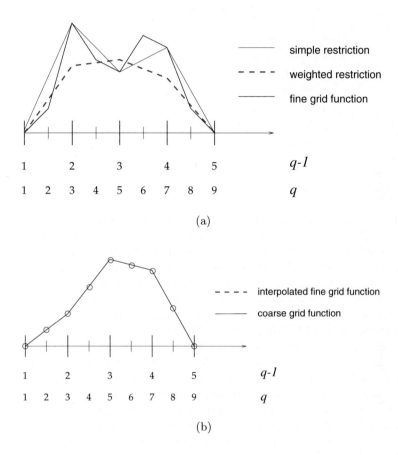

Fig. C.3. Example on (a) restriction and (b) prolongation on one-dimensional grids, where the cell size ratio of G^{q-1} and G^q equals two. The weighted restriction sets a coarse grid value u_j equal to $0.25u_{2j-1}^f + 0.5u_{2j}^f + 0.25u_{2j+1}^f$, where superscript f denotes a fine grid value and j is a coarse grid-point number in the figure. At the ends, the boundary conditions (here zero) must be fulfilled.

Algorithm C.4.

Multigrid method for linear systems.

routine LMG($\tilde{\boldsymbol{u}}^q, \boldsymbol{v}^q, \boldsymbol{f}^q, q$)
   if $q = 1$
      solve $\boldsymbol{A}^q \boldsymbol{v}^q = \boldsymbol{f}^q$ sufficiently accurately
   else
      $S(\tilde{\boldsymbol{u}}^q, \boldsymbol{v}^q, \boldsymbol{f}^q, \nu_q, q)$
      $\boldsymbol{r}^q = \boldsymbol{f}^q - \boldsymbol{A}^q \boldsymbol{v}^q$
      $\boldsymbol{f}^{q-1} = \boldsymbol{R}^q \boldsymbol{r}^q$
      $\tilde{\boldsymbol{u}}^{q-1} = \boldsymbol{0}$
      for $i = 1, \ldots, \gamma_q$
         LMG($\tilde{\boldsymbol{u}}^{q-1}, \boldsymbol{v}^{q-1}, \boldsymbol{f}^{q-1}, q-1$)
         $\tilde{\boldsymbol{u}}^{q-1} = \boldsymbol{v}^{q-1}$
      $\boldsymbol{v}^q \leftarrow \boldsymbol{v}^q + \boldsymbol{P}^{q-1} \boldsymbol{v}^{q-1}$
      $S(\boldsymbol{v}^q, \boldsymbol{v}^q, \boldsymbol{f}^q, \mu_q, q)$

Let us write out Algorithm C.4 in detail for the case of two grids, i.e., $K = 2$. We start the algorithm by calling LMG($\tilde{\boldsymbol{u}}^2, \boldsymbol{u}^2, \boldsymbol{b}, 2$), where $\tilde{\boldsymbol{u}}^2$ is an initial guess for the solution $\boldsymbol{u}^2$ of the discrete PDE problem on the fine grid. The algorithm then runs a smoother for ν_2 iterations and computes the residual $\boldsymbol{r}^2 = \boldsymbol{b} - \boldsymbol{A}^2 \boldsymbol{v}^2$, where now $\boldsymbol{v}^2$ is the result after ν_2 iterations of the smoother. This is called *pre-smoothing*. We then restrict $\boldsymbol{r}^2$ to $\boldsymbol{r}^1$. In a two-grid algorithm we branch into the $q = 1$ part of the LMG routine in the next recursive call LMG($\boldsymbol{0}, \boldsymbol{v}^1, \boldsymbol{f}^1, 1$). The system $\boldsymbol{A}^1 \boldsymbol{v}^1 = \boldsymbol{r}^1$ is then solved for the error, represented by the symbol $\boldsymbol{v}^1$ on this grid level (notice that we change the right-hand side from $\boldsymbol{b}$ on the finest level to the residual $\boldsymbol{r}^q$ on coarser levels, and the solution of linear systems with a residual on the right-hand side is then an error quantity, see (C.39)). In a two-grid algorithm it does not make sense to perform the inner call to LMG more than once, hence $\gamma_k = 1$. We then proceed with prolongating the error $\boldsymbol{v}^1$ to the fine grid and add this error as a correction to the result $\boldsymbol{v}^2$ obtained after the first smoothing process. Finally, we run the smoothing procedure μ_2 times, referred to as *post-smoothing*, to improve the $\boldsymbol{v}^2$ values.

The parameters ν_q, μ_q, and γ_q can be varied, yielding different versions of the multigrid strategy. Figure C.4 illustrates how the algorithm visits the various grid levels in the particular example of $K = 4$ and two choices of γ_q: 1 and 2. This results in the well-known V- and W-cycles. The reader is encouraged to work through the LMG algorithm in order to understand Figure C.4. To solve a given system of linear equations by multigrid, one runs a number of cycles, i.e., the LMG routine is embedded in a loop that runs until a suitable termination criterion is fulfilled.

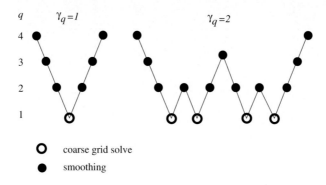

Fig. C.4. The LMG algorithm and the sequence of smoothing and coarse grids solves for the particular choice of $K = 4$; $\gamma_q = 1$ (V-cycle) and $\gamma_q = 2$ (W-cycle).

Multigrid can also be used as a preconditioner for Conjugate Gradient-like methods. Solving the system $\boldsymbol{My} = \boldsymbol{w}$ can then consist in running one multigrid cycle on $\boldsymbol{Ay} = \boldsymbol{w}$.

The perhaps most attractive feature of multigrid is that one can prove for many model problems that the complexity of the algorithm is $\mathcal{O}(n)$, where n is the number of unknowns in the system [125]. Moreover, the LMG algorithm can be slightly modified and then applied to nonlinear systems. We refer to [125] for intuitive analysis and extension of the concepts.

Diffpack offers extensive support for multigrid calculations, but the description of the software tools is beyond the scope of the present text. Readers can consult the introductory report [76]. In Chapter 6.4 we present a simulator for incompressible viscous fluid flow, where a special-purpose multigrid implementation is used for solving a 3D Poisson equation problem that arises in a solution algorithm for the Navier-Stokes equations.

Appendix D

Software Tools for Solving Linear Systems

This appendix presents various functionality in Diffpack for storing and solving linear systems. The importance of such tools is explained in the introduction to Appendix C. Appendix C also provides the necessary theoretical background for reading the present appendix. We also assume that the reader is familiar with the material in Chapter 1.

We start with describing various vector and matrix formats in Diffpack. Several examples demonstrate how to load a finite difference scheme into the matrix formats and how to create objects representing linear systems. We then outline in Appendix D.2 the syntax for using banded Gaussian elimination or the Conjugate Gradient method to solve the linear system. How to access basic iterative algorithms in general is treated in Appendices D.6–D.4, whereas preconditioning tools constitute the topic of Appendix D.5. Termination criteria and convergence monitors are covered in Appendix D.6. The final section D.7 combines various software tools discussed in this appendix with finite difference solvers from Chapter 1 for solving a transient 2D diffusion problem. This example is the major reference for programming implicit finite difference solvers in Diffpack.

Most of the material in Appendices D.1–D.2 is not needed when building finite element solvers, because the finite element toolbox in Diffpack offers an easy-to-use high-level interface to the construction and solution of linear systems. Nevertheless, Appendix D.2 contains many program examples which can be valuable for novice Diffpack programmers in general.

D.1 Storing and Initializing Linear Systems

As pointed out in Chapter 1.5.3, there are a number of different vector and matrix formats in Diffpack, see Figure 1.7. The reason for this diversity in data types is that different combinations of problem classes and solution algorithms lead to algebraic systems with different structural properties. Of particular importance is the sparsity structure of the coefficient matrix. Figure D.1 illustrates some common sparsity structures arising from discretization of partial differential equations. In this section we briefly review the available storage formats for vectors and matrices, before showing how Diffpack combines vector and matrix objects into a useful software abstraction for linear systems.

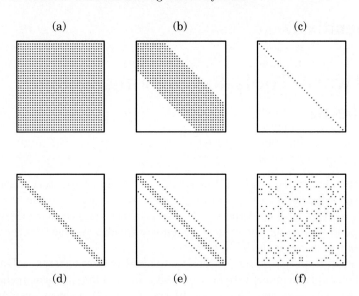

Fig. D.1. Different matrix formats supported by Diffpack. Each dot represents a nonzero matrix entry.

D.1.1   Vector and Matrix Formats

Vector Formats. In the context of algebraic systems, Diffpack offers three different data types representing vectors, `Vec`, `ArrayGen`, and `ArrayGenSel`. These classes are derived from the abstract base class `Vector`. Thus, you can allocate a `Handle(Vector)` variable in your program and at run time decide whether this should refer to a `Vec`, an `ArrayGen`, or an `ArrayGenSel` object.

In Chapter 1.5.3 it was demonstrated that these vector classes are based on lower-level array types like `VecSimplest` and `VecSimple`. While `Vec` offers a standard vector representation as a collection of entries v_i, $i = 1, 2, \ldots, n$, the classes `ArrayGen` and `ArrayGenSel` employ an index mapping that allows multiple indices, typically in two and three dimensions. The latter option is useful when the vector represents data sampled for the nodes in a box-shaped grid. For instance, this situation occurs for finite difference problems posed on rectangular domains.

For details of vector syntax we refer to the examples in Chapter 1.5.3 and to the man pages for the different vector classes.

Matrix Formats. Let $\boldsymbol{A}$ be an $m \times n$ matrix with entries $a_{i,j}$ for $i = 1, 2, \ldots, m$ and $j = 1, 2, \ldots, n$. If $\boldsymbol{A}$ is sparse, like the matrices shown in Figure D.1, it is important to utilize the sparsity in the implementation of numerical algorithms. This leads to the requirement of various matrix formats. The most widely applicable formats are dense matrices, banded matrices, general sparse matrices, structured sparse matrices, diagonal matrices, tridiagonal matrices,

and matrices represented in terms of finite difference stencils. Diffpack offers
the mentioned matrix formats. Each format is represented by a class. On the
top of the matrix hierarchy we have the abstract class `Matrix` that offers a
unified interface to all matrix formats. In the subclasses, the natural subscript operator function, `operator()`, is used for efficient inline indexing of
the matrix entries. Inline in C++ means that the function body is copied directly into the calling code, thus avoiding overhead related to function calls.
The `operator()` function is optimized for each format, and its arguments depend on the internal storage structure of the matrix entries. This ensures
user-friendly and highly efficient indexing syntax.

All matrix classes also offer a *virtual* function `elm(i,j)` for accessing entry
$a_{i,j}$. Virtual functions cannot be inlined, because it is not known at compile
time which version of a virtual function that will actually be called at run
time. A common indexing function for all matrix formats, like our `elm(i,j)`,
will always lead to a function call, which is much more costly than the simple
array look-up in the function body. A loop over the elements in a matrix, using `elm(i,j)` to access the entries, will therefore spend most of the CPU time
on calling the `elm(i,j)` function rather than processing the underlying array
structures. Nevertheless, there are situations where matrix-format independent access to array entries is more important than computational efficiency,
and `elm(i,j)` can be useful in such circumstances. Notice that assignment to
`elm(i,j)` might not always be possible; in a sparse matrix only some of the
index pairs `(i,j)` exist.

From a practical programming point of view, it turns out that efficient initialization of the matrix entries should be done in a code segment where the
matrix storage structure is known and where efficient `operator()` functions
can be used. In the rest of the code, including linear solvers, finite element
assembly etc., one can work with matrices through the generic `Matrix` interface only. This hides all storage-specific details and makes programming with
matrices more compact, easier, and less error-prone.

Here is a short list of Diffpack's matrix formats and their typical syntax
with respect to construction and entry assignment.

Class `Mat`: In dense $m \times n$ matrices[1], the values $a_{i,j}$ are stored for $i =$
1, 2, ..., m and $j = 1, 2, ..., n$, see Figure D.1a.
Example:

```
int m = 9; n = 6;
Mat(real) A(m,n);
A(3,4) = 3.14;          // 3rd row, 4th column
A.elm(3,4) = 3.14;      // 3rd row, 4th column
```

Here, `A(3,4)` and `A.elm(3,4)` are identical functions, except that the latter
is virtual and hence less efficient in a loop construct since it cannot be
inlined.

---

[1] Alternatively, dense matrices can be handled as `MatDense` objects. The name
`MatDense` is a synonym for `Mat`.

Class MatBand: In banded $n \times n$ matrices, the values $a_{i,j}$ are stored for indices (i, j) satisfying the inequality $|j - i| \leq q$ for $i, j = 1, 2, \ldots, n$. Here, $q + 1$ is the half of the total bandwidth.
Example:

```
int n = 9, q = 3;
MatBand(real) A(n,n,q+1);  // storing 2q+1 diags
A(3,4) = 3.14;             // 3rd row, 4th diag from below
A.elm(3,4) = 3.14;         // 3rd row, 4th column
```

`A(i,j)` employs indices that refer to the internal storage of a banded matrix. The storage structure makes use of a rectangular array, such that each diagonal is stored as a column. The transformation from a dense matrix index pair (i, j) to the actual storage location in the rectangular array is given by a function `denseIndex2bandIndex`.

Class MatDiag: In diagonal $n \times n$ matrices, the values $a_{i,j}$ are stored for $j = i$, $i = 1, 2, \ldots, n$, see Figure D.1c.
Example:

```
int n = 9;
MatDiag(real) A(n,n);
A(3,3) = 3.14;        // 3rd row, 3rd column
A.elm(3,3) = 3.14;    // 3rd row, 3rd column
```

We remark that `A(3,4)` is an illegal (out of range) subscript, while `s_o <<
A.elm(3,4)` is legal (a zero is printed), and `A.elm(3,4)=3.14` results in an error message. This natural behavior when indexing the matrix outside its sparsity pattern is common to all matrix classes.

Class MatTri: In tridiagonal $n \times n$ matrices, the values $a_{i,j}$ are stored for $j = i - 1, i, i + 1$, $i = 1, 2, \ldots, n$, see Figure D.1d.
Example:

```
int n = 9;
MatTri(real) A(n,n);
A(3,1) = 3.14;        // 3rd row, superdiagonal
A.elm(3,4) = 3.14;    // 3rd row, 4th column
```

`A(3,0)` denotes the main diagonal (3rd row), while `A(3,-1)` is the subdiagonal.

Class MatStructSparse: In structured sparse $n \times n$ matrices, only the main diagonal and the $2q$ sub- and super-diagonals are stored, see Figure D.1e. The diagonals are numbered from 1 to $2q + 1$, starting with the lowermost subdiagonal. This numbering is used directly in efficient indexing of `MatStructSparse` objects, as the example below demonstrates.

```
int n = 9; ndiag = 5;  // 5 diagonals
// set the offset from the main diagonal for each diagonal:
VecSimple(int) offset(ndiag);
offset(1) = -q; offset(2) = -1; offset(3) = 0;
offset(5) =  q; offset(4) =  1;
MatStructSparse(real) A(m,n,ndiag,offset);
```

```
A(3,1) = 1.14;              // 3rd row, 1st diagonal
A(3,2) = 1.14;              // 3rd row, 2nd diagonal
A(3,3) = 3.14;              // 3rd row, 3rd diagonal (main diag)
A(3,4) = 2.14;              // 3rd row, 4th diagonal
A(3,5) = 2.14;              // 3rd row, 5th diagonal
A.elm(3,3+q) = 2.14;        // 3rd row, column 3+q = 5th diag
```

If q_k is the offset of diagonal k, then $A(i,j)$ means $a_{i,i+q_j}$. In the current example, q_5 is $\mathtt{offset(5)=q}$, so the array entry $A(3,5)$ corresponds to $a_{3,3+q}$. A more comprehensive and instructive example is given on page 632.

Class MatSparse: In general sparse $n \times n$ matrices, the values $a_{i,j}$ are stored for indices $(i,j) \in \mathcal{P}$, where $\mathcal{P} \subset \{(i,j) : i,j = 1, 2, \ldots, n\}$ is a symmetric user-defined sparsity pattern, represented by a Diffpack object of class SparseDS. Typically, $\mathcal{P}$ consists of the matrix positions corresponding to the nonzero entries in $\boldsymbol{A}$, see Figure D.1f.
Example:

```
int n = 9, nnz = 14;    // nnz is max no of nonzero entries
SparseDS pattern(n,nnz);
// initialize sparsity pattern...

MatSparse(real) A(pattern);
A(3,4) = 3.14;              // 3rd row, 4th column
A.elm(3,4) = 3.14;         // 3rd row, 4th column
```

Here, we must remark that the assignments above are in general inefficient, because the syntax implies locating where in the sparse storage structure the entry $(3, 4)$ is actually stored. When implementing numerical algorithms that employ general sparse matrices, one should access the entries in the sequence they are stored in memory.

As illustrated in Figure 1.7, all matrix data types in Diffpack are derived from a common abstract base class Matrix. This encourages transparent use of matrix formats in the sense that the application can be based on the common Matrix interface,

```
Handle(Matrix) A;
```

At run time, the handle A can be rebound to a user-selected matrix representation, e.g.

```
A.rebind(new Mat(real)(m,n));
```

or

```
A.rebind(new MatBand(real)(n,bandwidth));
```

The Parameter Class for Initializing Matrix Objects. A common feature of many Diffpack class hierarchies is the existence of an associated *parameter class*. Such parameter classes are capable of holding any set of information needed for constructing objects of any class in the relevant hierarchy. Usually, such parameter classes are linked to the menu system as separate submenus, thus allowing the user to fill in the relevant data from input files or a graphical user interface.

The parameter class for the `Matrix` hierarchy is called `Matrix_prm`. If you browse the man page for `Matrix_prm`, you will see something like

```
class Matrix_prm(Type) : public HandleId
{
public:
  String   storage;       // name of subclass of Matrix(Type)
  bool     symm_storage;  // symmetry indicator (cheap storage)

  int      nrows;         // no of equations (no of rows in matrix)
  int      ncolumns;      // no of unknowns (no of columns in matrix)

  int      bandwidth;     // half-bandwidth (if banded matrix)
  bool     pivot_allowed; // true: allow pivoting (affects storage)

  int      ndiagonals;    // no of diags in struct. sparse matrices
  VecSort(int)  offset;   // offset values for each stored diagonal
  Handle(SparseDS) sparse_adrs;  // sparse matrix data structure
  real     threshold;     // for compression of sparsity patterns
  ...
};
```

The meaning of each data item should be pretty clear from the above comments. The parameter object is usually filled from the menu system (using `scan`, see Chapter 3.2). We can of course also initialize the members manually in a program. Here is an example concerning the creation of a structured sparse matrix capable of holding the five-point finite difference molecule for the 2D Laplacian on a $q \times q$ grid.

```
int q = 100, n = q*q;
Matrix_prm(real) pm;
pm.storage = "MatStructSparse";
pm.ncolumns = pm.nrows = n;  pm.diagonals = 5;
pm.offset.redim(pm.ndiagonals);
pm.offset(1) = -q;  pm.offset(1) = -1;  pm.offset(1) = 0;
pm.offset(1) =  1;  pm.offset(1) =  q;
Handle(Matrix(real)) A;
A.rebind(pm.create());
// *A has right size, but is not initialized
```

Calling `pm.createEmpty()` instead of `pm.create()` creates the right matrix type (provided `pm.storage` is initialized), but no other information from the parameter object is transferred to the matrix. You will then have to call the resulting matrix object's member function `redim` in order to complete the initialization.

Sparse Matrices. The class `MatSparse` is of special interest in the context of PDE problems. Discretization of PDEs usually leads to algebraic systems where the involved coefficient matrices are sparse, i.e., only a few nonzero entries occur in each row. Moreover, the count of nonzeros per row is usually independent of the matrix size, which means that for large-scale problems only a very small fraction of the matrix is populated with nonzeros. Clearly, this structural property should be taken advantage of. First, if it suffices to store only the nonzero entries, the memory requirement is significantly reduced. Second, and at least as important, computing time can be saved if the problem is solved with methods that can work on the nonzero entries alone. Iterative solvers are designed exactly for this purpose. Thus, the combination of compact storage schemes and efficient iterative solvers allows us to solve larger and more complicated problems.

When building a finite element application in Diffpack, you will not need any technical details on the construction of sparse matrices. This is automatically handled by the administrative class `LinEqAdmFE` whose usage is covered in Chapter 3. However, if your application is to take full and direct control of a `MatSparse` object, you will need some basic knowledge of the Compressed Row Storage (CRS) scheme. Consider the 5×5 sparse matrix

$$A = \begin{pmatrix} a_{1,1} & 0 & 0 & a_{1,4} & 0 \\ 0 & a_{2,2} & a_{2,3} & 0 & a_{2,5} \\ 0 & a_{3,2} & a_{3,3} & 0 & 0 \\ a_{4,1} & 0 & 0 & a_{4,4} & a_{4,5} \\ 0 & a_{5,2} & 0 & a_{5,4} & a_{5,5} \end{pmatrix}.$$

Using the CRS scheme, we need three arrays, respectively holding the nonzero matrix entries (A_s), references to the first stored value for each row (`irow`), and the column index for each stored entry (`jcol`). For the small example above, these arrays would have the following contents:

$$A_s = (a_{1,1}, a_{1,4}, a_{2,2}, a_{2,3}, a_{2,5}, a_{3,2}, a_{3,3}, a_{4,1}, a_{4,4}, a_{4,5}, a_{5,2}, a_{5,4}, a_{5,5}),$$
$$\texttt{irow} = (1, 3, 6, 8, 11, 14),$$
$$\texttt{jcol} = (1, 4, 2, 3, 5, 2, 3, 1, 4, 5, 2, 4, 5).$$

Here, notice that `irow` has length $n + 1$, whereas A_s and `jcol` have identical length $n_z = \texttt{irow}(n+1) - 1$. Observe that the last entry in `irow` "points" one position beyond the limit of A_s, i.e., $\texttt{irow}(n+1)$ should always have the value $n_z + 1$. The index information, i.e., the arrays `irow` and `jcol`, are part of the Diffpack class `SparseDS`. Objects of this type describes the sparsity pattern of a `MatSparse` matrix[2]. To see how the `SparseDS` structure should be filled, consult its man page, as well as the man page for class `MatSparse`.

---

[2] In fact, one `SparseDS` object can be shared by many `MatSparse` objects. In such cases, keep in mind that changes to a matrix sparsity pattern have influence on any matrix referencing the `SparseDS` object in question.

While `MatSparse` allows general sparsity patterns, `MatStructSparse` provides a more efficient representation of sparse matrices for which the nonzeros are located along matrix diagonals. Many PDE problems discretized on regular geometries lead to matrices with such structured patterns. As indicated above, the `MatStructSparse` constructor accepts a `VecSimple` object containing the offset of each diagonal to be stored. In this context, an offset value is the distance from the main diagonal, where negative (positive) values indicate a position in the lower (upper) triangular part of the matrix. That is, the matrix entry $a_{i,j}$ belongs to the diagonal with offset $j - i$. In particular, the main diagonal (zero offset) should always be stored.

Also the general sparse matrix format `MatSparse` accepts an offset vector. In this case, a sparsity pattern of type `SparseDS` is automatically constructed such that the same matrix positions are populated as for the corresponding `MatStructSparse` object. Moreover, the `MatSparse` class contains a constructor that accepts a dense matrix of type `Mat` and a drop tolerance. This constructor generates a sparsity pattern that corresponds to the dense matrix' entries that by magnitude are larger than the specified tolerance value. Only the entries fulfilling the drop tolerance test will then be copied to the sparse matrix. This approach can be a convenient way of constructing sparse matrices for test purposes, but due to the extremely low efficiency it is unsuitable for large- and medium-sized problems.

D.1.2   Detailed Matrix Examples

Consider the 3D Poisson problem

$$-\nabla^2 u = 1 \quad \text{on } \Omega = (0,1) \times (0,1) \times (0,1)$$

subject to the homogeneous Dirichlet condition $u = 0$ on the boundary $\partial\Omega$. Let this problem be discretized by centered differences on a uniform $q \times q \times q$ grid (not counting the boundary nodes). This procedure results in an $n \times n$ linear system $\boldsymbol{Ax} = \boldsymbol{b}$ where $n = q^3$ and $\boldsymbol{A}$ is a sparse matrix corresponding to the well-known seven-point point stencil in Figure D.2. Using our finite difference operator notation from Appendix A.3, the discrete problem can be written as

$$[\delta_x \delta_x u + \delta_y \delta_y u + \delta_z \delta_z u = -1]_{i,j,k} \, .$$

In the following examples we present four overloaded versions of a global function `genLaplace3D` that constructs $\boldsymbol{A}$ as a `Mat`, `MatBand`, `MatStructSparse`, or `MatSparse` object, respectively. The compilable source code for these examples is located in the directory `src/linalg/genLaplace3D`.

Using a Dense Matrix. When using a `Mat` object to represent the 3D Laplacian, the matrix indexing behaves as normal, i.e., the entries are addressed as `A(i,j)` where `i` and `j` are allowed to run from 1 to n. Consequently, the only tricky point is to treat the special cases where the stencil in Figure D.2

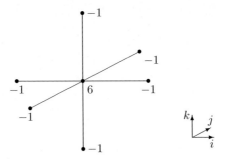

Fig. D.2. The seven-point stencil corresponding to a centered finite difference discretization of the 3D Laplace operator $\nabla^2 u$. This stencil can be compared to the 2D counterpart in Figure 1.5 on page 39.

meets the boundary. These situations lead to "holes" (single zero entries) in certain positions along the six sub- and superdiagonals in A, see Figure D.3.

```
void genLaplace3D (Mat(real)& A, int q)
{
    const int q2 = q*q;
    const int n  = q*q*q;   // using a q x q x q grid

    A.redim(n);  // square n x n matrix
    A.fill(0.0);

    int row = 0;  // row counter

    // run through all nodes in the q x q x q grid
    int i,j,k;
    for (k = 1; k <= q; ++k) {
      for (j = 1; j <= q; ++j) {
        for (i = 1; i <= q; ++i) {
          row++;  // treat next row
          if (k > 1) A(row,row-q2) = -1;
          if (j > 1) A(row,row-q)  = -1;
          if (i > 1) A(row,row-1)  = -1;
          A(row,row) = 6; // main diagonal present for all rows
          if (i < q) A(row,row+1)  = -1;
          if (j < q) A(row,row+q)  = -1;
          if (k < q) A(row,row+q2) = -1;
        }
      }
    }
}
```

We emphasize that this code is written for clarity, not for maximum computational efficiency. If the latter subject is in focus, we should avoid the if-tests inside the nested loops, but this requires expansion of the code segment (cf. Exercise D.1 and the solver associated with Appendix D.7).

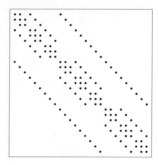

Fig. D.3. The sparsity pattern for the 3D Laplacian using a uniform $3 \times 3 \times 3$ discretization of the unit cube. Each dot represents a nonzero entry.

Using a Banded Matrix. Clearly, the use of a dense matrix is a waste of memory and computing time when all nonzero entries are located in a band centered around the main diagonal. This is the case even when the matrix will be subject to Gaussian elimination, since this process does not introduce new nonzero entries *outside* the band[3]. In contrast to the dense Mat representation, the class MatBand for banded matrices utilizes a different convention for indexing. In the present example the band consists of $2q^2 + 1$ diagonals and the half-bandwidth is $q^2 + 1$. This discussion assumes that we store the full band instead of exploiting the symmetry that is present for the Laplacian. Use of the symmetry option would need roughly half the storage space of the full band. The MatBand class can also set aside space for pivoting during the Gaussian elimination. Pivoting is not necessary when A stems from a finite difference (or finite element) discretization of the Laplace equation.

Internally in MatBand the diagonals are stored as columns in a $n \times (2q^2 + 1)$ array. The numbering convention is that the lower-most diagonal is stored in column 1, the next diagonal in column 2 etc. This scheme will place the main diagonal in column $q^2 + 1$ and the upper-most diagonal in column $2q^2 + 1$. In case of symmetric storage, the main diagonal is located in column 1 and the subdiagonals are not stored. All columns are adjusted such that the row index of the physical data structure matches the row index in the "logical matrix view". The operator(int,int) function works in the physical index space, while the member function elm assumes logical (dense matrix) indices $i, j = 1, 2, \ldots, n$. The member function denseIndex2bandIndex maps indices from the logical to the physical index space. All matrix formats in Diffpack have an elm function using logical indexing regardless of the physical data structure used, but the elm function is inefficient since it cannot be inlined.

---

[3] However, the Gaussian elimination will generate fill-in inside the band. (This fact makes the implementation of direct solvers for general sparse matrices rather intricate.)

The following piece of code builds the 3D Laplacian as a `MatBand` object using the physical indexing strategy:

```
void genLaplace3D (MatBand(real)& A, int q)
{
  const int n = q*q*q;        // using a q x q x q grid
  const int halfbw = q*q+1;   // half bandwidth (incl. main diag.)

  // column indices for the seven diagonals, counting from below
  const int d1=1, d2=q*(q-1)+1, d3=q*q, d4=q*q+1,
            d5=q*q+2, d6=q*(q+1)+1, d7=2*q*q+1;

  // don't use symmetric storage; MatBand is a suboptimal storage
  // structure for this problem anyway...
  A.redim (n, halfbw, false /*no symmetry*/, false /*no pivoting*/);
  A.fill(0.0);

  int row = 0;   // row counter

  // run through all nodes in the q x q x q grid
  int i,j,k;
  for (k = 1; k <= q; ++k) {
    for (j = 1; j <= q; ++j) {
      for (i = 1; i <= q; ++i) {
        row++;  // treat next row
        // NB! operator(int,int) uses data structure indexing
        if (k > 1) A(row,d1) = -1;
        if (j > 1) A(row,d2) = -1;
        if (i > 1) A(row,d3) = -1;
        // main diagonal present for all rows
        A(row,d4) = 6;
        if (i < q) A(row,d5) = -1;
        if (j < q) A(row,d6) = -1;
        if (k < q) A(row,d7) = -1;
      }
    }
  }
}
```

It might seem rather odd to present the internal data structure of the different matrix representations, and even encourage the use of highly specialized indexing functionality. After all, one of the most important issues in OOP is the use of data encapsulation which encourages the construction and use of black-box software where private details are hidden and only the public interface is known. However, the need for computational efficiency puts certain constraints on the use of object-oriented paradigms in numerical applications. For the present example this is evident since a general (logical) matrix indexing would be extremely expensive in terms of CPU time compared with the tailored indexing schemes utilizing knowledge of the data structure. Thus, this is a case where we deliberately break one fundamental rule of OOP. It is due to the same reason of efficiency that CPU-intensive functionality (e.g. matrix-vector products) is implemented as member functions in the various matrix classes, instead of relying on the general `elm`-type indexing. However, *this* design consideration is fully compatible with the concept of data encapsulation.

Using a Structured Sparse Matrix. The banded matrix approach used above is reasonable if we want to use Gaussian elimination to solve the associated system. However, if we instead turn to iterative methods where only the nonzero entries of A are needed, we can reduce memory cost and computing time by use of sparse matrices, see Appendix D.4. In the present case the matrix A has a regular sparsity pattern with only seven nonzero diagonals, see Figure D.3. Thus, we can use class `MatStructSparse` which only stores these diagonals. The underlying data structure is related to the one in `MatBand`, except that the columns representing the zero diagonals inside the band are removed. Thus the seven diagonals are stored as columns $1, 2, \ldots, 7$ in a rectangular array structure. The row index of this array structure matches the row index of the logical matrix indexing. In order to carry out operations on this data structure, there is need for an index vector `offsets` which tells how each stored diagonal is placed relatively to the main diagonal. That is, the entry located in row i of the kth diagonal corresponds to the entry `(i,i+offset(k))` in the matrix.

Here is a function that generates a `MatStructSparse` representation of the 3D Laplacian:

```
void genLaplace3D (MatStructSparse(real)& A, int q)
{
  const int ndiags = 7;    // 3D 7-point stencil
  const int n = q*q*q;     // using a q x q x q grid

  VecSimple(int) offsets(ndiags);
  offsets(1) = -q*q;
  offsets(2) = -q;
  offsets(3) = -1;
  offsets(4) =  0;   // main diagonal
  offsets(5) =  1;
  offsets(6) =  q;
  offsets(7) =  q*q;

  A.redim(n,ndiags,offsets);
  A.fill(0.0);

  int row = 0;   // row counter

  // run through all nodes in the q x q x q grid
  int i,j,k;
  for (k = 1; k <= q; ++k) {
    for (j = 1; j <= q; ++j) {
      for (i = 1; i <= q; ++i) {
        row++; // treat next row;
        // A(int,int) uses efficient data structure indexing
        if (k > 1) A(row,1) = -1;
        if (j > 1) A(row,2) = -1;
        if (i > 1) A(row,3) = -1;
        // main diagonal present for all rows
```

```
        A(row,4) = 6;
        if (i < q) A(row,5) = -1;
        if (j < q) A(row,6) = -1;
        if (k < q) A(row,7) = -1;
      }
    }
  }
}
```

Using a General Sparse Matrix. For the present case, the structured variant
of a sparse matrix format would be the best. However, in many applications
the resulting sparsity pattern is highly irregular, thus calling for the use of
the CRS scheme implemented in class `MatSparse`. The details of this format
are presented on page 627. The generic format requires slightly more book-
keeping than the structured one in order to keep track of the matrix indices.
Below you will find a `MatSparse` construction of the 3D Laplacian:

```
void genLaplace3D (MatSparse(real)& A, int q)
{
  const int q2  = q*q;
  const int n   = q*q*q;       // using a q x q x q grid
  const int nnz = 7*n-6*q2;    // nonzeros for 3D 7-point stencil

  Handle(SparseDS) pattern;    // allocate pattern storage
  pattern.rebind(new SparseDS(n,nnz));
  A.redim(*pattern);    // give A access to pattern
  A.fill(0.0);
  int row   = 0;  // row counter
  int entry = 0;  // entry counter (row by row)

  // run through all nodes in the q x q x q grid
  int i,j,k;
  for (k = 1; k <= q; ++k) {
    for (j = 1; j <= q; ++j) {
      for (i = 1; i <= q; ++i) {
        row++; // treat next row;
        pattern->irow(row) = entry+1;  // mark start of row
        // NB! operator(int) uses data structure indexing
        // (sliding down row by row)
        if (k > 1) { A(++entry) = -1; pattern->jcol(entry) = row-q2;}
        if (j > 1) { A(++entry) = -1; pattern->jcol(entry) = row-q;}
        if (i > 1) { A(++entry) = -1; pattern->jcol(entry) = row-1;}
        // main diagonal present for all rows
        A(++entry) = 6;
        pattern->jcol(entry) = row;
        if (i < q) { A(++entry) = -1; pattern->jcol(entry) = row+1;}
        if (j < q) { A(++entry) = -1; pattern->jcol(entry) = row+q;}
        if (k < q) { A(++entry) = -1; pattern->jcol(entry) = row+q2;}
      }
    }
  }
  pattern->irow(n+1) = nnz+1;
}
```

Exercise D.1. Looking at the `genLaplace3D` functions, we see that there are if-tests inside the nested loops over i, j, and k. This normally prevents compilers from optimizing the loops to the fullest extent. In this exercise we shall rewrite the loops to avoid the if-tests. This can be accomplished by letting the loops run from 2 to $q-1$ and then treat the indices 1 and q by special code segments. For example, we first set $k = 1$ and invoke a loop over i and j from 1 to q. All the points we visit in this loop are influenced by the boundary $z = 0$. Inside the loop we increment the `rows` counter as usual and fill all diagonals, except the lower-most one, with mathematical index (`row,row-q*q`). Thereafter we construct a loop over k from 2 to $q - 1$. Inside this loop we first make a special treatment of the $j = 1$ boundary by running through a loop over i from 1 to q and filling all diagonals, except the one corresponding to the mathematical index (`row,row-q`). Then we continue with the internal j points, but make special treatment of the first and last i index. Follow these ideas and implement an optimized version of `genLaplace3D` for the `MatStructSparse` matrix format. Choose $q = 300$ and compare the CPU time of the two versions of the routine.   ◇

D.1.3 Representation of Linear Systems

The general vector classes discussed in Chapter 1.5.3 and Appendix D.1.1 can serve as a basis for more abstract entities. The main focus of this chapter is to solve linear systems like $Ax = b$. It is then convenient to collect the entities A, x, and b in a separate class representing the linear system. If this new class refers to the abstract base classes `Matrix` and `Vector`, the representation is independent of technical details concerning the storage schemes.

As explained in Chapter 1.5.3 and Appendix D.1.1, the use of virtual functions is vital in the `Matrix` and `Vector` hierarchies. Numerical efficiency is achieved by implementing local versions of CPU-intensive algorithms (such as the LU factorization and the matrix-vector product) as virtual member functions for each matrix format. Thus, when a particular matrix instance is accessed by a `Matrix` base class pointer (handle) or reference, the tailored functionality is automatically invoked by the C++ run-time mechanisms for dynamic binding. The overhead of calling a virtual function is negligible when the function performs a few arithmetic operations. In a PDE context our virtual functions in the `Matrix` and `Vector` hierarchies normally process large array structures and often constitute the most time-consuming parts of a simulation program.

Block Matrices and Vectors. In many applications it would also be advisable to extend the concept of matrices and vectors into block structures. For instance, the discretized Laplacian posed on a $q \times q$ uniform, rectangular grid results in the block tridiagonal matrix structure

$$
A = \begin{pmatrix}
B & -I & & & \\
-I & B & -I & & \\
& \ddots & \ddots & \ddots & \\
& & -I & B & -I \\
& & & -I & B
\end{pmatrix} \in \mathbb{R}^{n,n} \text{ for } n = q^2,
$$

where

$$
I = \mathrm{diag}(1, 1, \ldots, 1) \in \mathbb{R}^{q,q}
$$

and

$$
B = \begin{pmatrix}
4 & -1 & & & \\
-1 & 4 & -1 & & \\
& \ddots & \ddots & \ddots & \\
& & -1 & 4 & -1 \\
& & & -1 & 4
\end{pmatrix} \in \mathbb{R}^{q,q}.
$$

A convenient representation of A can utilize the following block matrix:

```
class LinEqMatrix
{
protected:
  MatSimplest(Handle(Matrix)) matmat; // matrix of matrices
public:
  LinEqMatrix (int nblockrows, int nblockcolumns);
 ~LinEqMatrix ();

  const Matrix(NUMT)& operator(int i, int j) const;
  Matrix(NUMT)& operator(int i, int j);

  // matrix-vector product:
  void prod (const LinEqVector& x, LinEqVector& result) const;
  ...
};
```

The class `LinEqMatrix` organizes matrix blocks in a two-dimensional array structure with indices running from unity to `nblkrows` and `nblkcols`. Since each block is referenced by handles to `Matrix` objects, any matrix format may be used for the individual blocks. The only constraint in this respect is that individual blocks must have consistent sizes and that they match the dimensions and storage formats used for the corresponding blocks in other `LinEqMatrix` or `LinEqVector`[4] objects.

The `NUMT` parameter stands for NUMerical Type and is used in Diffpack to indicate `real` or `Complex`. One can usually think of `NUMT` as `real`, which is the default value. The presence of `NUMT` allows Diffpack to handle complex-valued

---

[4] Using a design quite similar to `LinEqMatrix`, the class `LinEqVector` represents block vectors.

linear systems and corresponding PDEs by simply compiling the application with a special option.

Returning to the example of the discretized Laplacian, the matrix A can be represented by a LinEqMatrix object with pointers to B and I, where B is a MatTri object and I is of type MatDiag. It is then sufficient to allocate *single* instances of B and I. This storage scheme will be conservative with respect to memory usage and will be an efficient alternative as long as the LinEqMatrix object has a fixed structure. We should mention that some numerical algorithms, e.g. in the field of mixed finite element methods, *require* the matrix and vectors in a linear system to be partitioned into blocks.

Having established a metaphor for block matrices and vectors, it is natural to express the linear system $Ax = b$ in terms of the LinEqMatrix and LinEqVector classes. It should be noticed that in many cases the matrix and vectors involved in a system will consist of single blocks, although the LinEqMatrix and LinEqVector classes are employed. These classes treat the single block case intelligently, thus resulting in negligible efficiency overhead. A standard linear system has the following representation in Diffpack.

```
class LinEqSystemStd : public LinEqSystem
{
protected:
  Handle(LinEqMatrix) Amat;   // coefficient matrix
  Handle(LinEqVector) sol;    // solution vector
  Handle(LinEqVector) rhs;    // right-hand side

public:
  LinEqSystemStd (LinEqMatrix& A, LinEqVector& x, LinEqVector& b);
  LinEqSystemStd (Matrix(NUMT)& A, Vector(NUMT)& x, Vector(NUMT)& b);
 ~LinEqSystemStd ();

  void residual (LinEqVector& r);  // r = A*x-b
  // matrix-vector product: result = A*y:
  void matvec (const LinEqVector& y, LinEqVector& result) const;
  ...
};
```

Here, the base class LinEqSystem is abstract and only serves the purpose of defining an interface to be obeyed by all possible representations of a linear system. The actual storage of data for the coefficient matrix, the solution vector, and the right-hand side is not addressed in class LinEqSystem, but rather in its subclass LinEqSystemStd. We remark that the code of class LinEqSystemStd is very short since the various functions, like matvec, are implemented by a call to LinEqMatrix::prod. Preconditioned linear systems are conveniently introduced as a subclass of LinEqSystemStd. A sketch of the LinEqSystem hierarchy is found in Figure D.5.

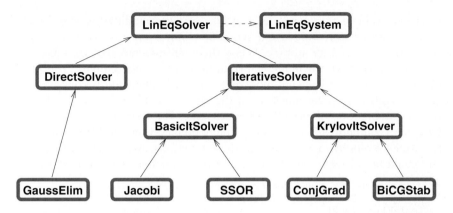

Fig. D.4. Linear solvers are organized in a class hierarchy, where the abstract base class `LinEqSolver` provides a transparent interface to all implemented methods, direct as well as iterative. The dotted line indicates a handle to the `LinEqSystem` object.

D.2   Programming with Linear Solvers

Regardless of the chosen algorithm for solving linear systems, the core operations that turn out to be CPU-intensive are implemented as member functions in the vector and matrix classes. In this way, the actual solver does not need to worry about efficiency considerations for a special matrix format since this is catered for by the low-level functionality, typically in virtual member functions like `prod` and `factLU` (in the `Matrix` hierarchy). In fact, the solver cannot see any of the details of matrix and vector storage formats. This allows a close relation between the solver's implementation and the corresponding mathematical algorithm without loss of efficiency.

As indicated in Figure D.4, all linear solvers are organized in a class hierarchy with base class `LinEqSolver`. At the next level it is natural to split this hierarchy into two subhierarchies, thus reflecting the division into direct (`DirectSolver`) and iterative (`IterativeSolver`) methods.

D.2.1   Gaussian Elimination

Direct methods are usually based on some variant of the Gaussian elimination procedure. Typically, such methods impose structural changes to the matrix through generation of *fill-in*. The term fill-in refers to nonzero values that occur during the elimination process in matrix positions that were initially occupied by zeroes. When employing a sparse matrix format where only nonzero entries are stored, we realize that direct methods will be closely tied to the technical details of each storage format. For instance, it may be advisable to renumber the unknowns of the original system in order to achieve certain sparsity patterns. It is also very common that sparse elimination algorithms start out with a symbolic elimination process that predicts the worst-case need for storage. When the resulting data structure has been built and initialized, the numerical computations are carried out. From these observations it is evident that most direct solvers will primarily consist of two or three calls to member functions of the `Matrix` hierarchy. Typically, this can be the combination of `factLU` and `forwBackLU`. For instance, the program in Chapter 1.2.3 addressing the two-point boundary-value problem, solves the resulting system by explicit calls to these functions. Alternatively, this problem could be solved by the use of the `LinEqSolver` derivative `GaussElim`:

```
// given Matrix A and Vector x, b
LinEqSystemStd system (A,x,b);
GaussElim gauss;
gauss.solve(system);
x.print(s_o,"The solution x");
```

As will be demonstrated in the examples below, the use of the `GaussElim` solver is more flexible than hardcoded calls to the functions performing the factorization and triangular solves, as it makes it fairly easy to switch to an iterative solution method. This is especially true when the solver is accessed through its base class interface `LinEqSolver`, such that we can select the particular solution method at run time.

D.2.2   A Simple Demo Program

Let us implement a class that can read a matrix system, i.e. A and b, from file and solve the system by means of Diffpack tools. The name of the class is `LinSys1` and the source code is found in the directory `src/linalg/LinSys1`.

```
class LinSys1
{
protected:
  Mat(real)                 A;         // coefficient matrix
  Vec(real)                 x, b;      // solution vector and rhs
  Handle(LinEqSystem)       system;    // linear system
  Handle(LinEqSolver_prm)   solv_prm;  // parameters for linear solver
  Handle(LinEqSolver)       solver;    // linear solver
  String                    filename;  // file with matrix data

public:
  LinSys1 () {}
  ~LinSys1 () {}
  void scan    ();
  void solveProblem ();
  void resultReport ();
};
```

In this class we require the coefficient matrix to be of type `Mat`, while the actual solver is accessed through a `LinEqSolver` handle named `solver`. The function `scan` loads the coefficient matrix and the right-hand side from the file specified by `filename`:

```
void LinSys1:: scan ()
{
  int n;
  initFromCommandLineArg("-n", n, 3);
  initFromCommandLineArg("-f", filename, "mat.dat");
  // instead of new LinEqSolver_prm we use the function construct:
  solv_prm.rebind (LinEqSolver_prm::construct());
  initFromCommandLineArg("-s", solv_prm->basic_method, "GaussElim");
  A.redim(n,n);  x.redim(n);  b.redim(n);

  // load matrix and right-hand side from file
  Is infile (filename, INFILE);  // Diffpack alternative to ifstream
  A.scan (infile);  A.print(s_o,"A");
  b.scan (infile);  b.print(s_o,"b");
}
```

The type of linear solver is read directly into the data member basic_method in the LinEqSolver_prm object. Class LinEqSolver_prm is the counter part to Matrix_prm in the Matrix hierarchy, that is, the parameter object contains all the parameters that are needed to select and initialize a subclass object in the LinEqSolver hierarchy. Because we have a dense matrix representation of A in this example, the other variables in the LinEqSolver_prm parameter object are not of interest. After reading the command-line options, the scan function loads the matrix A and the vector b from file. The following function solveProblem combines A, x, and b into a linear system, allocates the specified solver, and solves the system.

```
void LinSys1:: solveProblem ()
{
  system.rebind (new LinEqSystemStd (A, x, b));

  // make a solver based on the user's menu answers (we do not
  // know the solver type at compile time so we use a base class
  // pointer (or rather a handle), that can automatically deallocate
  // the solver when it is no longer in use)

  solver.rebind (solv_prm->create());

  x = 0.0;                    // remember to initialize a start vector!
  solver->solve (*system);   // compute the solution
}
```

Finally, the function resultReport asks the solver for statistics on the last solve operation. This information is stored in a LinEqStatBlk object before it is reported back to the user.

```
void LinSys1:: resultReport ()
{
  s_o << "\nHere is the computed solution:\n"; x.print(s_o,"x");

  // extract solver statistics
  LinEqStatBlk statistics; solver->performance (statistics);
  statistics.print (s_o);
}
```

The main program using the `LinSys1` class looks like this:

```
int main (int argc, const char* argv[])
{
  initDiffpack (argc, argv);
  LinSys1 problem;
  problem.scan();  problem.solveProblem();  problem.resultReport();
  return 0;
}
```

The resulting toy program can be run by typing `app`, thus using the default choices of solving the 3×3 system stored in the file `mat.dat` by the `GaussElim` method. You can specify the 5×5 system in `mat2.dat` by entering the command `./app -n 5 -f mat2.dat`. Alternatively, you can try the test problem in `mat3.dat` that contains the 27×27 matrix corresponding to the 3D Laplacian discretized on a $3 \times 3 \times 3$ grid.

D.2.3   A 3D Poisson Equation Solver

The purpose of the present section is to develop a Diffpack application for solving the three-dimensional Poisson equation on a unit cube, discretized by the common 7-point finite difference stencil:

$$[\delta_x\delta_x u + \delta_y\delta_y u + \delta_z\delta_z u = -1]_{i,j,k},$$

for $i, j, k = 0, 1, \ldots, q + 1$. The boundary points are left out of the linear system, i.e., the number of unknowns is $n = q^3$.

The solver, represented by class `LinSys2`, involves generation of the matrix entries and solution of the linear system by iterative methods. Appendix C presents algorithms and Diffpack tools for iterative solution of linear systems. Here, we merely show the syntax for calling some common iterative solution methods for linear systems, like the Conjugate Gradient (CG) method and the Successive Over-Relaxation (SOR) method. Iterative methods might contain several parameters that affect the computational performance. The relaxation parameter ω in SOR and SSOR constitute one example. Such parameters are collected in a `LinEqSolver_prm` object. This parameter object is most conveniently initialized using Diffpack's menu system, and the `LinSys2` class is therefore coupled hereto. Programming with and use of the menu system are thoroughly explained in Chapter 3.2. Here, we just outline the very basic steps in working with the menu system.

The `LinSys2` solver is not equipped with preconditioning. However, there are extended versions of `LinSys2`, implemented as the classes `LinSys3` and `LinSys4`, which offer preconditioning. These solvers are presented in Appendices D.5 and D.6.

The definition of class `LinSys2` becomes as follows.

```
class LinSys2 : public SimCase
{
protected:
  int                     q;       // q+1 grid points in each dir.
  MatStructSparse(real)   A;       // coefficient matrix
  Vec(real)               x, b;    // solution vector and rhs
  Handle(LinEqSystem)     system;  // linear system
  Handle(LinEqSolver_prm) solv_prm; // parameters for linear solver
  Handle(LinEqSolver)     solver;  // linear solver

public:
  LinSys2 () {}
  ~LinSys2 () {}
  virtual void adm     (MenuSystem& menu);
  virtual void define (MenuSystem& menu, int level = MAIN);
  virtual void scan    ();
  virtual void solveProblem ();
  virtual void resultReport ();
};
```

Deriving the simulator from class SimCase enables access to support for numerical experimentation, for instance, the multiple loop feature (see Chapter 3.4.2) and automatic report generation (see Appendix D.6 and Chapter 3.5.5).

As any module utilizing the menu system, the entry point adm is needed:

```
void LinSys2:: adm (MenuSystem& menu)
{
  SimCase::attach (menu); // enables later access to menu database
  define (menu);   // define/build the menu
  menu.prompt();   // prompt user, read menu answers
  scan ();         // read menu answers into class variables and init
}
```

The function adm relies on the presence of the functions define and scan:

```
void LinSys2:: define (MenuSystem& menu, int level)
{
  // define menu items
  menu.addItem (level,             // menu level
                "grid size q",     // menu command
                "grid has q*q*q points", // help
                "3");              // default answer

  // parameter objects are initialized by their construct function:
  solv_prm.rebind (LinEqSolver_prm::construct());
  // define solvers parameter menu as a submenu (level+1):
  solv_prm->define (menu,level+1);
}

void LinSys2:: scan ()
{
  // connect to the menu system database
  MenuSystem& menu = SimCase::getMenuSystem();

  q = menu.get("grid size q").getInt();  // read int (q)
```

```
    const int n = q*q*q;
    solv_prm->scan (menu);  // read menu answers into an object

    x.redim(n);  b.redim(n);
}
```

You should take notice of the first statement in scan, which is necessary in order to access the menu system object that was attached in adm.

The solveProblem function is almost identical to its counterpart in class LinSys1, but now we have an algorithm genLaplace3D, see page 632, for filling the MatStructSparse matrix object with entries according to the finite difference scheme.

```
void LinSys2:: solveProblem ()
{
    genLaplace3D (A,q);  // generate coefficient matrix
    b = 1.0/sqr(q+1);      // rhs in Poisson problem is f=1
    if (q < 4) {
      s_o->setRealFormat("%6.3f"); // change default output format
      A.printAscii(s_o,"A");
    }

    system.rebind (new LinEqSystemStd (A, x, b));

    // create solver according to user's menu choice
    solver.rebind (solv_prm->create());

    x = 0.0;                    // initialize a start vector!
    solver->solve (*system); // perform the linear solve
}
```

The present finite difference code is programmed at a fairly low level where the programmer is responsible for initializing the matrix entries. We remark that there exists a module in Diffpack which gives a flexible interface to finite difference programming, much like the finite element toolbox (see Chapter 3).

The function resultReport is slightly modified, now allowing a complete report in HTML format:

```
void LinSys2:: resultReport ()
{
    if (q < 4) {
      s_o << "\nHere is the solution computed by "
          << solv_prm->basic_method << ":\n"; x.print(s_o,"x");
    }
    // extract solver statistics
    LinEqStatBlk statistics; solver->performance (statistics);
    statistics.print (s_o);

    // write detailed convergence statistics, including plots
    HTMLReporter report (casename+".report");
    statistics.print (report, DETAILED /*print as much as possible*/);
}
```

Instead of an HTMLReporter, we could instead have declared a LaTeXReporter or a pure ASCIIReporter.

Due to the introduction of the menu system, also the `main` function has been changed. Now it reads:

```
#include <LinSys2.h>

int main (int argc, const char* argv[])
{
  initDiffpack (argc, argv);
  global_menu.init ("Poisson eq. FDM solver","LinSys2");
  LinSys2 problem;
  problem.adm (global_menu);    // define menu, prompt user, and scan
  problem.solveProblem();
  problem.resultReport();
  return 0;
}
```

The global menu object `global_menu` must be initialized. Moreover, the usual call to `scan` is now replaced by `adm`, which calls `scan` after the menu is defined.

Copy the directories `src/linalg/LinSys2` and `src/linalg/genLaplace3D` to your own directory tree, such that these two directories remain parallel[5] and compile the `LinSys2` application. You can now play with the SOR and CG methods, as well as with other solution methods. Here is an example on an input file `sor.i` for setting some of the menu items:

```
set grid size q = 3
sub LinEqSolver_prm
set basic method = SOR
set relaxation parameter = 1.8
set use default convergence criterion = ON
ok
ok
```

The first command sets the size of the linear system. The next command invokes the submenu for the `LinEqSolver_prm` parameters. The SOR method contains a relaxation parameter $\omega \in (0, 2)$ that influences the convergence speed. The current value of ω is set to 1.8. Finally, we specify the default termination (convergence) criterion for the iterative method. The default convergence test for the SOR method is that the Euclidean norm of the residual $r^k = b - Ax^k$ in the kth iteration is less than 10^{-4}. When this criterion is satisfied, the approximation x^k to x is taken as the solution (the vector x in the program). More sophisticated tools for monitoring the convergence of iterative methods are described in Appendix D.6.

Now run the program with the `sor.i` file as input to the menu system:

```
./app --casename sortest < Verify/sor.i
```

After completing the computation, use a web browser to view the HTML report named `sortest.report.html`. If you want to examine and adjust all the menu items through a graphical user interface, type

---

[5] The `LinSys2` application needs the file `../genLaplace3D/genLaplace3D.cpp` in the compilation.

```
./app --GUI --casename sortest --Default Verify/sor.i
```

provided that your program was compiled with the `Make` option `GUIMENU=tcl` (this option is often default – check the `$NOR/bt/src/MakeFlags` file).

Use `Verify/conjgrad.i` as input if you want to run the CG method. We remark that the default convergence criterion for the CG method is slightly different from the classical iterations: $||r^k||_2/||r^0||_2 \leq 10^{-4}$. Gaussian elimination is enabled in the `Verify/gauss.i` file.

Exercise D.2. Choose $q = 60$ and find the sensitivity of SOR to variations in the relaxation parameter ω. Aslo try to find, through experimentation, an ω that minimizes the number of SOR iterations. Compare the efficiency of SOR, with this optimal ω value, and the CG method. The efficiency measured in terms of "work cost" is listed in the HTML report.   ◇

D.3   Classical Iterative Methods

We recall from Appendix C.1.1 that the classical iterations (Jacobi, Gauss-Seidel, SOR, and SSOR) have a very simple structure on the form $Mx^k = Nx^{k-1} + b$, where $A = M - N$ and $k = 1, 2, \ldots$ is the iteration index. The implementation of these methods is therefore not much more than a loop supervised by some convergence test. Inside this loop we have to administer the two iterates x^k and x^{k-1}, and issue a call to low-level functionality that performs a given number of iterations. That is, in addition to the infrastructure offered by class `IterativeSolver` (see Figure D.4), the basic iterations need a vector object in order to store the previous iterate. This entity is a part of the derived class `BasicItSolver`, which serves as base class for the actual implementations of classical methods, exemplified by the classes `Jacobi` and `SSOR` in Figure D.4.

In order to play with the classical iterative methods, we may use the simple `LinSys2` program from Appendix D.2.3 or any of the finite element solvers from Chapters 3 or 4.2. To keep the numerics simple, we stick to the finite difference example as provided by the `LinSys2` solver, whose usage was explained in Appendix D.2.3. An appropriate input file for the `LinSys2` program, specifying Jacobi iteration as the solution method for linear systems, might look like this:

```
set grid size q = 3
sub LinEqSolver_prm
set basic method = Jacobi
set startvector mode = USER_START  ! Let user set start vector
set use default convergence criterion = ON
ok
ok
```

This input file explicitly tells the program to use the current value of the solution vector x as start vector when it enters the function `solver->solve`. Other alternatives to the default value `USER_START` are `RANDOM_START`, `ZERO_START`, and `RHS_START` which causes the start vector to be a random vector with entries in the interval $[-1, 1]$, the zero vector, or the right-hand side b. To view all possible menu items, run the program once and use the scripts `InputMenuFile` and `InputManualInLaTeX`, see Chapter 3.4.1.

SOR and SSOR iteration are enabled by changing `basic method` on the menu to `SOR` or `SSOR`. Recall that the Gauss-Seidel method corresponds to SOR with $\omega = 1$.

D.4   Conjugate Gradient-like Methods

In contrast to the classical iterations, the more sophisticated family of Conjugate Gradient-like methods, also known as Krylov subspace methods, puts more complicated demands on its surroundings. For instance, most Krylov iterations will recursively update the residual vector $r^k = b - Ax^k$, while others will also work on a *preconditioned* residual $s^k = M^{-1}(b - Ax^k) = M^{-1}r^k$. The framework from which the implementations of Krylov methods would be derived should take care of the allocation and initialization of the residual vectors needed by the given combination of a system and a solver. This also includes the detection of special cases, such as using an identity preconditioner $C = I$ which forces $s^k = r^k$ for all k. In this case the two entities r^k and s^k should actually share the same storage space. Such effects are easily obtained when accessing the residual vectors through handles.

To further complicate the internal picture of a flexible package for Krylov solvers, many of these methods compute certain values (residual norms, inner products etc.) that can be used outside the iteration for different purposes, such as monitoring of convergence, estimation of extreme eigenvalues, or tuning of some adaptive preconditioner. This type of information should be reused whenever needed in order to avoid recomputations. Diffpack takes care of this housekeeping by deriving Krylov methods from a specialized base class `KrylovItSolver`, see Figure D.4. Thus, the application programmer should not worry about these technical details.

D.4.1   Symmetric Systems

Many classes of PDEs lead to symmetric systems of algebraic equations. Krylov subspace methods can take advantage of the symmetry in the sense that the iterations will then be inexpensive in terms of computational cost and storage requirements.

The Conjugate Gradient Method. For problems where the coefficient matrix is symmetric positive definite, i.e., $A^T = A$ and $y^T A y > 0$ for all $y \neq 0$, the Conjugate Gradient method (see Appendix C.2) is the de facto choice of solver. In Diffpack this method is available as the class named `ConjGrad`. Since this solver does not require any user-specified parameters, assigning `ConjGrad` to the `basic method` menu item is the only initialization needed in the `LinEqSolver_prm` object.

Solving Symmetric Indefinite Problems. For symmetric problems lacking the property of positive definiteness, the class `Symmlq` provides an alternative strategy. To choose this method, simply set `basic method` on the menu equal to `Symmlq`.

D.4.2 Nonsymmetric Systems

In general, solving nonsymmetric systems is much harder than solving symmetric ones, and it is often impossible to predict what will be the most efficient solution strategy. Thus, it is very important that the software allows easy switching between different solvers and easy access to the corresponding performance statistics. As seen by the input files below, a Diffpack program like the one based on class `LinSys2` caters for such needs.

The Methods of Orthomin and GCR. The method called Orthomin is a popular member of the family of nonsymmetric Krylov solvers (see Appendix C.2). Usually this method has a name containing a parameter K, which specifies how many generations of basis vectors that should be kept for the update of the next iterate. In reports generated by Diffpack this *truncated* method is denoted T-OM(K). In general, the convergence rate and stability of the algorithm improves as K gets larger. However, large values of K also indicate a significant storage expense and a high computational cost per iteration, thus suggesting a tradeoff against convergence speed. The following input sets the K parameter to 5:

```
set basic method = Orthomin
set no of search vectors = 5
set restart = OFF
```

In the special case of $K = 0$, the Orthomin algorithm takes on a simpler form, and the implementation can be made more efficient. The resulting method is called Minres[6] and can be invoked as follows:

```
set basic method = MinRes
```

Finally, a close relative of Orthomin is the Generalized Conjugate Residual (CGR) method. Instead of the truncation to K basis vectors, the GCR algorithm uses *restarting*. That is, it runs for a specified number of iterations, K,

---

[6] This nonsymmetric solver should *not* be confused with the symmetric indefinite solver Minres, see [19]. This symmetric solver is not available in Diffpack.

before it starts over again using the last computed iterate as the new starting vector. For this reason the method is often denoted GCR(K). To emphasize that this algorithm is really a restarted Orthomin solver, reports generated by Diffpack refer to this method as R-OM(K). The following example uses $K = 7$:

```
set basic method = Orthomin
set no of search vectors = 7
set restart = ON
```

The GMRES Method. One of the best known (and also one of the most robust) nonsymmetric iterations is called the Generalized Minimal Residual method, or GMRES for short. This method uses restarting, just like GCR. A restart cycle of K iterations is therefore denoted by GMRES(K). The corresponding Diffpack class GMRES can be activated by the following input:

```
set basic method = GMRES
set no of search vectors = 5
set restart = ON
```

The CGS and BiCGStab Methods. In order to avoid the long recursions and high memory cost of the Orthomin, GCR, and GMRES methods, other Krylov methods like the Bi-Conjugate Gradient algorithm has been developed. However, this approach, which uses an auxiliary system to get rid of the expensive recursions, suffer from bad stability properties. In order to improve the stability without adding computational cost, the Conjugate Gradient Squared (CGS) algorithm was introduced. This method is self-contained and does not require any user-specified parameters. Its implementation in class CGS can be accessed by setting basic method equal to CGS on the menu. A newer variant with smoother convergence behavior is known as BiCGStab, with the class and menu name BiCGStab.

Transpose-Free QMR. The family of Quasi-Minimal Residual (QMR) methods constitutes yet another Krylov strategy based on short recurrences. In particular, there exist variants of the QMR algorithm that avoids the use of the transposed coefficient matrix $\boldsymbol{A}^T$. One such variant is the Transpose-Free QMR method available in class TFQMR.

D.5 Preconditioning Strategies

As explained in Appendix C.3.1, Krylov subspace solvers are usually combined with a *preconditioner* in order to speed up the convergence rate. That is, instead of solving $\boldsymbol{A}\boldsymbol{x} = \boldsymbol{b}$, we consider the equivalent system $\boldsymbol{M}^{-1}\boldsymbol{A}\boldsymbol{x} = \boldsymbol{M}^{-1}\boldsymbol{b}$. In each iteration of Krylov subspace methods we have solve a linear system with $\boldsymbol{M}$ as coefficient matrix. Alternatively, the preconditioner may

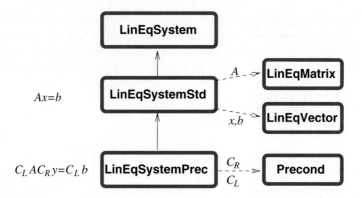

Fig. D.5. Combining matrix, vector, and preconditioner classes to form a linear system of equations.

be indirectly given in terms of a procedure implementing its action $\boldsymbol{M}^{-1}\boldsymbol{v}$ on the vector argument $\boldsymbol{v}$. We mention that Diffpack actually implements left and right preconditioning, $\boldsymbol{C}_L \boldsymbol{A} \boldsymbol{C}_R \boldsymbol{y} = \boldsymbol{C}_L \boldsymbol{b}$ for $\boldsymbol{y} = \boldsymbol{C}_R^{-1}\boldsymbol{x}$, see page 609.

Based on the `LinEqSystemStd` class introduced in Appendix D.1.3, an obvious extension for representation of preconditioned systems is

```
class LinEqSystemPrec : public LinEqSystemStd
{
  // A, x, and b structures are inherited from LinEqSystemStd.
protected:
  Handle(Precond) Cleft;   // left preconditioner
  Handle(Precond) Cright;  // right preconditioner
public:
  LinEqSystemPrec
    (Precond& C, LinEqMatrix& A,  LinEqVector& x, LinEqVector& b);
  LinEqSystemPrec
    (Precond& C, Matrix(NUMT)& A, Vector(NUMT)& x, Vector(NUMT)& b):
  void residual    (LinEqVector& r);  // r = b - A*x
  // matrix-vector product A*y:
  void matvec      (const LinEqVector& y, LinEqVector& result) const;
  // apply M: M^{-1}y (e.g. by solving M*result = y)
  void applyPrec   (const LinEqVector& y, LinEqVector& result) const;
  ...
};
```

We notice that the preconditioner is present in terms of a handle to `Precond`, which is the abstract base class in a hierarchy of different preconditioners, see Figure D.6. The most important change from the `LinEqSystem` base class, is the function `applyPrec` that actually computes the preconditioning step of applying $\boldsymbol{M}^{-1}$ to $\boldsymbol{y}$. The relation between `LinEqSystemStd` and `LinEqSystemPrec` is depicted in Figure D.5.

The `applyPrec` function is virtual in the `LinEqSystem` class hierarchy, despite the lack of a preconditioner in class `LinEqSystemStd`. However, we can

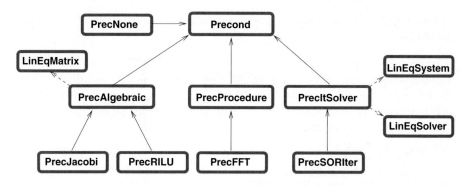

Fig. D.6. All preconditioners implemented in Diffpack are derived from the base class `Precond` as shown in this class hierarchy.

think of the standard linear system as a special case of the preconditioned system, where the preconditioner is the identity operator. This means that all iterative schemes in Diffpack are implemented in the presence of a preconditioner. Sending a `LinEqSystemStd` object to the solver, or a `LinEqSystemPrec` object with no preconditioner, simply results in a few calls to empty functions when the iterative algorithm executes statements involving the preconditioner.

A Sample Program with Preconditioning. By a few extra lines, the `LinSys2` code can be equipped with support for preconditioning. The resulting solver is called `LinSys3`. Two basic statements in the declaration of class `LinSys3` are new:

```
Handle(Precond_prm)     prec_prm; // parameters for preconditioner
Handle(LinEqSystemPrec) system;   // linear system
```

The preconditioning parameter object is put on the menu in `define` and initialized in `scan`, just as the `LinEqSolver_prm` object in class `LinSys2`. The creation of the linear system is slightly altered in the `solveProblem` function:

```
system.rebind (new LinEqSystemPrec(A, x, b));
// attach preconditioner (send in A since it is often demanded)
system->attach (*prec_prm, A);
```

Notice that the coefficient matrix A is used as argument to the function `attach` which initializes the preconditioner. Most preconditioners are constructed on basis of a matrix or differential operator. In the general case, the simulator itself can have `PrecProcedure` as base class and implement a proper `applyPrec` function.

The rest of the code is identical to what we had in class `LinSys2`. This example is typical when programming in Diffpack; as soon as you have a prototype solver for a problem up and going, more advanced algorithms and

software features can usually be added with small adjustments of the simulator class.

Classical Iterative Methods (Matrix Splittings). Well-known classical iterative methods, like Jacobi and SSOR, can be used as preconditioners (see
Sect C.3.2). The simplest choice is of course the Jacobi preconditioner, where
M is simply the diagonal of A. We can specify Jacobi preconditioning on
the `Precond_prm` submenu:

```
set grid size q = 5
sub LinEqSolver_prm
set basic method = ConjGrad
set startvector mode = USER_START
set use default convergence criterion = ON
set max iterations = 300
ok
sub Precond_prm
set preconditioning type = PrecJacobi
set left preconditioning = ON
set automatic init = ON
ok
ok
```

The SSOR preconditioner is also in frequent use, especially in combination
with the Conjugate Gradient method. Notice that the SOR preconditioner
is not applicable to iterative methods that require the coefficient matrix of
the preconditioned system to be symmetric and positive definite. In order to
change the Jacobi preconditioner used above to an SSOR preconditioner, the
following menu items must be set:

```
set preconditioning type = PrecSSOR
set (S)SOR relaxation parameter = 1.0
```

Inner Iterations. In addition to defining preconditioners in terms of matrix
splittings, classical iterations may be used for inner iterations. This consists
in performing *iterations* with a given classical method. Since most Krylov
methods require a fixed preconditioner, i.e., M is unchanged throughout the
Krylov iterations, such inner iterations should run for a fixed number of steps
and always use the same starting vector, e.g. the null vector. Numerical experiments indicate that this type of preconditioning can be efficient even for
complicated nonsymmetric systems. It has also been shown that inner iterations based on classical iterative schemes are intimately related to polynomial
preconditioners, see [19] and references therein.

Here are the necessary changes to the input file needed to run 10 SOR
iterations as an inner solver:

```
set preconditioning type = PrecSORIter
set (S)SOR relaxation parameter = 1.6
set inner steps = 10
```

Naturally, the SOR solver used for the inner iteration is itself implemented
as an instance of the corresponding `SOR` solver in the `LinEqSolver` hierarchy.

Incomplete Factorizations. Preconditioning by incomplete factorization is usually a simple and widely applicable technique for improving the convergence rates of iterative solvers. In Diffpack we have implemented the RILU preconditioner, where fill-in entries during the incomplete factorization process are multiplied by a factor ω before adding them to the main diagonal. The well-known ILU and MILU preconditioners are thus obtained by using RILU and $\omega = 0$ and $\omega = 1$, respectively. Specification of a RILU preconditioner with the all-round choice $\omega = 0.95$ takes the following form on the LinSys3 menu:

```
set preconditioning type = PrecRILU
set RILU relaxation parameter = 0.95
```

User-Defined Preconditioners. Any preconditioner in Diffpack must be derived from the base class Precond. The application programmer can extend this hierarchy by inheritance, thus bringing new preconditioners transparently into the existing framework. In many cases this can be done by implementing derivations of the base classes PrecAlgebraic and PrecProcedure. These classes are tailored to implementation of preconditioners expressed in terms of a matrix operator or as a procedure, respectively. For example, one can think of using a fast solver for the Laplace equation $\nabla^2 u = 0$ as preconditioner for an equation involving the variable coefficient Laplace operator $\nabla \cdot (\lambda(\boldsymbol{x})\nabla u)$. The simulator can then be derived from PrecProcedure and implement a fast constant-coefficient Laplace solver in the virtual applyPrec function. Attaching the this pointer (i.e. the simulator) as preconditioner in the linear system couples the tailored applyPrec function to the general linear solver and preconditioner library. In other words, the simulator becomes a part of the preconditioner hierarchy and provides a special-purpose preconditioner, closely connected to the PDE being solved. We refer to the man pages for the various preconditioner classes for further information.

Multigrid and Domain Decomposition. Finally, we mention that recent developments in Diffpack have resulted in iterative solvers and preconditioners based on multilevel techniques, including traditional multigrid algorithms as well as overlapping and non-overlapping (Schur complement) domain decomposition methods [25,76]. The programmer can develop ordinary solvers using simple and generic algorithms for linear systems, and as a post process extend the solver with support for optimal multigrid or domain decomposition methods.

D.6   Convergence History and Stopping Criteria

So far we have used iterative solvers without explicitly stating how to stop the iterations. Instead, we have relied on the (default) menu setting

```
set use default convergence criterion = ON
```

which uses the default convergence test associated with the solver of your choice. However, it is clear that different problems will need different measures of convergence. Moreover, it might be useful to combine several measures in a compound test, or even monitor certain entities without letting them influence the convergence test itself. For these purposes Diffpack offers the concept of *convergence monitors*. A conceptual discussion of convergence monitors and of how this abstraction fits into the Diffpack framework for iterative solvers can be found in [24].

Let us start by considering an example that illustrates two main points, the use of convergence monitors, and the possibility of providing a run-time choice of matrix formats[7]:

```
class LinSys4 : public SimCase
{
protected:
  Handle(Matrix_prm(real)) matx_prm; // parameters for coeff matrix
  Handle(Matrix(real))     A;        // coefficient matrix
  Vec(real)                x, b;     // solution vector, rhs
  Handle(Precond_prm)      prec_prm; // parameters for preconditioner
  ConvMonitorList_prm      conv_prm; // parameters for conv. monitor
  Handle(LinEqSystemPrec)  system;   // linear system
  Handle(LinEqSolver_prm)  solv_prm; // parameters for linear solver
  Handle(LinEqSolver)      solver;   // linear solver
  int                      n,q;      // grid has n=q*q*q points
  MultipleReporter         report;   // automatic report generation
  void makeSystem ();
  virtual void openReport();
  virtual void closeReport();
public:
  LinSys4 () {}
  ~LinSys4 () {}
  virtual void adm (MenuSystem& menu);
  virtual void define (MenuSystem& menu, int level = MAIN);
  virtual void scan ();
  virtual void solveProblem ();
  virtual void resultReport ();
};
```

Compared with the previous example, the new class LinSys4 has been equipped with a parameter class of type ConvMonitorList_prm. This class can hold the information needed to initialize one or more convergence monitors and arrange them in a linked list for compound evaluation. The new class has also a parameter class of type Matrix_prm needed for run-time selection of matrix formats. The MultipleReporter object, explained in Chapter 3.5.5, is a generalization of the HTMLReporter utility in the LinSys2 and LinSys3 solvers.

---

[7] Those using the Diffpack administrator LinEqAdmFE to solve the systems arising from finite element problems will have immediate access to these two facilities.

A Slight Digression: Choosing Matrix Formats. In the previous examples we have used an explicit instance of the dense matrix class `Mat` in the linear system. Now, we have replaced this matrix object by a `Handle(Matrix)`. Once the user has made his choice, the function `Matrix_prm::createEmpty` is called to set this handle,

```
A.rebind(matx_prm.createEmpty());
```

In this particular case we only pass the name of the matrix format (available as `matx_prm.storage`) instead of the complete `Matrix_prm` object. This is necessary since we want the relevant instance of the `genLaplace3D` function to set the matrix sizes and fill in the actual entries.

Since we still are solving the 3D Poisson problem, we want to utilize the functions developed in Appendix D.1.2. However, the C++ run-time system does not know how to automatically map a generic `Matrix` object to specific format like `MatSparse` or `MatBand`. This mapping must be done explicitly by the programmer. Such mapping between classes in the same hierarchy is referred to as *casting*. Upward casting, i.e., a mapping from a class back to its base class is always valid and is done automatically by the C++ run-time system. Downward casting must be stated explicitly and will cause fatal errors in situations where the casting is invalid. For further information see [14, pp. 342–343]. Since fatal errors will occur in the case that the object is of a different type than the one we assume, special care must be taken. In Diffpack there are two macros `CAST_REF` and `CAST_PTR` that can be used for safe downward casting of references and pointers, respectively. In the current example we have used the `CAST_REF` macro in combination with the macro `TYPEID_PTR` to cast the generic `Matrix` object entering `genLaplace3D` to its relevant subtype:

```
void LinSys4:: makeSystem ()
{
  // since the functions generating the 3D Laplacian allocates
  // the appropriate memory segments, we allocate empty matrices
  // by calling Matrix_prm(real)::createEmpty (which invokes a
  // constructor where no memory for the matrix entries is allocated)
  A.rebind(matx_prm->createEmpty());

  x.redim(n);  b.redim(n);
  b = 1.0/sqr(q+1);    // rhs in Poisson problem is f=1

  // we need to test for matrix types in order to do explicit casts;
  // downward casting is necessary to ensure optimal efficiency when
  // assigning the matrix entries

  if (TYPEID_PTR(A.getPtr(),Mat(real)))
    {
      Mat(real)& A_dense = CAST_REF(A(),Mat(real)); // A() is *A
      genLaplace3D (A_dense,q);  // global function
    }
  else if (TYPEID_PTR(A.getPtr(),MatBand(real)))
    {
```

```
      MatBand(real)& A_banded = CAST_REF(A(),MatBand(real));
      genLaplace3D (A_banded,q);
   }
  else if (TYPEID_PTR(A.getPtr(),MatStructSparse(real)))
   {
      MatStructSparse(real)& A_struct =
                          CAST_REF(A(),MatStructSparse(real));
      genLaplace3D (A_struct,q);
   }
  else if (TYPEID_PTR(A.getPtr(),MatSparse(real)))
   {
      MatSparse(real)& A_sparse = CAST_REF(A(),MatSparse(real));
      genLaplace3D (A_sparse,q);
   }
  else
    fatalerrorFP("LinSys4::genLaplace3D",
    "Cannot make Laplacian in %s format", TYPEID_NAME(A));
}
```

Illegal situations are caught by the last else clause in this function.

Adding Convergence Monitors to the Sample Program. Returning to this section's main issue of convergence monitors, we look at the bodies of the functions in class LinSys4[8]. The functions adm and resultReport are unchanged from the LinSys3 example, while the define and scan functions have minor changes in order to present and read the menu items concerning matrix formats and convergence monitors:

```
   #define MAX_CONV_MONITORS 2
   matx_prm.define (menu,level+1);
   conv_prm.define (menu,level+1,MAX_CONV_MONITORS);
```

and

```
   matx_prm.scan (menu);
   conv_prm.scan (menu);
```

Here, the value of MAX_CONV_MONITORS decides how many submenus will be created in which input data for different convergence monitors can be set. If this value is omitted in the call to conv_prm.define, a default value of 1 is used. You should notice that this parameter determines the *maximum* number of monitors that can be associated with your solver. For instance, the command

```
   set no of additional convergence monitors = 1
```

in the LinEqSolver submenu tells that this particular run will use only *one* of the monitor submenus (actually the first one listed).

   The most important change has taken place in solveProblem:

---

[8] The source code is located in the directory src/linalg/LinSys4.

```
// attach stopping criteria (convergence monitors), this can
// only be done if the solver is iterative

if (TYPEID_PTR(solver.getPtr(),GaussElim))
   warningFP("main","Your solver is not iterative (you have\n"
   "chosen the \"%s\" method - cannot attach convergence monitors)",
   solv_prm->basic_method.c_str());
else {
   // cast solver to the base class for iterative solvers and
   // attach the convergence monitor info:
   IterativeSolver* itsolver =
                   CAST_PTR(solver.getPtr(),IterativeSolver);
   itsolver->attach (conv_prm);
}
```

As seen above, the input concerning the convergence monitors is handed over to the solver which then will allocate and initialize the relevant `ConvMonitor` objects, before organizing them as a list structure.

Other changes in class `LinSys4`, compared with the `LinSys3` program, concern the `MultipleReporter` object. We simply refer to Chapter 3.5.5 and explanations in the source code of class `ReportPoisson2` for information regarding this subject. The purpose of the `LinSys4` solver is to utilize the automatic report generation tools in a simple way and thereby provide an environment for playing around with iterative methods for the Poisson equation problem discretized by finite differences.

Finally, we show the main program that has been slightly modified in order to allow automatic generation of loops over different set of input data (see Chapter 3.4.2).

```
#include <LinSys4.h>
int main (int argc, const char* argv[])
{
   initDiffpack (argc, argv);
   global_menu.init ("Solving the 3D Poisson problem","LinSys4");
   LinSys4 problem;

   // solve one or several problems, let the menu system function
   // multipleLoop take care of the adm, solveProblem, and
   // resultReport calls and allowing multiple answers to menu items
   global_menu.multipleLoop (problem);
   return 0;
}
```

The input file to be presented below shows how to input multiple values for certain menu items.

Residual-Based Monitors. All convergence monitors are implemented as subclasses in the `ConvMonitor` hierarchy. The two most commonly used tests are residual-based,

$$||\boldsymbol{r}^k|| \leq \varepsilon \qquad (D.1)$$

or

$$||\boldsymbol{r}^k||/||\boldsymbol{r}^0|| \leq \varepsilon \qquad (D.2)$$

where $r^k = b - Ax^k$. These tests are implemented as the classes `CMAbsResidual` and `CMRelResidual`, respectively. Consider the following input file[9]:

```
set grid size q = { 3 & 6 } ! try two values of q
sub Matrix_prm
! try different matrix formats:
set matrix type = {MatBand & Mat & MatStructSparse & MatSparse}
ok
sub LinEqSolver_prm
 set basic method = ConjGrad
 set use default convergence criterion = false
 set no of additional convergence monitors = 1
 set max iterations = 300
 ok
sub Precond_prm
 set preconditioning type = PrecJacobi
 ok
sub ConvMonitorList_prm
 sub Define ConvMonitor #1
  set #1: convergence monitor name = CMRelResidual
  set #1: residual type = ORIGINAL_RES
  set #1: convergence tolerance = 1.0e-6
  set #1: norm type = l2
  set #1: monitor mode = ON
  set #1: run time plot = OFF
  set #1: criterion mode = ON
  set #1: relative to rhs = OFF
 ok
ok
ok  ! leave menu and start simulation
```

One should note that the problem size and the matrix format has been subject to multiple choices. Consequently, this input file will cause the solution of $2 \times 4 = 8$ different problem scenarios, one for each set of input parameters, see Chapter 3.4.2 for more details.

Let us dissect the other items in the input file.

use default convergence criterion: This item is set to off in order to avoid automatic insertion of the solver's default convergence test. The default is false.

no of additional convergence monitors: This value (here equal to 1) decides how many convergence monitors that will be attached to our solver. This value must be in the range from 1 to the value we previously have called MAX_CONV_MONITORS.

max iterations: This value gives the maximum number of iterations. If the convergence test has not been satisfied before this limit is reached, the iterations will stop and the solver statistics will tell that we failed to find the solution with specified accuracy. In the current input file this value is set to 100. The default value is 300.

---

[9] This file and associated input files are located in the directory src/linalg/LinSys4/Verify.

#1: **convergence monitor name:** This is the name of the first convergence monitor in our list. The name specifies the type of convergence test, see Table D.1 for a list of available alternatives.

Table D.1. The convergence monitors available in Diffpack are listed showing their class names and the corresponding convergence measure. The "generic residual" v^k can be replaced by r^k, s^k, or z^k depending on the system and solver in question. The user can choose between different norms. For further information we refer to the relevant man pages.

Monitor class name	Convergence measure	Comment
CMAbsResidual	$\|\|v^k\|\|$	
CMAbsTrueResidual	$\|\|v^k\|\|$	computes v^k explicitly
CMRelResidual	$\|\|v^k\|\|/\|\|v^0\|\|$	
CMRelTrueResidual	$\|\|v^k\|\|/\|\|v^0\|\|$	computes v^k explicitly
CMRelResidualUB	$\|\|v^k\|\|/basevalue$	user-defined base value
CMRelMixResidual	$(r^k, v^k)/(r^0, v^0)$	Euclidean inner product
CMRelResSolution	$\|\|v^k\|\|/\|\|x^k\|\|$	
CMAbsSeqResidual	$\|\|v^k - v^{k-1}\|\|$	
CMAbsSeqSolution	$\|\|x^k - x^{k-1}\|\|$	
CMRelSeqResidual	$\|\|v^k - v^{k-1}\|\|/\|\|v^{k-1}\|\|$	
CMRelSeqSolution	$\|\|x^k - x^{k-1}\|\|/\|\|x^{k-1}\|\|$	
CMAbsRefSolution	$\|\|x^k - x_{\mathrm{ref}}\|\|$	needs reference solution
CMRelRefSolution	$\|\|x^k - x_{\mathrm{ref}}\|\|/\|\|x_{\mathrm{ref}}\|\|$	needs reference solution

#1: **residual type:** For the residual-based monitors, this item defines whether we are using the original residual r^k (ORIGINAL_RES), the preconditioned residual s^k (LEFTPREC_RES), or the pseudo-residual z^k (PSEUDO_RES). The default value is ORIGINAL_RES.

#1: **convergence tolerance:** This item defines the tolerance value ε needed in all types of monitors. The default value is $\varepsilon = 10^{-4}$.

#1: **norm type:** From this menu item the user can choose between the norms L^∞ (Linf), l^1 (l1), l^2 (l2), L^1 (L1), or L^2 (L2). The default value is l2.

#1: **monitor mode:** If this item is ON, the convergence history corresponding to the chosen measure is recorded for later presentation, analysis, plotting etc. The default value is ON.

#1: **run time plot:** If this item is ON, a convergence history plot is automatically displayed (using curveplot gnuplot). The default value is OFF.

#1: **criterion mode:** If this item is ON, the corresponding measure is used as a convergence test. That is, when the specified tolerance is reached, the current convergence test is flagged as satisfied. The default value is ON.

#1: **relative to rhs:** When this item is ON, the base value for the relative test uses the right-hand side rather than the initial residual. That is, when observing the original residual the base value is $||\boldsymbol{b}||$, but when observing any preconditioned residual it is $||\boldsymbol{M}^{-1}\boldsymbol{b}||$. The default value is OFF.

The input file listed above uses the absolute residual criterion. If we want to use the relative monitor (D.2) instead, the following statement,

```
set #1: convergence monitor name = CMRelResidual
```

must be inserted.

Special Residual-Based Monitors. It should also be mentioned that there are alternative implementations of the residual-based criteria discussed above, CMAbsTrueResidual and CMRelTrueResidual. The only difference from the two previously discussed monitors is that these two classes force the explicit computation of the residual. That is, at least the work of a matrix-vector product and a vector addition is spent extra. In the preconditioned case, this also involves the application of the left (and eventually the right) preconditioner(s). This might seem as a waste of computing time, but this type of criteria is needed for solvers that do not update the residuals themselves, e.g. as in Jacobi, SOR and SSOR. Moreover, even Krylov methods can make use of these "expensive" monitors for problems where the recursive residual updates become inaccurate due to poor conditioning.

There is a special version of the relative residual monitor, CMRelResidualUB that allows the user to set a specific base value instead of using the initial residual norm. In this case, the menu command

```
set #1: user base = 2.5
```

can be used. However, this particular criterion is usually explicitly coded to use data from another variable to generate the base value. When solving nonlinear PDEs, the base value might depend on parameters in the outer nonlinear iterations. For example,

```
// Handle(LinEqSolver)  solver;
IterativeSolver* itsolver =
    CAST_PTR(solver.getPtr(),IterativeSolver);
Handle(ConvMonitorList) cmlist; itsolver->getConvMonitors(cmlist);
const int nmonitors = cmlist.getNoEntries()
for (i = 1 ; i <= nmonitors; ++i) {
  // can monitor no. i set a user base value?
  if (cmlist()(i).userDefinedBaseValue())
      cmlist()(i).setBaseValue (mybasevalue);
}
```

Finally, the monitor CMRelMixResidual is based on

$$(\boldsymbol{r}^k, \boldsymbol{v}^k)/(\boldsymbol{r}^0, \boldsymbol{v}^0) \le \varepsilon, \tag{D.3}$$

where $\boldsymbol{v}^k$ is either $\boldsymbol{r}^k$, $\boldsymbol{s}^k$, or $\boldsymbol{z}^k$.

Solution-Based Monitors. Another type of convergence measures focuses on the computed solution. For instance, the monitor `CMAbsSeqSolution` observes the evolution of

$$||x^k - x^{k-1}|| \leq \varepsilon. \tag{D.4}$$

Similarly, there is a relative criterion

$$||x^k - x^{k-1}||/||x^{k-1}|| \leq \varepsilon \tag{D.5}$$

available as `CMRelSeqSolution`. Also, in cases where a reference solution is available, the variants

$$||x^k - x_{\text{ref}}|| \leq \varepsilon \tag{D.6}$$

and

$$||x^k - x_{\text{ref}}||/||x_{\text{ref}}|| \leq \varepsilon \tag{D.7}$$

are available as `CMAbsRefSolution` and `CMRelRefSolution`.

Presenting the Convergence History. As mentioned earlier, it is possible to request run-time plots of the convergence history. However, this information can also be extracted after execution, either by directly manipulating the generated `CurvePlot` files, or by generation of performance statistics in e.g. an HTML report. For instance, to view the convergence history of the `LinSys4` examples, try

```
curveplot gnuplot -f tmp.convhist_figures.pl.map -r '.' '.' '.'
```

This command will select all available convergence plots and show them in a Gnuplot window. If you add the option `-ps hist.ps`, you will get a PostScript file containing the selected curve plot(s). The plot in Figure D.7 is generated this way.

The sample program `LinSys4` uses the HTML approach as well, and you are encouraged to examine the generated report in order to find information on the linear solver's performance.

Compound Convergence Tests. In the example above we used only one monitor to test for convergence. However, Diffpack offers the possibility of combining several monitors into one compound test. Consider this excerpt from an input file:

```
sub LinEqSolver_prm
set basic method = ConjGrad
set use default convergence criterion = OFF
set no of additional convergence monitors = 2
...
sub ConvMonitorList_prm
 sub Define ConvMonitor #1
  set #1: convergence monitor name = CMAbsResidual
  set #1: residual type = ORIGINAL_RES
  set #1: max error = 1.0e-6
```

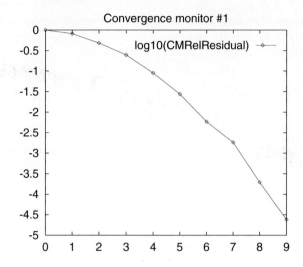

Fig. D.7. A sample curve plot showing the convergence history for a run of the LinSys4 program. The horizontal axis shows the iteration number, while the vertical axis shows the monitored value (here $||r^k||_2/||r^0||_2$) on a $\log_{10}$ scale. Here, the tolerance for the test was set to $\varepsilon = 10^{-4}$, thus reaching convergence in the 9th iteration.

```
  set #1: norm type = 12
  set #1: monitor mode = ON
  set #1: run time plot = OFF
  set #1: criterion mode = ON
  set #1: append at end of list = ON
  set #1: relational operator = CM_AND
  ok
 sub Define ConvMonitor #2
  set #2: convergence monitor name = CMRelSeqSolution
  set #2: max error = 1.0e-4
  set #2: norm type = 12
  set #2: monitor mode = ON
  set #2: run time plot = ON
  set #2: criterion mode = ON
  set #2: append at end of list = OFF
  ok
 ok
ok
```

This input will cause a combined convergence test consisting of (D.1) and (D.5). The residual based monitor will be put at the end of the list and both criteria need to be satisfied before the compound test is evaluated as true (using operator CM_AND instead of CM_OR). The evaluation of the list goes always from start to end, and both relational operators have the same precedence. Clearly, this scheme allows for the construction of quite complicated convergence tests, where all or some of the involved monitors contribute to the convergence test, while some or all also are used for pure reporting purposes.

D.7   Example: Implicit Methods for Transient Diffusion

We shall end our treatment of software tools for large sparse linear systems with an example on implementing implicit finite difference methods for a 2D time-dependent diffusion problem:

$$\frac{\partial u}{\partial t} = k \left(\frac{\partial^2 u}{\partial x^2} + \frac{\partial^2 u}{\partial y^2} \right), \quad (x, y) \in (0, a) \times (0, a), \ t > 0, \quad \text{(D.8)}$$

$$u = 0, \quad x = 0, a, \ y = 0, a, \quad \text{(D.9)}$$

$$u(x, y, 0) = I(x, y), \quad (x, y) \in [0, a] \times [0, a]. \quad \text{(D.10)}$$

The primary unknown $u(x, y, t)$ is the concentration of the diffusing specie, k is a diffusion coefficient, and $I(x, y)$ is a known initial distribution of the specie. The problem (D.8)–(D.10) can also be physically interpreted as heat conduction in a 2D square, u being a measure of the temperature.

First, we introduce a uniform grid on $(0, a) \times (0, a)$, with grid points (i, j), $i, j = 1, \ldots, m$. The grid increment $h = \Delta x = \Delta y$ then becomes $1/(m - 1)$. A quite flexible finite difference scheme for (D.8) is the θ-rule combined with centered spatial differences. Using the compact notation explained in Appendix A.3, and in particular in Example A.6 on page 502, the scheme for the current problem can be written as follows:

$$[\delta_t u]_{i,j}^{\ell - \frac{1}{2}} = \theta k [\delta_x \delta_x u + \delta_y \delta_y u]_i^\ell + (1 - \theta) k [\delta_x \delta_x u + \delta_y \delta_y u]_{i,j}^{\ell - 1}. \quad \text{(D.11)}$$

When $\theta = 0$, this is an explicit scheme where new $u_{i,j}^\ell$ values are computed by an explicit formula involving only known values $u_{i,j}^{\ell-1}$. The basic corresponding computational algorithm and implementation can then follow the ideas from Chapter 1.3, or from Chapter 1.6.3 if a more advanced Diffpack simulator is desired. Unfortunately, the results of Project 1.7.1 or Example A.19 on page 523 show that the explicit scheme has a severe stability requirement. When $\theta \geq 1/2$, there is no stability restriction on Δt, which can be convenient from an application point of view. However, $\theta > 0$ results in an *implicit scheme*, that is, the new $u_{i,j}^\ell$ values are coupled in a system of linear algebraic equations: $\boldsymbol{Au} = \boldsymbol{b}$, where $\boldsymbol{A}$ is a *sparse* $n \times n$ matrix, $\boldsymbol{u}$ is the vector of new $u_{i,j}^\ell$ values, and $\boldsymbol{b}$ is the right-hand side. The number of unknowns, n, equals m^2 in our grid. Furthermore, assuming that the grid points are numbered according to a double loop over j and i, with the fastest variation over i (i.e. the innermost loop runs over i), grid point (i, j) corresponds to the single index $(j - 1)m + i$ in the linear system $\boldsymbol{Au} = \boldsymbol{b}$.

The accuracy of the scheme (D.11) depends on θ. For $\theta \neq 1/2$ the truncation error becomes $(h^2, \Delta t)$, while $\theta = 1/2$, corresponding to the Crank-Nicolson scheme, leads to a more favorable truncation error: $\mathcal{O}(h^2, \Delta t^2)$. The choice $\theta = 1/2$ is therefore popular, since it appears to be the optimal combination of stability and accuracy.

Writing out (D.11) in detail and collecting the terms at time level ℓ on the left-hand side of the equation, gives

$$u_{i,j}^\ell - \theta\gamma\left(u_{i,j-1}^\ell + u_{i-1,j}^\ell - 4u_{i,j}^\ell + u_{i+1,j}^\ell + u_{i,j+1}^\ell\right) =$$
$$u_{i,j}^{\ell-1} + (1-\theta)\gamma\left(u_{i,j-1}^{\ell-1} + u_{i-1,j}^{\ell-1} - 4u_{i,j}^{\ell-1} + u_{i+1,j}^{\ell-1} + u_{i,j+1}^{\ell-1}\right), \quad \text{(D.12)}$$

where $\gamma = k\Delta t/h^2$. The coefficients on the left-hand side now determine the entries in row $(j-1)m + i$ in A, whereas the term on the right-hand side is to be placed in row $(j-1)m + i$ in b. The examples in Appendix D.1.2, combined with the suggested optimization in Exercise D.1, might be of great help when formulating a computational algorithm for generating the linear system at each time level.

Solution of the linear systems is definitely the most computationally intensive part of a finite difference-based simulator for (D.8)–(D.10). It is therefore crucial to apply efficient iterative solution methods, with preconditioning, as explained in Appendix C. A suitable solver for the present problem is the Conjugate Gradient method combined with MILU preconditioning. The most natural matrix storage format for finite difference equations on a square is the structured sparse matrix format, represented through class `MatStructSparse` in Diffpack.

The development of a simulator for (D.8)–(D.10) can start with the quite flexible `LinSys4` program from Appendix D.6, which is an extension of simpler demo programs from Appendices D.3–D.5. The `LinSys4` code must be extended with software tools from Chapters 1.5.5 and 1.6.3, that is, finite difference grids (`GridLattice`), scalar fields of type (`FieldLattice`) for u at the current and previous time level, time discretization parameters (`TimePrm`), and a time loop. In addition, we might want to have a flexible interface to visualization. The `SaveSimRes` class, covered in Chapters 3.3 and 3.10.2, is then a convenient tool that allows for interaction with many visualization systems (Plotmtv, Matlab, Vtk, AVS, IRIS Explorer). Although a simulator for (D.8)–(D.10) can be implemented in almost any computer language using direct array manipulation, we highly recommend to use the more high-level tools outlined above, because these provide greater flexibility when the simulator is used for investigating a physical problem through extensive numerical experimentation. Furthermore, if the shape of the domain changes and hence calls for finite element-based solution methods, the step from the currently suggested solver to a full finite element solver is conceptually small from an implementational point of view, despite the fact that the increase in numerical complexity is substantial. This is the advantage of using classes and high-level abstractions also in simple problems.

Readers having already studied classes `LinSys4`, `Wave1D1`, and perhaps `Wave2D1`, are strongly encouraged to create a new solver for (D.8)–(D.10). A suggested result is provided in the directory `src/linalg/Diffusion2D`.

Bibliography

[1] E. Acklam, D. Calhoun, and K.-A. Mardal. How to use Matlab in C++ programs. *Oslo Scientific Computing Archive*, University of Oslo, 1999. See URL *http://www.math.uio.no/OSCA*.

[2] E. Acklam, A. Jacobsen, and H. P. Langtangen. Optimizing C++ code for explicit finite difference schemes. *Oslo Scientific Computing Archive*, University of Oslo, 1999. See URL *http://www.math.uio.no/OSCA*.

[3] E. Acklam and H. P. Langtangen. Parallelization of explicit finite difference schemes via domain decomposition. *Oslo Scientific Computing Archive*, University of Oslo, 1999. See URL *http://www.math.uio.no/OSCA*.

[4] E. Acklam and H. P. Langtangen. Tools for simplified programming with staggered grids. *Oslo Scientific Computing Archive*, University of Oslo, 1999. See URL *http://www.math.uio.no/OSCA*.

[5] J. E. Akin. *Finite Elements for Analysis and Design*. Academic Press, 1994.

[6] M. B. Allen III, G. A. Behie, and J. A. Trangenstein. *Multiphase flow in porous media*. Lecture notes in engineering. Springer-Verlag, 1988.

[7] M. Alonso and E. J. Finn. *Physics*. Adison-Wesley, 1992.

[8] J. D. Anderson. *Computational Fluid Dynamics – The Basics with Applications*. McGraw-Hill, 1995.

[9] E. Arge, A. M. Bruaset, P. B. Calvin, J. F. Kanney, H. P. Langtangen, and C. T. Miller. On the efficiency of C++ in scientific computing. In M. Dæhlen and A. Tveito, editors, *Mathematical Models and Software Tools in Industrial Mathematics*, pages 91–118. Birkhäuser, 1997.

[10] E. Arge, A. M. Bruaset, and H. P. Langtangen, editors. *Modern Software Tools for Scientific Computing*. Birkhäuser, 1997.

[11] O. Axelsson. *Iterative Solution Methods*. Cambridge University Press, 1994.

[12] I. Babuska, T. Strouboulis, and C. S. Upadhyay. A model study of the quality of a posteriori error estimators for linear elliptic problems. *Comput. Methods Appl. Mech. Engrg.*, 114:307–378, 1994.

[13] R. Barrett, M. Berry, T. F. Chan, J. Demmel, J. Donato, J. Dongarra, V. Eijkhout, R. Pozo, C. Romine, and H. van der Vorst. *Templates for the Solution of Linear Systems: Building Blocks for Iterative Methods*. SIAM, 1993.

[14] J. J. Barton and L. R. Nackman. *Scientific and Engineering C++ – An Introduction with Advanced Techniques and Examples*. Addison-Wesley, 1994.

[15] K. J. Bathe. *Finite Element Procedures in Engineering Analysis*. Prentice-Hall, 1982.

[16] W. B. Bickford. *A First Course in the Finite Element Method*. Irwin, 2nd edition, 1994.

[17] S. C. Brenner and L. R. Scott. *The Mathematical Theory of Finite Element Methods*. Text in Applied Mathematics. Springer-Verlag, 1994.

[18] A. N. Brooks and T. J. R. Hughes. A streamline upwind/Petrov-Galerkin finite element formulation for advection domainated flows with particular empha sis on the incompressible Navier-Stokes equations. *Comput. Meth. Appl. Mech. Engrg.*, pages 199–259, 1982.

[19] A. M. Bruaset. *A Survey of Preconditioned Iterative Methods*. Addison-Wesley Pitman, 1995.

[20] A. M. Bruaset, X. Cai, H. P. Langtangen, and A. Tveito. Numerical solution of PDEs on parallel computers utilizing sequential simulators. In Y. Ishikawa, R. R. Oldehoeft, J. V. W. Reynders, and M. Tholburn, editors, *Scientific Computing in Object-Oriented Parallel Environments*, Lecture Notes in Computer Science, pages 161–168. Springer–Verlag, 1997.

[21] A. M. Bruaset, E. Holm, and H. P. Langtangen. Increasing the efficiency and reliability of software development for systems of PDEs. In E. Arge, A. M. Bruaset, and H. P. Langtangen, editors, *Modern Software Tools for Scientific Computing*. Birkhäuser, 1997.

[22] A. M. Bruaset and H. P. Langtangen. Basic tools for linear algebra. In M. Dæhlen and A. Tveito, editors, *Mathematical Models and Software Tools in Industrial Mathematics*, pages 27–44. Birkhäuser, 1997.

[23] A. M. Bruaset and H. P. Langtangen. A comprehensive set of tools for solving partial differential equations; Diffpack. In M. Dæhlen and A. Tveito, editors, *Mathematical Models and Software Tools in Industrial Mathematics*, pages 61–90. Birkhäuser, 1997.

[24] A. M. Bruaset and H. P. Langtangen. Object-oriented design of preconditioned iterative methods in Diffpack. *Transactions on Mathematical Software*, 23:50–80, 1997.

[25] A. M. Bruaset, H. P. Langtangen, and G. W. Zumbusch. Domain decomposition and multilevel methods in diffpack. In P. Bjørstad, M. Espedal, and D. Keyes, editors, *Proceedings of the 9th Conference on Domain Decomposition*. Wiley, 1997.

[26] X. Cai. *Numerical Methods for Partial Differential Equations and Their Object-Oriented Parallel Implementations*. Dr. scient (PhD) thesis, Department of Informatics, University of Oslo, 1998. See URL *http://www.ifi.uio.no/˜xingca/xingcai_dr.html*.

[27] A. J. Chorin. Numerical solution of the Navier-Stokes equations. *Math. Comp.*, 22:745–762, 1968.

[28] P. G. Ciarlet. *The Finite Element Method for Elliptic Problems*. North-Holland, 1978.

[29] P. G. Ciarlet. Finite element approximation theory. In H. Kardestuncer, editor, *Finite Element Handbook*. McGraw-Hill, 1987.

[30] R. D. Cook, D. S. Malkus, and M. E. Plesha. *Concepts and Applications of Finite Element Analysis*. Wiley, 3rd edition, 1989.

[31] C. T. Miller D. A. Barry and P. J. Culligan-Hensley. Temporal discretization errors in non-iterative split-operator approaches to solving chemical reaction/groundwater transport models. *J. of Contaminant Hydrology*, 22:1–17, 1996.

[32] G. Dahlquist and rA. Bjørk. *Numerical Methods*. Prentice-Hall, 1974.

[33] L. Demkowicz, J. T. Oden, W. Rachowicz, and O. Hardy. Toward a universal $h - p$ adaptive finite element strategy. Part 1. constrained approximation and data structure. *Comput. Methods Appl. Mech. Engrg.*, 77:79–112, 1989.

[34] Diffpack World Wide Web home page.
See URL *http://www.nobjects.com/Diffpack*.

[35] K. Eriksson, D. Estep, P. Hansbo, and C. Johnson. Introduction to adaptive methods for differential equations. *Acta Numerica*, pages 1–54, 1995.

[36] K. Eriksson, D. Estep, P. Hansbo, and C. Johnson. *Computational Differential Equations*. Cambridge University Press, 1996.

[37] M. J. Fagan. *Finite Element Analysis*. Longman Scientific & Technical, 1992.

[38] J. H. Ferziger and M. Perić. *Computational Methods for Fluid Dynamics*. Springer-Verlag, 1996.

[39] B. A. Finlayson. *Numerical Methods for Problems with Moving Fronts*. Ravenna Park Publishing, Seattle, USA, first edition, 1992.

[40] C. A. J. Fletcher. *Computational Techniques for Fluid Dynamics, Vol I and II*. Springer Series in Computational Physics. Springer-Verlag, 1988.

[41] N. D. Fowkes and J. J. Mahony. *An Introduction to Mathematical Modelling*. Wiley, 1994.

[42] J. E. F. Friedl. *Mastering Regular Expressions*. O'Reilly, 1997.

[43] R. Glowinski. *Numerical Methods for Nonlinear Variational Problems*. Springer-Verlag, 1984.

[44] P. M. Gresho and R. L. Sani. *Incompressible Flow and the Finite Element Method*. Wiley, 1998.

[45] M. Griebel, T. Dornseifer, and T. Neunhoeffer. *Numerical Simulation in Fluid Dynamics: A Practical Introduction*. SIAM, 1997.

[46] GRUMMP – Generation and Refinement of Unstructured, Mixed-Element Meshes in Parallel, World Wide Web home page.
See URL *http://tetra.mech.ubc.ca/GRUMMP*.

[47] M. D. Gunzburger and R. A. Nicolaides. *Incompressible computational fluid dynamics; Trends and advances*. Cambridge University Press, 1993.

[48] W. G. Habashi and M. M. Hafez. Compressible inviscid flow. In H. Kardestuncer, editor, *Finite Element Handbook*. McGraw-Hill, 1987.

[49] W. Hackbusch. *Iterative Solution of Large Sparse Systems of Equations*. Springer-Verlag, 1994.

[50] J. P. Den Hartog. *Advanced Strength of Materials*. Dover, 1987. (original edition published by McGraw-Hill, 1952).

[51] E. J. Holm and H. P. Langtangen. A uniform mesh refinement method with applications to porous media flow. *Int. J. Num. Meth. Fluids*, 28:679–702, 1998.

[52] Fastflo World Wide Web home page. See URL *http://www.nag.co.uk/simulation/Fastflo/fastflo.html*.

[53] FreeFEM World Wide Web home page. See URL *http://www.asci.fr/~prudhomm/gfem-html*.

[54] T. J. R. Hughes. Recent progress in the development and understanding of SUPG methods with special reference to the compressible Euler and Navier-Stokes equations. *Int. J. Num. Meth. Fluids*, 14(7):1261–1275, 1987.

[55] T. J. R. Hughes, W. K. Liu, and A. Brooks. Finite element analysis of incompressible viscous flows by the penalty function formulation. *J. Comp. Phys.*, 30:1–60, 1979.

[56] S. C. Hunter. *Mechanics of Continuous Media*. Wiley, 2nd edition, 1983.

[57] M. B. Allen III, I. Herrera, and G. F. Pinder. *Numerical Modeling in Science and Engineering*. Wiley, 1988.

[58] B. Joe. Geompack – a software package for the generation of meshes using geometric algorithms. *Adv. Eng. Software*, 13:325–331, 1991.

[59] C. Johnson. *Numerical Solutions of Partial Differential Equations by the Finite Element Method*. Cambridge University Press, 1987.

[60] L. D. Landau and E. M. Lifshitz. *Theory of Elasticity*. Course of Theoretical Physics, Volume 7. Pergamon Press, 2nd edition, 1970.

[61] H. P. Langtangen. Conjugate gradient methods and ILU preconditioning of non-symmetric matrix systems with arbitrary sparsity patterns. *Int. J. Num. Meth. Fluids*, 10:213–223, 1989.

[62] H. P. Langtangen. Implicit finite element methods for two-phase flow in oil reservoirs. *Int. J. Num. Meth. Fluids*, 20:651–681, 1990.

[63] H. P. Langtangen. Tips and frequently asked questions about Diffpack. Numerical Objects Report Series, Numerical Objects A.S., 1999. See URL *http://www.nobjects.com/Diffpack/FAQ*.

[64] H. P. Langtangen, N. Nunn, G. Pedersen, K. Samuelsson, H. Semb, and W. Shen. Finite element preprocessors in Diffpack. Numerical Objects Report Series, Numerical Objects A.S., 1999. See *http://www.nobjects.com/Reports*.

[65] H. P. Langtangen and G. Pedersen. Computational methods for weakly dispersive and nonlinear water waves. *Comput. Meth. Appl. Mech. Engrg.*, 1998.

[66] H. P. Langtangen and G. Pedersen. A Lagrangian model for run-up of shallow water waves. In M. Hafez and J. C. Heinrich, editors, *The proceedings of the Tenth International Conference on Finite Elements in Fluids*, 1998.

[67] R. J. LeVeque. *Numerical Methods for Conservation Laws*. Birkhäuser, 1992.

[68] J. A. Liggett. *Fluid Mechanics*. McGraw-Hill, 1994.

[69] C. C. Lin and L. A. Segel. *Mathematics Applied to Deterministic Problems in the Natural Sciences*. SIAM, 1988.

[70] W. K. Liu and T. Belytschko. Efficient linear and nonlinear heat conduction with a quadrilateral element. *Int. J. Num. Meth. Engng.*, 20:931–948, 1984.

[71] J. D. Logan. *Applied Mathematics; A Contemporary Approach*. Wiley, 1987.

[72] J. D. Logan. *An Introduction to Nonlinear Partial Differential Equations*. Wiley, 1994.

[73] M. Loukides and A. Oram. *Programming with GNU Software*. O'Reilly, 1997.

[74] B. Lucquin and O. Pironneau. *Introduction to Scientific Computing*. Wiley, 1998.

[75] L. E. Malvern. *Introduction to the Mechanics of a Continuous Medium*. Prentice-Hall, 1969.

[76] K.-A. Mardal and G. W. Zumbusch. Multigrid methods in Diffpack. *Oslo Scientific Computing Archive*, University of Oslo, 1999. See URL *http://www.math.uio.no/OSCA*.

[77] G. E. Mase. *Theory and Problems of Continuum Mechanics*. Schaum's Outline Series in Engineering. McGraw-Hill, 1970.

[78] Matlab World Wide Web home page. See URL *http://www.mathworks.com*.

[79] C. C. Mei. *The Applied Dynamcsi of Ocean Surface Waves*. World Scientific, 1989.

[80] C. T. Miller and A. J. Rabideau. Development of split-operator, petrov-galerkin methods to simulate transport and diffusion problems. *Water Resources Research*, 29:2227–2240, 1993.

[81] K. W. Morton. *Numerical Solution of Convection-Diffusion Problems*. Chapman & Hall, 1996.

[82] M. Nelson. *C++ Programmers Guide to the Standard Template Library*. IDG Books, 1995.

[83] C. D. Norton. *Object-Oriented Programming Paradigms in Scientific Computing*. PhD thesis, Rensselaer Polytechnic Institute, 1996.

[84] Object-Oriented Numerics Page. See URL *http://monet.uwaterloo.ca/blitz/oon.html*.

[85] J. T. Oden, L. Demkowicz, W. Rachowicz, and T. A. Westermann. Toward a universal $h-p$ adaptive finite element strategy. Part 2. a posteriori error estimates. *Comput. Methods Appl. Mech. Engrg.*, 77:113–180, 1989.

[86] H. Olsson. Object-oriented solvers for initial value problems. In E. Arge, A. M. Bruaset, and H. P. Langtangen, editors, *Modern Software Tools for Scientific Computing*. Birkhäuser, 1997.

[87] D. R. J. Owen and E. Hinton. *Finite Elements in Plasticity: Theory and Practice*. Pineridge Press Ltd., 1980.

[88] S. B. Palmer and M. S. Rogalski. *Advanced University Physics*. Gordon and Breach Publishers, 1996.

[89] D. H. Peregrine. Equations for water waves and the approximation behind them. In R. E. Meyer, editor, *Waves on beaches*, pages 357–412. Academic Press, New York, 1972.

[90] R. Peyret and T. D. Taylor. *Computational Methods for Fluid Flow*. Springer Series in Computational Physics. Spriver-Verlag, 1983.

[91] S. Prata. *C++ Primer Plus*. Waite Group Press, 2nd edition, 1995.

[92] W. H. Press, S. A. Teukolsky, W. T. Vetterling, and B. P. Flannery. *Numerical Recipes in C; The Art of Scientific Computing*. Cambridge University Press, 2nd edition, 1992.

[93] A. Quarteroni and A. Valli. *Numerical Approximation of Partial Differential Equations*. Springer Series in Computational Mathematics. Springer-Verlag, 1994.

[94] B. D. Reddy. *Functional analysis and boundary-value problems: an introductory treatment*. Longman Scientific & Technical, 1986.

[95] J. N. Reddy. On penalty function methods in the finite element analysis of flow problems. *Int. J. Num. Meth. Fluids*, 2:151–171, 1982.

[96] J. N. Reddy and D. K. Gartling. *The Finite Element Method in Heat Transfer and Fluid Dynamics*. CRC Press, 1994.

[97] G. Ren and T. Utnes. A finite element solution of the time-dependent incompressible Navier-Stokes equations using a modified velocity correction method. *Int. J. Num. Meth. Fluids*, 17:349–364, 1993.

[98] M. Renardy and R. C. Rogers. *An Introduction to Partial Differential Equations*. Springer-Verlag, 1993.

[99] R. D. Richtmyer and K. W. Morton. *Difference Methods for Initial-Value Problems*. Wiley, 1967.

[100] R. H. Sabersky, A. J. Acosta, and E. G. Hauptmann. *Fluid Flow: A First Course in Fluid Mechanics*. Maxwell Macmillian, 3 edition, 1989.

[101] R. Sampath and N. Zabaras. An object-oriented implementation of adjoint techniques for the design of complex continuum systems. *Int. J. Num. Meth. Engng.*, 1999.

[102] W. Schroeder, K. Martin, and B. Lorensen. *The Visualization Toolkit; an Object-Oriented Approach to 3D Graphics*. Prentice-Hall, 2nd edition, 1998.

[103] L. A. Segel. *Mathematics Applied to Continuum Mechanics*. Dover, 1987. (Original edition published by Macmillan, 1977).

[104] J. R. Shewchuk. Triangle: Engineering a 2D quality mesh generator and delaunay triangulator, 1996. See URL *http://www.cs.cmu.edu/~quake/tripaper/triangle0.html*.

[105] B. Smith, P. Bjørstad, and W. Gropp. *Domain Decomposition – Parallel Multilevel Methods for Elliptic Partial Differential Equations*. Cambridge University Press, 1996.

[106] I. M. Smith and D. V. Griffiths. *Programming the Finite Element Method*. Wiley, 3 edition, 1998.

[107] M. R. Spiegel. *Theory and Problems of Theoretical Mechanics.* Schaum's Outline Series in Science. McGraw-Hill, 1967.

[108] A. Srikanth and N. Zabaras. A computational model for the finite element analysis of thermoplasticity with ductile damage at finite strains. *Int. J. Num. Meth. Engng.*, 1999.

[109] S. Srinivasan. *Advanced Perl Programming.* O'Reilly, 1997.

[110] W. A. Strauss. *Partial Differential Equations: An Introduction.* Wiley, 1992.

[111] J. C. Strikwerda. *Finite Difference Schemes and Partial Differential Equations.* Wadsworth and Brooks/Cole, 1989.

[112] T. Strouboulis and K. A. Haque. Recent experiences with error estimation and adaptivity. Part I: Review of error estimators for scalar elliptic problems. *Comput. Methods Appl. Mech. Engrg.*, 97:399–436, 1992.

[113] T. Strouboulis and K. A. Haque. Recent experiences with error estimation and adaptivity. Part II: Error estimation for h-adaptive approximations on grids of triangles and quadrilaterals. *Comput. Methods Appl. Mech. Engrg.*, 100:359–430, 1992.

[114] B. Stroustrup. *The C++ Programming Language.* Addison-Wesley, 2nd edition, 1992.

[115] B. Szabo and I. Babuska. *Finite Element Analysis.* Wiley, 1991.

[116] M. Tabbara, T. Blacker, and T. Belytschko. Finite element derivative recovery by moving least squares interpolants. *Comput. Meth. Appl. Mech. Engrg.*, 117:211–223, 1994.

[117] R. Temam. Sur l'approximation de la solution des équations de Navier-Stokes par la méthode des pas fractionnaires. *Arc. Ration. Mech. Anal.*, 32:377–385, 1969.

[118] S. P. Timoshenko. *Theory of Elasticity.* McGraw-Hill, 3rd edition, 1982.

[119] D. L. Turcotte and G. Schubert. *Geodynamics – Applications of Continuum Physics to Geological Problems.* Wiley, 1982.

[120] A. Tveito and R. Winther. *Introduction to Partial Differential Equations; A Computational Approach.* Springer-Verlag, 1998.

[121] T. L. Veldhuizen and M. E. Jernigan. Will C++ be faster than Fortran? In Y. Ishikawa, R. R. Oldehoeft, J. V. W. Reynders, and M. Tholburn, editors, *Scientific Computing in Object-Oriented Parallel Environments*, Lecture Notes in Computer Science, pages 49–56. Springer-Verlag, 1997.

[122] R. Verfürth. *A Review of A Posteriori Error Estimation and Adaptive Mesh-Refinement Techniques.* Wiley Teubner, 1996.

[123] L. Wall, T. Christiansen, and R. L. Schwartz. *Programming Perl.* O'Reilly, 2nd edition, 1996.

[124] Z. U. A. Warsi. *Fluid Dynamics, Theoretical and Computational Approaches.* CRC Press, 1993.

[125] P. Wesseling. *An Introduction to Multigrid Methods.* Wiley, 1992.

[126] F. M. White. *Viscous Fluid Flow.* McGraw-Hill, 1991.

[127] N. Zabaras and A. Srikanth. An object oriented programming approach to the lagrangian fem analysis of large inelastic deformations and metal forming processes. *Int. J. Num. Meth. Engng.*, 1999.

[128] N. Zabaras and A. Srikanth. Using objects to model finite deformation plasticity. *Engineering with Computers*, 1999.

[129] E. Zauderer. *Partial Differential Equations of Applied Mathematics.* Wiley, 1989.

[130] O. C. Zienkiewicz and K. Morgan. *Finite Elements and Approximation.* Wiley, 1983.

[131] O. C. Zienkiewicz and R. L. Taylor. *The Finite Element Method, Vol I & II.* McGraw-Hill, 4 edition, 1989/91.

[132] O. C. Zienkiewicz and J. Z. Zhu. A simple error estimator and adaptive procedure for practical engineering analysis. *Int. J. Num. Meth. Engng.*, 24:337–357, 1987.

[133] O. C. Zienkiewicz and J. Z. Zhu. The super convergent patch recovery and a posteriori error estimates. Part 1: The recovery technique. *Int. J. Num. Meth. Engng.*, 33:1331–1364, 1992.

Index

$\nabla \cdot \lambda(\boldsymbol{x})\nabla$ operator, 37
$\partial u/\partial n$, 40

a posteriori error estimates, 207
a priori error estimates, 200
accuracy
– a posteriori estimates (FEM), 207
– a priori estimates (FEM), 200
– numerical dispersion relations, 516
– numerical experiments, 203, 245, 281, 308
– truncation error, 519
– wave equation (FDM), 516
– wave equation (FEM), 144
adaptivity, 205, 287
addBoIndNodes, 257, 538
addItem, 225, 283, 536
AddMakeSrc, 253
addMaterial, 260, 538
addSubMenu, 536
adm, 225, 285, 641
aform, 15
--allocreport, 550
animation
– of waves, 30, 328, 419
– using Gnuplot, 30
– using Matlab, 30, 419
– using Plotmtv, 315
– using Vtk, 316
app, 6
archiving simulation results, 30
array
– classes in Diffpack, 65
– plain C/C++, 59
ArrayGen, 65, 535, 622
ArrayGenSel, 82, 535
ArrayGenSimple, 70, 535
ArrayGenSimplest, 70, 535
artificial diffusion, 47, 177
assembly process (FEM)
– algorithm, 136

– Diffpack code, 565
– example in 1D, 136
– example in 2D, 165
assignEnum, 283
automatic report generation, 272, 312
automatic verification, 555
axisymmetric problems, 157, 284

backward Euler scheme, 103, 123, 346
.bashrc, 540
basis function, 110
BasisFuncGrid, 535
BiCGStab, 647
bilinear elements, 158, 228
binary storage of fields, 236
biquadratic elements, 166, 229
boNode, 257, 538
boundary conditions
– boundary indicators, 255
– Dirichlet, 218
– essential, 118
– essential (impl.), 218
– in weighted residual method, 116
– natural, 119
– natural (impl.), 260
– Neumann, 40, 260
– redefinition, 257
– Robin, 154, 260, 276, 311
boundary indicators, 255
boundary-layer problems, 47, 176
boundary-value problem, 6
Boussinesq wave equations, 423
Buckley-Leverett equation, 107
Burgers' equation, 107

calcElmMatVec, 261, 537, 566
calcTolerance, 549
--casedir, 31
casename, 31, 549

General Remarks

Lecture Notes are printed by photo-offset from the master-copy delivered in camera-ready form by the authors. For this purpose Springer-Verlag provides technical instructions for the preparation of manuscripts. See also *Editorial Policy*.

Careful preparation of manuscripts will help keep production time short and ensure a satisfactory appearance of the finished book. The actual production of a Lecture Notes volume normally takes approximately 12 weeks.

Authors receive 50 free copies of their book. No royalty is paid on Lecture Notes volumes.

For conference proceedings, editors receive a total of 50 free copies of their volume for distribution to the contributing authors.

Authors are entitled to purchase further copies of their book and other Springer mathematics books for their personal use, at a discount of 33,3 % directly from Springer-Verlag.

Commitment to publish is made by letter of intent rather than by signing a formal contract. Springer-Verlag secures the copyright for each volume.

Addresses:

Professor M. Griebel
Institut für Angewandte Mathematik
der Universität Bonn
Wegelerstr. 6
D-53115 Bonn, Germany
e-mail: griebel@iam.uni-bonn.de

Professor D. E. Keyes
Computer Science Department
Old Dominion University
Norfolk, VA 23529–0162, USA
e-mail: keyes@cs.odu.edu

Professor R. M. Nieminen
Laboratory of Physics
Helsinki University of Technology
02150 Espoo, Finland
e-mail: rniemine@csc.fi

Professor D. Roose
Department of Computer Science
Katholieke Universiteit Leuven
Celestijnenlaan 200A
3001 Leuven-Heverlee, Belgium
e-mail: dirk.roose@cs.kuleuven.ac.be

Professor T. Schlick
Department of Chemistry and
Courant Institute of Mathematical
Sciences
New York University
and Howard Hughes Medical Institute
251 Mercer Street, Rm 509
New York, NY 10012-1548, USA
e-mail: schlick@nyu.edu

Springer-Verlag, Mathematics Editorial
Tiergartenstrasse 17
D-69121 Heidelberg, Germany
Tel.: *49 (6221) 487-185
e-mail: peters@springer.de
http://www.springer.de/math/
peters.html

Editorial Policy

§1. Submissions are invited in the following categories:

i)   Research monographs
ii) Lecture and seminar notes
iii) Reports of meetings

Those considering a project which might be suitable for the series are strongly advised to contact the publisher or the series editors at an early stage.

§2. Categories i) and ii). These categories will be emphasized by Lecture Notes in Computational Science and Engineering. **Submissions by interdisciplinary teams of authors are encouraged.** The goal is to report new developments – quickly, informally, and in a way that will make them accessible to non-specialists. In the evaluation of submissions timeliness of the work is an important criterion. Texts should be well-rounded and reasonably self-contained. In most cases the work will contain results of others as well as those of the authors. In each case the author(s) should provide sufficient motivation, examples, and applications. In this respect, articles intended for a journal and Ph.D. theses will usually be deemed unsuitable for the Lecture Notes series. Proposals for volumes in this category should be submitted either to one of the series editors or to Springer-Verlag, Heidelberg, and will be refereed. A pro-visional judgment on the acceptability of a project can be based on partial information about the work: a detailed outline describing the contents of each chapter, the estimated length, a bibliography, and one or two sample chapters – or a first draft. A final decision whether to accept will rest on an evaluation of the completed work which should include

– at least 100 pages of text;
– a table of contents;
– an informative introduction perhaps with some historical remarks which should be
   accessible to readers unfamiliar with the topic treated;
– a subject index.

§3. Category iii). Reports of meetings will be considered for publication provided that they are both of exceptional interest and devoted to a single topic. In exceptional cases some other multi-authored volumes may be considered in this category. One (or more) expert participants will act as the scientific editor(s) of the volume. They select the papers which are suitable for inclusion and have them individually refereed as for a journal. Papers not closely related to the central topic are to be excluded. Organizers should contact Lecture Notes in Computational Science and Engineering at the planning stage.

§4. Format. Only works in English are considered. They should be submitted in camera-ready form according to Springer-Verlag's specifications. Electronic material can be included if appropriate. Please contact the publisher. Technical instructions and/or TEX macros are available on http://www.springer.de/author/tex/help-tex.html; the name of the macro package is "LNCSE – LaTEX2e class for Lecture Notes in Computational Science and Engineering". The macros can also be sent on request.

Lecture Notes
in Computational Science and Engineering

For further information on these books please have a look at our mathematics catalogue at the following URL: http://www.springer.de/math/index.html

Computer to plate: Mercedes Druck, Berlin
Binding: Buchbinderei Lüderitz & Bauer, Berlin